The 7th Guards Army from Kursk to Prague 1943 – 1945

A Day-by-Day Account of a Soviet Guards Combined Arms Infantry Army from 1943 to 1945

Dann Falk

Falken Books

2022

First Edition

You may contact the author at falkenbooks.com or on Facebook/Meta.

Published by Falken Books

ISBN: 978-1-7326074-2-2

To All Those Who Fought

All photographs are noted by source, and where necessary every reasonable effort has been made to trace copyright holders to obtain permissions for their use in this book. The author and publisher apologize for any errors.

The author sends a special thanks to Sherry Anderson for her continuing editing efforts and great support.

No work like this can be completed without years of research, then years of writing and further time gathering bits and pieces of information. It was a very long journey, but I was not alone on this journey; family and friends helped along the way, and many new friends from around the world freely gave valuable support and guidance as I needed it. Finally, once again, I owe a great deal to my wife Sarah for making this book possible. Without her unwavering assistance and tolerance of my continued jabbering about events from over 75 years ago, my efforts would have failed.

Also published by Falken Books:

The 64th Army at Stalingrad 1942-43

Table of Contents

Maps, Tables, Photos, Abbreviations

The author produced all the maps and takes full responsibility for any errors or omissions. They are of course not intended to be used as navigational charts, but as reference aids to assist in understanding the flow of battle and the movements of armies and units. Unless otherwise noted, the top of the map is always north.

Maps

Map Key

Rail Lines

Villages, Towns, Cities, RR Stations ●

Rivers

Capitals ★

Front Line Russian

Movement, Attacks Russian

Front Line German

Movement, Attacks German

Bridges (B)

Ferries (F)

Major Combat Units 81 GRD 25 GRC 711 ID

Army or Front Boundries

Corps Boundries

Armies Russian or Allied 7 GA 6 GTA

Armies German or Axis 6 Pz A 8 A

145.5

Prominet Peak ▲

Mountains ▲ ▲ ▲ ▲ ▲

Tables

Photos

All photographs are noted by source.

Abbreviations
Soviet and Soviet Allies

Stavka – Headquarters of the Supreme High Command
GKO – State Defense Committee
NKO – People's Commissariat for Defense
VGK – Supreme High Command - General Staff of the Red Army
RVGK – Reserve of the Supreme High Command
NKVD – Secret Police
A – Army – 57th Army – 57th A

GA – Guards Army – 7th GA
RC – Rifle Corps – 18 Rifle Corps – 18 RC
GRC – Guards Rifle Corps – 24 GRC
AAD – Anti Aircraft Division – 5 AAD
RD – Rifle Division – 422 Rifle Division - 422 RD
MtD – Mountain Division – 4 Ru MtD
RB – Rifle Brigade
MRB – Motorized Rifle Brigade
RR – Rifle Regiment
MRR – Motorized Rifle Regiment
CRR – Cadet Rifle Regiment
GAD – Guards Airborne Division
GRD – Guards Rifle Division
GRB – Guards Rifle Brigade
GRR – Guards Rifle Regiment
FR – Fortified Region

NKVD RD – NKVD Rifle Division (Secret Police)
NRB – Naval Rifle Brigade

TA – Tank Army – 5th Tank Army – 5th TA
TC – Tank Corps – 13 Tank Corps – 13 TC
TB – Tank Brigade
TR – Tank Regiment
SU – Self-Propelled AT Gun – SU Regiment - 1529 Self-Propelled Gun Regiment (SUR)
GMC - 4 Guards Mechanized Corps
GTB – Guards Tank Brigade
GTR – Guards Tank Regiment
CC – Cavalry Corps
CD – Cavalry Division
CR – Cavalry Regiment

AA – Anti Aircraft or Air Army – 5th AA
AT – Anti Tank or Armored Train – 38 Separate Armored Train Battalion – 38 ATB
A – Artillery – 265 Artillery Regiment – 265 AR
AAR – Antitank Artillery Regiment – 1669 AAR
GMR – Guards Mortar Regiment - Katyusha rocket launcher
VVS – (Voenno-Vozdushnye Sily) Red Army Air Force
ADD – Long-Range Aviation (subordinated directly to *Stavka*)

German and Axis

OKW – Armed Forces High Command
OKH – Army High Command (Command Authority for Eastern Front)
AG – Army Group - AG South

A – Army – 6th Army
Pz A – Panzer Army – 4th Panzer Army 4th Pz A
AF – Air Fleet – 4th Air Fleet
AC – Army Corps – XXV AC (25 Corps)
Pz C – Panzer Corps – XXXXVIII (48 Pz C)
Pz D –Panzer Division – 24 Pz D
MD – Motorized Division – 29 MD
SS Pz GD – SS Panzer Grenadier Division (These were *Waffen* or Armed SS units and not secret police)
Pz GD – Panzer Grenadier Division – Pz GD *Grossdeutschland*
VG – *Volksgrenadier* Division (peoples division)
ID – Infantry Division – 5 ID
Mt D – Mountain Division
IB – Infantry Brigade – 23 IB
GR – Grenadier Regiment
IR – Infantry Regiment – 215 IR

Ru A – Romanian Army – 4th Romanian Army – 4th Ru A
Ru RC – Romanian Rifle Corps – 7 Ru RC
Ru ID – Romanian Infantry Division – 20 Ru ID
Ru CD – Romanian Cavalry Division – 9 Ru CD
(For Romania as a Soviet ally it would be Ru RD)
Russian/German common usage on maps during WW II (Румынский = Ru) (Rumänisch = Ru)

It A – Italian Army – 8th Italian Army – 8th It A
It ID – Italian Infantry Division

Hu A – Hungarian Army – 3rd Hungarian Army – 3rd Hu A
Hu ID – Hungarian Infantry Division

Note: After their first introduction in the text, Soviet armies and smaller units will be abbreviated to simplify the narrative. For instance, the 5th Guards Tank Army will become 5th GTA and the 213 Rifle Division will become the 213 RD. Since they are the subject of the book, the exception will be the 7th Guards Army, which normally will be written out fully unless the shortened version, 7th GA, is appropriate. In turn, Axis Armies will be will be fully written out to more easily distinguish them from Soviet Armies with smaller Axis units abbreviated after their first introduction. For example, the 6 Panzer Division will become 6 Pz D.

Additionally, to help distinguish among the different armies and types of troops, all Soviet or Russian/Allied infantry units will be referred to as Rifle units and armored units will be called Tank units. German/Axis infantry units will be referred to as Infantry/Grenadier and armored units as Panzer units.

Within the text, Eastern Front and Russian Front denote the same thing. However, Eastern Front is used when discussing the German point of view of the war against the Soviet Union. Conversely, Russian Front is used when looking at events from the Soviet point of view.

The map style and scale will change slightly from chapter to chapter to accommodate the changing historical situation and to reduce complexity. Every effort has been made to keep the maps free of clutter to better illustrate the story.

Note: Throughout the book the author uses TsAMO ID numbers (#) to identify and to recall known Soviet/Russian historical TsAMO records. This method is faster than writing out the Fund, Inventory, and File numbers for each record. (See Appendix 1.)

Preface

This volume is a sequel to my first book, *The 64th Army at Stalingrad 1942-43*. The story begins right after the end of the fighting at Stalingrad and continues to the end of World War II with the newly designated 7th Guards Army.

Writing this book was made possible through a long struggle by many authors and researchers to reveal the true history (objective truth) of the Soviet/German conflict during World War II.

In the beginning, historians were limited to few sources of information, and authors produced early works based mostly upon personal accounts and officially approved views of the fighting. German and Russian language titles dominated the research field, with a growing number of titles being translated into English.

After many decades of hard work writing and researching largely alone, authors at last had access to the newly developed Internet which began a worldwide revolutionary exchange of information and ideas that would eventually allow growing access to previously restricted German and Soviet records and now the Russian TsAMO World War II archive.

Without the unending efforts of many previous authors, researchers, historians and numerous other dedicated individuals, my efforts would have utterly failed to produce anything of value. Finally, I am pleased once again to add a small piece to the historical puzzle that was the Soviet/German war on the Russian/Eastern Front.

Dann Falk

The 7th Guards Army from Kursk to Prague 1943 – 1945

A Day-by-Day account of a Soviet Guards Combined Arms Infantry Army

Introduction

After Stalingrad - The Transfer to a New Front

"A victory at Kursk would shine like a beacon to the world."

"Whenever I think of this attack [the battle for Kursk], my stomach turns over."

"We will not capitulate – no, never! We may be destroyed, but if we are, we shall drag a world with us – a world in flames."

"If the war is lost, the nation will also perish. This fate is inevitable. There is no necessity to take into consideration the basis, which the people will need to continue even a most primitive existence. On the contrary, it will be better to destroy these things ourselves, because this nation will have proved to be the weaker one and the future will belong solely to the stronger eastern nation. Besides, those who will remain after the battle are only the inferior ones, for the good ones have all been killed."

Adolf Hitler

"The death of one man is a tragedy. The death of millions is a statistic."

"Great Britain provided time, the United States provided money, and Soviet Russia provided blood."

"In the Soviet army it takes more courage to retreat than advance."

"I trust no one, not even myself."

"The reason why there is now no communist government in Paris is because in the circumstances of 1945, the Soviet army was not able to reach French soil."

Joseph Stalin

These comments give us some idea of Hitler's impression of what the attack at Kursk meant to him and what the consequences were for failing to destroy the Soviet Union, thus losing the war.

In turn, Stalin makes his own thoughts clear; during the war the individual did not really matter. In the long run, after the successful conclusion of the war, the remaining powerful states would decide the future.

Great Victory at Stalingrad

The end of 1942 and the start of 1943 saw the Red Army victorious over Axis forces at Stalingrad. With this victory, the strategic situation dramatically shifted in favor of the Red Army. As the new year developed, the entire world watched as Soviet forces advanced along a broad front toward the west, driving Axis armies before them in full retreat.

German troops retreat under snow and mud conditions.

Meanwhile, the depleted Soviet Armies that remained in and around Stalingrad were busy policing the battlefield, gathering large numbers of POWs, and generally taking stock of the situation. This Stalingrad Group of Forces was composed of the 62nd, 64th, and 66th Armies. Many of these troops must have been in a state of shock having survived this great battle. The truth is, these armies were exhausted and badly needed rest, reinforcements, and time to recover.

Despite this, the 64th Army was doing its part south of Stalingrad by trying to deal with more than 50,000 POWs in improvised camps. Without adequate food, medical support or even basic housing, thousands of these prisoners soon died of the severe cold, rampant disease or starvation. The situation for all involved was grim.

The overall success of several different winter offensives made the Red Army appear to be unstoppable, but in late February 1943 the situation changed yet again. Field Marshal Erich von Manstein had finally gained freedom of action from Adolf Hitler, which allowed him to launch a German winter counteroffensive in southern Russia. Using regrouped formations from Army Group South, von Manstein was able to stop the Soviet offensive by using mainly flanking attacks against dispersed Russian Armies. Red Army units of the Voronezh and Southwest Fronts were unable to counter this unexpected attack, and in turn they lost ground, were driven back toward the east, and suffered heavy losses along the way.

Field Marshal Erich von Manstein in 1943.

Eventually Joseph Stalin and *Stavka* would be forced to draw upon distant reserve units to counter this new German attack. The Red Army needed many days and weeks to gather the forces that would eventually bring an end to the winter offensive season and restore stability to the front lines. One of the distant formations called upon to counter this threat was the 64th Army currently stationed at Stalingrad. This Army would soon be on its way back to the active front.

Ultimately the 64th Army would be given a great reward for its heroic actions during the battle for Stalingrad. For its mass bravery and success during this momentous battle, it would be awarded the designation of Guards (for Guardians) of the Soviet Union, thus becoming the elite 7th Guards Army. During the war only 11 infantry armies and 6 tank armies were awarded Guards status. The 64th Army would join the ranks of the Guards, but first it had to reach a new and distant battlefield. Their story continues…

28 February 1943

On this date, the 64th Army had a reported combat strength of 31,198 men with an estimated support strength of 12,000, for a total ration strength of 43,198 men and women. [1]

With the changing situation at the active front, the 64th Army's time in the rear area was over. *Stavka* needed to bring up fresh forces. The 64th Army was sent to the Voronezh Front area to help stabilize the front and to prevent the Germans from gaining more territory. The first of two Orders/Directives concerning the transfer of the 64th Army was NKO Order #36799 on 28 February 1943 at 05:30 hours. This NKO order should be considered an alert order, which provided detailed instructions about the transfer of the Army by rail to the Voronezh Front. The order outlined which units were being moved, how many trains were allocated, and which areas would serve as loading and unloading stations. (See Appendix 2.) The official *Stavka* Directive #46063 transferring the 64th Army, soon followed:

Stavka Directive #46063

Authorized Stavka, to the command of the Voronezh Front and the 64th Army about reassignment of the Army

28 February 1943 17 h 00 min GHQ orders:

1. The 64th army of six infantry divisions (15 GR, 36 GR, 29, 38, 204 and 422 RD); 156 and 1111 Gun Artillery Regiment; 186, 493, 496, and 500 Antitank Artillery Regiment; 838 Artillery Observation Battalion; 27 Guards Tank Brigade; 224 and 245 Tank Regiments [the two tank regiments are coming from Gorki] *relocate to the area of Valuiki, at the disposal of the commander of the Voronezh Front.*

2. Redeployment to proceed by rail. Start sending on March 1, with the transport of the army into the new area to finish by March 15.

3. Resupply all infantry divisions with personnel, horses, arms, and in the new areas of deployment, bringing the number of men in each Infantry Division, including the Guards, to 8,000.

Reply upon Execution. GHQ Stalin

This Directive officially removed the 173 RD, 214 RD, and the 233 RD from the 64th Army and substituted the 29 RD, 38 RD, and the 422 RD. The reason for this change of rifle divisions assigned to the 64th Army is unknown. These three newly assigned rifle divisions (29, 38, 422 RDs), and the original three rifle divisions (15, 36 GRD and 204 RD) had the following strengths:

Unit	Men	Horses	Vehicles	Tractors	Guns (45-122mm)	Mortars (50-120 mm)
29 RD	3,083	446	56	5	27	32
38 RD	3,534	659	114	13	29	72
422 RD	4,307	451	117	11	50	79
15 GRD	4,889	1,033	149	12	60	139
36 GRD	3,483	248	161	27	65	59
204 RD	2,612	294	128	17	34	75
Totals	**21,908**	**3,131**	**725**	**85**	**265**	**456**

Table #1: February 1943 Rifle Division Strengths.

As can be seen, the final six rifle divisions had a combined combat strength of 21,908 men; 3,131 horses; 725 vehicles; 85 tractors; 456 mortars; and 265 guns. This was a powerful force but far below the authorized strength of 8,000 men for each division. The Soviets command hoped this new deployment would allow enough time to bring these units back to full fighting strength as ordered in #3 of the Directive. However, the Special Commission sent from the GKO to investigate the situation at Stalingrad stated that 50-60 percent of the weapons and equipment (within the Stalingrad Group of Forces) needed repair and that it was impossible to carry out these repairs under the conditions at Stalingrad.

Directive #46063 also was hopelessly optimistic. To comply with this order, the 64th Army would have needed to:

- Gather the required combat units, medical units, and support formations and then provide them with supplies for the journey (food, water, fodder, wood, fuel and ammunition).
- Assemble the different sub-formations at the designated loading stations at designated times.
- Load the troops and equipment as quickly as possible, but only if the required number of trains and types of cars arrived at the correct stations.
- Set off on the long journey, but only if the rail lines were clear and the steam engines were filled to capacity with fuel and water.

Only then could the 64th Army begin its long, slow trip along overloaded rail lines toward the front. Also, the train crews needed to be changed periodically and the steam engines serviced over this lengthy journey.

To complicate matters, due to logistical issues, units of the 64th Army were sent via four different routes from Stalingrad and Gorki to their destinations in the Valuiki area. (See Map #1, Rail Map showing the four routes from Stalingrad and Gorki to the Valuiki area.) (See Appendix 3 about steam engine use.)

1. The first route covered approximately 730 miles: Stalingrad – Povorino – Gryazi –Yelets – Kastornoe – to the Valuiki area. Taking this route were the 64th Army Headquarters, 15 GRD, 38 RD, 29 RD and 27 GTB and most if not all of the first group of supply, medical and rear support units.
2. The second route covered 869 miles and would transport the 36 GRD, 204 RD and the 422 RD. From Stalingrad these units crossed to the east side of the frozen Volga on foot and entrained at Zaplavnoe. Then they continued via rail to Baskunchak, turning north to Urbach and then moving west, crossing back to the west side of the Volga to Saratov. Finally, they traveled to Rtishchevo - Balashov - Povorino - Taloyaya - Liski, to arrive in the Valuiki area.
3. The third route covered the remaining units of the Army at Stalingrad, mostly artillery and AT units who would take a shorter route that covered approximately 400 miles. This route was Stalingrad - Povorino - Taloyaya - Liski - Valuiki.
4. Additionally, the two newly assigned tank regiments, 224 and 245 TRs, needed to travel all the way from Gorki, some 720 miles away, to reach the Valuiki area.

Note: While the 224 and 245 TRs were assigned to the 64th Army, they never went into combat with the Army. As noted above in Directive #46063, the 64th Army and the two TRs were "*at the disposal of the commander of the Voronezh Front.*" The 224 TR supported the 8th GA and the 245 fought with the 6th GA during the Kursk battle.

After traveling for many days, the slow-moving troop trains eventually would arrive at a series of vaguely defined and dispersed locations for detraining. Finally, as these sub-formations detrained, they would march through the snow to assembly areas in order to reform their combat units.

All this was to be accomplished in only 15 days during the middle of the Russian winter. Furthermore, this transfer was intended to help block a powerful German counteroffensive that was currently advancing into the same general region where the 64th Army was scheduled to detrain. In a situation like this, a slight miscalculation on the part of *Stavka* or the Voronezh Front headquarters could be disastrous for the 64th Army.

Regardless of the difficulties in carrying out this order, the 64th Army was duty-bound to begin its journey back to the active front, leaving behind Stalingrad and all its horrors. With a new battlefront calling them forward, the Army gathered its strength and prepared to leave. But its leaders and men were no longer part of a new and inexperienced Army that first fought at Stalingrad, but members of an Army that had faced a great trial by fire and had survived with honor!

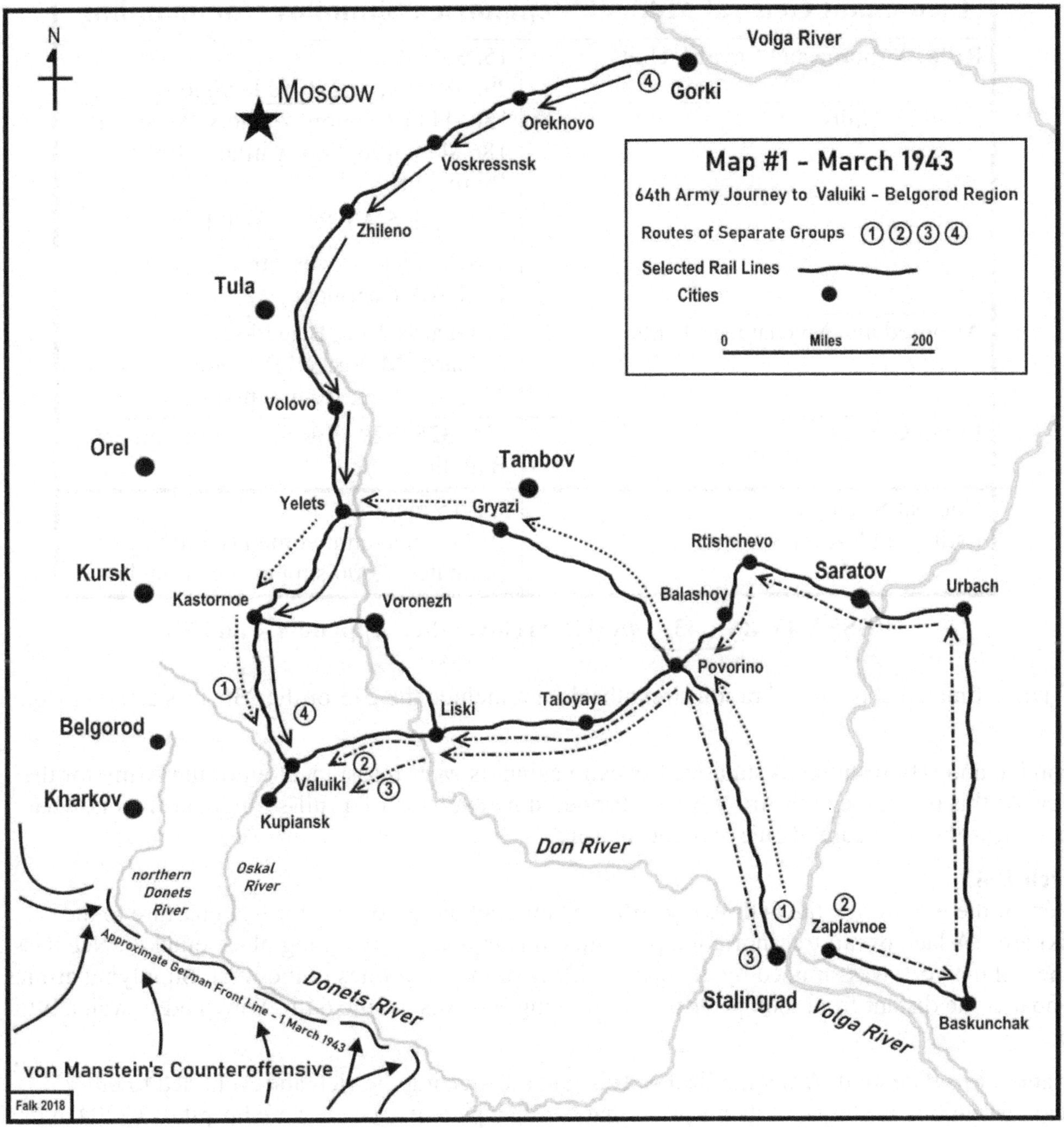

1 March 1943	
64th Army	
Assigned – Voronezh Front	
Lieutenant General Mikhail Stepanovich Shumilov Commanding	
Rifle, Airborne and Cavalry Units	15, 36 GRD 29, 38, 204, 422 Rifle Divisions
Artillery Units	156, 1111 Cannon Artillery Regiments 186, 493, 496, 500 Antitank Artillery Regiments 1, 2, 3, 4, 5 AT Rifle Battalions 838 Artillery Observation Battalion 1261 AA Regiment
Armored and Mechanized Units	27 Guards Tank Brigade 7 Guards Motorized Brigade 224, 245 Tank Regiments
Engineer Units	175, 328, 329, 330 Separate Engineer Battalions
Combat Strength Estimated Ration Strength	32,115 men [2] 39,115 men and women (including an estimated 7,000 support personnel)

(BSSA 43, BSA 43, TsAMO Archive) (See Appendix 4 and 5.)

Note: The estimated combat and ration strengths do not include the 245 or the 230 TRs arriving from Gorki.

The combat and ration strengths stabilized as extra subunits were reassigned out of the Army for the transfer. At this point, the 64th Army was a stripped-down version of a full-strength army, but what remained was the hard core of an experienced army.

1 March 1943

The start of the month saw the 64th Army still at Stalingrad but preparing for its departure to the Valuiki area. A lack of rail transportation prevented this transfer from taking place quickly. Needless to say, the rail network was strained by the rapid shifting of the front lines to the west. Supplying armies in combat at the distant front took priority over moving reserves, so the 64th Army had to wait a bit longer.

On or about this date, 64th Army medical staff began transferring the sick and wounded to other medical institutions from two of its hospitals that had few patients. These were hospitals #2204 and #2207, known as HPPG or Mobile Surgical Field Hospitals. The hospitals were emptied so that when transportation did become available, these two hospitals would be free for an early transfer to the new combat area. An army on the move would need medical support at both the start and at the end of the transfer. No one really knew what the conditions would be at the detraining locations or if they would have any type of support at all. It was best to be prepared for anything.

Map #1 shows the three planned transfer routes from Stalingrad and the one from Gorki to the Valuiki area. Obviously, the three routes from Stalingrad were not the most direct routes to the new deployment location. The transfer from Stalingrad consisted of a long trip to the north using three different routes and then looping to the west and south. Of course, the Germans were responsible for this odd routing of the army. During their retreats over the previous few months, German forces took the time to destroy key rail lines and major bridges to the west of Stalingrad and between the Don and Donets Rivers. Because of this, the direct rail line west from Stalingrad was not opened to traffic until 24 April 1943, which was far too late for transferring the 64th Army to the front.

More importantly, the vital rail line from Laski to Valuiki and on to Kupiansk received special treatment from the Germans. They used a so-called *Schienenwolf* (Rail Wolf), which was a large steel hook mounted on a flat car pulled by one or two steam engines. The Rail Wolf not only plowed up the rail bed but also ripped the sleepers (railroad ties) apart and twisted the rails. Moving along at about 6 miles per hour, a crew of 10-12 men could devastate a rail line with ease. These demolition experts also took the time to damage or to destroy major bridges and signal equipment along the way with explosives.

A German solider prepares a Rail Wolf to destroy a rail line.

Despite their best efforts, the Russian 19 Railway Brigade (RR repair brigade), the only one working in this area, was overwhelmed by the amount of damage. Even with the conscripted help of the local population, they only were able to restore the rail line between Laski and Valuiki on 5 February and then the line from Valuiki to Kupiansk on 18 February 1943. Even after this, the newly repaired single-track line only could handle 10-12 pairs of trains per day. The VOSO (Military Transportation Services) was blamed for allowing the Germans to destroy these rail lines and for other logistical failures during this period. Adding to these problems, the Germans already had converted the Kupiansk Station and connecting rail lines to the narrower Western European gauge tracks, so for the Russian repair crews, proceeding beyond Kupiansk would require much more time and effort.

Because of these disruptions of the rail net, the resulting backup in this general rear area was monumental. On the line from Rtishchevo to Balashov to Povorino to Taloyaya was a backlog of over 6,300 railroad cars with military supplies destined for the front. Over these same rail lines, an additional 156 full trains, 85 of them carrying troops, some of which were units of the 64th Army, were trying to move to the front. They were forced to sit idle for days at a time. With conditions like these a "timely advance of troop trains was not kept." What would normally be a trip of 4-10 days was taking 10-70 days and train speeds were down to 40 miles/day in some places. [3]

Because the Voronezh Front Headquarters complained about the supply and reinforcement situation, Stalin became directly involved. Stalin gave the Deputy Commissar of the Railways (NKPS) A. A. Kovalev the task to expedite the transfer of supplies and troops, including the 64th Army, and to transfer railway brigades working in the rear closer to the front to clear the way. Ultimately the head of the TSUPVOSO (Central Department of Military Transport Services) decided to send trains carrying the 64th Army on a one-way trip into the Valuiki area. Empty trains were not allowed back out, thus effectively doubling the number of full trains that could use the limited rail capacity to 20-24 trains per day. This plan, of course, produced a backlog of empty trains accumulating on sidings and in stations along the way, but the combat troops were able to reach their detraining locations much faster. This somewhat extreme action was needed because *Stavka* was worried the Germans were going to develop their offensive further to the north and east, threatening Kharkov, Belgorod, and the weakened Soviet armies holding the Donets River line.

Also, on 1 March 1943, the NKO issued Order #104 promoting four rifle divisions within the 64th Army to Guards status. They were the 29 RD into the 72 GRD, 38 RD into 73 GRD, 204 RD into 78 GRD, and 422 RD into the 81 GRD. Even with this order, some time would pass before these new designations appeared on charts, tables and maps used by various commanding headquarters.

A typical GRD at this time contained 18 distinct units or formations with the following personnel:

Headquarters – 116 men and women (all units listed at full strength)
Three Rifle Regiments – 2,714 each
Artillery Regiment – 991
Separate Engineer Battalion – 164
Separate AT Battalion – 204
Separate Reconnaissance Company – 79
Separate Signals Company – 154
Separate Chemical Defense Company – 34
Separate Transport Company – 83
Field Mobile Bakery – 54 (able to produce 12-15 metric tons of bread in 24 hours)
Medical Sanitation Battalion – 90
Veterinarian Hospital – 9
Divisional Workshop – 11
SMERSH – Counterintelligence Department – 22
Field Pay office of the State Bank, *Gosbank* – 3
Separate Training Battalion – 392

Thus, a typical GRD, at this time, should have contained 10,548 troops, but most divisions did not reach full *Shtat* strength.[4] The planned strength for a unit would depend upon which *Shtatniy* (*Shtat* for short) that was used. Each *Shtat,* which were frequently revised, provided an official authorized count of troops, vehicles and armament for each type of unit. A Russian *Shtat* was equivalent to the American

TO&E (Table of Organization and Equipment) or the German *Kriegsstaerkenachweisung (K.St.N.)*, (Organization, Composition and Armament). [5]

In Germany, Hitler appointed General Heinz Guderian to the post of Inspector General of Armored Troops. In this position, Guderian had the authority to rebuild and to prepare the depleted Panzer force, both tanks and troops, for future battles. In essence Guderian's efforts in this area made the future German attack at Kursk possible.

2 March 1943

Even with urgent orders to move to a new combat location, the Army's paperwork continued. On this date the staff of the 64th Army and the Stalingrad Group of Forces issued this secret message:

Memorandum of Staff Operations Division 64th Army Chiefs of Staff Stalingrad Group of Forces and instill a course of cleaning the area from corpses.
Stalingrad active army № 0121 Secret
March 2, 1943

During the cleaning work of the bodies of the army allotted territory of parts of the army that are attached, between 23 February to 02 March 1943 [during 8 days] *was collected and buried the following corpses:*

Clean-Up Unit	Soviet Dead	Enemy Dead	Totals
36 GRD	42	339	381 and 11 horses
204 RD	162	2,840	3,002 and 80 horses
38 MRB	313	2,884	3,197
173 RD	1,663	4,330	5,993
214 RD	290	1,324	1,614
233 RD	280	980	1,260
Totals	**2,750**	**12,697**	**15,447 and 91 horses**

Table #2: March 1943 Corpse Collection and Burial.

Cleaned area of 548 km².

Acts of completion of works are presented only in part for the Members of the Army (36 GRD and 204 RD).

In the area of the Voroponovo Railroad Station, on the site of the former prisoner of war camp for Red Army troops, during the excavation of trenches and dugouts discovered the bodies of soldiers and commanders of the Red Army who were tortured by the German occupiers.
Staff Operations Division 64th Army

Chiefs of Staff Stalingrad group of forces [6]

As can be seen by this report, even at this late date, the rifle divisions of the 64th Army still were able to recover frozen corpses from the battlefield at a rate of almost 2,000 per day. The 62nd and 66th Armies also were engaged in these same types of tasks. Despite hundreds of thousands of the Red Army dead that had already been recovered from around the city, recovery troops always found a few more. This recovery effort continues to this day. We can assume that all the 64th Army troops were eager to move away from this city of death; anything would be better than this endless labor as undertakers.

3 March 1943

The NKO issued a second Order #36861 about the transportation of the 64th Army to the Valuiki area. The Deputy Chief of the General Staff A.G. Karponosov used a telegram to report this plan of the Army's transfer from the Stalingrad area on 3 March 1943. This order restated the earlier transfer order issued on 28 February and altered it somewhat. Units of the 64th Army should have started major loading efforts on 2 March; however, they still were mostly in the Stalingrad area awaiting transport.

Order #36861

Commander-In-Chief of Forces of Voronezh Front

About the Plan of the Transportation of the Connections 64th Army

March 3, 1943 19 h of 45 min

1. Report the plan of the transportation by railroad. Joining connections of 64th Army:

(This transportation plan goes on to list the number of trains needed to move the 64th Army to its new area of operations.)

HQs of the 64th Army with only necessary units from the army's rear support - 20 trains (loading at Beketovka Station) (This would have included basic supply and medical units)
38 Rifle Division – 7 trains (loading at Beketovka Station)
1111 Cannon Artillery Regiment – 2 trains (loading at Beketovka Station)
15 Guards Rifle Division – 10 trains (loading at Sarepta Station)
838 Artillery Observation Battalion – 1 train (loading at Sarepta Station)
29 Rifle Division – 6 trains (loading at Voroponovo Station)
493 AT Artillery Regiment – 1 train (loading at Voroponovo Station)
496 AT Artillery Regiment – 1 train (loading at Voroponovo Station)
27 Guards Tank Brigade – 3 trains (loading at Gumrak Station)
156 Cannon Artillery Regiment – 2 trains (loading at Gumrak Station)
186 AT Artillery Regiment – 1 train (loading at Gumrak Station)
500 AT Artillery Regiment – 1 train (loading at Gumrak Station)
36 Guards Rifle Division – 7 trains (loading at Zaplavnoe Station, left bank of the Volga)
204 Rifle Division – 6 trains (loading at Zaplavnoe Station, left bank of the Volga)
422 Rifle Division – 6 trains (loading at Zaplavnoe Station, left bank of the Volga)
230 Tank Regiment – 2 trains (loading at Gorki Station)
245 Tank regiment – 2 trains (loading at Gorki Station)

These orders called for a total of 78 trains, with an average of 50 cars in each train, or approximately 3,900 rail cars.

As the order outlines, the 64th Army was entraining at five major stations around Stalingrad: Beketovka, Sarepta, Voroponovo, Gumrak, and the Zaplavnoe Station, which was located on the far-left bank of the Volga River, some 31 miles away from the main grouping of the Army. The three rifle divisions assigned to the Zaplavnoe Station, 36 GRD, 204 RD, 422 RD, would need to march these 31 miles, crossing the frozen Volga and Akhtuba Rivers at several places just to reach their 19 trains. [7] Once loaded, they traveled north to Urbach, west to Saratov and Rtishchevo, then via Balashov to Povorino, and on toward their detraining stations. Just to reach Povorino was a journey of 500 miles. These units were ordered to take this long indirect route to reduce the heavy demand on the rail line directly north from Stalingrad to Povorino.

Traveling by rail during the war was not a pleasant experience. The standard 40 or 8 (40 men or 8 horses) was used to fill a boxcar, but rail cars were routinely overloaded. During the winter, boxcars should have been heated, but that was not always possible, and they contained only rudimentary sanitary facilities: a hole in the floor or a bucket. The troops would have taken turns sleeping on the floor due to lack of space, with limited food and water being distributed during brief steam engine refueling/watering stops at stations. Overall, the troops suffered much during the slow, dirty, and cold journey to the front.

Compare these train and rail car totals with the movement of the 64th Army when it first moved to Stalingrad in July 1942. At that time, the Army's combat elements required 96 trains with some 4,800 cars. So, a smaller but much more experienced army now was moving back to the front. Once again, the 64th Army was leaving behind many of their rear area logistical and support units. They would have to carry on without many of their supply, medical and administrative formations until further rail transportation became available.

So, on or about this date, the first few trains carrying the leading elements of the 64th Army left the Stalingrad area. These first trains almost certainly carried local security forces, various command and planning staffs, and communication troops as well as engineering support.[8] The planned six to seven-day trip eventually took longer than expected for most of the Army, but the good news was that the detraining locations around the city of Valuiki still were clear of the enemy.

At the distant front south of Kharkov, the already in position 3rd TA finally launched its attack to the south to support the Southwest Front. This attack quickly ran into advancing units of the 1, 2 and 3 SS Panzer Grenadier Divisions, in other words, the entire II SS Panzer Corps. By the end of the day, the 12 and 15 Tank Corps along with supporting units (comprising most of the attacking units of the 3rd TA) were surrounded. With the surrounded Russian units fighting courageously, the Germans needed several days to reduce this pocket.

4 March 1943

Back at Stalingrad, the medical staff of the 64th Army planned ahead when they organized medical inspections at the stations where troops soon would load onto military trains. "Special attention was paid to the counter epidemic measures" since the imminent lengthy rail passage and the congestion of people in the railroad cars would prevent regular sanitary/medical activities. To prevent the outbreak of diseases required "the special vigilance of commanders and medical workers." [9] So before loading, all personnel, equipment and clothing underwent extensive medical and sanitary examinations. This precautionary measure was most effective and prevented any major outbreak of typhus or other diseases during or immediately after the transfer.

5 March 1943

The Germans finally reduced the pocket containing the remnants of the 3rd TA, but many of these Red Army tank men broke out and reached friendly lines. The problem was that these survivors were forced to leave behind most of their equipment, so they would not be of much help in the coming battles. To counter this problem, the Voronezh Front sent three tank brigades and a Guards rifle division as reinforcements to hold the current front line south of Kharkov. However, the Voronezh Front did not know if these units would be enough to stop the attacking German mobile divisions.

6 March 1943

The Germans renewed their attacks and quickly broke through the front, forcing the Russian units to fall back north toward Kharkov.

While the Army's leading organizational elements entrained from Stalingrad starting 3 March, on or about this date, the first combat units of the 64th Army started to leave the Stalingrad area. Most likely these would have been the leading elements of the 15 and 72 Guards Rifle Divisions. *Stavka* also sent even more units from its reserves to help bolster the defense of Kharkov. The 19 RD and the 86 TB soon joined the fight west of Kharkov and helped slow the German advance in this area.

To further disrupt the smooth transfer of Russian reinforcements and supplies to the front, the Germans continued to interfere with Russian rail operations. Luftwaffe reconnaissance aircraft closely watched the repair progress on the major 2,132-foot-long Don River Bridge west of Stalingrad. The Russians had been working day and night on this key bridge and were almost finished rebuilding it when Luftwaffe bombers struck. They dropped 350 incendiary bombs, which destroyed a 500-foot section of the bridge and killed 102 workers including two important engineers. This attack delayed the opening of the bridge and rail line until 24 April 1943. The 64th Army, fortunately, was taking the long northern route to the front. [10] Also, on this date the Voronezh Front's Military Council issued order No. 095, which in part talked about conducting a mobilization of Soviet civilians in the newly liberated territory, collecting servicemen freed from captivity or from behind the lines, and calling up civilians liable for military service. At this time, these measures were not vigorously implemented, resulting in a lack of replacements for the reserve army regiments. [11] This would not be the last time the issue of replacements for the fighting forces became a problem.

7 March 1943

Concentrated attacks by the German II SS Pz C continued toward Kharkov with the Panzer Grenadier Division *Grossdeutschland* (Greater Germany) joining in with supporting attacks on their left flank. The 3rd TA and the 69th Army largely were helpless to stop these powerful assaults. As the advance continued, the 40th A, positioned west of Kharkov, soon was threatened too.

8 March 1943

Exactly when General Shumilov left the Stalingrad area for the Voronezh Front is unknown, but it was most likely early during the transfer. Shumilov might have flown ahead to await the arrival of his Army at the detraining locations (see the following photo) or perhaps he took an early train. A TsAMO map from 15 March 1943, Archive ID #100210818, *Map of the decision of the commander of 64 A on the deployment of formations arriving by rail on 15.3.43,* is a strong indicator that General Shumilov was on site by this date making critical decisions. Army HQ staff officers were more than able to coordinate the movements and loading of units onto the trains at Stalingrad. The real decisions were going to be made along the way and at the detraining stations. This also was the last scheduled day to load trains carrying the Army; however, loading likely took several days longer due to various unforeseen situations.

(Front row, R to L) Commander of the 64th Army Lieutenant General M.S. Shumilov, member of the Military Council Konstantin Kirikovic Abramov (with his arm in a sling), member of the Military Council S.T. Serdyuk, and head of the Special Division of State Security Major I.I. Clay (half image) near Stalingrad in early 1943. Note the PS-84/Il-2 transport (DC-3/C-47) in the background.

The German II SS Pz C and the Grossdeutschland Division continued their attacks northward, splitting the 40th A away from the 69th A and beginning the envelopment of Kharkov from the west.

9 March 1943

German forces continued their attacks, isolating defending Russian forces from each other and destroying them piecemeal. The II SS Pz C moved to further envelop Kharkov and even to penetrate into the suburbs of the city.

Before the war, Kharkov was the fourth largest city in the Soviet Union with some 900,000 citizens. After the war broke out in 1941, the population swelled to over a million due to refugees and then decreased again, as German forces advanced to capture Kharkov for the first time in October 1941. In February 1943, the second battle for Kharkov resulted in the Red Army recapturing the city. But now in March 1943, the third battle for Kharkov was underway and the population was further reduced to approximately 350,000 mostly sick and starving people. After two years of total war, the accumulated damage to the city and to its population was considerable.

10 March 1943

To put even more pressure on the Voronezh Front, German forces launched a two corps attack (VII and LII Corps with seven infantry divisions), with the *Grossdeutschland* Division and another infantry division in support. This attack was directed against elements of the Russian 40th and 69th Armies to the west of Kharkov. Also on this date, Hitler once again visited Field Marshal von Manstein at his headquarters. They talked about the current situation and how operations should be finished before the muddy season (mid-March-late April) set in. Hitler must have been very pleased with the results of von Manstein's counteroffensive.

Leading elements of headquarters, logistical, and medical support units of the 64th Army began to arrive at stations from Volokonovka to Valuiki. These staff officers hurriedly prepared for the arrival of the first combat units of the army. The active front line was moving northward and was now only about

89 miles from Valuiki. The Voronezh Front officers quickly realized that due to the complex and dislocated nature of the 64th Army transfer, they needed to implement temporary measures to deal with the situation and to support the army in this new location.

From the medical point of view, Soviet leaders decided to form an Operations Group of Medical Units to support the army as it arrived. The 69th A provided some of these medical units, while others came from the small medical units that traveled with the combat troops themselves.

Logistically, until its own supply and transportation units arrived, the 64th Army needed to draw supplies and support from nearby Red Army formations, most notably from the depots of the 69th A and the Voronezh Front itself. These supplies arrived via trucks (or horse-drawn wagons) organized into motor regiments, with battalion- sized units used for bulk cargo and smaller formations like companies for fuel delivery or for other special items.

As for supplies, a Combined Arms Infantry Army required a wide variety of provisions. A short list included:

POL/Fuel

Gasoline, motor oil, kerosene, diesel, aviation fuel, aviation oil, and lubricants.

Rations and Fodder

Bread, flour, biscuits, potatoes, grain for flour, concentrates/canned goods, pasta, single grain cereals, dry rations, livestock, meat, meat concentrates, concentrated fish, butter, fat, sugar, tea, salt, vegetables, tobacco, raw tobacco, vodka, barley oats, oats, and local and imported fodder for horses. [12]

Ammunition/Equipment

Ammunition (thousands of tons of ammunition for every type of weapon from pistols to artillery), cleaning supplies (gun oil, rags, brushes, spare parts and weaponry repair equipment) to service the thousands of weapons within the army.

Radios, field telephones, telegraph equipment, miles of wire and other command and control equipment.

Clothing and Kit

Hats, helmets, shirts, pants, boots, jackets, gloves, underwear, socks/foot wraps, belts, packs and other types of web gear. Most of this clothing was provided in both winter and summer varieties, as needed.

Spoons, knives, forks, pots, pans, plates, canteens, lamps, matches, shovels, saws, picks, buckets, and other assorted tools and utensils.

All these supplies and more were necessary every day, regardless of the state of affairs. The army might be attacking or retreating, resting or advancing, or even being transported by rail; providing supplies was a never-ending struggle to keep the Army alive and functioning.

11 March 1943

Stavka ordered the 1st Tank Army and the 21st Army into position north of Belgorod to stabilize the situation in that area. Accordingly, the 64th Army could not arrive soon enough to assist at the front.

General Mikhail Shumilov and his driver ride in their Lend Lease jeep in early 1943.

12 March 1943

The Head of the Signals Service of the 64th Army, Lt. Colonel Borisenko, presented a plan for setting up wire communication lines (land lines) covering the 26-mile area between Volokonovka and Valuiki. This river valley region was chosen as the detraining and assembly location for units of the 64th Army coming from Stalingrad. See Map #2 for detraining and assembly locations. [13] The individual assembly locations changed somewhat when the units detrained and started to deploy. At this time the 64th Army's detraining and assembly areas were only about 70 miles from the fighting around Kharkov, so *Stavka* and General Shumilov certainly watched these developments with great care. One account stated that

> The echelons of our division [29 RD] were moving westward [by rail]. The closer to the front, the more often the fascist planes appeared in the sky. A lot of abandoned enemy equipment lay at the stations and railway sidings. At the end of March, following Valuyki [Valuiki], we reached the Solod [Solotti railway] junction. Here [we were] ordered to disembark.

Moving westward from the train stations, the sub-formations of the 29 RD (soon to be 72 GRD) had a difficult time.

> … movement is slowed down by impassable [muddy roads]. Rushing [vehicles], cars skid [off the road], horses [lose] strength pulling artillery and carts, although they are assisted by gun crews, scouts, signalmen, everyone who can. Only on April 5, the gunners reached the Shebekino, Krapivnoye, Rzhavets concentration area. Everyone is exhausted, tumbled down and dirty as devils. [14]

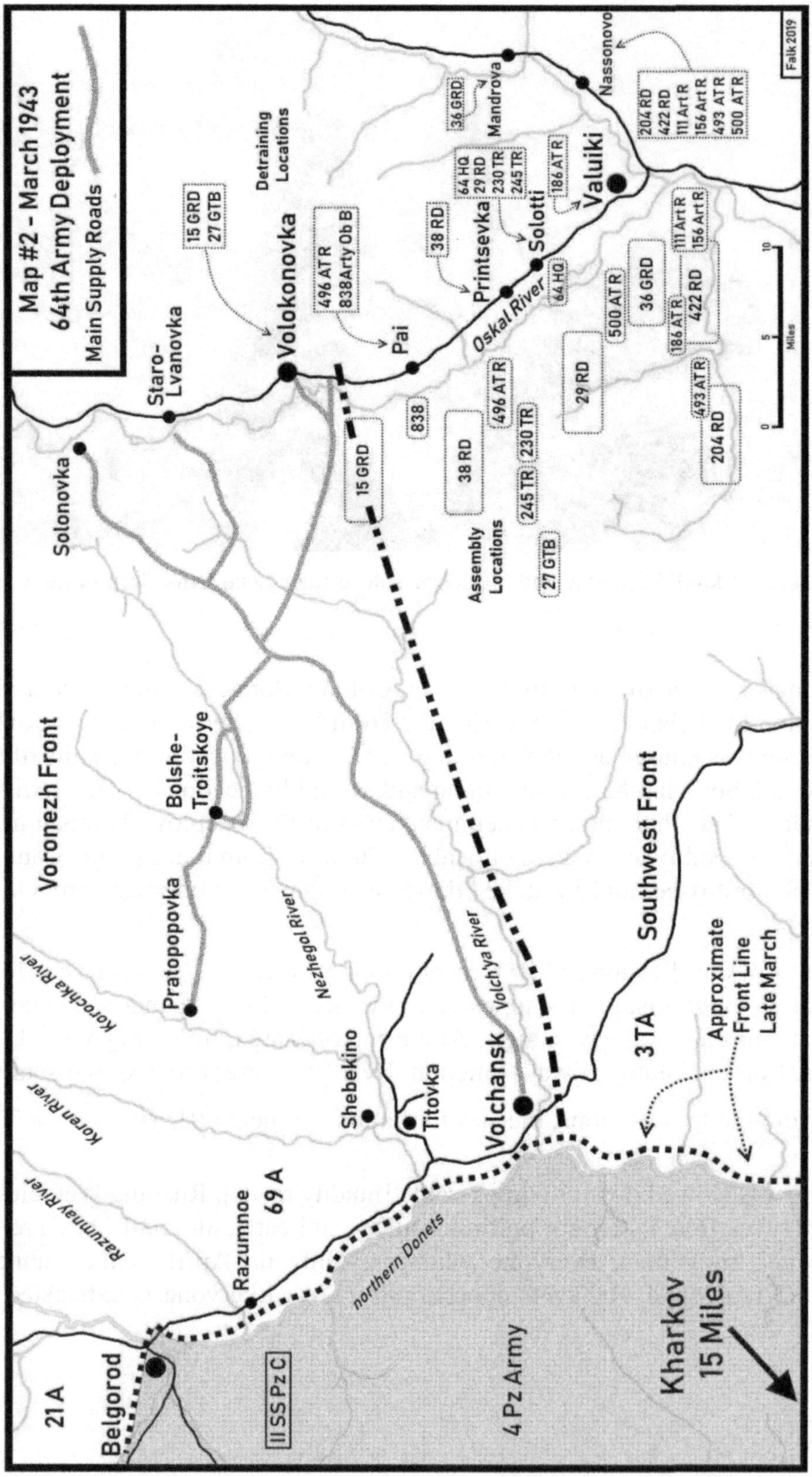

Map #2 - March 1943
64th Army Deployment
Main Supply Roads
Detraining Locations
Assembly Locations
15 GRD
27 GTB
Volokonovka
Staro-Lvanovka
Solonovka
496 AT R
838 Arty Ob B
Pai
38 RD
Printsevka
64 HQ
29 RD
230 TR
245 TR
Solotti
186 AT R
Valuiki
36 GRD
Mandrova
Nassonovo
204 RD
422 RD
111 Art R
156 Art R
493 AT R
500 AT R
Oskal River
64 HQ
500 AT R
36 GRD
186 AT R
422 RD
111 Art R
156 Art R
29 RD
493 AT R
204 RD
838
15 GRD
38 RD
496 AT R
230 TR
245 TR
27 GTB
0
5
10
Miles
Falk 2019
Voronezh Front
Bolshe-Troitskoye
Pratopopovka
Nezhegol River
Korochka River
Koren River
Razumnay River
Volch'ya River
Southwest Front
Approximate Front Line Late March
3 TA
Shebekino
Titovka
Volchansk
69 A
Razumnoe
northern Donets
21 A
Belgorod
II SS Pz C
4 Pz Army
Kharkov 15 Miles

13 March 1943

The 3 SS Pz GD cut the lines of communication for the 3rd TA and then isolated the remnants of the 3rd TA within Kharkov. Intense fighting took place within the city itself.

In Berlin, Adolf Hitler and the OKH (Army High Command) issued preliminary Operational Order #5, "Instructions for the conduct of combat operations during the next months." The focus for Army Group Center and South was upon the newly forming Kursk Bulge. This order envisioned a mid-April German attack in the Kursk area right after the end of the muddy period and before any Russian offensive. [15] This was a surprisingly far-sighted planning order. The general idea for attacking Kursk was to use a double envelopment attack to cut off the bulge at the base, (a distance of approximately 110 miles) and to destroy, it was hoped, a considerable number of surrounded Soviet troops. A successful offensive also would have the added benefit of reducing the overall length of the front line, which would release German forces for use elsewhere.

By this stage of the war, not everyone in the Third Reich was willing to follow Hitler blindly to the inevitable destruction of Germany. The first organized attempt by a small group of German military officers to assassinate Adolf Hitler occurred this day. They planted a bomb on his private aircraft, but the bomb failed to explode.

14 March 1943

Heavy fighting continued to take place within and around Kharkov, but the isolated Russian forces had no chance of holding out. *Stavka* had no fresh forces to commit to the battle for Kharkov.

15 March 1943

In the Valuiki area, a newly completed and more refined detraining map showed assembly locations for the major units of the 64th Army. With 64th Army HQ located near the Solotti Station, General Shumilov and most of the army staff officers were now on site coordinating the placement of the combat formations as they came off the trains. These adjustments of detraining and assembly locations were normal, but they required army level command decisions. This map also now shows the Army controlling the 230 and 245 TRs plus the 27 GTB.

One account stated that leading elements of the 15 GRD had already reached the northern Donets River and began digging in with parts of two regiments. After sending out scouts, the division eventually [possibly on 15 March or later] reported back to General Shumilov that "the SS Panzer Corps, with the support of the infantry, pursuing on the heels of the remnants of our cavalry corps and the 3rd Tank Army, in all likelihood intends to cross the northern Donets on the move and seize a bridgehead on the other side." This information most likely came from Red Army units fleeing to the north from the Kharkov area. General Shumilov thanked the troops for taking up the defense of the river line and then said, "Please note, the Nazis must not pass in any case, otherwise they will [reach the] rear of a large group of our troops." [16] Note, these reported events might have taken place a few days later as the situation at the front was extremely disorganized, with some units retreating in front of the German offensive as newly arrived units from the 64th Army were just arriving piecemeal at the front.

Kharkov fell to concentrated German attacks mainly by units of the II SS Pz Corps. So ended the third battle for Kharkov, which was now firmly back in German hands.

16 March 1943

German forces fully occupied Kharkov and the surrounding areas and continued their advance to the north and northeast. The covering 69th A and the 3rd TA conducted a fighting withdrawal, but the troops mostly were trying just to save themselves. Their strength was gone. The men were exhausted and their weapons, tanks and artillery were strewn across the battlefield. The lack of supplies and reinforcements provided only two options to these armies: retreat and live, or stand and die.

In Moscow Josef Stalin called Marshal Georgy Zhukov on the phone. Zhukov, who was then near Leningrad, was told by Stalin, "You may have to go to the Kharkov area." Two days later, after stopping in Moscow to talk with Stalin, Zhukov arrived at the Voronezh Front HQ, but the damage was done. Kharkov was lost once again, and the Germans continued to advance northward. [17]

Also of note, Soviet secret agent "Dora" (Alexander Rado), operating out of Switzerland, transmitted to Moscow information about the 13 March OKH Operational Order #5. This was the first solid information Moscow received about German summer plans in the Kursk sector. [18]

17 March 1943

Sometime around this date, Field Marshal von Manstein decided to call off further major attacks due to bad weather, mud, and the growing exhaustion of German forces. But before the front line stabilized, German units continued their attack toward Belgorod on the northern Donets River. The front line now was within 50 miles of the detraining locations of the 64th Army. The danger was that the enemy suddenly might arrive before the 64th Army could properly assemble.

18 March 1943

With one last effort, Germans units from the II SS Pz C captured the city of Belgorod and even crossed the northern Donets River, [19] capturing the large village of Mikhailovka on the east side of the river. They also sent numerous reconnaissance units further to the east. The Russian command reacted swiftly to this development and ordered hastily assembled rifle and cavalry divisions into the area to block further German operations on the east side of the river.

While this was taking place, further subunits of the 15 GRD and the 29 RD (new 72 GRD) of the 64th Army begun to move out of their deployment and assembly areas, heading west toward Belgorod and the northern Donets River.

They were currently too far away to help the Soviet forces at Belgorod. These two rifle divisions were almost certainly the first major combat units to arrive at their detraining locations, after a rail trip of five-eight days. The 15 GRD began to arrive on 10 March, with the new 72 GRD on 13 March. This was a very fast transfer under adverse conditions, but most of the army would require twice as long as originally planned, taking 10-15 days or more to arrive at their detraining stations.

The start of the spring thaw, which brought on the muddy period or *rasputitsa,* now forced a halt to major mobile offensive operations along the front because of the thick mud. Any significant German attack east of Belgorod would be delayed for some time.

For *Stavka,* the Voronezh Front Commander General Filipp Golikov, and General Shumilov the spring muddy period, which typically lasts four to six weeks, came at a perfect time. The change in the weather forced an end to major combat operations and imposed a recovery period on both armies. With the entire 64th Army deploying into an active combat zone, imminent major combat operations were inadvisable. The 64th Army was moving and vulnerable. Its units were strung out along the rail lines and scattered about at various train stations.

Also of consideration was a limited logistical and medical support structure. The mud was a hindrance for everyone, but the spring thaw might have saved the 64th Army from being committed to battle piecemeal and destroyed trying to stop von Manstein's ongoing counteroffensive. At this date, German forces were not firmly established along the northern Donets River as they were just approaching the river line. Eventually, German forces finally consolidated their defensive positions along the western side of the northern Donets River, thus leaving the 64th Army free to assemble its forces in peace some 50 miles behind the front line.

19 March 1943

After a long fighting withdrawal, the 69th A retreated to the east side of the northern Donets River and dug in. The 69th A now covered a 26-mile section of the front from just north of Belgorod south to the Koren River near the cities of Shebekino and Titovka. This position was also the far-left flank of the Voronezh Front. At this point, the 69th A was severely weakened, having lost large numbers of men and equipment trying to fend off German attacks for the past month. Just to the south of the 69th A, the 3rd TA was holding its section of the front. Having broken through German encirclements east of Kharkov, its scattered and understrength units were now in position on the east side the northern Donets River south of Volchansk. The remnants of the 3rd TA occupied the far-right flank of the Southwest Front and also were severely weakened. (See Map #2.)

Meanwhile, units of the 64th Army finally started to reach the frontline area in growing strength. The situation was confused, but it appeared that units of the 64th Army were beginning to occupy the weak gap in the line between Titovka and Volchansk. Another account stated that on this date, the leading elements of the 72 GRD were just reaching the northern Donets River area with supporting artillery far behind to the rear. [20]

A small but significant fact about the detraining and assembly area of the 64th Army is that they were in the river valley between Volokonovka and Valuiki. This valley was located directly to the rear of the 3rd Tank Army and directly behind the neighboring Southwest Front. So, when the time came to move the bulk of the army to the front, units of the 64th Army first would need to move north and then west along the Volch'ya River line in order to reach their new position behind and south of the 69th A in the Voronezh Front's section of the newly established front line.

20 March 1943

Voronezh Front Commander General Filipp Golikov apparently already had allocated a section of the front to the 64th Army. This is denoted by changes in army boundary lines on historical maps of this area. The 64th Army would soon take over the southern section of the front line held by the 69th A. Apparently, General Golikov was worried about continued German attacks across the northern Donets River. To counter this possibility, during the previous several days, he started to move in Voronezh Front subordinate units to reinforce and to strengthen the river line held by the 69th A. This soon resulted in a confusing mixture of formations from the 64th Army, 69th A and Voronezh Front along this section of the front line. Most likely these Voronezh Front subordinate units that moved into place along the river were officially transferred to the 69th A, but they would not remain within that army for long. Events at the front were changing too quickly.

Also, on or about this date, the 213 RD and the 6 GCC were assigned to the 64th Army. The 213 RD was at full strength with a listing of 10,538 men. The 6 GCC, containing the 8 and 13 Guards Cavalry Divisions, was listed as having 6,353 men in total on 30 March 1943. The 213 RD had recently been part of the *Stavka* reserve, and the 6 GCC was one of the Voronezh Front subordinate units. These and many other units were sent to the Belgorod/Kharkov area to stabilize the front and to stop the German offensive. Additionally, the 64th Army had several additional units under its control on this date: the 27 GTB, the 201 TB, and the 230, 245, and 262 TRs. Any added tank support units were always well received by an infantry army.

With the addition of the 213 RD, and the 6 GCC, the 64th Army now controlled six Guards' rifle divisions, one regular rifle division, two Guards' cavalry divisions, two tank brigades and three tank regiments. With these new units the 64th Army had increased their strength and fighting ability, but this order of battle would not last long. The transfer of the 6 GCC to the 64th Army mostly was done to facilitate the transfer of these two cavalry divisions from the front line. This same procedure would be repeated again in the coming days as the 64th Army took its place at the front in sections then covered by the 69th A.

Over the next several days the 21st A occupied positions to block any German movement to the north from Belgorod. The front lines were stabilizing as the muddy season limited movement and as more reinforcements arrived to strengthen the defense lines on both sides. However, Russians forces did not yet firmly hold their front line, as some troops occupied isolated strong points with the land between them not even covered.

At the same time, units of the German II SS Pz C firmly aligned themselves along the northern Donets River south of Belgorod. They were arranged in the following order from north to south: 1 SS Panzer Grenadier Division *Leibstandarte* Adolf Hitler (Hitler's Bodyguard), 2 SS Panzer Grenadier Division *Das Reich* (The Empire as in Third Reich), and finally the 3 SS Panzer Grenadier Division *Totenkopf* (Death's Head). Interestingly, the Voronezh Front's 313 Radio Battalion had also located (triangulated) the positions of all three SS divisions along the front. [21]

Of special note for this day, a period Russian map displays units of the 6 GCC and the 72 GRD engaging and pushing back German advanced positions on the east side of the Donets River. One of these German positions was actually east of the main north-south rail line. This coordinated Russian attack was launched at 13:00 hours and evidently was directed against reconnaissance units of the 3 SS Pz GD. This first attack by the 64th Army, in this new area of operations, was successful and cleared most of the east bank of the river. With this setback, the next time German forces attacked along this section of the front, they would be forced to conduct an assault crossing of the northern Donets River against dug-in defenders. Scattered fighting in this area continued until at least 25 March.

A group of patriotic German military officers once again attempted to assassinate Adolf Hitler. One of their members planned to act as a suicide bomber in order to kill Hitler, but Hitler left the area before the timed bomb could explode, so the conspirators deactivated the bomb. There would be many other assassination attempts against Hitler, but they all would fail.

21 March 1943

By 21 March the 69th A, 64th Army, and 3rd TA were aligned along most of the northern Donets River effectively screening the detraining and assembly areas of the remaining 64th Army units still arriving from Stalingrad. Voronezh Front subordinate units also took their place in the front line. Both the 15 GRD and 72 GRD apparently had reached the Donets River line and were in the process of taking over a small section of the front line. Deploying these divisions first made sense because the 15 GRD was the strongest rifle division within the 64th Army with 7,407 men, and the 72 GRD had 6,729 men. Both of these divisions had been located behind the front, with the 72 GRD to the north and the 15 GRD at the very southern end of the 69th A's line.

The 64th Army now was set to move in with substantial forces between 69th A units to the north and 3rd TA units to the south. With the spring thaw, General Shumilov and Voronezh Front Commander General Golikov had time to deploy the 64th Army to the front in an organized manner.

Meanwhile, the *Stavka* High Command also had time to devise a strategy for the coming spring and summer battles.

On this same day, von Manstein suggested to General Hermann Hoth, commander of the 4th Panzer Army, that he cross the northern Donets (in strength) and straighten out his front line, but Hoth refused, saying that "his troops were worn out and the river provided a defensive advantage over a shorter line in the open steppe." [22] General Hoth also knew his advanced reconnaissance troops had just been pushed back over the river by the 64th Army. That forced retreat also might have influenced his decision.

22 March 1943

The 69th A continued to cover sections of the front from north of Belgorod toward the south with newly assigned Voronezh Front formations filling in along the line. Also, on or about this date, the 15

GRD and the 72 GRD had completely replaced the 6 GCC along the northern Donets River. These first two rifle divisions of the 64th Army had moved into their assigned areas and fully took over sections of the front from the 6 GCC during the previous few days and nights.

At this time, two complete surgical hospitals #2204 and #2207 of the 64th Army arrived. They detrained at the Volokonovka Station and within a few days began to support the forward troops who had moved into position along the upper Donets River. The Therapeutic and Infectious Hospital #2712, along with a Field Army Evacuation Point #24, also arrived and set up in Volokonovka. Other medical units from Stalingrad would arrive soon, but not all of them would follow the 64th Army in its new location. Volokonovka itself quickly became the main supply center of the entire 64th Army and its main medical evacuation center for wounded troops.

Another secret message arrived in Moscow from Switzerland. This message provided further details about German plans to attack the Kursk Bulge in the spring. [23] For Stalin and *Stavka*, the evidence supporting an impending attack was adding up.

23 March 1943

As 64th Army units began to replace existing formations along the Donets River, the exchange of units along this section of the front took place as quietly and discreetly as possible. To achieve this, much emphasis was placed upon using *maskirovka* (deception) methods. These deception methods were used to mislead and to deceive the Germans about what was taking place on the Russian side of the line. These methods included the use of radio silence, camouflage, moving units in or out of locations only at night or during bad weather, and finally having the new units occupy the pre-existing positions and fortifications of the former occupants. [24]

Also, over the previous few days, formations of the 73 GRD moved into supporting positions behind the 15 GRD and the 72 GRD. Additionally, the 27 GTB moved up and deployed into a supporting position behind the front. So now the 64th Army had at least parts of three rifle divisions and a tank brigade holding its section of the front line.

Despite their best efforts to remain concealed, on this day the 64th Army appeared on German situation maps for the first time. Everyone was watching for clues as to what was going on behind the front lines. Aircraft were taking air photos and specialists were listening for that one betraying radio transmission or phone message, not to mention receiving information from deserters crossing the lines to give themselves up to the enemy. Apparently, the new frontline activities of the 64th Army had only remained secret for about three days. The Germans now knew the 64th Army had arrived from Stalingrad.

The German 320 Infantry Division likewise moved in to replace the 2 SS Panzer Grenadier Division on the river line. The Germans too were exchanging units at the front.

24 March 1943

The combat strength of the 64th Army now stood at 54,327 men. [25]

As more and more units of the 64th Army arrived by train and assembled, they were sent off toward their future positions along the front line. These movements were not completed too quickly, due to the mud and general disorder of the rail transfer, but they involved a steady flow of men, equipment, and supplies to the front.

Stavka acted to clarify the command and control of the new front line on the north side of the Kursk Bulge. It issued Directives #30081 and #30082, which disbanded the old Kursk Front and created a new Orel Front. This change did not last very long as *Stavka* was working on additional plans for future operations.

25 March 1943

By this date, the 6 GCC had fully given up the 12-mile section of the front it had been holding and moved to the rear of the 64th Army. This same 12-mile section of the front, from Razumnoe in the north to Shebekino in the south, was taken over completely by the 15 GRD and the 72 GRD of the 64th Army with the 73 GRD and the 27 GTB in support. This exchange of units probably occurred over the previous few days and mostly at night. So, on or about this date, the 64th Army held at least a 12-mile section of the front line. Across the river, the German 320 Infantry Division held the same section of front with elements of the II SS Pz C supporting. However, neither side was looking for a fight. Both were more interested in shifting divisions and corps around, disengaging units from the front, and trying to organize the newly established front line.

A preliminary plan already was in place for the 64th Army to take over the remaining sections of the front held by the 69th A and by any additional units from the Voronezh Front. This plan apparently was designed to replace 69th A formations working from the south to the north. As units of the 64th Army relieved the different formations of the 69th A, the 69th A formations withdrew to the rear for rest and refitting. From this date forward, the 64th Army HQ Staff began working with the Voronezh Front to produce a series of planning maps/diagrams detailing several possible deployments of the entire 64th Army to the front. These plans would not be finalized until Stalin and *Stavka* decided the overall strategy for the summer season.

26 March 1943

Since the start of the muddy period, heavy fighting had all but disappeared. The normal daily routine became one of sniper fire, sudden artillery and mortar harassing bombardments, and nightly scouting raids. Both sides were exhausted, and the troops took full advantage of this pause in combat. But despite the lull in the fighting, German and Russian troops still launched probing attacks and reconnaissance in force missions along the line. They were looking for weakness in defenses and gathering intelligence about their adversaries. The front line actually was alive with activity, but most of this action was hidden below ground level as troops on both sides dug in, or the activities were covered by the dark of night.

The spring mud determined not only the level of fighting but also the number of supplies that were able to reach the front. One Russian artillery unit reported on the situation,

> Very tight with the delivery of food and ammunition. A hard limit was imposed on the expenditure of shells and cartridges. The commanders of all units, political agencies, mobilized personnel, free from carrying out combat service, were to deliver food, shells, mines and cartridges on foot. The distance to…where our rear services were located, was about 20 kilometers [12 miles]. And the columns of one hundred or more people walked back and forth toward each other. From the warehouses were carried in bags of cereal, salt, meat, canned goods, cartridges, and on specially made sleds – shells and mines. Only in this way did we succeed in providing the regiment with a minimum of products and ammunition. [26]

27 March 1943

With wounded men returning from the hospital and new units being assigned to the army, the 64th Army became larger and stronger. With the addition of local conscripts drafted into the army, the combat strength of the 64th Army reached about 69,000 men. This was a considerable improvement in strength from just a few months before.

28 March 1943

On 28 March 1943, General N.F. Vatutin took over command of the Voronezh Front. He quickly signed Order #0087, which formally took over control of the five field armies of the front. They were the 21st, 38th, 40th, 64th and 69th Armies. [27]

The KHPPG #4344 (Mobile Surgical Field Hospital), on loan from the Voronezh Front, began to support the 64th Army. This hospital at first worked as an intermediate hospital, used to evacuate patients to the other rear area hospitals. Hospital #4344 would remain with the 64th Army at least until August 1943.

On left, General Nikolai Vatutin and Member of the Military Council Nikita Khrushchev at the front.

29 March 1943

Stavka issued Directive #30085 renaming the Orel Front to the Bryansk Front. This finalized the reorganization of commands on the north side of the Kursk Bulge. The reorganization in other areas would happen later.

Sometime around this date, the last combat elements of the 72 GRD finally loaded onto trains at Stalingrad. This is noteworthy because the leading elements of this division had already been deployed along the Donets River and were engaged with the enemy. The Soviet logistics system appeared to be overwhelmed and straining under the load of supporting a newly established front line in the middle of winter.

Working together, the command staff (including Lieutenant General Shumilov, Major General Laskin, and Colonel Georgy Serdyuk), produce a 31-page defensive plan for the 64th Army along the new front. This plan was very close to the final plan that they would use to defeat the forthcoming German attack against what would become the 7th Guards Army.

30 March 1943

As further units of the 64th Army were ready to leave their detraining and assembly locations, they were deployed toward the new front south of Belgorod. But these units did not automatically take over frontline positions; they were held back behind the lines until a smooth changeover could take place. On this day the 167 RD and 183 RD from the 69th A appeared on strength tables as being assigned to the 64th Army. This was the start of the next phase in the final changeover between the 69th and 64th Armies, which finished in early April.

31 March 1943

By the end of the month, all of the remaining medical units assigned to the 64th Army had arrived in the new area of operations. Because of this, the Voronezh Front disbanded the supporting Operations Group of medical units. As more and more hospitals of the 64th Army became operational, the loaned medical units gradually returned to the control of the 69th A and the Voronezh Front. [28] This should have been an indication that most if not all combat and support units of the army had arrived from Stalingrad, but this was not the case. Looking ahead somewhat, the Deputy Defense Commissar of the State Defense Committee, E. A. Shchadenko, issued memorandum #130721 on 28 May 1943 about the transfer of military units from the Stalingrad Group of Forces to new locations. This was in regards to an earlier GKO Order #2911 from 18 February 1943. Shchadenko stated in part:

- Sending combat formations and units 62nd, 64th and 66th armies, [the] 7 RC, ten Rifle [divisions] - finished on April 14.
- Shipping tank, artillery, anti-aircraft and mortar units and formations RGK - finished on 10 May.
- Sending portions of chemical and engineering, as well as the front and rear parts of the Army - finished on 22 May.

Total sent 662 echelons (trains). [29]

So, the armies that had been located at Stalingrad and currently were deployed along the front line would need to wait another seven weeks or so to receive their remaining combat, artillery and support units. Despite this, the combat strength of the 64th Army now stood at about 72,000 men.

With the front stabilized and the muddy period preventing major combat actions, General Shumilov focused his attention on bringing his army back up to strength, absorbing new weapons and equipment, and training raw recruits for future battles. Sooner or later the front lines once again would erupt into flames, and the 64th Army would need to stand ready to defeat the enemy. (See Map #3 – Front Line End of March.)

A historical document from the TsAMO record ID #132636511 shows that during all of March, the traveling 64th Army suffered the following losses: Killed – 188, wounded – 424, gone missing – 42, totaling 654 men and women. Apparently, traveling by rail from Stalingrad to the new front was not without its dangers. The “gone missing” total of 42 is interesting and could indicate deserters or just soldiers that were searching for food or water and missed their train.

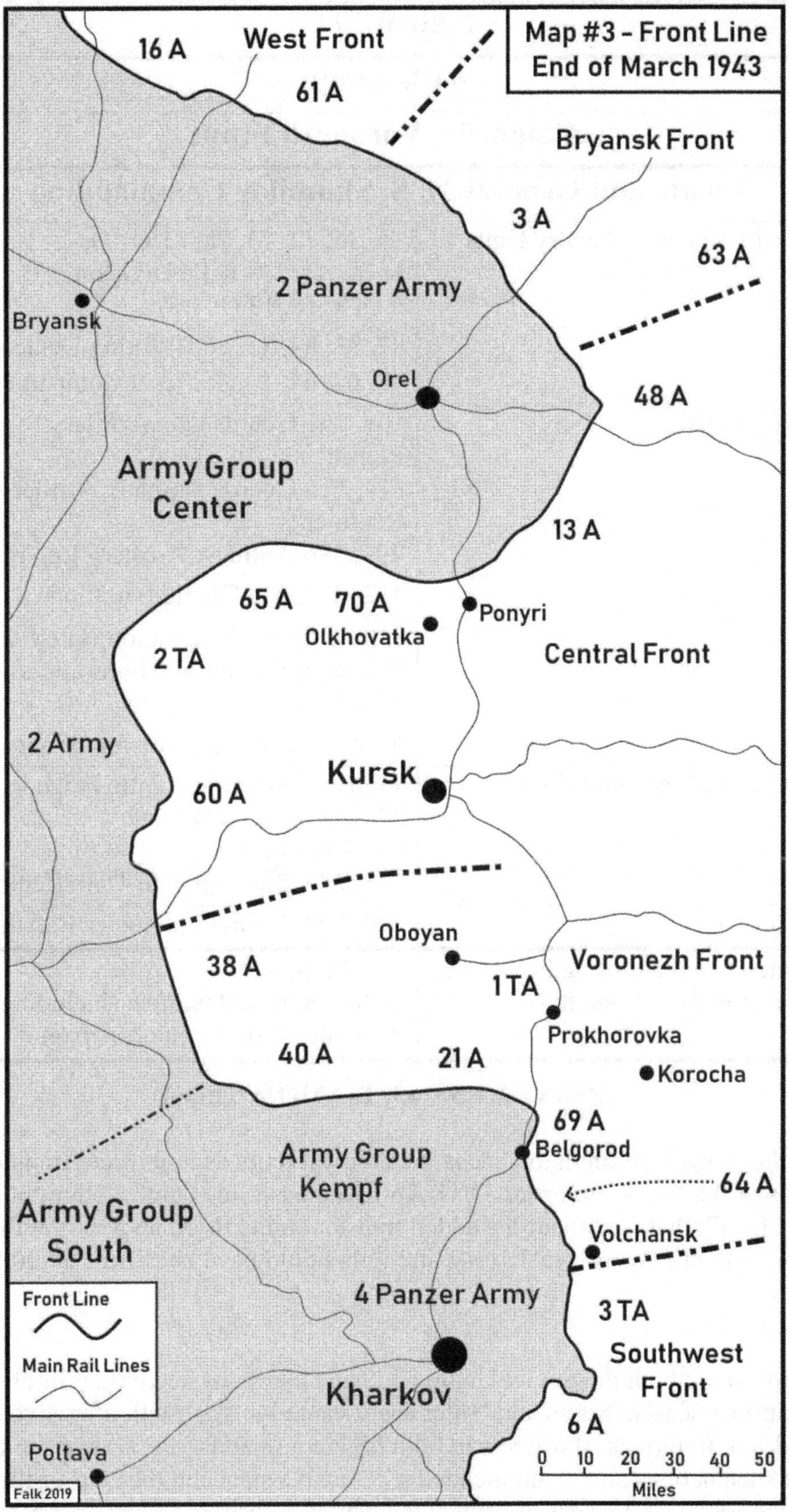
Map #3 - Front Line
End of March 1943
16 A
West Front
61 A
Bryansk Front
3 A
63 A
2 Panzer Army
Bryansk
Orel
48 A
Army Group
Center
13 A
65 A
70 A
Ponyri
Olkhovatka
2 TA
Central Front
2 Army
Kursk
60 A
Oboyan
38 A
Voronezh Front
1 TA
Prokhorovka
40 A
21 A
Korocha
69 A
Belgorod
Army Group
Kempf
64 A
Army Group
South
Volchansk
Front Line
Main Rail Lines
4 Panzer Army
3 TA
Southwest
Front
Kharkov
6 A
Poltava
0 10 20 30 40 50
Miles
Falk 2019

1 April 1943	
64th Army	
Assigned – Voronezh Front	
Lieutenant General M. S. Shumilov Commanding	
Rifle, Airborne and Cavalry Units	15, 36, 72, 73, 78, 81 GRDs 167, 183, 213 Rifle Divisions 6 GCC (8, 13 GCDs) 82, 83, 84, 85, 86 Blocking Detachments 62, 63, 64, 15, 66 Penal Companies
Artillery Units	109, 161 Guards Gun Artillery Regiments 114, 115 Guards Antitank Artillery Regiments 493, 496 Antitank Artillery Regiments 1, 2, 3, 4, 5 AT Rifle Battalions 838 Artillery Observation Battalion 662 Separate Aerial Observation/Air Raid Warning Company Mortar Regiment (with no number)
Armored and Mechanized Units	7 Guards Motorized Rifle Brigade 27 Guards Tank Brigade 201 Tank Brigade 148, 245, 262 Separate Tank Regiments
Engineer Units	175, 329 Separate Engineer Battalions
Estimated Combat Strength Estimated Ration Strength	71,942 men [30] 78,942 men and women (including an estimated 7,000 support personnel)

(BSSA 43, BSA 43, TsAMO Archive)

Note: During April the Army's strength totals changed almost daily as units were transferred in response to the needs of the Voronezh Front. On 1 April, the base-line combat strength of the 64th Army was listed as 51,007; that is without the added transient units listed above.[31] With about 7,000 people in supporting units, the Army's real ration strength would have been near 58,000 people, both men and women.

1 April 1943

The start of the month found both the Soviet Union and Nazi Germany searching for an overall strategy for the spring and summer season. Stalin and Hitler each wanted to start with offensive operations as soon as the ground dried. In turn, staff officers in both high commands, *Stavka* and the OKH, advised their leaders against such actions and recommended a defensive stance at the start of the season. The debate would continue for several weeks as each side knew this spring and summer offensive season could very well decide the war on the Eastern Front/Russian Front.

With the transfer of both the 167 and 183 RDs to 64th Army control in March, the stage was set for further changes at the front. General Shumilov considered several defensive plans for the 64th Army, and now the Voronezh Front HQ approved one. This plan called for four rifle divisions to hold the first line of the defensive, with three more rifle divisions occupying the second defensive line directly behind them. So, the first defensive line, from north to south, would be the 73 GRD, 213 RD, 72 GRD, and 15 GRD. The second defensive line, from north to south, contained the 81 GRD, 78 GRD, and 36 GRD. Independent tank, antitank, artillery and engineer units were assigned as needed to increase the strength of the front. (See Map #4 for the April deployment.) The 6 GCC still was assigned to the 64th Army and was positioned to the rear as a reserve force.

On the German side, Major General Dr. Hans Speidel, chief of staff of Army Detachment Kempf drafted an outline for the southern section of proposed Operation *Zitadelle* (Citadel), the German summer offensive plan to surround and to destroy Soviet armies within the Kursk salient. His plan called for three Panzer Corps from 4th Panzer Army and AD Kempf to attack the southern side of the Kursk Bulge. The 4th Panzer Army would be deployed north and west of Belgorod with Army Detachment Kempf on the right flank, located between Belgorod and Kharkov. [32]

Lieutenant General Warner Kempf Commander of Army Detachment Kempf

2 April 1943

On or about this date the 213 RD started to deploy into the positions held by the 183 RD with the 73 GRD likewise moving up to relieve the 167 RD. Once again, this process took place mostly at night and required several days to complete. The relieved units withdrew toward the 69th A for rest, replenishment and reassignment.

The Chief of the General Staff (*Stavka*) General A. M. Vasilevsky issued a directive to all fronts and armies. This directive ordered all commanders to "use the period of the spring *rasputitsa* [muddy period] to organize better defensive positions, particularly antitank, to develop defensives, … to create reserves on the main axes, and to improve force combat training." He then said that they also should pay attention to offensive matters. Vasilevsky's statements clearly indicated that *Stavka* was undecided upon which strategy should be implemented at the start of the summer season, attack or defend. [33]

In Berlin, Hitler considered two other spring attacks in Russia other than just Operation Citadel against the Kursk Bulge. These other attacks were Operation *Habicht* (Hawk) and Operation Panther; both of these more limited operations would be directed against the Russian Southwest Front, directly east of Kharkov. Note: Overall, all three of these offensives should be considered spoiling attacks, designed to destroy or to weaken enemy forces that were in the process of forming or assembling for an attack of their own. German planners hoped that a sudden powerful Panzer assault, from the north and south, would weaken Soviet forces enough to reduce or to delay any major Soviet offensive operations during the summer months. At best this was wishful thinking on the part of Hitler and the OKH.

From the German point of view, the weather was still a big factor for any operation. The ground needed to be dry enough for extensive mobile operations on both the north and south sides of the Kursk Bulge. Therefore, Hitler ordered that "if Operation *Habicht* could not start by 17 April, it would be superseded by Panther which would have to be ready by 1 May." If neither of these operations could be launched, "then Army Group Center and South would go straight to Citadel." [34]

3 April 1943

To continue implementing the defensive plan for the 64th Army, the 78 GRD and the 36 GRD moved into position along the second defensive line behind the left flank of the Army, backing up the 72 and 15 GRDs.

With the nearly simultaneous movement of four or more rifle divisions toward the front and two more divisions moving to the rear, *maskirovka* methods were used to cover and to disguise this transfer of units. One employed method restricted the use of radio traffic. During this time,

> To conceal the deployment of the 7th Guards Army [64th Army] into the area east of Belgorod and to ensure radio security throughout the Voronezh Front, all radio transmissions were prohibited [from lower HQs]. All sets could receive transmissions, but these were sent from armies and corps only in short 10-15 second bursts. All acknowledgements were then sent by wire communications. To maintain communications security all radio call signs were changed daily, and frequencies changed five times a month. When the 64th Army was deployed into the 69th A's sector, all orders to that army passed through 69th A communications. [35]

Note: The 64th Army would officially become the 7th Guards Army next month.

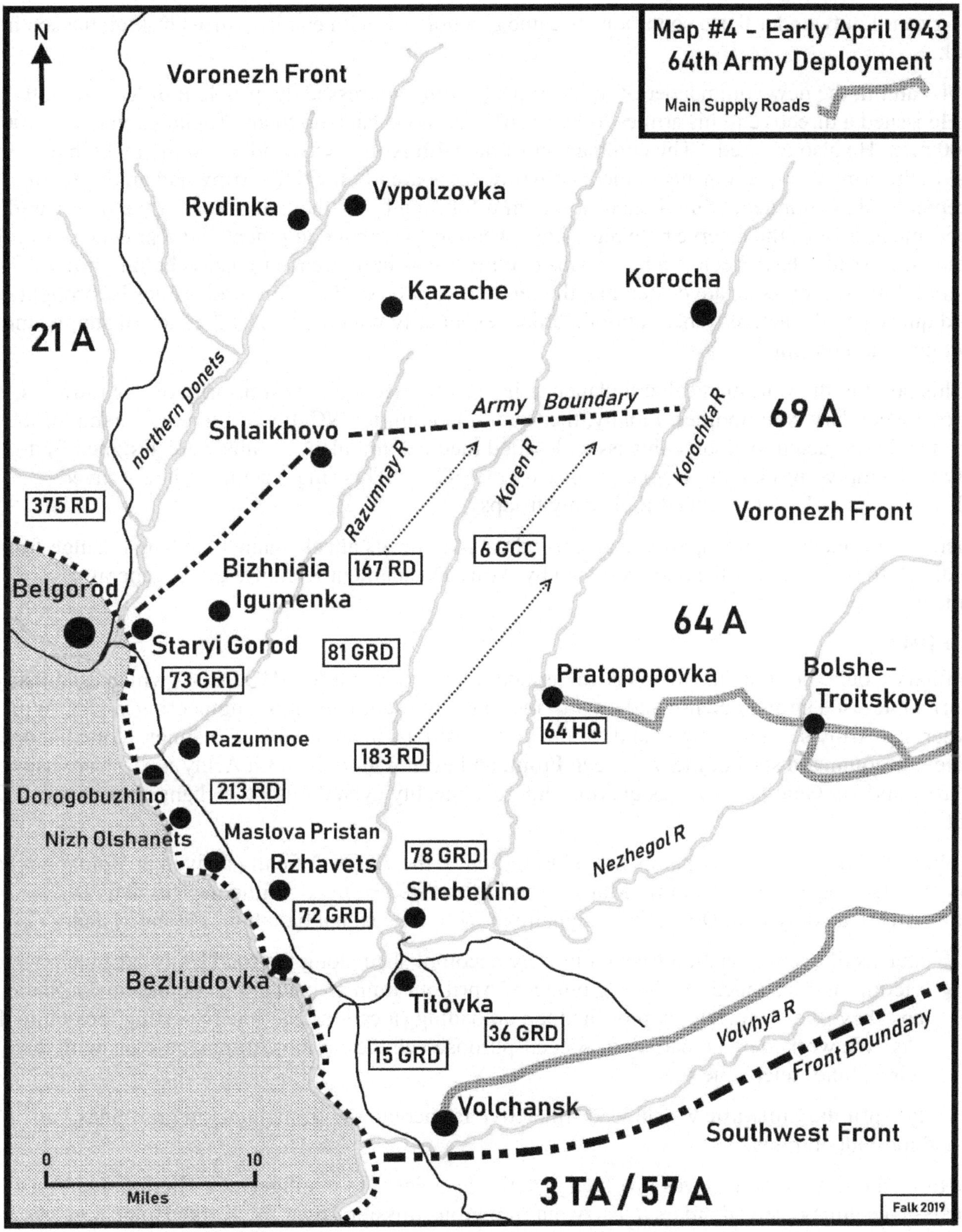

The spring mud still dominated the overall situation. For Red Army units at the front, bogged down supplies meant rations were reduced to a minimum and "even aircraft were used to deliver hardtack and salt, which were air-dropped." Along with the hunger, the lack of vitamin A caused troops of the 64th Army to suffer "a mass outbreak of nocturnal blindness [nyctalopia]. A soldier suffering from this condition could no longer see well in the evening and at nighttime became practically blind." [36] Even standing guard duty under these conditions became a major risk with enemy patrols lurking unseen in the dark.

General Vatutin, the new commander of the Voronezh Front, addressed the problem of low strength units. He issued a directive to his armies to bring rifle divisions back up to an official strength of 7,000 to 8,000 men. He also ordered, "The commander of the 64th Army is to send 6,000 men who have arrived in the army as replacements to the 21st Army." Apparently the 64th Army had an excess of replacements. He continued, "Simultaneously, I am informing [you] that measures for the most rapid transport movement of the reserve rifle regiments, weapons, combat equipment and rear echelons of the 21st and 64th Armies have been taken." [37] This comment was in reference to rear echelons still at Stalingrad. Finally, someone advocated that the remaining parts of the Stalingrad armies be brought forward quickly to the active front. Vatutin's orders eventually would alleviate the lack of supply and support units at the front.

Up to this point in the war, medical/ambulance trains had the second lowest priority on the railways, just above secondary commodities. Finally, the Central Committee NCO (executive leadership of the Communist Party) acted to resolve this issue. Loaded medical/ambulance trains were assigned first priority with empty ones having second priority on the rails. [38] This single act most likely saved countless lives of sick and wounded Red Army troops.

The German 4th Panzer Army approved an early draft plan for its attack against the Kursk Bulge from the south. Both this plan and the one presented by Army Detachment Kempf were in general agreement.

4 April 1943

The Military Council of the 7th Guards Army issued to the troops Order #NQ 00147 which established basic principles of a strong defense and a set of measures for deployment of engineering works. With this order, the Army started to dig in and to fortify its positions in an organized manner. Note that on this date, the command staff of the Voronezh Front, and even within the 64th Army itself, were using 64th Army and 7th Guards Army designations interchangeably even though the change was not yet official.

On or about this date, the 101 Separate Signal Battalion of the 72 GRD finally arrived at Volokonovka Station after leaving Stalingrad in mid-March. The Luftwaffe continued to impede Russian rail operations, as related by Ivan Drozhcha, a signalman within the signal battalion. He stated that,

> attacks of enemy aviation frequently were accomplished along the road [railroad;] there were victims and wounded. At the beginning of April our train arrived at the Volokonovka Station and they [the train crew] approached the unloading [location] cautiously casting looks into the sky. We worked on unloading of special-purpose vehicles, radio equipment, coils with the wires, [and] telephone sets. [39]

So, in early April the Luftwaffe was indeed still active far behind the front, attacking rail lines, rail yards and individual trains.

At this time, the Luftwaffe also was involved in activities related to intelligence gathering. Photo reconnaissance units received orders to carry out numerous missions over the Kursk Bulge area. As a result, about 7,200 square miles of territory were photographed at a scale of 1/25,000. These images greatly helped the Germans plan their attack in the Kursk area. [40]

5 April 1943

The 64th Army's reported combat strength stood at 55,854 men and women. [41] Due to recent transfers out of the Army, the 64th was back to its core units that had arrived from Stalingrad. [42]

On or about this date, the 81 GRD started to arrive at the front. This was the final rifle division of the 64th Army to move into a forward position. The 81 GRD took its place in the second defensive line at the far-right flank of the Army, backing up the 73 GRD and the 213 RD in the front line. Once again, this movement took several days. Also, the last combat elements of the 72 GRD arrived at the front after a 65-mile march from Valuiki. Note that numerous sub-formations of all the rifle divisions continued to arrive at the front for weeks to come. The divisions were in place at the front, but they were not yet fully operational as they still lacked key units.

At a meeting in Kharkov, Field Marshal von Manstein discussed with his commanders the three separate but related operations planned by Hitler for the summer: Operation Citadel, Operation Hawk and Operation Panther. Eventually von Manstein "told his subordinate commanders that the final choice would probably be Citadel." [43] Nevertheless, von Manstein prepared his assault forces for any attack at the end of April or by mid-May. He knew his armies needed to strike first before the Russians had completed their own reorganization.

6 April 1943

The 167 RD and 183 RD that had been transferred to the 64th Army no longer were listed on strength tables as being assigned to the 64th Army. This implies the troop exchange at the front was completed. The 183 RD then was reassigned back to the 69th A and it withdrew to a new position to the rear of the 64th Army. The 167 RD most likely was assigned to the Voronezh Front reserve at first and later was transferred to the 38th A. Also, the 6 GCC was no longer listed with the 64th Army. The 6 GCC withdrew into *Stavka* reserve, and in June 1943 was sent north to the Western Front. Needless to say, the 64th Army total ration strength fell as these units were reassigned, resulting in an estimated ration strength of 55,000 men and women. [44]

Additionally, the 1181 AA Regiment from the 5 AA Division, the first of many new units, appeared on the Army's strength report.

Stavka issued a directive creating a reserve group in the Kursk area called the Steppe Military District. This group eventually would be renamed as the Steppe Front.

Additionally, the Voronezh Front issued order 003/op "For the purpose of raising troop's training to a higher level." The order commanded that "exercise in combat and political training are to be organized with the rank and file in all of the front's units and formations." [45]

7 April 1943

A historical planning map from this date displayed battle schemes for the 64th Army showing how it would counterattack against potential enemy attacks at the flanks or along boundary lines between armies. This map showed attacks toward the north to assist the 21st A and to the south to help the 3rd TA. The map also revealed all seven rifle divisions of the 64th Army in position along the front, but these divisions most likely did not fully occupy these positions at this time. Additionally, it indicated several tank units in position to support these possible attacks. Staff officers routinely produced plans like these to counter possible enemy attacks as the overall situation changed at the front.

8 April 1943

Per a request from Stalin, General Zhukov, Deputy Commander in Chief of the General Staff and member of *Stavka*, sent Stalin his assessment on what should be done about the Kursk Bulge and the summer season. Zhukov advocated staying on the defensive at Kursk and only launching powerful counterattacks after German forces had been weakened attacking entrenched Russian forces. General Aleksandr Vasilevsky, the Commander in Chief of the General Staff, supported Zhukov's view, but Stalin was not sure about staying on the defensive and letting the Germans attack at will. Stalin really wanted to strike first to disrupt German plans. Undecided, Stalin finally chose to ask for the opinion of his front commanders, General Vatutin, the Voronezh Front commander, and General Rokossovsky, the

Central Front commander. Stalin gave them until 12 April to present their assessment of the situation. [46] This request appears to have been sent to them by order or communication #11990 from Stalin/*Stavka*. [47]

On the other side of the front line, General Otto M. W. Model, commander of the German 9th Army (which was designated to be the northern arm of the Kursk attack) sent Army Group Center a warning message about his army. Model stated that his army was too weak to attack any time soon. [48] The 9th Army had played a large role in defeating the Russian Operation Mars offensive (November-December 1942) against Army Group Center. In doing so, the 9th Army suffered heavy casualties and had not yet recovered its strength. The weakness of the 9th Army was known to the OKH and AG Center, but Model was being proactive in pointing this out in the hope the offensive would be canceled. Ultimately, Model failed to convince Hitler to give up the Kursk attack.

9 April 1943

With all its rifle divisions occupying position along the front line, the 64th Army now held the area it would defend when the Germans eventually launched their attack. (See Map #5.) For the 64th Army, the first defensive line covered 32 miles along the front. The second defensive line extended for about 35 miles. The third defensive line ranged for nearly 75 miles and was manned mostly by units of the 69th A. Behind these positions, the Voronezh Front developed even more defensive lines. Needless to say, within the first three army-level defense lines, all available troops were hard at work digging in and improving their new positions. Behind the 64th Army, thousands of civilians also dug trenches and antitank ditches and worked on improving the roads. Note: Map #5 provides a general layout of the three main army-level defense lines, but on the ground, positions and emplacements are not that precise. Fortified towns and villages, backup positions, fake earthworks, real and fake minefields, and various other emplacements were scattered throughout the area. The Red Army had plenty of time to prepare for the coming battle.

10 April 1943

On this day the 64th Army's reported combat strength was 59,383 men.

The Voronezh Front assigned the 1438 SU Regiment to the 64th Army. [49] This was a new unit equipped with a new weapon - a powerful 122 mm Howitzer mounted on a T-34 tank chassis. These SU-122s, as they were called, were similar to the German *Sturmgeschutz* (assault gun). With the addition of this and other units, the 64th Army received the best and newest equipment available. However, the 1438 SU Regiment would take some time to actually arrive in the 64th Army' zone of operations. (See Appendix 6.)

General Rokossovsky and his Central Front staff submitted their views to Stalin about the Kursk Bulge. Their view was mostly in line with the one presented by Zhukov: the Germans would attack at Kursk with a powerful force from the north and south, and the Central Front should be prepared to counter their attacks. They even went on to advocate an early large-scale attack against the Orel area to disrupt German plans. [50] (See Map #3 for Orel area layout.)

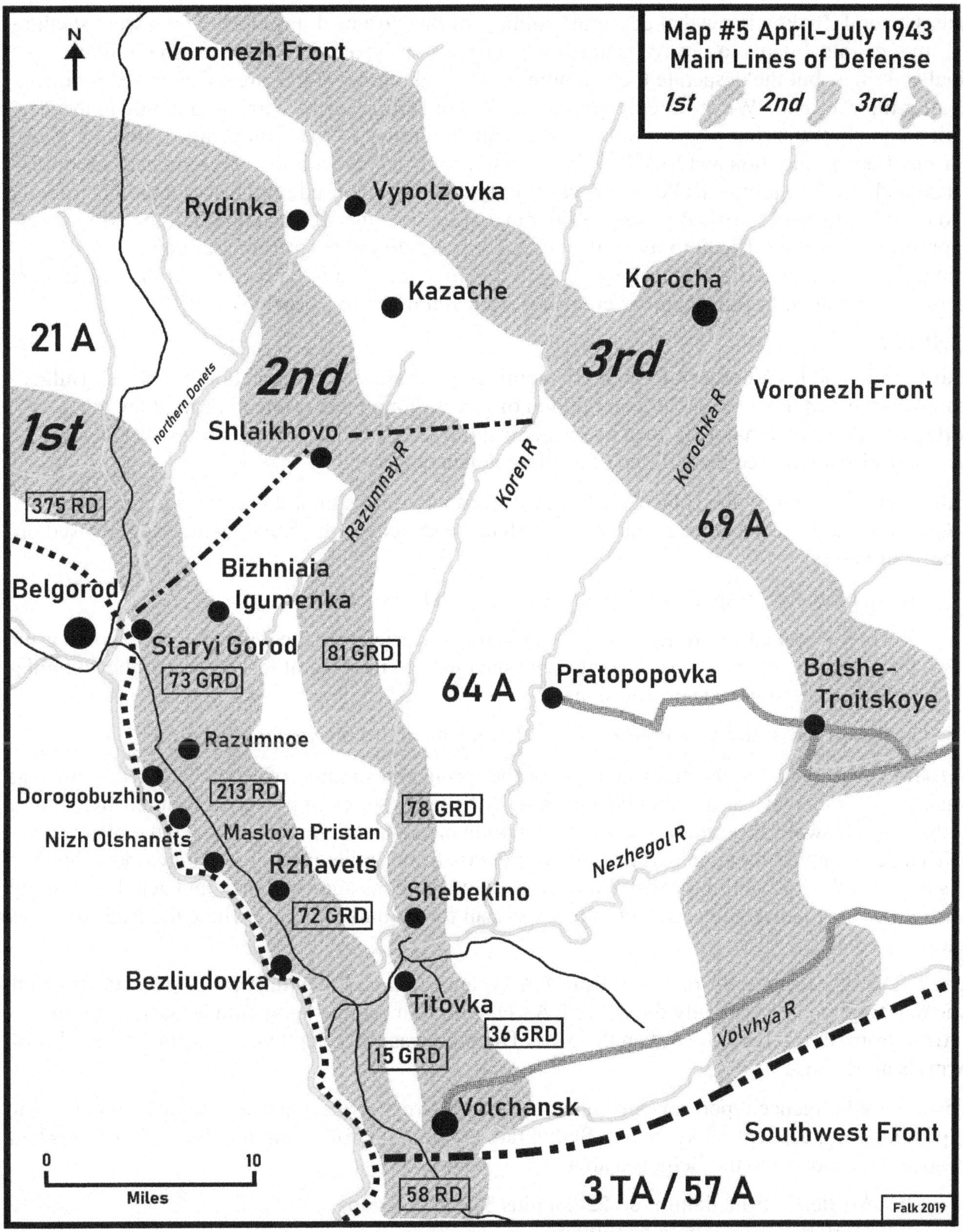

11 April 1943

During the day, General Zhukov conducted a hurried inspection of the frontline units of the Voronezh Front. He made several suggestions, and then he flew to Moscow. Stalin had ordered him to be back by 11 April to discuss plans for the summer offensive season. He also needed to attend the *Stavka* meeting in Moscow on 12 April.

Whether General Zhukov knew that desperate soldiers of the Voronezh Front even then were stealing food from local inhabitants to survive is unknown. The army was quick to crack down on these criminal activities, but the desperate thefts continued throughout the Kursk area. General Shumilov wrote after the war that, "We are deeply grateful to the population...who came to our aide in these difficult days...In the spring *rasputitsa*, in the difficult conditions of roads made impassible by mud, they brought up ammunition and food." [51] The overall problem was with the rear logistical services, both road and rail. The armies lacked their rear supply units, which used both trucks and horse-drawn carts to transfer supplies to rifle divisions. With most of these units still at Stalingrad, the army-level transport effort found it difficult to move the needed supplies to the front from rear depots. Additionally, the rail network was lacking capacity to move the required amounts of rations and other stores to the front area. Improvements were being made but much too slowly.

12 April 1943

General Vatutin and his Voronezh Front staff submitted their views to Stalin about the Kursk Bulge. After making note of a powerful German buildup of forces along the Voronezh Front, they too supported the idea that a major German attack against the Kursk Bulge was pending within a few weeks. They also advocated the buildup of a strong defensive force. [52]

Later that same evening, Stalin, Zhukov, and other *Stavka* members finally decided against a pre-emptive attack and instead adopted a deliberate defensive stance for the Kursk area. They focused on three basic premises:

1. An offensive to spoil the German attack would be pointless.

2. The Central and Voronezh Fronts would remain strictly on the defensive, and would grind down the enemy in defensive battles, but the moment the assault lost its power they would go over to a decisive counteroffensive.

3. The Bryansk and Western Fronts would prepare an offensive aimed at Orel. [53]

So, *Stavka* finally agreed on the main strategy for the spring and summer of 1943. The Red Army was to remain on the defensive at first and then eventually launch a series of powerful counteroffensives all along the front. Now all that *Stavka* and the front commanders had to do was to make the plan work. Note: This decision was based in part upon secret reports from Soviet intelligence networks in the west. By this time *Stavka* was well informed about German attack plans against Kursk and acted accordingly. They would use the Kursk Bulge as bait to lure German forces into attacking where the Red Army was strongest. [54]

Also on 12 April, a second AA regiment from 5 AAD appeared on the 64th Army's strength report; this was the 670 Regiment. Eventually the entire 5 AAD, with four regiments, would be assigned to the 64th Army from *Stavka* reserves. When the 5 AA Division was fully deployed, it would greatly boost the Army's air defense.

Moreover, an intelligence report revealed how many 64th Army artillery pieces and tanks remained in firing positions and concentration areas at Stalingrad. Overall, the following numbers of weapons had not yet been transported to the Belgorod area:

263 Artillery - from 45mm to 152mm plus 37mm AA guns.
256 Mortars – from 82mm to 120mm.
124 Tanks – 10 - KV 1, 99 - T 34, 15 - T 70. [55]

As can be seen, a significant amount of the Army's firepower had not yet arrived at the distant front line. The Soviet defenders were fortunate that the Germans continued to delay their planned spring attack.

13 April 1943

At this time General Shumilov and his staff were very busy. The 64th Army had just occupied frontline positions, so they had to keep an eye on enemy activities. They had to decide upon the placement of combat units and supporting combat elements of the army. They also needed to deal with the problem of minimal logistical arrangements. Additionally, all the armies of the Voronezh Front needed to build or to improve roads in their areas, as well as to establish proper supply depots for all types of provisions and materials. Furthermore, they needed medical facilities for both humans and horses. Finally, a priority was given to establishing a robust wire and radio communications network throughout their individual areas of operations as well as to all neighboring armies and higher headquarters. Time was short because one never knew when the enemy might strike.

14 April 1943

Russian armies all along the front in the Kursk area dug in and tried to improve their fighting positions. *Stavka*, in turn, attempted to deliver the maximum number of men, ammunition, equipment and general supplies to the forces within the Kursk Bulge. The majority of these provisions were moving via rail, so during this same time frame, both the Luftwaffe and VVS actively attacked the enemy's rear areas with emphasis on rail line interdiction and logistical attacks. To support this logistical buildup within the Kursk Bulge and the Voronezh Front in particular, the Russians began the construction of a new 58-mile rail line between Stary Oskol and Rzhava. This urgently needed connector rail line finally was opened on 17 July and could carry 12 pairs of trains per day. [56] This opening came after the defensive battle for Kursk had been decided, but regardless of the date completed, this new rail line relieved some of the congestion within the Kursk area and supported the Red Army efforts to go over to the offensive during the summer.

15 April 1943

The 64th Army listed a combat strength of 54,468 men.

Adolf Hitler decided the Kursk attack would be the first offensive launched that campaigning season; therefore, he issued Operations Order #6 for Operation Citadel, the German code name for its proposed large spring offensive against the Kursk Bulge. Hitler wanted this attack launched as quickly as possible with an estimated start date no earlier than 3 May 1943. With this decision, the German forces facing the Kursk Bulge began an almost frantic preparation effort. They did not have much time to get ready for such a major attack.

Once again General Model informed Army Group Center, his higher headquarters, that his 9th Army was not ready to take part in the coming offensive.

16 April 1943

On this day the 64th Army was designated as the 7th Guards Army (Guards for "guardians of the Soviet Union"). This honorific title was awarded to the entire 64th Army for its distinguished service during the fighting in and around Stalingrad. The rifle divisions within the 64th Army had been awarded Guards status earlier on 1 March 1943. In general, the Guards distinction was originally initiated by the order of People's Commissar for Defense of USSR Order #303 on 18 September 1941.

All Guards units or formations were considered to have elite status within the military. They received more pay; privates and junior NCOs received double pay, while sergeants and officers received 50% more pay. Normally, Guards units were given better equipment, personnel, and rations. [57] They also had the honor of leading major attacks and were expected to achieve their objectives. A Distinctive Guards badge was later introduced into use on 21 May 1943 and was proudly displayed by soldiers on their uniforms or painted on military equipment. The Guards title only was awarded to 11 combined arms infantry armies and 6 tank armies during World War II. [58] However, starting in 1943, a Guard's

Infantry Army was intended to have two rifle corps and an assigned or subordinate tank corps. At this time, the 7th Guards Army contained none of these formations. [59]

The Guards title also was awarded to several hundred smaller combat formations or separate units during the war. The significance of receiving this title on boosting morale cannot be understated. For instance, infantryman/mortar crewmember Mansur Abdulin related,

> We learned that our 293 RD was to become the 66 GRD while still near Stalingrad: now we were impatient to receive the Guards badge. We wanted to get it as soon as possible, so we could go around looking smart. In June, after we received the banner [Guards Flag], the Guards badges were at last sent to our regiment. Finally, each of us had one! In those days soldiers rarely had medals, let alone orders. For many Guardsmen the badge was the only decoration on their blouse, and it often seemed to a soldier that he did not need anything else.... Men started painting Guards signs on tanks, guns, cars, towing trucks and even on two-wheel carts. [60]

Of course, the awarding of Guards status to the Army was good for troop morale, and the story would have been widely promoted by the Army paper, "For the Motherland".

Editorial Staff for the 64th/7th Guards Army Paper "For the Motherland".

Of special note, Lieutenant General M.S. Shumilov, who had taken over command of the 64th Army in July 1942, continued to command the new 7th Guards Army throughout the rest of the war. He was eventually promoted to Colonel General in October 1943 for his outstanding service. (See Appendix 7.)

The Soviet intelligence network in London was hard at work and provided further information about German attack plans against the Kursk Bulge, including the German code name of the attack, *Unternehmen Zitadelle* (Operation Citadel). [61]

The 73 GRD launched a final attack against the small German Mikhailovka bridgehead on the east side of the northern Donets, directly across the river from Belgorod. This attack failed to drive the Germans back over the river. [62] The Germans would hold this bridgehead throughout the battle for Kursk but were never able to use it offensively.

17 April 1943

In a somewhat strange move, Stalin ordered General Zhukov and several other officers from the high command to fly 450 miles south of Kursk to the North Caucasus Front. They were sent to assist in the elimination of the German 17th Army bridgehead on the Taman Peninsula, which was protecting the Crimea from Russian invasion. After several weeks of effort, they were unable to destroy the German bridgehead, so Zhukov and the others returned to Moscow on 13 May.

For a third time, General Model informed Army Group Center HQ that his 9th Army lacked the forces needed to successfully attack at Kursk by the required date. "Army Group Center paid no attention" to his appeals. [63]

On this day the 7th Guards Army's combat strength stood at 63,222 men and women. Women did serve with combat units throughout the Red Army during the war, in vital roles such as surgeons, combat medics, snipers, pilots and other dangerous jobs. Approximately 200,000 women were decorated during the war, with 89 of them earning the Soviet Union's highest award, the Hero of the Soviet Union. [64] An unknown number of these heroic women served with the 7th Guards Army, but on "average 2,000 to 3,000 women soldiers [served] per field army, with around 20,000 women per front." [65] Of special note, at the start of the war the Soviet Union possessed 140,796 medical doctors with 96,000 (68%) of them being women. Of course, not all these doctors were military doctors, which lead to "an acute shortage of [military] medical doctors and other medical personnel" during the war.[66]

Red Army female snipers take aim at the enemy.

18 April 1943

The Russian buildup in and around the Kursk Bulge continued unabated. By this date, the 69th A was deployed with its five rifle divisions along the third defensive line for its entire 75-mile length. This position not only provided a defense in depth for the Voronezh Front, but they also effectively covered and protected the boundary line between the new 6th Guards Army (former 21st Army) and the 7th Guards Army. From this position the 69th A was able to support either or both armies with fresh units as needed. The Red Army had learned the hard way to pay close attention to these "who is in control" areas between armies. The enemy had an annoying habit of attacking these vulnerable spots where effective coverage of the front line might be lacking and command and control issues might develop during battle.

Fearing that the starting date for Operation Citadel might slip, von Manstein sent a letter to General Kurt Zeitzler, Chief of Staff OKH, and through him to Hitler. Von Manstein restated that Citadel should be launched as soon as possible because he knew the Russians were reinforcing their armies faster than he was, and every day of delay allowed the enemy to fortify the Kursk Bulge further.

19 April 1943

The 7th Guards Army combat strength stood at 64,411 men.

The new 7th Guards Army retained most but not all of the previous medical units that were assigned to the 64th Army at Stalingrad. A few new units were added at this new location. Despite some confusion within the historical record, the known assigned medical units, on or about this date, are listed below:

One Field Army Evacuation Point – PEP #24, and the HQ for the Field Evacuation Point – GOPEP also numbered #24 (used to transfer wounded to the deep rear).
One Evacuation Hospital – EG #2670.
Four Mobile Surgical Hospitals – HPPG #869, 2204, 2208, 2409.
Two Mobile Field Hospitals – PPG #220, 4344.
Two Infectious Hospitals – IPPG #2241, 2242.
Two Hospitals for the Lightly Wounded – GLR #3247 and 3255.
One Separate Ambulance Company (Auto) – #97, equipped with 48 GAZ-55 ambulances each capable of carrying 10 sitting or 4 stretchers (litter) wounded.
One Ambulance Company (Horse-Drawn) – #251.
One Disinfection Company – #111.
Various medical support facilities like medical depots, laundry detachments, bathhouses, and other sanitation services.

Also, to care for the Army's many horses and other livestock, the following veterinary units were assigned:

Two Army Veterinary Hospitals – AVL #958, 960.
One Evacuation Veterinary Hospital – #959.
One Mobile Field Veterinary Laboratory – #16.

Medical support for any wounded soldiers started in each rifle division, right at the front, with a Field Sanitary Post. Wounded troops then were move in succession to a Battalion Medical Station, to the Regimental Medical Point, to the Medical Battalion, and finally to the nearest Field Hospital. In theory, the badly wounded then were moved to the Field Evacuation Point and from there to a front level hospital, if there was room and if transportation were available. Others could be moved directly to the deep rear for long-term recovery.

The men and women of the 7th Guards Army should have received outstanding medical support from these experienced medical personnel. The lessons learned from their work at Stalingrad would further improve their care and handling of wounded soldiers on this new battlefield. However, only time would tell if these improvements in skill and training would be sufficient to overcome the ever-present problem of poorly equipped and under supplied medical units. Note: These assigned medical units would change once again before the battle for Kursk began.

20 April 1943

By this time, the final combat formations of the 7th Guards Army should have arrived from Stalingrad to their detraining locations. [67] The remaining support units would take until mid-May to arrive. Also arriving under 7th Guards Army control were the 26 and 34 Separate Armored Train Battalions (ATB), with two armored trains each. The command staff certainly welcomed these armored trains because of the important support that similar units at Stalingrad gave to the Army.

The 26 ATB was assigned to the Army on 10 April and contained trains #683 *Alexander Nevsky* and #707 *Alexander Suvorov*. The 26 ATB stayed with the Army only 15 days before leaving on 25 April for rearmament in Moscow. The 34 ATB contained trains #667 *Death to Fascism* and #712 *Krasnoufimsky* (Railroad Worker). The 34 ATB arrived in early April and train #712 stayed with the Army until it needed repairs after being bombed on 29 May.

To address the question of the understrength armies of the Voronezh Front, General Vatutin looked to locally raised replacements as a partial answer when he assigned different civilian districts in the surrounding area to his armies. The 7th Guards Army was given the Shebekino and Prispeshenskoe Districts (not displayed on a map), "where they were supposed to conduct a house-to-house mobilization of the male population." This was forced mobilization of any and all males who could be pressed into service. The plan was to add some 84,000 conscripts to Voronezh Front units; 39,000 from march replacements (newly trained men arriving from the rear); 11,000 men from local hospital returns; and 34,000 men from the newly liberated territory. [68] Integrating all these raw or partially trained men into combat formations would take time and effort. Every day of delay in the German attack was used to rebuild the depleted armies.

21 April 1943

After *Stavka* completed the preliminary planning for the defensive phase of the coming battle, they issued several directives implementing this plan for the summer season. These directives were sent to all front commanders that were involved: Voronezh, Central, Southwestern, Bryansk and the soon-to-be formed Steppe Front. These early directives covered two basic issues:

1. Clear the entire civilian population from a zone starting at the front and extending 25 Km (15 miles) to the rear.
2. Start construction of two to three military defensive lines immediately, one behind each other in the frontal [combat] zone. Prepare all towns and population centers for defense. [69]

This was truly the start of the battle for the Kursk Bulge. Hitler and Stalin had both made their decisions, but which side would complete their preparations first, the German attack plan or the Russian defensive one?

Also, on this day the Voronezh Front commanders sent a "Top Secret: Special Importance" report directly to Stalin. This multi-page report described the Front's commanders' perception of enemy forces arrayed against them, the status of the armies under their command, and the Front's future plans.

The report went on to say that all first echelon armies (38th A, 40th A, 6th GA, 7th GA) would be refitted by 25 April. The report outlined that for future offensive operations the "7th Guards Army will be in reserve in the *Khar'kov* region." This implies that if the 7th Guards Army were not attacked by the German offensive, then it would be held in reserve as a counterattack force.

Four days later *Stavka* sent an order stating, "The commander of the Voronezh Front will place the 7th Guards Army in front reserve by 20 May." This order indicated that *Stavka* believed (at this time) the expected German attack would be launched directly north toward Kursk only from the section of the front west of Belgorod, and not also to the east where the 7th Guards Army was positioned. [70] This would be the Belgorod-Oboyan-Kursk direction (See Map #3 for Kursk Bulge layout.) *Stavka* and the Voronezh Front would further narrow and define the most likely German attack zones in the future.

22 April 1943

In addition to intensive planning efforts by both high commands for the ground battle, the battle in the air continued. The VVS used its superior numbers to "launch preemptive air strikes wherever possible into German rear areas to create confusion and [to] slow the flow of reinforcements to the front…" For example, "on April 22, VVS bombers attacked Luftwaffe reconnaissance bases at Orsha and Orsha South [Orel combat zone], inflicting severe losses and leaving the enemy with only one strategic reconnaissance air unit operational in the Kursk area." [71] The Luftwaffe in turn continued with attacks mostly against logistical targets while trying to preserve its strength for the main battle to come.

23 April 1943

Simultaneously with the defensive planning, *Stavka* created a new Steppe Military District (soon to be called the Steppe Front) behind the entire Kursk Bulge area. General I.S. Konev commanded this Steppe Military District, and he was tasked with amassing a huge strategic reserve that would be used both during the defensive phase and then the following offensive phase of the overall Kursk battle. [72]

On the German side of the line, both the 4th Panzer Army and Army Detachment Kempf published their schedules for deploying their attacking forces into position along the south side of the Kursk Bulge.

During this day, when visiting the HQ of the 72 GRD, in a powerfully moving ceremony, General Shumilov personally presented the Division its Guard's Banner. Other awards like the Order of Lenin or the Order of the Red Banner were awarded to individual soldiers and separate units of the Division. Everyone was very pleased with these awards for their actions at Stalingrad. [73] General Shumilov, said, "I believe that under this glorious Banner you will perform new heroic deeds for the glory of our great Motherland." [74]

The 7th Guards Army ration strength now stood at 62,303 men and women. [75]

24 April 1943

On or about this date, within Army Group Center, General Model, commander of the German 9th Army, stated for the fourth time that his army would not be ready for the Kursk attack on 3 May, but perhaps by 15 May. Hitler and the OKH remained undecided about the attack date.

At this same time, the 57th Army was reformed in the Southwest Front's area of operation. After being disbanded at Stalingrad in February 1943, this new 57th Army was reformed from the remnants of the 3rd TA. This new army soon began operating along the front-line south of the 7th Guards Army and was in effect taking over the section of the front previously held by the dissolved 3rd TA. This process would not be complete until May 1943.The Germans were giving *Stavka* plenty of time to reorganize and to regroup the Red Army in and around Kursk.

25 April 1943

The 7th Guards Army listed its combat strength as 62,592 men.

German Army Group South sent its assessment to OKW [Wehrmacht/Armed Forces High Command] for the Citadel offensive. "We must expect him [the enemy] to be completely ready defensively and capable of strong counteroffensives. But his offensive capabilities will probably not forestall the launching of Citadel." Army Group Center also gave a detailed report on the enemy situation as well as its own actions in support of both offensive and defensive plans. [76]

The army boundary between the 4th Panzer Army and Army Detachment Kempf was fixed in place, so commanders at different levels could better coordinate their efforts and control their areas of responsibility.

Within the 7th Guards Army, the flow of information was much more limited. For example, a Soviet solider stated: A "modern reader who is interested in the history of the Great Patriotic War certainly knows about the events of spring and summer of 1943 much more than we, officers and soldiers [who] were direct participants in the events." The writer goes on to say:

> Our horizons were limited, figuratively speaking, to the limits of the two-mile field map, on which the layout of the units and divisions of the division was plotted. The boundaries of the division's operational range were marked on this map by red lines. Behind the left was placed the division number, which was the left neighbor, for the right was the division number, which was the right neighbor. And that's all. No unnecessary details… The details were known only to the divisional commanders and, possibly, to the commanders of the regiments. But only those that had to do with the nearest neighbors, to the band of the army. The rest was hidden even from the commanders of the divisions. Therefore, the officers and soldiers of the 72nd Guards, although they heard the expression "Kursk Bulge," although they knew that they were on the southern tip of it, could not even imagine that the main military events of the summer campaign of the forty-third year [1943] would unfold precisely here. [77]

This type of remark is the same in all armies; those on the front line know the least and risk the most.

26 April 1943

General Model sent a written report to Hitler warning him of "the growing strength of the Soviet defensives facing him, now more than 20 Km in depth." [78] This document must have gotten Hitler's attention because the next day General Model flew to southern Germany to meet with Hitler.

OKH agreed to postpone Operation Citadel by two days, to 5 May, most likely due to wet weather on the battlefield.

27 April 1943

General Model flew to Berchtesgaden to meet with Hitler and to express his views and reservations about the proposed attack.

Within the Kursk Bulge, Russian forces continued to entrench and to strengthen their positions.

28 April 1943

The 7th Guards Army listed a combat strength of 72,623 men.

During his meeting with Hitler, General Model voiced his concerns about the ability of his army to carry out the northern attack against the Kursk Bulge in a timely manner. After reviewing air photos of the attack zone with Hitler, he reiterated his reluctance to launch the attack at all. Hitler listened to Model's doubts but made no decision about Citadel.

29 April 1943

At this stage of Soviet and German preparations for upcoming battle, we have contrasting views from each high command as well as from both Stalin and Hitler. Stalin at first wanted to launch a preemptive attack but was persuaded to choose a defensive stance by his high command. *Stavka* in turn was unanimous in their belief that the Germans would attack the Kursk Bulge in force. They felt the best strategy to defeat such an attack was to use a deeply echeloned defensive position to slow down and to degrade the enemy's forces before launching a strategic counteroffensive. Hitler in turn always wanted to attack and only briefly considered a defensive stance to begin the summer season. At the same time, the OKH was at first supportive of an early offensive, but as time passed and the odds of success diminished with each passing week, their views changed to one of opposition. Adding to this mix, German field commanders forcibly voiced their opinions against a significantly delayed attack. Ironically, in the end, both sides were kept waiting for Adolf Hitler to make his final pronouncement on Operation Citadel.

30 April 1943

By the end of the month von Manstein, Army Group South, 4th Panzer Army and Army Detachment Kempf were almost ready to launch the attack against the southern side of the Kursk Bulge. They were certain to be ready by the new 5 May deadline set by Hitler, but what about the 9th Army and the northern arm of the offensive? Would they be ready in time?

The German OKH issued another postponement order about Operation Citadel, due to heavy rains in the Kursk area. [79] This delay moved the starting date of the offensive to 9 May. But, even after this change, General Model stated he could not meet this deadline; the northern arm of the offensive was not ready and would not be for some time. More postponements would follow in the coming weeks.

The 7th Guards Army issued Intelligence Document #225 about losses suffered during the month of April. The major combat units lost 683 killed; 1,147 wounded; and 114 missing, for a total of 1,944 casualties. Within this total, the artillery lost 14 killed, 35 wounded, and 2 missing. Tank units lost 6 killed, 6 wounded, and zero missing. This was during a quiet time with no major combat operations taking place. These totals reflected over 64 casualties per day. [80]

The 7th Guards Army listed a combat strength of 69,513 men. This decrease of 3,110 troops from the 28 April report was most likely due to normal losses and transfers.

1 May 1943	
7th Guards Army	
Assigned – Voronezh Front	
Lieutenant General M.S. Shumilov Commanding	
Rifle, Airborne and Cavalry Units	24 Guards Rifle Corps (15, 36, 72 GRDs) 25 Guards Rifle Corps (73, 81 GRDs, 213 RD) 78 GRD (Penal units most likely still were attached to the Army but documentation is lacking.)
Artillery Units	8 Separate Destroyer Brigade 109, 161 Guards Army Artillery Regiments 114, 115 Guards Antitank Artillery Regiments 1, 2 Antitank Artillery Regiments 8 Antitank Artillery Brigade 1, 2, 3, 4, 5 AT Rifle Battalions 5 Antiaircraft Artillery Division (670, 743, 1119, 1181 Antiaircraft Artillery Regiments) 162 Guards AA Regiment 662 Separate Aerial Observation/Air Raid Warning Company
Armored and Mechanized Units	27 Guards Tank Brigade 1438 SU Regiment 148, 167, 262 Separate Tank Regiments 34 Separate Armored Train Battalion
Engineer Units	60 Engineer - Sapper Brigade 175, 329 - Separate Engineer Battalions
Reported Combat Strength Estimated Ration Strength	69,335 men [81] 76,335 men and women (including an estimated 7,000 support personnel)

(BSSA 43, BSA 43, TsAMO Archive)

1 May 1943

The 7th Guards Army listed a combat strength of 69,335 men. Note: All the following reported strength reports come from TsAMO historical documents.

New combat units and subunits transferred into or out of the Army as the Army command staff and Front HQ made adjustments to the force structure. These alterations reflected changes in the defensive plan and the availability of new formations and equipment.

Due to the break in combat activities, intensive training within the 7th Guards Army began at all levels. Officers and staffs worked on troop control and writing combat documents, and they followed up with realistic training sessions and war games to demonstrate their new skills. Finally, instructors gave a series of lectures on different themes like "supporting force regrouping" or "organizing the antitank defense."

The emphasis was placed on organizational skills, defensive preparations, maneuvering of forces, and the control of all types of fire. [82] The 7th Guards Army had extra time to prepare, and General Shumilov put it to good use.

2 May 1943

The air war in and around the Kursk Bulge intensified as both the VVS and the Luftwaffe conducted various operations to disrupt their enemy's activities. The VVS concentrated on rail and airfield attacks, while the Luftwaffe focused mostly on the disruption of Russian logistical efforts by road and rail. The VVS also introduced free-roaming hunter tactics behind German lines. Small groups of aircraft (2-4) hunted for and attacked trains, motorized columns and infantry within their assigned areas. "During the two months prior to the Battle of Kursk, these free hunters flew 2,000 sorties." [83] These and other types of attacks kept the pressure on the German logistical system and forced AA units and the Luftwaffe to respond to attacks all over the rear area.

Within the 7th Guards Army, the entire 5 Antiaircraft Artillery Division at last was deployed along with another separate AA regiment. The five AA regiments were the 670, 743, 1119, 1181 and the independent regiment 162. The Army needed these AA units because of the increase in German air activity.

Stavka issued the first of three alerts during May warning of an imminent German attack. Hitler's constant changing of the attack date kept Stalin and his entire high command guessing.

3 May 1943

Field Marshal Model once again met with Hitler about the Citadel attack and "present[ed] Hitler with an extremely gloomy report of his army's chances" of success; Model pushed for the offensive to be postponed for at least a month so he could reinforce his army. Despite opposition from all the other senior commanders who were present at this meeting, "Reluctantly Hitler ordered the offensive postponed until mid-June." [84] Now the Red Army would have even more time to prepare for this decisive summer battle.

Starting this night and for the next nine nights in a row, the Soviet ADD (Long-Range Aviation) branch attacked German rail installations at Bryansk, with great effect. [85]

On this day the 7th Guards Army's ration strength stood at 70,026 men and women.

4 May 1943

During the nights of 4-5 May, the 73 GRD moved to the rear and occupied the second defense line as the 81 GRD moved up to take its place at the front along the northern Donets. This was a normal rotation of units which allowed the 73 GRD to safely rest and train its men behind the front. This adjustment also was made because the 73 GRD was, at this time, the weakest division in the Army with only 6,824 men. [86]

The VVS launched heavy night attacks against the rail junctions of Minsk and Orsha; both received heavy damage with the loss of many rail cars full of supplies. [87]

Back on the ground, Russian forces within the Kursk Bulge continued to improve their skills. David M. Glantz and Harold S. Orenstein, in their book, *The Battle for Kursk 1943 The Soviet General Staff Study,* p. 45, stated:

> While exploiting the temporary lull in combat operations, together with the defensive work needed to fortify their positions, the Voronezh and Central Front's forces conducted intensive training. The principal training task was to teach infantry, artillery, tank, and engineer troops to fight against German tanks, mainly the T-VI [Tiger] and against self-propelled "Ferdinand" guns to overcome their fear of tanks and to install into them an iron steadfastness in the defense.

Over the next several weeks, troops trained on specially designed fields, three to four miles behind the front, where infantry companies learned the use of all sorts of weapons like AT rifles, grenades and Molotov cocktails (hand thrown bottles of incendiary liquid) for combat against tanks. Trainers stressed different themes like repelling an enemy tank attack, conducting battalion withdrawals and occupying intermediate defensive positions and cut-off positions, destroying enemy infantry supporting tanks, launching counterattacks, and even conducting "combat in encirclement." All this was undertaken on a daily basis with units rotating into and out of these training areas.[88] Veterans of earlier battles could not help but be impressed by this organized training effort, but the raw recruits were the ones who gained the most. With this intensive training, maybe more of these new recruits would survive the battle and live to fight another day.

In Germany Hitler oversaw yet another high-level conference where he and his generals discussed Operation Citadel. After four long hours, Hitler dismissed his generals and said he would consider the timing of the attack once again. [89] Following this conference, Hitler made no announcement concerning the timing of the offensive. His indecisiveness would continue for months which would allow the Soviets to amass an enormous superiority in defensive and offensive forces in the Kursk area.

5 May 1943

The VVS and ADD units continued night attacks against German rear area facilities while the Luftwaffe conducted its own air operations.

With the air war raging above them, ground troops with the help of conscripted civilians built new fortified positions all over the front. The 7th Guards Army continued to conscript new recruits from the surrounding territory, but they encountered problems with this effort. By this stage of the war there just were not that many able-bodied men left in the civilian population. Most of the ones that they found were suffering from starvation and/or were sick from various infectious diseases. The solution was prompt medical attention, which included giving them "high-caloric food [which] was required in order to bring the conscripts up to specified physical standards." [90] These new recruits also needed enough time to heal and to recover. Simultaneously the Army conducted a massive fortification effort, together with a vigorous training program of all troops and officers. All these efforts continued despite the fact that the Army itself suffered from a shortage of food, clothing and various types of weapons and equipment.

6 May 1943

The VVS launched heavy day attacks against Luftwaffe airfields causing substantial damage and the destruction of 22 German aircraft. The Germans were surprised by this attack and shot down only 21 Russian aircraft all day.

7 May 1943

On this day and then again on 8 May, the VVS launched heavy attacks against Luftwaffe airfields, but this time the Germans were better prepared. Some 23 German aircraft were shot down or put out of operation on the ground, but the VVS lost in total about 101 aircraft.

During the night, Luftwaffe bombers attacked the rail junction and marshalling yards at Kursk. They had correctly identified the Kursk rail hub as one of the key logistical areas for the Russian buildup throughout the Kursk area.

8 May 1943

ADD night attacks continued to pound German rail lines, disrupting both the transportation of supplies and the movement of troops.

Stalin issued another warning alert. "Stalin insisted that the entire region be brought to a complete state of readiness in anticipation of a German offensive by 10 May." [91] The Germans did not attack, so nothing came of this warning.

9 May 1943

Keeping to the 7th Guard's Army training routine, trainers sent artillery batteries to special ranges to practice direct firing against tanks. This in itself was unusual, as artillery pieces are indirect fire weapons. But the defenders of the Kursk Bulge were taking no chances; they planned to deploy every weapon against the most dangerous element of the enemy's formations, their dreaded Panzers. These firing ranges used moving mock-up tanks, and the batteries were forced to shift to different firing position several times. Soldiers also reviewed conventional artillery training with an emphasis on opening fire quickly, learning different signals, working as part of a combined arms group, and learning fire cooperation with AT guns. Also during May, "Red Army Main Artillery Directorate representatives carefully inspected the equipment of all artillery units. A significant number of the guns which required factory repair were replaced by serviceable ones." [92] The Red Army was using this extra training time to increase effectiveness at all levels. Everything was being reviewed and adjusted for the impending defensive battle, as well as for the subsequent offensive battles to come.

10 May 1943

On or about this date, two Guards Rifle Corps (GRC) HQs were added to the 7th Guards Army's order of battle; they were the 24 and 25 GRC. These Corps HQ had an immediate impact on the operations of the Army and basically consisted of the HQ staff with a few added support units. This change relieved the 7th Guards Army command staff from needing to directly control every combat formation under its command. Historically, back at Stalingrad, the 64th Army had been one of the first armies to gain one of the newly reintroduced corps HQ when it was assigned the 7 RC with three rifle brigades.

The low point for active Rifle Corps HQs within the Red Army was on 1 January 1942 when there were only 6 in the entire Army, but by 1 January 1943 the total had risen to 34, and by 1 July 1943 the number of Corps HQs had reached 82. [93] Now before the battle of Kursk, a revitalized Red Army assigned several rifle Corps HQs to each army, and each corps normally controlled two to three rifle divisions. The addition of these two new headquarters greatly improved the command-and-control arrangement within the Army. They also helped to make a Combined Arms Infantry Army, like the 7th Guards Army, more flexible in combat and a much more dangerous opponent.

The 7th Guards Army ration strength stood at 66,954 men and women. [94]

11 May 1943

The Soviet Long-Range Aviation (ADD) concluded their night attacks against the German rail installations at Bryansk. These attacks ended successfully, destroying some 1,200 tons of ammunition. Both sides would continue their air efforts for the rest of the month. [95]

Once again, concerning Operation Citadel, "Hitler decided not to decide, although in practical terms that meant a postponement." [96] He set a new date for 12 June, but this decision did not please army leadership in the high command or with the armies at the front. At some level, Hitler must have known or sensed the great gamble of the Kursk attack. This attack would be his last chance to decisively influence the outcome of the war in the east; therefore, he hesitated. Hitler was being influenced by his gut feelings instead of relying upon logic, logistics, or the cold hard military calculations that his generals were trained to use.

12 May 1943

Training efforts within the Red Army also extended to engineering units, especially ones that were to install land mines or to create obstructions on the battle field. Once again officers set up special training areas to allow engineers, combat engineers (Sappers) and tank units to train together.

Also, chemical defense units were active, issuing various types of equipment, such as gas masks, and manning chemical observation points. Their efforts included helping medical units construct decontamination points and train all troops in the use of gas masks. [97] Despite a slim chance that chemical weapons would be used by the enemy, nothing was left to chance.

13 May 1943

With the surrender of the last Axis troops in Tunisia, North Africa, which yielded 275,000 Axis POWs, Hitler once again postponed Operation Citadel. This Allied victory caused Hitler to shift the start date of the Kursk attack to the end of June in order to prepare German forces for the defense of Italy.

14 May 1943

For the first time in the Army's history, the 7th Guards had a sizeable armored force assigned to it, but not a full Tank Corps, as was originally proposed to provide support to Guard Combined Arms Armies. The available armored support included both tanks and self-propelled guns. Training efforts were extended to include these vital mobile units. Army staff officers, using war games, planned numerous tank force responses to various enemy attacks. Training for these planned attacks required tank commanders to conduct personal reconnaissance of the terrain including all march routes and start lines. These routes were marked for both day and night use, and then engineering elements improved the roads and bridges along the way. Local minefields, radio signal tables and so on were marked on maps, using one set of maps for each attack variant. Once commanders finalized these plans, different units trained in the field, with troops actually driving their tanks to staging areas to support the different attack or defensive efforts. Because of this rigorous training across the Army, each person knew his or her part of the plan, whether at the front or working behind the lines.

Command staffs up to corps level also conducted their own exercises to verify all aspects of the different plans and to modify them as needed. [98] Once again, the gift of extra time was used to produce extensive battle plans. Then these plans were reviewed and revised and finally put to use to train combat formations on the actual future battlefield. This prelude to the battle for Kursk was becoming one huge training exercise for the Red Army.

Additionally, the 38 Separate Armored Train Battalion, with two trains #730 *Krasnaya Zvezda* (A Red Star) and #754 *Pravda* (Truth), joined the 34 ATB within the Army. The 7th Guards now had four armored trains to support its left flank.

Russian infantry troops train with live ammunition.

15 May 1943

The 7th Guards Army ration strength stood at 72,848 men and women.

The 258 Separate AA Regiment appeared on strength reports. So now the Army had six full-strength AA regiments deployed within the Army's area of operations. These regiments deployed 16 – 76 or 85-mm AA guns and 81 – 20, 25, or 37 mm AA guns. Additionally, the Soviet army employed numerous infantry weapons in the anti-air role.

On or about this date, the newly reformed 57th A fully occupied the frontline positions south of the 7th Guards Army, which was the right flank of the Southwest Front. The 57th A contained some 60,000 men and 100 tanks and was destined to take an active part in the fighting during the Russian counteroffensive after the Red Army defeated the German Kursk attack.

Also of note, the Germans launched Operation *Zigeunerbaron* (Gypsy Baron- May-June). This operation was one of several powerful German anti-partisan operation intended to clear the rear area behind Army Group Center of Russian partisans.

Soviet partisan operations were a resistance movement made up of ex-soldiers and conscripted civilians who launched guerrilla attacks behind German lines. These irregular forces carried out attacks with growing strength as different bands of partisans coordinated their raids against logistical targets and received increased material support from Moscow. The Germans, in turn, killed any and all partisans on sight and ruthlessly hunted them down through swamps and woods where they gathered.

16 May 1943

To control the actions of the entire 7th Guards Army during a major battle required the army command staff to stay in communication with its own units, friendly armies on each flank, and higher headquarters. To accomplish this required extensive wire and radio links among various command and observation posts as well as communication with combat formations. Tight troop control was vital for a successful defensive operation. The 7th Guards Army constructed its main HQ underground along with one reserve command post and three or four main observation posts. This same type of arrangement also existed at the corps, division, and regimental levels. The construction and location of these vital control elements used the strongest bunkers and trenches as well as any natural cover and concealment. *Maskirovka* methods were used throughout the planning and placement of these command-and-control centers, including the use of the following guidelines:

- Place all facilities outside of and away from population centers.
- Disperse and carefully camouflage all facilities.
- Restrict motor vehicle movement or use within 1-2 miles from the command centers.
- Control entry to the area by passes and protect all facilities with local security forces. [99]

The use of numerous wire lines, radio transmitters (both fixed and mobile), and vehicles and motorcycles for couriers ensured a largely uninterrupted communication network. Having months to prepare meant the Red Army at all levels developed a sound command and control plan. The methodical implementation of that plan meant the command staffs were ready for whatever assault strategy the Germans used. This allowed them to maintain not only control of their forces but also control of themselves and HQ staff during the confusion of battle.

Russian soldiers man an observation post in a trench near Kursk.

Unfortunately, by this date, the 7th Guards Army had only placed 22.4% of its planned emplacement of 65,000 antitank mines and only 16.9% of its planned 84,000 anti-infantry mines. The spring thaw and disrupted deliveries of the allocated explosives slowed the work. Unit commanders took note of this situation so that by 5 June the mine emplacement totals had greatly increased with the 7th Guards Army laying 46.2% of its planned antitank mines and 35.7% of its anti-infantry mines. [100]

17 May 1943

The 7th Guards Army had a rear area organization which controlled the three main types of replacements flowing into the Army. First were new recruits arriving from trainings units in the rear. Second were men returning from the hospital, and third were civilian men found and conscripted within newly liberated territories. These civilian recruits were called "Booty Troops" as in spoils taken from an enemy. Replacements from the rear were formed into march companies or battalions and were sent to field replacement regiments at the front or army level. The army level replacement regiment also was used "as a processing center for returning wounded, conscripted civilians, and stragglers." From here the newly acquired manpower would flow to divisional level replacement battalions. "Each division incorporated a replacement battalion, which was also called a school battalion, for training newly arrived replacements and for holding recuperating sick and wounded." [101] Overall, this was an effective system, and it ensured a steady flow of men and women into the army's organization. But this system only worked if it was controlled properly, and that was not always done.

Speaking of women, as the war progressed some 2 million women were drafted into the Red Army to perform different jobs and also to take over non-combat tasks so men could fight at the front. A combined arms army like the 7th Guards most likely contained hundreds of women serving in various jobs like military police, communications personnel and within medical units. Women volunteers also were able to serve in combat positions as snipers, combat pilots, and in other hazardous duties like medical aids caring for the wounded right at the front. Those frontline positions were far more dangerous than work in hospitals which were located far enough to the rear to be clear of the day-to-day combat dangers of the front.

18 May 1943

By this time, the 7th Guards Army had assumed an overall layout of forward units being supported by additional units in rear positions. For example, normally each rifle division had two regiments in the front line and one supporting regiment in the rear. The Army itself had four rifle divisions on the front line with three rifle divisions in support located along the second main defensive line. Intermingled with these major rifle formations were numerous smaller units strengthening critical areas. The command staff constantly made adjustments to this general deployment plan as the Army added new supporting arms like artillery, antitank and antiaircraft units. The work to improve defensive positions never stopped as the weeks went by.

The historical record reveals an interesting bit of information about the Army during this time. Every other night, armored trains #754 and #730 from the 38 Separate Armored Train Battalion made fire raids on various German targets in support of frontline units of the 7th Guards Army. The Germans were not yet successful in stopping these harassing attacks. [102] [103] Due to the limited layout of active, but safe, rail lines in the frontline area, these armored trains were only able to support units on the far-left flank of the army. They supported the 24 GRC and the 15, 36 and 72 GRDs, which were holding that section of the front.

19 May 1943

Once again *Stavka* issued a warning alert of an impending German attack, and once again the anticipated German attack never materialized. Stalin and others within and outside of the High Command were beginning to ask themselves, Did they make the correct choice by remaining on the defensive so long? Only time would tell.

20 May 1943

The problem with army replacements in general and with the number of men reaching forward units in particular did not go away, so the Voronezh Front leadership once again took action. The Voronezh Front's Deputy Chief of Staff issued order No. 0025/OU, dated 20 May 1943, about the lack of organized efforts in the rear areas of the 7th Guards and the 40th A. Apparently, both armies had lost thousands of replacements who wandered around in the rear areas without being assigned to frontline units. [104] Despite this warning, slack control of the rear areas and disregard of orders continued.

Even with these problems, the 7th Guards Army ration strength now stood at 74,138 men and women.

At 23:00 hours, "on the night of 20 to 21 May, two platoons of the 15th Guards Rifle Division conducted reconnaissance to determine the defense system [and] the presence of enemy forces and weapons in the village of Ogurtsovo," [105] which was on the west side of the northern Donets River. With the assistance of the 34 Separate Armored Train Battalion, armored trains #712 and 667 were allowed to fire 100 rounds of high-explosive shells (76mm) in support. This small reconnaissance raid was successful. This example serves to reinforce the understanding of what all combat troops know to be true: the front was always an active and dangerous place to be.

On the other hand, sometimes Red Army soldiers were ordered to undertake less effective reconnaissance activities. The regimental intelligence officer from the 72 GRD stated:

> And several times we were entrusted with "stupid assignments", on behalf of the political department. [We] went into the German rear and scattered leaflets and appeals to German soldiers in the trenches. The risk associated with the execution of such "party orders" was many times higher than the benefits that these leaflets could bring. But in the army, you can't argue with orders... [106]

21 May 1943

With the assignment to the Army of the 5 AAD with four AA regiments, and two separate AA regiments, all from the high command's reserve, the Army could protect several vital areas at once. One of these regiments was deployed to cover the main supply station at Volokonovka, while the others covered infantry, tank and artillery concentration areas. Even with all these dedicated AA guns, more infantry weapons soon were added to boost the Army's AA defense. They would certainly be needed for the imminent battle.

22 May 1943

The Luftwaffe launched a large daylight attack with 111 aircraft against the Kursk Railroad Station area; it was an effective attack. "The whole rail junction area was turned into an inferno." [107] In response to this attack, the Red Army strengthened the AA defensives of the Kursk rail junction.

23 May 1943

At this time the 2nd and 17th Air Armies were active over the Voronezh Front intercepting Luftwaffe aircraft and attacking selected ground targets. In the coming weeks these air armies would shift their activities to directly support the armies on the ground, but that was still in the future.

An additional indication of what was taking place behind the Russian front lines came to light when careless handling of 2,700 captured German T-35 antitank mines near Voronezh caused an explosion which killed 7 military personnel and 15-20 civilians and destroyed 2 trucks. These mines were destined to be employed within the Kursk defensive zone. This incident illustrated both how Russian discipline in the rear area was lax and how the Russians employed any available material to strengthen the minefields. [108]

Also, on this day an unknown number of German aircraft found and attacked armored train #754. The engine was hit and one armored car was wrecked. Train crew members suffered 9 killed and 15 wounded. Train #754 withdrew for repairs and only returned to duty in July, apparently too late to assist in the defensive of the Kursk Bulge.

24 May 1943

Far to the south of Kursk, major Russian attacks against the German Kuban bridgehead failed once again. German troops repelled all attacks in this area and inflicted significant casualties to forces of the North Caucasus Front. Russian leaders temporarily called off further offensive operations in that area. These attacks had little impact on the rest of the front.

25 May 1943

Russian attacks along the lower Donets River in the Stalino area also failed to breach strong German defensive positions. (See Map #21 for the location of Stalino.) Once again, these attacks had little impact on the rest of the front, but they reminded German commanders that unlike them, the Red Army was able to attack in force at more than one location at a time. The balance of power had decisively shifted in favor of the Red Army.

The 7th Guards Army ration strength now stood at 73,814 men and women. As the days went by, the Soviet Forces in the Kursk area only grew stronger. General Shumilov and his command staff must have started to wonder, When will the enemy attack? This quiet time can't last much longer!

26 May 1943

Due to the late nature of the German offensive and the several false warnings by *Stavka*, an "impatient Vatutin [Commander of the Voronezh Front] and his political Member of the Military Council Nikita Sergeevich Khrushchev, urged a preemptive attack on Kharkov, arguing that the German offensive had been canceled." [109] This proposed offensive was supported by a detailed planning map, [110] which required the Voronezh Front to shift armies to new locations, including the 7th Guards Army. Only due to vigorous efforts by *Stavka* members was Stalin persuaded to keep to the original plan and to stay on the defensive. The leaders on both sides were becoming increasingly impatient; they must have asked themselves over and over again, When will the battle start? The answer, of course, was when Adolf Hitler decided it would start.

27 May 1943

Behind the front lines a logistical battle was fought as trains and truck drivers continued to deliver needed supplies. Each front and each army needed thousands of tons of ammunition, fuel, lubricants, fodder for the horses, and food for the soldiers. Slowly, the logistical support units were able to stockpile reserves of all items, ultimately enough for both the defensive phase of the battle and even the follow up offensives phase. Once again German air units attempted to interfere with this supply effort like they had at Stalingrad, but they were unable to block the flow of supplies and reinforcements for long. The delayed German offensive allowed the Russians sufficient time to logistically prepare for the coming battle despite critical problems and bottlenecks in the supply line. The Red Army and VVS forces would not want for bombs and bullets in the days ahead.

28 May 1943

The Deputy Defense Commissar of the State Defense Committee (GKO), E. A. Shchadenko issued Memorandum #130721 on 28 May 1943 about the transfer of military units from the Stalingrad Group of Forces to new locations. This was in regards to an earlier GKO Order #2911 from 18 February 1943.

This memorandum covered in general the composition of the Stalingrad Group of Forces, but interestingly it made note of 7,560 people, "Red Army soldiers who were in captivity," and suggested that they all were checked. This implies they were being sent back to their military units because the NKVD had checked and cleared these former POWs. The unlucky ones that failed this security review either were shot, sent to penal units at the front, or maybe even to a *gulag* forced-labor camp.

The common solider also had to be on guard against the NKVD because they had numerous informants within the ranks. These internal spies would read personal letters or the diaries of their fellow soldiers and report any incriminating information to political officers. This internal spying was so bad that the common solider thought that "in reality, the Germans were not the only enemy; it could equally well prove to be the person immediately next to you." [111]

Memorandum #130721 went on to list formations in transit and how long it would take for them to reach their parent units:

- The transfer of combat formations and units of the 62nd, 64th [7th Guards Army] and 66th armies, 7 RC, and ten RDs would be complete only by 14 April.
- Tank, artillery, anti-aircraft and mortar units and formations VGK would be complete only on 10 May.
- Chemical, Engineering and rear parts of the armies would be completed only on 22 May. For a total of 662 echelons (trains). [112]

These transfers amounted to approximately 29,000 rail cars' worth of men and equipment. As can be seen, many of these vital support units only now were reaching the frontline armies in late May, showing delays from the previously proposed arrival dates in April and early May. The Russians were fortunate that the Germans had delayed their attack for so long.

29 May 1943

The 7th Guards Army HQ issued a summary of material support for units of the Army. Overall, the food situation had improved with ample stocks of flour, pasta, and vodka, but the army had almost no reserves of meat, fish and tobacco. Stocks of fuel for all types of vehicles were adequate. Horses had enough hay but almost no oats. All types of ammunition were sufficient for a minor battle, but inadequate for a major one. The supply situation was moving in the right direction; all that was needed now was a bit more time to fulfill all the needs of the Army. [113]

However, on this day 28 German aircraft found and attacked Armored Train #712, of the 34 Separate Armored Train Battalion, which was working in support of the 7th Guards Army. The air raid lasted almost 30 minutes, with the Germans losing only one aircraft. The train was hit and two cars were wrecked; crew members suffered 5 killed and 16 wounded. After dark, the Russians restored the damaged rail line and removed the train to the rear for repairs.

Due to the delay in their offensive, high level German field commanders actively conducted map exercises (war games) to fine tune the Kursk attack plan. General Kempf held the first one on 29 May, with Erhard Raus, who commanded an infantry corps under Kempf, coordinating the game. Field Marshal von Manstein also attended this game. [114]

30 May 1943

The 7th Guards Army ration strength stood at 74,197 men and women.

Field Marshal von Richthofen was busy preparing his 4th Air Fleet to support von Manstein's southern attack against the Kursk Bulge. The 4th Air Fleet had supported the 4th Pz Army and General Kempf's XXXXVIII Pz C at Stalingrad, and von Richthofen was looking forward to doing so once again, but soon his efforts would be directed elsewhere. [115]

For the VVS and the Partisan movement, the end of May revealed a victory of sorts. Within the Army Group Center's area of operations, Soviet air and land operations caused the loss of 35 locomotives and 106 rail cars, along with significant blockages of rail lines, [116] not to mention the loss of considerable amounts of supplies and equipment.

31 May 1943

While Field Marshal Model was away from the front on leave, he left General Walther Weiss in command of his 9th Army. Over the next several weeks, General Model continued in his attempts to convince Hitler to cancel Operation Citadel. June would be the decisive month to formulate a different strategy for the summer season. Model must have thought that if only Hitler would listen to reason and accept the fact of overwhelming Russian strength within the Kursk Bulge area, that a different, more realistic plan would be chosen. Field Marshal von Manstein had proposed just such a plan, the so-called Backhand Blow. Basically, let the Soviets attack first, and then launch a powerful counterattack, like the one von Manstein had undertaken to recapture Kharkov and Belgorod in the early spring of 1943. Hitler rejected von Manstein's plan and stuck with his desire to attack first at all costs with Kursk as his chosen target.

On the other side of the front line, Stalin and *Stavka* continued to transfer a steady stream of combat formations into not only the Kursk Bulge, but also into the surrounding fronts as well. With each passing day Russian defensives grew more robust and the strength of armies destined for early offensive operations become an overwhelming force. The opportunity for a German victory at Kursk had passed. Soon Stalin would be the one dictating the tempo of battle during the summer and fall of 1943.

The 7th Guards Army issued an intelligence document about losses for May 1943. The major combat units lost 371 killed; 1,124 wounded; 33 missing; 611 sick; and 112 for other reasons, for a total of 2,251 casualties. Within this total, the artillery lost 18 killed, 49 wounded, and none missing. Tank units lost 11 killed, 29 wounded and none missing. This was during a quiet time with no major combat operations taking place. These figures revealed over 72 casualties per day. Losses like this due to skirmishes, air raids, accidents and illness continued to accumulate even when fighting was restrained, but how much longer could this calm period last? [117]

Seven tank or SU units now were assigned to the 7th Guards Army. These units were the ones the Army would take into battle. They contained the following number of tanks:

27 GTB – 50 – T 34, 4 – T 70
201 TB – 11 – MK II (Matilda), 23 – MK III (Valentine) (British tanks)
148 Separate TR – 33 – T 34. 8 – T 70
167 Separate TR – 32 – T 34, 7 – T 70, 3 – BA 64 (light armored cars)
262 Separate TR – 24 – KV 1
1438 SUR – 12 – SU 122, 9 – SU 76
1529 SUR – 12 – SU 152, 1 – KV 1

These units had a total of 193 tanks, 33 SUs, and 3 armored cars. [118] By 4 July the Army contained a total of 224 tanks with only 14 of them under repair, and 33 SUs assigned with 4 under repair. These in-service rates displayed a high state of readiness of the armored force. [119]

The Luftwaffe, of course, was watching the arrival of all these troops and equipment into the area via reconnaissance aircraft. To counter this buildup, during May the Luftwaffe conducted 22 air raids against rail targets behind the 7th Guards Army. This effort slowed but did not stop the Red Army buildup. [120]

But the fighter units of the VVS were striking back against the Luftwaffe's all-seeing eye in the sky. During May, the VVS shot down 29 German reconnaissance aircraft; in June that number jumped to 54 aircraft. A Red Army squadron commander stated, "We organized small groups of aircraft (2-4) to ambush the enemy reconnaissance planes. These were stationed at airfields located only [8-15] miles from the front. The interception of enemy reconnaissance aircraft through ambushes was wide-spread in the spring and summer of 1943 before the battle for Kursk." [121]

1 June 1943	
7th Guards Army	
Assigned – Voronezh Front	
Lieutenant General M.S. Shumilov Commanding	
Rifle, Airborne and Cavalry Units	24 Guards Rifle Corps (15, 36, 72 GRDs) 25 Guards Rifle Corps (73, 78, 81 GRDs) 213 RD (army reserve)
Artillery Units	8 Separate Destroyer Brigade 109, 161 Guards Gun Artillery Regiments 1112 Gun Artillery Regiment 114, 115 Guards Antitank Artillery Regiments 1, 2 Antitank Artillery Regiments 290 Mortar Regiment 1, 2, 3, 4, 5 AT Rifle Battalions 5 Antiaircraft Artillery Division (670, 743, 1119, 1181 Antiaircraft Artillery Regiments) 162, 258 Guards AA Regiments 662 Separate Aerial Observation/Air Raid Warning Company
Armored and Mechanized Units	27 Guards Tank Brigade 201 Tank Brigade 148, 167, 262 Separate Tank Regiments 1438, 1529 Self-propelled Gun Regiments 34, 38 Separate Armored Train Battalions (three trains active)
Engineer Units	60 Engineer-Sapper Brigade 175, 329 Separate Engineer Battalions
Estimated Combat Strength Estimated Ration Strength	73,190 men [122] 80,455 men and women [123] (including an estimated 7,000 support personnel)

(BSSA 43, BSA 43, TsAMO Archive)

1 June 1943

The medical establishment had changed in May, so at this point the known medical and related units within the 7th Guards Army are listed below:

One HQ for the Field Evacuation Point - #24.
One Field Evacuation Point - #24, (used to transfer wounded to the deep rear).
Nine Mobile Surgical Hospitals – HPPG #699, 869, 2204, 2207, 2208, 2209, 2409, 2670, 4344.
Two Therapeutic Hospitals – TPPG #34, 2712.
Two Hospitals for the Lightly Wounded – GIR #3247 and 3255.
One Infectious Hospital – XG #2241.

One Separate Ambulance Company (Auto) - #97, equipped with 48 GAZ-55 ambulances each capable of carrying 10 sitting or 4 stretcher (litter) wounded.
One Ambulance Company (Horse-Drawn) - #251.
One Sanitary Epidemiological Unit - #115.
One Disinfection Company - #111.
One Separate Medical Reinforcement Company - #51.
One Pathological and Anatomical Laboratory - #112.
Three Field Laundry Squads - #116, 117, 118.
Four Laundry Bathhouses - #1, 2, 3, 4.

These units represented approximately 5,100 hospital beds for the entire Army.

The veterinary services remained unchanged with

Two Army Veterinary Hospitals - #958, 960.
One Evacuation Veterinary Hospital - #959.
One Mobile Field Veterinary Laboratory - #16.

Note that the number of beds per hospital and/or medical unit changed over time, so the number of beds available to the Army, as indicated throughout this book, is an approximation.

Additionally, the 7th Guards Army had 29 other rear area service units assigned to it.

Clearly all the command authorities were fine-tuning the Order of Battle of the armies around the Kursk Bulge area. The expected German attack had not come yet, so preparations continued as more combat and support units, supplies and equipment flowed into the battle area. Along with these efforts, the Russians further developed the entrenchments, making them deeper and more extensive, and the minefields grew thicker and more widespread. They used camouflage and a full range of deception methods to hide, as much as possible, these improvements from the Germans. The Russians were very rigorous in their efforts to prepare a suitable killing ground for their enemy.

Soviet deception methods allowed each front to develop its own plan. These plans included creating false tank concentration areas, fake radio nets and tanks, phony airfields equipped with their own simulated aircraft, and mock minefields, trenches, artillery positions and command posts. All this was designed to be seen and attacked by the enemy. The deception worked well enough. In the coming weeks, the Luftwaffe would bomb and strafe these replica positions while leaving real combat positions unharmed.

The deployment of the Army's major combat units hardly varied from the previous weeks. The only notable exception was the movement of the 213 RD from its frontline position to a backup position in the second defensive line. The 78 GRD moved into the 25 GRC position as the 213 RD moved out. The reason for this switch is not known, but it might be related to the directive about grouping all Guards Units together, or possibly it was because the 213 RD was untested in combat.

The addition of the 1529 Self-Propelled Gun Regiment (SUR) with its massive 152 mm Howitzer added to the firepower of the Army. This was one of the first new SU-152 units to join the Red Army. The 1529 SUR was formed from March to May 1943 and then quickly assigned to the 7th Guards Army in June. The design for the SU-152 was rushed into production to counter new German heavy tanks like the Tiger and Panther. [124] Eventually the SU-152 became known as *Zveroboy* (Beast Killer) because it could destroy these new tanks. [125]

At the start of June, the 7th Guards Army had become a fearsome combined arms army. A summation of its men and equipment displays this power:

73,190 men in combat units - with 30,296 in the Infantry; 16,807 in the Artillery; and about 7,000 men and women in supporting roles.

Machine Guns – 802 Heavy MGs; 113 AA MGs; 2,444 Light MGs; and 5,205 Sub-Machine Guns.
Mortars – 319 - 50mm, 520 - 82mm and 191 - 120mm.
Artillery – 359 - 45mm, 341 - 76mm, 92 - 122mm, 12 - 152mm (SU-152) and 93 AA Guns.
Small Arms – 45,229 rifles and 2,052 AT rifles, with an additional 252 flamethrowers.
Armored Vehicles – 194 tanks and 6 armored cars. [126]

With the postponement of the German attack, the 7th Guards Army had even more time to train troops, deploy men and weapons in ideal positions, and prepare the battlefield even further.

A Red Army soldier loads/fires a 120 mm mortar.

2 June 1943

Despite the delay in the ground attack, the Germans were not inactive. On this day the Luftwaffe launched a particularly intensive day and night bombardment on the Kursk railway junction. Dubbed Operation Carmen, "According to instructions, the marshalling yards at Kursk were to be attacked relentlessly during the hours of daylight on June 2 and the following night." [127] German aircraft flew 824 sorties in total, attacking for hours in groups of 20, 40 and even 70 aircraft. They made their bombing runs from various directions and bombed the station area over a three to five-hour period at random intervals. Overall, they dropped about 2,000 bombs, including a number with delayed action fuses, some lasting up to 72 hours.

The damage was extensive. Some "7 miles of the rail line…, 12 turnouts, 300 communication lanes, 154 loaded cars, 8 locomotives, a water reservoir, a pumping station, a distribution network, a depot for five stalls, mechanical workshops, a car repair center, a coal trestle, a wooden bridge, an overpass on the Voronezh detour and other objects" were destroyed or damaged. [128] Repair crews needed 11 hours to open a through line and a full week to repair or replace damaged equipment or facilities. Unfortunately, the human toll for this attack is unknown.

This powerful attack against the city of Kursk once again caught the Russians by surprise. Because of this they greatly improved their air defensive system in the area with more AA artillery units and fighter squadrons patrolling the rail network and especially rail junctions and stations. Note: Several Russian commanders reported to *Stavka* that during these attacks anti-aircraft guns and fighters had shot down 123 German aircraft, which was a blatant exaggeration. Only about 31 German aircraft actually were shot down with the VVS losing about 27 fighters. Because of this obvious lie, *Stavka* rebuked Colonel Gavrilov, chief of staff of the Voronezh-Belogrebsk Anti-aircraft Divisional Area, for giving false information. [129] This was a case of putting a positive spin on bad news which was then passed up the chain of command.

In turn the VVS launched a series of heavy attacks against the German-held city of Orel and inflected heavy damage on the city and rail lines. These VVS attacks also included raids on Luftwaffe airfields in the area.

German commanders continued with their map exercises to support planning for the Kursk attack. On this day, General Hoth held a war game for the 4th Panzer Army with key corps and divisional commanders attending, including Luftwaffe officers. The same day, the III Pz C also held a separate war game for its officers. Several days later (date unknown) General Paul Hausser, commander of the II SS Pz C, and General Otto von Knobelsdorff, commander of the XXXXVIII Pz C, conducted their own war game. These two corps were assigned to the 4th Panzer Army. Both General Hoth and von Manstein also attended this exercise. [130] The planning never really stopped, but the clock also kept running. Time now clearly was working against the Germans.

3 June 1943

Using the battle for Stalingrad as recent example, commanders and medical personnel gave much thought to the evacuation of the sick and wounded from this new battlefield. They planned to move military casualties from the front during the Battle of Kursk toward the rear more rapidly, carried out mostly "by railway transport according to the following principle:

1. From army field hospitals to army hospital bases for a distance of 30 to 100 km [18-62 miles] via ambulances of front formation.
2. From army to frontline hospital bases for a distance of 100 to 400 km [62-248 miles] - by temporary military-sanitary trains (VVSP).
3. From frontline hospital bases to deep rear [hospitals] by permanent military sanitary trains (VSP)."

"By early March 1943 there were 293 VSP, 130 VVSP [medical trains] and up to 250 sanitary [aircraft] flights in operation." During the battle "the transport aviation of General N. S. Skripko, provided ammunition supplies…, canned blood and blood substitute fluid…, and [flew out] more than 21,000 wounded…[on] return flights." One such unit, "the Sanitary Squadron of the 62nd Independent Guards

Regiment flew for 5,150 hours in July-August, while 4,284 wounded were evacuated; hospitals received 27 tons of canned blood and 2 tons of medicines." [131]

The use of numerous aircraft to transport supplies to the front and wounded to the rear was a great improvement in medical-related support. These aircraft also provided direct links between headquarters by delivering "directly to the frontline staff officers urgent military documents." This type of effort clearly showed how the Red Army had learned its lessons from the Stalingrad battle and was much better prepared to take care of wounded soldiers during the upcoming battle.

On this day, the 7th Guards Army issued another summary of material support for units and formations of the Army. Overall food stocks were adequate, but most formations were out of vodka, naphtha (lamp and cooking fuel) and had low supplies of meat, soap and hay. [132] Ammunition supplies also were sufficient for several days of combat but not much more. The Red Army needed to stockpile additional ammunition before the German attack.

Nurses load a wounded solider into a U-2 (Po-2) air ambulance aircraft (U-2S1 variant).

4 June 1943

Much has been written about the extensive fortification efforts to defeat the expected German attack all along the Kursk Bulge. The northern side of the bulge was covered by the Central Front. The southern side of the bulge was covered by the Voronezh Front where the 6th and 7th Guards Armies were deployed. Activities of these two armies are the main focus of this southern part of the battle for Kursk.

The 6th and 7th Guards Armies each created three separate defensive lines within their area of the front. In turn, the Voronezh Front itself developed three more defensive lines behind the frontline armies. Also, by June the civilian workforce had swelled to over 300,000 people. These civilian labors were kept busy digging trenches and AT ditches, building bunkers, and improving roads and airfields behind the frontline area. [133] Despite all this effort, due to an apparent planning error in laying out the defensive positions for the 7th Guards Army, the fortifications were weaker than expected. For some reason, units of the 7th Guards Army produced too few cutoff positions between each of its three main defensive lines. These cutoff positions were used to link the three defensive lines together, at right angles, thus preventing enemy forces from freely maneuvering between the main defensive lines. History shows this oversight would cost the 7th Guards Army much blood and threaten the entire defensive position of the Voronezh Front. [134]

Despite the omission of cutoff positions, and after more than three months of preparations, the defensive works had reached staggering levels of completeness. Just for the Voronezh Front's section, the Russians built approximately 2,600 miles of trenches and 310 miles of AT obstacles. For the 13th Army (Central Front), 6th and 7th Guards Army, some 2,400 AT mines and 2,700 antipersonnel mines were laid per each mile of the front line. Also, within the 7th Guards' area, soldiers built 27 special AT strong points. Numerous bunkers, wire and engineered obstacles supported these defensive positions. All these preparations were designed to force German tanks into "kill zones" where every type of weapon could be brought to bear against them. [135] By 5 July, the 7th Guards Army had deployed at least 66,814 antitank mines and 85,140 anti-personnel mines. During the battle, engineer units laid an additional 13,309 mines, [136] mostly in front of advancing German units. Note: A large percentage of these mines were laid in the first army-level defense line, with lesser amounts in the second and third lines.

One new weapon of special note was the use of numerous flame or fire mines. These fire mines were deployed mostly within the positions of the 6th and 7th Guards Armies. These simple-to-build fire mines consisted of a wooden box filled with bottles of flammable material and a small explosive charge. When the small explosive charge was detonated, it would "in turn detonate the incendiary bottles. The resulting flames, fragments and shock wave effectively destroyed infantry and any vehicles within a 40-meter [43 yard] radius. The flames would reach 30 to 40 meters [32-43 yards] in height and then cascade downward upon targets in the area of effect." [137] These flame mines or MOF mines each contained up to 19 bottles of flammable material.

By the start of the German attack, a reported 300,000 bottles of flammable material were deployed. They were arranged in a checkerboard fashion in their own flame minefields. One Russian account disclosed that "in the area of Kreida Station, in the 7th Guards Army's defensive sector, eight enemy tanks burned up on the flame minefields. For the entire period of the defense, the enemy lost more than 70 tanks on the flame minefields laid by the Voronezh Front." [138] These reported flame minefield tank losses appear to be excessive, but they might in fact be describing losses for all types of German vehicles. Interestingly, a German tank crew member confirmed the use of these flame mines. Talking about Russian minefields, he said, "There were even dug-in containers full of incinerating oil also triggered by tripwires. These had much the same effect as flame-throwers." [139] In any case, the Germans had no idea of the extent of Red Army preparations, and that lack of knowledge would cost them the battle.

But the 7th Guards Army suffered its own troop losses to these extensive minefields. In the area of the 24 GRC from 15 May to 1 June 1943, the Corps suffered 7 killed and 39 wounded men. [140] Minefields are indiscriminate killers.

During this night, the Luftwaffe bombed the GAZ #1 tank factory at Gorki. Their bombs inflicted heavy damage to buildings producing T-70 light tanks with lesser damage to other structures. This was the start of a German nocturnal strategic bomber offensive against Soviet rear area industrial targets. [141] This air offensive only was possible because of the pause in ground combat during the buildup to the Kursk offensive.

5 June 1943

OKH and most German frontline headquarters personnel agreed that they had to launch Operation Citadel by mid-June at the latest if they were to have any hope of success. Hitler's indecision, of course, prevented offensive operations from starting in a timely manner.

6 June 1943

Operation *Zigeunerbaron* (Gypsy Baron) was concluded on this day with mixed results. The Germans wanted to clear a large pocket of partisans south of Bryansk, west of the Orel Bulge. The Axis troops achieved some success, but the partisans mostly just dispersed and returned later to this same area before the end of July. Altogether German troops launched five separate anti-partisan operations before the Citadel offensive to relieve the pressure on critical rail lines on the north side of the Kursk Bulge. [142]

7 June 1943

The German attack plan, as finally outlined, contained two concentric attacks against the north and south sides of the Kursk Bulge. These attacks were designed to cut off the bulge and then destroy the substantial Soviet forces that would be surrounded. The northern attack, launched directly south from Orel, was at first aimed at the cities of Olkbovatka and Ponyriy, then straight south toward the city of Kursk. The southern attack would be launched from each side of Belgorod aimed directly to the northeast toward the area between Oboyan, Prokhorovka and Korocha. Once reaching these positions, the attack then would turn north toward Kursk. (See Map #6 – German attack plan.)

The northern attack would be an exclusive 9th Army operation, commanded by Colonel General Walter Model. The 9th Army contained five corps with 5 Panzer Divisions and 13 Infantry Divisions. These units would be attacking along roughly a 24-mile front and would be supported by the 6th Air Fleet.

The more powerful southern attack was split between the 4th Panzer Army commanded by Colonel General Hoth and Army Detachment Kempf commanded by Lieutenant General Warner Kempf. The 4th Panzer Army contained three corps with two Panzer Divisions, four Panzer Grenadier Divisions and four Infantry Divisions. Army Detachment Kempf contained three corps with three Panzer Divisions and six Infantry Divisions. This combined force would attack along approximately a 44-mile front and would be supported by the 4th Air Fleet commanded by Colonel General von Richthofen.

(L-R) Colonel General Wolfram von Richthofen, Commander of Luftflotte 4; Lieutenant General Warner Kempf, Commander of Army Detachment Kempf and Colonel General Hermann Hoth, Commander of the 4th Panzer Army.

Note: The Soviets incorrectly believed the northern attack was to be the most powerful one and they planned accordingly. They strengthened the defending Central Front armies with substantially more artillery units, containing more than 2,000 guns, over what was assigned to the Voronezh Front.

With these substantial German forces gathered for the Kursk attack, the Red Army needed to act cautiously. On the attack, German Panzer formations were known to quickly overwhelm defenders, then surround and destroy entire armies. Only time would tell if the tried-and-true *Blitzkrieg* (Lightning War) tactics would once again win the day and the battle.

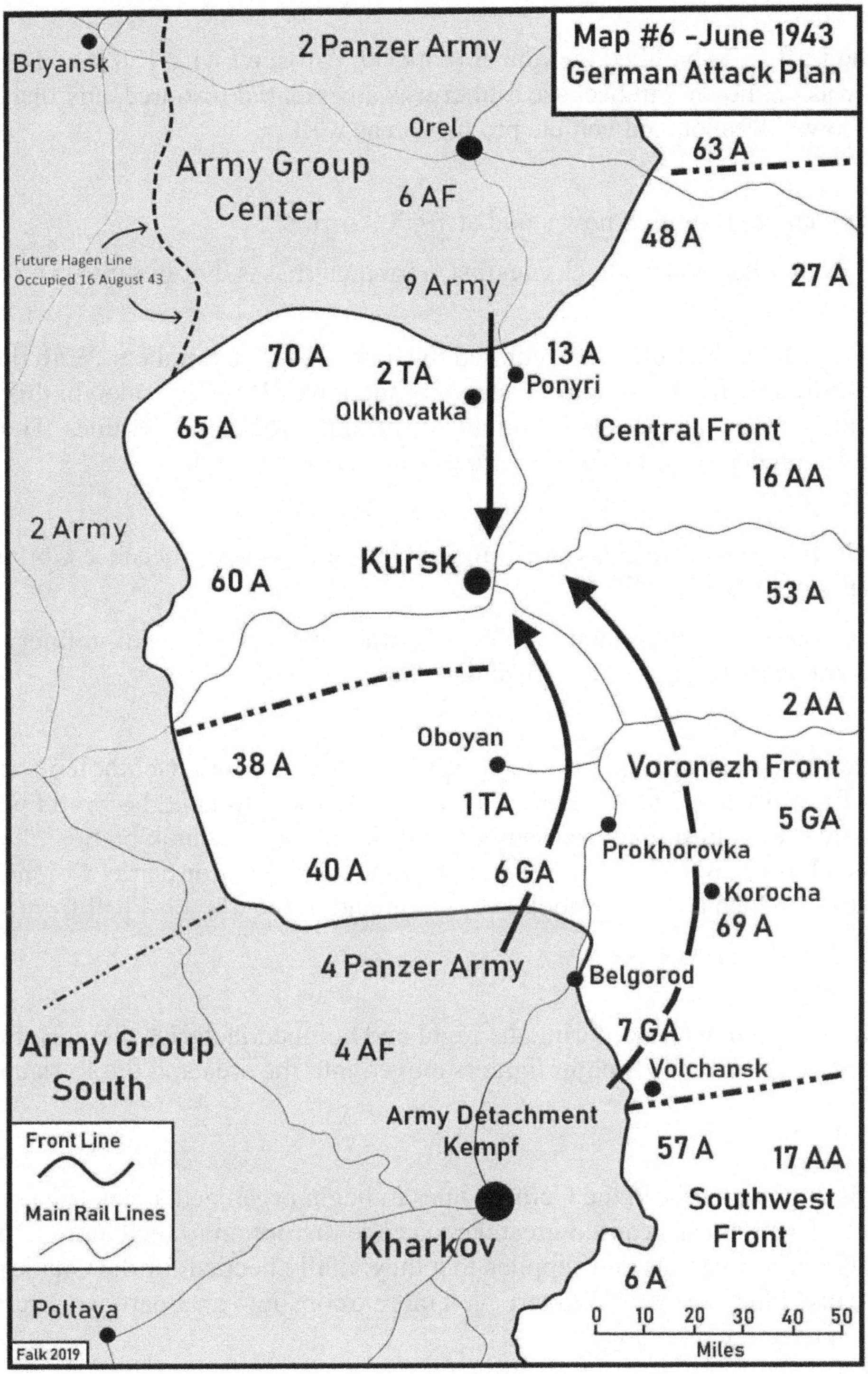

8 June 1943

On or about this date, Field Marshal von Richthofen left for Berlin to attend a series of meetings. As a result of these talks, von Richthofen was given command of the 2nd Air Fleet in Italy. "Von Richthofen's hope was that he would only take over air operations in Italy for four to eight weeks in order to meet the expected Allied Attack" [143], but that was not to be. He would die of a brain tumor on 12 July 1945 in Austria, never seeing Russia again.

The VVS launched a series of heavy attacks against German airfields over the next three days, but the German air defense system worked well and the attackers suffered considerable losses.

9 June 1943

During the night, the Luftwaffe bombed the rubber works at Yaroslavl with 138 He 111 bombers. The amount of damage was unknown, but because rubber was an essential resource, any disruption in rubber production slowed weapon and vehicle production as well.

10 June 1943

The 7th Guards Army combat strength now stood at 79,009 men. [144]

The VVS launched further heavy day attacks against German airbases, but in turn suffered heavy losses themselves.

Once again, during the night, the Luftwaffe returned to Gorki with 126 bombers. With these attacks in the Gorki area, the Luftwaffe tried to stop or to slow production of Russian tanks. In this they were partly successful, but Germany already had lost the industrial battle behind the lines. The Red Army and Air Force would soon demonstrate their growing strength.

11 June 1943

Field Marshal von Richthofen left for Italy at Hermann Göring's instance, because Göring saw him as a rival and wanted him out of Berlin. [145]

Once again, German bombers launched a nighttime raid, this time against the oil refinery at Saratov. They attacked this same target again on 13, 15, and 16 June.

12 June 1943

The German's master of tactical air support operations Field Marshal von Richthofen arrived in Italy to assume command of Luftwaffe's 2nd Air Fleet. Due to this new assignment, he would not be available to guide the Luftwaffe efforts to support the southern attack during the Kursk battle. [146] Von Manstein was not pleased he had lost von Richthofen's expertise just before a major battle. Colonel General (*Generaloberst*) Otto Dessloch (*Deßloch*) took over command of the 4th Air Fleet from von Richthofen.

13 June 1943

Again, the Luftwaffe returned to Gorki during the night and bombed factories there. But German losses were starting to increase as more VVS night fighters moved into the area and the Russians improved AA defensives.

14 June 1943

Stavka ordered partisan groups behind the German lines to begin organized attacks against the entire German rail network. These attacks caused great damage and disruption to the German rail network and slowed the accumulation of weapons and supplies that they vitally needed for the Citadel offensive. [147] These attacks also caused the Germans to divert even more troops into anti-partisan sweeps to clear the rail lines before the start of the offensive.

At the Voronezh Front, General Vatutin issued order No. 00151/OU which stated in part that the 7th Guards Army left 4,470 trained men lingering six days in the reserve rifle regiment rather than reassigning them to strengthen rifle divisions. In other cases, some 2,500 and then 700 men also were not reassigned to the front in a timely manner. To correct these errors, General Vatutin ordered that reserve regiments move closer to the army headquarters, no more than 15 miles away, and finally Vatutin stated, "The chief of staff of 7th Guards Army, Major General Lukin (not to be confused with the former chief of staff General Laskin), is rebuked for failing to take steps to uncover inherent resources and the timely dispatch of replacements for the troops." [148]

General Vatutin also issued order No. 00152/OU on the same day about understrength rifle divisions and stated, "Of all the armies, the 7th Guards Army stands out in the worst light on this matter." He then went on with the resulting punishment:

> I am announcing a rebuke to the [following] commanders: 72nd Guards Rifle Division — Major General A.P. Losev; 15th Guards Rifle Division — Major General E. I. Vasilenko; 36th Guards Rifle Division — Major General M. I. Denisenko for showing complacency toward my order regarding the increased staffing of the rifle companies, and I am warning that if the rifle companies haven't been brought up to the strength that I have set by 20 June, harsher measures of punishment will be taken toward them. [149]

So, the front commander General Vatutin directly rebuked the 7th Guards Army chief of staff and three divisional commanders, but spared General Shumilov from any reprimands. One wonders if General Shumilov might have been out sick or even away from the front during this time.

15 June 1943

The 7th Guards Army combat strength stood at 78,820 men.

During inspection tours of the front during May and June, General Vatutin (and others) emphasized the need to strengthen the boundaries between his armies. Within the 7th Guards Army, the far-right flank was covered by the 81 GRD and this area required special attention. This area needed to be strongly defended for two reasons: first, boundaries between armies or divisions always were vulnerable, and second, Vatutin had a plan to use the position east of Belgorod, the Staryi Gorod area, to split the expected German assault force into two groups. Two detached attacking groups would have a hard time supporting each other, thus making the Germans' objectives harder to attain. To achieve that level of disruption of the German attack plan, the Russians needed to convert the area covered by the 81 GRD into a fortress. General Shumilov would turn Vatutin's plan into reality by massively reinforcing the 81 GRD with firepower. By the start of the battle, the 81 GRD had the following reinforcements at its disposal:

> 161, 265 Cannon Artillery Regiments
> 153 Guards Artillery Regiment
> 290 Mortar Regiment
> 114 Guards Destroyer Antitank Artillery Regiment
> 97, 315 Guards Mortar Regiments (Katyusha rockets)
> 262 Separate Tank Regiment (24 – KV 1 heavy tanks)
> 2 AT Rifle Battalion [150]

Of course, Generals Vatutin and Shumilov did not know the Germans would attack this specific area, but they could hope.

16 June 1943

General Heinz Guderian asked Hitler to postpone Citadel so he could continue to rebuild the Panzer forces and to fix the teething problems with the Panther tanks. In turn, von Manstein and von Kluge pressed for an immediate attack before the Soviet defensives became even stronger. Meanwhile Hitler believed German air reconnaissance had exaggerated Soviet defensives and that the offensive could proceed as planned. [151]

17 June 1943

Once again, the extended delay in combat operations worked in favor of the Soviet defender. For example, "Much attention was paid to the medical service. First the wounded [existing wounded from all armies] were to be evacuated into the rear before the fighting started." This made available an estimated 90,000 hospital beds overall, that had been deployed for the battle. Also, "A great deal of effort went into the organization of subterranean operating and pre- and post-operating rooms at first-line hospitals and medical battalions. These structures usually had four or five layers of overhead logs banked with two or three meters of earth." [152] Additionally, the delay in major fighting allowed extensive training of medical personnel and the stockpiling of all types of medical equipment and supplies. This level of preparation of medical support was unheard-of at Stalingrad. At this time the 7th Guards Army had about 5,100 beds in its own medical facilities.

18 June 1943

The OKW, High Command of the German Armed Forces, "recommended to Hitler that he abandon Citadel in order to free strategic reserves for the defense in both East and West." Hitler rejected this advice and set the Operation Citadel start date to 3 July 1943. [153]

19 June 1943

Logistically, the Central and Voronezh Fronts had adequate ammunition supplies. The two fronts each had truck carrying capacity of some 6,000 tons, which would theoretically allow each front to receive about 6,000 tons of ammunition, POL, and rations every day. At least that was the plan. However not all of these trucks were operational, which reduced the lift capability. Despite these problems, ammunition levels reached "from 2.5 to 3 fire units [Unit of Fire] for weapons of basic calibers at the two fronts and up to 5 fire units for weapons of heavy caliber. Automobile fuel was sufficient to refuel each vehicle almost five times...and six times on the Voronezh Front. As for food, there were enough rations for 20-25 days [for the Central Front]." [154] The logistics chief of the Central Front General Nikolai Antipenko reported these estimates; however, his report seems to be overly optimistic. Sadly, supply problems and shortfalls still existed for the Voronezh Front and throughout other frontline armies.

The Soviet logistical system used three main categories of provisions to plan operations. A basic fuel load was called a *zapravki* at approximately 402 tons, a munitions load (Unit of Fire) was a *boekomplekt* at 2,168 tons and a food/fodder load was called *s/dacha* which contained approximately 409 tons. For example, a Soviet Army preparing for an offensive might amass 5 *zapravki* (2,010 tons of fuel), 2.5 *boekomplekt* (5,420 tons of munitions) and 13 *s/dacha* (5,317 tons of food/fodder) before being ready to attack. [155] The logisticians would calculate the requirements for an army to undertake a specific operation, then they would analyze how long it would take to accumulate the required supplies and where to store the supplies until needed. Obviously, during this planning and accumulation phase, *Stavka* or the Front commander might order the army to attack regardless of the supply situation, or the enemy might attack; either of these situations would greatly complicate logistical planning.

20 June 1943

The appalling basic supply situation affecting frontline armies continued. For example, "M. S. Shumilov on 20 June 1943 personally requested emergency aid from Lieutenant General S. P. Ivanov, the Voronezh Front's Chief of Staff:

> ... the Army presently lacks sufficient flour or hardtack, tinned meat or meat, and sugar in its stockpiles; I can feed the divisions a full assortment [of products] only tomorrow, but for 21 June [the] 36th Guards and 81st Guards Rifle Divisions will not have meat, while 36th, 73rd and 78th Rifle Divisions will not have flour or hardtack, and there is still nothing on the way. I ask that you take urgent measures.

The overall supply problem was compounded by laziness or dereliction of duty by commanders at the front and within the rear supply organizations. [Front line veterans] often called those who served in the rear echelon units "rear rats." Because of their negligence, tens of thousands of men each day who were under the stress of mortal danger were unable to receive the most basic needs — a clean uniform and a decent meal. [156]

Another night attack, this time against Gorki, cost the Germans six bombers. These bombers would soon be needed for other targets.

21 June 1943

Hitler confirmed the opening date for Operation Citadel as 3 July, and then quickly changed it to 5 July 1943. Everyone concerned must have wondered if Hitler ever would launch this operation.

This night, a final raid against Gorki cost the Luftwaffe two bombers shot down by VVS night fighters. "Thus ended the relatively short-lived German strategic bomber offensive on the Eastern Front." [157]

22 June 1943

Toward the end of June, Lieutenant-General of Technical Troops, I. V. Kovalev summed up the supply situation. "At the end of June, the supply of troops, military equipment and supplies of goods by rail from the deep rear was basically completed." [158] General Shumilov and other frontline commanders might disagree with this opinion as their troops were still short of all types of supplies. It is one thing to have full warehouses and depots in the deep rear area, and another for the supplies to actually reach the front. Despite shortages, at this point the Red Army was more than ready for whatever the Germans could do. The Germans, on the other hand, were seriously short of fuel both for vehicles and for aircraft.

23 June 1943

Another German anti-partisan operation named *Tannhaeuser* (named after a 13th century German poet), was concluded on this day with no contact being made with partisan units. The partisans, who were very familiar with their local areas, just slipped away into heavily wooded or swampy terrain and returned when the Germans were occupied elsewhere.

24 June 1943

NKO issued Order #38 that announced incentive payments for soldiers that disabled or burned-out enemy armor.

- AT Rifle teams could earn 500 rubles for the team leader and 250 for his assistant.
- Tank crew commander, driver and gunner could earn 500 rubles each with the loader earning 200 rubles.
- Artillery gun commander and gun layer (person who aims the gun) could earn 500 rubles with the remaining crew members earning 200 rubles each.
- For destroying a tank using individual AT weapons, each fighter or commander would earn 1,000 rubles with smaller amounts awarded to each individual of a larger tank hunting team.

This order went into effect on 1 July 1943 and was intended to inspire fighters and commanders to increase their effectiveness against enemy tanks. [159]

But there were a few problems with this plan. Soviet soldiers out in the field had nothing to purchase with rubles, so most of them sent any pay home to their parents or families. But, due to the massive war effort, (Guns! Tanks! Aircraft!) civilian goods hardly were produced. So, civilian workers putting in 10-12-14 hour workdays also found little to buy. To illustrate this, during the time of the Soviet Union, a Russian folk saying went, "They pretend to pay us, and we pretend to work. Some people have millions [of rubles] stashed away, [but] the joke is, there's nothing to spend it on." This pretty much sums up the situation.

In terms of foreign exchange during the war, the situation was much the same. Rubles were almost worthless outside the Soviet Union. That is why Stalin began sending gold to the west as partial payment for Lend Lease deliveries. One of these shipments, containing 465 gold bars (23 lbs. each) worth approximately 50 million dollars during World War II, were on-board the British cruiser HMS Edinburgh when it sunk in 1942. This lost treasure partly was recovered, but not until 1981. [160]

In the far off Obersalzberg area of the Bavarian Alps, Hitler was at his Berghof home with Dr. Paul Joseph Goebbels, his propaganda minister. Hitler admitted to Goebbels that his grand designs in the east were now lost dreams. He went on to say the Kursk attack would only be a minor correction of the front line[161] and not a great victory. But Hitler was wrong! It would be a great victory, but only for the Red Army, and soon after the front line would experience a major correction as it began to move rapidly to the west.

25 June 1943

At last, Hitler confirmed the final date for launching Operation Citadel as 5 July 1943.

The 7th Guards Army, at this time, had a combined combat strength of 79,615 men. Its seven rifle divisions had the following strengths:

15 GRD - 8,250
36 GRD - 8,260
72 GRD - 8,208
73 GRD - 8,295
78 GRD - 7,850
81 GRD - 8,073
213 RD - 8,669

The continued inflow of new troops had boosted these rifle divisions to near the desired strength.

26 June 1943

General Hoth stated that chances for success at Kursk decreased each day. The truth was even grimmer than that; the possibility for any successful German summer offensive had in fact long passed. During this summer offensive season, Stalin would be dictating the pace of combat operations, with Hitler and the German Army struggling to parry the increasingly powerful thrusts of the restored Red Army.

27 June 1943

By late June, the OKH and army commanders had lost much of their confidence in the attack. They briefly considered a last-minute change in the overall plan, but Hitler wanted to stay with the original strategy. In the end "the original attack plan was maintained despite the doubts that had arisen in many places." But it was noted that "during the rest period prior to the offensive, the state of training, equipment, and morale of the units earmarked to participate reached the highest peak ever attained heretofore and thereafter during the entire Russian campaign. The assault divisions and army troops moved quickly and efficiently into their assembly areas." [162] This was the last calm period before the breaking storm of battle.

28 June 1943

By this date, on the southern side of the Kursk Bulge, General Vatutin's Voronezh Front had assembled a vast force to oppose Field Marshal von Manstein's Army Group South. These two groups contained approximately the following men and armaments:

Formation	Men	Guns and Mortars	Tanks	Aircraft
Voronezh Front	625,591	8,718	1,704	881 (2nd AA)
AG South	397,000	3,600	1,617	1,100 (4th AF)

Table #3: June 1943 Southern Side of the Kursk Bulge Strength Comparisons.

The exact number of troops and equipment that actually were committed to the battle varied, but as can be seen by these totals, the odds were stacked in the Soviets' favor. Common rule-of-thumb knowledge states that normally the attacking army needed a 2 or 3 to 1 advantage over the defender to be successful. In this case, the defender heavily outnumbered the attacker in manpower, guns and mortars. This fact alone did not bode well for German chances at Kursk. Stalin and *Stavka*, of course, had even more armies in reserve. The overall Soviet plan was not only to defeat the German Kursk attack, but also to launch massive counterattacks all along the front after the Red Army won the defensive battle. In the southern section of the front, they planned to drive the Germans back toward the far-off Dnepr River and to recapture Kiev, the capital of Ukraine.

29 June 1943

The sustained Russian attacks against rail targets both by partisans and by the VVS during June were successful. The Germans reported the loss of 298 locomotives; 1,222 rail cars; and 44 bridges just in the Army Group Center's area of operations alone. Additionally, "between 1 April and 4 July 1943, the ADD (Long Range Aviation) performed 9,400 sorties against rail targets." [163] This effort was taking its toll on German ground and air forces as well as on their supplies.

The 7th Guards Army combat strength now stood at 77,224 men. Its seven rifle divisions had further increased their fighting power to the following strengths:

15 GRD - 8,472
36 GRD - 8,535
72 GRD - 8,455
73 GRD - 8,452
78 GRD - 8,223
81 GRD - 8,461
213 RD - 8,756

As can be seen, the 7th Guards Army had finally brought its rifle divisions up to the authorized strength of at least 8,000 men each. The Army now fully was ready for battle.

30 June 1943

All the air activity during June not only caused losses in German aircraft and flight crews, but the fighting also consumed vast amounts of scarce fuel. Stocks of B4 [89 octane] fuel dropped to 4,886 tons, with daily consumption running at 287 tons. "During June, against a total consumption of 8,643 tons of B4, only 5,733 tons had been delivered. During the same month, the total consumption of C3 [94-octane aviation fuel] used by our Fw 190s had been 1,079 tons, as compared to 441 tons delivered."[164] One wonders if this situation would have been different had Field Marshal von Richthofen not been sent to Italy earlier in the month. The consequences of having limited fuel stocks would force the Luftwaffe to ration its ground support effort to only critical situations and/or during the climax of a particular battle. The persistent efforts by the VVS in May and June had indeed won a victory over the German Luftwaffe even before the coming ground battle had begun.

At the same time, the VVS benefitted from increasing amounts of high-octane (99 octane) gasoline received from the U.S. and UK via Lend Lease. The U.S. also sent Russia large amounts of octane blending agents, which allowed the Russians to boost the octane levels of locally produced gasoline. The octane level for aircraft fuel is critical because higher octane equals more power, which allowed aircraft to fly further, carry heavier loads, and/or fly faster.

> As the result of the fierce struggle for strategic air supremacy in the spring and early summer 1943, the power of the German aviation was sapped away…[The Luftwaffe] was already losing the strategic initiative but continued to maintain its grasp on it with all forces. The struggle was resumed with new forces at the beginning of July 1943, when the Germans went over to the offensive in the Kursk sector. [165]

On or about this date, General Reinhard Gehlen, head of the *Fremde Heere Ost* (Foreign Armies East, *FHO*, intelligence unit), [166] concluded that the operation (Citadel) "could no longer be justified on any grounds, since both essential preconditions-numerical superiority at the decisive point, and the element of surprise, were no longer met." He went on to state: "The Russians have been expecting our attack in the relevant sections of the front for weeks and have done everything, with characteristic energy, to absorb it at an early stage both by building several defensive positions and by an appropriate deployment of forces." In closing he stated, "I consider the envisaged operation a critical error of the highest order that will have very serious consequences." [167] Ultimately, despite Gehlen's critical assessment of the situation, Hitler ignored this accurate intelligence report and decided to continue with Operation Citadel as planned.

On the eve of Operation Citadel, a review of the ages of key leaders and commanders might be insightful:

- Adolf Hitler was 54, while Joseph Stalin was 63.
- Army Group South Commander Erich von Manstein was 55, while the Voronezh Front Commander Nikolai Vatutin was 41.
- 4th Panzer Army Commander Herman Hoth was 58 while the 5th Guards Tank Army Commander Pavel Rotmistrov turned 42 on 6 July.
- Army Detachment Kempf Commander Warner Kempf was 57 while 7th Guards Army Commander Mikhail Shumilov was 47.

No one knows if the age difference among these individuals had any effect on the battle, but overall, Red Army commanders were far younger than their Wehrmacht counterparts. [168] With the Germans' age came patience, experience and wisdom, but the Soviet's relative youth may have offered more energy, resilience and optimism for the ensuing battle.

Despite the 7th Guards Army not being actively engaged in combat, life along the Kursk front was not without its dangers. A partial casualty report (TsAMO ID #133373291), covering only ten days, shows the Army suffered 860 casualties, killed, wounded, sick/injured, or missing. If this loss rate is extrapolated out for the rest of the month, then the Army most likely suffered about 2,560 casualties for the entire month.

The stage was set, forces were assembled, attack and defense plans were completed, and weapons and supplies were gathered in vast quantities. Little chance remained to alter what would take place. The last few days ticked by as millions of troops only could wait for the first signs of the coming battle to appear!

1 July 1943	
7th Guards Army	
Assigned – Voronezh Front	
Lieutenant General M.S. Shumilov Commanding	
Rifle, Airborne and Cavalry Units	24 Guards Rifle Corps (15, 36, 72 GRDs) 25 Guards Rifle Corps (73, 78, 81 GRDs) 213 RD (army reserve) 82, 83, 84, 85, 86 Blocking Detachments 62, 63, 64, 65, 66 Penal Companies
Artillery Units	109, 161, 265 Guards Gun Artillery Regiments 30 Antitank Artillery Brigade 114, 115 Guards Antitank Artillery Regiments 1669, 1670 Antitank Artillery Regiments 1, 2, 3, 4, 5 AT Rifle Battalions [169] One Molotov Cocktail Company (unknown #) Five Flamethrower Companies, (176 Backpack, 191 Separate and three other unknown companies) 290 Mortar Regiment 5 Antiaircraft Artillery Division (670, 743, 1119, 1181 AA Regiments) 162, 258 Guards AA Regiments 662 Separate Aerial Observation/Air Raid Warning Company
Armored and Mechanized Units	27 Guards Tank Brigade 201 Tank Brigade 148, 167, 262 Separate Tank Regiments 1438, 1529 Self-propelled Gun Regiments 34, 38 Separate Armored Train Battalions (three trains active)
Engineer Units	60 Engineer-Sapper Brigade 175, 329 Separate Engineer Battalions
Reported Combat Strength Estimated Ration Strength	81,850 men [170] 88,850 men and women (including an estimated 7,000 support personnel)

(BSSA 43, BSA 43, TsAMO Archive) [171]

1 July 1943

Weather conditions leading up to and during the battle were normal for summer months in central Russia. Random scattered rain showers and even thunderstorms occurred throughout the area followed by hot, dry conditions. Noteworthy rains occurred for the Voronezh Front on 1, 3, 4, 7, 9, 10, 11 and 16 July. [172]

A typical account related, "The weather was not always favorable. I recall a few awful thunderstorms causing the soil to turn into something remotely akin to liquid concrete making any progress with tracked vehicles next to impossible." Another recalled that after a rain, "As soon as the black soil had dried out it was as hard a surface as tarmac [black top]. But after a thunderstorm it softened up again and turned to black mud within 15 minutes, liable to take one's boots off." [173] While weather conditions affected both sides, bad weather had a greater impact upon the attacker, who was forced to advance and attain objectives to succeed.

By this date all Russian armies within the Kursk Bulge area were on constant alert. Intelligence reports indicated the Germans would attack soon, but the exact date, time and location were unknown. Large reserve forces were ready to directly intervene in the upcoming battle if needed. After months of preparation, additional armies were prepared to launch major offensives of their own when *Stavka* felt the time was right. The Red Army was ready for anything the Germans would do.

Meanwhile, the 7th Guards Army had settled into its final positions. From north to south, the 25 GRC had the 81 GRD and 78 GRD in the front line with the 73 GRD supporting in the second line.

The 24 GRC had the 72 GRD and 36 GRD in the front line with the 15 GRD supporting in the second line. The 24 GRC also had both the 34 and 38 Separate Armored Train Battalions with three active trains (with one under repair) in support throughout the battle. Additionally, the 213 RD was under the 7th Guards Army control; it too was placed in a supporting position in the second defensive line. (See Map #7. Note: Not all units are displayed on the maps.)

In the Army's rear area, the 30 AT Artillery Brigade (with three AT regiments) and the 5 AA Artillery Division (with four AA regiments) were deployed in the second army-level defense line, along with two extra AA regiments. Of the six AA regiments available, five covered infantry units while the sixth one defended the Army's main supply center at Volokonovka. Volokonovka was far to the rear and one of the original detraining locations when the 64th Army arrived at the front in March. Along the way, the Army also was assigned a dedicated Molotov Cocktail company for tank hunting purposes. Additionally, the Army had five Flamethrower Companies, most likely spread out with rifle or engineer units along the front. (The deployment location of these small units is unknown). Not to be forgotten, the deputy commander of the 7th Guards Army, Lieutenant General V. S. Petrov, arrived at the headquarters of the two armored train battalions and ordered them to support frontline units with their fire to prevent the enemy from breaking through the Army's defenses. [174] Every weapon was being prepared for the coming battle.

The 7th Guards Army issued a report on food and ammunition supplies held by the Army. Ammunition stores were adequate, but there were no stocks of naphtha or vodka, and low amounts of meat and hay. In other words, the Army was adequately but not overly supplied for the coming battle. [175] Of note, the strength of the Army's rear area support (non-combat) units had stabilized at about 7,000 men and women. This total would vary somewhat depending on conditions and daily casualties, but it basically remained unchanged for the rest of the year.

2 July 1943

"German radio news service reported in the morning that *Generalfeldmarschall* von Manstein would fly to Bucharest in order to award the Crimean Shield in Gold to the Romanian Chief of State, Marshal Antonescu. He would, it was spread around, remain several days in the Romanian capital." However,

"*Stavka* recognized this as a German deception operation." Meanwhile, after photos were taken at the ceremony, von Manstein flew back to the front that same day and then traveled in his command train to the Belgorod area to be ready for the coming battle. [176]

Stavka alerted the front commanders that the German offensive would start sometime between 3-6 July.

3 July 1943

For the following combat actions, this account will cover mostly the operations of the Voronezh Front, and the 6th and 7th Guards Armies along with the 69th A. On the German side, the main focus will be on the activities of Army Group Kampf but also will include relevant actions from the nearby 4th Panzer Army on the south side of the Kursk Bulge. Events on the north side of the Kursk Bulge will be discussed briefly as needed.

A great deal of confusion still exists in the literature about different starting times of the battle for Kursk. The Germans used Berlin time and the Russians used "Decreed Time," [177] which meant a one-hour difference between the two. So, if the Germans documented their own attack as starting at 05:00 hours in German records, the Russians would record this attack in Soviet records as starting at 04:00 hours. This discrepancy makes it all the more confusing for historians and readers to follow the battle as it unfolded. (Note: The author has made no attempt to alter the stated times reported in different historical records or battle accounts, so the reader should take into account this time shift between German and Soviet records.)

Rainy weather on this day restricted the movements and actions of both the Germans and Russians.

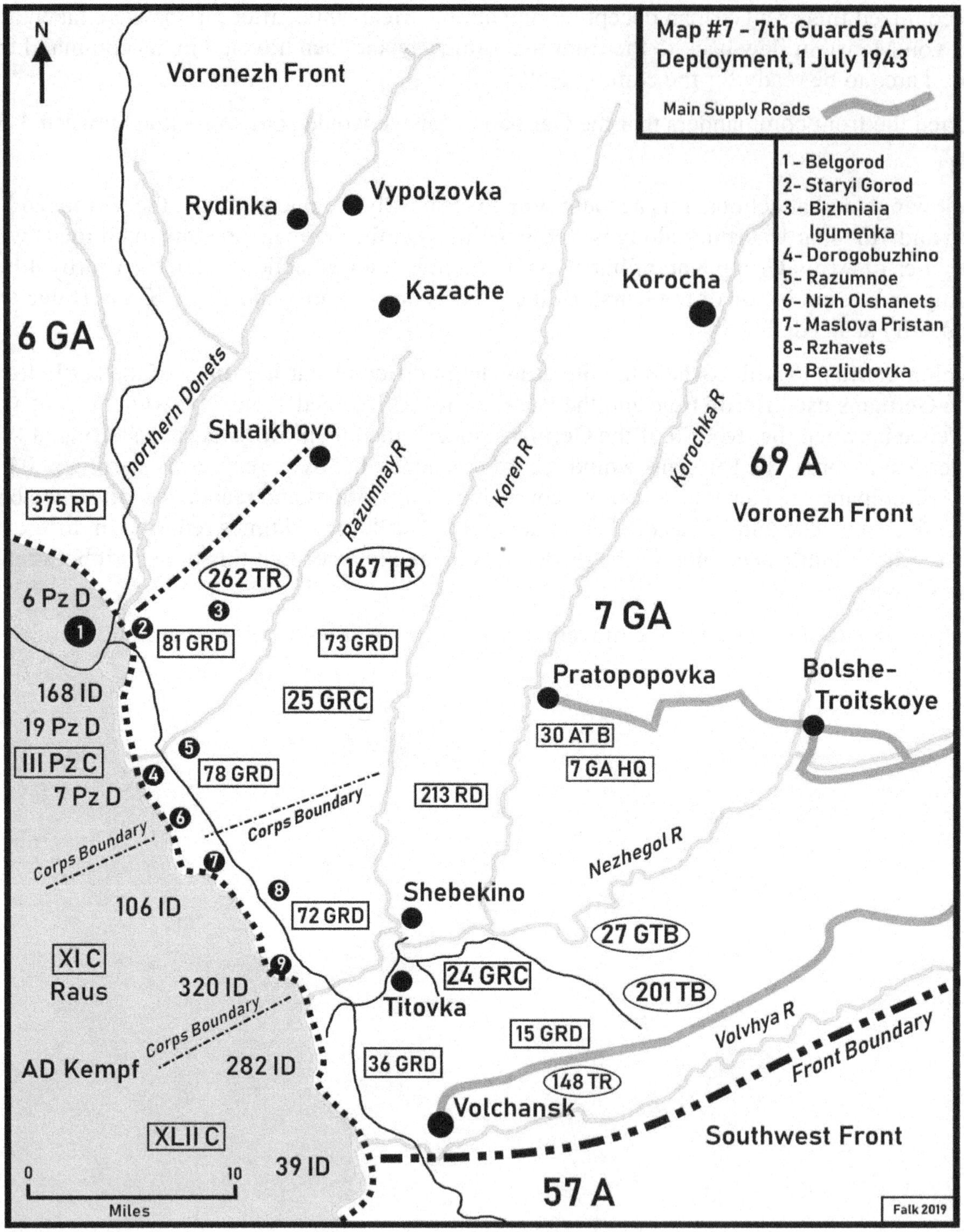

4 July 1943

The Germans made their first preparatory move of Operation Citadel in the southern attack area. "Shortly after 16:00 hours [15:00 hours Russian time] on 4 July 1943, as was their custom prior to a major offensive, German forces along the southern face of the Kursk Bulge launched a reconnaissance-in-force." [178] This initial attack by the 4th Panzer Army was directed against positions of the 6th Guards Army. The attack was used to remove any enemy forward combat outposts and to determine the forward edge of the main enemy defensive line. This reconnaissance-in-force also allowed the Germans to establish good artillery observation positions to support the main attack that they would launch the next day. Units of the 7th Guards Army naturally noticed these actions and reported that in their area "the enemy conducted reconnaissance by small groups and in some areas tried to cross the eastern bank of the [northern Donets] river." [179]

The reconnaissance-in-force lasted most of the night and was supported by artillery and air attacks in the 4th Panzer Army's sector. Most frontline troops, on both sides, knew the start of the main attack must be only hours away. The night passed slowly as the tension increased on both sides of the front line. Just when would the real fighting start? Meanwhile both in the north and south, German assault forces had already moved into their deployment locations, but most had not yet been released to their assault positions. Once again, heavy rains fell in the area.

At 22:00 hours, [23:00 hours German time] 7th Guards Army Commander General Shumilov (code name Stepnoi) reported via teleprinter machine to General Vatutin (code name Nikolaev), Commander of the Voronezh Front, about the situation at the front: "I take the attacks by fire that the enemy is conducting as cover for the rotation of units. I've observed no large increase in infantry in this area, just the movement of small enemy groups from platoon-size to rarely company-size to and from the forward edge." He then went on to talk about Germans cutting down trees for defensive works. He added, "No enemy reinforcements have been detected." [180] This exchange appears to indicate General Shumilov was unaware the German assault was about to begin. Shumilov's message also indicated that Army Group Kampf largely was successful in concealing from the Soviets the massing of assault forces.

At about 22:30 hours, the commander of the 6th Guards Army ordered a short five-minute artillery strike against forward German positions of the 4th Panzer Army.

Sometime between 22:30 and 24:00 hours, General Vatutin ordered the commanders of the 40th Army and the 6th and 7th Guards Armies to conduct the pre-planned artillery counter preparation fire, but they were to concentrate it mostly in front of the 6th Guards Army sector. Whatever time this first large bombardment might have been ordered to start; it was a 30-minute-long barrage fired from hundreds of artillery pieces all along the front. This bombardment was intended to disrupt German troop deployment as units moved into their attack positions.

5 July 1943 – Day One of Operation Citadel

The 7th Guards Army combat strength now stood at 76,831 men, a slight reduction from the start of the month. With the addition of 7,000 support troops, the Army's ration strength had reached 83,831 people. These totals changed significantly, on a day-by-day basis, as the battle unfolded.

Early in the morning, the VVS launched a massive surprise attack against local German airfields filled with fighters and bombers. The 2nd AA from the Voronezh Front and the 17th AA from the Southwest Front participated. But the Luftwaffe had two advantages: mobile radar stations gave some warning of the approaching air attack, and most German aircraft were already airborne flying assigned missions to support the ground attack. Overall, the airfield attack mission failed, resulting in substantial losses for the VVS. During 5 July in the south, the Germans lost about 45 aircraft with VVS losses amounting to more than 150 aircraft. [181] Some Russian fighter losses for the 17th AA were not recorded.

During the early morning hours of 5 July, before the Germans launched their attack, General Vatutin again ordered the 6th and 7th Guards Armies to conduct a second planned artillery counter preparation fire, starting on or about 03:00 hours. This time General Shumilov and the 7th Guards Army got in the first shots against German assault forces concentrated along its own front. General Shumilov decided to conduct the ordered artillery counter-fire using options 3, 5 and 6, according to a previously developed plan. This concentrated the Army's artillery fire against targets directly in front of the 81, 78, and 72 Guards Rifle Divisions, in other words the 25 GRC holding the right flank of the Army. This barrage consisted of a five-minute fire raid which passed without interference from the enemy, then a 15-minute methodical fire against identified concentration positions. Then a last ten-minute fire raid at

03:20 hours [Russian time]. During this last phase, the enemy began their own artillery preparation fire to support their attack.

In total, the 7th Guards Army used 0.5 unit of fire worth of ammunition of all calibers for this artillery counter bombardment.[182] During the first earlier barrage ordered by General Vatutin, something close to another 0.5 unit of fire had been consumed. General Vatutin had put restrictions on the number of rounds that could be fired. So, the 7th Guards Army had fired about half of its on-hand artillery ammunition during these two separate artillery bombardments. The supply trucks and wagons must have been busy for the rest of the day hauling more ammunition from the rear ammo dumps to restock the guns. (Note: A Unit of Fire [UoF] is a logistical planning term [different for each army] for a standard, arbitrary amount of ammunition for a specific type of weapon. For instance, a UoF for a single Russian 122 mm heavy artillery piece is 40 rounds weighing 2,200 lbs. while a UoF for a single light 76 mm gun consists of 140 rounds and weighs 1,960 lbs.) [183]

Supply trucks drop off ammunition for 76 mm Zis-3 divisional guns.

Due to new information from German records, we now known that these preemptive bombardments all along the front delivered mixed results. In some areas they were not very effective as most German troops had not yet moved into their assault positions and were still in bunkers and trenches. In other areas, the Red Army scored direct hits on assembly areas with heavy loses of both men and equipment. These initial Russian artillery attacks mostly came too early to catch German units out in the open. Overall, Germans forces in the south suffered relatively few casualties and lost only minor amounts of equipment.

For General Shumilov, luck was on his side. These early barrages had their greatest impact against units of Army Group Kampf, which was deployed directly in front of the 7th Guards Army. AD Kempf had to cross the river and this section of the northern Donets was some 230-260 feet wide (70-80 meters) and the river banks were 3-5 feet high (1-1.5 meters). In this case the restrictive nature of possible river crossing points over the northern Donets, and good intelligence work, allowed the 7th Guards Army to identify the few possible crossing points before the battle. [184] This information then permitted the bombardment to achieve some disruption of the German timetable and to force a delay in the crossing of the river by some German units, most notably the 6 Panzer Division [185] and 19 Panzer Division. The losses of key bridges, bridging equipment, and engineering troops prevented the 6 Panzer Division from crossing into the shallow bridgehead at Mikhailovka and forced it to deploy further south to cross the river, which took the entire day, wasting much valuable time.

The early artillery fire also achieved significant disruption of the 19 Panzer Division during its crossing of the northern Donets River. Overall, German casualties would have been heavy in some areas and much lighter in others. Formations of Army Group Kampf had moved up early because of the added time needed to build bridges over the northern Donets. Because of this, they were hit the hardest.

Then on or about 03:30, German artillery units started their 30-minute bombardment of Russian positions. The German main ground assault, which was on time more or less, began at approximately 04:15 hours. [186] Operation Citadel was underway! Different German units along the south side of the Kursk Bulge were forced to stagger their opening artillery attacks and ground assaults, depending upon local conditions. Rain during the night made roads muddy which slowed movement, and the repeated Russian artillery bombardments caused various levels of disruption all along the front.

Fierce German attacks by two corps from AD Kempf, III Panzer and Corps Raus (XI Corps), aimed directly into the defenses of the 7th Guards Army, were able to force several crossings over the northern Donets and to establish bridgeheads on the east side. These attacks were met by powerful counterattacks from tanks and infantry along with large amounts of artillery and mortar fire from the rifle divisions and supporting units within the first army-level defense line.

At the end of the day, the 81 GRD had withstood repeated German assaults by the 168 ID, which failed to clear the way for the 6 Pz D to enter the Mikhailovka bridgehead. The 81 GRD gave up minor amounts of ground and repeatedly pounded German units, engineers, and bridging sites with strong artillery and rocket attacks as well as with flanking fire. As planned, on the first day the 81 GRD had completely prevented the German 6 Pz D from crossing the river and taking part in the assault. Its neighbor to the south, the 78 GRD, was forced away from the river but held a continuous front line that limited German gains during the first day. Likewise, the 72 GRD also lost some ground but firmly blocked further German advances. (See Map #8.)

General Shumilov committed his reserve 213 RD to a local counterattack toward Rzhavets. Attacking with tank support, the 213 RD struck advancing German units, mostly the 320 ID, thus limiting their progress for the day.

So, after an entire day of combat against the 7th Guards Army, the German offensive had only gained a small bridgehead about 7 miles wide and 2 miles deep on the east side of the river. German forces were firmly stuck in the first defensive line, but the deep thrust by the 7 Pz D was just able to pierce it in one place. Note: The Germans did not directly attack the sector of the front held by the 36 GRD and the supporting 15 GRD.

An indication of just how brutal the fighting was occurred in the German 320 ID of Corps Raus. They too had advanced over the Donets River on the far-right flank of the German attack, and during the day had gained only minor amounts of ground. A powerful Russian counterattack with tanks quickly stopped their advance. During this counterattack, units of the 72 GRD had captured some 150 German infantry from the 320 ID. Later that day, while monitoring Russian communications, the Germans heard what they assumed was a regimental commander reporting to the divisional commander, "I have 150 Fritzes. What shall I do with them?"

The reply was, "Keep a few for interrogation, and have the others liquidated."

Later that night the Russian regimental commander reported, "The order was executed and … the majority of the Fritzes had been killed immediately, the remainder after they had been interrogated." [187] With incidents like this as motivation, most troops on both sides preferred to fight to the death rather than to surrender. Of special note, the Russians were worried about the careless use of radio and telephone landline communication systems. They stated the Germans were able to eavesdrop from 5 kilometers away. Maybe this is what happened in this case with the 72 GRD.

To the north, the German 4th Panzer Army made significant gains against the defending 6th Guards Army but were unable to achieve a major breakthrough.

The German attack against the 7th Guards Army made only minor gains at a heavy cost. The northern Donets River and Russian defensive positions proved to be a considerable obstacle on this first day of Operation Citadel. (See Map #8.)

Meanwhile, overnight General Shumilov shifted his troops around to counter the next series of attacks that were sure to come in the morning. This effort was greatly helped when the Voronezh Front ordered the 69th A to transfer control of the 111 RD and the 270 RD (49 RC) to the 7th Guards Army. The Voronezh Front also hurried to move the 92 and 94 GRDs (35 GRC) south to assist the 7th Guards Army. Note: The overnight shifting of units by General Shumilov makes it difficult to map their movements, so in most cases, their final positions are displayed instead. Also, the 35 GRC was in fact the recently promoted 7 RC, which fought with the 64th Army at Stalingrad. The 7 RC had been promoted to Guards status on 18 April 1943.

The fighting above the battlefield was just as brutal as on the ground. During the day, along the southern section of the Kursk Bulge, the VVS flew 1,688 sorties while the Luftwaffe flew 2,387 (1,071 by Stukas, 536 by bombers, 335 by ground-attack planes and 371 by fighters with the final 74 sorties most likely being reconnaissance missions). Just over the northern Donets River area, the VVS lost 55 Il-2 (Sturmovik) trying to attack the river crossings being used by AD Kempf Soviet sources state that the VVS lost 186 aircraft in the south during this first day of battle. German sources list Luftwaffe losses as 27 aircraft due to hostile action with another 18 suffered severe to medium damage, for a total of 45 aircraft. [188] Thus, this battle produced a Soviet aircraft loss ratio of 186 to 45 (4.13:1).

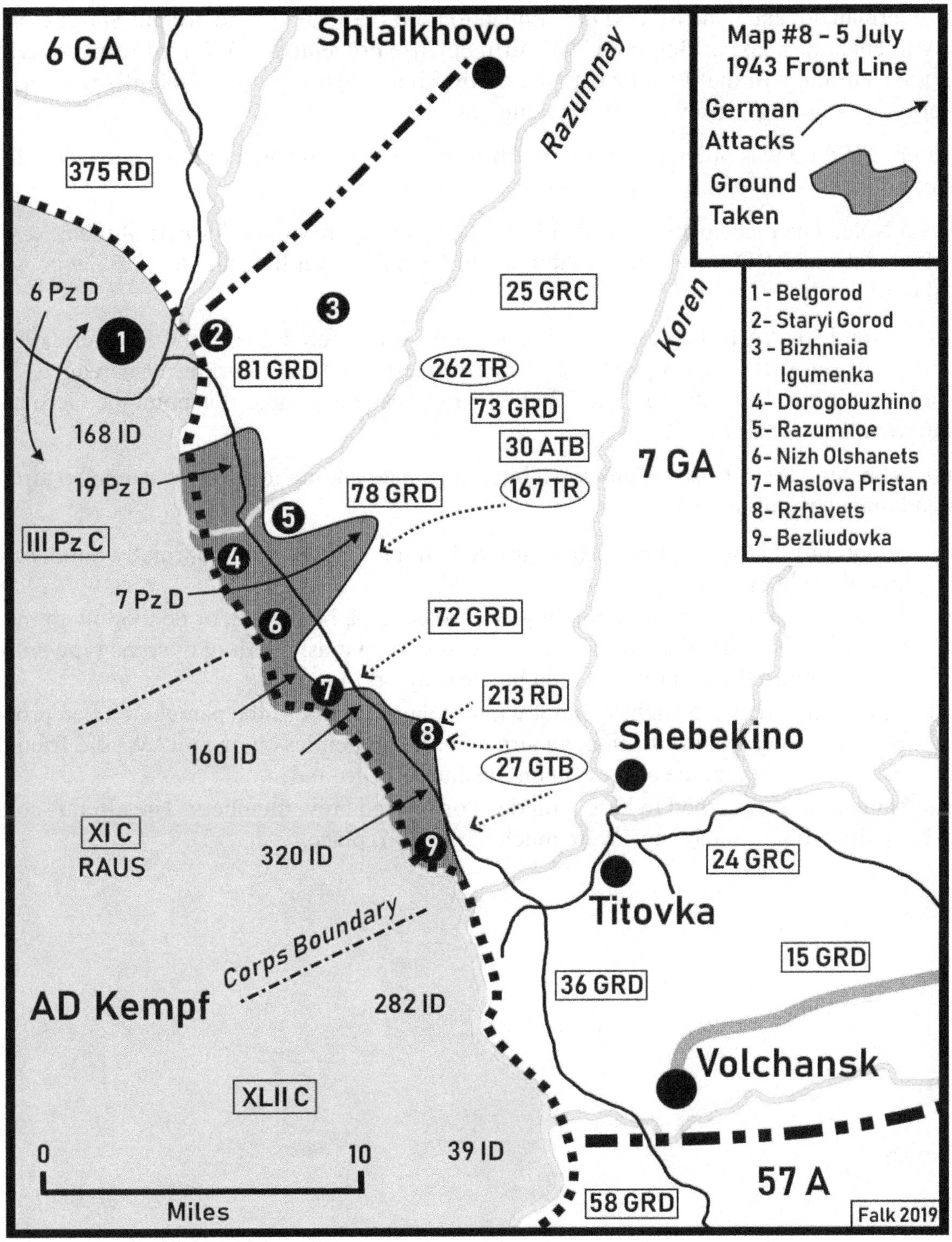

6 July 1943 - Day Two of Operation Citadel

Corps Raus, 320 ID began to abandon the small bridgehead on the east side of the river south of Bezlidovka. Constant attacks by the 72 GRD with the support of tanks and heavy artillery fire made the bridgehead too costly to hold.

The three German Panzer divisions of AD Kempf pushed forward and greatly expanded their bridgehead, breaching the first army-level defense line and reaching the second-level defense line in force. The 6 Pz D was forced to cross the northern Donets over bridges established by the 7 Pz D, which took much time and overloaded the road network in that area.

Despite this German advance, the 81 GRD continued to hold the vital ground around Staryi Gorod and Bizhniaia Igumenka, just east of Belgorod. This firm defense prevented AD Kempf from covering and supporting the right flank of the II SS Pz C to the north. That support was AD Kempf's main job, and General Kempf was failing to provide this flanking cover.

Also, the reserve 198 ID was ordered to move north from the Kharkov area to reinforce Corps Raus. This transfer would take several days to complete.

(See Map #9.) Note: The placement of the 213 RD on maps is approximate for part of the next week. The 72 GRD and the 213 RD were shifting their positions in the front line, and that exchange would not be finalized until 9 July.

During the day, the 17th AA lost 28 aircraft while the 2nd AA suffered the loss of 51 aircraft. The opposing *Fliegerkorps* VIII (Air Corps 8) lost 11 aircraft destroyed and a further 11 damaged, [189] producing a VVS aircraft loss ratio of 79 to 22 (3.59:1). As can be seen, at this point the Germans continued to dominate the skies.

The human cost behind these losses is unknown but a fortunate ending to a lost or damaged aircraft would depend upon several factors:

- An aircraft that was damaged or destroyed on the ground would normally not involve the loss of flight crews.
- A damaged aircraft in flight might go out of control, be on fire, or be too low to the ground for the crews to use their parachutes. In that case, a crash landing of some type with possible injuries or fatalities would be the only option.
- Exiting an aircraft in flight was not without danger: Would the parachute open properly? Would an enemy aircraft shoot an airman hanging helpless in the air? Would friendly or enemy troops greet the aircrew upon reaching the ground?
- Many aircraft returned to base with dead or injured crew members. The aircraft could be repaired, but trained crews were much harder to replace.

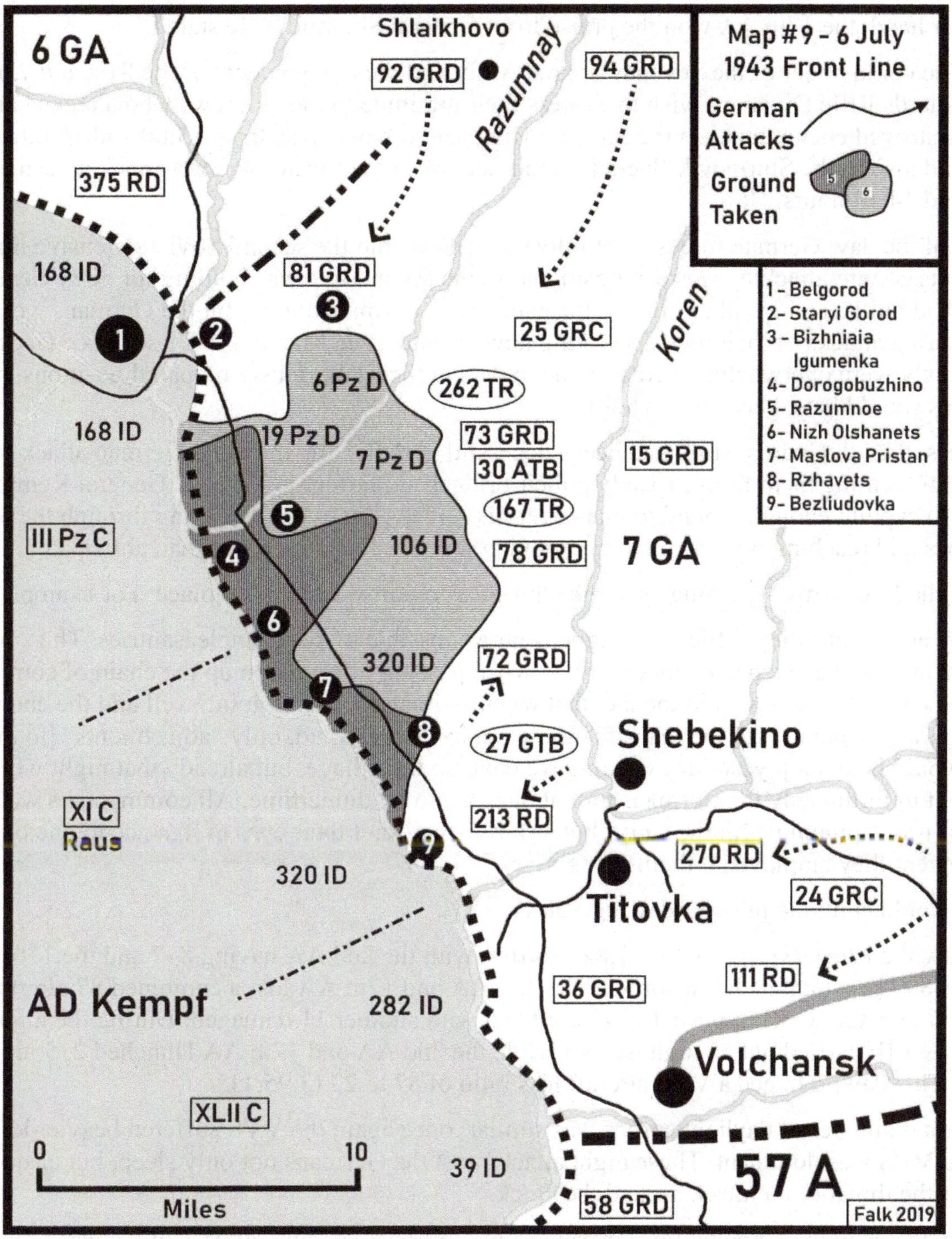

7 July 1943 - Day Three of Operation Citadel

The 92 and 94 GRD, along with the 31 ATB, joined the 7th Guards Army in the front line. These reserve units were subordinate to the Voronezh Front. General Vatutin sent the 7th Guards Army a steady flow of new units to build a solid barrier in front of AD Kempf's attack. The 49 RC came forward from the 69th A to fill the gap between the 24 GRC to the south and the 25 GRC in the north. The 111 RD and 270 RD also shifted north to hold the Koren River line. As the front shifted and opened up, the Red Army had plenty of new units to plug frontline holes and to launch counterattacks.

Despite the rains in the area, General Shumilov noted that pilots from the 2nd AA, using Sturmovik aircraft, bombed and strafed his 148 TR, among other Red Army units of the Voronezh Front. [190] He was not pleased, as friendly fire could be just as lethal as enemy fire.

On the other hand, the 17th AA won the praise from General Shumilov. He stated,

> The commander of the Seventh Guards Army and the commanders of the 81st and 73rd Guards Rifle Divisions wish to express their gratitude to the air crews who attacked and destroyed enemy tanks in the Melekhov and Chyornaya areas from 15:00 and 16:00 hours, and also to the Sturmovik flyers for their actions in the Maslov Pristan area between 13:25 and 14:13 hours. [191]

At the end of the day, German forces were fully engaged within the second Soviet defensive line with Russian units counterattacking German positions, so the second line was holding for now. General Shumilov had built a solid wall of units in the path of the German attack. But the Germans would not stop conducting attacks of their own and taking new ground. (See Map #10.) These fierce German assaults slowly were overwhelming Russian defenders in their laboriously prepared positions, positions the Russians should have been able to hold.

Nevertheless, the 7th Guards Army had been successful in deflecting the main German attack to the north, thus preventing them from expanding their bridgehead farther to the east. General Kempf was struggling to capture enough ground to maneuver his forces effectively. Breaking through the Russian defense lines and reaching open ground was proving much harder to achieve than anticipated.

Of note, in the Red Army "spinning" a combat progress report was commonplace. For example:

> by any means they [officers] sought as far as possible to avoid unpleasantries. Thus, in the Red Army there existed a practice of lying when passing information up the chain of command. The essence of it consisted in the fact that when a situation wasn't going well and the enemy was gaining ground, then the real situation wouldn't be reported, only "adjustments" [to the report would be made]: yesterday evening we gave up the village, but already that night we won it back, but unfortunately the Germans took it back again by dinnertime. All commanders were well aware of this "spinning", [of the story] but as a rule they shut their eyes to it, since to one or another extent they employed it themselves. [192]

The battle would continue the next day regardless.

This day saw the *Fliegerkorps* VIII fly 1,829 sorties, with the 2nd AA having 847 and the 17th AA producing 689, for a total of 1,536 sorties. The 2nd AA and 17th AA lost a combined 87 aircraft. Once again, the *Fliegerkorps* VIII reported 11 aircraft lost with another 11 damaged. During the night, *Fliegerkorps* VIII carried out 52 night sorties while the 2nd AA and 17th AA launched 275 nighttime sorties. [193] This day produced a VVS aircraft loss ratio of 87 to 22 (3.95:1).

While the total number of daylight sorties was similar, once again the VVS suffered heavier losses. But at night the VVS was dominant. These night attacks cost the Germans not only sleep, but also kept them under the threat of air attack around the clock.

After heavy losses in the air and with German ground forces advancing, after 7 July, "when the German troops expanded their bridgeheads, [air] attacks against the crossings [over the Donets] began to taper off." This shift in focus of the VVS also was due to heavy losses from ground fire. "During the battle [in the area of the 7th Guards Army], the formations of the 17th Air Army suffered very heavy losses, primarily from anti-aircraft artillery fire." The 17th Air Army had in fact lost 56 IL-2 aircraft in the first three days. [194]

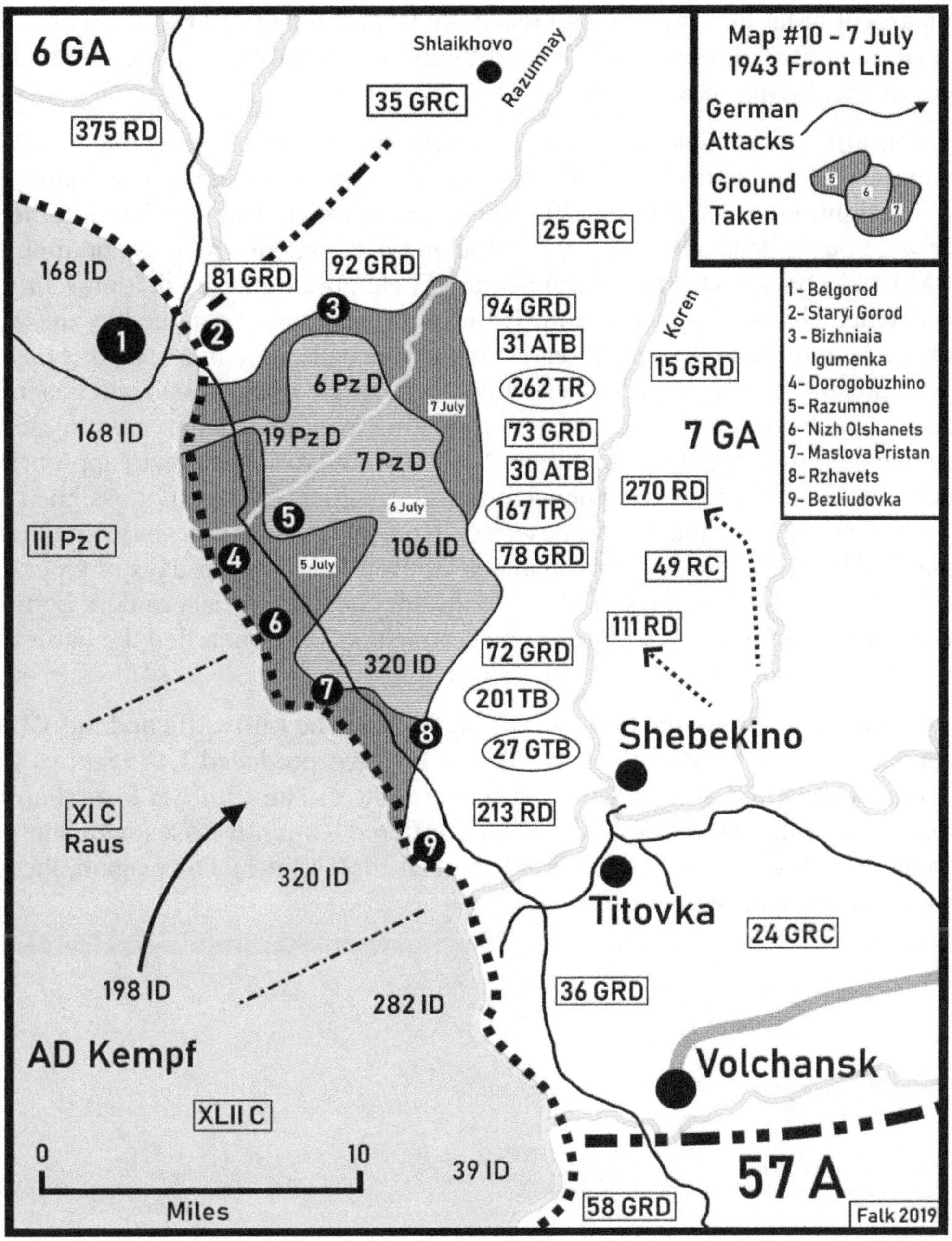

8 July 1943 - Day Four of Operation Citadel

The III Pz C launched a powerful attack to the north, between the first and second Soviet defense lines, and reached the small village of Shlaikhovo. The Germans had found the weakness in the design of Russian fortifications.

To counter German attacks in this area, the 15, 73 and 94 GRDs launched a series of flank attacks against the 7 Pz D, starting late in the evening of 7 July and then again at 10:30 hours on 8 July. These regimental size attacks were disorganized, only supported by the 30 ATB with little artillery. After fighting most of the day, they had gained little ground and were soon forced back by strong German counterattacks supported by tanks and artillery. [195] Losses appeared to be heavy for both sides.

With the continued German advance to the north, the 81 and 92 GRD were forced away from the 7th Guards Army, which resulted in supply and command problems. As the battlefield opened up, units were shifted around in response to enemy actions and become disorganized. To fill the growing gap between armies, General Vatutin urgently ordered the 305 RD (69th A) to move south toward

Shlaikhovo. This urgent order took six hours to reach the HQ of the 305 RD and its late arrival "created a real threat of a German breakthrough." [196] General Shumilov was not the only commander having command and control problems. (See Map #11.)

Of note, Valeriy Zamulin, in his book *The Forgotten Battle of the Kursk Salient*, page 399-400, stated that, "Signs of fatigue and the unsteadiness of units were also appearing in the 7th Guards Army." That during "the heaviest fighting on 5-6 July, 15-20 soldiers and commanders who had left the battlefield without an order were being detained each day by blocking detachments, pickets and mobile groups of NKVD troops. [Most likely OO NKVD/Special Section.]" He goes on to say that on 7 July the number of men detained reached almost 200 men… and "[o]n 8 July the number detained by the 92 NKVD RR [197] alone multiplied. According to the operational summary of its headquarters, on this day 758 military servicemen were detained." Numerous Army checkpoints, traffic control points and general rear security units also would have added to the barriers retreating troops needed to pass to escape the front. The fate of these detained men is unknown, but the NKVD would deal with them appropriately. Most would be returned quickly to their units, some were sent for medical attention or assigned to penal units, and a few were executed on the spot as deserters. Honestly, most of these fleeing men came from rifle divisions that had been hit hard by German attacks during the first four days of Operation Citadel, the 73, 78, and 81 GRDs. After four days of almost constant combat, artillery and air bombardment, along with little sleep or food, it should come as no surprise that some men fled the battlefield. Combat soldiers, on both sides, can only endure so much before breaking.

Due to losses and high usage during the first days of battle, both the Luftwaffe and the VVS recorded a decrease in sortie rates. During the day, the 2nd AA and 17th AA produced 1,185 sorties while Fliegerkorps VIII launched 1,686 sorties. Losses also were lower. The 17th AA lost 14 aircraft, with the 2nd AA losing 47 aircraft, while *Fliegerkorps* VIII suffered 9 aircraft losses with another 12 damaged, [198] producing a VVS to Luftwaffe loss ratio of 61 to 21 (2.9:1). Once again, the Luftwaffe was able to maintain an advantage over the VVS.

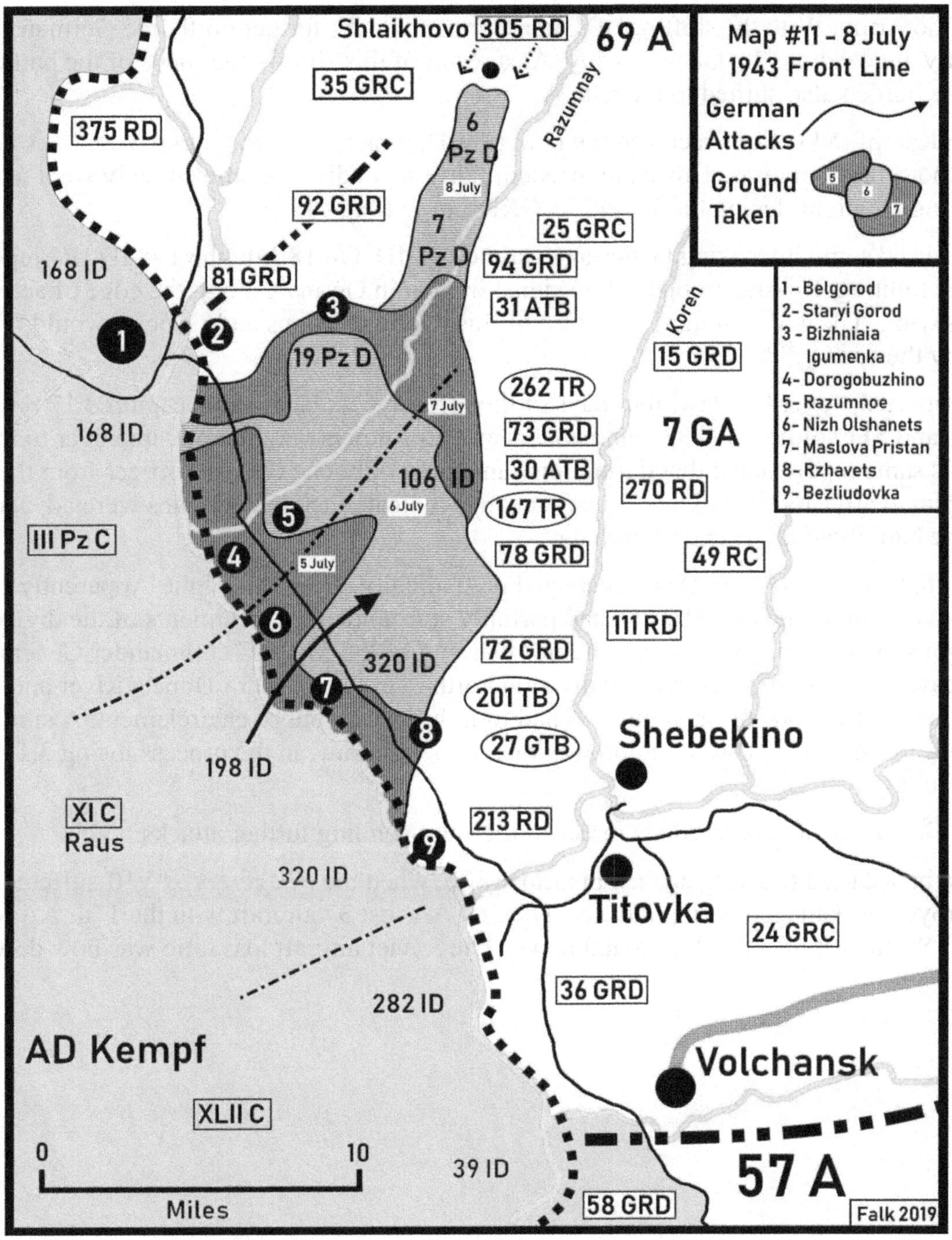

9 July 1943 – Day Five of Operation Citadel

This day was the first of a three-day stretch of substantial rain.

The Voronezh Front's Military Council directed the 69th A to assume control of the section of the front between the 6th Guards Army and the 7th Guards Army. So, the 81 GRD, 92 GRD, 94 GRD, 305 RD, 375 RD and 31 ATB (and other units) were assigned to and supplied by the 69th A. Along with this action, General Vatutin also changed the boundary lines between his armies to improve command and control arrangements. With this change, the 69th A was charged with stopping the advance of the German III Pz C toward the north. Likewise, General Shumilov issued Order #00285/op outlining corps sectors for the 24, 25 GRC and 49 RC of the 7th Guards Army.

Meanwhile the 7th Guards Army was to conduct an active defense (limited attacks and counterattacks) against the flank of AD Kempf, XI C Raus. The idea was to engage and to hold in position German units facing the Army. With the shifting of major German attacks further north, the German III Pz C no longer directly attacked the 7th Guards Army. As a result of this shift in the focus of the battle, the major defense burden also shifted to the 69th A.

Army Group Kampf, XI C Raus received the new 198 ID, which was brought up from the south near Kharkov in motorized columns. This badly needed, fairly fresh division immediately went into the frontline during the night, facing the 15 and 73 GRDs. (See Map #12.)

The Russians used some interesting tactics against the 198 ID. On 18 July the 198 ID HQ reported that during earlier fighting, Russian troops put up signs, written in German, along the edge of active minefields, saying "Cleared of mines." Because of this, German troops and vehicles would "fall into the trap set by the enemy." [199]

The 7 Pz D reported, "over [the last] four days of the offensive its units alone captured 17 Soviet commanders and 511 soldiers and 33 commanders and 50 soldiers voluntarily came over to the German side." [200] This same report stated that during this same time only one German officer from the division had been captured alive by the Soviets. These statistics indicate that the Germans were advancing and occupying the battlefield at the end of each day.

The situation for the 81 and 92 GRD had changed dramatically during the night. Apparently German forces were overrunning the 81 GRD and had partially surrounded two regiments of the division. Then later that night someone, maybe General Shumilov or maybe the 35 GRC commander General S. G. Goriachev, gave the order to withdraw the division north over the northern Donets River and out of the forming pocket. At least one of the regiments had to fight its way out of encirclement. A similar situation arose for the 92 GRD when it too was forced to withdraw, in the process losing 3,131 men during 9-10 July.

In the north, General Model ordered his forces to regroup, pending further attacks.

The air battle had started to swing against the Luftwaffe, when the *Fliegerkorps* VIII suffered 28 aircraft destroyed or damaged during the day. The 2nd AA lost 37 aircraft with the 17th AA losing another 9. [201] Sortie rates for the day are unknown. The Soviet aircraft loss ratio was now down to 46 to 28 (1.6:1).

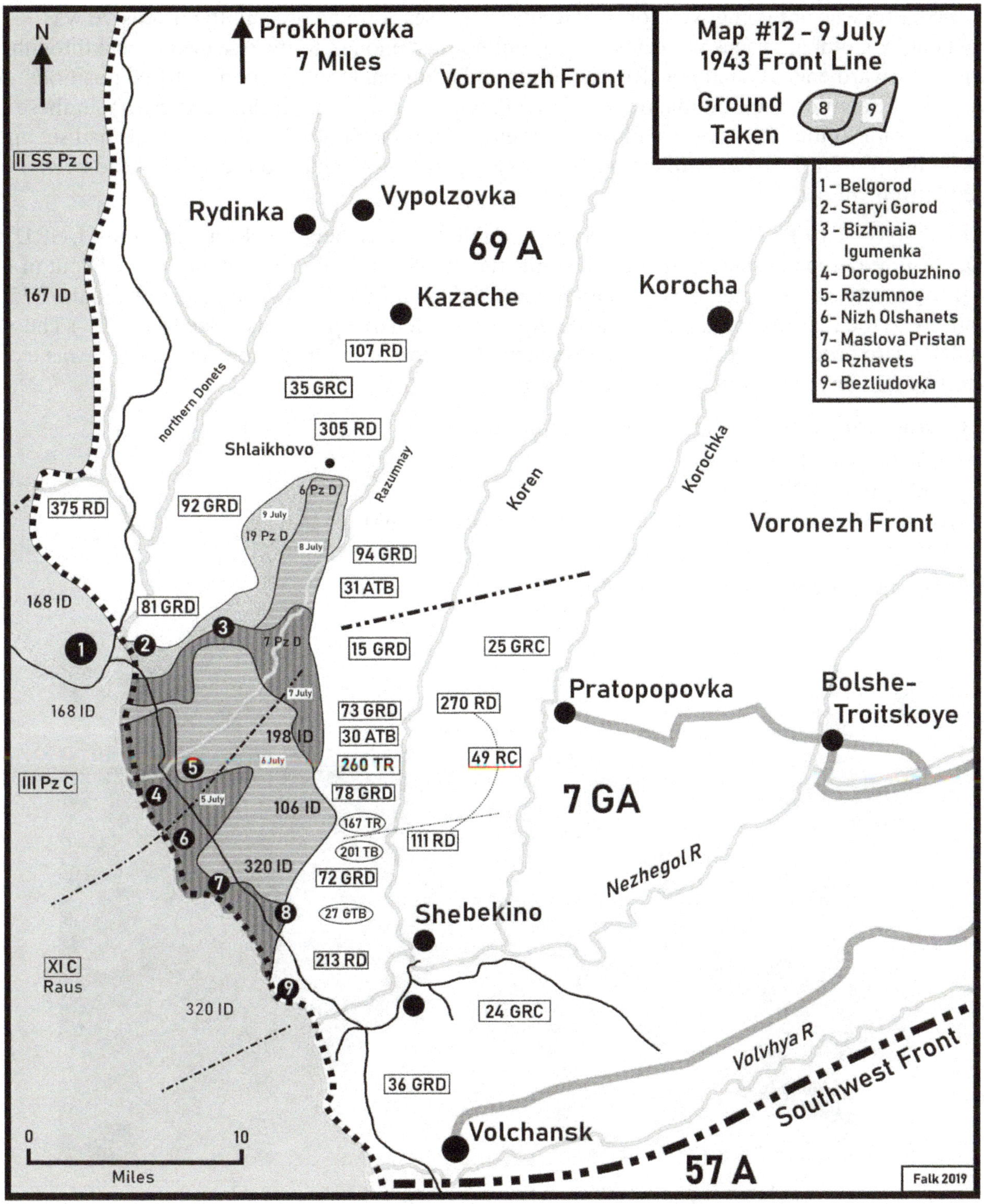

10 July 1943 - Day Six of Operation Citadel

On the north side of the Kursk Bulge, General Model launched two well-planned attacks on 10 and 11 July. The troops did not achieve the expected breakthrough. The 9th Army was unable to make any headway.

In the south, the attacking German 198 ID gained some ground, but by the end of the day the 7th Guards Army had all but stopped the forward advance of XI Corps Raus to the east. Corps Raus then went over to the defense in an attempt to hold the newly captured ground. AD Kempf had shifted its attention to the III Pz C and its attack to the north in an attempt to support the advance of the II SS Pz C.

Overall, the battle was not going well for the Germans. Losses were heavy and the offensive was behind schedule. Combat groups were advancing, but not fast enough to make a clean breakthrough. Every move forward encountered new Russian units and additional unidentified fortified positions. As General Rudolf Schmidt, the Commander of the 19 Pz D, related, "What little we knew about these fortifications prior to the offensive, we didn't even guess one-fourth of what the Russians had set up here. Each clump of bushes, each collective farm, every patch of woods and the heights had been converted into fortresses." [202]

The German 168 ID, 19 Pz D and 6 Pz D were at last able to focus their attacks against the 81 GRD and 92 GRD. Despite fierce resistance, by the end of the day they had finally forced the 81 GRD out of its fortress like positions and captured Staryi Gorod and Bizhniaia Igumenka. The 81 GRD and the 375 RD withdrew to the rear to reorganize and to occupy new defensive positions. (See Map #13.) The 89 and 93 GRD were ordered south to fill in the weakened section of the front line. The 69th A was trying desperately to halt any further northward advance by the III Pz C.

The 7th Guards Army reported the following combat strengths:

Rifle units - 54,983 men
Tank units - 3,323 men
Artillery units - 7,079 (including the 97 GMR-Katyusha)
Anti-Aircraft Units - 2,144 men

These troops amounted to a combat total of 68,529 men. [203]

A late war photo of Russian troops loading Katyusha rocket launchers.

Interestingly the 7th Guards Army still retained Penal Companies (PC - criminals) and Blocking Detachments (BD - to backup less reliable troops). During the first six days of combat, Penal Companies #5, 62, 63, 65, and 66 lost the following troops:

170 killed
508 wounded
260 missing in action
2 in hospital with illness

These figures total 940 men. [204]

These penal companies were actually over strength because normally each would contain between 150-200 men. During this time the 5 PC only lost 5 men wounded, so the other four PCs suffered the remaining 935 casualties. In other words, they were wiped out.

The three Blocking Detachments (sometimes known as Barrier units), #82, 85 and 86 only lost three killed, six wounded and three missing during the same time period. (See Appendix 8.)

Due mostly to bad weather, *Fliegerkorps* VIII flew only 682 sorties with the VVS flying 524. In turn, the *Fliegerkorps* VIII lost 8 aircraft while the 2nd AA lost 25 aircraft and the 17th AA lost none. [205] This day's air battles produced an increase in Soviet losses with a loss ratio of 25 to 8 (3.125:1).

In another theater of this World War, more than 150,000 Allied soldiers within the British 8th Army and the American 7th Army conducted an amphibious landing on the Italian island of Sicily, Operation Husky, in the Mediterranean Sea. The weak Axis defenders were taken by surprise and were unprepared for a long battle. Nevertheless, Axis forces held out on the island until 17 August, which delayed any Allied landing on the Italian mainland. Hitler, worried about the vulnerability of Fortress Europa in the west, soon reacted to this new threat, which in turn had a decisive influence upon Operation Citadel.

From this point forward, the interaction between Allied offensive operations in the East and West would greatly shape Hitler's military options for the rest of the war. Hitler was now engaged in a two-front land war in Europe, and he knew it was impossible to win a two-front war. Nazi German also had to contend with two other major fighting fronts, namely the naval war in the Atlantic Ocean and the massive air war being fought over occupied Europe.

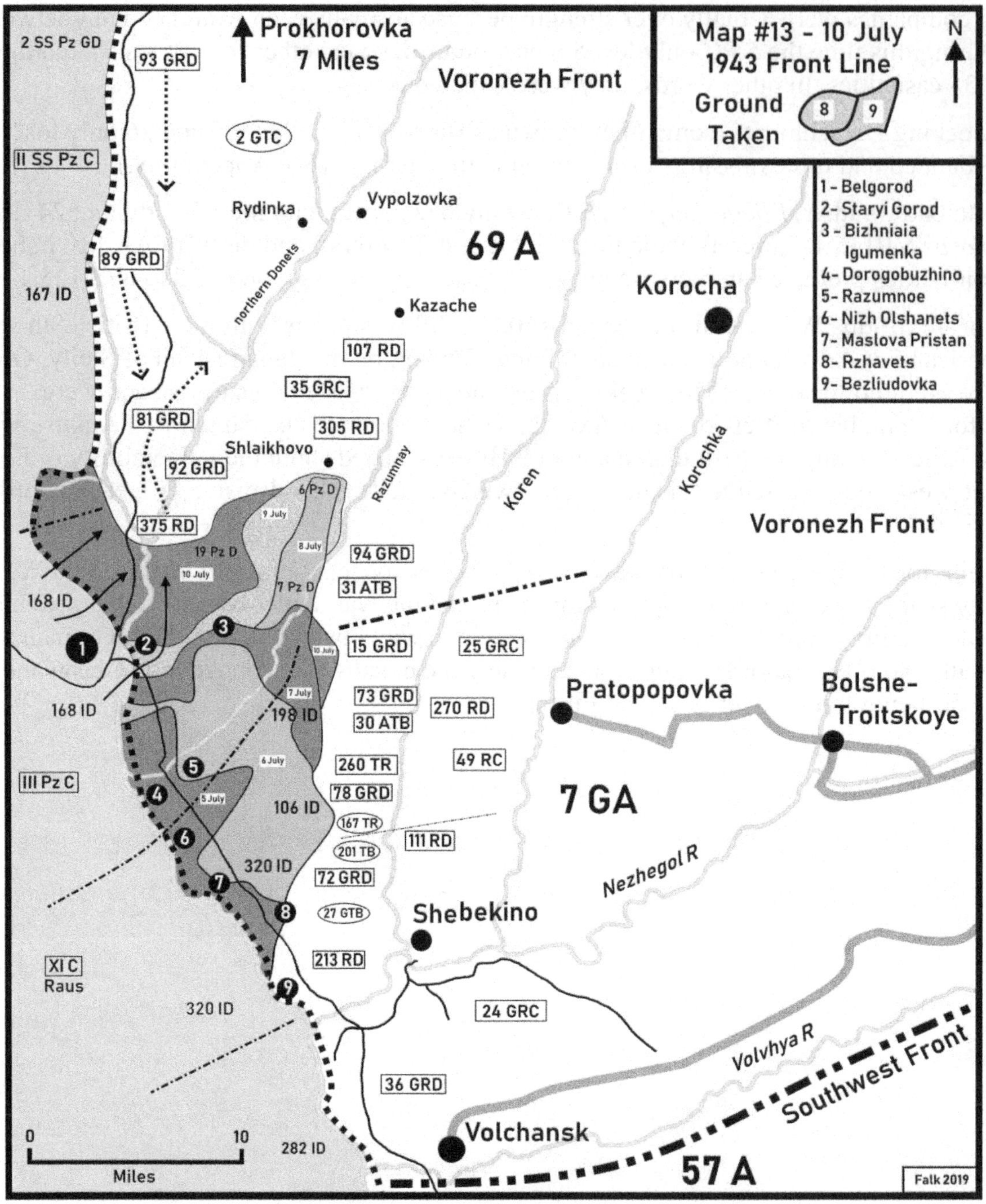

11 July 1943.- Day Seven of Operation Citadel

AD Kempf launched an all-out attack to the north in an attempt to support the II SS Pz C attack in the Prokhorovka area. The III Pz C attack shattered Russian positions of the 69th A, surrounded several units, and gained at least five miles in one day.

The XI Corps Raus held its positions and defended against increasingly powerful assaults from the 7th Guards Army during the day. These active defense attacks were made at various locations and with varying strengths. These attacks were intended to hold German troops in this area and not allow them to move northward. The 69th A in turn was shifting forces around in an attempt to block the advance of the III Pz C.

A bit further to the north, the 4th Panzer Army expanded its attack over the small Psel River and advanced along a tank friendly corridor directly toward the town and rail station of Prokhorovka.

Russian forces were fiercely defending this entire area, as it was the third and last army-level line of defense.

In his command train HQ, located at Dolbino Station on the rail line between Belgorod and Kharkov, von Manstein gathered his commanders for a conference. He wanted to know "[w]hat to do next, and was it possible at all for both formations – II SS Panzer Corps and III Panzer Corps – to reach Prokhorovka?" Warner Kempf could not say that the III Pz C could break through in time. In turn, General Friedrich Fangohr (Chief of Staff, Fourth Panzer Army) stated, "It would be good if the II SS Panzer Corps continued its offensive to the northeast [toward Prokhorovka], since everything previously planned was based on this calculation." [206] The battle would continue for now, for stopping attacks in the south would be the end of Operation Citadel, and it would hand the initiative over to the Soviets.

Meanwhile, *Stavka* and the Voronezh Front commanders planned two coordinated sets of attacks for the next day. Voronezh Front troops would launch a frontal attack on 12 July, all along the front line to halt Army Group South's advance toward Kursk. The second attack would be against the Orel Bulge.

In the air the battle continued as before, but the loss ratio had shifted further in favor of the VVS. During the day, the Fliegerkorps VIII flew over 1,000 sorties with the 2nd AA and 17th AA flying 595. The *Fliegerkorps* VIII lost 16 aircraft while the 2nd AA and 17th AA only lost 19 aircraft. During the night, the Luftwaffe launched at total of 110 aircraft on night bombing and harassment missions with the VVS launching 240 of their own. [207] The Soviet aircraft loss ratio decreased once again to 19 to 16 (1.18:1).

12 July 1943 - Day Eight of Operation Citadel

During the early morning hours, both the West and Bryansk Fronts launched Operation *Kutuzov* (named after General Mikhail Kutuzov, who fought against Napoleon in 1812). [208] This was a major attack against the northern and eastern side of the Orel Bulge. In essence this offensive was designed to collapse the Orel Bulge, capture Orel, then strike into the rear area of the northern wing of the German Citadel attack.

This offensive was triggered when after two major attacks on 10 and 11 July, General Model's forces were stopped from advancing by strong Russian defenses. *Stavka* correctly understood that Model's troops were now fully engaged. Stalin and *Stavka* waited for this sign of weakness before striking. Fighting on the north and east side of the Orel Bulge was intense, and it would increase over the coming days as the Soviet armies struggled to break through the defending German forces. With the start of this Soviet offensive, General Model ordered his 9th Army to cease its attack toward Kursk and to switch to the defense, effectively ending Operation Citadel's northern attack against the Kursk Bulge. After eight days of intensive combat, the 9th Army had advanced only about nine miles, suffering 20,273 casualties in the process but had lost only moderate amounts of equipment. [209]

On the south side of the Kursk Bulge, a decisive battle began when the II SS Pz C continued their direct advance toward the city of Prokhorovka. The Germans were about to break through the third army-level defensive line and reach open ground, but the entire Soviet 5th GTA with more than 500 tanks, and the 5th GA, along with numerous supporting formations, stood in their way. In some ways the fighting around Prokhorovka was the highpoint of the southern German attack. At the end of the day, the German II SS Pz C was unable to break the Russian defense line and gain the freedom of movement it desperately needed. Based upon the latest research, the losses for this single day of fighting are telling:

German units of the II SS Pz C, in and around Prokhorovka lost 48 tanks destroyed, and suffered 937 casualties.

Russian forces of the 33 GRC, 18 TC, 29 TC, 2 TC and 2 GTC lost 309 tanks destroyed, and suffered 5,051 casualties. [210]

This single battle truly was a great clash of armored forces, but in retrospect, these totals are far more accurate and believable than earlier exaggerated accounts covering the battle for Prokhorovka.

The Russians also were having problems with friendly fire from their VVS air support. Over 11-12 July, the 69th A and the Voronezh Front personnel complained about a sharp increase "of fratricidal bombing and strafing attacks by [air] crews of the 2nd and 17th Air Armies." For example, the commander of the 69th A Lieutenant General Vasilli Dmtrievich stated, "It is one thing if *Sturmoviki* were pounding the rear areas flat, [but is it possible for troops at the front to wave off attacking VVS pilots] when the enemy is firing at you from every barrel?" He went on to say, "There wasn't a day when the 'Red Star Falcons' didn't pounce on their own troops several times. The attacks struck both the rear areas and the front." [211] Sadly, friendly fire air attacks occurred on both sides. Men were killed or wounded, equipment was destroyed, but the battle would continue regardless of these misdirected attacks.

AD Kempf continued its all-out attack with the III Pz C to the north in an attempt to support the main attack against Prokhorovka. But at the end of the day, they were about 12 miles away from this key battle. The III Pz C threatened to encircle the 48 RC of the 69th A but were unable to close the gap. The 7th Guards Army and the 69th A had done their jobs well. For eight days AD Kempf had been prevented from joining the II SS Pz C, thus depriving von Manstein of the advantage of forming a combined overwhelming attack force. The great German attack against the Kursk Bulge has almost run its course.

As part of the overall major counterattack plan of the Voronezh Front, the 7th Guards Army launched a heavy attack against the boundary between the 198 and 106 IDs with converging attacks toward the village of Krutoi Log. The right flank of the attack contained the 73 GRD and 270 RD with the supporting 201 TB. The left flank attack included the 111 RD with the 27 GTB in support. All these units were controlled by the 49 GRC.

Against the long and exposed right flank of AG Kampf, XI Corps Raus, this attack should have caused significant damage. But these attacks gained little ground due to the low combat strength of the units involved and poor coordination between units. The 111 and 270 RD only had 3,915 infantry between them with the 73 GRD having an additional 4,593 men. To boost the attack strength of the 49 RC, General Shumilov added the 265 AR, 1669 AAR, 309 GMR, 132 AT Rifle Battalion and the 2/47 Separate Rifle Battalion to the left flank attack. For the 73 GRD, he added 97 GMR, 161 AR and the 131 AT Rifle Battalion. [212]

The main attack failed to gain any useful ground, but the secondary attack further south with the 24 GRC, 72 GRD and 213 RD captured the area around Rzhavets and to some extent advanced along the northern Donets River. These attacks continued during the night and into the next day. (See Map #14.)

Von Manstein ordered the XXIV Pz C, with the SS Pz GD Viking and 23 Pz D, to move north from Kharkov during the night, and to concentrate in the area around Belgorod. These two divisions had only 104 tanks and 7 assault guns between them. [213] Ultimately Hitler refused to release this Corps to von Manstein, so Army Group South fought on with the units on hand.

The 2nd AA had 472 serviceable aircraft with the 17th AA having just over 300 aircraft. *Fliegerkorps* VIII also had been badly depleted. During the day, the 2nd AA and 17th AA produced a total of 893 sorties, with *Fliegerkorps* VIII flying less, with 654 sorties. The VVS deployed most of its aircraft to the flanks leaving the area over Prokhorovka for the Luftwaffe to dominate. The *Fliegerkorps* VIII suffered 19 losses, with only one of them due to air-to-air combat and claimed to have shot down 16

Soviet aircraft. Overall, the VVS lost 31 aircraft for the day with most of them shot down by ground fire. [214] The Soviet aircraft loss ratio of 31 to 19 (1.63:1) remained low.

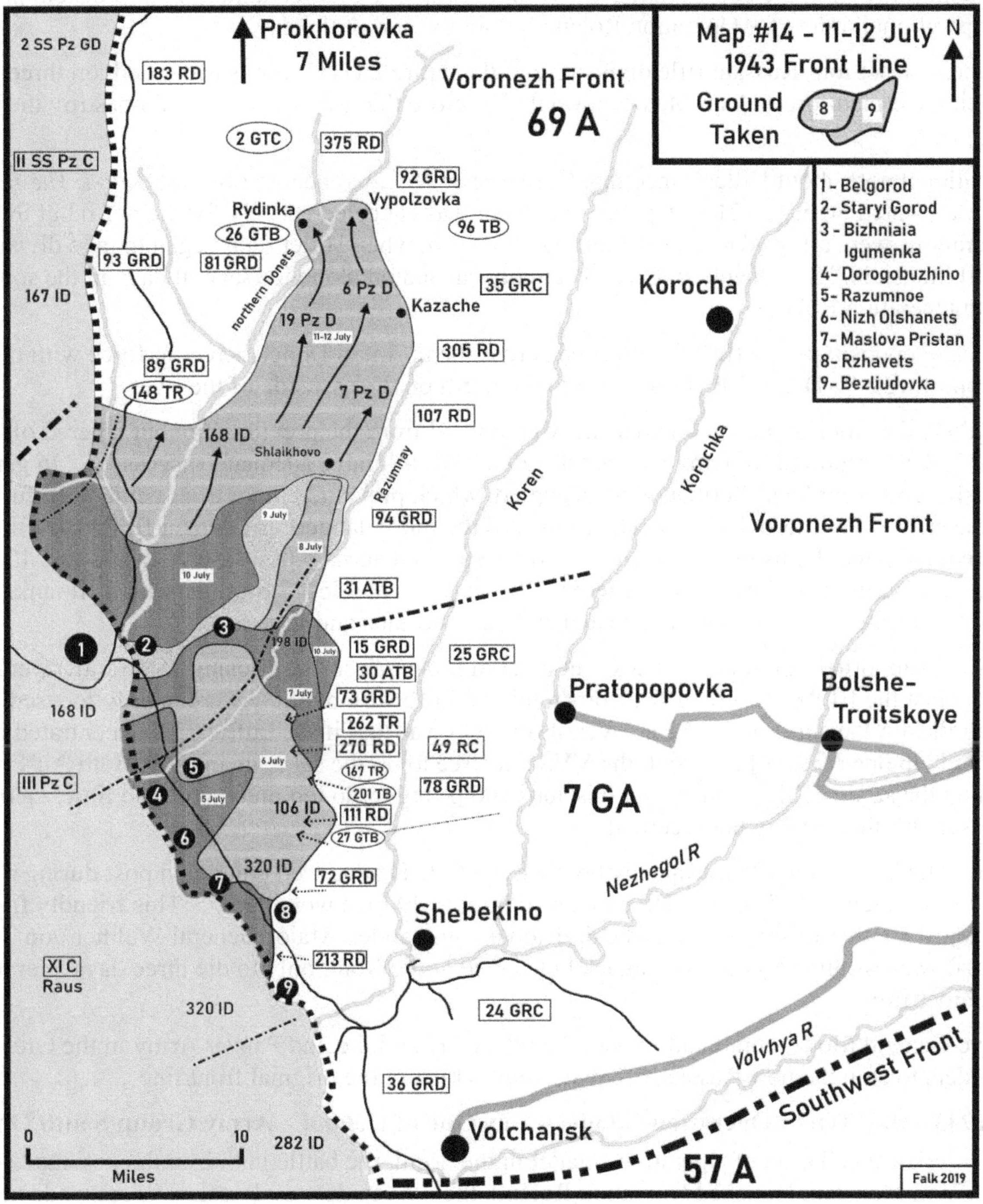

13 July 1943 - Day Nine of Operation Citadel – Army Group South Only

Calling von Manstein and General Gunther von Kluge, commander of Army Group Center, to his headquarters at (Wolf's Lair) near the town of Rastenburg in East Prussia (now called Gierloz, Poland), Hitler canceled Operation Citadel and ordered major forces transferred from the Eastern Front to the West and to Italy. Von Kluge readily agreed with Hitler, but von Manstein was able to win approval to continue his attacks in the south for several more days to destroy Russian operational reserves. The idea was to join the forward elements of the 4th Panzer Army and AD Kempf, crush any Russian troops caught between these two forces, and finally form a shortened continuous front line. The commanders hoped this operation would reduce Russian reserves in the area and delay any major Russian attack in the south that summer. This new limited German effort was code named Operation Roland. [215] (Roland was a Paladin [knight] in the court of the Holy Roman Emperor Charlemagne.) Von Manstein promptly sent out operational orders for Operation Roland before the end of the day.

At this time, at least four Russian rifle divisions and the entire 2 GTC were surrounded, on three sides, by advancing German forces. Von Manstein wanted to close the last gap so he could destroy these forces.

As an additional note about Hitler's meeting. Christopher A. Lawrence, in his book *Kursk The Battle of Prokhorovka*, stated on page 1219, "Operation Citadel was canceled entirely for the north but in the south operations were allowed to proceed through the 17th, when Hitler then began to take divisions from [von] Manstein." So, Operation Citadel actually carried on alone for several days in the south along with Operation Roland.

Against heavy opposition, the III Pz C attacked across the northern Donets River in force with the 19 Pz D; the supporting 6 Pz D and 7 Pz D attacked to the north on the east side of the river.

Fighting along the front of the 7th Guards Army continued from the previous day but tapered off over time. The 25 GRC reported to Army HQ that the attack "didn't have adequate success." [216] In general, this offensive only kept local German units occupied, which prevented them from being used further north at the critical Prokhorovka battle. Operations of the more dangerous German III Pz C further north were unaffected. Units of the 7th Guards Army suffered about 5,000 casualties and lost 45 tanks during this attack. In turn, Corps Raus suffered 728 casualties and lost 7 Sturmgeschutz III tank destroyers. [217] (Note: These values are debated among historians and researchers.)

The VVS did supported Shumilov's attack, but most of their 777 sorties, (losing 28 aircraft) were directed against the III Pz C. Luftwaffe efforts totaled 656 sorties, but they were mainly focused on supporting the 4th Panzer Army. [218] The overall problem was that if the Luftwaffe concentrated its airpower to help one sector of the front, the VVS was free to do the same in another. Both Air Forces were feeling the strain of flying incessant missions and trying to do too much with too little. German aircraft losses for this day are not recorded.

Of special note: The Luftwaffe mistakenly bombed the 6 Pz D advanced command post during a group meeting. Losses included 15 men killed with an additional 49 men wounded. [219] This friendly fire attack was just another mishap of war. The divisional commander, Major General Walther von Hunersdorff, was wounded during this attack but stayed at the front, only to die three days later to Russian sniper fire.

Hitler gave General Model command of both the 9th Army and the 2nd Panzer Army in the Orel area. He had orders to stop the new Russian offensive and to restore the original front line.

14 July 1943 - Day Ten of Operation Citadel – Day One of Roland – Army Group South Only

Despite Hitler's cancellation of Operation Citadel in the north, the battle in the south continued and expanded. Von Manstein launched Operation Roland, the plan to destroy Russian forces wedged between the 4th Panzer Army and AD Kempf. The III Pz C continued its attack toward the north with both the 19 Pz D and the supporting 7 Pz D. They attacked on the west side of the northern Donets River, greatly expanding their bridgehead there. General Kempf certainly was following orders from von Manstein as he tried to maneuver his forces to support the goals of Operation Roland. (See Map #15.)

Meanwhile, the II SS Pz C finished their attacks in the Prokhorovka area. The Russian defenders were much too strong to dislodge. The advance on Prokhorovka had now ended. But the 2 SS Pz GD attacked to the southeast in an attempt to link up with the III Pz C. That night, Russian units began to withdraw from between these two converging forces. They basically were surrounded at this point.

During the day, *Fliegerkorps* VIII made a maximum effort flying 1,452 sorties. The Luftwaffe lost nine aircraft destroyed and nine damaged. Matching them, units of the 2nd AA and 17th AA flew 1,033 sorties but lost 30 aircraft in the process. [220] This day produced a Soviet aircraft loss ratio of 30 to 18 (1.66:1). While the ratio remained low, the fact remained that the Russians had 30 less aircraft to put into the air the next day. The Luftwaffe in turn lost fewer aircraft, but they also had fewer replacement aircraft and crews to call upon. Repairs to the damaged but salvageable aircraft and replacement of lost air crew must have been a continuous and exhausting task for both sides. A massive war of attrition was taking place in the air above the Kursk battlefield.

The 7th Guards Army front line was mostly quiet, with both sides resting their forces.

15 July 1943 - Day Eleven of Operation Citadel – Day Two of Operation Roland – Army Group South Only

AD Kempf, the III Pz C attacked and advanced against scattered Russian resistance. The Russian 69th A, 48 RC, had ordered the 81 GRD, 89 GRD, 93 GRD and the 375 RD to withdraw to the northeast in the morning, out of the forming pocket. This withdrawal also included parts of the 2 GTC. German troops still tried to firmly close the gap between the two largely separated groups.

The 213 RD of the 7th Guards Army launched a surprise, battalion size attack against the German 320 ID south of Rzhavets. They captured no ground, but the 213 RD suffered 21 men killed and 75 wounded. German losses were not reported. [221]

Christer Bergstrom, in his book, *Black Cross Red Star Air War Over the Eastern Front Vol. 5*, page 291 stated that the 2nd AA and 17th AA lost 11 aircraft during the day, mostly due to ground fire because Luftwaffe air units were being called away to fight battles in other areas. German aircraft losses for this day are not recorded.

In the Orel Bulge area, the Russian Central Front opened its own attack toward Orel. Major Russian forces attacked the Orel Bulge from the north, east and now the south. The Luftwaffe responded to this offensive by transferring air units from the south of Kursk to the north.

16 July 1943 - Day Twelve of Operation Citadel – Day Three of Operation Roland – Army Group South Only

Most of the escaping troops of the 48 RC were clear of the pocked by early morning, but losses of troops, animals, and equipment were heavy. The 94 alone reported these losses for 15-16 July:

3,426 – men and women
200 – horses
273 – machineguns
106 – AT rifles
47 – 50 mm mortars
76 – 82 mm mortars
6 – 120 mm mortars
6 – 45 mm AT guns
3 – 76 mm regimental guns
5 – cars [or trucks] [222]

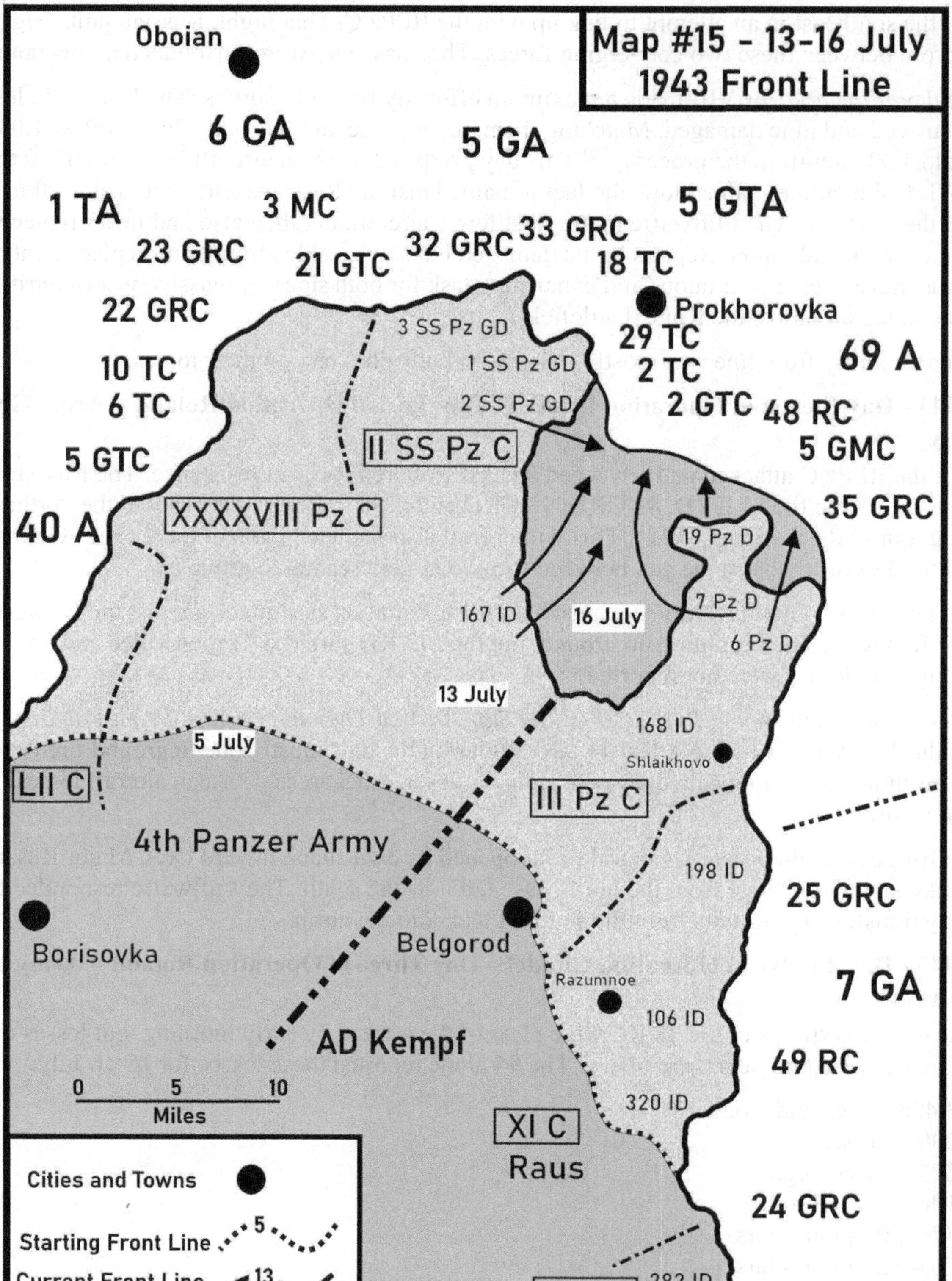

During this time period, historians lack reliable data for the other three partially surrounded Russian divisions of the 48 RC. But, using the 89 GRD as an example for these divisions would lead us to the conclusion that they also had suffered significant losses, perhaps over 10,000 men and much equipment. Christopher Lawrence, in his book *Kursk*, on page 1114 stated, "It is clear that while the Soviet decision to withdraw was a fundamentally necessary one, its execution was flawed and improvised. They needed to get some transports inside the Donets triangle that could have pulled out some of the heavy equipment ahead of the withdrawing units." He went on to say, "Overall, the withdrawal was poorly planned and conducted."

The German 4th Panzer Army and Army Detachment Kempf linked to form a continuous front. With this, Operation Roland achieved a partial success, but this success would be fleeting. Despite inflicting heavy losses upon Russian forces in the area, depleted Russian rifle and tank units were able to escape from the pocket and eventually would be rebuilt. German forces still needed to clear pockets of Russian troops from behind the lines, but they had ended major combat actives. Operation Roland was now all but over.

Von Manstein knew that German Panzer and infantry formations were likewise greatly weakened during Citadel and Roland. The question remained, Would von Manstein have enough time to rebuild his formations to meet the expected Russian offensives that he knew were coming that summer?

The 7th Guards Army mostly was quiet along the front with only small, company-sized attacks against the German 198 ID. Army units improved their unchanged positions and conducted harassing fire against the enemy. The Army reported losses of 40 killed and 120 wounded for the day. [223]

During the day 2nd AA and 17th AA flew 926 sorties while *Fliegerkorps* VIII conducted 499. Once again, the 2nd AA and 17th AA lost 33 aircraft during the day, mostly due to ground fire. [224] German aircraft losses along the southern Kursk area were not recorded.

In the Orel area, General Model ordered the construction of the Hagen Line, a fortified defensive line across the base of the Orel Bulge. Model hoped he could stabilize the front along this line. (See Map #6.)

To meet the changing strategic situation, early in the morning *Stavka* issued two separate Directives #30150 and #30151, which altered the structure of the Fronts.

Stavka Directive #30150 to the commander of the Steppe Front was about changes of fighting strength. This was sent on 16 July 1943, 01 h 00 min.

> *GHQ orders:* [in part]
>
> *1. From July 18 23.00 hours. In the Steppe Front include 69 Army ... and 7th Guards [A]rmy with all the reinforcements, the army and the rear part of the institutions by transferring the army in areas occupied by them from the Voronezh Front.*
> *2. Then the Steppe Front gives up the 52 Army, the 3, 5 and 7 Guards cavalry corps, and transfers them to the reserve of the Supreme Command in the areas of concentration of the present.*
> *3. After this the composition of the Steppe Front will have 69 Army .... with attached reinforcements; 7th Guards Army ... with attached reinforcements; 53rd Army, 47th Army, 3 Guards and 4 Guards Army, and 3 Guards Mechanized Corps.*
>
> *GHQ J. Stalin G. Zhukov*

Then *Stavka* Directive number #30151 arrived for the commander of the Voronezh Front about the transfer of the 69th Army and 7th Guards Armies to the Steppe Front effective on July 18 at 23:00 hours.

> *July 16, 1943 01 h 00 min*
> *GHQ orders:*
>
> *1. From July 18 to 23.00, the transfer of the Steppe Front 69th Army and 7th Guards Army*

with all reinforcements, the army and the rear parts of the institutions. Army to transfer the areas occupied by them.

2. The line of demarcation between the Voronezh and Steppe fronts from 23:00 07.18.1943 to install: Stary Oskol, Skorodnoye Far leash, Lucky (and south). All items, except for Stary Oskol, to the Steppe Front inclusive.

3. To entrust responsibility for the junction between Voronezh and Steppe Fronts to the commander of the Voronezh Front, for which the latter to have the 5th Guards Tank Army in the region of Prizrachnoye.

4. On receipt and execution to deliver.

GHQ J. STALIN Zhukov [225]

This rearrangement of armies and change in power centers was necessary to allow the Voronezh and Steppe Fronts to go over to the offensive toward Belgorod and Kharkov.

The Voronezh Front released a casualty list for the 6th Guards Army, 7th Guards Army and the 69th Armies between 5-16 July. This was based upon data received from the armies on 23 July. During 12 days of combat, these armies suffered the following casualties:

6th Guards Army – 24,921
7th Guards Army – 23,390
69th Army – 29,267

These statistics total 77,578 men. [226]

In comparison, the III Pz C (6 Pz D, 7 Pz D, 19 Pz D, and 168 ID) suffered the following casualties between 4-20 July. Initially they were fighting against the 7th Guards Army and later the 69th A as well:

6 Pz D – 1,821
7 Pz D – 1,487
19 Pz D – 2,118
168 ID – 2,623
Corps units – 167

These figures total 8,216 men. [227]

The total reported casualties for AD Kempf, III Pz C and XI Corps Raus, during the German attack (approximate date range, 5-15 July), are as follows:

2,450 killed
12,482 wounded
709 missing

These numbers total 15,641 casualties. [228]

This would imply the three divisions of XI Corps Raus - 106, 168 and 320 IDs -- suffered about 7,425 of these casualties during this same time period. Note: Another account puts this number at 8,754 men for 5-20 July. [229] This value could be true if the AD Kempf totals covered fewer days, such as 5-15 July, as noted above.

17 July 1943 - Day Thirteen of Operation Citadel – Day Four of Operation Roland – Army Group South Only

The OKH ordered the II SS Pz C, shortly after 13:00 hours, to halt Operation Roland and to withdraw to an assembly area around Belgorod, in preparation for its transfer to the west. [230] With this order, Hitler and the German High Command concluded Operation Citadel and the related Operation Roland. The II SS Pz C began withdrawing during the night. Without this powerful Panzer Corps at his disposal, von Manstein knew Operation Citadel had truly ended. At this point, maintaining the extended front line was worthless, as it required extensive forces to hold. Therefore, sometime on or about this date, von Manstein ordered the gradual withdraw of the 4th Panzer Army and Army Detachment Kempf to the south. As German forces withdrew to their start line, they gave up the territory that they had captured at such a heavy cost.

After thirteen days of battle, the Germans had neither cut off the Kursk Bulge nor destroyed the assembled Russian armies, but they did claim to have captured some 34,000 prisoners (southern attack only). [231] Thus, Operation Citadel had decisively failed to achieve its objectives. The Red Army had won a great defensive victory, but Red Army losses were serious both in terms of troops and of equipment. Meanwhile, perfect summer weather all along the Russian Front gave *Stavka* the opportunity for further offensive operations.

Once again, the 7th Guards Army remained quiet with no major activity. Army units continued to improve their unchanged positions and to conduct artillery-harassing fire against the enemy. The Army reported losses of 24 killed and 70 wounded for the day. [232] This was just another typical day at the front.

The *Fliegerkorps* VIII flew only 138 sorties over the southern Kursk Bulge area losing 3 Ju-87s, while the 2nd AA flew 484 missions losing 16 aircraft, [233] producing a Soviet loss ratio of 16 to 3 (6.33:1). Most of the Luftwaffe's strength had been shifted north to support the defenders of Orel or south of the Kursk Bulge to counter new Russian attacks in that section of the front. Both the Luftwaffe and the VVS could and did intervene decisively in the ground fighting, but their influence was not lasting as the battle moved on to new areas and new situations.

On this very day, far to the south, the Southwestern and Southern Fronts opened their attack from the city of Izyun, then south along the Mius River to the Sea of Azov. These powerful attacks were directed against the German 1st Panzer and 6th Army of Army Group South and were designed as a diversion to draw away German reserves from the Belgorod/Kharkov area. [234] Once again *Stavka* had waited for the opportune moment to begin another offensive to distract German attention from a more critical area. The initiative had now clearly passed to the Russian side.

18 July 1943

Stavka activated the Steppe Front, and it began moving its units into positions opposite Belgorod. The 7th Guards and 69th Armies, along with other forces, were transferred to the Steppe Front for the upcoming operation.

Of special note, from 4 July to 18 July, the 7th Guards Army consumed 2,368 tons (short 2,000 lbs. tons) of ammunition of all types (214,821 Kg) while the Army received from supply only1,692 tons. The resupply effort was not fully able to replace the expenditure, but at no time was the 7th Guards Army lacking for ammunition. [235]

The 7th Guards Army became active again. The 15, 73 GRD, 270 and 111 RD attacked the German 106 and 198 ID. Losses for the Army were heavy, numbering 699 men. German losses were approximately 60 men.

Across the entire southern section of the Kursk Bulge, total casualties for the day were 888 German troops and 4,150 Russian troops.

Both sides knew the German forces were withdrawing, so why die trying to defend or capture ground that would be given up anyway? This was the last day of any substantial combat during the German retreat. [236]

Despite diligent research by various historians, accurate day-to-day aircraft loss totals for both the VVS and the Luftwaffe during the Kursk battle, 5-18 July, are difficult to compile. Christer Bergstrom, in his series of books on the Kursk battle, comes the closest. He states that aviation losses during the battle, covering the southern Kursk Bulge attack area, amounted to approximately 677 Russian aircraft and 220 German aircraft lost, [237] which produced an overall loss ratio of 3.07:1 in Germany's favor. Despite inflicting more damage and losses to the enemy, both on the ground and in the air, the German Kursk attack had failed to achieve its objectives. For the remainder of the war, the Red Army and VVS now would have the strategic advantage.

19 July 1943

Due to the Allied invasion of Sicily, Hitler summoned Benito Mussolini, the Fascists leader and Prime Minister of Italy, to a meeting in Germany to discuss future actions. Mussolini was clearly a broken man. As Axis powers began to lose battles and ground, Mussolini also began to lose support of the Italian population and the government. The truth was, the Italian Army was not capable of defending the island of Sicily without help, let alone the Italian mainland, and people were beginning to realize it. Because of this, Hitler and the German High Command already were thinking about militarily taking over Italy, Operation *Achse* (Axis), to prevent Italy from leaving the Axis Alliance and surrendering to the Allies.

On the Eastern Front, German forces in the south continued their planned withdraw, finishing around 24 July. Advancing Russian troops did not press the Germans too forcefully, but sporadic fighting did break out in different areas.

20 July 1943

The 7th Guards Army reported a combat strength of 61,509 men with 48,679 in the main combat units.

About Orel, "Hitler issued a *Fuehrer* order [to General Model] forbidding further withdraws by the Ninth and Second Panzer Armies." [238] Model knew that obeying this order would put his two armies in grave danger.

Field Marshal von Manstein made a public statement to the troops. He thanked the forces of 4th Panzer Army, Army Detachment Kempf and the 4th Air Fleet for their efforts during the battle for Kursk. He went on to list their accomplishments and the damage inflicted upon the enemy. He concluded, "I know that you will meet the challenges that now confront you just as you always have in the past, in a manner worthy of our fallen heroes." [239] He was clearly trying to boost the morale of defeated troops.

21 July 1943

After a bit of fighting, the Germans once again were able to disengage and to pull back to new positions. As the front line shortened, both sides withdrew units to the rear for refitting and replenishment. Weary units were in no shape for a major battle.

As German units retreated, they destroyed any and all damaged equipment or supplies that they could not evacuate. Certainly, they also left a few stray land mines around for the Russians to find.

22 July 1943

On or about this date, *Stavka* issued instructions for Operation *Rumyantsev* (named after the 18th century Field Marshal Perter Rumyantsev). This plan called for the total destruction of German forces of Army Group South, from Belgorod to Kharkov and beyond.

Hitler relented on the Orel issue and allowed General Model to conduct an "elastic defense." [240] Now Model had a chance to save his armies from the overwhelming Soviet Orel attack.

On this date, a historical OOB chart from the TsAMO Archive showed the 7th Guards Army gained two independent squadrons of aircraft. The #50 OKE was an independent spotting air squadron, most likely used as artillery spotters, and the #578 OE was an independent communications air squadron, for reconnaissance or transportation duties. Both of these squadrons most likely were equipped with 12

PO-2/U2 biplane aircraft. These two squadrons would provide useful support during upcoming offensive operations.

23 July 1943

German forces south of Kursk had retreated almost back to their stating positions. Nearly all the gains of Operation Citadel had been erased. The ambitious goals of Operation Citadel had ended in total failure. Soon it would be the turn of the Red Army to carry out their plans for the summer months.

Stalin, of course, wanted to launch Operation *Rumyantsev* immediately, but General Zhukov and others persuaded him to wait until logistical preparations were complete, estimated to be in eight days' time. Meanwhile, rest and replenishment were the main focus of depleted units.

24 July 1943

Stalin issued an order of the day to commanders about the victory at Kursk. In part he stated, "Thus the legend that in a summer offensive the Germans are always successful, and that the Soviet troops are compelled to retreat, is refuted." In closing he said, "I express my gratitude to all men, commanders and political workers of the troops under your command for their excellent operations. Immortal glory to the heroes who fell in the fight for the freedom and honor of our Motherland!" [241]

Retreating German forces north of Belgorod stopped a few miles short of their original start line because they had found better defensive positions. South of Belgorod, they did fall back over the northern Donets River as that was the best position overall. The new shortened front line only required 11 German divisions for a firm defense, thus releasing other units for use elsewhere.

General Zhukov blamed the less than aggressive pursuit on the exhaustion of the 6th Guards Army, 7th Guards Army, 1st Tank Army and the lack of action on the part of the 5th GTA. [242]

The 7th Guards Army had a reported combat strength of 63,186 troops with 49,925 in the main combat units. The Army slowly regained its strength and reorganized as it prepared for further combat.

Christopher A. Lawrence, in his book, *Kursk the Battle of Prokhorovka*, page 1207-1236, presents a very nice "post mortem" of the battle. A few bits of information are presented below from this massive tome:

> Personnel Losses 4-18 July, (south battle only):
>
> German – 5,612 killed; 27,627 wounded; 1,142 missing – Total 34,381
> Russian – 27,046 killed; 66,961 wounded; 32,801 missing – Total 126,808
>
> Tank Losses: Destroyed, Damaged or Broken Down:
>
> German – 1,536
> Russian – 2,471

During the main German attack, the Germans normally controlled the battlefield at the end of the day, so many of their tanks were repaired and returned to the fight. So, their actual destroyed tank totals would be much lower than this value. Likewise, damaged Russian tanks found on the battlefield were routinely destroyed by German troops, thus preventing their repair.

Finally, a Russian Loss Report covering the entire Voronezh Front, 4-22 July, is summarized below:

> 4-16 July: 18,097 killed; 47,272 wounded; 24,851 missing; 29 captured – Total 90,249

One can see the losses during the defensive battle were quite heavy for the Russians.

In turn, as the Germans withdrew from the Kursk bulge, Russian losses dropped significantly.

> 16-22 July: 2,481 killed; 7,155 wounded; 1,047 missing – Total 10,683

Russian grand total (4-22 July 1943)

20,578 killed; 54,387 wounded; 25,898 missing; 29 captured – Total 100,892

Note: These and earlier personnel loss figures do not agree with each other (which is typical), but they are based on the best current available information and subject to change if new records from the Russian archives are released someday. (See Appendix 9.)

Regardless of the battle's outcome, both sides suffered terrible losses, so now we need to speak of the dead. Both the Red Army and the *Wehrmacht* (unified armed forces of Nazi Germany) had established procedures to deal with the dead upon the battlefield. The German practices were officially controlled at several different levels and most accounts, both written and in historical photos, showed that they were effective. The Red Army, on the other hand, dealt with the dead in a more haphazard manner.

German efforts to deal with the battlefield dead contained three levels:

1. The unit that the dead belonged to was responsible for the burial. For that purpose, every unit from company/battery level upwards was supposed to appoint a *Gräberoffizier or Gräberunteroffizier* (Grave Officer or NCO) from its ranks and to organize a burial group and report the grave locations. This was most appropriate as the soldiers in the field would be the ones to take care of their dead comrades.
2. A *Wehrmachtgräberoffizier* (*Wehrmacht* grave officer or W.G.O.), who was attached to every army and moved around with the army, supervised these lower-level Grave Officers or NCOs. This officer also acted as a central source of information regarding the dead. The W.G.O. likewise was responsible for dead enemy soldiers and was ordered to treat them in a manner similar to German casualties in accordance with the Geneva Convention. However, the burial groups for the Red Army dead were formed from enemy POWs and more mass graves were used as opposed to individual burials.
3. Then behind the combat zone were *Bodenständige Wehrmachtgräberoffiziere* (unmovable *Wehrmacht* Grave Officers who had a specific territory and did not move around with the army). The *Bodenständige's* main responsibility was to register and to secure all graves in its area and to organize the care for the graves, rebury the dead if necessary, and so on. [243]

Each German soldier was issued a perforated metal Identification Tag or *Erkennungsmarke* (Dog Tag in the U.S. Military). Upon death, the tag/disk would be broken in half so that one part remained with the body and the Grave officers retained the other half for official recordkeeping. An account from the Pz GD *Grossdeutschland* details this process:

> The fallen are brought to us at the *Tross* [rear services] in the provision's lorries [trucks]. It is my job to 'prepare' these poor boys. First, I break off the ID Tag for collection by the *Spiess* [also known as der Spiess "The Spear" a *Hauptfeldwebel* or Sergeant Major or First Sergeant], then I go through all the pockets. I pack in[to] the fallen soldier's washbag private belongings such as purses, other papers, pipe, photos, letters, harmonica. I remove still usable boots. Finally, the body is sewn into a [woolen] blanket or groundsheet and later removed to an easily recognizable place near some salient feature and then interred. [244]

The Soviet Union and the Red Army often treated their dead in a less than official manner. It was one thing to establish regulations in Moscow and then send out orders to the army for implementation, but another thing entirely to put burial orders into effect out in the field, during or after a battle, and have the fallen buried.

On December 22, 1941, the Directive of the Main Political Directorate of the Red Army "On the organization of the burial of the corpses of the dead Red Army soldiers" was published. Whether this was the first official directive concerning the dead is unknown. Statements attest to a less than organized process dealing with the dead, especially during the early part of the war. A report from August 1943 included the following evidence:

> There are shameful cases of mass abandonment of unburied corpses on the battlefield.
> The burial teams are few in number and do not cope with their work, as a result of which the corpses are not buried for several days or even left on the battlefield, where they decompose. Special graves are not [dug for the dead]. For graves, trenches, cracks, bomb craters, road ditches are used or [bodies even are buried in the forest]. The graves are not filled up; there are no grave [markers] indicating the names of the victims. …at the edge of the forest, the bodies of 9 soldiers were found, not buried for 5 days, and in the forest – the bodies of 8 dead Red Army soldiers, of which two bodies were not buried, and six were carelessly covered with a thin layer of earth, from under which the limbs of the legs are visible. The bodies of the victims were not identified, since they did not have any documents with them.
> Often, there are no schemes [maps] for the geographical location of mass and individual graves, or fictitious burial places are indicated. Notices for the dead servicemen are sent to the district military enlistment offices with a delay, sometimes up to 4 months. The registration of personal belongings of the deceased is extremely unsatisfactory [I]n most cases they are not taken into account at all and are not sent to relatives, and in fact they represent not only material value, but are mainly a precious memory of a Red Army soldier, a loved one who died in the battle for our Homeland. The listed facts indicate that the issue of the correct registration of irrecoverable losses of personnel by the headquarters of all levels is not given due attention, and individual commanders have not yet understood the entire political and universal importance of timeliness and complete personal accounting of irrecoverable losses and proper burial of the dead. Such a boorish attitude towards the [bodies] of soldiers who gave their lives for the honor, freedom and independence of our Motherland, and their memory, cannot be tolerated.

This report goes on to say that dedicated regimental burial teams did exist but only numbered 3-6 people and that looting on the battlefield was revealed. The dead were, at times, buried without boots, jackets, watches, and other valuables and belongings. [245] The burial teams, once organized, set to work with their gruesome task and did their duty. By all accounts, the enemy dead were treated in a similar manner with maybe the exception of enemy dead being interred in a different/separate trench or shell crater.

As in other armies, Soviet soldiers also were issued identification metal medallions, or a pencil case or ebony capsule containing a paper note with personal data of the dead soldier. [246] But too many times, all documents were collected from the dead during burial, leaving little or no individual identifying information behind.

In the end, it would be up to the troops on both sides of the front to take matters in hand. The stench of hundreds or perhaps thousands of dead bodies, both human and animal, would spur them to perform quick burials. The dead also presented the threat of disease and the psychological stress of watching the decomposition of a friend or even an enemy solider. The soldiers knew that it was best not to dwell upon death, and to put the dead underground and out of sight.

Starting on 24 July and lasting until 2 August, both the British RAF and U.S. Air Forces launched six raids, totaling 2,600 heavy bomber sorties, against the northern German city of Hamburg. The city was devastated as the raids killed at least 46,000 and injured another 37,000, mostly civilians. Joseph Goebbels, Nazi Germany's propaganda minister, wrote in his diary that these air raids and their devastation were "a catastrophe, the extent of which simply staggers the imagination." [247]

25 July 1943

In the Mediterranean, the dictator Benito Mussolini was deposed and arrested on orders of the king of Italy, Victor Emanuel III. The fall of Mussolini came as a surprise to Hitler; he did not take the news well.

By this date the 7th Guards Army had reoccupied almost all the territory it had lost to the German attack. By doing so, it achieved a great victory! But the price had been high to defeat the massive German attack and then force the Germans back to their starting point. For the 7th Guards Army, a historical report records the total bloodbath for both the defense phase and counterattack phase (5 July to 25 July 1943):

> Killed - 449 Officers: 1,915 NCOs: and 4,250 Enlisted (total killed 6,614)
>
> Wounded/shell shocked/burned - 1,221 Officers: 5,769 NOCs: and 13,128 Enlisted (total wounded 20,118)
>
> Sick - 31 Officers: 63 NCOs: and 315 Enlisted (total sick 409)
>
> Missing - 218 Officers: 1,078 NCOs: and 2,589 Enlisted (total missing 3,885) and 31 other losses, for a total of 31,063 casualties. [248]

These losses are equivalent to three full strength rifle divisions!

Obviously, the hospitals were overflowing with wounded. Russian reports maintain that sanitary losses (that is wounded who died after reaching medical support,) were about 6% of the wounded. If true, this would add an additional 1,200 dead to the final death totals. Note: Most historical casualty reports from this period do not agree with each other, so all these values should be considered approximations.

Of special note: During the battle for Kursk, no information was found to indicate medical units of the 7th Guards Army suffered any major losses. This implies that medical and other support units were withdrawn in a timely manner before the advancing Germans reached their positions. The lack of losses and timely movement of medical units shows good staff work at Army and Corps HQs.

26 July 1943

In the weeks following the combat in and around Kursk, General Shumilov recommended Major General N.A Vasil'ev, commander of the 24 GRC; Major General G.B Safiulin, commander of the 25 GRC; and five rifle division commanders for the Hero of the Soviet Union award. A number of other commanders also received the Hero of the Soviet Union award. For Shumilov himself, in a few months' time, he would be awarded Hero of the Soviet Union, Order of Lenin and the Gold Star Medal. Note: The Hero of the Soviet Union awarded came with the Gold Star Medal that soldiers wore on their tunic. Also, the Order of Lenin was normally awarded to Hero of the Soviet Union recipients at the same time.

Hitler suddenly ordered the entire II SS Panzer Corps transferred to Italy, but then he changed his mind and sent only the 1 SS Panzer Grenadier Division *Leibstandarte* to Italy, which allowed the 2 and 3 SS Pz GDs to remain on the Eastern Front. This decision was vital to AG South as these two divisions were needed to first attack the Russian bridgehead on the Mius River and then later on, to counterattack the Soviets offensive towards Kharkov, which would begin on 3 August 1943.

As a side note, the mood/morale of victorious Soviet soldiers often varied. A mortar gunner of the *Grossdeutschland* Division fighting to the west of Orel gave this account during a Russian artillery bombardment:

> Raising my nose [from his mortar pit] to assess our plight, I saw two Ivans running about distractedly and in a loud voice I shouted to them [in Russian] "Ivan, come here quickly, hands up and weapons down!" They came to us at once in the mortar pit. [After searching them and giving them a cigarette, one of the Russians said,] "If Hitler and Stalin are dead,

> then we shall have peace." [The mortar gunner then commented,] I had heard the same thing before often enough from prisoners…and they believed it. [249]

27 July 1943

The 7th Guards Army had a reported combat strength of 59,458 men.

Fighting a losing battle, General Model was able to temporarily stabilize his defensive line around Orel, but he knew he could not hold the vital city for long.

28 July 1943

The 7th Guards Army has a reported combat strength of 58,627 men with 45,743 in the main combat units. These totals had fallen as small units continued to transfer out of the Army.

In a total change of mind, Hitler issued orders for Operation *Herbstreise* (Autumn Journey) to start as soon as possible. This operation called for the total and speedy withdrawal of German forces from the Orel Bulge. [250] Now Model was free to withdraw his 9th Army and 2nd Panzer Army in a timely and organized manner, thus "shortening the front" or *Frontverkürzung* and freeing up 19 divisions to be used elsewhere or to be rebuilt behind the front. [251]

For his determination and leadership during the Kursk battle, General Paul Hausser, Commander of the II SS Pz C, was awarded the Oak Leaves to his Knight's Cross. [252] Hitler was pleased with the performance of his SS Panzer Commander.

Sometime around this date, the 7th Guards Army had cleared the entire east side of the northern Donets River except for a final bridgehead, the old German Mikhailovka bridgehead near Belgorod, on the east side of the river. The Army would eliminate this obstacle in due course. A bit further north, the 69th A now occupied positions right up to the new German defense line. With these actions, both armies were ready for future offensive operations.

29 July 1943

Combat actions still took place at the front but at low levels. The 7th Guards Army issued a combat report #193 that listed the casualties for the day – 60 killed and 106 wounded. The next day Army HQ issued an operational report #391 updating the losses for 29 July to 61 killed and 138 wounded. Compiling, issuing, and updating reports/paperwork is a never-ending job for the Headquarters staff.

The Army issued a summary of material support for units within the Army:

- Low ammunition supplies (all types)
  No naphtha listed for combat units (in summer weather, troops had little need for heating fuel)
- Good supplies of sugar, biscuits, bread flour, hay, canned meat, and salt
- Low to moderate amounts of gasoline, diesel, kerosene, fish, fish concentrates, oats, crushed oats, and macaroni
- Low levels of concentrates, meat, vodka, tobacco and soap. [253]

This report shows a fairly well-supplied Army, but they still needed additional stocks of ammunition before starting the next round of combat.

The Army also processed orders (from *Stavka*) to transport by rail the 270 RD into *Stavka* reserve as a separate unit. This was Intelligence Document #590. The 270 RD only contained 2,681 men (combat troops), so withdrawing it to the rear for rebuilding was the correct decision. [254]

Hitler sent for his favorite commando, Otto Skorzeny. During their meeting, Hitler ordered Skorzeny to find and then to rescue Mussolini from his imprisonment. Intelligence sources thought that Mussolini was being held somewhere in Italy. This rescue effort would eventually become Operation *Eiche* (Oak).

30 July 1943

Far to the south in the Donbass, German forces counterattacked the Russian bridgehead over the Mius River. This attack was possible because of the transfer of the II SS Pz C from the Kursk area. The counterattacked lasted from 30 July to 2 August and was successful. But it also kept major German forces away from the Belgorod/Kharkov sector, which was the Russians' intent all along.

Colonel General Ivan Stepanovich Konev. He would become Marshal Konev on 20 February 1944.

Colonel General Ivan Stepanovich Konev commanded the Steppe Front, which now held a section of the front line. The Steppe Front included the following armies (arrayed south to north): 7th Guards Army, 69th A, 53rd A, along with the 1st Mechanized Corps supported by the 5th Air Army. Konev stated the 7th Guards Army had 50 tanks, the 69th A contained 88 tanks, the 53rd A had 53 tanks, and the 1st Mechanized Corps had 200 tanks. This was not a huge force to conduct a major attack. [255]

The Voronezh Front, still commanded by General Vatutin, contained the following armies (arrayed east to west): 5th GTA, 5th GA, 6th GA, 1st TA, 40th A and the 38th A, supported by the 2nd Air Army. These troops would be the main strike force for Operation *Rumyantsev*.

31 July 1943

The 7th Guards Army had a reported combat strength of 55,825 men with 42,682 in the main combat units.

Operation *Herbstreise* began. General Model started to abandon the entire Orel Bulge in an orderly manner. During this withdraw, "50,000 tons of supplies and 20,000 wounded soldiers" were evacuated from Orel itself. "Subsequently, while heavy fighting raged, the Germans executed a scorched earth policy [destroying anything of use to the enemy] throughout the salient, burning the rye harvest and forcing 250,000 civilians, with their cattle and possessions, toward the west." [256]

Along the south side of the Kursk Bulge, final arrangements for Operation *Rumyantsev* were underway. Planning for Operation *Rumyantsev* had produced three different variants.

- The first plan would have required major regrouping of armies, which would have taken much time and effort. This plan also would have delayed the offensive, and that was something Stalin would not allow.
- The second plan also would have required major regrouping of armies and significant amounts of bridging equipment to cross river obstacles, which in turn would have slowed the tempo of the attack. Stalin rejected this plan, too.
- Therefore, the third variation was just what Stalin wanted. The Voronezh and Steppe Fronts would begin their offensive as soon as possible with only minor regrouping. This entailed a more direct frontal assault with key outflanking maneuvers along the northern Donets River and an envelopment of Kharkov from the west by tank units. The initial focus of this offensive would be against the boundary line between the 4th Panzer Army to the west and Army Detachment Kempf to the southeast. Split these two German forces apart, *Stavka* thought, and the victorious Red Army could quickly liberate Kharkov.

At least that was their thinking.

Overall, the losses (casualties) suffered by the 7th Guards Army during all of July 1944 are reported to have totaled 38,300 men and women, and of these, more than 4,000 were missing in action. [257] Note that these missing troops might be dead and lost on the battlefield, captured POWs, POWs who voluntarily surrendered to the Germans or soldiers who deserted to the rear and somehow avoided being detained by Blocking Detachments or NKVD units.

1 August 1943	
7th Guards Army	
Assigned – Steppe Front	
Lieutenant General M.S. Shumilov Commanding	
Rifle, Airborne and Cavalry Units	24 Guards Rifle Corps (36, 72 GRDs, 213 RD) 25 Guards Rifle Corps (73, 78, 81 GRDs) 49 Rifle Corps (15 GRD, 111 RD) 82, 83, 84, 85, 86 Blocking Detachments 2, 5, 62, 63, 64, 65, 66 Penal Companies
Artillery Units	109, 161, 265 Guards Gun Artillery Regiments 30 AT Brigade 115 Guards Tank Destroyer Regiment 1669 Tank Destroyer Regiment 1, 2, 3, 4, 5, 131, 132 AT Rifle Battalions 5 Antiaircraft Artillery Division (670, 743, 1119, 1181 AA Regiments) 162, 258 Guards AA Regiments 662 Separate Aerial Observation/Air Raid Warning Company
Armored and Mechanized Units	27 Guards Tank Brigade 201 Tank Brigade 167, 262 Separate Tank Regiments 1438 SU Regiment 34, 38 Separate Armored Train Battalions (four trains active)
Engineer Units	60 Engineer Sapper Battalion 175, 329 Separate Engineer Battalions
Reported Combat Strength Estimated Ration Strength	54,602 men 61,602 men and women (including an estimated 7,000 support personnel)

(BSSA 43, BSA 43, TsAMO Archive)

1 August 1943

The start of August found the 7th Guards Army with eight rifle divisions organized into three corps along with good tank and artillery support. This firepower would be needed because the Voronezh, Steppe, and Southwest Fronts were going to attack on 3 August, launching Operation *Rumyantsev* (named after the 18th-century Russian Field Marshal Peter Rumyantsev). [258] The operation was planned to recapture Kharkov and the surrounding area. Despite the replenishment of men and equipment over the previous month, the 7th Guards Army still was understrength, having 27,000 fewer men than from its high point in early July. Note: For some unknown reason, the Armored Train Battalions 34 and 38 were removed temporarily from the control of the 7th Guards Army.

Logistically, the Army was in good shape for the coming offensive in regards to amassed ammunition, fuel and food/fodder, but it lacked transportation vehicles. H.G.W. Davie writes in *The Journal of*

Slavic Military Studies, 2020, Vol. 33, No 4, p. 595, an article called "*The Logistics of the Combined-Arms-Army-the Rear: High Mobility Through Limited Means*" lays out a disturbing picture of the state of the Army's mobility:

> In August, the [7th Guards Army] possessed 71 percent of its [vehicle] establishment with 2,160 Östic, 632 captured, and 445 imported [British and American Lend Lease] vehicles. However, only 76 percent of these were running, less than half the establishment figure. Non-domestic manufactured vehicles remained around 30 percent of the fleet, and it can be assumed that, as before, imported vehicles were concentrated in artillery and tank units and trophy vehicles in rear units. So on the eve of the offensive, [the 7th Guards Army] mustered 62,000 personnel, 987 artillery pieces, 79 operational tanks [and SUs], 5,885 horses, and 2,158 vehicles, which gave a ratio of 11 men per horse and 24 men per vehicle, which indicates that [the Army] was seriously short of transport by even the meager Soviet standards.

Due to the reduced number of serviceable logistical vehicles and limited rail service in the operational area, the 7th Guards Army now was supplied and equipped to support an advance of some 45 miles (70 Km) before it would encounter supply issues. Logistically this meant the Army would be able to advance to the gates of Kharkov before they would need a pause in operations to bring forward supplies from Steppe Front depots and Army level supply bases.

The medical support of the Army still was focused on the just completed defensive operation. That is until 21 July when the first information, of a general nature, concerning a new offensive was passed to medical officers of the 7th Guards Army. Working with vague information, the medical staff began planning for the new offensive on 25 July and finished on 28 July; Army HQ approved the plan on 29 July, all without medical staff officers even having knowledge of the military plan! They had to essentially guess what the Army was going to be doing over the next few weeks. Everything was a secret even to their own medical staff. Only on 2 August, the day before the attack, was the medical department informed and given detailed information. Fortunately, they had deduced the general plan correctly as a Belgorod/Kharkov operation.

Lacking the time to make major adjustments, the medical staff offered support for this new offensive as best they could. Because the hurried evacuation of wounded out of the Army, over the previous few days, to the Steppe Front and beyond did not go as planned, 5,951 wounded and sick remained in Army hospitals, (these hospitals contained 5,100 beds) with a further 600 wounded in divisional level medical evacuation points. These men were the remaining wounded and sick from the defensive, then offensive, Kursk fighting in July. So, the 7th Guards Army would start this new offensive with its medical facilities holding over 6,551 wounded and sick. The medical staff only had been able to clear the wounded from three HPPGs (Mobile Surgical Field Hospitals) and one PPG (Mobile Field Hospital) that were to directly support the attack. This effort could hardly be called excessive medical support when they were expecting 1,300 patients per day during the upcoming operation. [259]

Once again, the medical establishment for the Army had changed. The known medical and related units within the 7th Guards Army are listed below:

> One HQ for the Field Evacuation Point - #24.
> One Field Evacuation Point - #24, (used to transfer wounded to the deep rear).
> Eleven Mobile Surgical Hospitals – HPPG/PPG #869, 899, 2204, 2207, 2208, 2209, 2409, 4344, 4611, 5217, 5226.
> Two Therapeutic Hospitals – TPPG #34, 2712.
> Two Hospitals for the Lightly Wounded – GIR #3247 and 3255.
> Two Infectious Hospitals – XG #2241, 2242.

One Separate Ambulance Company (Auto) - #97, equipped with 48 GAZ-55 ambulances each capable of carrying 10 sitting or 4 stretchers (litter) wounded.
One Ambulance Company (Horse-Drawn) - #251.
One Sanitary Epidemiological Unit - #115.
One Disinfection Company - #111.
One Separate Medical Reinforcement Company - #51.
One Pathological and Anatomical Laboratory - #112.
Three Field Laundry Squads - #116, 117, 118.
Four Laundry Bathhouses - #1, 2, 3, 4.

This represented approximately 5,100 hospital beds for the entire Army.

The veterinary services remained unchanged with:

Two Army Veterinary Hospitals - #958, 960.
One Evacuation Veterinary Hospital - #959.
One Mobile Field Veterinary Laboratory - #16.

Stavka representative Marshal Georgy Zhukov was on hand at the command post of the 5th Guards Army when he issued the final instructions, from *Stavka*, to army and corps commanders about the imminent offensive. [260] The assembled forces needed to complete only minor final preparations; the date was set, the armies were ready, the countdown had begun.

The U.S. Army Air Force launched a major long-range heavy bomber attack from bases in North Africa against the Ploesti oil refineries in Romania. This attack destroyed about 40 percent of Romania's refining capacity. This damage soon would reduce fuel supplies for the Germans on the Eastern Front.

During the night, several companies from the 72 GRD (24 GRC) crossed the northern Donets River and attempted to form a bridgehead on the western bank; this was near the town of Toplinka. After German counterattacks, the front line was restored, but more fighting would continue in this same general area during the next several days. [261] This action was almost certainly a diversionary attack meant to keep German attention focused on the wrong area of the front.

The 7th Guards Army had no plans to force a major crossing of the northern Donets in the Toplinka area. The 49 RC and the 25 GRC were tightly concentrated right next to Belgorod for that purpose. These two corps were waiting for the 69th A to advance on Belgorod from the north and to support their crossing of the river. This support would allow the 7th Guards Army to directly assault Belgorod from the east. Ultimately, the Steppe Front plan called for the 69th and 7th Guards Armies to advance south side by side, conducting a flanking attack against German positions along the west side of the river. This maneuver would eliminate the need to launch an unsupported attack across the river directly into prepared German positions.

2 August 1943

The 7th Guards Army had a reported combat strength of 54,994 men with a further 7,000 men and women in supporting roles, totaling 61,994 men and women. Approximately 1,473 women worked with the rear support troops [262] and many more women served in forward positions within combat units and divisional level medical units.

The Army issued a summary of material support for combat units. The report showed a modest increase in most ammunition types, with a good increase in some basic food supplies. Overall, the logistical needs of the Army were being met with a small surplus building up. [263]

Additionally, Army personnel produced a report listing the daily (non-combat) requirement of food and fuel/lubricants for the Army: 198 tons of food and 334 tons of fuel/lubricants. [264] Note: At this time the

Army was not active in combat, so fuel use would have been low and any ammunition expenditure only would increase the required number of tons per day.

3 August 1943 – Day One of Operation *Rumyantsev*

(Note: Soviet records report Operation *Rumyantsev* taking place in three distinct stages: Stage One, 3-5 August; Stage Two, 6-10 August; and Stage Three, 11-23 August; with the operation ending with the fall of Kharkov.)

On the first day of the Operation *Rumyantsev* attack, the 7th Guards Army had a reported combat strength of 54,757 men. (Note: These reported combat strengths were most likely recorded at the end of the day, and all these reports were retrieved from the Russian TsAMO Archive).

For Operation *Rumyantsev,* the Voronezh and Steppe Fronts contained 1,144,000 men; 2,418 tanks and SUs; 12,866 artillery; and 767 rocket launchers. Also available were two Air Armies, the 5th with 769 aircraft and the 2nd with 753 aircraft, yielding a combined total of 1,522 aircraft, with 1,311 of them operational. [265] [266] *Stavka* could commit additional forces if needed.

On the German side, the 4th Panzer Army and AD Kempf had been denuded, containing some 210,000 men and approximately 250 tanks. [267] German forces had support from the 4th Air Fleet, which contained 1,138 aircraft with 796 of them operational. [268] (See Map #16.) As can be seen, the defending German forces both on land and in the air were badly outnumbered.

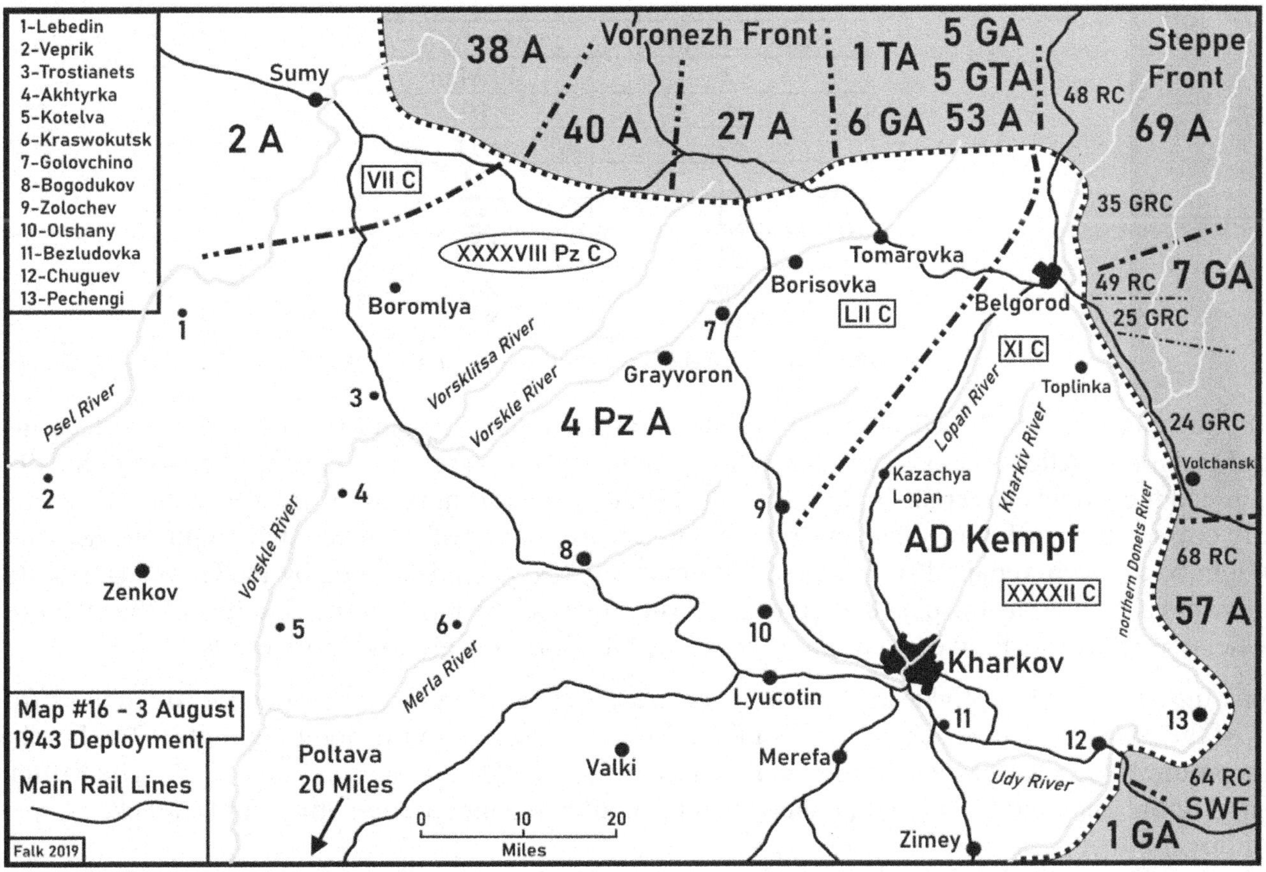

Starting at 05:00 hours (Russian time), the attack opened with a powerful 170-minute-long artillery bombardment. At 07:55 hours the ground attack began with the support of the 2nd and 5th AAs as well as long-range aircraft from the Southwest Front's 17th AA. In the first six days of the offensive, these Air Armies flew over 13,000 sorties; the Luftwaffe could not match this effort, which allowed the VVS to gain air superiority over the battlefield. [269]

The Voronezh Front conducted the main attack to the west of Belgorod. On the right flank of the Steppe Front, the 53rd and 69th Armies launched supporting attacks by assaulting the well-fortified German defenses directly around Belgorod. By 13:00 hours, the 1st TA and 5th GTA were breaking through the German defensive lines. German forces simply were overwhelmed by the power of this attack. Apparently, the 7th Guards Army initially focused its main attention upon reducing the German Mikhailovka bridgehead (east of Belgorod) and only provided artillery fire support to the other armies on this first day. But in the Toplinka area, the Soviets launched further small attacks, some employing Russian troops that spoke German and wore German uniforms. 270 These attacks failed to secure a significant bridgehead. As author Boris Ilyich Mutovin stated in Chapter 5 of his book Through All the Trials, "From Belgorod to Kharkov, two hours by train. We [78 GRD] went this way in 16 days. [We] passed with heavy fights." Securing Kharkov would be a long hard road indeed as the German defenders fought for every position, delaying the Soviet advance as long as possible. (See Maps #17, 18, 19.) At the end of Day One of the offensive, TsAMO record ID #133390890, p. 13, records that the 7th Guards Army's rifle divisions admitted 798 sick or wounded troops to sanitary/medical services. The number of dead was not reported in this or in any of the following daily reports.

4 August 1943 – Day Two of Operation *Rumyantsev*

The 7th Guards Army had a reported combat strength of 54,590 men. Interestingly, the data also included the reported strength of the Army's penal companies. The strengths vary wildly and one can see which units had been in combat recently. (See Table #4).

Penal Company	Strength
2	30 Men
5	19
62	221
63	54
64	33
65	260
66	277
Total	**894**

Table #4: Penal Company Strengths -- 4 August 1943 [271]

During the morning, the 7th Guards Army finally eliminated the German Mikhailovka bridgehead just east of Belgorod. After this long overdue victory, in the afternoon forward units of the Army, the 49 RC, forced the Donets River along an eight-mile section of the front south of Belgorod. [272] The Army finally crossed the northern Donets in strength with the assistance of engineers who built one heavy and two light bridges and repaired six bridges that previously had been destroyed by the Germans. But this victory was a day late as the plan called for the Army to cross the river on the first day of the attack. This delay might have allowed German forces to withdraw from Belgorod more easily.

Troops fighting in the Toplinka area also were active, and they even received effective air support from units of the 5th AA. General Shumilov sent the following telegram to the bombing units, "The bomber strike produced by your groups during the period 07:15–07:25 [hours] gave a good result. The Military Council of the 7th Guards Army announced thanks to all personnel participating in the attack. Shumilov." [273]

At the end of Day Two of the offensive, the 7th Guards Army's rifle divisions admitted 534 sick or wounded troops to sanitary services.

5 August 1943 – Day Three of Operation *Rumyantsev*

The 7th Guards Army had an estimated combat strength of 53,857 men, only a slight decrease from the previous day.

After intense fighting, Belgorod fell to units of the 69th A and the 7th Guards Army. The 93 GRD and the 111 RD of the 7th Guards Army lead the way into Belgorod from the east. Simultaneously, north of Kursk, Orel fell to units of the 3rd and 63rd Armies. (See Map #17.)

Later in the day, in far off Moscow, "To celebrate the twin victories [of capturing both Belgorod and Orel on the same day], Stalin ordered an artillery salute of twelve volleys from 120 guns. It was the first time in the war that such a salute had been fired, and in Moscow some citizens, thinking it was an air raid, took to their cellars." [274]

More of these salutes would take place in the near future. (See Appendix 10.)

Early in the morning, once again fighting around Toplinka intensified. Units of the 7th Guards Army somehow were able to quickly bridge the river and, to the Germans' surprise, brought tanks over to the western side of the Donets. Heavy fighting broke out which lasted into the night. Only after recovering their old positions did the Germans discover how the Russians were able to build a bridge so quickly.

Two rows of undamaged tanks were discovered, which had been driven into the water one behind the other and which served as supports for [the] improvised submarine bridge. Planking had been hastily placed on top of them and attached with ropes [by engineers]. [275]

The top of the bridge was about three feet under the surface of the river, making it all but invisible to German observers. This ingenious engineering effort produced a usable crossing of the river, but in the end the Germans were able to destroy all the tanks and regain their original positions.

Units of the Voronezh Front broke through the German positions between Tomarovka and Belgorod and advanced against moderate resistance. Tomarovka itself was a heavily fortified city. Despite being almost completely encircled by advancing Russian tank units, this German position delayed the Russian advance. After fighting for three days and nights, the Russians finally took Tomarovka at the end of the day on 6 August.

At the end of Day Three of the offensive, the 7th Guards Army's rifle divisions admitted 757 sick or wounded troops to sanitary services.

The German III Pz C headquarters arrived, near Zolochev, to reinforce AD Kempf, but its first combat unit, the 3 Pz D, was still south of Kharkov. (See Map #17.)

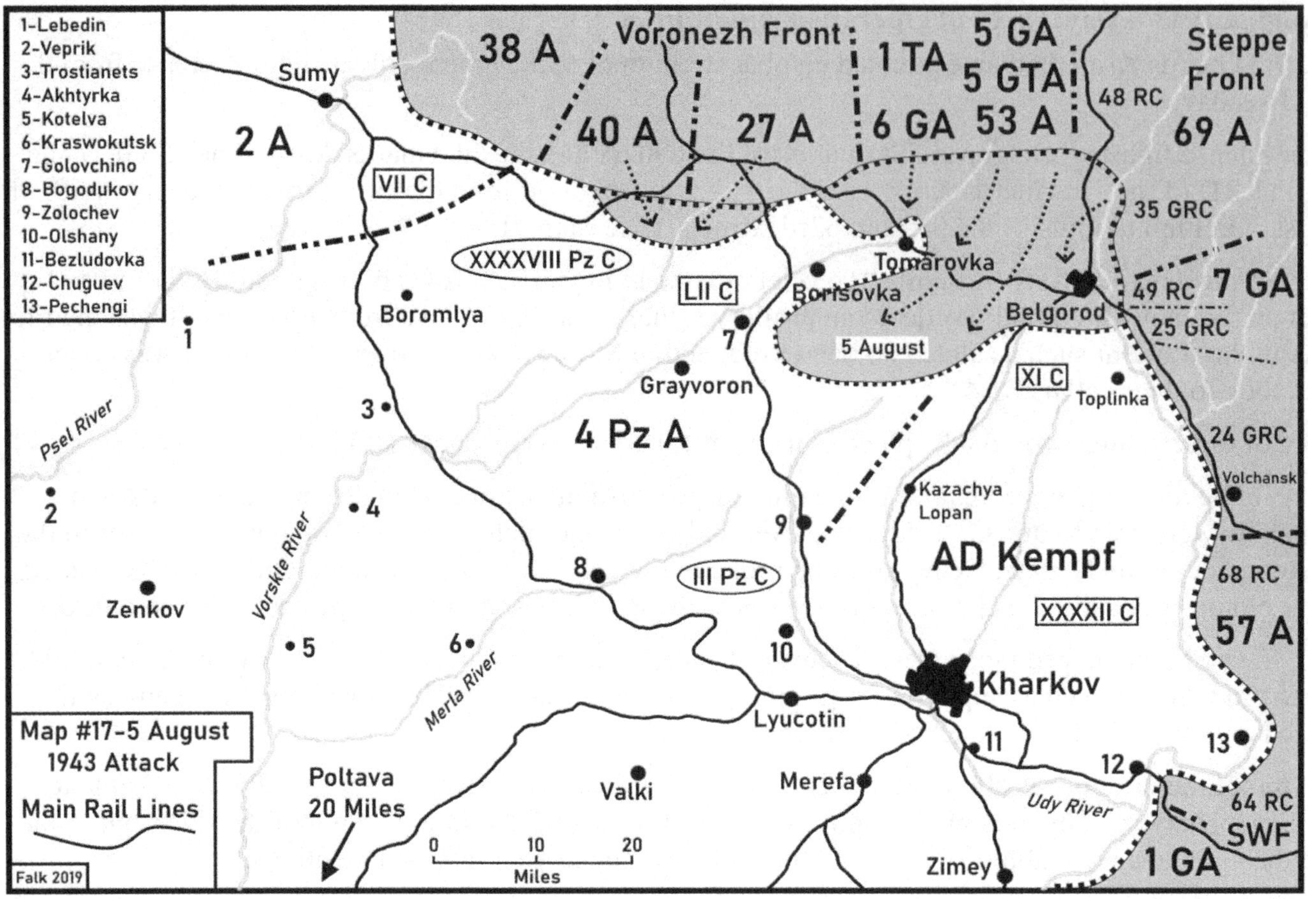

6 August 1943 – Day Four of Operation *Rumyantsev* - Stage Two of the Offensive

The 7th Guards Army's overall reported combat strength was 53,529 men, [276], but a second report, listed in Table #5, separately displayed a more detailed headcount within the rifle divisions.

Division	Total # of Men	Officers	Sergeants	Privates
15 GRD	4,518	667	1,240	2,611
36 GRD	8,021	857	2,363	4,801
72 GRD	3,237	520	622	2,095
73 GRD	3,892	565	863	2,464
78 GRD	4,674	669	972	3,033
81 GRD	6,163	800	1,473	3,890
111 RD	4,252	576	1,035	2,641
213 RD	4,028	595	1,049	2,384
Totals	**38,785***	**5,249**	**9,617**	**23,919**

*** Not including attached combat support units.** [277]

Table #5: Headcount Report for the 7th Guards Army's Rifle Divisions.

As can be seen, only the 36 GRD could be considered a full-strength unit. The remainder fell between 40-50 percent with the 81 GRD at about 75 percent of full strength. These figures also indicated the Army had approximately 14,744 men in supporting tank, artillery, AA, self-propelled AT guns and other antitank units.

The 69th and 7th Guards Armies continued their advance south of Belgorod making slow but steady progress against determined German resistance. Both the 49 RC and the 25 GRC had fully crossed the Donets River. The 25 GRC waited on the east side of the river until the other corps could unhinge and outflank the formidable German defenses along the west side of the Donets. Units under the command of General Warner Kempf (AD Kempf) now began a difficult but superbly executed phased withdraw to the south in order to avoid being surrounded by flanking Russian tank armies. (See Map #18.) General Kempf also intended to slow the Russian attack as much as possible and to inflict maximum damage to the advancing Red Army. Once again, fighting around Toplinka flared up, which lasted until the main battle on the western side of the northern Donets River reached that area.

Stavka representative General Zhukov, Commander of the Steppe Front General Konev, and Chief of Staff of the Steppe Front General Zakharov produced a report about the developments in the Belgorod/Kharkov sector. This report was sent to Comrade Ivanov (Stalin).

August 6, 1943

We report:
In connection with the successful breakthrough of the front of the enemy and the development of the offensive in the Belgorod-Kharkov direction [,] operation will continue to pursue the following plan: (a synopsis of the relevant parts of this long report are listed below)

1. The 53rd A will continue its advance along the Belgorod-Kharkov highway. The supporting 69th A needs to be replenished with 20,000 men as soon as possible. The 7th Guards Army will continue its advance toward Kharkov and attack toward the east side of the city to help the 57th A force the River Seversky Donets (northern Donets River).
2. It would be advisable for the 57th A of the Southwest Front to be handed over and made subordinate to the Steppe Front. This will allow coordinated attacks against Kharkov.
3. In the second stage of the Kharkov operation, the 5th GTA should also be made subordinate to the Steppe Front.
   We suggest that the Kharkov operation should tentatively be arranged according the following plan:
   a) The 53rd Army and the 5th Guards Tank Army will assault Kharkov from the south and southwest.
   b) Shumilov's Army (7th Guards Army) will attack from the north.
   c) The 57th Army will advance from the east and envelop Kharkov from the south.
   d) The 69th Army, (if it has been replenished by this time) will continue its advance to the south.
4. *To conduct the Kharkov operation, besides the 20,000 extra men, we need another 15,000 men for the divisions of the 53rd and 7th Guards Armies; 200 T-34* [medium] *and 100 T-70* [light] *tanks besides 36 – KV* [heavy] *tanks to bring the tank units' front line up to full complement. Four self-propelled artillery regiments and two engineer brigades need to be moved up. The air forces on the Front should be brought up to full complement with assault and combat aircraft and bombers in the following numbers: 90 fighters, 40 PE-3 and 60 IL-2 planes.*

Please endorse
Zhukov, Konev, Zakharov [278]

This report clearly indicated the failing strength of the assaulting armies and the need to regroup in order to deliver a coordinated attack to finally capture Kharkov. Whether Stalin and *Stavka* sent the requested reinforcements is unknown, but the battle for Kharkov continued just the same.

The German Pz GD *Grossdeutschland* arrived back in the 4th Panzer Army but during two days of combat was only able to slow the advance of the Russian 40th A. However, von Manstein was assembling other forces, under the control of the newly arrive III Pz C, to deal with this new Russian offensive.

With the fall of Belgorod, General Konev, Commander of the Steppe Front, asked/ordered the People's Commissariat of Railways (NKPS) to restore the rail line on the east side of the northern Donets River in the section covered by the 7th Guards Army. This was a 31-mile section from Volchansk to Belgorod. The time allocated for the completion of this line was 7-8 days. The restoration of this section of track would allow trains to cross the northern Donets River at Belgorod and supply the armies of the Steppe Front advancing toward Kharkov on the west side of the river. Despite some unwelcome interference from Nikita Khrushchev, Military Council member of the VF, the 14 Bridge and 9 Track Battalions from the 19 Railway Brigade restored the rail line in the allocated time. [279]

As the Steppe Front advanced south toward Kharkov, the 7 Railway Brigade worked to restore rail service just behind the front line. Due to the stubborn resistance put up by the defending German troops, the repairs to the railway line were able to keep pace with the advancing Red Army. Because of this, when Kharkov was finally liberated on 23 August, the first train arrived at the Kharkov passenger station only five days later, on 28 August. The restoration of the Kharkov railway network was critical to supply the advance of both the Voronezh and Steppe Fronts. The Germans had, of course, destroyed the entire network and railway support structures within Kharkov before withdrawing from the city. By 30 August, enough damage had been repaired within the city to allow through traffic to support the advance of the fronts. The total restoration of the rail facilities would take many months. [280]

On or about this date, the 7th Guards Army issued an updated, more current supply consumption report. The report stated the Army needed 120 tons of food, 55 tons of fuel and lubricants and 600 tons of ammunition, for a total of 775 tons per day. These numbers illustrate why an operational local rail line was so important. [281]

At the end of Day Four of the offensive, the 7th Guards Army's rifle divisions admitted 914 sick or wounded troops to sanitary services.

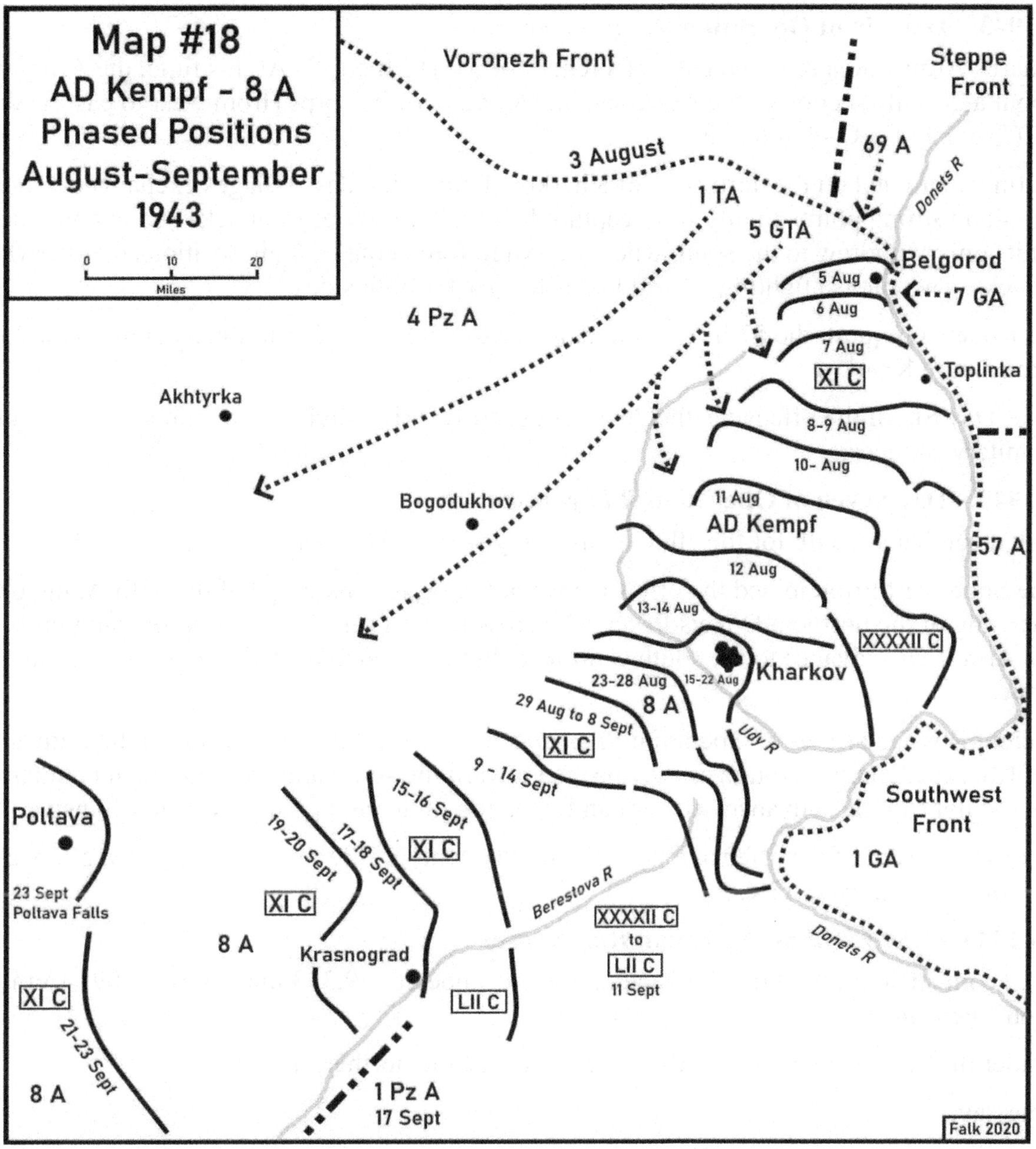

7 August 1943 – Day Five of Operation *Rumyantsev*

The 7th Guards Army had a reported combat strength of 52,667 men. [282]

By this time the medical system was in full operation taking care of thousands of wounded troops. One of the main problems was the lack of ambulances to transport the wounded to the rear. The few ambulances (truck or wagon) only transported about 25% of the men, with the remaining 75% evacuated by other means such as empty supply trucks, supply wagons, railroad cars, and maybe a few by air. [283]

The 69th and 7th Guards Armies continued their advance south toward Kharkov, but they had to repeatedly bring up artillery to blast their way through fiercely defended German positions. All this effort cost time and much sacrifice on the part of the combat units. The German plan was to blunt Russian attacks with prepared positions and to inflict the maximum number of casualties on attacking Russian units before withdrawing.

At the end of Day Five of the offensive, the 7th Guards Army's rifle divisions admitted 570 sick or wounded troops to sanitary services.

8 August 1943 – Day Six of Operation *Rumyantsev*

The 7th Guards Army had a reported combat strength of 51,413 men. [284] At this time, the Army was holding about a 20-mile section of the front, and had all three of its corps (from west to east) -- the 49 RC, 25 GRC, and 24 GRC -- on the line.

Heavy fighting continued on the north side of Kharkov. During the day, villages exchanged hands several times and strong points finally were captured only after a series of attacks and counterattacks. German units only withdrew to the south when they were forced out of their positions by overwhelming Russian attacks. The intense fighting caused losses to soar for both sides.

At General Konev's request, the 57th A was transferred to the Steppe Front, thus adding to the forces deployed to capture Kharkov.

At the end of Day Six of the offensive, the 7th Guards Army's rifle divisions admitted 423 sick or wounded troops to sanitary services.

9 August 1943 – Day Seven of Operation *Rumyantsev*

The estimated combat strength for the 7th Guards Army was 50,313 men.

Units of the Southwest Front joined the offensive when the right flank corps of the 57th A, the 68 RC, forced a crossing of the northern Donets River. The crossing was eased by the approaching units of the 7th Guards Army. The advance to the south continued, but the attacking Red Armies lost strength at an alarming rate.

The other eight Russian armies of Operation *Rumyantsev*, Voronezh Front, continued their advance to the west of Kharkov, gaining ground as they pushed toward the south and west. German counterattacks slowed but did not stop this advance, as German forces fought to contain the Russian offensive.

At the end of Day Seven of the offensive, the 7th Guards Army's rifle divisions admitted 525 sick or wounded troops to sanitary services.

10 August 1943 – Day Eight of Operation *Rumyantsev*

The reported combat strength of the 7th Guards Army dropped to 49,303 men, down 5,691 from the start of Operation *Rumyantsev.*

The remainder of the 57th A crossed to the western side of the northern Donets.

On the same day,

> two groups of [VVS] dive bombers bombed a cluster of [German] manpower and artillery and mortar batteries in the Tavrovo area [northeast of Kharkov]. How effective these blows were can be seen from the telegram of the commander of the 7th Guards Army sent to the commander of the air corps, Colonel I. Polbin: Your bombers were exceptionally well on their first flight. Thanks to this, the Tavrovo point was taken, and after the bombing… ground units occupied the northern edge of the point and conducted battles for mastering the [village of] Brodok. Shumilov. [285]

As related, airpower, if applied correctly, could be vital in breaking an enemy position and clearing the way for further advances.

By this date, three HPPGs (surgical hospitals #2207, 2409, 4344) were in place on the west side of the northern Donets Rive and directly supported the Army. At this point in the battle, medical units of the Army were dealing with an average of 633 casualties per day, which translates into over 5,000 losses over the previous eight days of combat. [286]

At the end of Day Eight of the offensive, the 7th Guards Army's rifle divisions admitted 264 sick or wounded troops to sanitary services.

11 August 1943 – Day Nine of Operation *Rumyantsev* - Stage Three of the Offensive

The 7th Guards Army had a reported combat strength of 48,912 men. (See Map #19.)

On or about 11 August, the 57th A was officially transferred to the Steppe Front.

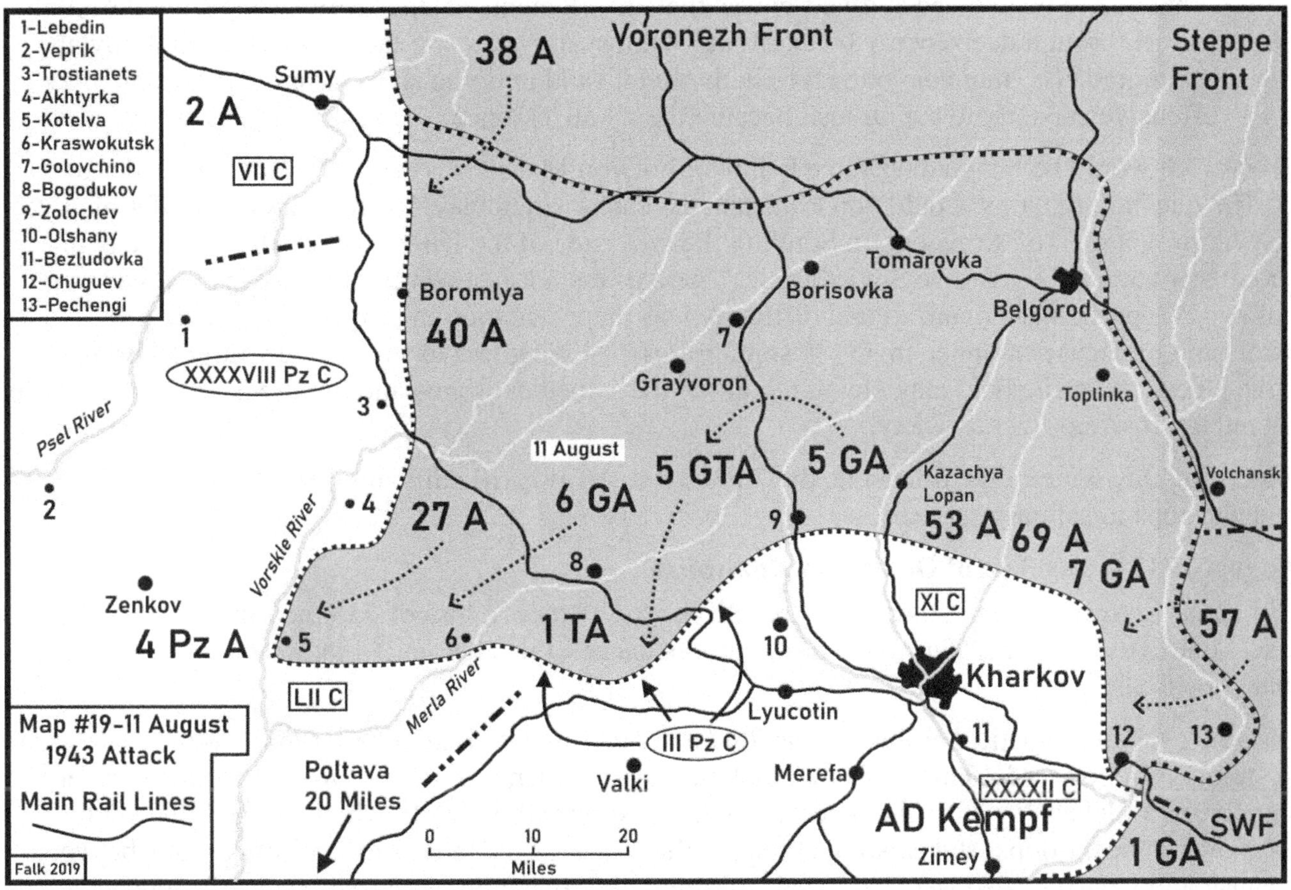

By the end of the day, the 53rd, 69th, and 7th Guards Armies all had reached the outermost Kharkov defensive positions. The 7th Guards Army led with its 25 GRC and the other two corps provided flank protection. The 57th A still was short of this defense line. The entire city was encircled by strong points, mine fields, barbed wire obstacles, and even antitank ditches.[287] The main battle for Kharkov was about to begin. The Germans were not going to give up the city without a prolonged struggle. In addition to being a major city, Kharkov was also a supply, rail and support center. Forcing the Germans out of Kharkov would not only be a propaganda victory for the Red Army, but also would greatly complicate German logistical efforts for Army Group South.

To expedite the attack upon Kharkov, the Steppe Front reinforced their attacking armies with an additional 4,234 guns and mortars. [288] But these reinforcements took several days to arrive, so the main attack upon the city was delayed until 19 August. General Konev understood that after nine days of constant combat, his forces needed some time to rest and to regroup before finally capturing the big prize. This massive reinforcement of firepower soon would become the standard tactic of future Russian offensive operations. The Red Army had learned it was easier to kill Germans with a Howitzer than with a rifle. Around the city of Bogodukhov (#8 on Map 18) the 2 SS Panzer Grenadier Division *Das Reich* and the 3 SS Panzer Grenadier Division *Totenkopf,* of the III Pz C, engaged the leading formations of the 1st Tank Army. David Glantz and Jonathan House, in their book *When Titans Clashed*, summed up this battle nicely on page170, "Initially, the German forces halted the Soviets and severely mauled the leading three brigades of the 1st Tank Army. The next day, however, 5th GTA sent reinforcements, and the Germans were fought to a draw between 13 and 17 August."

These battles outflanking Kharkov from the west cut road and rail access on that side of the city and slowly forced von Manstein to realize he would not be able to hold Kharkov for much longer. Von Manstein, in his book, *Lost Victories*, on page 450, brings into focus the overall German strategy at this point in time:

> What had to be avoided at all costs was that any elements of the Army Group should become cut off through deep enemy breakthroughs and suffer the same fate as [the] Sixth Army at Stalingrad. To "maintain ourselves in the field," and in doing so to wear down the enemy's offensive capacity to the utmost, became the whole essence of this struggle.

This approach would be the guiding force behind all of von Manstein's efforts for the remainder of 1943. The question remains, Could von Manstein execute the defensive performance of a lifetime by safely withdrawing all of Army Group South to the west side of the Dnepr River, all the while being under constant attack by massive Soviet armies? And it was not just combat formations that had to be withdrawn. "A particular organizational difficulty was the evacuation of 200,000 wounded, plus hospitals and medical personnel, in 387 hospital trains." [289] Needless to say, the few railway bridges over the Dnepr had to be held long enough to allow these medical trains and hundreds of other trains to reach and to cross the river to safety.

At the end of Day Nine of the offensive, the 7th Guards Army's rifle divisions admitted 519 sick or wounded troops to sanitary services.

12 August 1943 – Day Ten of Operation *Rumyantsev*

The 7th Guards Army reported a combat strength of 48,179 men, a loss of 733 men from the day before. Attacks by the Army during the day were not successful. They failed to breach the tough German positions.

To ease the growing logistical problems, the 7th Guards Army HQ issued Order #006 to move the Army's main supply base forward 45 miles, from Volokonovka to Volchansk on the east side of the northern Donets River. Volchansk was well connected to the existing rail network, which allowed the swift transfer of supplies, troops, and existing depots to the new area by 15 August. Military roads between Volchansk and Kharkov also were extended and improved to handle increased levels of supplies moving to the fighting front. [290]

The depleted infantry units of AD Kempf struggled to hold onto Kharkov. General Kempf, fearing a flank attack by the 57th A, proposed to von Manstein to withdraw his troops from Kharkov and to form a shorter line to the south. Von Manstein did not object, but Hitler overruled von Manstein and ordered Kempf to hold Kharkov at all costs.

After much urging on the part of the high command, Hitler issued Fuhrer Order No. 10, about the immediate construction of the *Ostwall* (East Wall) defensive line along the western side of the Dnepr River. As it would turn out, this order came much too late to be effective.

When the exhausted troops of AG South eventually arrived on the western bank of the Dnepr, they found that almost nothing had been prepared. "The troops, influenced by talk of an 'East Wall,' were dismayed to find upon crossing that nothing had been built and that much of the proposed front had not even been surveyed. Later, one of the army chiefs of staff was to warn that troop morale could not again withstand such a shock." [291]

At the end of Day Ten of the offensive, the 7th Guards Army's rifle divisions admitted 362 sick or wounded troops to sanitary services.

13 August 1943 – Day Eleven of Operation *Rumyantsev*

The 7th Guards Army had an estimated combat strength of 47,279 men, a further loss of 900 men. General Shumilov shifted the Army's attack to the left flank, striking Kharkov more from the northeast, making progress and once again easing the way for the 57th A to advance.

By this date the 7th Guards Army carried out HQ Order #006 as it moved its main logistical supply center from Volokonovka to Volchansk on the northern Donets River, some 60-70 miles closer to the front. Besides a rail line, this new location also had a good supply road to the rear. Medical support also moved as they too had to adjust to these new logistical arrangements as they already had several hospitals at work on the west side of the northern Donets River. (See Map #2 for supply base location.)

The 1st Guards Army of the Southwest Front began crossing the northern Donets River on the left flank of the 57th A. This action was another threat to the German position at Kharkov.

At the end of Day 11 of the offensive, the 7th Guards Army's rifle divisions admitted 503 sick or wounded troops to sanitary services.

14 August 1943 – Day Twelve of Operation *Rumyantsev*

The 7th Guards Army had a reported combat strength of 46,379 men, another loss of 900 men. The Army conducted low level combat along the front.

General Kempf would not relent in his efforts to save his troops from being surrounded at Kharkov, so at Hitler's direction, von Manstein relieved Kempf of his command, and appointed General Otto Wöehler, von Manstein's Chief-of-Staff, in his place. Kempf was then transferred to *Führerreserve* (leaders reserve) until 1944 when he was finally given another command.

To the west of Kharkov, fighting still raged between opposing tank and Panzer formations around Bogodukhov. To the southeast, the 1st Guards Army expanded its position on the west side of the Donets River and deployed several divisions against the city of Zimey. The noose was closing around Kharkov.

At the end of Day 12 of the offensive, the 7th Guards Army's rifle divisions admitted 719 sick or wounded troops to sanitary services.

15 August 1943 – Day Thirteen of Operation *Rumyantsev*

The 7th Guards Army had a reported combat strength of 44,640 men, down 1,739 from the day before. These heavy losses included those killed, wounded, sick, or missing.

The 53rd, 69th, 7th Guards and 57th Armies continued low level attacks into the several outer defensive lines around Kharkov. These positions gave the relatively weak German defenders a great advantage as the attacking Russian armies were forced to fight through layer after layer of fortified positions, minefields and obstacles.

At the end of Day 13 of the offensive, the 7th Guards Army's rifle divisions admitted 694 sick or wounded troops to sanitary services.

16 August 1943 – Day Fourteen of Operation *Rumyantsev*

The 7th Guards Army had a reported combat strength of 43,352 men, down another 1,288 men. The Army prepared for what was hoped to be the final attack against Kharkov.

In the north, German armies finished their withdraw from the Orel Bulge and occupied the prepared defensive Hagen Line. (See Map #6 for the location of the Hagen Line.) Advancing Russian forces repeatedly attacked this defense line and failed to break through, suffering heavy losses in the process. Eventually *Stavka* called off the attacks and ended Operation *Kutuzov*. [292]

Due to the change of command, AD Kempf was renamed as the 8th Army.

The 1st Guards Army expanded its position further, capturing the city of Zimey and cutting one of the rail lines south of Kharkov.

At the end of Day 14 of the offensive, the 7th Guards Army's rifle divisions admitted 1,130 sick or wounded troops to sanitary services.

17 August 1943 – Day Fifteen of Operation *Rumyantsev*

Reported combat strength of the 7th Guards Army was down to 42,328 men, a further loss of 1,047 men from the day before. Little or no combat activity was reported, but losses, for whatever reason, continued to accumulate.

The 7th Guards Army HQ issued a summary of material support for units of the Army. The report shows the Army held moderate amounts of ammunition for all weapons, and large stocks of biscuits and bread flour along with grains and macaroni. The Army had low amounts of sugar, fish, meat and vodka, but had good amounts of canned meat and hay for the horses. So, the Army was being adequately supplied under current combat conditions. [293]

At the same time, engineer and road maintenance troops were busy behind the lines building and repairing existing bridges and constantly working on the roads. The Separate Road Construction Battalions #62 and 165 and the Separate Bridge Building Battalion #156 worked nonstop to keep the lines of supply and communication open. The Army at this time was being supplied over a single main road on the west side of the Donets River with no rail access at all. The road maintenance troops also were developing a second, more direct supply road to Kharkov. That road was 48 miles long and would be in use before the end of the month. For the Army to fight and to win, supplies must flow. The Separate Road Operation Battalion #22 faced a constant battle against nature and the enemy to provide the needed logistical and infrastructure support. [294]

At the end of Day 15 of the offensive, the 7th Guards Army's rifle divisions admitted 551 sick or wounded troops to sanitary services.

18 August 1943 – Day Sixteen of Operation *Rumyantsev*

The 7th Guards Army had a reported combat strength of 41,281 men, a loss of another 1,047 combat troops (same as the previous day). These reported losses, having the exact same amount several days in the row, suggests that someone in the Army HQ probably was just estimating these daily losses until the real values could reach the HQ from the combat units.

On this day, a greatly weakened Red Army finished occupying the entire Orel Bulge area. German troops had successfully withdrawn, under heavy rain clouds and constant air attacks by the VVS, to the newly fortified Hagen Line. [295] The Russians had successfully completed Operation *Kutuzov*, but Red Army forces in that area now were exhausted and depleted of their offensive strength.

In the south, strong German positions and the weakness of its own rifle divisions prevented the 7th Guards Army from advancing further toward Kharkov. The Army at this time only had some 20 tanks remaining to support further attacks.

Far to the west of Kharkov, the Germans launched a two-pronged attack to close the open gap in the front line between the 4th Panzer Army and the new 8th Army. The Germans would close this gap between Akhtyrka and Krasnokutsk (numbers 4 and 6 on the Map #19) on 20 August. During this operation several Russian units were surrounded, but most were able to break through the thin German lines to reach safety. This counterattack finally stopped the Russian offensive in this area.

Adolf Hitler, realizing Kharkov was doomed, gave General Wöehler permission to abandon Kharkov but only "in the event of extreme emergency." [296] General Wöehler would do just that in a few days' time.

Back in the Third Reich,

> On August 18, 1943, the Luftwaffe's Chief of Staff, Colonel General [Hans] Jeschonnek, placed a gun to his temple and blew his brains out. His suicide was the direct result of the two massive blows [the western] Allied bombers had launched the previous day and evening (Schweinfurt/Regensburg and Peenemunde). Jeschonnek had ignored the industrial, logistical and technical basis on which modern war between industrialized states…[was] fought. That curious blindness…now led his air force and nation into a hopeless situation. In [July and August] the Luftwaffe fought three great air battles and on each one of the three fronts [Germany, Italy, Russia] the Germans lost more than 1,000 aircraft. The impact of the pressure exerted by three different fronts forced the Germans to shut down the air war somewhere. There was no other alternative but to defend the Reich. [297] (See Appendix 11.)

As a result of this crisis, the air war in the east and in the Mediterranean became secondary theaters of war for the Luftwaffe. Despite this change of focus, the Luftwaffe never disappeared from the skies over the Eastern Front, but it gradually lost its ability to intervene decisively in the fighting.

By the end of Day 16 of the offensive, the 7th Guards Army's rifle divisions admitted 421 sick or wounded to sanitary services.

19 August 1943 – Day Seventeen of Operation *Rumyantsev*

The 7th Guards Army had a reported combat strength of 41,125 men. The Army also lost the support of the 50 OKE (Independent Spotting Air Squadron), leaving only the 578 OE (Independent Communications Air Squadron) at its disposal.

Following strong artillery and air attacks, the final assault against German positions around Kharkov began with all armies exerting a maximum effort. Progress was slow and costly. General Konev stated in his book *Front Commander's Notes*, Chapter 2, that in order to save the Kharkov Tractor Plant undue damage, "It is better if the army of General M.S. Shumilov will storm individual buildings of the plant and conduct street battles." In other words, save your ammunition, expend the lives of your soldiers instead, and do not unduly harm the factory.

At the end of Day 17 of the offensive, the 7th Guards Army's rifle divisions admitted 506 sick or wounded troops the sanitary services.

20 August 1943 – Day Eighteen of Operation *Rumyantsev*

The 7th Guards Army reported a combat strength of 41,383 men. This was the first time since the start of Operation *Rumyantsev* that the Army showed a slight increase in combat strength. Unfortunately, this increase was due only to the 34 and 38 Armored Train Battalions once again being assigned to the Army. These two brigades added 392 men to the Army's combat strength totals.

Note that at this time there was no active rail bridge or rail line on the west side of the northern Donets River for the 7th Guards Army to use. So, any available armored trains were most likely positioned to defend the main supply base at Volchansk from air attack.

The attack against Kharkov continued with full force as the surrounding armies made slow but steady progress against enemy positions.

The weakened German divisions that were defending Kharkov depended upon artillery support to compensate for the lack of infantry, but now the artillery was running out of ammunition. Talking with General Otto Wöehler, Commander of the new 8th Army, von Manstein stated that "twenty-four trainloads of ammunition were on the way from Germany, but [von Manstein] had to agree that they would hardly come in time" to hold Kharkov. [298]

The 57th and 1st Guards Armies made good progress south of Kharkov against determined German resistance by several infantry divisions and part of the 6 Pz D.

Stavka released the 5th GTA from the Voronezh Front to the Steppe Front in order to support their attack upon Kharkov.

At the end of Day 18 of the offensive, the 7th Guards Army's rifle divisions admitted 190 sick or wounded troops the sanitary services.

21 August 1943 – Day Nineteen of Operation *Rumyantsev*

Reported combat strength for the 7th Guards Army was 41,248 men.

The 53rd, 69th, 7th Guards and 57th Armies pressed in tighter around the city. Defending German forces were now pushed back against the built-up area of Kharkov. At this point, any quick flanking movement by the Russians would block the last escape route out of the city. The 5th GTA moved in from the west to add its support to the attack.

At the end of Day 19 of the offensive, the 7th Guards Army's rifle divisions admitted 444 sick or wounded troops to sanitary services.

22 August 1943 – Day Twenty of Operation *Rumyantsev*

The 7th Guards Army now had a reported combat strength of 40,580 men. Once again, the Army's strength was melting away day-after-day.

During the night, and after consulting with von Manstein, General Wöehler began withdrawing his 8th Army from Kharkov. He had no choice. Advancing Russian troops had almost completely surrounded Kharkov and already controlled most of the exits from the city.

After learning the Germans were beginning their withdraw, General Konev ordered a night assault against Kharkov to prevent the enemy from escaping unscathed.

> All night on 23 August [and during the early morning], street fighting was going on in the city, fires were burning, [and] strong explosions were heard. The soldiers of the 53rd, 69th, 7th Guards, 57th Armies and 5th GTA, showing courage and bravery, skillfully circumvented the enemy strongholds, seeping into his defenses, [and attacking] his garrisons from the rear. Step by step Soviet soldiers cleared Kharkov of fascist invaders. [299]

South of Kharkov, a counterattack by parts of the 6 Pz D with support from the 161 ID forced back the advanced units of the 1st Guards Army. The only route out of Kharkov now was clear, but this was only a temporary reprieve.

Much further south, once again Russian armies attacked over the Mius River and formed a bridgehead that could not be reduced. The German 6th Army simply lacked the strength to restore the situation.

At the end of Day 20 of the offensive, the 7th Guards Army's rifle divisions admitted 394 sick or wounded troops to sanitary services.

23 August 1943 – Day Twenty-One of Operation *Rumyantsev*

The 7th Guards Army had a reported combat strength of 35,894 men. This represented a one-day loss of 4,686 men from the previous day. Street fighting within a major city was very costly.

Of special note, at this time the advancing 69th A to the right and the 57th A to the left had squeezed the 7th Guards Army between them so much that the Army was only holding a tiny section of the front just a few miles wide. The benefit of this arrangement was that most of the Army's units were essentially in the rear, where they could get some rest and replenishment.

By 12:00 hours, Kharkov had completely fallen to the advancing Red Army. During the entire war, this was the fourth and final battle for control of Kharkov. Overnight the active front line had shifted several miles to the south of the city. Of course, German infantry units once again occupied prepared positions just south of the Udy River. Both forces were exhausted as Operation *Rumyantsev* sputtered out in the ruins of Kharkov. (See Map #20.)

Alexey Grigorievich, an artillery gun technician serving in the artillery of 7th Guards Army said, "Since Kharkov was already surrounded on both sides by our troops, the Germans retreated in front of us. They always did it smartly and accurately. In the morning we woke up, but they are not in front of us."

Grigorievich was wounded during the fighting for Kharkov and his commander sent him to the hospital. He said,

> At first I was in Stary Oskol - I lay in the front-line hospital for several days, then an ambulance train arrived, and without any further ado I was taken to an evacuation hospital in Baku - to a hospital in the deep rear. I stayed there for 4 months, and they nevertheless cured me. The medical commission [then] decided, [that] 'After a short 15-day vacation, [he would be] fit for combat service, returns to his unit!' [300]

Officially, Operation *Rumyantsev* ended with the fall of Kharkov. During this 21-day offensive, a Soviet 1983 *Military-Medical Journal #8* account of the fighting stated that the 7th Guards Army suffered 17,199 total casualties, with 11,105 wounded and 2,184 sick, which leaves approximately 3,910 dead and missing on the battlefield (these values are almost certainly under-reported, see below). In some hospitals the medical load reached 200 percent of capacity and for the Army overall it reached 144 percent of standard capacity. [301]

But, historical TsAMO records ID #133373339 and 133373377 tell a slightly different story. They show that from 5-10 August the Army suffered 13,262 causalities and then between 10-31 August the Army suffered a further 10,923 causalities for a total of 24,185 causalities over 27 days of combat. If losses for the remaining four days of the month are estimated at 250 men and women per day, then the total increases by a further 1,000 people, bringing the grand total to approximately 25,185 causalities for the month.

Then to further clarify the issue, TsAMO record ID #133390890, p. 13, *The number of wounded and sick admitted to sanitary facilities by stage of the operation,* records that the Army's rifle divisions admitted 2,189 sick or wounded troops for Stage One of the operation; 2,696 for Stage Two; and 6,591 for Stage Three; totaling 11,476 men and women. If we use these wounded and sick values that were actually admitted to hospital with the approximate 25,185 total causalities listed above, we can calculate a value for the killed, missing and captured. Total causalities of 25,185 minus admitted causalities of 11,476 equals approximately 13,709 killed, missing, or POWs during the entire month. These loss values, taken from historical documents, are far more believable and higher than what the 1983 Soviet *Military-Medical Journal #8* would lead us to believe.

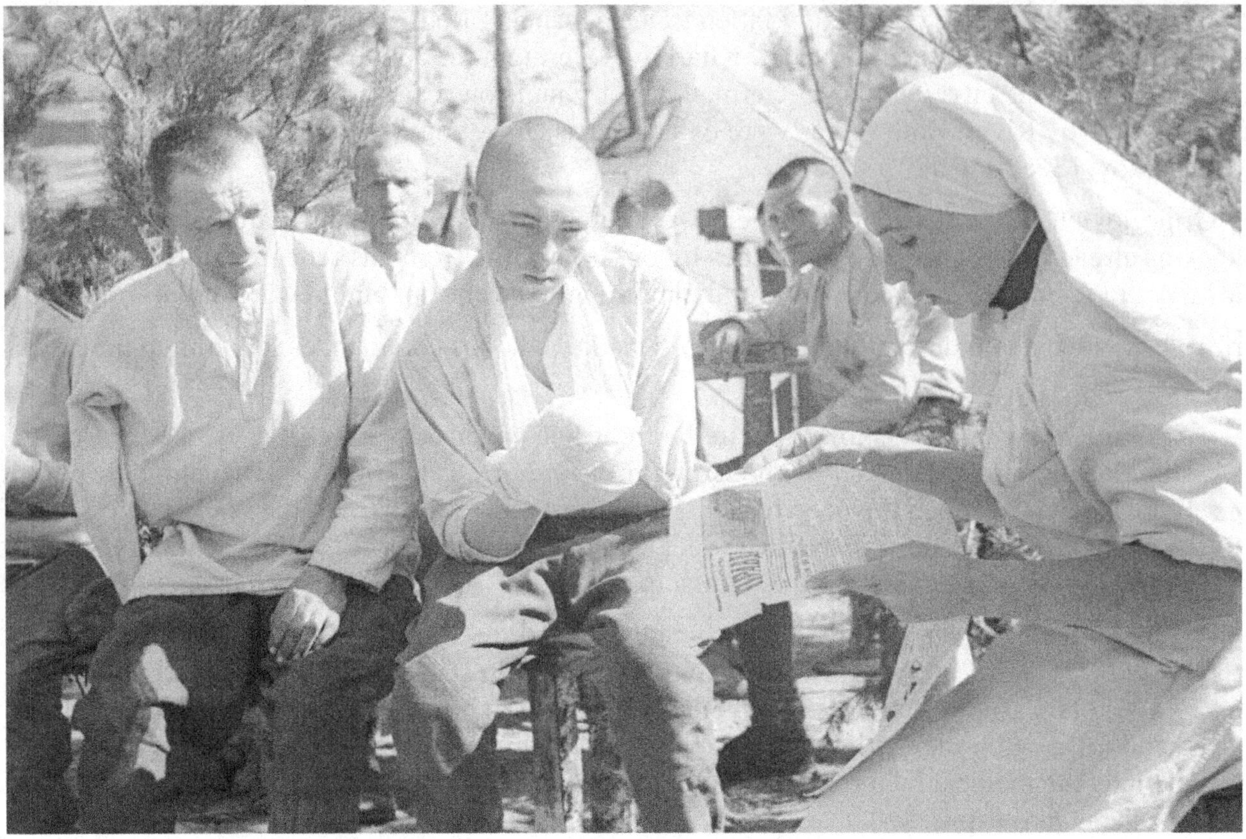

A Russian nurse reads to wounded soldiers behind the lines.

For Operation *Rumyantsev* overall, the Russians lost 71,611 killed or missing, with 183,955 sick and wounded, along with the loss of 1,864 tanks, 423 artillery and 942 aircraft. (Note: These losses are based upon the best available information.) German losses were reported as 8,933 killed or missing, with 16,135 wounded, along with the loss of about 240 tanks and 147 aircraft. [302] [303] In any case, retaking Kharkov for the last time was a long battle for the Russians, both at the front and behind the lines. (See Appendix 12.)

Of note, the use of the railroad to evacuate wounded from the east side of the northern Donets River to medical facilities outside the 7th Guards Army did function well during the offensive, totaling some 8,040 wounded transported between 3-23 August. This effort did ease the burden on army level medical support. [304]

Also on this date, five HPPGs (surgical hospitals #869, 2207, 2208, 2409, 4344) and one XG hospital (infectious hospital #2242) were located on the west side of the northern Donets River in direct support of the Army. The new, fully operational main supply and support center at Volchansk was now some 35 miles behind the front, which was quite adequate. Additionally, a new railway supply line for the 7th Guards and other armies was opened over the Donets River just east of Kharkov at Chuguev (#12 on Map 20). The 57th A captured this area on 11 August, so the rail bridge was repaired very quickly.

In Moscow, a happy Stalin praised the conquerors of Kharkov.

> Ten divisions were given the honorary title of '*Khar'kov.*' A number of units, a large number of generals, officers, sergeants, and Red Army enlisted men, received government medals. The capital of our motherland, in the name of the motherland, saluted the valorous forces that had liberated *Khar'kov*, with 20 artillery salvos from 224 guns. [305]

Another aspect of the Red Army on the offensive was that large numbers of civilians were liberated from German occupation, some 190,000 in Kharkov alone. [306] These liberated people had to be investigated for their conduct during the German occupation because spies and collaborators were everywhere. The NKVD took care of this aspect of liberation. The liberated population also had to be fed, given medical care, incorporated into the ranks of the Red Army or put to work supporting the war effort. The medical establishment of the 7th Guards Army took steps to reduce contact between civilians and the troops because they learned that 46.9 percent of treated civilians were suffering from malaria with fewer numbers of civilians having dysentery, syphilis and typhoid. Due to therapeutic measures taken earlier, few Army troops came down with theses debilitating diseases. [307]

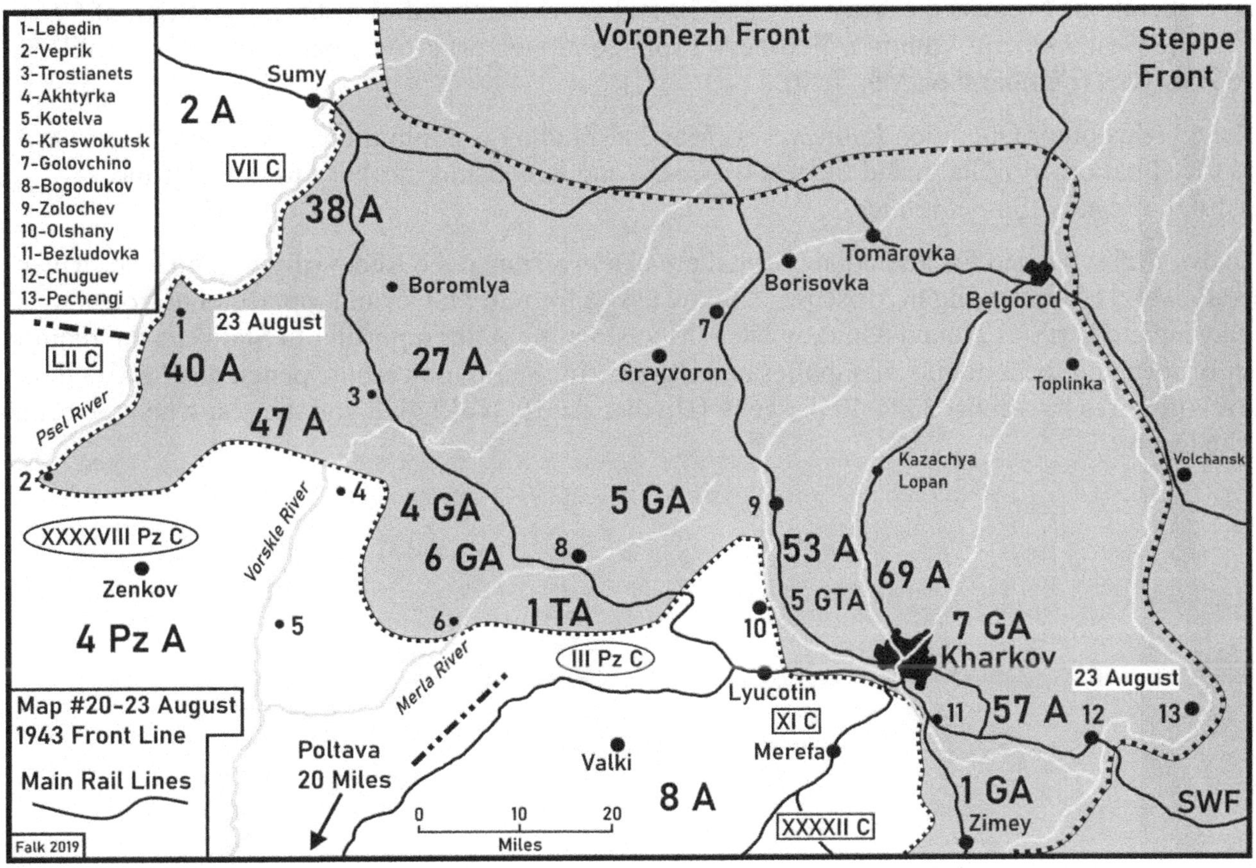

Yielding to pressure from five advancing Russian armies, the German 6th Army and 1st Panzer Army abandoned the long-held Mius River defensive line. Both armies now began a long withdraw to the west while conducting an active defense along the way. They were headed for the mighty Dnepr River, which offered the hope of safety behind its waters, which were between 2,300 to 3,000 feet wide and over 30 feet deep. Soon, the rest of Army Group South followed them towards the Dnepr and its presumed safety. (See Map #21.) Note: Construction of fortified positions along the west side of the Dnepr River had already begun. The building of this so called *Ostwall* (East Wall) was ordered on 12 August 1943 and had hardly started by the time Army Group South urgently needed a secure fallback position.

At the end of Day 21 of the offensive, the 7th Guards Army's rifle divisions admitted 209 sick or wounded troops to sanitary services.

24 August 1943

The 7th Guards Army had a reported combat strength of 35,747 men.

Despite the fall of Kharkov, heavy fighting continued for several days in and around the city. The combat strength of rifle divisions assigned to the 7th Guards Army continued to fall and hovered around 3,000 effective troops in each division.

25 August 1943

The 7th Guards Army had a reported combat strength of 35,401 men. Interestingly, a HQ operational report on this same day showed the 7th Guards Army had recovered some of its tank strength. The report showed the following tanks in its five armored units: 1 – KV 1, 26 – T 34s, 3 – T 70s, 4 – MK IIs (Matilda), 6 – MK IIIs (Valentine) (British tanks), and 1 – SU 122, for a total of 41 tanks and assault guns. [308]

All along the Steppe Front, scattered fighting continued. By this time the 5th GTA ceased major offensive operations because the army only had 50 tanks remaining out of a starting strength of 503 tanks. To the west, General Vatutin's Voronezh Front continued offensive operations and was able to capture Akhtyrka (number 4 on Map #20).

With the conclusion of Operation *Rumyantsev*, Marshal Zhukov was summoned back to Moscow to discuss the situation with Stalin and the rest of *Stavka* and to consider further missions for the successful, but weary, Soviet armies.

In Moscow, Stalin pushed for immediate frontal attacks to advance the Red Army to the Dnepr River. Understandably, after arriving in Moscow, Zhukov asked for more tanks and replacement troops to bring the depleted armies around Kharkov back up to strength. After reporting to Stalin on the required number of men and the amounts of supplies and equipment, Stalin took a blue pencil and "slashed my [Zhukov's] figures by almost 30 to 40 per cent. [He then said] 'We'll give you the rest when the Fronts reach the Dnepr.'" [309]

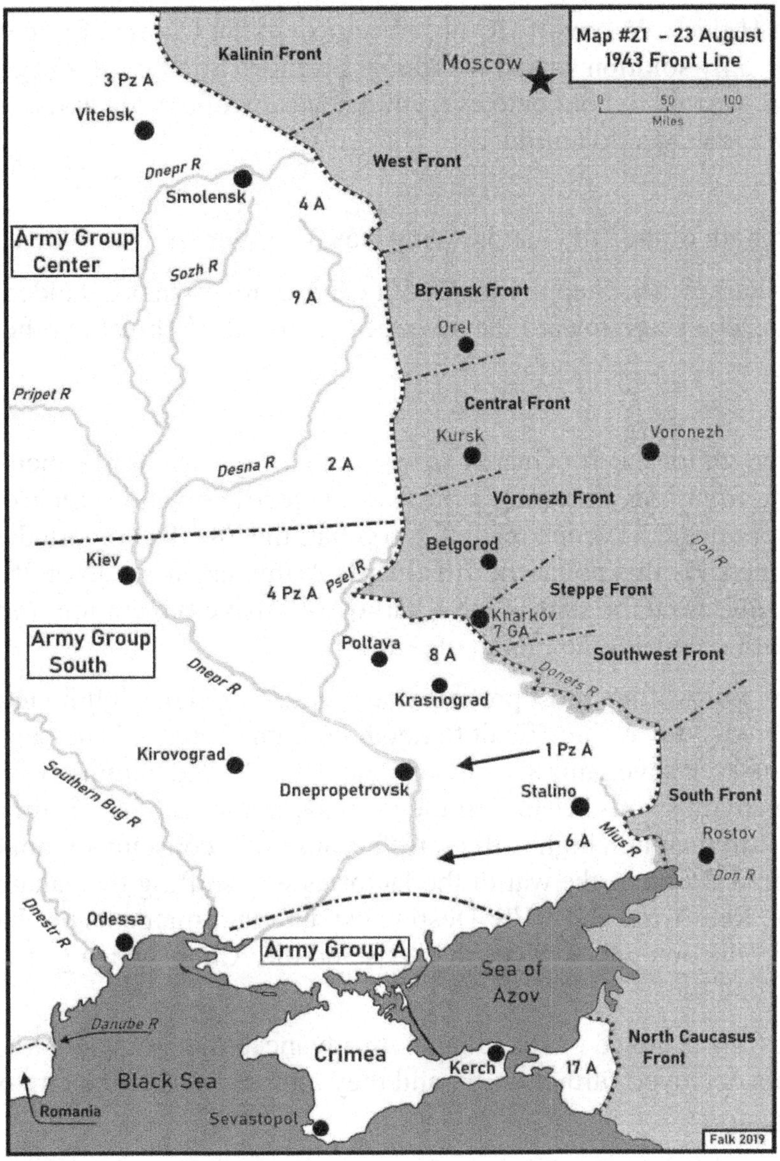

26 August 1943

The 7th Guards Army had a reported combat strength of 35,319 men.

Fighting in and around Kharkov tapered off as frontline units dug in and rested.

One hundred fifty miles to the north, the Russian Central Front launched a new offensive with 579,000 troops. The defending German 2nd Army was unable to resist this attack and was forced to withdraw.

27 August 1943

The 7th Guards Army had a reported combat strength of 34,791 men.

The 4th Panzer Army and the 8th Army were able to stabilize a weak continuous front around Kharkov, thus bringing to an end the final offensive efforts of Operation *Rumyantsev*. But von Manstein and other German commanders knew further withdraws were needed to save their armies. Army Group South was spread out and too weak to hold the existing extended front line. The next obvious natural defense position would be the Dnepr River far to the west, but would Hitler, Stalin, *Stavka* and the Red Army let them reach that safe haven over 100 miles away?

Units of the Voronezh Front continued to attack far to the west of Kharkov, threatening to break through the thin German line.

Von Manstein arrived at Hitler's Werewolf HQ near Vinnitsa in the Ukraine for a meeting. Von Manstein argued that the only solution was to withdraw the entire Southern Front quickly to the Dnepr River line. Hitler, in turn, wanted to hold onto everything. Meanwhile, back at the front, Russian armies now were dictating how events would unfold.

28 August 1943

The reported combat strength of the 7th Guards Army was 31,449 men.

Directly to the west of Kharkov, the Steppe Front's 5th GTA, with 50 tanks, suddenly came to life, attacking with three corps due south toward the town of Merefa, easily breaking the fragile German line.

29 August 1943

The reported combat strength for the 7th Guards Army on this day was 29,818 men. To illustrate what this number meant on the front line, the Army's three weakest rifle divisions reported the following strengths: the 73 GRD reported 2,815 men ready for combat; the 78 GRD reported 2,922 men; and the 213 RD reported 2,957 men. By this point, the 7th Guards Army has lost at least 25,176 combat troops killed, wounded, or missing, from the start of the Kharkov offensive (Operation *Rumyantsev*). This is the lowest combat strength report for the month of August.

Guenther Rall, one of the Luftwaffe's most proficient air aces, scored his 200th victory on this day in the Kharkov area. He stated, "It wasn't difficult to notch up high scores…" He went on to say, "We often went on five, six, or even seven missions in a single day." [310] The outnumbered Luftwaffe fighter pilots needed to fly so many missions because of the growing numbers of VVS aircraft. The VVS air armies, in turn, were able to maintain high sortie rates because of a constant stream of new pilots and aircraft to the front. Behind the lines, the war of the factories was shifting the balance of power decisively in favor of the Red Army and VVS. Despite expanding armaments production, Germany was never able to close the production gap between the Axis and the Allies, which guaranteed a losing strategic position in the long run.

Elements of the 5 Guards Mechanized Corps (5th GTA) advanced to a point south of Kharkov, outflanking German units deployed further north, and they almost reached the city of Merefa. The German line was coming apart.

30 August 1943

The reported combat strength for the 7th Guards Army was 30,013 men, up slightly from the day before.

The III Pz C engaged the 5th GTA south of Kharkov and tried to restore the situation. But unknown to von Manstein, the fighting around Kharkov was not an isolated affair, as David Glantz states in his book, *Forgotten Battles*, Vol V, Part One, pages 8-10. Glantz said that these attacks were

> [t]he second stage of the summer-fall campaign [which] involved a wholesale expansion of the Red Army's Kursk strategic offensive to encompass all *fronts* attacking along virtually every strategic axis. The Stavka's intent was to drive German forces from all regions east of the [Dnepr] River, reach and cross the river, and secure strategic bridgeheads on the west (far) banks of the river.

The Steppe Front and the soon-to-be-committed 7th Guards Army were part of the Poltava Offensive (24 August to 30 September). Ultimately, the aim of this offensive was to liberate the cities of Poltava and Krasnograd, then reach and cross the Dnepr River in force. Only now was *Stavka's* long-range plan unfolding for all to see.

31 August 1943

The 7th Guards Army had a reported combat strength of 30,458 men. This was roughly equivalent to three full strength Guard rifle divisions.

Despite the 7th Guards Army having a larger number of vehicles assigned to it than a non-Guards army, its *Shtat* (order of battle) listed over 3,500 vehicles as authorized, yet at the start of August only about 3,000 vehicles were operational. Because of this, the Army continued to depend upon huge numbers of horses for transport. During the August offensive, the 7th Guards Army started with 5,885 horses but lost 875 horses killed and suffered another 803 wounded and 329 sick; that was about 34 percent of the starting total. So, humans were not the only ones who suffered and died during an attack.

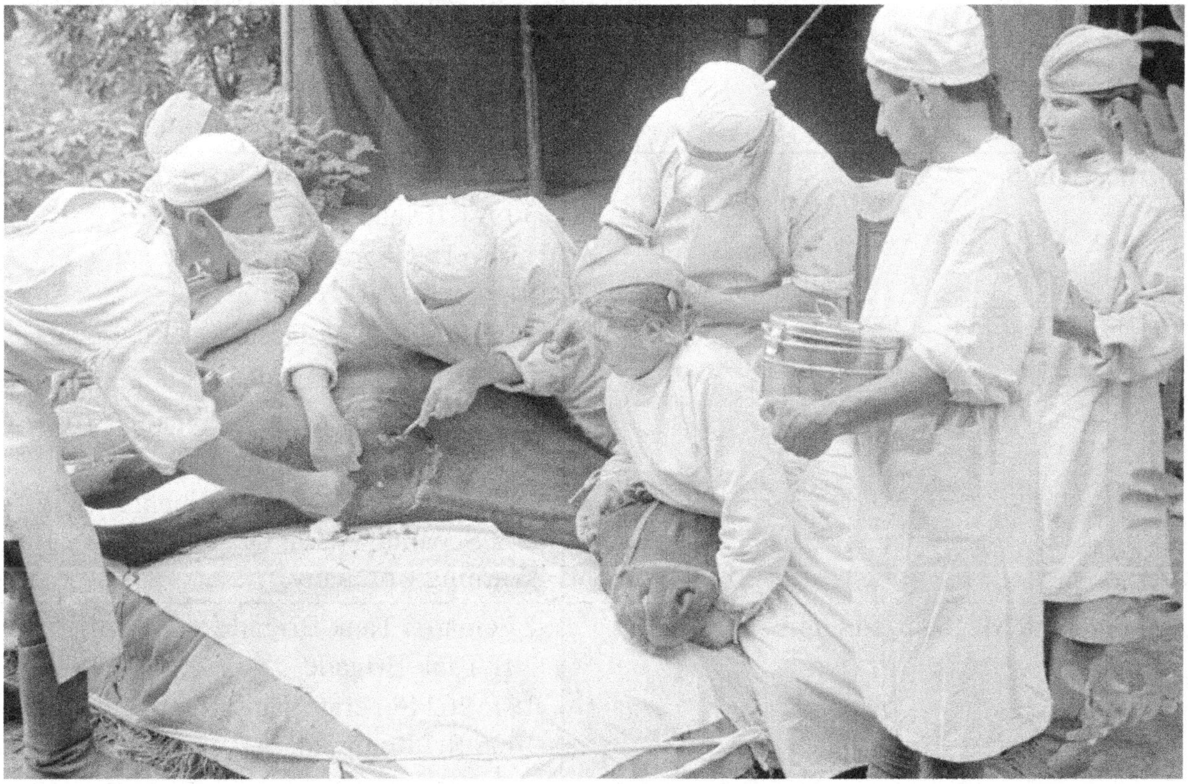

Russian Veterinarians operate on a wounded horse.

Supply vehicles also were damaged or destroyed during Operation *Rumyantsev*. The Steppe Front in total lost 863 vehicles, of which 32 percent were lost due to enemy air raids, 39 percent from artillery and mortar fire, and 29 percent from minefields. [311] This number included total losses for each army and reserves of the Steppe Front. Despite its losses in terms of humans, horses and equipment, the 7th Guards Army performed well and played a key part in the final liberation of Kharkov. [312] After the 7th Guards Army's short rest around Kharkov, the Army was about to take part in a new operation, one that would unfold to the west and reach the Dnepr River.

Von Manstein reported to OKH that Army Group South suffered 133,000 casualties in August and only received 33,000 replacements in turn. The message was clear: without substantial reinforcements, the only thing von Manstein could do was withdraw his forces to the Dnepr River line as quickly as possible to avoid the total destruction of his armies.

1 September 1943	
7th Guards Army	
Assigned – Steppe Front	
Lieutenant General M.S. Shumilov Commanding	
Rifle, Airborne and Cavalry Units	24 Guards Rifle Corps (73 GRD, 213 RD) 25 Guards Rifle Corps (78, 81 GRDs) 49 Rifle Corps (15 GRD, 111 RD) 1, 2, 3, 4, 5, 131, 132 Separate AT Rifle Battalions 82, 83, 84, 85, 86 Blocking Detachments 62, 63, 64, 65, 66 Penal Companies
Artillery Units	109, 161, Guards Gun Artillery Regiments 30 AT Brigade 114, 115 Guards Tank Destroyer Regiments 1669, 1670 Tank Destroyer Regiments 290 Mortar Regiment 5 Antiaircraft Artillery Division (670, 743, 1119, 1181 AA Regiments) 162, 258 Guards AA Regiments 662 Separate Aerial Observation/Air Raid Warning Company
Armored and Mechanized Units	27 Guards Tank Brigade 201 Tank Brigade 167, 262 Separate Tank Regiments 1438 SU Regiment 34, 38 Separate Armored Train Battalions (four trains active)
Engineer Units	60 Engineer Sapper Battalion 175, 329 Separate Engineer Battalions
Reported Combat Strength Estimated Ration Strength	30,698 men 37,698 men and women (including an estimated 7,000 support personnel)

(BSSA 43, BSA 43, TsAMO Archive)

1 September 1943

The Army started the month with only six rifle divisions assigned to their three-rifle corps; the remaining divisions appeared to be out of the front line, rebuilding in the rear area. The Army also still had most of its artillery and tank support units from the previous month.

The 15 GRD moved up to the Udy River in preparation for an attack the next day.

2 September 1943

The Steppe Front and 7th Guards Army units attacked southward from Kharkov. The 15 GRD made a dawn attack over the swampy shores of the Udy River. The Army seized a bridgehead on the south side of the river as they broke the German defense line.

The Voronezh Front captured Sumy as it advanced to the west.

3 September 1943

The 7th Guards Army continued its attacks toward Merefa in the face of stubborn resistance from German positions in fortified buildings. General Konev talked with General Shumilov about the slowness of the advance and offered to organize a "massive strike by artillery and aircraft to suppress the enemy artillery and mortar system." [313] The effort proved to be successful as Shumilov was able to break through the tough German fortifications.

Von Manstein traveled to East Prussia to meet with Hitler at his HQ. The meeting did not go well, as von Manstein failed to resolve with Hitler the persistent command and control issues as well as the force distribution problem on the Eastern Front. The only bright spot was Hitler's approval of withdrawing the 17th Army from the Taman Peninsula and deploying it to the Crimea. [314] This move not only saved the 17th Army from imminent destruction but also bolstered the Axis defense of the Crimea which denied Soviet naval and air forces the use of this centrally located Black Sea peninsula.

On this day, King Victor Emmanuel III of Italy signed the secret Armistice of Cassibile (Sicily) with the Western Allies. Apparently, a Soviet representative was not present for the signing of the Armistice. The King (and other leaders) realized the Axis Powers were losing the war and wanted to break from Hitler and remove Italy from the war. Five days later, on 8 September, the Italian government released the news to the public causing panic and joy within the Italian population. The war was over for Italy! Hitler, of course, would not stand for this. He launched Operation *Achse,* which directed German military units to begin taking control of Italy.

The British 8th Army, commanded by General Bernard Montgomery, crossed the straights of Messina (between Sicily and Italy) and landed at the southern tip of Italy. Sadly for the Italian population, their involvement in the war was far from over. The invasion of Italy proper now had begun.

4 September 1943

The 7th Guards Army continued its attack upon Merefa, finally making some progress.

A historical chart shows three Army corps, in the first echelon, contained the following rifle divisions:

24 GRC – 72, 73 GRDs and 213 RD
25 GRC – 78, 81 GRDs
49 RC – 15 GRD and 111 RD

The Army chart also shows only five AT Rifle Battalions (1, 4, 5, 131 and 132) in action, along with only two Penal Companies (63 and 64), but all five Blocking Detachments (82, 83, 84, 85, 86) were active. [315] This illustrates how quickly the active unit lineup could change. We can assume the units that should be assigned, but are missing from this chart, were still in reserve being rebuilt behind the lines.

5 September 1943

The 7th Guards Army captured the town of Merefa and its railway junction directly south of Kharkov. The capture of this position unhinged the German defense along the Donets River. After the seizure of Merefa, the rate of advance of the 7th Guards Army increased to about 18 miles per day. Assessing the events of those days, the commander of the Steppe Front, General I. S. Konev, subsequently noted:

> The path was opened for a faster offensive of the 57th and 7th Guards Armies to the Dnepr. But there was still a stubborn struggle with the strong Poltava grouping of the 8th German

> Army. In the direction of Poltava and Kremenchug, under the attacks of the Steppe Front, the largest group of the 8th German Army retreated, consisting of the 3rd, 47th Tank and 11th Army Corps. [316]

Poltava would not fall to the Steppe Front until 23 September.

A historical document shows that on this date the 23 mile section of railroad from Kharkov to outside Merefa had been restored to a usable condition. This indicated the Soviet Railroad Brigades had continued to work right up to the front line. This meant the same type of support also might continue all the way to Krasnograd, a major rail center, some 47 miles closer to the Dnepr River. [317] For an advancing Soviet army, the logistical importance of a nearby operational railroad cannot be overstated. Army level rear area logistical units can only complete the delivery of bulk goods that the railroad delivers. After all, Napoleon Bonaparte stated, "An army marches on its stomach." If the beans and bullets stop arriving, an army will soon lack the means to carry on.

6 September 1943

Stavka ordered a general change to offensive operations in Ukraine. The Voronezh Front was ordered to advance directly toward Kiev while the Steppe Front directed its attack toward Poltava and Kremenchug with the aim to cross the Dnepr River between Kremenchug and Dnepropetrovsk. [318] The plan called for the armies of the Steppe Front to reach the Dnepr along an 80-mile front.

After a successful pre-planned artillery barrage, units of the 7th Guards Army attacked south out of their positions around Merefa. German forces withdrew to the southwest to avoid being surrounded.

7 September 1943

A Soviet book about the 15 GRD shows that fighting during this time was intense and unrelenting. "The formations of the 7th Guards Army of General M. S. Shumilov, overcoming fierce resistance at a number of intermediate lines of defense, did not stop the attack, day or night, getting closer and closer to the Dnepr on a wide front." [319] This statement was basically true, but the attack stalled from time to time along the way. On this date, leading units of the 15, 81 and 78 GRDs were trying to break through one of these German intermediate positions about 40 miles northeast of Krasnograd. Determined German resistance in this area and the next German defense line halted the advance for seven days until 14 September. Of note, the 7th Guards Army troops had the sole responsibility to capture Krasnograd and in their weakened state, that would take time.

8 September 1943

On this day, a historical chart shows only two Army corps in the first echelon. They contained the following rifle divisions:

> 25 GRC – 78 and 81 GRDs
> 49 RC – 15 GRD and 111 RD

Of note, all these rifle divisions were operating with combat strength of only 2,000 - 3,000 troops, or about 20 to 25 percent of normal strength. The 7th Guards Army was an exhausted, depleted Army.

Hitler flew to von Manstein's HQ at Zaporozhe, Ukraine, for an Army Group meeting. Hitler promised reinforcements for Army Group South (that would never arrive). Von Manstein soon realized that Hitler's promises were empty and resigned himself to withdraw Army Group South over the Dnepr River to safety. Most of his armies were already withdrawing toward the west anyway under heavy pressure from the advancing Red Army. This was to be Hitler's last visit to the Eastern Front. In essence, for operational reasons, it did not matter as *Stavka* was arranging for the Eastern Front to come to him.

The desperate efforts in Russia were not the only concern for Hitler. Closer to the German homeland, the US 1st Army, with attached British units, made an amphibious landing at Salerno in southern Italy. Now two Allied armies were positioned to advance up the boot of Italy.

9 September 1943

The German 17th Army began withdrawing its troops to the Crimea. Most of these troops would be used to safeguard and to hold the Crimea, thus preventing the VVS from bombing the vital Romanian oil fields.

As the 7th Guards Army advanced, troops paid special attention to gather supplies and equipment that would be useful for the upcoming river crossing. Different units of the Army started to plan for this major river crossing using what was called "improvised means":

> ...direct preparations for the crossing were widely held. Fighters of a separate combat engineer battalion under the command of the Guard Major Polyank still carried several boats with a carrying capacity of five to twenty people from the Vorskla River. Using every opportunity, they tried to replenish the arsenal of crossing facilities. With the help of the local population, old fishing longboats were found and repaired, [and] planks, logs, [and] cables were accumulated. At the initiative of [various] soldiers... rafts and home-made ferries were built in all divisions. Experienced swimmers practiced with beginners, [and] showed [them] how to stay on the water using a log, a barrel, [or] an inflatable chamber. [320]

10 September 1943

On this day, a historical record shows the 7th Guards Army having a combat strength of only 20,128 men. The four divisions assigned to the front line had the following troop strengths:

15 GRD – 3,064
72 GRD – 2,806
81 GRD – 2,972
111 RD – 3,064 [321]

These troop strength numbers were far below the 8,000-man full-strength totals for rifle divisions. The strength of the Army clearly was wasting away under the burden of constant combat.

Also on this day, the Army issued a logistical report of the weight of one day's supply of food, fuel and lubricants for the Army. The Army needed 162.2 tons of food and 271.5 tons of all types of fuel and lubricants, for 433.7 tons in total. Ammunition requirements only would add to the needed tonnage. With the Army's main supply base currently 50-60 miles behind the front, the logistical support units must have been strained to supply these basic needs of the Army. [322]

Further south in the German Kuban bridgehead, the North Caucasus Front attacked with approximately 250,000 troops against positions of the 17th Army. The 17th Army continued to withdraw its forces to the Crimea.

11 September 1943

During late August and early September, the autumn rains began to fall in the south, causing all sorts of problems for both the Germans and the Red Army. As the autumn muddy period or *rasputitsa* took hold of the landscape, it actually helped the German defensive effort. Von Manstein directed his forces to firmly hold cities and major road centers, thus forcing the advancing Red Army to either fight for these centers or to try to bypass them through the deep mud; either way it caused a delay upon the attackers. Tenacious German resistance, the inclement weather, and the resulting logistical disruptions became more intense the further the Red Army advanced. What began as a fast pursuit of a defeated enemy developed into a battle to overcome one fortified position after another. Recognizing that a great victory was so close, army commanders, front commanders and even *Stavka* urged their troops forward.

However, nature had a say too, as cold, rain, and mud, along with little food and little rest, hampered a weakened and weary Red Army as it forged ahead toward the Dnepr River.

To further impede Soviet efforts, von Manstein issued a separate order implementing the Scorched Earth policy for his command. This was part of the German sequential ARLZ Measures (*Auflockerung*-Dispersal, *Räumung*-Removal, *Lähmung*-Paralysis, and *Zerstörung*-Destruction) concerning the clearing of equipment, food, and people from occupied areas so that the advancing Red Army would be denied their use. [323]

12 September 1943

General Raus, commander of the German XI Corps, 8th Army, highlighted the overall German defense plan, which took into account the weather situation:

> Above all the enemy wanted to capture major cities commanding the road net needed for speedy maneuvers. In view of the rainy weather, the possession of hard-surface highways became a decisive factor to both sides since the mud prevented any movement off the roads. We took this factor into account and concentrated our antitank defenses in and around important towns. During the phase, therefore, cities became far more important than during the fighting in the area north of Kharkov. [324]

In far off Italy, Captain Otto Skorzeny, while not in command of the Mussolini rescue mission, was the commander of a Special Forces unit of the *Waffen* SS (Armed SS) that took part. The main force was comprised of German *Fallschirmjäger* (Luftwaffe Paratroops). The mission to free Mussolini from his Campo Imperatore mountaintop jail was successful. Then Skorzeny personally took control of Mussolini and flew off the mountaintop in a daring escape aboard a Fi 156 *Storch* (Stork) light aircraft. The *Storch* was badly overloaded with the pilot, Mussolini, and the 6-foot 4-inch Skorzeny crammed into the two-seat aircraft. Skorzeny personally delivered Mussolini to Hitler on 14 September and was ultimately awarded the Iron Cross 1st Class for his successful mission.

13 September 1943

To hurry the attack along, General Shumilov started issuing verbal orders to regimental commanders of the 81 GRD. They were ordered to assemble battalion-sized combat groups and to launch non-stop attacks toward Krasnograd. [325] The 7th Guards Army was breaking through the current German defense line. General Shumilov had the German 8th Army on the run, and he did not want to give them a chance to rest or to properly fortify another position. The Dnepr River was only about 65 miles away.

14 September 1943

The 7th Guards Army reported a combat strength of 20,162 men.

With these few men, the 7th Guards Army broke the German line defending the approaches to Krasnograd. German units fell back toward the city.

15 September 1943

The 7th Guards Army reported a combat strength of 21,093 men as the Army continued to advance toward Krasnograd. In turn, the Germans briefly blocked their advance for two days in the area of Pervomaiskoe village.

Once again, von Manstein flew to East Prussia to meet with Hitler at his HQ. After heated talks between these two stubborn leaders, von Manstein finally won approval for Army Group South to begin a general withdrawal to the Dnepr. The new proposed defense line (*Ostwall*) would begin at Melitopol (north of the Crimea), then north to the Dnepr and then on to Kiev and Army Group Center. The race was now on to see who would reach the Dnepr River first, a resurgent Red Army or the battered forces of Army Group South. (See Map #22.)

As Army Group South withdrew, German units started to use six main crossing points over the Dnepr River. Most of the 4th Panzer Army headed to Kiev to cross, and its XXIV Pz C headed to Kanev. The 8th Army planned to use Cherkassy and Kremenchug to cross while the 1st Panzer Army would use Dnepropetrovsk and Zaporozhye. The 6th Army would be transferred to Army Group A on 17 September and would remain east of the southern part of the Dnepr River.

To give his forces the best possible chance to reach and to cross the Dnepr successfully, von Manstein implemented the use of a Scorched Earth policy, within 15 miles of the Dnepr, to slow and to hinder the enemies' advance. This order involved the destruction or evacuation of anything that could be of use to the enemy. Civilians, horses, cattle and machinery were taken to the west. Surplus food supplies, buildings and anything that would help the Russians cross the Dnepr were destroyed. "On instructions of the Supreme Command, the civil population was also evacuated in order to prevent the Soviets from conscripting all able-bodied males and making the remaining population to carry out military works." [326] A regular German soldier described the results of this order:

> We are having to evacuate all the territory east of the Dnepr and form a new front line to the rear of it. Furthermore, all grain, cattle, horses, everything of value has to be brought back behind the river, even the remaining civilians. An emigration of nations! I looked on as crops in the fields were harvested, threshed, loaded aboard lorries and driven off. The Russians will arrive in a land bereft of everything. All the large haystacks, often the size of a house, were set afire, and we watched the flames eat their way through the stubble still standing in the fields.... even the railway lines were torn up by the "rail ripper" [Rail Wolf]. [327]

Also, in another prudent move, von Manstein relocated his AG South HQ to Kirovograd on the western side of the Dnepr River. Relocating an entire Army Group HQ was no easy feat. At this time, von Manstein's HQ most likely contained over 2,000 personnel, several hundred vehicles, liaison aircraft, a mass of communication equipment, security troops, food and medical support, and any other supplies and equipment necessary to keep the HQ running. The old HQ would remain operational until the new HQ location was fully functional. This transfer process could take several weeks to complete, depending upon the weather and enemy interference.

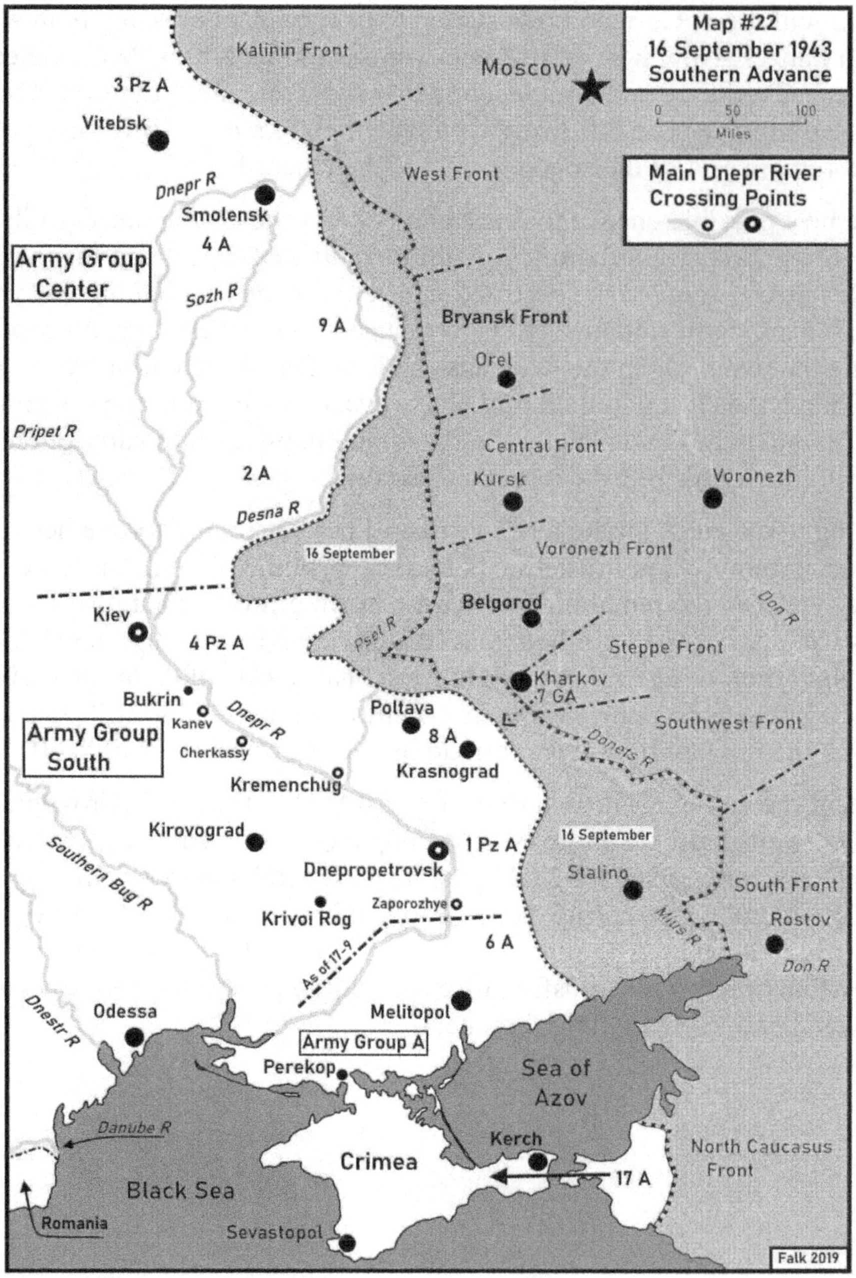

16 September 1943

Along the temporary defensive line in front of Krasnograd, German troops had orders to slowly withdraw, when needed, first to Krasnograd, then to the Dnepr River. As German formations began withdrawing to the southwest, units of the 7th Guards Army followed them, keeping the pressure on the retreating Germans.

17 September 1943

The Germans continued to retreat, with Russian formations advancing as quickly as possible.

18 September 1943

The battle for Krasnograd began. The 72, 78 and 81 GRDs of the 7th Guards Army attacked from the north and east. Supporting formations from the 57th A, who were approaching from the east, attacked over the small Berestova River and into the city.

19 September 1943

Units of the 7th Guards Army participated in the liberation of Krasnograd with the 57th A. By the end of the day, Krasnograd had fallen into Soviet hands. The way to the Dnepr River now was open. The 24 GRC went into the second echelon for rest and replenishment.

20 September 1943

Once again, after the liberation of Krasnograd, the pace of the advance accelerated for units of the 7th Guards Army. The 81 GRD alone was able to clear and/or capture at least seven settlements in one day. Later this same night, under orders from General Shumilov, the 81 GRD continued to advance and to attack German units found in various villages. In essence, at this point in the battle, the advance to the Dnepr River only slowed when German rear guard units deployed, stood their ground, and fought. After a quick battle, German rear guard units then would withdraw to escape being surrounded.

On this day, the 7th Guards Army only deployed four rifle divisions to the front, the 72 and 81 GRD on the left flank and the 15 GRD and 111 RD on the right. They were attacking with a total frontage of approximately 14 miles and were about 56 miles from the Dnepr River.

A retreating German *Tross* [rear services] unit from the Pz GD *Grossdeutschland* had crossed the Dnepr after some delay and reached Kirovograd. The soldiers found that a "peacetime economy was in full swing" within Kirovograd and maybe more importantly, that "German girls were working as auxiliaries on the staff, in signals units and as nurses." [328]

21 September 1943

Units of the 3rd Guards Tank Army, Voronezh Front, reached the east side of the Dnepr River south of Kiev.

In turn, as units of Army Group South withdrew to the Dnepr, they were directed toward six main crossing points. They also were ordered to hold three substantial bridgeheads on the east side of the river. These bridgeheads were located at Kremenchug, Dnepropetrovsk and Zaporozhe. Hitler had delusions of launching counterattacks against the Red Army from these bridgeheads.

As German combat formations headed to the west, they took with them vast amounts of supplies and equipment. General Raus stated his forces of the XI Corps, fighting in and around Poltava (with an estimated population of 60,000), evacuated 30,000 civilian vehicles (carts and wagons); 64,000 horses; 80,000 head of cattle and thousands of civilians. This massive evacuation effort did impede a speedy withdraw, but it was necessary to prevented the Red Army from gaining the use of these materials and provisions. In a daring move, Raus was able to clear the rail line out of Poltava sufficiently to evacuate "[t]welve freight trains loaded with tanks and other valuable equipment" from the city, allowing them to reach the Dnepr crossing at Kremenchug and to move safely away from the battle. [329]

On or about this date, *Stavka* withdrew the 111 RD from the 7th Guards Army and sent it into *Stavka* reserve for rebuilding. Eventually it would be transferred to the 5th GA.

22 September 1943

Units of the 3rd GTA begin crossing the Dnepr River near the village of Bukrin, and they formed a small bridgehead on the west side of the river. Due to its location and difficult terrain, this bridgehead would prove to be difficult to use and only helped to divert German forces to this area.

The leading units of the 7th Guards Army, 78 GRD, were now only 24 miles from the river.

Units of the Central Front reached and crossed the Dnepr River north of Kiev. German resistance was fierce, but the defenders could not hold the river line.

23 September 1943

After heavy fighting, advancing Steppe Front troops took control of the city of Poltava. This victory allowed the Russians to open a major logistical route to support the offensive. Once the Soviets restored the rail lines, supplies eventually would reach all the way to the Dnepr.

During the night, two brigades of Russian paratroopers (4,575 men) dropped on the west side of the Dnepr near the Bukrin bridgehead. This bold plan was intended to seize the area until the arrival of friendly forces from the east side of the river. Unfortunately, the airborne drop was a disaster as the scattered units came down among several German divisions that were in the area. After the first day, about 1,000 paratroopers had already been killed or captured. Most of the surviving paratroopers fought in small groups as they tried to achieve something behind German lines. Ultimately about 60 percent of these paratroopers were killed, captured or went missing. This would be the last major Soviet airborne operation during the war. [330]

The 7th Guards Army HQ released a report on ammunition supplies within the Army. The report broke down the totals as "With Troops" or "In Stock". The Army was holding good stocks of small arms ammunition (1.1 to 3.9 units of fire), moderate amounts of mortar and up to 76mm guns (.6 to 3.5 units of fire), and low amounts of heavy artillery ammunition (.26 to .45 units of fire). In all, the Army had reasonable amounts of ammunition but not enough for a major battle.

Regrettably, the 7th Guards Army was at the end of a long, mostly road bound, supply line. The major rivers in the area were of no use for transporting supplies because they flowed mostly north/south. To adequately support the troops on the west side of the Dnepr River required an active rail line, and the nearest railroad crossing of the Dnepr was at Kremenchug, 24 miles away to the northwest. History shows that the retreating Germans would soon blow up both the road and rail bridges at Kremenchug. The next nearest railroad bridge at Dnepropetrovsk was 60 miles to the southeast, and the Germans still held the city and the rail bridge. Clearly if the Red Army wished to advance past the Dnepr River, they would need temporary railroad bridges to support that advance.

Until then, with the fall of Poltava, the Red Army was expected to repair the rail line out of Poltava to the west in a reasonable amount of time; this was on the east side of the Dnepr River. That implied the closest usable rail line and supply road for the 7th Guards Army eventually would be about 30 miles to the north around the village of Beliki on the Poltava-Kremenchug rail line. Sooner or later, that would be the logical place for Shumilov to locate the main supply base for the 7th Guards Army, but the historical record is mute on this point. Additionally, logistical planners would have been aware that other armies in the area also were seeking supply routes, and they would make an effort not to hinder their supply traffic if possible. Most likely at this time, the 7th Guards Army supply base was located somewhere near Kharkov, far to the rear.

"Within the four Ukrainian fronts there were eight extra-class railway bridges across the Dnepr - two each at Kiev and Dnepropetrovsk and one each at Kanev, Kremenchug, Cherkassy and Zaporozhye." [331] The Germans would ensure that they destroyed these bridges before the Red Army and railroad engineer units could capture them.

Pushing far ahead of the rest of the Army, at 23:00 hours, units of the 78 GRD, 7th Guards Army, seized a bridgehead on the island of Glinsk-Borodaevska in the middle of the Dnepr. Fighting raged all night as the Germans had placed a defending infantry unit on the island.

The Dnepr River was a wide and meandering river with many islands and side channels that offered both challenges and concealment. Trees and bushes would hide from view an approaching force. Finding a ford allowed an easy crossing to an island, a short swim would gain access to another, and then came the main channel, which was the largest obstacle to an advancing army. Crossing a major river with a few hundred men was not the problem; the greater problem was the thousands that would follow and the many tons of supplies and equipment that would be needed to gain and then to defend a bridgehead on the far shore. The crossing point topography also needed to be assessed to allow for easy use of small rafts, then pontoon rafts, pontoon bridges, and finally road bridges. (See Map #23.)

With the fall of Poltava, the 7th Guards Army had the 69th A to the north, covering its right flank, and the 57th A to the south, covering its left flank. But General Konev, commander of the Steppe Front, had secretly brought up the replenished 37th Army behind the 69th A. Because the Russians were using *maskirovka* methods, the Germans were unaware this strong army was rapidly approaching the Dnepr. At the opportune moment, Konev planned to relieve the 69th A at the front with the 37th A, then force the Dnepr and form a large bridgehead with this fresh army.

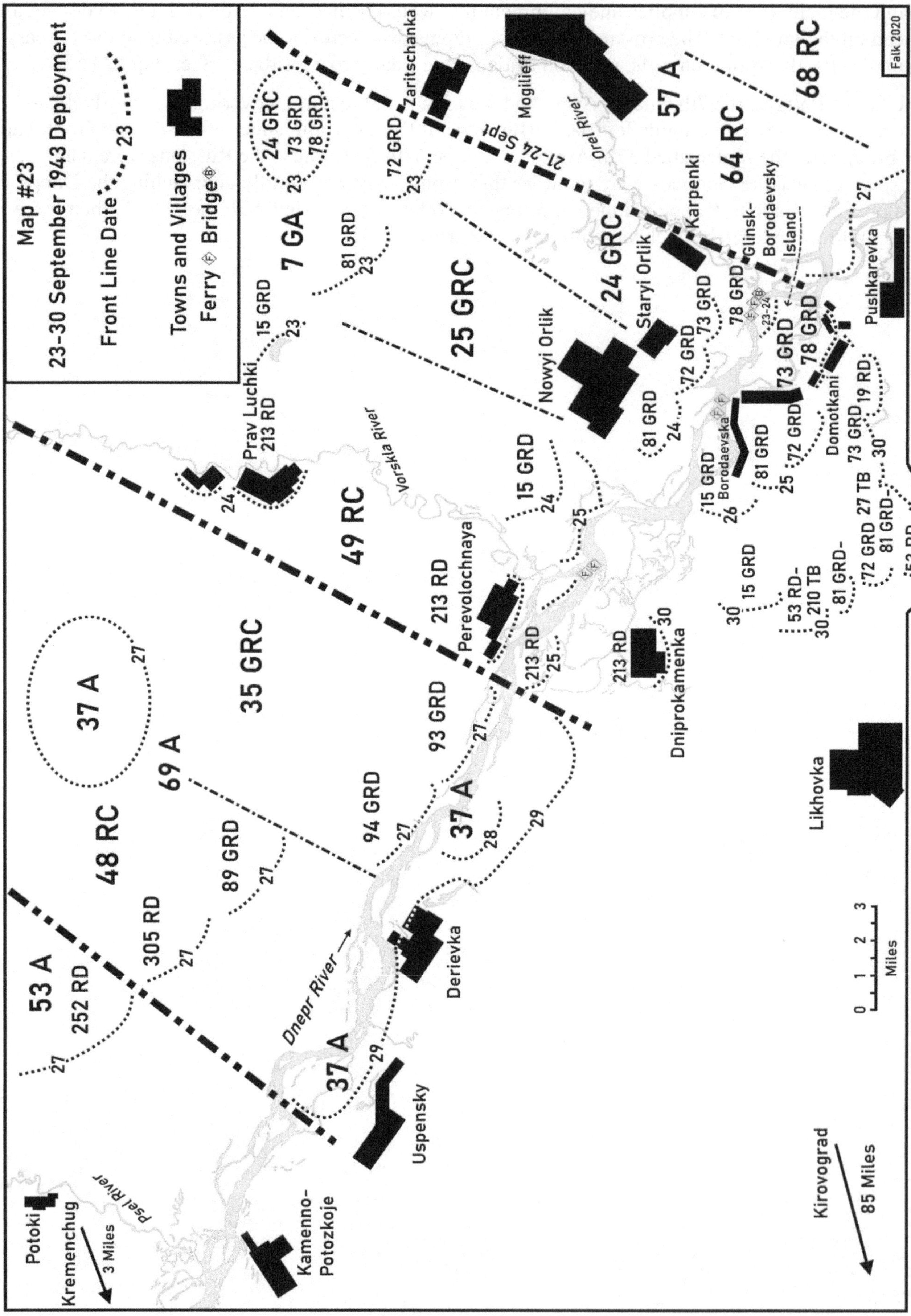

Map #23
23-30 September 1943 Deployment
Front Line Date 23
Towns and Villages
Ferry F Bridge B
Falk 2020
53 A
252 RD
48 RC
305 RD
89 GRD
69 A
37 A
35 GRC
94 GRD
93 GRD
49 RC
213 RD
Prav Luchki
213 RD
Vorskla River
15 GRD
7 GA
81 GRD
72 GRD
24 GRC
73 GRD
78 GRD
Zaritschanka
21-24 Sept
Mogilieff
Orel River
57 A
68 RC
64 RC
Karpenki
Glinsk-Borodaevsky Island
24 GRC
25 GRC
Nowyi Orlik
Staryi Orlik
73 GRD
78 GRD
72 GRD
Pushkarevka
Domotkani
19 RD
73 GRD
72 GRD
81 GRD
15 GRD
Borodaevska
27 TB
72 GRD
81 GRD
53 RD
15 GRD
53 RD-
210 TB
81 GRD-
Perevolochnaya
Dniprokamenka
213 RD
Likhovka
Derievka
Dnepr River
Uspensky
37 A
Kamenno-Potozkoje
Potoki
Kremenchug
3 Miles
Psel River
Kirovograd
85 Miles
0 1 2 3
Miles

24 September 1943

Fighting on Glinsk-Borodaevska Island continued as the Army prepared to cross the main channel in strength and to gain the right bank of the Dnepr. German artillery and mortar fire constantly rained down as a German "frame" (Fw 189) reconnaissance aircraft flew overhead directing the fire. Later during the day, "General M.S. Shumilov summoned all division commanders, corps, and political unit chiefs to his observation post and ordered [them] to force the Dnepr [that is cross to the right bank of the river] on the night of September 25." [332] This main crossing attack would be in the area of the 24 and 25 GRC along a section of the river with a frontage of about four miles wide, which included the Glinsk-Borodaevska Island area.

Some four to five miles upstream, units of the 15 GRD were ordered to create the appearance of a crossing attempt to distract the Germans from the main attack area. Note: The main force of the 15 GRD would move south from the 49 RC sector and cross the river in the Borodaevska area (i.e. in the sector of the 25 GRC). Overall, this crossing attack was planned to clear the Germans from the Glinsk-Borodaevska Island and to land on the far shore between the villages of Borodaevska and Domotkani. If successful, parts of five rifle divisions then would be on the far shore. They would be the 15, 81, 72, 73 and 78 GRDs, and they would hold an initial position on the right bank of the river.

The VVS also was active when 40 bombers successfully attacked the German bridgehead in the Kremenchug area and partly destroyed the road bridge over the river. This attack greatly delayed German efforts to cross the Dnepr as heavy weapons were forced to use the one rail bridge.

General Konev ordered the 37th A to advance, by forced march, to relieve the 69th A at the front by 25 September. Then, after relieving the 69th A, it was to force a crossing of the Dnepr, from the march, by 27 September. General Konev was trying to take advantage of the success of the 7th Guards Army that had already reached the river and engaged with defending German troops. (See Map #23.)

During the evening, units from the 15, 72, 73 and 81 GRDs, 7th Guards Army, conducted their own forced march and reached the Dnepr in some strength. At this point along the river, the far western bank was approximately 300 feet higher than the eastern bank. The cliffs and the river itself would be formidable obstacles, but only if they were defended firmly.

At the far north end of the Dnepr River, after intense fighting, German forces of Army Group Center began withdrawing from Smolensk after almost being surrounded by armies from the Kalinin and Western Fronts. Soviet forces liberated Smolensk the next day. Smolensk, a major population and industrial center, had been left devastated after nearly two years of German occupation.

25 September 1943

Major units of the 7th Guards Army crossed the Dnepr River south of Kremenchug. The battle for the Borodaevska bridgehead began. In a night attack, the 73 GRD crossed the river using improvised means, helping to capture the large island of Borodaevska along the way. For this feat, the divisional commander, Colonel S. A. Kozak, was awarded the title of the Hero of the Soviet Union. Units of the 7th Guards Army formed a connected bridgehead on the west (right side) of the river between the villages of Borodaevska and Domotkani. [333] During the day, the bridgehead grew to about 7 miles wide and 3 miles deep and was held by subunits of the 81, 72, 73, and 78 GRDs (arrayed from north to south). The 7th Guards Army had successfully crossed the mighty Dnepr River in force!

If crossing the Dnepr wasn't enough good news, even the weather began to improve. From 25 September to 15 October, steady dry weather developed along the Dnepr. [334] As the mud dried out, mobility improved and logistical support became easier.

General Konev, Commander of the Steppe Front, made the following statement about this crossing in his book, *Front Commander's Notes*, Militera.lib.ru, 1972, Chapter 2.

> The experience of forcing the Dnepr by the 7th Guards Army M.S. Shumilov is noteworthy. This army captured right-bank bridgeheads before anyone else.
>
> During the approach to the Dnepr in the area of Perevolnaya, [at] the mouth of Vorskla [river], [at Nowyi] Orlik … it seems it was September 22 or 23, I met with commander M.S. Shumilov and army engineer V. Ya. Plyaskin at [Nowyi] Orlik. We walked along the coast, examined the approaches, roughly determined the width of the river at 700-800 m [760-870 yards], examined the opposite bank with binoculars and found that the enemy's defense on that bank was not yet complete. All this prompted us, without wasting time, to begin forcing the river and [to] act decisively. But, unfortunately, there were almost no crossing facilities, not counting several inflatable boats. The area is steppe, there is no forest and, therefore, there is nowhere to take auxiliary materials for the construction of rafts and bridges. There was nothing on the shore except the trap willow. True, some prudent commanders on the way collected boats, barrels, boards from ruined houses, and everything, of course, came in handy. But this was not enough for the army. We decided everything that is in the nearest settlements (wooden sheds, doors, roofs, barrels, wattle), to use for rafts.
>
> We must pay tribute to M.S. Shumilov and the chief of the army's engineering service, General V. Ya. Plyaskin; they immediately set tasks for the divisions and, under their own guidance, began to prepare rafts for transporting infantry with machine guns, individual guns and sappers [combat engineers] to that shore. Of course, at the same time they took all measures to ensure the crossing of artillery fire and air defense.

On the night of September 25, units of the 7th Guards Army crossed the Dnepr and captured a small piece of land near the village of Domotkan. This was the beginning of the forced crossing of the Dnepr by the troops of the Steppe Front.

Konev continued,

> The difficulties in this matter are undeniably great, especially when there aren't enough effective means of crossing at hand. I ordered the unit commanders to cross…the Dnepr with the advanced detachments. This greatly contributed to the forcing of the river and ensuring the seizure of bridgeheads, as commanders led the fighters by their example, took the initiative, and were creative in solving problems.

By this date, the 7th Guards Army also was able to mobilize a significant number of newly liberated men from local populations into the ranks of the Army. A historical document shows the Army mobilized 5,789 men, with 3,131 being equipped (with weapons and uniforms) and 2,748 being unequipped. [335] The Germans were of course aware of these actives, hence the orders to evacuate the civilian population during the retreat. "Experience during recent withdrawals has shown that in all regained territories, the Red Army immediately mobilized all able-bodied men between sixteen and sixty and incorporated them into the fighting units." General Nehring, commander of the XXIV Pz C, observed, "These men were not 100 per cent soldiers, but they ran along with the rest, they burst into the gaps in the line and flooded us." [336]

26 September 1943

Units of the 7th Guards Army continued to cross the Dnepr using any and all means, reinforcing and expanding the small bridgehead on the west bank, but due to the lack of proper crossing equipment, the movement of men and supplies was very slow. At this time, units of the 15 GRD also were able to cross the Dnepr and to help capture the village of Borodaevska.

Despite this success, the new bridgehead was weakly held. Taking advantage of this weakness, the Germans launched continuous strong attacks against the bridgehead with tanks and aircraft in support that caused heavy losses to the defenders. With the survival of the bridgehead in doubt, General Shumilov contemplated withdrawing his troops from the right side of the river, but General Konev convinced him to keep fighting.

Part of the problem was the complete lack of air support from the VVS. The Luftwaffe was freely roaming over the small foothold on the west side of the river, bombing at will. With the help of General Konev, proper air support finally was organized and massed artillery fire was brought to bear upon the attacking enemy. With that, the enemy attacks began to wane and the bridgehead was saved.

Note that the opposing Germans forces in the area were comprised of only depleted remnants from the 306 and 355 IDs, supported by a few tanks. German tactical doctrine called for the formation of *ad hoc Kampfgruppe KG* (Battle Groups) from the remains of worn-out units, thus giving them some type of offensive capability. Ultimately, the High Command would disband the 355 ID due to the heavy losses it had suffered during the retreat to the Dnepr.

Just south of the 7th Guards Army, the 57th A also began to cross to islands in the Dnepr.

27 September 1943

On this day, a historical chart shows all three 7th Guards Army corps aligned in the first echelon. They now contained the following rifle divisions:

24 GRC – 73, 78 GRDs and the newly arrived 19 RD
25 GRC – 72, 81 GRDs
49 RC – 15 GRD and 213 RD

General Shumilov was switching units in and out of the front line as the combat situation demanded. The overall intent was to direct the best-prepared and strongest units across the river. Also on this day, the 89 Guards Engineer Battalion, from the 78 GRD, assembled the first pontoon bridge crossing the main channel of the Dnepr River, thus linking Glinsk-Borodaevska Island to the left bank. Now men, equipment and supplies could more easily cross the river. The 158 Guards Artillery Regiment, also from the 78 GRD, immediately crossed to the right side of the river and joined in the fighting for Domotkani.

During the night, the Germans launched a counterattack against Glinsk-Borodaevska Island in an attempt to cut off the main Russian force on the right bank of the Dnepr. Fighting continued all night, but by morning the defending Russian troops had repelled the German attack. Once again, the stout defense by the 7th Guards Army saved the bridgehead.

At this time, the 37th A struggled with an extreme logistical problem. The Army's supply base was 110 miles to the rear at the small village of Kovyagi on the Kharkov-Poltava rail line. This village was located about 27 miles west of Kharkov at the end of a restored section of rail line. At this distance, the entire auto transport of the 37th A could only deliver 200 tons per day to the front, with the stated need of 500 tons of ammunition, 75 tons of fuel and lubricants and 50 tons of food, for a requirement of 625 tons per day. Due to this shortfall of supply, Steppe Front logistical planners gave priority to ammunition and fuel deliveries because combat troops urgently needed those supplies to cross the Dnepr. [337] During this period, the historical record does not reveal the location of the 7th Guards Armies supply base, but likewise, it must have been far to the rear, which would explain the similar supply shortages within the 7th Guards Army.

28 September 1943

After quickly replacing units of the 69th A, units of the 37th A advanced to the river and began crossing the Dnepr during the night, just to the north of the 7th Guards Army. This assault was a day late from what General Konev had ordered. Two divisions lead the way, the 62 and 92 GRDs. Due to a strong and deadly response by the Germans, the 92 GRD could not successfully cross the river, but the 62

GRD was able to reach and to hold a section on the west side of the river. By the end of the day, the 37th A had formed two small bridgeheads. General Konev's plan was to quickly link the bridgeheads of the 37th A with those of the 7th Guards Army, thus creating a large continuous lodgment area. These two armies ultimately would accomplish this objective, but doing so would take time and much effort.

The Germans, on the other hand, wanted to prevent any linkup between bridgeheads on the right side of the river. For example, on this day the 7th Guards Army reported a German attack toward the village of Borodaevska with 500 troops and 21 tanks. Needless to say, fighting in and around the bridgehead was constant.

On or about this date, General Shumilov ordered the newly arrived 19 RD to cross to the west side of the Dnepr River. Additionally, General Konev transferred the 53 RD to the 7th Guards Army. When it arrived, General Shumilov moved the fresh 53 RD over the Dnepr and into the bridgehead to support the main attack.

29 September 1943

On this day, a historical chart continued to show all three 7th Guards Army corps in the front line. They contained the following rifle divisions:

24 GRC – 73, 78 GRDs and 19 RD
25 GRC – 72, 81 GRDs and 53 RD (Note: The newly-assigned 53 RD was not counted on this day's combat strength report.)
49 RC – 15 GRD and 213 RD [338]

The Army had a reported combat strength of 34,116 men and women.

By this date, the 7th Guards Army had transported to the west side of the Dnepr River the following:

Combat Soldiers – 9,234 (about 30 percent of the Army's combat strength)

Machine Guns – 104
Sub Machine Guns – 396
82 mm Mortars – 182
120 mm Mortars – 30
[76 mm] Regimental Guns – 42
45 mm AT – 123
[76 mm] Divisional Guns – 56 [339]

The 7th Guards Army's Borodaevska bridgehead now measured 12.5 miles wide and 5 miles deep. While an improvement, the bridgehead was still too small to permit a major breakthrough. During the day, the first 14 tanks of the 201 TB crossed the river via the newly installed pontoon bridge and engaged the enemy in the Domotkani village area. The tank support was greatly appreciated by the infantry, and with their help the bridgehead expanded, but fighting within Domotkani was at times hand-to-hand, with parts of the village changing hands several times during the day. By the end of the month, the 27 GTB also would join the fighting on the west side of the Dnepr River.

Powerful German attacks by units of the I Panzer Army were directed against the 37th A and 7th Guards Army for the next ten days. Von Manstein knew these Russian positions were a major threat to the stability of the Dnepr River defense line, and everything had to be done to drive the Red Army back over the river. Due to these attacks, the 7th Guards Army and the 37th A were forced to give up some ground, but they securely held both of their bridgeheads.

These attacks were headed by parts of the 3 SS Panzer Grenadier Division *Totenkopf* (Death's Head) along with the 6, 9 and 23 Pz Ds. The *Kampfgruppe* from the *Grossdeutschland* Panzer Grenadier Division, and KGs from the 306 and 355 IDs also joined the fight.

Units of the 5th Guards and 53rd As eliminated the German bridgehead on the east side of the river, at Kremenchug and captured the city of Kremenchug. Most of the defending German units safely withdrew to the west bank of the Dnepr, blowing up 6 of the 11 spans of the railway bridge and additional facilities along the way.

Von Manstein issued an order of the day to AG South:

> Soldiers of Army Group South!
>
> After weeks of difficult attacks and defensive battles, the armies of the army group have withdrawn across the Dnepr to again face the enemy in more suitable positions. The enemy, who has been defeated by you whether in the defense or in the counterattack, could not stop this maneuver. He gained territory, but paid a high price in losses. That this withdrawal succeeded on a wide front, over few crossings sites, with our most important equipment and while destroying the most important war installations, is the result of the courage of all the troops that were well lead during a critical situation by their commanders, under conditions reflecting an enormous numerical superiority in favor of the enemy. The execution of this especially difficult operation is yet another testimony to the brilliance of the leadership, especially the work of the staff. My thanks to both the leadership and the troops for their magnificent efforts in this summer's great defensive battle. That the enemy, in spite of his strength, could not stop us from executing our mission, gives us the opportunity to master new missions.
>
> Signed von Manstein Field Marshal. [340]

While AG South was forced to retreat in the face of superior Soviet forces, von Manstein's message reflected German propaganda and bravado at its finest.

The Soviet summer offensive, otherwise known as the Lower Dnepr Strategic Offensive, cost the Red Army dearly. David Glantz states in his book, *Forgotten Battles of the German-Soviet War, Vol V, Summer-Fall Campaign (Part Two)* on page 817, Figure 30. "Between 28 September – 20 December 1943, the 2nd, 3rd, and 4th Ukrainian Fronts (currently the Steppe, Southwest and South Fronts) suffered the following losses:

Front	Killed, Missing, POW	Wounded and Sick	Total
2 UF	77,400	226,217	303,617
3 UF	34,821	132,248	167,069
4 UF	60,980	222,726	283,706
Total	**173,201**	**581,191**	**754,392**

Table #6: Front Losses between 28 September – 20 December 1943.

So, from a starting total strength of 1,506,400 men and women for these three fronts, they collectively suffered 754,392 losses from all causes during these 84 days of offensive operations, which was a casualty rate of about 50%. This equates to an average loss rate of 8,980 casualties per day.

30 September 1943

Between 1-30 September, the 7th Guards Army reported suffering the following total casualties:

Officers – 1,034 men and women
Sergeants – 1,952
Privates – 8,697
Total – 11,683

Of these casualties, 1,925 were killed; 8,925 were wounded; and 833 were sick or missing. [341]

The 7th Guards Army ended the month with a combat strength of 43,392 men and women. This sharp increase in combat strength was due mostly to the 53 RD being added to the Army at nearly full strength with 8,163 men. In turn, the newly assigned but weakened 19 RD could claim only 3,707 men and women within its ranks.

Partial Army rifle units on the west side of the Dnepr, or on islands, included elements from at least the following major units:

49 RC – 15 GRD, 213 RD
25 GRC – 72, 81 GRD, 53 RD
24 GRC – 73, 78 GRD, 19 RD

The strength of these incomplete units on the west side of the river totaled about 12,195 men.

Supporting units on islands or on the west side of the river included:

27 GTB, 201 Tank Brigade
30 AT Brigade
161 Guards Gun Artillery Regiment
115 Guards Tank Destroyer Regiment
265 Guards, 290 Mortar Regiments

The estimated strength of supporting units totaled about 4,359 men.

These incomplete Army units totaled 16,554 men or about 38 percent of the 7th Guards Army's total combat strength. The remaining subunits would be transferred to the west side of the river when possible.

As can be seen on Map #23, General Shumilov had pushed across the river as much of his army that would fit into the bridgehead. What remained on the eastern shore were the remaining units of infantry, tank, and artillery that were waiting to cross. Additionally, until the bridgehead could be enlarged sufficiently, the Army and Corps HQs, medical, supply and other support units would remain on the east side of the river indefinitely. Note: Due to the confused nature of the river crossing and the needs of combat, the 81 GRD and the 53 RD were divided, with each division fighting in two separated distinct groups along the front line.

By this date, the 49 RC had two ferries in operation on the river in its sector. The 25 GRC also had two ferries in operation, and the 24 GRC had two ferries and one pontoon bridge connecting to Glinsk-Borodaevska Island. In total, the engineers of the 7th Guards Army were operating three separate crossing points over the Dnepr using seven sets of crossing equipment.

During September, the supporting VVS 5th Air Army flew 7,917 sorties in the area of the Steppe Front. This was far more than the Luftwaffe could produce with only about 3,250 sorties. [342] This disparity was not due to lack of effort on the part of the Luftwaffe. During 1943, the Luftwaffe devoted 80 percent of its total sorties on the Eastern Front to the tactical support of the army. [343] The problem was lack of aircraft, spare parts, fuel and demands from other fronts. In essence, the German air force was fighting too many powerful enemies in too many locations to achieve a decisive result in any one theatre of operation.

To increase the pressure on the Borodaevska bridgehead, the Germans moved the LII Corps and several more *Kampfgruppe* into the area, including the late-arriving KG from the Pz GD *Grossdeutschland* (*GD*). Clashes with units of the Pz GD *Grossdeutschland* were already occurring and more soon would follow. These additional forces sound impressive, but all these units were far below strength and only a desperate attempt to hold the Dnepr River line compelled German commanders to throw these depleted units back into combat time after time.

Interestingly, the Germans did comment upon Soviet success and failures along the Dnepr.

> During the few remaining days of September, the Soviets achieved twenty-three river crossings along the 440 miles [of the Dnepr] from Loyev [north of Kiev] to Zaporozhye. But the Soviets nevertheless made one mistake in their calculations. True enough, they got their companies, battalions, and even regiments across the river quickly and established bridgeheads on the other side; but they found it difficult to enlarge these bridgeheads on the scale, which would have allowed a major offensive to be launched from them. They had enormous difficulties in ferrying tanks, heavy equipment, and ammunition across the river. For that they would have needed solid bridges, but these could not be built from their minute bridgeheads. [344]

This German observation accurately described the situation in which General Shumilov's 7th Guards Army and Lieutenant General Mikhail Sharokhin's 37th A found themselves. They were tightly constrained in several small, shallow bridgeheads under constant air, artillery and ground attack.

The end of the month saw the victorious Red Army firmly up against the Dnepr River along most of its length and across it in a few areas. They had not achieved their decisive breakthrough yet, but that would come in time. (See Map #24.)

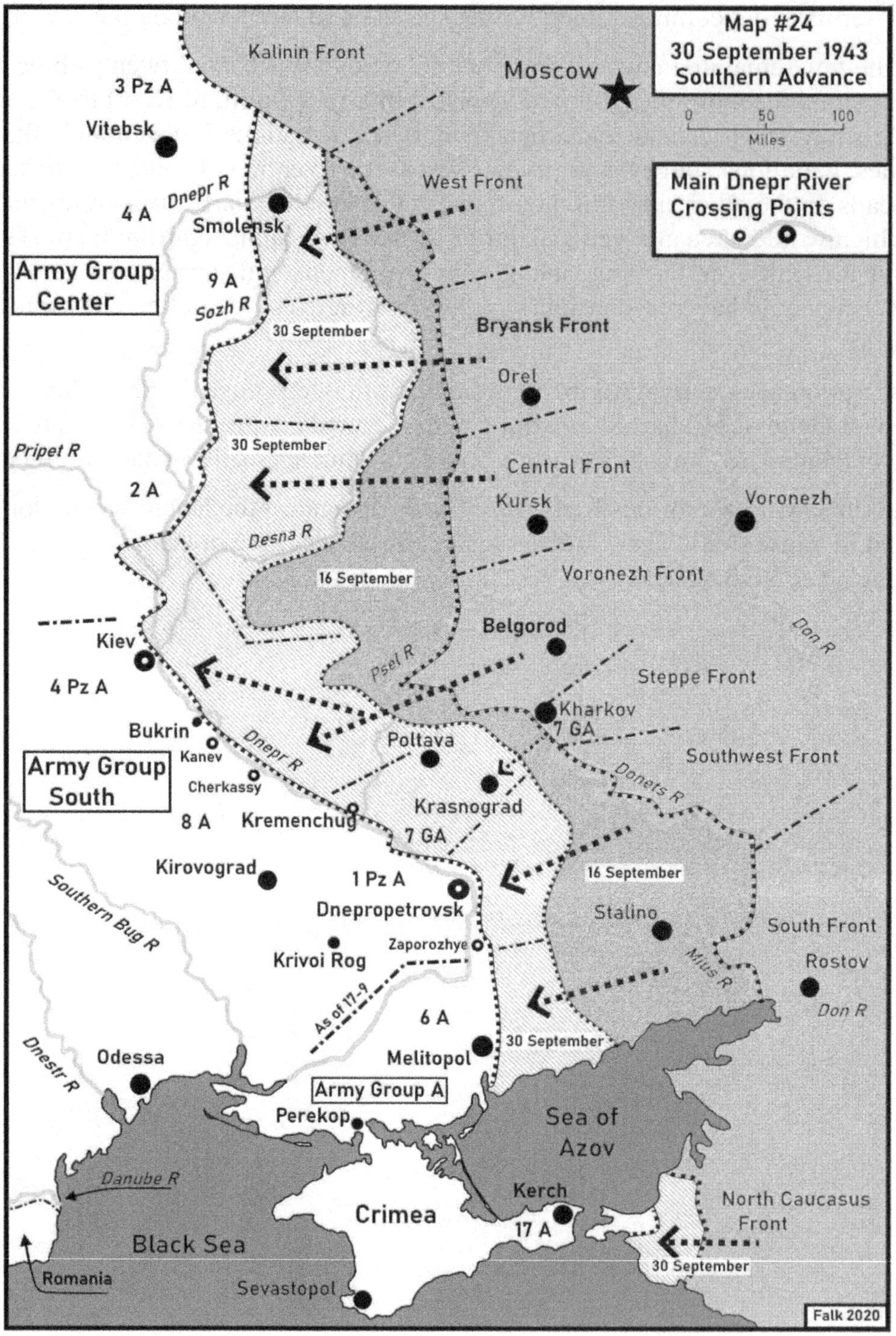
Map #24
30 September 1943
Southern Advance
0 50 100
Miles
Main Dnepr River
Crossing Points
Kalinin Front
Moscow
3 Pz A
Vitebsk
West Front
4 A
Dnepr R
Smolensk
Army Group
Center
9 A
Sozh R
30 September
Bryansk Front
Orel
30 September
Pripet R
Central Front
2 A
Kursk
Voronezh
Desna R
16 September
Voronezh Front
Belgorod
Don R
Kiev
Psel R
Steppe Front
4 Pz A
Kharkov
7 GA
Bukrin
Kanev
Poltava
Southwest Front
Army Group
South
Dnepr R
Cherkassy
Donets R
Krasnograd
8 A
Kremenchug
7 GA
Kirovograd
1 Pz A
16 September
Southern Bug R
Dnepropetrovsk
Stalino
South Front
Zaporozhye
Krivoi Rog
Mius R
Rostov
As of 17-9
Don R
6 A
Dnestr R
Odessa
Melitopol
30 September
Army Group A
Perekop
Sea of
Azov
Danube R
Kerch
North Caucasus
Front
Crimea
17 A
Black Sea
30 September
Romania
Sevastopol
Falk 2020

1 October 1943	
7th Guards Army	
Assigned – Steppe Front (renamed 2nd Ukrainian Front)	
Colonel General M.S. Shumilov Commanding	
Rifle, Airborne and Cavalry Units	24 Guards Rifle Corps (73, 78 GRDs, 19 RD) 25 Guards Rifle Corps (72, 81 GRDs, 53 RD) 49 Rifle Corps (15 GRD, 213 RD) 2, 3 Separate AT Rifle Battalions 82, 83, 84, 85, 86 Blocking Detachments 62, 63, 64, 65, 66 Penal Companies
Artillery Units	11 Artillery Division (40 Howitzer Artillery Brigade, 45 Gun Artillery Brigade, 31 Gun Artillery Brigade) 161 Guards Gun Artillery Regiment 114, 115 Guards, 1669 Tank Destroyer Regiments 30 Antitank Artillery Brigade 265 Guards, 263, 290 Mortar Regiments 5 Antiaircraft Artillery Division (670, 743, 1119, 1181 AA Regiments) 162 Guards AA Regiment 662 Separate Aerial Observation/Air Raid Warning Company
Armored and Mechanized Units	27, 96 Guards Tank Brigades 201, 262 Separate Tank Regiments 34, 38 Separate Armored Train Battalions (four trains active)
Engineer Units	60 Engineer Sapper Battalion 175, 329 Separate Engineer Battalions 40 Pontoon Bridge Battalion
Estimated Combat Strength Estimated Ration Strength	30,698 men 37,698 men and women (including an estimated 7,000 support personnel)

(BSSA 43, BSA 43, TsAMO Archive)

The 7th Guards Army was the first army of the Steppe Front to reach and to cross the Dnepr River in strength and to seize a sizeable bridgehead on the western shore. The start of October saw additional bridgeheads established within the zone of the Steppe Front. As the battle on the west side of the river raged, German forces launched nearly constant attacks in their attempts to eliminate or to reduce these Soviet gains.

1 October 1943

For outstanding service during the battle for Stalingrad and Kursk, General Shumilov was promoted to Colonel General.

The 78 GRD withdrew into Army reserve for rest and replenishment; the newly arrived 19 RD took over its place in the front line. Also, sometime within 1-5 October, the 48 GRD was assigned to the 7th Guards Army, showing up within its ranks on 5 October.

To give the reader some idea of the greater scope of the war, the following table displays the estimated size and distribution of German forces throughout Europe in October, and the corresponding Soviet strength deployed on the Eastern/Russian Front as well as throughout the Soviet Union. These values should be considered ration strengths.

Location	German Forces [345]	Soviet Forces [346]
Eastern/Russian Front	3,900,000	6,700,000
Additional Soviet Areas Behind the Main Front	-	4,400,000
Finland	180,000	-
Norway	315,000	-
Denmark	110,000	-
Western Europe	1,370,000	-
Italy	330,000	-
Southeastern Europe	610,000	-
Totals	**6,815,000**	**11,100,000**

Table #7: Distribution of German and Soviet Forces in October 1943.

Note that despite Germany having its forces scattered all over Europe and fighting on many fronts, German Axis allies would be able periodically to add several hundred thousand troops to these totals. This was especially true in Italy and when the fighting on the Eastern Front reached the Balkans. Additionally, Germany was logistically in a difficult position by being forced to transport supplies, weapons and troops to all points of the compass by land, sea and air while being under constant attack by Western naval and air forces.

On the other hand, the Soviet Union had the advantage of dealing with only one, but arguably huge, fighting front. Furthermore, the Soviet Union could focus on pushing its armies to the west via multiple overland routes, and on drawing strength and resources from its vast interior, much of which was untouched by war.

During the day, the Luftwaffe repeatedly bombed the 7th Guards Army within the bridgehead area. When the aircraft withdrew, German artillery units renewed their bombardment and pounded Russian positions. Russian commanders took this to be an effort to soften up the defenders for further ground attacks.

In the area of the 37th A, German forces launched a major attack with tanks in support, but after making some progress, Russian troops pushed back the attackers.

On this day, Hitler's attention again was diverted from Russia because Allied forces captured the city of Naples as they advanced up the boot of Italy.

2 October 1943

A historic document shows the Army now had a reported combat strength of 47,672 men and women.

Overhead, the 5th Air Army had 712 aircraft to support and to protect the Steppe Front bridgeheads.

As additional units arrived in the area, German attacks against the Borodaevska and other bridgeheads continued to increase in intensity, both on the ground and in the air. The 1st Panzer Army and the Luftwaffe were making a major effort to reduce or to destroy all these small bridgeheads in the Steppe Front area.

Newly promoted Colonel General M.S. Shumilov, Commander of the 7th Guards Army.

3 October 1943

Sometime in early October, Field Marshal von Manstein once again moved his Army Group South HQ. The headquarters moved from Kirovograd to Vinnitsa (Hitler's former headquarters) located about 125 miles southwest of Kiev. The HQ accommodations were much improved.

German forces launched a powerful attack supported by aircraft and artillery against the 7th Guards Army. Fighting continued for most of the day as KG from Pz GD *Grossdeutschland* and KG from the 306 and 355 ID strove to crush the Russian bridgehead. The 7th Guards Army's front line was forced back as German forces reached the Dnepr River in the center of the bridgehead, near Hill 147.4, but the stubborn defenders held out in other areas.

Almost simultaneous with these German efforts on 3 October, General Konev issued the following order:

> [To] the 37th commander in the section Uspenskoye, Derievka, collective farm. Voroshilov temporarily [will] go on the defensive. On the left flank, using three divisions, the entire bulk of the army artillery from the right and left banks of the Dnepr River, [and] all the tanks that crossed in the morning of October 3, 1943 …[will] deliver a decisive blow in the direction of Annovka, 177.0 [Hill 177.0, See Map #25] and, together with units of the 7th Guards Army, defeat and destroy the advancing enemy group in front of the 7th Guards Army. [347]

The plan was for the 37th A and 7th Guards Army to link their bridgeheads and to destroy the German forces blocking the advance of the 7th Guards Army. The counterattack did not go as planned as the ongoing German attack was too strong to force back. Not until 15 October would the real breakout and linkup of these separate bridgeheads take place.

About 16 miles south of the 7th Guards Army, the 57th A seized a small position on the right bank of the Dnepr. The German reaction to this intrusion simply was to seal off this new foothold and to continue their attacks against the Borodaevska bridgehead unimpeded.

Key commanders submitted to *Stavka* for approval a general plan for the Steppe Front's offensive to the southwest. However, the Soviet leaders still needed to complete much detailed planning work before they could execute this operation.

4 October 1943

The 5th GTA continued its attack toward Krivoi Rog but was hampered by heavy rains and mud.

The 7th Guards Army had a reported combat strength of 49,901 men and women. Commanders made an effort to replenish the divisions on the right side of the river with men and equipment as supplies arrived at the front.

The German advance continued as the pressure increased against the Borodaevska bridgehead. During the day, German troops reached the Dnepr in the center of the bridgehead and seized the western half of the Borodaevska village. This advance allowed the Germans to bring fire against any remaining positions along the right side of the river in the area of hill 147.4 and therefore effectively split the 7th Guards Army position into two parts. But could the Germans hold these newly captured positions overnight?

In other news, in the Mediterranean Sea, the Allies recaptured the island of Corsica, located west of the Italian mainland. Attacks against German positions throughout the Mediterranean theatre of war continued.

5 October 1943

The 7th Guards Army had a reported combat strength of 53,787 men and women with ten rifle divisions assigned: 15, 28, 48, 72, 73, 78, and 81 Guards, 19, 53 and 213 RDs. The replenishment of divisions on the right side of the river continued.

By 5 October, the German counterattack had forced back the front line to the river, occupied the remaining Russian positions around Hill 147.4 in the center of the bridgehead, and decisively splitting the 7th Guards Army's bridgehead into two parts. On the far-right flank, the 213 and 15 GRD of the 49 RC were in limited contact with units of the 37th A but were cut off from the main bridgehead around Borodaevska. At Borodaevska, the 24 and 25 GRC containing parts of seven rifle divisions held a reduced, but secure, bridgehead. The Germans now were firmly wedged between the two parts of the Army and actually held a four-mile section of the riverfront. The newly arrived 9 Pz D was leading the attack against the left flank of the 7th Guards Army. (See Map #25.)

Most likely due to supply issues, the 5th Air Army was able to fly only 200 sorties in support of the Steppe Front, while the Luftwaffe was able to fly some 400 sorties in support of their attack against the Borodaevska bridgehead.

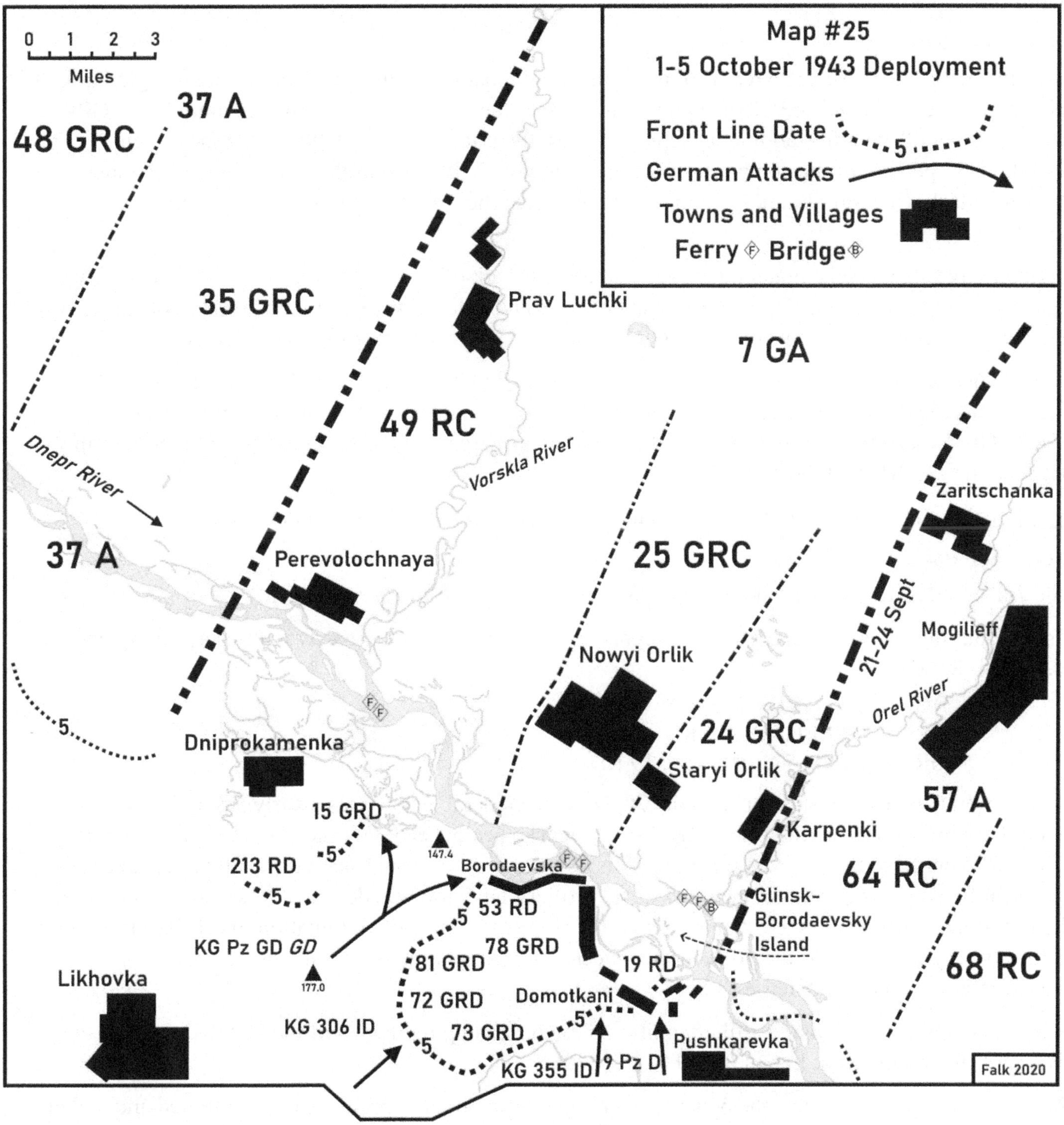

6 October 1943

The 7th Guards Army had a reported combat strength of 54,175 men and women.

Low-level combat occurred during the day in the Borodaevska bridgehead. Both sides gathered their forces and assessed the situation.

The 5th Air Army flew 270 sorties in support of the Steppe Front, while left flank units of the 37th A made repeated attacks in their attempt to link up with the 7th Guards Army and to secure both of their flanks.

7 October 1943

The 7th Guards Army issued a report of the combat composition of the Army on the south/right bank of the Dnepr. The 49 RC had 6,930 men on the right bank; the 24 GRC had 5,856 men; and the 25 GRC had 6,322 men. With a few additional support troops, the total came to 20,495 men and women. [348] This was about 37 percent of the combat strength of the Army. With the Borodaevska bridgehead under constant attack, General Shumilov continued to reinforce the right bank with as many combat troops as possible.

German forces launched a major attack against the left flank of the 7th Guards Army. During heavy fighting they pushed deeply into the Russian defenses and reached the eastern end of Borodaevska. A late day counterattack by units of the 78 GRD force the Germans back and cleared Borodaevska.

The 28 GRD arrived at the left bank of the Dnepr and officially joined the 49 RC. Interestingly, the 28 GRD already had orders to cross the Dnepr near Perevolochnaya to join the 49 RC on the other side. The 28 GRD was a fresh rifle division and was ordered to cross the river and to fill a dangerous gap between the 213 RD and the 15 GRD.

Additionally, the 5th GTA was transferred to the Steppe Front. The buildup continued.

8 October 1943

The 7th Guards Army had a reported combat strength of 55,826 men and women

On or about this date, the 41 GRD was assigned to the Army. After their arrival, the corps assignments were as follows:

49 RC – 15, 28 GRDs, 213 RD
25 GRC – 72, 81 GRDs, 53 RD
24 GRC – 41, 48, 73, 78 GRDs, 19 RD

With eleven rifle divisions assigned to the 7th Guards, General Konev was clearly strengthening the Army for a breakout from the Dnepr River Borodaevska bridgehead. Of note, the 11 Artillery Division also was assigned to the Army at this time. This new type of divisional size unit would provide powerful fire support during future attacks. An Artillery Division was designed to control and to direct the use of mass artillery assets. "Quite simply, the concentration and coordination of artillery firepower in the artillery division was the difference between a grinding battle of attrition and a decisive breakthrough of German defenses." [349]

During the night, rifle division artillery units and the remaining regiments of the 28 GRD crossed the river and entered the front line of the 49 RC.

Regardless of reinforcements, the powerful German attack from the previous day continued and ended once again with the Red Army defenders tenaciously holding their current positions.

9 October 1943

The 7th Guards Army had a reported combat strength of 54,849 men and women.

The powerful German attack continued unabated, supported by lavish air and artillery support. The Luftwaffe also bombed the far shore of the Dnepr to disrupt the transfer of men, equipment and supplies into the bridgehead.

Further south, the remaining units of the German 17th Army completed their withdraw to the Crimea. The battle for the Caucasus had ended in victory for the Red Army as the Taman Peninsula was now fully under the control of the North Caucasus Front. Thus, the 17th Army was ordered to release eight divisions for use further north and along the Dnepr.

10 October 1943

The 7th Guards Army had a reported combat strength of 57,677 men and women. Despite the relentless German attacks, new formations arrived and replacements kept increasing the Army's overall strength.

For air coverage of the entire Army Group South and Army Group A area, the German 4th Air Fleet had 834 aircraft on hand for support, but only 562 of these aircraft were operational. Despite the small number of aircraft available, on this day, the Germans committed the bulk of these aircraft against Red Army bridgeheads over the Dnepr River and against railroad targets in the rear.

However, the VVS did not sit idle. The 5th Air Army intervened in ground combat in and around the Borodaevska bridgehead, helping the 7th Guards Army resist the ongoing German ground attack. Later in the day, combat activities decreased along the front.

By the end of the day, left flank units of the 37th A firmly linked up with the isolated 213 RD, the 15, 28 GRDs, and the 49 RC of the 7th Guards Army.

11 October 1943

The 7th Guards Army had a reported combat strength of 56,619 men and women, a loss of just over 1,000 troops from the day before.

Combat operations around the Borodaevska bridgehead continued to subside as both sides gathered their strength and planned further assaults.

General Konev, the Steppe Front Commander, considered the battle to capture, expand and then hold the Borodaevska bridgehead had taken 15 days, from 27 September to 11 October. Fighting in the bridgehead area had certainly taken place before and after these dates, but during this time, the Red Army has seized a large lodgment area on the right-hand bank of the Dnepr River, and they then held it long enough to stabilize the newly captured territory.

With Marshal Zhukov's support, *Stavka* approved the submitted plan for an offensive in the direction of Krivoi Rog. By advancing to the southwest, this offensive would outflank and threaten to cut off any enemy forces retreating from the Dnepropetrovsk area, specifically the 1st Panzer Army.

To simplify the plan, the Red Army would use the existing bridgeheads over the Dnepr to launch the offensive. For this effort, commanders brought forward and deployed several additional armies in and behind the existing bridgeheads currently held by the 37th A, 7th Guards and the 57th A of the Steppe Front. The 37th A, commanded by General Mikhail Sharokhin, would make the main attack so its bridgehead received the 5th GA. The 5th GTA and the 7th Guards Mechanized Corps were deployed to the rear, ready to exploit the anticipated breakthrough. General Shumilov's 7th Guards Army and 57th A, commanded by Lieutenant General Nikolai Aleksandrovich Gagen, would make their own attacks at the appropriate time.

The movements of these newly assigned armies were conducted using strict *maskirovka* methods. General Konev stated, "Measures were used to conduct a regrouping of forces of the front and material to support them during a short period and [to conceal it] from the enemy." [350] The Germans failed to detect the regrouping of these forces. The 37th A received the most engineering support from the Steppe Front, which included numerous pontoons, ferries and boats. With four ferry crossing points and 11 active ferries, the 37th A's bridgehead had the capability to support the main attack. [351]

An interesting side note to all this buildup activity was that General Konev prohibited the use of any river crossing equipment within the 37th and 5th GA bridgehead area during daylight hours. Crossings could only take place at night to conceal the buildup of forces in this area. But in the 7th Guards Army bridgehead area, a bit further south, Konev ordered Shumilov to keep all bridges and ferries in sight during the day and to move all his reinforcements into the bridgehead. All this visible activity would attract the attention of the Luftwaffe and indicated a buildup of Red Army forces in this southern area and not in the adjacent 37th A's bridgehead. In essence, Konev was using the 7th Guards Army as a decoy to divert the Germans' attention away from the main attack area. [352]

12 October 1943

The 7th Guards Army had a reported combat strength of 62,237 men and women. No major fighting took place along the front line.

13 October 1943

The 7th Guards Army had a reported combat strength of 65,505 men and women. The increase in strength from the previous day showed that reinforcements and/or returns from medical units continued to strengthen the Army.

Again, the front line was quiet as Russian commanders shifted their forces around and replenished supplies in preparation of the coming offensive. German forces were, of course, doing the same thing.

In Western Europe, Marshal Pietro Badoglio, the head of the Italian government, declared war on Germany. This declaration sent a clear message to the Italian people and to the Allies just where Italy stood in regard to their previous Axis allies.

14 October 1943

As commanders repositioned and reassigned forces, the 7th Guards Army had a reported combat strength of 66,374 men and women. This value changed daily due to ongoing casualties and arriving replacements.

Once again, German units attacked in strength trying to expand the corridor between the 49th and 25th Corps. After two assaults, they failed to gain much ground and later in the day the attacks subsided. This would prove to be the last major German assault against the Borodaevska bridgehead. Next it would be General Shumilov's turn to strike.

At Zaporozhye on the lower Dnepr River, the Southwest Front crushed the German bridgehead on the east side of the river and captured the city. Before giving up the area, German engineers destroyed all the river bridges, industrial facilities and the local Dnepr hydroelectric dam, which would remain out of operation until 1947.

15 October 1943

The 7th Guards Army reported an increased combat strength of 67,191 men and women.

At 10:00 hours, the neighboring 37th and 5th GAs went on the offensive. This was the beginning of the Steppe Front's Krivoi Rog Operation (15 October-3 November), which was yet another attack in a series of Red Army offensives collectively called the Winter Strategic Offensive. On this first day, only

105 tanks from the 5th GTA were able to cross the Dnepr to support the attack. A small bridgehead and limited road network prevented the full power of the 5th GTA and other mobile units from being employed en masse.

At the appropriate moment, the 15 GRD, 49 RC and the 78 GRD, 25 GRC attacked toward each other and finally drove the defending German units from the western end of Borodaevska. They then went on to capture Hill 147.4, thus closing the gap between the two separated bridgeheads. Turning to the southwest, the attack continued as the remaining units of the Army pushed the defending German formations away from the river and further expanded the size of the bridgehead, capturing the critical German observation point located on Hill 177.0.

VVS units of the 5th Air Army flew 470 sorties to support this attack; the enemy flew some 250 sorties. As a result of this unexpected offensive, German units were forced to fall back in some disorder. By the end of the day, the now combined bridgehead of the 5th GA, 37th A and 7th Guards Army reach a depth of about 7 miles and a width of 27 miles. (See Map #26.)

General Konev, General Shumilov, General Sharokhin of the 37th A and General Gagen of the 57th A were now ready to exploit this victory and to use the additional forces that General Konev had moved into the area. Together, they planned to fully breach the German front line to achieve the decisive breakout for which they had all been fighting.

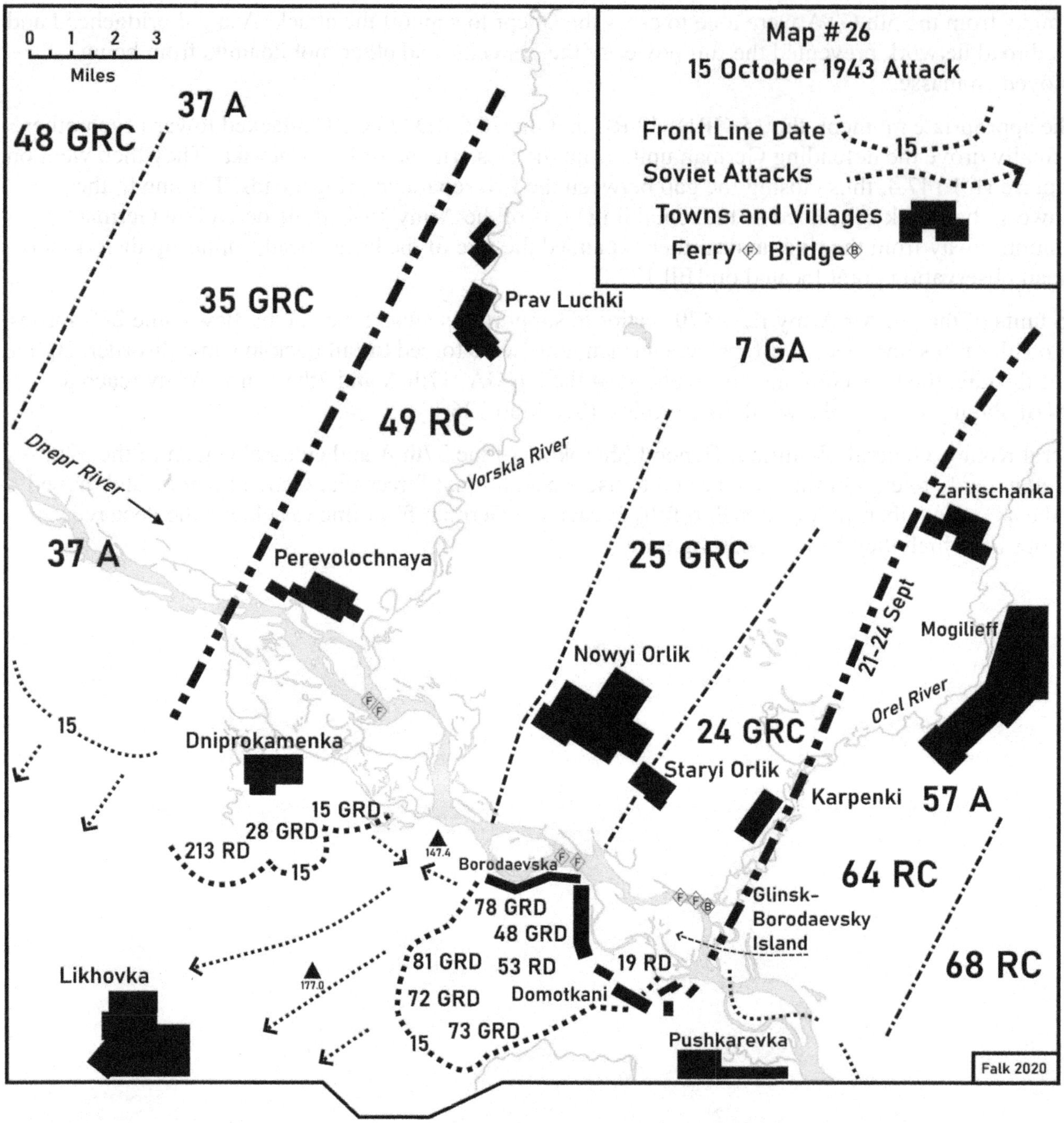

16 October 1943

The 7th Guards Army had a reported combat strength of 67,842 men and women.

The offensive continued to advance with growing force as the Germans offered determined resistance. On this second day of the offensive, congestion at the limited crossing points allowed only 109 tanks from the 5th GTA to cross the Dnepr.

At this point in the offensive, the 7th Guards Army was aiming toward Krivoi Rog, some 65 miles to the southwest.

17 October 1943

The offensive continued as the armies of the Steppe Front advanced to the west and southwest. This assault was near the boundary line between the German 8th Army to the north and the 1st Panzer Army to the south. This location, between armies, might have caused some command-and-control problems for the Germans as their combat formations fell back. As von Manstein stated in his book *Lost Victories,* page 481, "A wide gap opened between the two armies, [8th A and 1st Pz A] leaving the enemy free to drive deep into the [Dnepr] bend towards Krivoi Rog and Nikopol, the retention of which Hitler considered essential to the German war effort [due to their mineral deposits]."

The bridgehead had been pushed out a further 11 miles. With this new space and with room to maneuver, tank and mobile units that had not yet crossed the Dnepr began to cross to the right side of the river, expanding the breakout. [353]

Von Manstein, knowing his forces were badly weakened by months of fighting, called upon the high command for additional units to reinforce his faltering front.

For once, though somewhat belatedly, the OKH could offer effective help. Earlier in the month the Operations Branch, OKW, had decided that the danger of an invasion in the West had passed for the time being and that the Eastern Front had to be reinforced even if it meant taking risks in other theaters. Subsequently the OKW released the 14th and 24th Panzer Divisions. Then on 20 October it also offered the 1st and 25th Panzer Divisions, the 1 SS Panzer Grenadier Division *Leibstandarte* Adolf Hitler (LSSAH), and the 384th Infantry Division. These five Panzer/Grenadier divisions constituted a powerful mobile reserve, but the question was whether they could arrive in Russia in time to stabilize the front. [354]

18 October 1943

The 7th Guards Army had a reported combat strength of 63,540 men and women. The offensive had started to take its toll of the Army's strength as casualties mounted.

The offensive of the Steppe Front continued, but supply issues slowed the advance even as the German front line became somewhat disorganized.

The Pz GD *Grossdeutschland* was defending east of Likhovka, but as the Red Army offensive expanded, the *Grossdeutschland* began further withdraws south toward Krivoi Rog on 19 October.

19 October 1943

The 7th Guards Army again reported a decline in combat strength to 62,486 men and women.

At its furthest point of advance, the 7th Guards Army had penetrated 15 miles into the positions of Army Group South. To hold that area, the Army had ten rifle divisions in the front line. Because the 57th A had not yet advanced over the Dnepr in strength, the left flank of the 7th Guards Army was still fixed at the river near Borodaevska. The 53rd A moved south from the Kremenchug area and over the river, joining the 5th GA, 5th GTA, 7th GA, 37th A and the 7 Mechanized Corps within the expanding bridgehead. (See Map #27.)

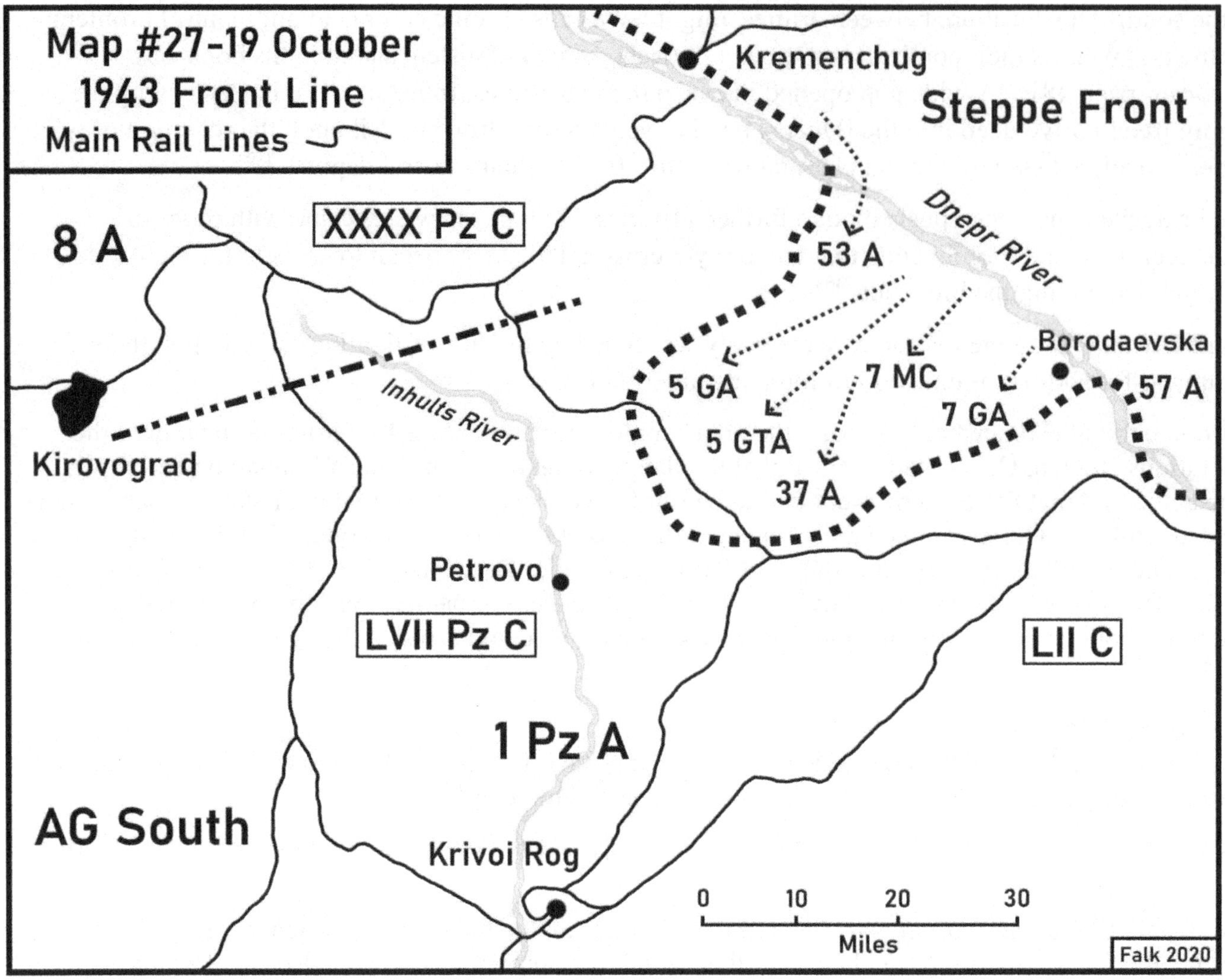

20 October 1943

To mark recent changes at the front and to highlight recent victories, *Stavka* began renaming their front commands to more correctly fit the current situation and location. The Voronezh Front was renamed the 1st Ukrainian Front (1 UF). The Steppe Front, in which the 7th Guards Army fought, became the 2nd Ukrainian Front (2 UF). The Southwest Front became the 3rd Ukrainian Front (3 UF) and the South Front became the 4th Ukrainian Front (4 UF). The North Caucasus Front remained the same at this time, but on 15 November 1943, the North Caucasus Front would be reformed into the Independent Coastal Army, for operational use along the Black Sea coast.

Despite the name changes, units of the new 2 UF continued their advance on the western side of the Dnepr River. In turn, Luftwaffe forces conducted powerful attacks against this ongoing Red Army offensive.

21 October 1943

The Borodaevska bridgehead now had expanded to approximately 30 miles wide by 40 miles deep. With increasing German resistance, the 2 UF's advance slowed.

The 57th A began to cross the Dnepr, filling in on the left flank of the 7th Guards Army south of Borodaevska.

22 October 1943

General Konev stated in his book *Front Commander's Notes*, the "Battle of the Dnepr" chapter, that at this point the 3rd

> …experienced great difficulties in replenishing troops and in supplying the army with equipment and ammunition. Taking all measures for the development of the offensive, we used Po-2 aircraft [small biplanes] to transfer fuel. In addition, Headquarters assigned us 10 Douglas transport aircraft [C-47]. Animal transport then played a large role. All means were used; everything was done so that the troops of the front went forward….

The Russians needed all this effort because they had no rail line in operation over the Dnepr. Simultaneously with this forward logistical effort, General Konev redirected troops, civilians, and rear area transport personnel to harvest and then to transport grain from behind the lines to the deep rear, specifically Moscow and Leningrad. One wonders how they could effectively accomplish both of these divergent goals.

South of the Dnepr, due to heavy pressure, German forces began withdrawing from the critical city of Melitopol. Its loss would endanger the entire front of Army Group A by allowing the Red Army an unobstructed advance to the lower Dnepr River. The loss of Melitopol also would allow the Soviets to cut off land access to the Axis position and troops in the Crimea.

In far off Germany, the operational headquarters of the SS (SS-FHA) ordered the 1, 2, 3 and 5 SS Panzer Grenadier Divisions to be renamed and reorganized as full Panzer Divisions with one Pz Regiment containing two Pz Battalions. These two battalions were to be equipped with mostly Pz IVs and Pz V (Panther) tanks as soon as they could be rebuilt or equipped. [355] At this time, the best medium tank in the Red Army was the T-34/85, but it had just entered full production and would not become abundant until the following year. The T-34/85 was equal to or slightly superior to the German Pz IV, but it was totally outclassed by the Panther. Despite these changes, the massive Red Army tank force would continue to lead the advance to the west as the increasingly outnumbered German *Panzer Truppen* (Tank Troops) extracted a heavy price along the way.

23 October 1943

Units of the 4 UF, (former South Front), liberated the city of Melitopol from the enemy. This was serious news for the German 17th Army in the Crimea. If the Germans could not stop the Soviet offensive, the 17th Army soon would be cut off from the mainland and trapped within the Crimea.

24 October 1943

The 5th GTA continued its attack toward Krivoi Rog but was hampered by heavy rains and mud.

25 October 1943

The 3 UF, with the assistance from the 2 UF, liberated Dnepropetrovsk. Before the Germans lost Dnepropetrovsk, they destroyed an extra-class metal bridge across the Dnepr with demolition charges, dropping 12 of 15 spans. They also blew up the large reinforced concrete railway bridge on the Nizhnedneprovsk-Apostolovo rail line section. [356] The 3 UF was now free to cross the Dnepr in strength, once their engineers developed temporary crossing points. With the loss of Dnepropetrovsk and its bridgehead, the only AG South position remaining on the east side of the lower Dnepr was the Nikopol bridgehead. Due to Hitler 's orders, this bridgehead was to be strongly held in an attempt to secure critical supplies of manganese ore, which was being mined and stored in that area.

26 October 1943

To acknowledge his war achievements,

> By a decree of the Presidium of the Supreme Soviet of the USSR [on 26 October 1943], for successfully crossing the [Dnepr] River, firmly securing and expanding the bridgehead on the western bank of the [Dnepr] River, and showing courage and heroism of the Guard, Colonel General Mikhail Stepanovich Shumilov was awarded the title Hero of the Soviet Union with the [accompanying] Order of Lenin and the Gold Star Medals. [357]

These awards were a well-deserved recognition of a great general.

On or about this date, General Konev determined that the 7th Guards Army would be more useful if it were realigned with the anticipated direction of future attacks, which would be more toward the Kremenchug area. So, the 2 UF began transferring the 7th Guards from its current position on the east side of the expanding Borodaevska bridgehead to the west side of the bridgehead 25 miles away. Guard units were thought of as elite combat units, so they were expected to spearhead major attacks, if possible. As the 7th Guards Army departed, the 57th A moved south to cover that section of the front line. Note: This shifting of the 7th Guards Army to the west also may have been in response to a confusing series of mobile battles with units of the newly arrived German XXXX Pz C, that were currently taking place west of the Inhults River. (See Maps #27, #28 and #29 for general location).

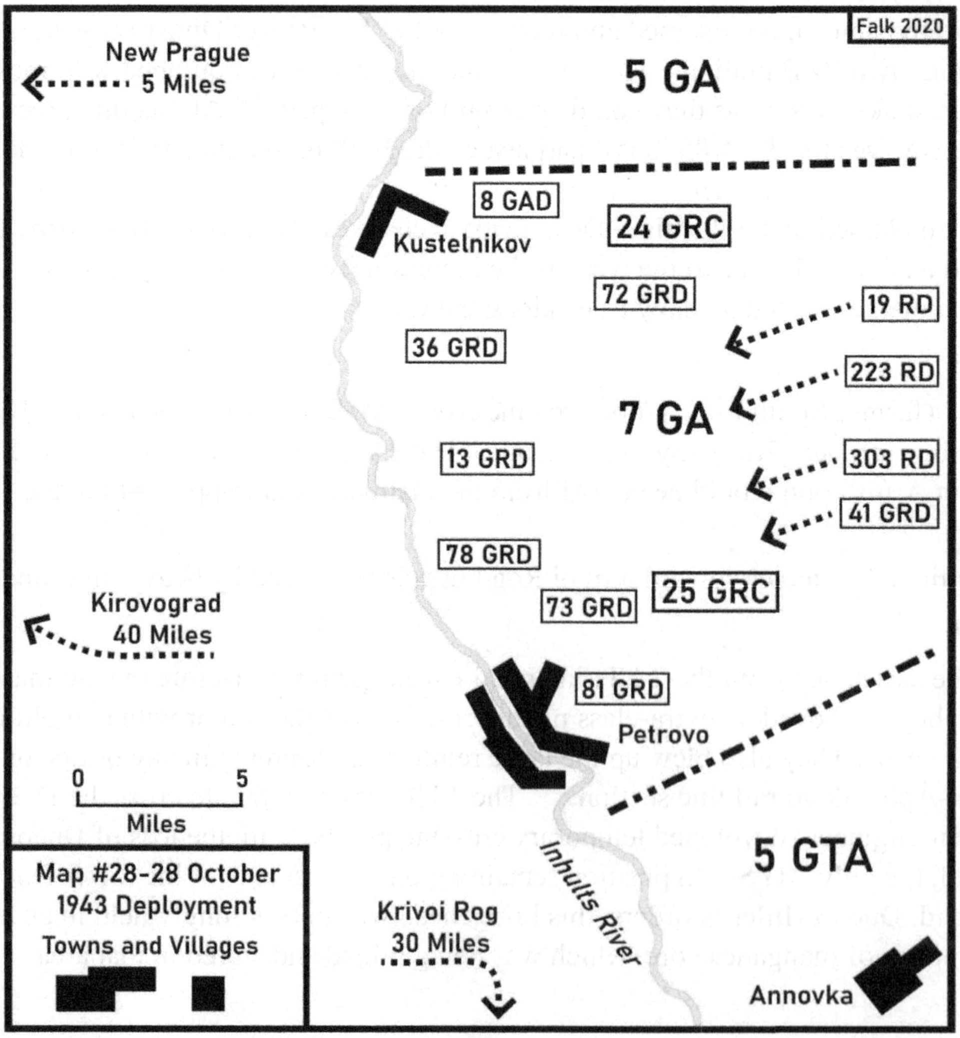

27 October 1943

The7th Guards Army continued to shift its forces to the west.

On the south side of the lower Dnepr River, German forces conducted weak attacks against advancing Red Army troops of the 4 UF. This was a futile attempt to prevent the Red Army from cutting off the Crimea from the mainland.

28 October 1943

Units of the 7th Guards Army began to arrive at their new position on the west side of the Borodaevska bridgehead. The Army was aligned along the Inhults River, starting at the town of Petrovo and extending northward about 17 miles. The 24 GRC was deployed on the right flank and the 25 GRC on the left flank with the 49 RC in reserve. (See Map #28.)

The arrival of the 7th Guards Army was fortunate because units of the 5 GMC, from the 5th GTA, were engaged with German Panzer units on the far side of the Inhults River. Some sources indicate that advanced rifle formations of the 7th Guards Army even then were supporting the 5 GMC on the west side of the river. Enemy forces confronting the 5 GMC were the leading elements of the newly arrived XXXX Pz C (14 Pz D, 24 Pz D, and the 3 SS Pz D *Totenkopf*.) Over the next several days, the 5 GMC gradually retreated to the Inhults River and withdrew behind the now deployed 7th Guards Army to safety.

By 31 October, the 3 SS Pz D *Totenkopf* was positioned along the Inhults, screening the front of the 7th Guards Army. This maneuver allowed the fresh 14 and 24 Pz Ds to slip further south and to attack into the right flank of the 5th GTA, which had almost reached Krivoi Rog. This series of Panzer counterattacks were all part of von Manstein's effort to prevent Krivoi Rog, a vital rail center, from falling to the Red Army.

Further south, other formations of the 5th GTA approached Krivoi Rog from the north and northwest. German resistance was building as von Manstein moved reinforcements into the area to blunt the Soviet attack.

29 October 1943

Between 1-29 October, the 7th Guards Army reported the following casualties:

Killed – 4,212 men and women
Wounded – 20,668
Sick – 1,033
Gone Missing – 3,578
For other reasons – 53
Captured – 3

These numbers totaled – 29, 547 casualties. [358]

Note: The number of soldiers captured would have been far higher than three. Many of the "Gone Missing" were in fact captured by the Germans during battle, and a sizeable number would have gone over to the other side, willingly giving themselves up.

The reasons why so many troops surrendered to the enemy are varied and occurred in both the German and Soviet armies. During most of the war, the average Red Army soldier feared being shot by his or her own officers or *politruk* (political commissar). One soldier stated, "[If] I'm going to desert this place, it's better to flee to the Germans; you can stay alive with them, but if you flee to the rear, you know you'll be shot." [359] At times, soldiers on both sides became tired of the fighting, preferring to take their chances as non-combative POWs. As they saw it, surrender was their only option if they wanted to stay alive. Fear, cold, hunger, and a sense of being on the losing side at times drove soldiers/warriors to become just men, ready to give-up. This is why the Red Army and increasingly the German *Heer* (Army) spent so much time and effort bolstering morale and indoctrinating soldiers with Soviet or Nazi propaganda. Either ideology left little room for personal choice.

The battle for Krivoi Rog was reaching a critical point. The depleted 5th GTA continued to advance toward the city against growing German resistance.

30 October 1943

The 7th Guards Army had a reported combat strength of 43,310 men and women, as it settled into its new position holding a section of the front line, generally facing west toward Kirovograd some 40 miles away. A historical map shows the 25 GRC, 81, 73, 87 and 41 GRDs holding a small bridgehead on the west side of the Inhults River across from Petrovo. This small bridgehead, about 4 miles wide and 2.5 miles deep, must have been seized after major units of the 3 SS Pz D *Totenkopf* moved further to the south. To support the 7th GA, the front commander ordered the 1 MC forward and positioned it south of Petrovo on the east side of the river. The 1 MC would cross the river soon and form its own bridgehead. (See Map #29 for location.)

Having brought up mobile reserves from both the 8th Army and 1st Panzer Army, (XXXXVIII, XXXX Pz Cs, and the supporting LVII Pz C), von Manstein continued the ongoing series of powerful counterattacks against the 2 UF spearhead (5th GTA) that was approaching Krivoi Rog. Krivoi Rog had to be saved because it was a vital supply and rail center for the 1st Panzer Army operating along the lower Dnepr River. With the 11 and 23 Pz Ds leading the way from the south, their spearhead struck to the north of Krivoi Rog against the left flank of the 5th GTA. From the north, the 14 and 24 Pz Ds likewise drove into the exposed right flank of the 5th GTA.

31 October 1943

In October, VVS units of the 5th Air Army flew 10,165 sorties covering the Red Army troops of the 2 UF. A weakened Luftwaffe could not match these totals.

The German counterattacks to drive Russian forces back to the north and away from Krivoi Rog continued without letup.

Further south, forces of the 4 UF captured the town of Chaplinka, which was located 18 miles north of the city of Perekop. Perekop was the key to the Crimea. Not only was it a rail center, but also Perekop occupied, like a cork in a bottle, the narrow land bridge that connected the Crimea to the mainland. The failure to control this vital area meant German and Romanian forces within the Crimea soon would be isolated.

1 November 1943	
7th Guards Army	
Assigned – 2nd Ukrainian Front	
Colonel General M.S. Shumilov Commanding	
Rifle, Airborne and Cavalry Units	24 Guards Rifle Corps (36, 72 GRDs, 8 GAD, 223 RD) 25 Guards Rifle Corps (13, 41, 73, 78, 81 GRDs) 19, 303 RDs 2, 3 Separate AT Rifle Battalions 82, 83, 84, 85, 86 Blocking Detachments 62, 63, 64, 65, 66 Penal Companies
Artillery Units	11 Artillery Division (40 Howitzer Artillery Brigade, 45 Gun Artillery Brigade, 31 Gun Artillery Brigade) 161, 265 Guards Gun Artillery Regiments 114 Guards, 1669 Tank Destroyer Regiments 30 Antitank Artillery Brigade 263, 290 Mortar Regiments 5 Antiaircraft Artillery Division (670, 743, 1119, 1181 AA Regiments) 162 Guards AA Regiment 662 Separate Aerial Observation/Air Raid Warning Company
Armored and Mechanized Units	27 Guards Tank Brigade 121, 262 Tank Brigades 34, 38 Separate Armored Train Battalions
Engineer Units	60 Engineer Sapper Battalion 175, 329 Separate Engineer Battalions 40 Pontoon Bridge Battalion
Estimated Combat Strength Estimated Ration Strength	66,000 men 73,000 men and women (including an estimated 7,000 support personnel)

(BSSA 43, BSA 43, TsAMO Archive)

1 November 1943

Advancing units of the 51st A, 4 UF, captured the city of Perekop, thus isolating the German 17th Army and supporting Romanian units within the Crimea. Hitler refused to withdraw the 17th Army from the Crimea in a timely manner because he was worried the Red Army Air Force would use the Crimea as a base to bomb the vital Ploiesti oil fields and refineries in Romania.

Hitler needed to be worried, as the Soviets were the first to bomb Ploiesti starting on 13 July 1941, losing four of six bombers, but this small force managed to destroy 9,000 tons of crude oil. During the following weeks, the VVS had continued to bomb Romanian oil targets including the port city of Constanta, which suffered more than 50 small raids. These nuisance raids only ended after German and Romanian forces captured most of the Crimea in mid-November 1941. The potential loss of the Crimea once again would put Ploiesti in danger from the East. [360]

At this time, Ploiesti was supplying about 30-50 percent of all Axis fuel and crude oil needs. The oil fields and refineries had recovered from the damage sustained during the 1 August 1943 USAF bombing raid, and Hitler was in no mood to permit a repeat attack, this time from the east by the Long-Range bomber force (ADD) of the Red Army Air Force.

At the eastern end of the Crimea, units of the Russian 56th and 18th Armies, North Caucasus Front, crossed the Kerch Strait and landed on each side of the city of Kerch. The 56th A was to the north, and the 18th A was to the south. Fighting was heavy as these armies formed two new bridgeheads.

A bit further north of the Crimea, von Manstein continued a series of counterattacks against Russian forces of the 2 UF that were advancing against Krivoi Rog. These attacks successfully halted and then began to push back the advance of the Red Army in this area.

2 November 1943

The 7th Guards Army had a reported combat strength of 66,052 men and women. The 7th Guards Army held its front with little or no activity as the 2 UF prepared for further operations. The fresh German 384 ID was deployed now along the west side of the Inhults River facing the 7th Guards Army.

The German 4th Air Fleet had a reported strength of 610 aircraft, [361] which implied that only about 415 of them would have been operational. That number of air-worthy aircraft was far too few to cover both AG South and AG A effectively.

The attacking northern and southern German Panzer spearheads met north of Krivoi Rog. The Germans had closed the gap between the 8th Army to the north and the 1st Panzer Army to the south. Krivoi Rog was saved from capture by the advancing Soviet units.

3 November 1943

The 7th Guards Army had a reported combat strength of 67,437 men and women.

The 37th A now was fighting on the 7th Guards Army's left flank, and the 5th GA was on the right flank. General Konev shifted armies around as needed to respond to the threat of the German attack from the south and west.

Interestingly, on or about this date, the 7th Guards Army, with the help of the 1 MC, continued to hold a small and slightly expanded bridgehead on the western side of the Inhults River at the town of Petrovo. To achieve this, they pushed back parts of the German 384 ID, the 3 SS Pz D and the 11 Pz D. A historical TsAMO Archive map, ID #100250135, shows the 7th Guards Army, with the 81, 73, 78 and 41 GRDs, firmly holding a bridgehead about 4 miles wide and 2.5 miles deep. The 19 and 303 RDs were in reserve on the east side of the river. On the left flank of this bridgehead, the 1 MK had crossed the river, and was holding its own adjacent and connected bridgehead, which was about 5 miles wide and 3.5 miles deep, with the 5 GMC in reserve. (See Map #29.). Note: These two MCs had been inserted into the front line between the 7th Guards Army to the north and the 37th A to the south. Overall, powerful German combat formations had been diverted to the north and south to deal with equally powerful attacks by the 2 UF. This diversion weakened the German center, thus allowing the 7th Guards Army to seize and to hold this small bridgehead over the Inhults River.

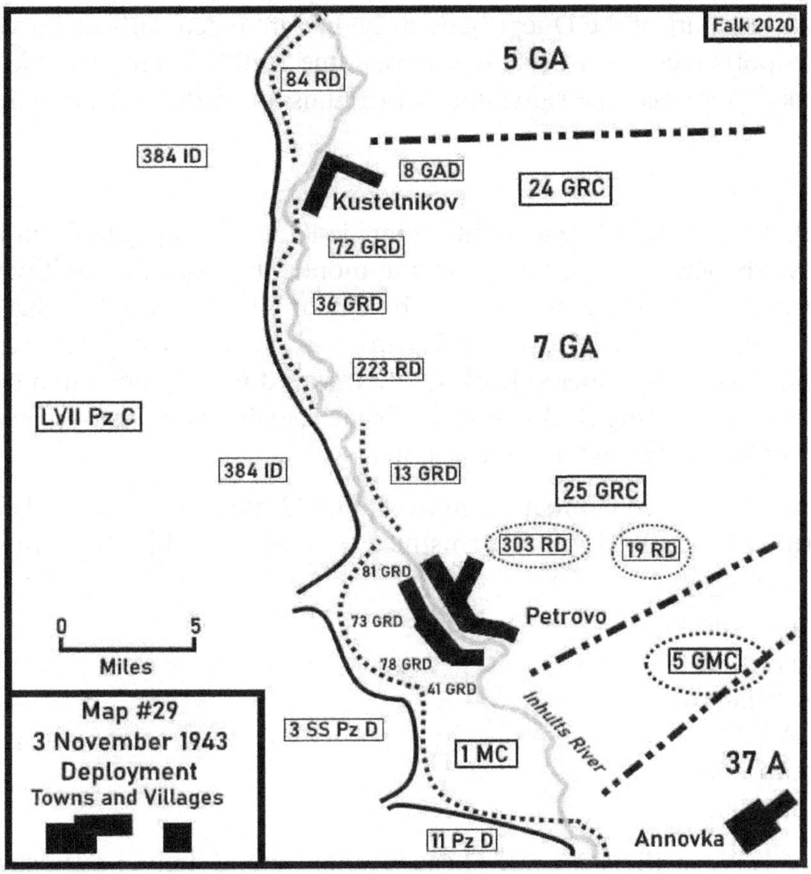

Discussing the overall strategic situation, Hitler stated in his Directive #51:

> …but a greater danger now appears in the west: an Anglo-American landing! In the east, the vast extent of the territory makes it possible for us to lose ground, even on a large scale, without a fatal blow being dealt to the nervous system of Germany. It is very different in the west! Should the enemy succeed in breaching our defenses on a wide front here, the immediate consequences would be unpredictable. Everything indicates that the enemy will launch an offensive against the western front of Europe at the latest in the spring, perhaps even earlier. [362]

Obviously, the successful Allied invasion of Italy weighed heavily on Hitler's mind. But what Hitler feared most was the very real threat of a massive Allied invasion of northern France. He felt this invasion must somehow be prevented or repelled at all cost. As this threat from the west grew, all other areas of the German war effort suffered. Gradually, growing numbers of troops, equipment and supplies were diverted to France to counter this danger. The negative impact of that choice would increasingly be felt on Germany's Eastern Front.

4 November 1943

With eleven rifle divisions assigned, the 7th Guards Army had a reported combat strength of 73,089 men and women. Newly assigned units and reinforcements continued to swell the ranks of the Army.

The German counterattack to defend Krivoi Rog ended with complete success, driving Red Army forces 19 miles to the north and away from the city and its critical rail line. Von Manstein commented on this operation:

> Thanks to the exemplary manner in which the two armies cooperated [8th A and 1st Pz A], the counterblow delivered at the end of October north of Krivoi Rog – which already had the enemy at its gates – ended in a neat success. The enemy, instead of cutting off First Panzer

Army in the eastern part of the Dnepr bend as he had intended, suffered a severe setback. According to reports received from the armies, some 10,000 enemy were killed, apart from which 350 tanks, more than the same number of guns and 5,000 prisoners fell into our hands. [363]

5 November 43

To the north, the 38th Army of the 1 UF was fighting inside Kiev, the capital of Ukraine. The capture of Kiev would not only give the Red Army a huge boost in morale, but logistically it would also make accessible a major road and rail crossing points over the Dnepr River. From the Black Sea in the south to Smolensk in the north, all along the line, the Red Army was making steady progress to the west along a 1,200-mile front. No one on either side of that line would fail to notice that one critical position after another was lost to the advancing Red Army. Their inescapable conclusion: Germany was losing the war. The Red Army could be slowed but not stopped.

The 7th Guards Army had a reported combat strength of 75,322 men and women. The Army also issued a report, TsAMO record ID #13130891, on firearms/heavy weapons within the Army. The total artillery listed for this day was:

103 – 152 mm Artillery
137 – 122 mm artillery
316 – 76 mm artillery/AT
230 – 45 mm AT
183 – 120 mm mortars
431 – 82 mm mortars

This list shows a total of 1,400 artillery pieces. These weapons could deliver a substantial amount of firepower that General Shumilov could bring down upon the enemy at will. Of this total, the 11 Artillery Division (11 AD), on its own, supplied 35-152 mm, 39-122 mm and 21-76 mm artillery pieces, for a total of 95 weapons.

A few days after von Manstein defeated the 2 UF's first attack against Krivoi Rog, *Stavka* issued Directive #30238 ordering General Konev to postpone his planned offensive against Kirovograd and to once again attack south toward Krivoi Rog with three armies, the 37th, 57th and the 5th GTA. Konev had orders to envelop Krivoi Rog from the west while the 3 UF attacked from the east. This new offensive was set to begin sometime between 12-14 November; that allowed only seven to nine days for planning and regrouping of forces.

After further discussions with *Stavka*, General Konev decided to include the 7th Guards Army as flank support for the main attack. From its position around Petrovo, the 7th Guards Army needed to break through the German line directly to its front, then turn its attack 90° to the left, before advancing due south at least 30 miles in an attempt to protect the right flank of the main attack. This bold maneuver, by an infantry army, would in turn expose the 7th Guards Army's extended right flank to attacks from the substantial German forces assembling around Kirovograd, the 1 Pz D, 7 Pz D (just arriving by rail) and the 1 SS Pz D *Leibstandarte* Adolf Hitler. At first glance, this appears to be an overwhelming task for the 7th Guards Army to accomplish with only the 1 MC in support. But General Konev might have planned, from the start, to use the 7th Guards Army's attack as a diversion, drawing in German reserves from the Kirovograd area, involving them in a running battle with a veteran guard's infantry army, and thus preventing them from interfering with the main attack against Krivoi Rog. General Konev's writings are silent in this regard.

In the Crimea, the attacking 56th A was able to enlarge its bridgehead toward Kerch, but the small bridgehead of the 18th A, south of Kerch, was isolated and ultimately destroyed on 11 December.

6 November 1943

The 7th Guards Army had a reported combat strength of 76,072 men and women.

The Soviet 38th A completed the capture of Kiev, driving out all defending German forces. The armies of the 1 UF continued the expansion of their bridgehead on the west side of the Dnepr.

7 November 1943

The 7th Guards Army had a reported combat strength of 76,313 men and women.

Von Manstein flew to meet with Hitler at his HQ, (location unknown but most likely at *Wolfsschanze,* Wolf's Lair, in east Prussia) to discuss the situation of AG South. During this meeting Hitler decided nothing about the growing Soviet attacks in the Kiev area, the status of the Crimea, fighting in the Dnepr River bend, or the Nikopol bridgehead on the east side of the Dnepr. Without clear guidance, von Manstein was unable to use effectively the forces available to him.

8 November 1943

The 7th Guards Army had a reported combat strength of 77,531 men and women as they rested and replenished combat formations for the next battle.

Facing the 7th Guards Army bridgehead, the German 76 ID replaced the 3 SS Pz D and the 11 Pz D, which allowed these two Panzer Divisions to move southeast, further protecting Krivoi Rog.

9 November 1943

The 7th Guards Army had a reported combat strength of 78,962 men and women. The 49 RC was displayed on a planning map as wedged between the 25 GRC to the south and the 24 GRC to the north. This planning map, TsAMO record ID #100232766, also displayed the Army's attack plan to the west of Petrovo.

10 November 1943

The 7th Guards Army had a reported combat strength of 65,699 men and women. With the removal of the 13 GRD, which most likely occurred late on 9 November, the combat strength of the Army had fallen noticeably. Of note, on or about this date, the 41 GRD had a reported combat strength of only 2,845 men, indicating the depleted nature of most of the units assigned to the Army.

The 49 RC once again was assigned to the 7th Guards Army. The rifle corps almost certainly was in reserve, refitting and training reinforcements for the Army. The corps assignments were as follows:

24 GRC – 36, 72 GRDs, 8 GAD
49 RC – 81 GRD, 223 RD
25 GRC – 41, 73, 78 GRDs
19, 303 RDs - in reserve

Additionally, it appears the 1 MC was assigned directly to the 7th Guards Army to act in a supporting role for the upcoming offensive. The 1 MC contained the 19, 35 and 37 Mechanized Brigades (MB) and the 219 Tank Brigade (TB).

11 November 1943

The 7th Guards Army had a reported combat strength of 66,442 men and women.

General Konev continued planning for the new offensive to capture Krivoi Rog, and the 7th Guards Army was ready to play its part.

In the Crimea, once again the attacking 56th A enlarged its bridgehead and reached the outskirts of Kerch. The limited German forces in the area were unable to prevent this advance.

12 November 1943

The 7th Guards Army had a reported combat strength of 66,445 men and women. The Luftwaffe conducted a minor attack against the Army, totaling 14 sorties, dropping 36 high explosive (HE) and 3 incendiaries bombs. The results of this attack are unrecorded in known historical archives.

13 November 1943

The 7th Guards Army had a reported combat strength of 66,391 men and women. Thus, the German air attack from the previous day seemed to have had little impact on the Army's personnel strength, which would be expected from the deployment of only 14 sorties and the use of only 39 bombs.

During the early morning hours, the 52nd A on the far-right flank of the 2 UF attacked across the Dnepr River without artillery preparation to insure surprise. The Army's objectives were twofold: divert the Germans' attention to that area and capture the city of Cherkassy. Logistical provisioning for the offensive amounted to 2.5 combat loads of ammunition and 8-10 days' worth of rations. The Army expected no further logistical support as the 2 UF diverted all further supplies to the upcoming Krivoi Rog offensive further south. This diversion of resources also included the support of the VVS, which would be reserved for the 2 UF's main attack. The 52nd A did have the support of eight hospitals with a total of 4,300 beds/cots, which was adequate. [364]

On this day, at 12:00 hours, a half a day earlier than the fixed deadline, a Railroad (construction) Brigade completed a short-term floating RR bridge over the Dnepr at Dnepropetrovsk.[365] This railroad bridge would ease the logistical support effort for all the armies operating on the right side of the Dnepr River. The floating RR bridge was used for 28 days until 11 December 1943 when ice tore off the floating bridge from the anchors, taking it downstream to a temporary road bridge. At that time, a train of 88 wagons with a total weight of 1,850 tons was located on the floating bridge. The fate of the crew and the supply train is unknown, but most likely the Red Army would have conducted a rescue effort.

14 November 1943

The 7th Guards Army had a reported combat strength of 67,500 men and women.

General Konev once again turned the armies of the 2 UF southward, including the 7th Guards Army, to attack toward Krivoi Rog from the northwest and the west. As before, the persistent mud slowed the advance, and once again, defending German forces delayed the attack long enough so that eventually the 2 UF called it off on 21 November.

The Luftwaffe conducted a major attack against the Army totaling 64 sorties, dropping 376 HE bombs. This was clearly in response to the new offensive of the 2 UF. Once again, the effect of this aerial bombardment upon the Army was not recorded.

15 November 1943

The 7th Guards Army had a reported combat strength of 67,741 men and women. Once more, the German air attack from the previous day seemed to have had little impact on the Army's personnel strength despite 64 sorties and the dropping of 376 HE bombs. Continuing with their air effort, the Luftwaffe paid another social call, conducting a minor attack against the Army, totaling 11 sorties and dropping 60 HE bombs. Again, the results of these gifts from above are unknown.

The 7th Guards Army advanced to the west and enlarged its bridgehead in heavy fighting. Supporting units of the 57th A, on the left flank, also advanced to the south. In response, the Germans returned the 11 Pz D and part of the 3 SS Pz D to the area to block any further loss of ground. The 7th Guards Army was threatening to break through the German line and advance upon Kirovograd from the south or Krivoi Rog from the north.

In the Crimea area, *Stavka* ordered the North Caucasus Front and the 56th A to be reformed into the Independent Coastal Army; this was the second formation of the Independent Coastal Army.

16 November 1943

The 7th Guards Army had a reported combat strength of 65,783 men and women. With the loss of 1,958 troops from the previous day as an indication, heavy fighting continued. The German line held firm, making further progress costly.

17 November 1943

The 7th Guards Army had a reported combat strength of 65,232 men and women. A further loss of 551 troops from the previous day suggested combat activity had started to wane.

18 November 1943

The 7th Guards Army continued combat operations, but they expected no immediate breakthrough. The Germans withdrew the 11 Pz D from the line, replacing it with the units of the 76 ID.

During the night, German units launched a successful night attack against units of the 25 GRC and suffered no losses. This attack was a raid or a reconnaissance in force mission.

19 November 1943

The 7th Guards Army had a reported combat strength of 64,170 men and women. The loss of only 1,062 men and women over two days indicates that the intensity of combat had diminished. As a further indication of shifting priorities, the 11 Pz D moved to the north in response to attacks by the 5th GA near the Dnepr River.

20 November 1943

Minor shifting of the front line around the tiny Hill 194, on the right flank of the Army, was the big gain during the day. The rest of the front line remained static. The 7th Guards Army clearly had failed to break the German line.

21 November 1943

This second 2 UF Krivoi Rog offensive once again failed to capture the city or even to make significant progress anywhere. Lacking any real hope of reaching Krivoi Rog, General Konev suspended the offensive. General Konev then regrouped the 5th GTA and transferred it to the north to support a northern attack against Kirovograd. This attack would take place from 22 November to 9 December.

Once again, at 00:15 hours, the Germans launched a night attack against the 7th Guards Army, with 197 troops and 20 armored personnel carriers. They successfully captured a local hill (height and position unknown). During this action they suffered only one wounded. The failure of his troops to effectively counter these night actions bothered General Shumilov and he would, in time, take action to correct these tactical and operational deficiencies.

22 November 1943

German forces north of Krivoi Rog made minor adjustments to their front line. While in front of the 7th Guards Army, but behind the front line, the German 14 Pz D moved into position.

In far off North Africa, the Cairo Conference (codenamed Sextant, 22-26 November) took place in Cairo, Egypt. American President Franklin Roosevelt, British Prime Minister Winston Churchill, and Generalissimo Chiang Kai-shek of the Republic of China attended. During this conference, the Allied leaders outlined their future positions concerning the war with Japan.

23 November 1943

The 7th Guards Army had a reported combat strength of 61,698 men and women. This represented a loss over the previous four days of 2,472 people, or 618 men and women per day. Not all of these losses were combat related as winter weather, mud, rain and short rations would account for numerous people becoming ill.

The 57th A now was deployed on the left flank of the 7th Guards Army, and the 5th GA held the line to the right.

General Konev, in his book, *Front Commander's Notes*, Militera.lib.ru, 1972, Chapter 2, stated:

> On November 23, I reported to the Supreme High Commander-in-Chief about the battles beyond the Dnieper [Dnepr River], about the large strategic bridgehead that had been won, that the troops were fighting well, and had high morale. But, being continuously in the battles for about four months, the fighters were physically tired; the troops needed rest and replenishment. I asked for permission to temporarily switch to defense at the occupied line. J V Stalin expressed complete satisfaction with the actions of the front troops and agreed with my proposal.

While half of the armies of 2 UF were allowed to rest and to resupply, specifically the 7th Guards, 57th and 37th Armies, Konev informed Stalin the 53rd A, 5th GTA and 5th GA would continue their attack in the north along the Dnepr River and in the area northeast of Kremenchug. This was an attempt to clear the entire right bank of the Dnepr River of the enemy and to outflank Kirovograd from the north.

24 November 1943

The 7th Guards Army had a reported combat strength of 61,211 men and women. The small reduction in combat strength from the day before implied a reduced level of combat that took place along the front.

25 November 1943

The 7th Guards Army had a reported combat strength of 61,132 men and women. Hardly any change in combat strength from the previous day indicated the front was quiet. The Germans were content to maintain a stable front in this area.

Along the central Dnepr, the 53rd A cleared the right bank of the river across from Kremenchug. The advance to the northwest continued at a slow but steady rate.

Further south along the lower Dnepr River, the Red Army now occupied the entire left bank of the river all the way to the mouth of the Dnepr. The only exception was the German Kremenchug bridgehead, which tenaciously held out against repeated attacks.

26 November 1943

The 7th Guards Army had a reported combat strength of 60,669 men and women. Once again, the 7th Guards Army became active with attacks along its right flank. They gained a trivial amount of ground. Minor attacks like this one were routine. The front-line zone was never a safe place to be. One side or another was always trying to gain an advantage by capturing a new position or seizing an opportunity to inflict losses upon the enemy.

Stalin left the Soviet Union for the Tehran Conference (codenamed Eureka – 28 November to 1 December). Stalin and his entourage flew from Baku in southern Russia to Tehran, Iran, in four Russian PS-84/Il-2 transports. These were license-built, Soviet versions of the American Douglas DC-3/C-47aircraft. Twenty-seven VVS fighters escorted Stalin's aircraft. President Franklin Roosevelt and Prime Minister Winston Churchill joined Stalin in Tehran, having arrived from Cairo. This was the first "Big Three" conference among Stalin, Roosevelt and Churchill. The leaders discussed the Allies' future positions concerning the war in Europe, the second front, Japan, and post-war relations. Meanwhile, all around the world, the war continued in full force.

27 November 1943

The 7th Guards Army had a reported combat strength of 59,958 men and women. The Army continued its right flank attack against units of the German 384 ID, capturing more ground.

The 7th Guards Army had the 5th GA on its right flank to the north and the 57th A on its left flank to the south. Minor combat took place within the bridgehead over the Inhults River.

28 November 1943

The 7th Guards Army had an overall reported combat strength of 59,121 men and women. A listing of the current combat strength of major rifle formations, men in the front line, illustrates the weakened condition of the infantry units within the Army:

8 GAD – 4,657 men and women
36 GRD – 3,602
41 GRD – 2,881
72 GRD – 3,787
73 GRD – 4,274
78 GRD – 3,323
81 GRD – 3,040
213 RD – 3,669
303 RD – 2,906

Total 32,139 men and women. [366]

Note: At this stage of the war, a full-strength rifle division would have contained about 8,500 men and women.

Despite this weakness, on this day, the Army's right flank slightly pushed back the German line, suffering heavy casualties for little gain.

29 November 1943

The 7th Guards Army had a reported combat strength of 48,607 men and women. Most of this reduction in strength from the previous day was due to the shifting of combat units out of the Army rather than from combat losses. The transferred troops included units like the 1 MC, which had a reduced strength of approximately 7,000-10,000 people, along with support units like the 40 Pontoon Bridge Battalion. Losses from the previous day's attack would have been in the range of 300-500 people. The 7th Guards Army captured little or no ground this day.

The Luftwaffe conducted a minor attack against the Army, totaling 16 sorties and dropping 65 HE bombs. The result of this attack is unknown.

During a special ceremony at the Soviet embassy in Tehran, Winston Churchill presented Joseph Stalin with the Sword of Stalingrad. This sword was a bejeweled ceremonial long sword created to honor the defenders of Stalingrad. During the presentation, Churchill declared, "I am commanded [by King George VI] to present this sword of honor as a token of homage of the British people." Kissing the sword, Stalin accepted it on the behalf of the defenders of Stalingrad and thanked the British people. After the ceremony, Stalin handed the sword to Marshal Kliment Voroshilov who promptly allowed the sword to slip out of it scabbard, falling noisily to the floor. It appears Marshal Voroshilov suffered no punishment for this slip of the sword.

30 November 1943

As can be seen on Map #30, the strategic situation had changed significantly from the middle of October. While suffering substantial losses both in troops and equipment, the Red Army, 2 UF, and the 7th Guards Army made major advances to the west, but they failed to achieve the distant goals set by Stalin and *Stavka*.

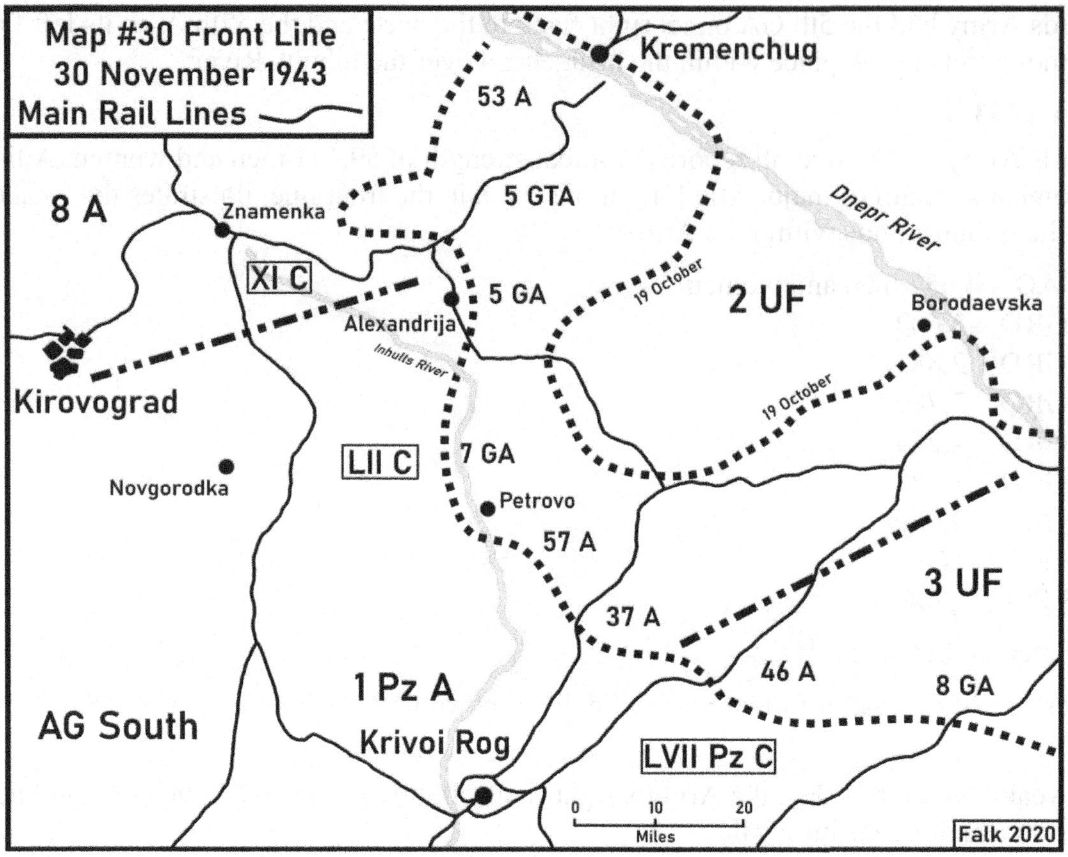

The 7th Guards Army had a reported combat strength of 48,560 men and women. The Army's troops were tired, and for now General Shumilov was satisfied to just hold its slightly enlarged bridgehead over the Inhults River. The 223 RD rejoined the 49 RC as the 41 GRD withdrew to the rear for refitting.

An incomplete losses report from TsAMO record ID #131308926, covering only 10-28 November (19 days) shows the Army suffered 10,127 casualties. If we use a conservative estimate of 250 casualties per day for the remaining 11 days of the month, we can add a further 2,750 casualties, which equals a more realistic grand total of 12,877.

The Luftwaffe once again conducted a minor attack against the Army, totaling 12 sorties and dropping 17 HE bombs with unrecorded results.

Facing the realities of the failing strength of the 2 UF, General Konev issued special Order #917 to the 37th and 57th Armies to shift to a "strong defense." [367] In other words, the entire left flank of the 2 UF now assumed a defensive stance. This implied building defensive positions, laying minefields, and generally getting ready to defend their current positions without taking any further offensive actions.

Despite the Luftwaffe's failing strength in the east, combat formations continued to make an effort to support the ground forces. For example, a historical TsAMO record ID #454130105 "*Report on the actions of aviation in the 7th Guards Army Area,*" presents an interesting picture. During all of November, Luftwaffe aircraft flew over the 7th Guards Army area on only 19 days, totaling 313 aircraft over flights. During these same 19 days, the Luftwaffe only conducted 141 combat sorties directly against the Army, dropping 638 high explosive (HE) bombs and 9 incendiaries. In total, one major Luftwaffe raid, of 64 sorties, was launched on 14 November, with only four minor raids (more than 10 sorties) and eight harassing raids (less than 10 sorties) delivered during the rest of the month. The results of these attacks upon the 7th Guards Army are unrecorded, but the lack of noteworthy damage reports suggests the overall effect was minimal.

Note: The men and women of the 662 Separate Aerial Observation/Air Warning Company were most likely the troops who produced these historical reports of German aircraft activity over the Army in November.

The end of the month saw the 2 UF and the 7th Guards Army achieve a genuine victory by greatly expanding their bridgehead and liberating a huge section of the Ukraine on the west side of the Dnepr River. This effort soundly defeated the German attempt to hold and to fortify the Dnepr River line.

In turn, the Germans won a defensive victory by twice thwarting the Red Army's efforts to capture the city of Krivoi Rog and to cut off the 1st Panzer Army in the great bend of the Dnepr. German troops of AG South also prevented the Red Army from expanding their bridgehead further to the west and capturing Kirovograd. In the end, both sides had unfinished business at the start of a new month.

1 December 1943	
7th Guards Army	
Assigned – 2 Ukrainian Front	
Colonel General M.S. Shumilov Commanding	
Rifle, Airborne and Cavalry Units	24 Guards Rifle Corps (36, 72, 81 GRDs, 8 GAD) 25 Guards Rifle Corps (41, 73, 78 GRDs) 49 Rifle Corps (19, 223, 303 RDs) 2, 3 Separate AT Rifle Battalions 82, 83, 84, 85, 86 Blocking Detachments 62, 63, 64, 65, 66 Penal Companies
Artillery Units	11 Artillery Division (40 Howitzer Artillery Brigade, 45 Gun Artillery Brigade, 31 Gun Artillery Brigade) 161 Guards Gun Artillery Regiment 114, 115 Guards, 1317, 1669 Tank Destroyer Regiments 30, 33 Antitank Artillery Brigades 263, 290 Mortar Regiments 97, 309 Guards Mortar Regiments 5 Antiaircraft Artillery Division (670, 743, 1119, 1181 AA Regiments) 162 Guards AA Regiment
Armored and Mechanized Units	27 Guards Tank Brigade 34, 38 Separate Armored Train Battalions
Engineer Units	60 Engineer Sapper Battalion 175, 329 Separate Engineer Battalions
Reported Combat Strength Estimated Ration Strength	48,558 men 58,298 men and women (including an estimated 7,000 support personnel)

(BSSA 43, BSA 43, TsAMO Archive)

1 December 1943

The 7th Guards Army had a reported combat strength of 48,558 men and women. The start of the month saw the Army holding its expanded bridgehead over the Inhults River and facing the German 76 and 384 IDs, with the 3 SS and 23 Pz Ds nearby. The front was quiet.

2 December 1943

The 7th Guards Army had a reported combat strength of 49,739 men and women.

The front continued to be quiet, but General Konev was about to launch the 7th Guards Army, along with the supporting 1 MC, on a new offensive.

3 December 1943

The 7th Guards Army had a reported combat strength of 49,444 men and women.

The latest 2 UF offensive began early in the morning and was aimed in the general direction of

Kirovograd. The initial goal of the attack was to capture the city of Alexandrija, which was split by the front line. (See Map #30 for location.) Progress was slow at first with rain and wet snow turning firm soil into mud, which caused a delay of the main 7th Guards Army attack by 24-48 hours.

4 December 1943

The 7th Guards Army had a reported combat strength of 49,745 men and women.

The 7th Guards Army's minor opening effort started with attacks from the Army's right flank to the northwest, in support of the neighboring 5th GA. The 53rd A and the 5th GTA continued their own attacks further north.

5 December 1943

The 7th Guards Army had a reported combat strength of 49,853 men and women. Fighting was renewed and expanded along the Army's right flak with a major attack supported by the 1 MC. The entire 2 UF front north of the 7th Guards Army, all the way to the Dnepr River, was conducting offensive operations.

6 December 1943

The 7th Guards Army had a reported combat strength of 49,578 men and women.

Units of the 5th GA, 7th Guards Army and 1 MC pushed back the German 376 and 384 IDs, gaining some ground. The 5th GA captured Alexandrija after two days of heavy fighting, with the Germans withdrawing to the west after the battle.

7 December 1943

The 7th Guards Army had a reported combat strength of 49,283 men and women. Due to the slow start of the fighting for the 7th Guards Army, the Army suffered minimal casualties during these initial battles.

Elements of the 3 SS Pz D launched strong counterattacks against the right flank of the 7th Guards Army. The 5th GA also suffered major counterattacks. Heavy fighting delayed the advance of both armies.

From north to south, the 53rd A, 5th GTA, 5th GA and 7th Guards Army pushed back the German line further to the west as weakened German units began to fall back toward Kirovograd.

Due to a perceived weakness in the overall German situation of AG South, *Stavka*, sensing an opportunity, ordered General Konev to continue his attack toward Kirovograd while simultaneously launching a new offensive against Krivoi Rog, all without proper planning or logistical support. *Stavka* issued Directive #30258 on 8 December, committing General Konev and the armies of the 2 UF to attack to the west while also advancing south to Krivoi Rog. [368] What would persuade *Stavka* to think these two diverging attacks were possible or even prudent under the prevailing circumstances? Such an effort was clearly beyond sound operational doctrine and the forces at hand. Once again, this was a case of *Stavka* attempting to achieve too much too quickly with too few troops. *Stavka* could have succeeded in capturing either objective, but Stalin wanted both at the same time, and that was not to be.

8 December 1943

The 7th Guards Army had a reported combat strength of 58,837 men and women. This sudden increase was most likely due to General Konev assigning the 7 Mechanized Corps (MC) to the Army. This Corps contained the 16, 63, 64 Mechanized Brigades (MB), the 41 Guards Tank Brigade (GTB), the 1449 SU Regiment (SUR) with Su-76 assault guns, and the 1821 SU Regiment with Su-85 assault guns. The 7 MC was a powerful mobile force that greatly added to the 7th Guards Army's offensive strength.

At this time, the exact location of the 7 MC was unknown, but it most likely was moving up from the rear.

9 December 1943

The 7th Guards Army had a reported combat strength of 58,324 men and women. At this time the 7th Guards Army was about 30 miles from Kirovograd with adjacent Red Army units even closer.

After seven days of continuous combat, with minimal gains and heavy fighting, General Konev once again launched the armies of the 2 UF toward Kirovograd. The battle continued.

Meanwhile, *Stavka* issued Directive #30262 to clarifying its intent to Konev. This directive reinforced the 2 UFs left flank with additional units, taken from the 3 UF. Konev's other armies were ordered to capture Znamenka, northeast of Kirovograd, and to firmly hold the Inhults River line, thus allowing the 7th Guards, 57th and 37th Armies freedom of action to advance upon Krivoi Rog. [369] General Konev was ordered to submit his operational plan to *Stavka* by the next day, 10 December, which was an absurdly short deadline!

10 December 1943

The 7th Guards Army had a reported combat strength of 64,443 men and women.

The right flank armies of the 2 UF continue to push German units back toward Kirovograd, but progress was slowed by determined German defensive efforts and the changeable weather conditions.

On the left flank of the 2 UF, the reinforcing troops from the 3 UF would not be in position until 17 December so apparently, around this date, the 57th and 37th As began their offensive south toward Krivoi Rog on their own. The 7th Guards Army, in turn, was fully occupied trying to support the advance along its front and therefore could not disengage and turn to the south.

The 662 Separate Aerial Observation/Air Raid Warning Company once again was active within the Army. The observation company had disappeared from the Army's rolls for several days for unknown reasons.

The GAU or Main Artillery Directorate, which provided technical support for various branch services like artillery, armor, engineer, and signal [370] became involved in railroad support of the 2 UF. In this case, the GAU organized the repair of a vital rail line and bridge over the Dnepr for the 2 UF. At the end of September 1943, Red Army troops had liberated Kremenchug and crushed the enemy bridgehead over the Dnepr River. Unfortunately, the Germans retained the right bank, west side, of the river, which prevented any bridging work. When retreating beyond the Dnepr, the Germans blew up 6 out of 11 spans of the railway bridge and destroyed other bridges also. On 25 November, the 53rd A finally had cleared the Germans from this section of the right bank of the Dnepr; this victory allowed the restoration of this vital railway bridge to begin on 10 December. After 33 days of engineering work, the rail bridge would be completed on 12 January 1944. [371] With the restoration of this vital rail line and rail bridge, the 2 UF had significantly more logistical support on the west side of the Dnepr.

11 December 1943

The 7th Guards Army had a reported combat strength of 64,006 men and women.

While units of the 7th Guards Army were supporting the advance of the 1 and 7 MCs, the 303 RD was withdrawn into reserve (second echelon), most likely for refitting.

12 December 1943

The 7th Guards Army had a reported combat strength of 63,744 men and women.

The 1 and 7 MCs, with support from the 7th Guards Army and the 35 GRC from the 5th GA, continued to advance with heavy fighting. The German front line was stabilizing as new Panzer units were brought into the area. Of note, during this offensive, the 35 GRC, which contained the 93, 94 GRDs and the 84 RD, might have been assigned temporally to the 7th Guards Army. Records are unclear on this point.

13 December 1943

The 7th Guards Army had a reported combat strength of 63,151 men and women.

Units of the 1 and 7 MCs penetrated into a gap in the German line, southeast of Kirovograd, trying to force a breakthrough in this area. This gap in the German line was centered upon the village of Novgorodka, which would be the focus of heavy fighting for the next nine days.

14 December 1943

The 7th Guards Army had a reported combat strength of 62,600 men and women.

German Panzer divisions counterattacked to the east of Kirovograd, pushing the forward units of the Red Army back and threatening to surround units of the 1 MC, 7 MC and other Red Army formations. The Germans almost had restored a continuous front line to the east of Kirovograd. The German 11 Pz D and 2 *Fallschirmjäger* (Parachute) Division (2 FJD) attacked from the north with the 13 and 17 Pz Ds advancing from the south. They aimed to recapture the tactically important city of Novgorodka, and to surround various Red Army units in the area. (See Map #31.)

Further north along the Dnepr, the 52nd A finally completed the capture/liberation of the city of Cherkassy, a commanding German strongpoint on the Dnepr.

The attacks by the unsupported 57th and 37th As made little or no progress toward Krivoi Rog. Historical German situation maps show no change in their position north of Krivoi Rog.

15 December 1943

The 7th Guards Army had a reported combat strength of 62,088 men and women with 11 rifle divisions on the front line and with the 7 MC in support. The 303 RD was in reserve.

The 7 MC was fighting far forward of the right flank of the Army in the sector of the 24 GRC with the 53 and 409 RDs in support. The 1 MC and 35 GRC were acting as right flank support (not shown on any map). The exact placement or makeup of all the units engaged in this battle is difficult to determine due to the confused nature of the fighting. German Panzer counterattacks threatened to totally cut off many Red Army formations of this advanced force.

16 December 1943

The 7th Guards Army had a reported combat strength of 61,664 men and women.

German divisions almost fully surrounded Red Army units fighting around the city of Novgorodka. Meanwhile, the 7th Guards Army attempted to advance and to clear a path to these surrounded units.

17 December 1943

Heavy fighting continued as the 7th Guards Army attempted to free the semi-encircled units. This was a very confusing battlefield as formations from the 5th GA, 35 GRC, 1 MC, 7 MC and the 7th Guards Army all were fighting in a restricted area. Red Army units continued to hold the village of Novgorodka.

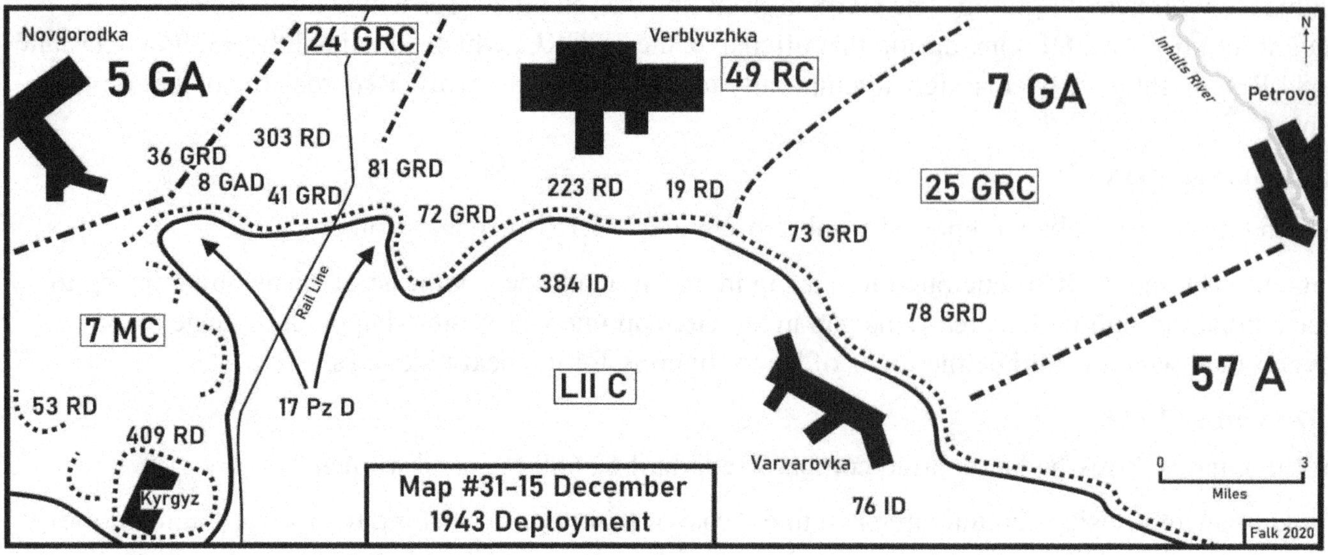

18 December 1943

Due to the heavy fighting, the 7th Guards Army reported a combat strength of 50,544 men and women. This drop in strength of over 11,000 troops over the previous two days of fighting was due most likely to one or more major units, like one of the mechanized corps, being transferred out of the Army. The 78 GRD also was transferred to the 57th A on this day. Within the Red Army it was normal for front commanders to shift different units and formation in or out of an army on the front line. Even large formations like an entire rifle, mechanized or tank corps would be moved around as needed. Some of these formations were withdrawn to the rear for rest or refitting, while others were moved to reinforce a major attack or to defend against an enemy advance.

Units from the German 2 FJD and supporting assault guns captured two key hills, 170.3 and 163.7, which dominated the area, located just south of Novgorodka (not shown on Map #31).

19 December1943

The fighting continued as the 7th Guards Army reported a combat strength of 42,148 men and women, indicating a drop of over 8,000 troops in one day. Once again, this almost certainly indicated the transfer of one or more major units out of the Army. In any event, on this day, the 1 and 2 MCs no longer were assigned to the Army. The 7th Guards Army could not sustain combat losses or transfers of this magnitude and remain a viable offensive fighting force. On its own, the Army lacked the power to launch and to sustain a major offensive operation.

General Konev called off the offensive. This latest offensive by the 2 UF did capture the cities of Alexandrija and Znamenka (a large railway junction) and pushed back the German front line, but it failed to secure Kirovograd and utterly failed to make any progress toward Krivoi Rog. The territory the Russians had gained did not justify the heavy losses of both troops and equipment to the units that had been partially surrounded.

20 December 1943

The 7th Guards Army had a reported combat strength of 42,464 men and women. On or about this date, the 303 RD returned to active duty with the 25 GRC.

Once again, General Konev asked *Stavka* for permission to temporarily go on the defensive to rest and to regroup his 2 UF forces. *Stavka* agreed and set 5-7 January 1944 as the start date for a new offensive. [372] This order indicated *Stavka* expected a resumption of offensive operations in only 16 to 18 days.

German forces that had been fighting both the 5th and 7th Guards Armies had restored their defensive line around Kirovograd. The village of Novgorodka still needed to be cleared of lingering Red Army troops, a task that would require a few more days of low-level combat. Once again, AG South had defeated a major attack by the 2 UF launched toward Kirovograd, repelling it with heavy losses. As for the simultaneous and the badly planned attack against Krivoi Rog, it hardly even started and was never a threat to the German control of that city.

On 20 December, General Alfred Jodl, chief of the Operations Staff of the German armed forces (OKW), and Hitler talked about the overall strategy for the coming year. Hitler told Jodl

> There is no doubt at all...that the attack in the west will come in the spring...When they attack in the west, then that attack will decide the war...When that attack is beaten off, the story is over. Then we can take forces away again [from France and the west] without delay. [373]

In other words, delay in the East as much as possible, then defeat the Allied invasion of Western Europe. When the decisive Western Front battle had been successfully concluded, then transfer large forces to the east and defeat the Soviet Union. If this degree of high-level strategy was even possible at this stage of the war remained to be seen. Even if Germany was victorious against the Western invasion, thinking that Nazi German could then go on to defeat the Soviet Union was just self-deception.

21 December 1943

An exhausted 7th Guards Army reported a combat strength of 42,653 men and women. Active combat operations almost were finished for troops that badly needed to rest and to resupply.

22 December 1943

The 7th Guards Army had a reported combat strength of 42,669 men and women. The front remained quiet as both sides tried to recover from the latest fighting. Before the next attack, the battered remnants of the 2 UFs tank and mechanized forces needed rest and replenishment to recover their strength.

23 December 1943

The 7th Guards Army had a reported combat strength of 43,810 men and women.

The Army issued a report of troop losses from 19-23 December (five days). This was recorded in TsAMO record ID #131308951: Killed - 268, wounded - 719, gone missing - 44, (and 49 others) for a total of 1,080 men and women. These losses covered the time period after General Konev cancelled the offensive.

On or about this date, the victorious defending German units completely cleared the final Red Army formations from the village of Novgorodka. Once again, this village was firmly in their hands.

At this point, the front line remained stable until the next series of offensive operations began in early January 1944. See Map #32, which displays the territory gained by the Red Army during four months, September-December 1943, of nearly constant combat. During these four months, the 1 UF had advanced about 270 miles to the west, the 2 UF advanced about 180 miles, the 3 UF advanced about 220 miles, and the 4 UF advanced about 310 miles. Note: The overall Red Army losses for this four-month period are listed in the 29 September 1943 entry.

24 December 1943

The 7th Guards Army had a reported combat strength of 44,024 men and women.

A historical document from the TsAMO Archive ID #131308953 reported that rear area support troops for the 7th Guards Army numbered 6,952 people. This value of around 7,000 support personnel had remained steady for months. This suggested the medical, engineering, communication and logistical support these troops provided satisfied the needs of the Army.

25 December 1943

The 7th Guards Army had a reported combat strength of 44,671 men and women, a slight increase from the previous day, suggesting little or no fighting took place on this Christmas day.

A chart from the TsAMO Archive ID #1312344482 displayed a new corps arrangement within the Army. The corps assignments were as follows:

24 GRC – 36 GRD, 8 GAD, 53 RD
49 RC – 81, 72 GRDs, 19, 223 RDs
25 GRC – 41 GRD, 303, 409 RDs

As can be seen, the 73 and the 78 GRDs were withdrawn to the rear to be rebuilt.

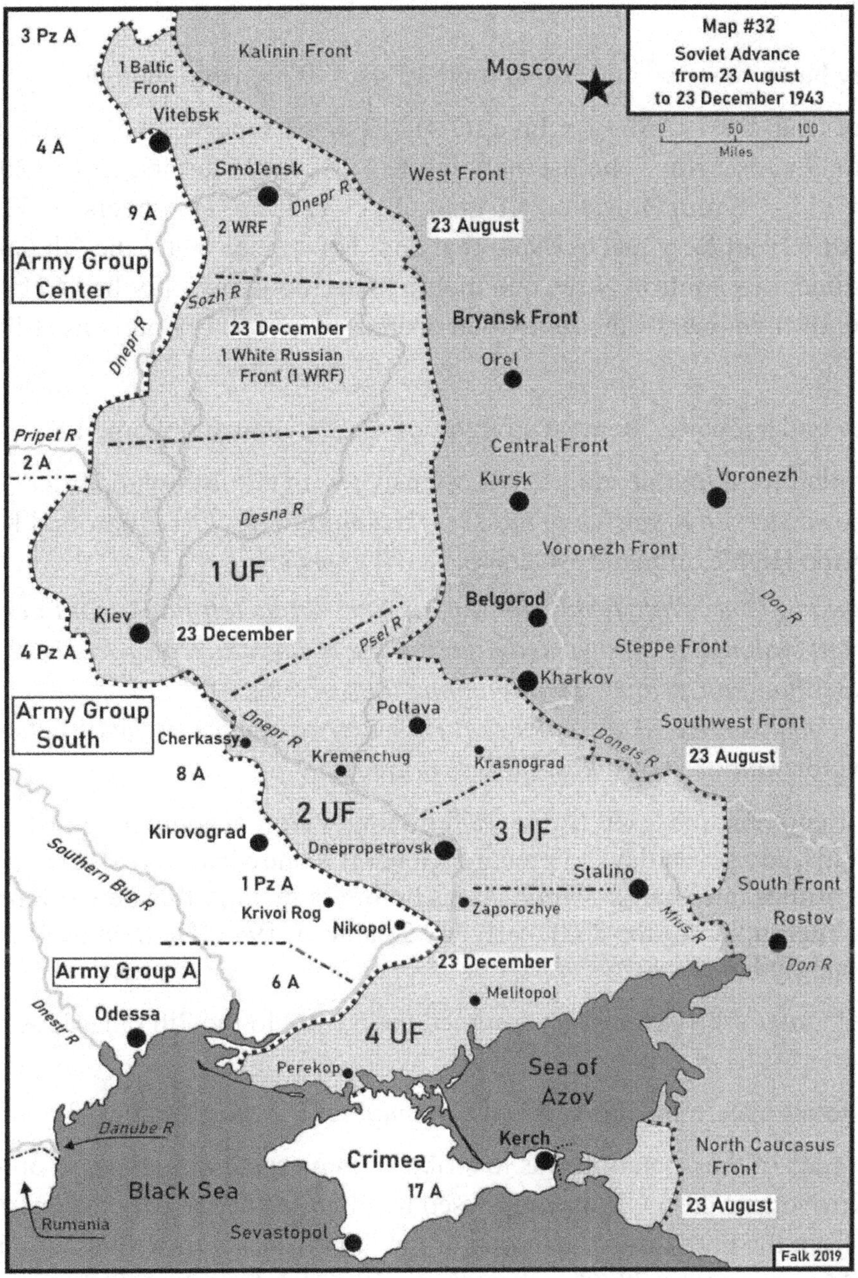

26 December 1943

The 7th Guards Army had a reported combat strength of 44,575 men and women.

The overall weather conditions were sporadic, as cold with snow was followed by warmer weather and rain. Different historical accounts talk of an early spring, but the historical record provides little detailed weather data during this period. Obviously, with the rain came muddy road conditions, but even if a road at first was frozen solid and covered with thick snow, those conditions would not last long. After thousands of men, hundreds of horses, wagons and trucks marched or drove along it, the surface of any frozen road soon would turn into a thick layer of sticky mud.

27 December 1943

German units defending directly east of Kirovograd decreased their combat activities and changed to a defensive posture along the newly established line. Units of the 7th Guards Army were doing the same, resting, resupplying, bringing up reinforcements, and waiting for the next round of combat to begin.

28 December 1943

The 7th Guards Army had a reported combat strength of 44,489 men and women.

A historical document from the TsAMO Archive ID #131308961 reported on the quantity of artillery and mortars within the 7th Guards Army. The list included 131 - 82 mm mortars, 129 - 120 mm mortars, 75 - 45 mm AT guns, 111 - 76 mm Howitzer/AT guns, 124 - 122 mm Howitzers, and 51 - 152 mm Howitzers for a total of 621 artillery pieces. Note that on 5 November 1943, the 7th Guards Army listed 1,400 artillery pieces under its control. Now, one month later, the Army only had 621 artillery pieces remaining, so the overall firepower of the Army had decreased noticeably but was still substantial.

29 December 1943

The 7th Guards Army had a reported combat strength of 41,273 men and women.

Stavka changed their plan for the impending 2 UF offensive with the following order:
In connection with the successful offensive of the 1st Ukrainian Front forces [around Kiev], *the STAVKA VGK* [Supreme High Command – General Staff] *...orders:*

1. *2 Ukrainian Front, while firmly holding its lines on its left flank, not later than 5 January* [1944] *will resume the offensive, delivering the main attack on Kirovograd with the force of not less than four armies, including one tank army....*
2. *Simultaneously deliver a secondary attack with two armies in the direction of Shpola and Khristinovka station* [northwest of Kirovograd]. [374]

With this change in orientation, the 2 UF had to shift its forces from a southern attack focus to a more northwesterly one. With only seven days to prepare, General Konev had to rush the resupply effort and hurriedly regroup his armies into their assembly areas before launching this new offensive. General Konev completed the regrouping of the 2 UF early, by 2 January 1944. He then used the remaining time to rest and to resupply his troops.

General Konev issued Order #001020, which assigned the 33 RC to the 7th Guards Army, effective on 30 December 1943.

Konev restated the overall plan in his book, *Front Commander's Notes*, Chapter 3:

> The general idea of the operation was to strike at convergent directions in order to surround the entire Kirovograd group. It was supposed to encircle and destroy the enemy group by covering it from the north with the troops of the 5th Guards Army together with the 7th Mechanized Corps and the 7th Guards with the 5th GTA from the south.

30 December 1943

The 7th Guards Army had a reported combat strength of 41,878 men and women.

As both Axis and Soviet troops rested and regrouped for further combat on the Eastern Front, Hitler and the German people had more immediate concerns. During the previous six weeks, in the night skies over Berlin, the British RAF Bomber Command assaulted Berlin with over 14,000 tons of bombs. Many more attacks by British and American bombers would follow in the coming months and years, as Berlin incrementally was pounded into rubble. The Luftwaffe had, of course, been trying to do the same thing to London and other major cities of Europe since 1939.

31 December 1943

The 7th Guards Army had a reported combat strength of 41,666 men and women.

The Army issued a report, *The Composition of the Artillery 7th Guards Army,* TsAMO record ID #455185274. The total artillery listed for this day was:

152 mm Artillery – 68
122 mm Artillery – 136
76 mm Artillery/AT – 285
45 mm AT – (none)
120 mm Mortars – 154
82 mm Mortars – 310

Totaling: 953 pieces

This was a slight reduction from the previous month's total, but a significant increase from the 28 December listing of 621 pieces. The reason for this discrepancy is unknown. Regardless, 953 artillery tubes represented a substantial amount of firepower that General Shumilov would soon be using to hammer the enemy. After all, the Soviets did call artillery the "God of War."

An incomplete losses report, recorded in TsAMO record ID #131308951, covering only 19-23 December (five days) shows the Army suffered 1,080 casualties. If we use a conservative estimate of 200 casualties per day for the remaining 26 days of the month, we can add a further 5,200 casualties, which equals a more realistic grand total of 6,280 casualties for the month.

A New Year's Eve surprise arrived at 23:45 hours, as the Luftwaffe paid a visit to the high-water railway bridge that was under construction near Kiev. Three waves of aircraft bombed the bridge area, killing and wounding numerous workers and damaging the construction site. Despite this attack, the high-water bridge was ready to receive trains by 20 February 1944.

When the spring thaw came, the temporary low-water bridge over the Dnepr at Kiev would be destroyed by rising water and drifting ice, so it was vital a high-water bridge be in place and be ready to carry heavy rail traffic to the west. [375] Planning ahead, railway engineers designed this Kiev high-water bridge to last ten years, so that it would be long after the war before they would need to build a permanent replacement bridge.

While the Luftwaffe still was able to hit some targets hard, the total number of sorties the Luftwaffe generated in one region did not match what the VVS produced. For instance, during December 1943, the 5th AA aviators flew 8,045 sorties covering 2 UF operations. Unfortunately, the historical record fails to record the Luftwaffe's response in this area, but undoubtedly, due to lack of aircraft and aviation fuel, the Luftwaffe total was much lower.

Overall, aircraft numbers were not favorable for the Luftwaffe. On the entire Eastern Front, the Luftwaffe could only muster 1,633 combat aircraft while the VVS deployed 8,818 aircraft with another 313 in reserve. [376]

Soviet production of aircraft and all types of military equipment was surging ahead, far outpacing even the growing German production. The coming new year would not improve this situation for Germany.

1 January 1944	
7th Guards Army	
Assigned – 2 Ukrainian Front	
Colonel General M.S. Shumilov Commanding	
Rifle, Airborne and Cavalry Units	24 Guards Rifle Corps (36, 41, GRDs, 8 GAD) 25 Guards Rifle Corps (72, 81 GRDs, 409 RD) 33 Rifle Corps (50, 297, 303 RDs) 35 Guards Rifle Corps (93, 94 GRDs, 78 RD) 3 Separate AT Rifle Battalion 62, 63, 64, 65, 66, 94 Penal Companies
Artillery Units	11 Artillery Division (40 Howitzer Artillery Brigade, 45 Gun Artillery Brigade, 31 Gun Artillery Brigade) 161 Guards Gun Artillery Regiment 1110 Gun Artillery Regiment 114, 115 Guards Tank Destroyer Regiments 1661, 1669 Tank Destroyer Regiments 30 Antitank Artillery Brigade 263, 290 Mortar Regiments 97, 302, 309 Guards Mortar Regiments 5 Antiaircraft Artillery Division (670, 743, 1119, 1181 AA Regiments) 162 Guards AA Regiment 662 Separate Aerial Observation/Air Raid Warning Company
Armored and Mechanized Units	27 Guards Tank Brigade 34, 38 Separate Armored Train Battalions
Engineer Units	27 Separate Flamethrower Battalion 175, 329 Separate Engineer Battalions 46, 49, 528 Engineering Battalions
Estimated Combat Strength Estimated Ration Strength	56,519 men * 63,519 men and women * (including an estimated 7,000 support personnel)

(BSSA 44, BSA 44, TsAMO Archive)

*** TsAMO Archive document ID #131310108, page 4, provides strength data for the 12 RDs of the 7th Guards Army, at 47,892 men. The total combat and ration strength for the Army is missing, so the author is forced to estimate these.**

BSA data shows the 27 Separate Flamethrower Battalion was newly assigned to the 7th Guards Army. At full strength, this unit would have contained 390 men with 240 ROKS-2 flamethrowers.

Due to the lack of operational rail lines on the west side of the Dnepr River, front level motorized transports had to move a significant volume of cargo under harsh off-road conditions. This was during the autumn mud and winter snows of 1943/44. So, of the 1,272 vehicles that were at the disposal of the 2 UF at the beginning of 1944, most were worn-out, with only 640 being serviceable. [377] Each vehicle (truck) could carry about 1.5 tons of cargo, so during the muddy period, these 640 serviceable trucks could move about 960 tons of cargo around 40-50 miles per day on well-maintained dirt roads. In the most difficult sections of deep mud, these trucks could move supplies only about 10 miles per day. Winter snow and ice provided a firmer driving surface, but deep blowing snow could clog a road and block all traffic as easily as mud. A German account from Russia nicely illustrates the problem about driving in winter snow conditions:

> The snow, which fell in great abundance, never melted, and in places was already over three feet deep.... We shoveled away the snow so that our trucks could move forward fifteen or twenty miles in the morning, only to find our efforts covered over again the same day....The road leaving the city was well ploughed, and we rolled along at a brisk pace. There must have been road gangs working around the clock. The snow banks on either side of the road were nearly twelve feet high. [378]

This German account went on to say that this truck convoy took eight days to travel 250 miles, or about 31 miles per day.

1 January 1944

The 7th Guards Army had been reinforced substantially. With four rifle corps containing 12 rifle divisions, the Army was once again ready for offensive operations. Sadly, despite being reinforced, these divisions were below strength, from the low of 2,079 men for the 41 RD to the high of 4,921 men for the 50 RD (estimated strengths based upon earlier reports).

During January, Axis naval forces operating in the Black Sea were able to transfer 33,000 tons of supplies to the trapped troops of the 17th Army holding the Crimea. [379] Soviet naval and air interference was minimal.

2 January 1944

The latest 2 UF offensive against Kirovograd almost was ready. General Konev, 2 UF Commander, stated in his book *Front Commander's Notes,* Chapter 3:

> The 7th Guards Army (commander General M. S. Shumilov) was ordered to deliver the main blow in close cooperation with the 5th Guards Tank Army (commander General P. A. Rotmistrov) in the general direction of Plavin, Pokrovskoye, bypassing Kirovograd from the southwest. By the end of the second day of the operation, they were supposed to take control of Kirovograd and reach the line of Fedorovka, Yuryevka, Ingulo-Kamenka. (See Map #33.)

In terms of local weather conditions, General Konev's new offensive was in luck. He went on to say:

> Meteorological conditions were favorable. Dry weather, slight frost, slight snow cover, which reached only 20 cm [8 inches], the absence of snow drifts, good roads - all this contributed to the maneuver of our troops and [to] the supply of all necessary supplies. Only thick clouds and fogs limited the combat operations of aviation and impeded the work of artillery.

Facing the 7th Guards Army, on the other side of the front line, from north to south, was part of the German 376 ID, the entire 2 FJD with 3,200 men, and part of the 13 Pz D deployed around Novgorodka.

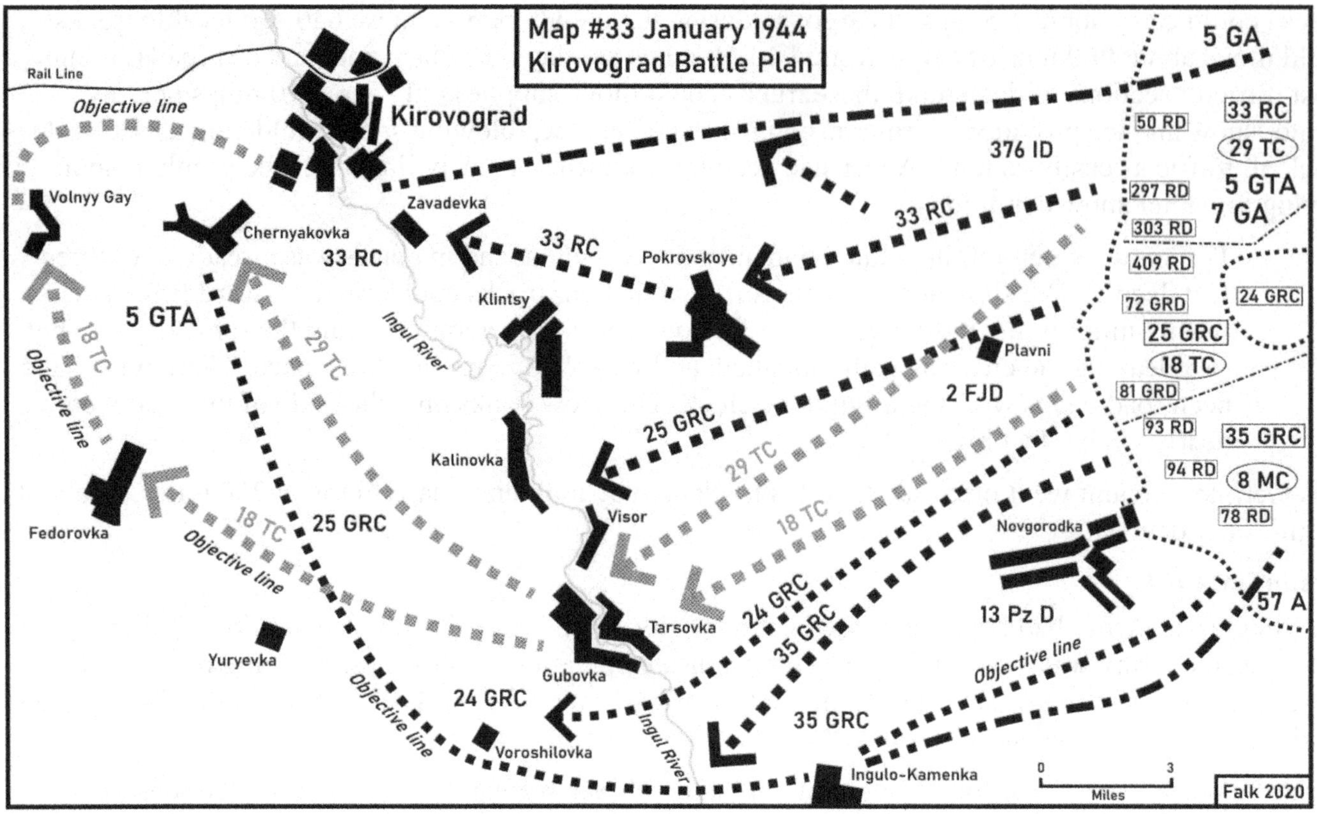

3 January 1944

At this time, the 7th Guards Army's command staff consisted of Colonel General M.S. Shumilov, Commander; Major General A. Mukhin and Colonel F.K. Shcherbak, Members of the Military Council; and Major General G.S. Lukin, Chief of Staff. [380] These officers had been working together for many months as a dedicated, harmonious team, and would continue together to coordinate the Kirovograd battle plan.

4 January 1944

The 7th Guards Army released a report, TsAMO record ID #136074325, listing AT and artillery support for the upcoming offensive: 300 – 45 mm AT guns, 632 – Mortars (82 and 122 mm), 380 – 76 mm Artillery/AT guns, 130 – 122 mm Howitzers and 83 – 152 mm Howitzers, for a total of 1,225 artillery tubes and 300 dedicated AT guns. Of this total mass of artillery, some 1,080 tubes would be directed against German forces along a five-mile section of the right flank. The intent, of course, was to blast a hole in the German defense line, thus allowing for a swift breakthrough into their rear area with mobile forces from the 5th GTA. At least that was the plan. Note that on Map #33 the two tank corps had different objective lines from the three rifle corps.

To further enhance coordination between the 5 GTA and the 7th Guards Army, the staffs of the two armies worked closely together.

> Due to the fact that the 7th Guards Army did not have its own tanks and self-propelled [guns], two tank brigades had to be allocated to it for direct support of the infantry – [the] 32nd and the 181st. Before the offensive, the headquarters of rifle, tank formations and artillery units exchanged communication and radio signal tables, prepared general fire schemes, [and] outlined common landmarks and target designation procedures. The officers of the operational department of the headquarters of the 7th Guards Army, as well as representatives of the artillery groups, were sent to the tank corps. There were spotter officers in the battle formations of the tank companies. [381]

Von Manstein flew to Hitler's HQ. (Once again, the location is unknown, but it was most likely at *Wolfsschanze,* Wolf's Lair, in East Prussia). The meeting was intended to convince Hitler of the vital need to give up the Nikopol bridgehead and the Crimea in order to redeploy these forces and to shorten Army Group South and AG A's front line by at least 560 miles. The newly released forces could then be used to strengthen AG South's left flank around Kiev. Hitler refused to authorize any withdraws along the southern section of the front and once again rejected von Manstein's third attempt to have a Commander in Chief (of the Eastern Front) appointed. This person would have had complete independence within the framework of the overall grand strategy set by Hitler. Von Manstein returned to his HQ after his futile effort to reintroduce military professionalism to the eastern war effort and to harmonize the overall strategy on the Eastern Front. [382]

Pz GD *Grossdeutschland* received orders to disengage from its position east of Krivoi Rog and to move north, most likely by rail, to new positions west of Kirovograd. This move probably was due to German intelligence sources spotting indications of a new offensive in the Kirovograd area.

At the start of January, the Soviet 5th Air Army had 702 operational aircraft [383] while the German 4th Air Fleet had 849 aircraft on hand with 625 of them operational. [384] Apparently neither side would enjoy air superiority during the upcoming battles.

5 January 1944

At 08:10 hours, artillery and VVS aircraft began their 50-minute-long bombardment of German positions. German forces estimated the Soviet artillery expended 188,000 rounds of ammunition during this initial bombardment. Then at 09:00 (Russian time), the Red Army launched the ground assault. The Kirovograd operation of the 2 UF had begun.

The 7th Guards Army held a noticeable superiority over the enemy in the attack area of 3:1 in infantry, 8:1 in artillery, and 7:1 in mortars. General Shumilov must have launched this offensive with confidence in his troops and weaponry.

Of the 7th Guards Army's total frontage of 13 miles, the first phase of the attack was focused against just a 5-mile section of the front on the Army's right flank. From north to south, the 33 RC and 25 GRC led the offensive toward Kirovograd. The 35 GRC on the far-left flank maintained its position during the day, making only minor holding attacks to keep local German forces in place. Shumilov had the 24 GRC in reserve with three RDs, and the 5th GTA also remained in reserve until the 7th Guards Army made the expected breakthrough.

> With the beginning of the attack on the front edge of the enemy's defenses, the rifle formations of the 7th Guards Army met stubborn resistance from the Nazis and in two hours of battle advanced only two kilometers [1.2 miles]. From the reports received, it followed that the enemy had a large number of tanks, 88 mm guns, mortars and other [artillery] weapons that were not suppressed during the artillery preparation. In addition, in the way of [the Soviet's] advancing troops, dense minefields and multi-row barbed wire obstacles were discovered. [385]

After another hour of heavy fighting, the 7th Guards Army still was unable to make a clean breakthrough of the German defense line. This lack of progress delayed the advance of the 5th GTA, the main strike force of the entire offensive. Because the 7th Guards Army was clearly unable to achieve its objectives due to strong German resistance, Colonel General Pavel Rotmistrov, the commander of the 5th GTA, stated,

> Having understood the current situation and having talked with the commander of the 7th Guards, MS Shumilov, I decided at twelve o'clock to send the main forces of my army into battle to complete the breakthrough of the enemy defense. [Marshal] Konev agreed, and two of our tank corps [the 18 and 29 TCs] rushed into battle. [386]

With the released 5th GTA to assist in forcing a breakthrough, Marshal Konev had changed the tempo of the offensive, but the true impact of this change would not be felt until late the next day.

The stout German defense was no accident. German intelligence efforts had detected a buildup of forces in the 7th Guards Army area and had planned accordingly. Expecting an offensive, German forces just happened to have large stocks of artillery ammunition on hand. So much ammunition was available that they were able to fire off approximately 177,000 artillery rounds this first day. [387]

Advancing Soviet infantry move past an abandoned German 88 mm AA/AT Gun.

At the end of the day, only the 7th Guard's Army center and right flank units made any progress at all with the left flank holding their original positions. This was a bad showing for a Guards Army. Due to this lackluster progress and the inability of even the 5th GTA to force a decisive Russian breakthrough, General Konev stated, "Thus, the 7th Guards Army in its offensive had only partial success on the right flank and in the center." Because of this lack of success, Konev decided to shift the 8 MC (the 5th GTA reserve corps) north to the area of the 5th GA, which had made a clean breakthrough of the German front. In this new location, east of Kirovograd, the 8 MC joined the 7 MC on 6 January to further exploit the 5th GA early success. At the end of the day, units of the 7th Guards and 5th GTA had advanced only 2-3 miles. (See Map #34.)

Sergei Andreevich Otroschenkov, a Russian tank crew member of the 5 GTA, stated, "Our tanks attacked and after a few kilometers the Germans met us with well-organized fire near the village of Plavni. All day we tried to break through the German defense. The battalion and the brigade suffered heavy losses."

Later, during the night, the tanks had moved on to another village "where a real friend, a front-line comrade…Ivan I. Rodin" was shot and killed by Germans. Otroschenkov said,

> His death upset me greatly. The next day our battalion captured some prisoners. This never happened to me again, and I forbade my subordinates from acting like this, but on that day I

broke. I shot the captured Germans myself. I got a good dressing down from the brigade commander, but there was no serious punishment. [388]

The book *The Anvil of War,* edited by Peter G. Tsouras, p. 199, states that on the first day of the offensive, German sources claim to have destroyed 153 tanks and caused heavy casualties to the attacking Red Army infantry. In comparison, the 10 Pz GD alone lost 620 men during the day. Other German units must have suffered similar losses.

6 January 1944

The offensive continued with 2 UF armies facing determined German resistance and counterattacks lead by small groups of Panzers. Of note, elements of the 13 Pz D launched a counterattack into the left flank of the 7th Guards Army just west of Novgorodka. This attack was not very powerful, but it was enough to prevent the Russian offensive from spreading further south. German sources also noted these small attacks "of active defense deceived the enemy as to German strength." As a side benefit, the Germans noted that the advancing Russian mobile forces "[p]aid too much attention to flank cover." [389]

The 5th and 7th Guards Infantry Armies continued their attacks and advanced westward in the face of German counterattacks. The two tank corps of the 5th GTA continued to support the advance of the 7th Guards Army. The overall breakthrough zone of the 2 UF increased to 43 miles wide by 18 miles deep. In the 7th Guards Army area, after two days of effort, the advance only amounted to about 10 miles at the end of the day. A frustrated General Shumilov committed his reserve corps, the 24 GRC, which attacked with two of its three divisions on the left flank of the 7th Guards Army. Shumilov wanted the 24 GRC to provide flank protection for the 25 GRC and 33 RC so they could advance faster. Attacks continued all along the front during the night.

The weather conditions remained favorable for the offensive. But the Germans also benefited from good mobility, which allowed them to shift mobile forces around, reinforce threatened sectors of the front and launch quick, powerful counterattacks.

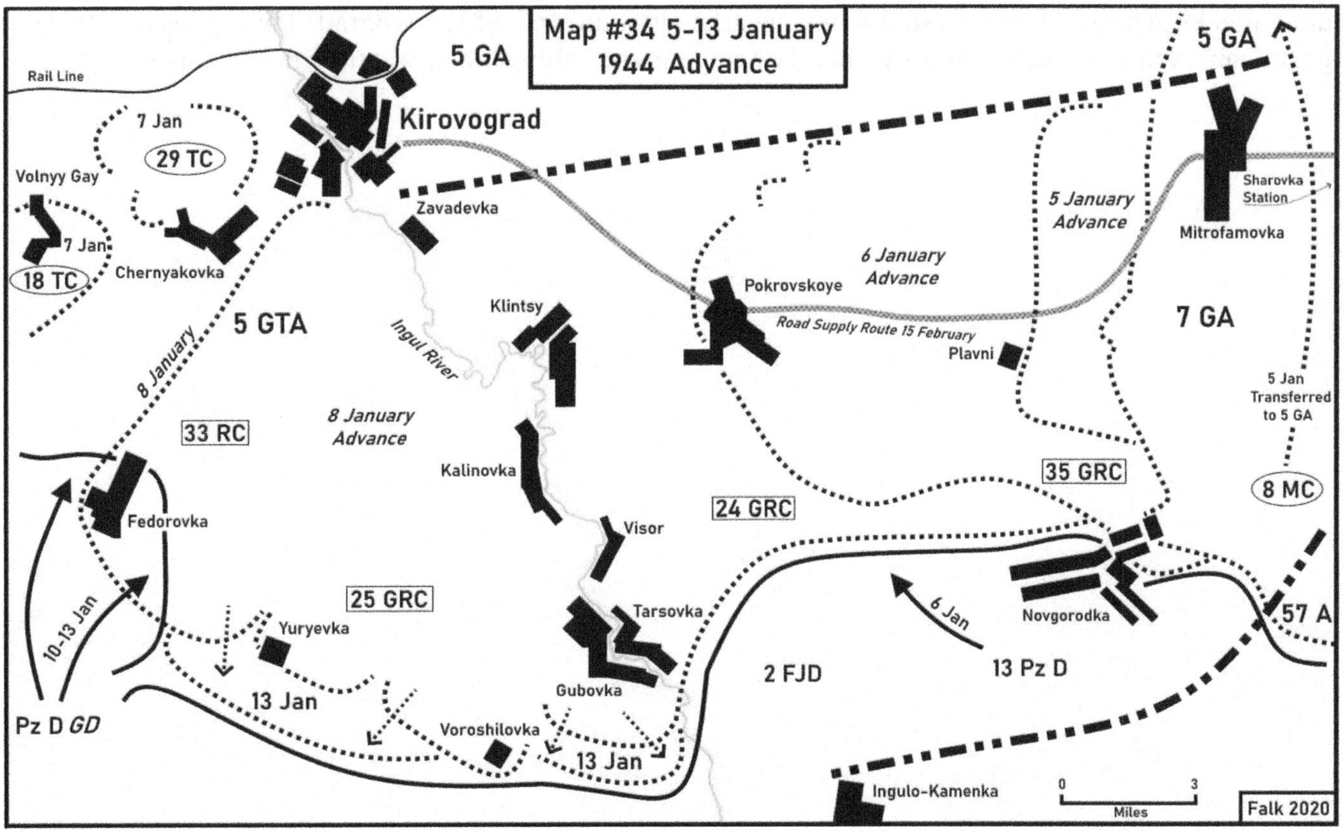

Map #34 5-13 January 1944 Advance

7 January 1944

The rifle divisions of the 7th Guards Army continued to advance and to gain ground south of Kirovograd, but the 18 and 29 TCs fought past German positions, turned to the north, and reached the area west of Kirovograd. With support from the 53rd A and 5th GA to the north and east, major elements of three German divisions, 10 Pz GD, 14 Pz D and 376 ID, now were surrounded in and around Kirovograd. A fourth division, the 3 Pz D, launched a successful breakout attack to the northwest, linking up with the leading elements of the 11 Pz D the next day.

8 January 1944

After house-to-house fighting, units of the 5th GTA, 5th GA, 7th GA and 53rd A liberated the city of Kirovograd. The surrounded German divisions quickly were forced out of the city to the north, where they occupied the hill town of Lelekovka and the surrounding area. (This area is off Map #34 to the north). Due to the fast takeover, combat damage to the city apparently was minimal with the added benefit of the Germans only having time for a few deliberate demolitions. In far off Moscow, Stalin ordered a salute to be fired from 224 guns to mark the liberation of the city.

Kirovograd itself was a fairly large city that functioned as a German supply and rail center. As German and Soviet troops fought over the city, they paid little attention to any remaining civilians. The civilians knew that if they wanted to live, they needed to keep their heads down and to stay out of the way. For the Red Army, Kirovograd eventually would become a rail and logistical center, but it also was a symbol of liberation, with several participating units receiving the honorific title of Kirovograd!

One of the units assigned to the 7th Guards Army, the 41 GRD, 24 GRC, felt left out of the fighting for Kirovograd. "The Guards really wanted to participate in the liberation of Kirovograd, but the division was advancing a little south of it." [390] The 41 GRD had its hands full on the left flank of the advance as it was facing the remains of the German 2 FJD and the 13 Pz D in and around Novgorodka. In this area, the German line held firm, blocking the planned capture of Novgorodka and any further advance to the south by the 24 and 35 GRCs. Newly arrived formations of Pz GD *Grossdeutschland* launched counterattacks against advancing Red Army spearheads southwest of Kirovograd. This series of attacks engaged tank and rifle units of the 7th Guards Army and possibly some from the 5th GTA as well.

A German *Feldgendarmerie*, Military Police/Traffic Control, signals vehicles while a motorcycle dispatch rider awaits his turn.

9 January 1944

The 33 RC and 25 GRC dug in along the Kirovograd, Fedorovka, Yuryevka, Voroshilovka line, while the 24 GRC and 35 GRC held the remaining left flank of the Army. The 5th GA covered the Army's right flank.

During the evening, after making small local attacks to mask the withdraw, formations of the 10 Pz GD, 14 Pz D and 376 ID successfully withdrew from the Lelekovka pocket, reaching safety 8 miles to the west at the town of Gruzkoe. (Off of Map #34 to the north).

The Pz GD *Grossdeutschland* continued its flanking attacks south of Kirovograd with some support from the leading elements of the newly arrived 3 SS Pz D *Totenkopf.* These attacks drove back toward Kirovograd the most advanced Red Army formations operating in the south.

10 January 1944

Units of the 25 GRC become active, attacking along the front between Yuryevka and Voroshilovka, capturing some ground in the process. This action most likely was meant to improve their local defensive positions.

A historical document from the TsAMO Archive ID #131310108 lists the strengths of the three rifle divisions of the 35 GRC on 1 January, and another document, TsAMO record ID #132344486, lists the strength of the same rifle divisions as of 10 January, that is after five days of combat during this offensive. (See Table #8.)

Rifle Division's	1 January Strength	10 January Strength	Gain/Loss
93 RD	4,429 men	3,206 men	-1,223 men
94 RD	4,700	3,697	-1,003 men
78 RD	3,309	4,382	+1,074 men

Table #8: 35 GRC Strength Change during Five Days of Combat -- 5-10 January 1944.

Despite these two documents being an incomplete record for the Army, they do show that over the previous five days of battle, these divisions showed a loss of about 200 men/day, and the overall low strengths of the 93 and 94 RDs might explain the lack of progress on the ground by the 35 GRC. Clearly the 78 RD did not take part in the attack but held its position as it began to recover its strength.

11 January 1944

Units of the 25 GRC continued their low-level attacks to the south, as the remainder of the Army's front was quiet. The front line in the south stabilized as German forces began to close the gap in their line. German forces also destroyed a small pocket of Russian troops southwest of Kirovograd.

12 January 1944

As the 25 GRC continued to adjust its positions, German forces became active, launching an attack in the Fedorovka area. Overall, German forces were attempting to close the gap in their front line and to drive back any outlying Red Army outposts and scouting parties.

Seeing a new opportunity, the Soviet VGK (Supreme High Command) issued Directive #220006 outlining a coordinated attack by the 1 and 2 UFs south of Kiev. German forces in this area were holding an exposed bulge in the line and *Stavka* wanted to surround and to destroy the German forces in the Korsun-Shevchenkovskii (Korsun) area. The Soviet plan was for the 2 UF to attack from the south on 25 January with the 1 UF attacking from the north on 26 January.

13 January 1944

Counterattacking German units, most likely from Pz GD *Grossdeutschland* (*GD*), recaptured most of Fedorovka. This German success blocked any further advance by units of the 7th Guards Army in that area. Meanwhile, the 25 GRC ceased their attacks and moved to a defensive posture to hold the ground they had recently captured.

In response to *Stavka's* new Korsun offensive directive, General Konev began scaling back 2 UF offensive operations in and around Kirovograd. Preparations for the Korsun attack had to be pushed forward quickly to meet the tight schedule, and this would have occupied all of Konev's attention. General Shumilov was now free to operate more independently, to resupply and reinforce his army, and to make preparations for the inevitable resumption of offensive operations south of Kirovograd.

14 January 1944

Fighting was at times heavy north of Kirovograd as Russian tank units attempted to breach the German line. As fighting south of Kirovograd declined, due to exhaustion, lack of troops and tanks, and the return of mixed weather (which caused snowy/muddy conditions), both sides took stock of their situation.

In the far north at Leningrad, a new Red Army offensive began against the besieging German forces.

15 January 1944

The Russians launched another major attack north of Kirovograd with some success, opening a seven-mile gap in the German line and driving forward three miles. South of Kirovograd, units of General Shumilov's 7th Guards Army held their current positions. The supporting units of the 5th GTA, 18 and 29 TC had suffered heavy tank losses when fighting south and west of Kirovograd and were most likely withdrawn by this time for refitting and resupply.

16 January 1944

As Russian tank and infantry forces gathered to continue their attack north of Kirovograd, a German counterattack lead by the newly arrived 3 SS Pz D *Totenkopf* along with three supporting infantry divisions spoiled the pending Russian attack and, at last, closed the gap in the German line. With the end of this final battle, combat around Kirovograd dropped off considerably.

Units of the 2 UF did achieve the liberation of Kirovograd, but they did so at a heavy cost in men and equipment. In turn, German forces lost their positions in and around Kirovograd, but they ultimately were able to prevent a major breakthrough. Both sides could claim a partial triumph of sorts, but neither side had achieved a decisive victory.

During the operation to liberate Kirovograd, the 5th Air Army flew 2,485 sorties in support. The number of Luftwaffe sorties is unknown, but should have been similar. Bad weather hampered both air forces.

Further south, forces of the 3 UF regrouped their armies over the next three nights in preparation for a further attempt to capture Krivoi Rog.

17 January 1944

In and around Kirovograd, both sides took stock of the battle. Russian formations had only 120 tanks operational and rifle divisions were severely short of infantry. The Germans in turn suffered similar losses with Panzer and infantry units critically understrength. The estimated combat strength of the 7th Guards Army at this time was 44,000. This represented a loss of some 12,000 men and women during the offensive; however, some of these losses were due to transfers out of the Army.

18 January 1944

The front around Kirovograd was quiet, but the 5th and 7th Guards Armies regrouped their troops and conducted reconnaissance operations all along the front. Two TsAMO Archive Documents ID #132309772 and #13230977 on 18 January listed the strength of the 27 GTB: 21 – T 34s and 3 - SU 122s. The next day the totals were 23 – T 34s and 3 - SU 122s. Most likely the two T-34 tanks added on 19 January were tanks that were repaired by either the 37 or 62 Mobile Repair bases.

Local German commanders took notice of these increased activities in and around Kirovograd.

19 January 1944

The 5th GA began simulating the concentration of the 5th GTA to the west of Kirovograd by deploying mock tanks and troops and by sending false radio messages to indicate increased activity in this area. The 5th GTA, in reality, was moving away to the northwest in preparation for the next offensive. At first, German intelligence units were deceived by these *maskirovka* methods.

20 January 1944

Stavka and Konev continued their preparations for the real attack further north while the 5th and 7th Guards Armies' activities provided a useful distraction.

21 January 1944

General Konev, in his book *Front Commander's Notes*, Chapter 4, talking about the upcoming Korsun-Shevchenkovsky operation, stated:

> It was tempting to use the most experienced armies of our front, which took part in the defeat of the Stalingrad enemy grouping, the 5th Guards Army of General A.S. Zhadov and the 7th Guards Army of General M.S. Shumilov in the direction of the main attack. But, unfortunately, it was impossible to regroup these armies from the left wing to the direction of the main attack.

The 5th and 7th Guard Armies were at this time facing west and southwest and the new offensive mandated by *Stavka* was directed to the northwest on the right flank of the 2 UF. So, it was not practical to shift these two armies to the right flank of the 2 UF. But the 5th and 7th Guard Armies would be called upon to provide some assistance to the Korsun offensive. Konev had a special plan for them also.

General Konev was able to redeploy and reequip the 5th GTA from the left flank of the 2 UF to the right flank in time to participate in the Korsun attack. Ultimately, the 1 and 2 UFs received 400 new tanks from *Stavka* reserve to rebuild their depleted tank units.

The German 8th Army, using radio intelligence, detected and identified the dummy tank activity west of Kirovograd and noted that the HQ staff and other units of the 5th GTA had moved northward.

22 January 1944

Stavka reinforced the 2 UF with the 5 Guards Cavalry Corps (GCC). Konev's plan was to attack the base of the Korsun salient with the 4th Guards and 53rd Armies along an 11-mile sector of the front and then to introduce the 5th GTA into the breach to achieve the desired breakthrough. All troop movements were, of course, conducted mostly at night and as secretly as possible.

The 5th and 7th Guards Armies, in and around Kirovograd, continued to use *maskirovka* methods to simulate attack preparations by the 5th GTA in that area. Per orders from Konev, they also would launch a very real attack the next day. This would be a diversionary attack, intended to keep German attention focused upon the Kirovograd area and not further north where the main Korsun offensive would take place on 24 January.

After recognizing the regrouping of Russian forces behind the lines, the German 8th Army began transferring the 14 Pz D northward with the 11 Pz D following the next day. This was mostly in response to the movement of the 5th GTA to the north. But, the use of *maskirovka* methods had slowed the German response, which ultimately lead to a decisive breakthrough by the Russians in the Korsun area.

In Italy, Allied forces conducted Operation Shingle, an amphibious landing at Anzio, on the coast south of Rome. The landing was successful, but a fast response by German forces prevented an Allied breakout into the Italian interior and the capture of Rome.

23 January 1944

The 5th and 7th Guards Armies began their diversionary attacks northwest and south of Kirovograd.

The 5th GA conducted their attack northwest of Kirovograd with the 116 RD, 85 GRD, 111 RD and 97 GRD, aligned from north to south. The 5th GA's plan was to advance about three miles into the German defensive positions along a nine-mile front. This advance would be sufficient to force a German response.

For the 7th Guards Army, planning maps are not available, but a TsAMO record ID #132653901 can be used to identify the attack location. The 25 GRC would attack with the 81 GRD, and the 24 GRC would attack with the 36, 41 GRDs with artillery fire support coming from adjacent uncommitted divisions. The 24 GRC's (36, 41 GRDs) diversionary attack was directed to the south between the villages of Tarsovka and Novgorodka along a five-mile front. The 25 GRC (81 GRD) would attack eastward into the flank of German positions, crossing the Ingual River in the process. (See Map #34 for the general location.) The overall objective was to seize control of four prominent heights: Hills 163.4 and 179.6 for the 36 and 41 GRDs with Hills 166.2 and 168.6 assigned to the 81 GRD.

Of special note: The 7th Guards Army must have captured documents or interrogated a prisoner because a TsAMO Archive map ID #131310324 shows the Army command staff knew the German positions that were under attack were being held by the Von Oppen Battle Group of the 13 Pz D. *Oberstleutnant* (Lieutenant Colonel) Alexander von Oppen was most likely commanding Panzer Grenadier Regiment 66 at this time.

24 January 1944

The 2 UF launched the southern arm of the Korsun offensive. The 4th G and 53rd As opened the attack while the 5th GTA, now reinforced with 218 tanks, stood ready to exploit the opportunity.

At the beginning of the new offensive, the 1 and 2 UFs deployed 336,700 men; 524 tanks and assault guns; 5,300 artillery and mortars; and 1,054 combat aircraft. [391] The defending Germans were vastly outnumbered.

The 5th and 7th Guards Armies diversionary attacks probably continued, but the historical record is silent in this regard.

Of special note, the two Armored Train Battalions #34 and 38 that were assigned to the 7th Guards Army were hard at work behind the front. Battalion #34 operated in and around the Krasnograd supply base with two trains. Battalion #38 operated in and around the Poltava supply base, also with two trains. The cities of Krasnograd and Poltava were located on the two main railway supply lines to the 2 and 3 UFs and were subject to sporadic Luftwaffe attack, so the armored trains and their AA Guns added to the air defense of these supply and rail areas. [392]

25 January 1944

The 1 UF joined the Korsun offensive, attacking from the north.

By this date, the 5th and 7th Guards Armies almost certainly had suspended their diversionary attacks. The amount of ground they gained or lost is unknown, but a historical map from February shows the 7th Guards Army's front line in the same place. By the very nature of the two widely separated attack areas, it appears the 5th GA attack was the real/main diversionary attack. The 7th Guards Army attack seems to have been launched to clear up the Army's left flank and to improve its overall position. Maybe these two attacks did create some type of diversion or doubt in the minds of the opposing German commanders, but all eyes were definitely focused upon events further north and the new Red Army offensive.

The 2 UF transferred the 94 GRD, with a combat strength of 4,339 men, out of the 7th Guards Army and into the front reserve. The 94 GRD would never return to the 7th Guards Army, and it eventually fought in Berlin at the end of the war. This transfer was part of General Konev's effort to shift forces to areas that were more active and in need of reinforcements.

26 January 1944

The weather and ground conditions started to change noticeably as a January thaw set in, with the melting snow turning the roads into paths of bottomless mud. The mud made any type of movement difficult and major attacks extremely challenging.

The unyielding *rasputitsa* conditions forced the 5th and 7th Guards Armies to hold their positions. Troops manning the front were occupied with routine jobs like digging in and keeping dry. The rear area supply troops worked hard to resupply the front-line troops with all manner of provisions and efforts were made to bring reinforcements up to the front for more advanced training. In addition, the armies undertook an occasional reconnaissance mission or delivered an artillery or mortar bombardment to engage the enemy. The Germans, of course, were compelled to return the favor.

27 January 1944

General Konev stated in his book, *Front Commander's Notes*, Chapter 4, about the hardships the weather imposed on his armies during the Korsun attack:

> The weather and terrain conditions for the preparation of the operation were extremely unfavorable. The sudden thaw … made it more difficult for the troops to move and to supply them with fuel and ammunition. According to weather forecasters, it means that from January 27 to February 18, it rained for 10 days [with] sleet, the rest - snow. Only 5 days were without precipitation. The average daily temperature ranged from -5.5 to + 4.9° [22 to 41° F].

With massive battles taking place all along the Eastern Front, von Manstein, in his book *Lost Victories*, on page 511, recounts how Adolf Hitler

> summoned all army group and army commanders on the Eastern Front, in addition to a large number of other senior officers, to General Headquarters. He [Hitler] wished to address us in person on the need for National-Socialist education inside the army. The more difficult the military situation became, the greater importance he attached to 'faith' as a guarantee of victory.

During the talk, Hitler insulted all the gathered officers to which von Manstein responded to uphold the honor of the officer corps. Afterward, von Manstein and Hitler had a few choice words in private. Von Manstein recounted, "I knew, of course, that he [Hitler] would not forgive me for the retort I had made earlier. But I had much more important things to worry about than my personal relations with the Supreme Commander." This testy exchange with Hitler would come back to haunt von Manstein in the future.

By 27 January, the Moscow/Leningrad rail line was cleared of enemy forces. After 872 days (2 years and three months) and the loss of some 1,500,000 lives, Moscow announced the German siege of Leningrad was over. German forces of Army Group North withdrew south to the Narva River line.

28 January 1944

The spearheads of the 1 UF and 2 UF met at the city of Zvenigorodka, thus achieving a double envelopment of two German Corps, XI and XXXXII, and forming the Korsun pocket. Approximately 59,000 German troops now were cut off within the newly formed pocket. The next phase of the Russian plan was for the 1 and 2 UFs to quickly form an inner and outer ring around the surrounded German forces. The inner ring was needed to keep the trapped Germans within the pocket while the outer ring, of course, would be used to defeat the expected German relieving force coming from somewhere outside the pocket. Which side would be ready first, offense or defense?

29 January 1944

Around the Korsun pocket, fighting continued unabated despite the mud, as Russian forces sought to expand their breakthrough area as they repelled small but dangerous German counterattacks. Increased pressure by the surrounding Red Army forces began to reduce the area within the Korsun pocket.

The 7th Guards Army held its current position as it prepared for further operations.

30 January 1944

The 3 UF began its initial diversionary attacks toward Krivoi Rog. This was the start of yet another offensive to capture Krivoi Rog and then to cut off and to destroy German forces along the lower Dnepr River.

31 January 1944

A TsAMO record ID #132309786, on 31 January 1944, listed the current strength of the 27 GTB as having 27 – T 34s and 3 – SU 122s. Which *Shtat* (by date) was being used for the 27 GTB is unknown, but a November 1943 *Shtat* for a tank brigade included 63 medium tanks, which would make the 27 GTB significantly understrength. The three SU 122s is an oddity, but they were most likely remnants held over from the battle of Kursk and added to the 27 GTB to increase high-explosive fire support. An assault gun regiment of 21 vehicles (Su-76, 85,100 or 152) would routinely be added to tank brigades later in 1944 to increase anti-tank firepower.

The 3 UF began its main attack toward Krivoi Rog with the 4 UF launching its own attack against the German Nikopol bridgehead position.

So ended a critical month with the following highlights:

- Soviet forces achieved a victory when they regained control of the long-sought city of Kirovograd, but the armies of the 2 UF suffered substantial losses doing so.
- Further north, armies of the 1 and 2 UF successfully surrounded considerable enemy forces in and around the city of Korsun, but that battle was as of yet undecided.
- To the south, the 3 and 4 UFs launched their own offensives toward German positions around Krivoi Rog and along the lower Dnepr. These battles were as well undecided.
- In the far south, the German 17th Army continued to hold the Crimea, but this was a position of little benefit other than to protect the Romanian oil fields. For the 17th Army itself and the German and Romanian troops involved, it invited eventual disaster.
- In the center of this vast battlefield, the 7th Guards Army and others waited. These armies were occupied with recovering their strength, amassing supplies, training new troops and planning further advances to the west and toward the end of the war.

During the offensive to capture and to hold Kirovograd between 5-9 January, (five days), the 7th Guards Army suffered the following reported losses: 954 killed and 4,605 wounded, or 1,111 casualties per day. If we use a conservative estimate of 250 casualties per day for the remaining 26 days of the month, we can add a further 6,500 casualties, which equals a more realistic grand total of 12,059 casualties for the month. Additionally, the Army lost 1 - 152 mm, 3 - 122 mm, 11 - 76 mm artillery pieces and 2 - 82 mm mortars. Tank losses amounted to 10 - T 34s. This data is from the TsAMO record ID #113983041, p. 33.

1 February 1944	
7th Guards Army	
Assigned – 2 Ukrainian Front	
Colonel General M.S. Shumilov Commanding	
Rifle, Airborne and Cavalry Units	24 Guards Rifle Corps (36, 41 GRDs, 8 GAD) 25 Guards Rifle Corps (72, 81 GRDs, 53 RD) 33 Rifle Corps (297, 303 RD) 35 Guards Rifle Corps (93 GRD, 78 RD) 409 RD 3 Separate AT Rifle Battalion 62, 63, 64, 65, 66, 94 Penal Companies
Artillery Units	11 Artillery Division (40 Howitzer Artillery Brigade, 45 Gun Artillery Brigade) 161 Guards Gun Artillery Regiment 1110 Gun Artillery Regiment 114, 115 Guards Tank Destroyer Regiments 11, 30 Antitank Artillery Brigade 263, 290 Mortar Regiments 5 Antiaircraft Artillery Division (670, 743, 1119, 1181 AA Regiments) 162, 225 Guards AA Regiments 662 Separate Aerial Observation/Air Raid Warning Company
Armored and Mechanized Units	27 Guards Tank Brigade 34, 38 Separate Armored Train Battalions
Engineer Units	27 Separate Flamethrower Battalion 5 Engineer Sapper Squad 175, 329 Separate Engineer Battalions 46, 49, 528 Engineering Battalions
Estimated Combat Strength Reported Ration Strength	42,855 men * 49,855 men and women * (including an estimated 7,000 support personnel)

(BSSA 44, BSA 44, TsAMO Archive)

*** A TsAMO Archive report for the 2 UF, ID #130357744, lists the ration strength of the 7th Guards Army on 1 February 1944 as 49,855 people along with 8,283 horses.**

At the start of the month, the 7th Guards Army had the following units transferred out of the Army: 97, 302, 309 Guards Mortar Regiments (rockets), the 1661 and 1669 Tank Destroyer Regiments, and the 31 Gun Artillery Brigade. Most if not all the transferred units were sent north to reinforce the Korsun pocket forces.

The Army also gained the following units: the 225 Guards AA Regiment, 11 Antitank Artillery Brigade, and 5 Engineer Sapper Squad.

1 February 1944

In a TsAMO record "Operations Log" ID #113985362, the 2 UF made note of the actions of the 5th and 7th Guards Armies along with the 57th A; they "fought and [performed] reconnaissance from [their] previous position[s]." In other words, the 2 UF was busy with operations around the Korsun pocket and nothing important was taking place along the 7th Guards Army's particular section of the front. As a result, these three armies held their assigned sections of the front and engaged themselves with maintaining and rebuilding their depleted forces.

The German 11 and 13 Pz Ds, XXXXVII Pz C, launched the first rescue attack toward the Korsun pocket. They were attacking directly toward the southernmost part of the pocket. The two Pz Ds were able to advance 17-21 miles, capturing ground along the way, but they could not reach the pocket due to powerful Russian counterattacks. The attacking forces were unable to cross the Shpolka River in force. With the pocket still about 11 miles away, heavy losses on both sides put an end to the fighting.

During February, Axis naval forces operating in the Black Sea transferred 4,053 men and 52,455 tons of supplies to the 17th Army in the Crimea. On the way back, these same naval forces transported 12,000 tons of supplies and equipment and over 10,000 wounded and on leave soldiers from the Crimea. All convoys got through without loss from Soviet naval or air forces. [393] Additionally, the Germans moved the entire 73 ID to the Crimea, mostly by air, with all the equipment coming by sea. At this time, approximately 200,000 Axis troops were stationed in the Crimea.

Large numbers of Axis forces continued to be deployed to the Crimea for three main reasons:

- To control the north coast of the Black Sea and to disrupt Soviet operations on land, sea and air.
- To retain the powerful political influence the Crimea had upon the Axis Balkan states of Romania, Bulgaria, and even neutral Turkey.
- To keep the Red Army Air Force (VVS) and more importantly the Soviet long-range aviation (ADD) out of range of the absolutely vital Romania oil fields and refineries in and around the city of Ploesti.

The Romanian oil fields were the most important reason Hitler felt Germany needed to control the Crimea at all costs.

2 February 1944

Once again, TsAMO record "Operations Log," ID #113985362, from the 2 UF stated that the 5th GA, 7th Guards and 57th A "fought and [performed] reconnaissance from [their] previous position[s]." Note: The 2 UF had to report something to *Stavka*, and so they did. This type of report is an example of Soviet military language for "nothing much was going on, but the armies were occupied doing something."

The 7th Guards Army had nine rifle divisions in the front line, so two divisions must have been in the second echelon resting and/or rebuilding.

3 February 1944

The 7th Guards Army continued to hold its position. Little or no fighting took place.

Further north, German forces within the Korsun pocket began to move toward the south, hoping to reach any outside relieving force.

4 February 1944

General Konev ordered the 7th Guards Army to transfer the 41 GRD (estimated combat strength 2-3,000), sending it north in a forced march to join the 4th Guards Army. The 2 UF needed additional forces to block German rescue efforts toward the Axis troops trapped in the Korsun pocket. As the division moved off, weather and movement conditions were bad.

> The commanders and political workers explained the task to the soldiers, although in general terms we are going to finish off the encircled enemy grouping; we must hurry. The march was incredibly difficult. The distance is more than 200 km [125 miles], and after rain the dirt roads turned into a mess. One after another, huge potholes filled with liquid mud. It was impossible to walk on foot along the road, and the soldiers moved along the curbs, with difficulty [walking with] mud-clad boots. Cars [trucks] skidded, many of them had to be left [behind], and shells, cartridges, [and other] products [were transferred] to carts. The horses were exhausted and stopped. Then the soldiers took in duffel bags all that could [be carried]. [They] walked day and night, walked to exhaustion. Guns and howitzers were dragged with horses in [shifts]: they transported one gun for 1-1.5 km [1 mile], then returned for the second. [394]

The 41 GRD would never return to the 7th Guards Army, but it fought valiantly, assigned to other armies until the end of the war.

The second German rescue attempt toward the Korsun pocket was launched (code name Wanda). This attack set out about 15 miles west of the first attack. The 16 and 17 Pz Ds, with supporting infantry divisions of the III Pz C, advanced quickly but were only able to advance some 12 miles before reaching the Gniloi Tikitsch River, where they were stopped by Russian forces. The encircled troops were still 17 miles away.

5 February 1944

The 5th GA, 7th Guards Army and 57th A again held their positions.

Armies of the 3 UF approached Krivoi Rog. The defending German forces were under tremendous strain trying to hold their positions.

6 February 1944

The weather in the area of the 2 UF shifted between light rain during the day to frost at night. This mix insured difficult cross-country movement. The 7th Guards Army had the advantage of being fixed in place so it could focus upon restoring its combat strength and not conducting active operations.

Bombers of the ADD (Long Range Aviation of the VVS) paid a visit to Helsinki, Finland, on 6 February and then again on 27 February. These heavy attacks caused a great deal of damage to the capital and reminded the Finnish government, in a blunt way, that the war with the Soviet Union was unfinished business.

7 February 1944

On the lower Dnepr River, armies from the 3 and 4 UFs captured the city of Nikopol and the manganese ore mines in that area. The alloying (blending) of steel and manganese produces high-strength steel, which was critical for the manufacture of vehicles, weapons and munitions during the war. So, the loss of these mines was another blow to German war production. Just how long would reserve stocks of manganese ore last? The clock was ticking.

8 February 1944

The Soviet command presented a (surrender or die) ultimatum to the encircled German troops in the Korsun pocket. General Wilhelm Stemmermann, commander of all the surrounded German troops, (also call Group Stemmermann) rejected the ultimatum. The fighting and slaughter continued.

During the night of 8-9 February, the Luftwaffe delivered to the airfield at Korsun 150 tons of ammunition, and 60 cubic meters of fuel (15,850 gallons). They flew out 566 wounded soldiers on the return flights. [395] Everyone was thinking this was a little Stalingrad, and they were right! Only this time the battle would turn out somewhat differently.

9 February 1944

The rear supply troops of the 7th Guards Army experienced a small tragedy. At 15:30 hours an abandoned cigarette started a fire that spread to a stockpile of 650 captured German 105 mm artillery shells. The resulting explosion killed one officer and one solider and wounded seven other people. [396]

10 February 1944

The 7th Guards Army issued Headquarters Combat Report #41, according to TsAMO record ID #131310433. "Because the Army had almost completely relocated [moved forward during recent fighting], new positions and defensive structures were erected. Reconnaissance [was carried out] and [the] positions of Army formations [were] unchanged." The report continued that on 9 February "the Army suffered four casualties and on… 10 February 30 soldiers and one officer were [casualties]." The cause of these casualties is unknown.

Additional German units, most notably the 1 SS Pz D *Leibstandarte*, were brought up to help in the attack toward the Korsun pocket.

11 February 1944

General Konev ordered the 7th Guards Army to transfer its only tank unit, the 27 GTB, with 28 - T 34s and 2 - SU 122s, to the Korsun pocket area as the fighting intensified. They were ordered to leave on 12 February and to hurry north, most likely by rail. This is another example of a front commander shifting around units and formations under his command like pieces on a chessboard. Reinforcements were needed further north, so armies along quiet sections of the front were stripped of whatever that was required.

12 February 1944

Due to the breakout of some German troops from the pocket, *Stavka* ordered that all troops fighting around the Korsun pocket would be under of the command of the 2 UF (i.e., General Konev). Both General Vatutin, 1 UF commander, and Marshal Zhukov, *Stavka* representative who was coordinating the offensive, were not pleased with this change in command. *Stavka's* order positioned General Konev to be able to claim all the glory for the Korsun battle. [397]

General Konev stated in his book, *Front Commander's Notes*, Chapter 4:

> The 53rd, 5th Guards, 7th Guards, [and] 57th Armies conducted reconnaissance and occupied their former position. The 5th, 7th, and 57th armies defended the occupied lines on the left wing of the front [;] they were ordered to have reserves in case of maneuvers both in areas of the encircled group, and for launching an offensive according to the general plan of the front.

He went on to say that the VVS was flying 163 sorties per day in support of the Korsun battle while the enemy was flying 150 sorties and another 60 transport sorties.

Due mostly to the 7th Guards Army's inactivity, a TsAMO record, *Summary of Material Support,* ID #132344489, showed the Army as having good supplies of food, a large quantity of vodka and tobacco, and large stores of oats and hay for the horses. Overall, the Army was well supplied with foodstuffs, which helped both man and beast stay healthy and strong.

13 February 1944

Weather conditions began to change; a storm approached the area of the 2 UF, which reduced visibility. The resulting fog, in turn, limited air support from both the VVS and the Luftwaffe.

14 February 1944

Troops of the 2 UF captured the city of Korsun. The overall pocket was greatly compressed by this time, as German forces within the pocket slowly fought their way south. Every step closer to freedom left a trail of death in its wake. Time was running out for the encircled troops.

Temperatures in the Korsun pocket area dropped, and snow began to fall.

15 February 1944

A TsAMO record ID #136672985, *Transportation Unit map for the 7th Guards Army*, showed a supply road linking Kirovograd with the rail line 30 miles to the east behind the Army's front. The route followed Kirovograd, Pokrovskoye, Plavin, Mitrofanovka to Sharovka Station. (See Map #34 for road supply route.) This map also implied the Sharovka Station and rail line were active and receiving cargo. With this transportation layout, even under the worst mud or snow conditions, Army level transportation should have been able to keep divisional supply dumps fully stocked.

With extra time to prepare his army before the next big battle, General Shumilov instituted, on this day, a broad base training schedule. The titles of three TsAMO Archive Documents, ID #133391694, 133391707 and 133391710, illustrate this effort:

- *The training program for the sapper training of rifle units in the formations of 7th Guards Army*
- *Schedule for combat training of officers of the chemical department 7th Guards Army*
- *Training schedule with the officers of the staff of engineering troops 7th Guards Army*

16 February 1944

With the German III Pz C attacking from outside the pocket, General Stemmermann launched his breakout attempt at 23:00 hours. Weather conditions deteriorated as the ongoing snowstorm increased in strength. Drifting snow blocked roads and made overall travel conditions difficult. German forces, mostly advancing on foot, still were able to move out of the pocket under heavy Russian fire. At first, the Russian response to the breakout was slow due to the reduced visibility. Furthermore, the movement of additional Red Army forces into the breakout area was delayed due to poor road conditions.

17 February 1944

During the night, the last rear-guard soldiers within the Korsun pocket reached German lines over the Gniloi Tikitsch River. German sources relate the grim ending of the pocket:

> The veil of silence fell over the fate of the wounded left behind as it did over those who entered the hopeless situation of captivity…. The prisoners [who] returned after many years, mostly starting in 1949…, reported what they had seen – that the Soviets had, as was their custom, immediately shot the wounded who were incapable of walking. [398]

The remaining troops, including lightly wounded, were taken prisoner, but many would not survive a brutal winter's march into captivity.

The victorious forces of the 2 UF completed the destruction of the German forces within the Korsun pocket. Historical Soviet sources claim they captured 18,000 prisoners from the Korsun pocket and killed 55,000 others, but these numbers were exaggerated wartime propaganda.

German sources claim that

> 27,703 German soldiers and 1,063 *Hiwis* [Russian helpers] had broken out unscathed. In addition, 7,496 wounded and injured had managed to reach III Panzer Corps... Also, 4,161 wounded and sick had been evacuated prior to the breakout aboard aircraft... Thus, a total of 40,423 men had survived the encirclement without becoming prisoners. [399]

An estimated 59,000 men had been initially trapped within the pocket, which meant some 19,000 men either died or were taken prisoner by the Russians.

Within the last group of German soldiers trying to escape from the pocket, General Wilhelm Stemmermann was killed. Russian troops found his body, stating "a shell splinter got him in the back... We buried him decently. We can afford to bury a general decently." A later comment stated, "Dead generals aren't all that frequent, so we gave him a proper burial." [400]

The one thing both sides did agree upon was that the escaping German troops left behind almost all their heavy equipment. Anything that would slow a person down was abandoned upon the battlefield during the last confusing night and day of the breakout. Russian troops could pick through the debris of the routed German force at leisure. Eventually, the surviving troops were re-formed into new combat units, but it would take some time for Germany to replace the combat equipment of two complete infantry corps. Despite heavy losses, both inside and outside the pocket, the Germans achieved their main objective: they rescued most of the trapped troops. In this way, there would be no Stalingrad on the Dnepr; the men were not left to their fate as was the German 6th Army at Stalingrad. For the Red Army and General Konev, another victory was added to their growing list of triumphs.

18 February 1944

General Konev recounted statements made by Major Kampov about the last group of German troops in the Korsun pocket:

> He described, somewhat gleefully, Germany's latest catastrophe: some eighty, or even a hundred thousand Germans had refused to surrender and had been forced into a narrow space [;] then the tanks smashed their heavy equipment and machine-gun posts while the Cossack cavalry finally finished them off. 'We let the Cossacks cut them up for as long as they wished. They even hacked off the hands of those who raised them to surrender!' the Marshal said with a smile. [401]

Alexander Werth, in his book *Russia at War,* p. 780-781, added a slightly different Russian account of the final bloodbath:

> They [German troops] had now emerged from the ravines and reached open country. Then it happened. It was about six o'clock in the morning. Our tanks and our cavalry suddenly appeared and rushed straight into the thick of two columns. What happened then is hard to describe. The Germans ran in all directions. And for the next four hours our tanks raced up and down the plain, crushing them by the hundreds.

For his achievements in the Ukraine and for the Korsun battle, Stalin promoted General Konev to Marshal of the Soviet Union on 20 February 1944. Some believed that Stalin himself admired Konev for his ruthlessness.

On 18 February, on behalf of the Motherland, Moscow saluted the Korsun victory of the Soviet Armed Forces with 20 artillery volleys of 224 guns.

The 2 UF received orders from *Stavka* to prepare for further offensive operations to the west, in the Uman direction and the Southern Bug River. Note: The Bug River and the Southern Bug River (also called the Vuzhni Bug) are two different rivers. The Bug River is in the north and mostly in Poland, while the Southern Bug is in Ukraine.

During the war the Vuzhni Bug was commonly called the Southern Bug to avoid confusion with the Bug River further north.

After participating in the capture of Nikopol on 7 February, the 3 UF continued its advance to the west while clearing the great bend in the Dnepr River along the way. Hitler now had forever lost the supposedly vital manganese mines that were supplying the Nazi war economy. Bit by bit, Hitler's empire was slipping away.

19 February 1944

Once again, General Konev stated in his book, *Front Commander's Notes*, Chapter 4:

> The shock groups [special assault groups] were continuously strengthened by troops operating on less active sections of the front. Corps, divisions of the 6th, 7th Guards and 57th armies, were transferred to more important sections. A large number of tanks, artillery and engineering units were also deployed from other armies. [General Konev also recounted,] We must never forget the feat of women, children, the elderly, when they, along with the soldiers, sometimes [with] swollen [feet], in poor clothes, half-starved, walked dozens of kilometers through the mud to bring ammunition to the troops.

Having saved all possible stragglers from the pocket, the III Pz C began withdrawing to a less exposed position. The battle of the Korsun pocket was over.

20 February 1944

The 7th Guards Army continued to rest and to rebuild the strength of its units.

General Konev stated,

> After delivering a short report by telephone to Stavka Headquarters about the end of the battle of Korsun-Shevchenkovsky, Stalin said, "Congratulations on your success. The government has an opinion to award you the title of Marshal of the Soviet Union. How do you look at it, do you mind? May I congratulate you?"
>
> I could only answer, "Thank you, Comrade Stalin."
>
> In this way, General Konev became the twelfth Marshal of the Soviet Union. Two days later, *Stavka* delivered by plane the marshal epaulettes sent by Marshal of the Soviet Union G.K. Zhukov. Marshal Konev said, "It was attention, congratulations, and a priceless gift."

21 February 1944

The 8th Guards and 37th Armies of the 3 UF fought their way into Krivoi Rog. The weakened defending German forces at this point barely could resist.

The 57th A was transferred from the 2 UF to the 3 UF.

22 February 1944

General Shumilov issued Directive #0090 to the Army about training to fight at night. The key point was, "During December 1943 and January 1944, the Germans used nightly offensive operations against units of our army… [and at times] with impunity and without loss." The Army needed training on how to prevent this from occurring again. He went on to list seven reasons the Germans were successful in their night attacks, and then seven measures to prevent them from doing so again. [402] This directive further expanded the 7th Guards Army's training schedule. General Shumilov was being very attentive to the necessities of keeping his army in fighting shape.

Several historical maps show that on this date the 57th A took over a six-mile section of the front from the 7th Guards Army directly in front of the village of Novgorodka. This transfer allowed the 7th Guards to shift its own forces further north, amassing troops near Kirovograd.

Further south, after prolonged and heavy fighting, troops of the 3 UF finally liberated Krivoi Rog from German occupation. This victory unhinged the German defensive position along the lower Dnepr, ultimately leading to the loss of the Nikopol bridgehead.

23 February 1944

On the orders of Stalin, as reprisals for real or imagined acts against Soviet authority or for "cooperating with the Nazi invaders," over 100,000 NKVD officers and soldiers began the roundup and deportation of Chechens and Ingush peoples from the North Caucasus region to Central Asia. The plan was to deport some 450,000 people, later revised to 495,000, but some Chechen sources put the total at 650,000. The total death count during and after the deportation was put at over 220,000, but the true total will never be known. [403] In 2004, the European Parliament acknowledged this campaign by the Soviet Union, against its own citizens, as an act of genocide.

24 February 1944

The 7th Guards Army maintained its position and continued to prepare for the impending offensive.

On the diplomatic front, Soviet and Finnish diplomats met in Stockholm, Sweden, to discuss a separate peace between Finland and the Soviet Union. The Finnish diplomats rejected the Soviet terms for peace, but they agreed that further negotiations would take place in Moscow on 26 March. [404]

25 February 1944

On or about this date, Hitler ordered the OKW to refurbish its plan for the military occupation of Hungary, Operation *Margarethe* (Pearl). Hungary was now only a reluctant Axis ally. As the fortunes of war had turned against Germany, so too had Hungary's support slipped away.

26 February 1944

The 7th Guards Army began regrouping troops and artillery units to its right flank for the upcoming offensive. They used standard *maskirovka* methods to confuse, as much as possible, German intelligence officers, but no major deception effort was made. [405] However, the Germans did spot the concentration of 7th Guards Army forces for the offensive toward the end of the month. The upshot of this discovery was two major Luftwaffe attacks on 26 and 27 February. With a major new 2 UF offensive only days away, it was doubtful AG South would be able to respond adequately to this new threat.

The Luftwaffe had been quiet in the area of the 7th Guards Army during the month. Up to this point, only 80 aircraft had overflown the Army with no attacks, but that was going to change this day. In waves of up to 25 aircraft, German bombers attacked the rifle division headquarters of the 25 GRC. The 53 RD HQ was bombed in the village of Markov, and the 72 GRD HQ was hit in Yuryevka. Attacking aircraft numbered 100: 65 - Ju 87, 13 - Ju 88, 10 - Me 109, 3 - He 111, 1 - Fw 189 and 8 - Fw 190. During this attack, the Luftwaffe dropped 765 bombs. The result of this attack was not recorded, but this sizable effort must have caused losses to both troops and equipment. [406] The inclusion of large numbers of Ju 87 aircraft (Stuka dive bombers) suggested a precision attack was intended against pinpoint targets.

27 February 1944

In a repeat performance, the Luftwaffe returned to attack the same general area, but this time they focused their attention upon the 81 GRD HQ in Gubovka. Attacking aircraft numbered 112: 3 - Ju 88, 2 - Me 109, 77 - He 111 and 30 - Fw 190. This attack dropped 120 bombs. The large number of He 111 aircraft (horizontal bombers) might be related to the village of Gubovka being larger than most and located on the Ingual River, which contained several bridges. The inclusion of large numbers of horizontal bombers suggested an area attack against the entire village. Once again, losses on the ground were unrecorded.

Stavka approved Marshal Konev's plan for the 2 UF offensive. Forces of the 2 UF included 480,000 troops directly involved in the offensive; 8,890 pieces of artillery and mortars; 670 serviceable tanks; and 551 aircraft. The opposing German forces were far less abundant, with the exception of the Luftwaffe having some 500 aircraft on hand.

The 57th A launched an attack on its left flank, making a four-mile advance. This action was most likely a diversionary attack to support the start of the 2 UFs Uman/[Southern] Bug River offensive. Note: The 57th A had been transferred to the 3 UF on 21 February, so this attack was also part of the 3 UF's ongoing offensive.

28 February 1944

David Glantz, in his book *Fallen Soviet Commanders*, on page 203, relates how General N. F Vatutin, commander of the 1 UF (former commander of the Voronezh Front during the battle for Kursk) was traveling between HQs far behind the front lines when he and his escort were ambushed by Ukrainian nationalists of the Ukrainian Insurgent Army (UPA). Vatutin was seriously wounded in the leg and was quickly taken to a hospital in Kiev. Vatutin eventually developed septicemia (blood infection), which led to a condition called sepsis, which killed him on 15 April 1944.

The UPA was a force to be reckoned with as it controlled large sections of Ukrainian territory. In response to this attack upon Vatutin and the destruction of (a reported) NKVD battalion, the Soviets deployed 30,000 troops on an anti-partisan sweep against the UPA throughout the Volyn province in northwest Ukraine. [407] Losses were heavy on both sides during this fighting, but the UPA survived to fight on for independence long after the war.

David Glantz expands upon officer losses for the Red Army in his book, *Colossus Reborn*, with a table on page 535. This table shows that 631,008 Red Army Officers died and a further 392,085 were missing in action during the war (1941-1945), for a total of 1,023,093 killed or missing. Far from being exempt from the horrendous casualties at the front, these numbers illustrate the extent of the overall carnage of the Red Army officer corps.

A historical map shows the 7th Guards Army holding the section of the front directly southwest of Kirovograd, with the 5th GA to the north. With the 57th A now part of the 3 UF, the 7th Guards Army had become the left flank army for the entire 2 UF. This was an important position because the flank army, either left or right flank of the front, is responsible for maintaining contact with units of the adjacent front and for guarding the vulnerable boundary between fronts.

The new offensive plans were in place with Soviet commanders settling the final details. The 5th GA and the 7th Guards Army were slated to conduct a supporting attack, on the left flank of the 2 UF, toward the city of Novo-Ukrainka, while the right flank armies of the 2 UF attacked toward the German supply and rail center of Uman. Marshal Konev and other commanders hoped these attacks would lead to an advance by the entire 2 UF as far as the Southern Bug River.

29 February 1944 (Leap Day/Year)

The 2 UF issued a report, TsAMO record ID #130357746, on the amount of ammunition and fuel each army of the front had available for the upcoming offensive. The 7th Guards Army had listed:

> Small arms ammunition - between 1.2 to 2.1 units of fire
> Artillery and mortars - between 1.3 to 2.7 units of fire
> Fuel - gasoline 2.6, diesel 2.7, naphtha 8.3, and kerosene 2.6 standard loads (refills)

These were adequate amounts of ammunition and fuels to conduct offensive operations.

With the front mostly inactive up to this point with no major operations, the 7th Guards Army should have suffered few combat casualties, but the weather, sickness, and accidents would have added additional casualties. Unfortunately, there are no historical Soviet casualty reports for the 7th Guards Army during this month, so the rough estimate of 35 casualties per day is used, which equals a total of 980 losses for the month. This conservative estimate is based upon past casualty reports for a non-combat month.

1 March 1944	
7th Guards Army	
Assigned – 2 Ukrainian Front	
Colonel General M.S. Shumilov Commanding	
Rifle, Airborne and Cavalry Units	24 Guards Rifle Corps (36, 81, GRDs, 8 GAD) 25 Guards Rifle Corps (72 GRD, 53, 297 RDs) 27 Guards Rifle Corps (50 RD) 409 RD 3 Separate AT Rifle Battalion 62, 63, 64, 65, 66, 94 Penal Companies
Artillery Units	11 Artillery Division (40 Howitzer Artillery Brigade, 45 Gun Artillery Brigade) 161 Guards Gun Artillery Regiment 114, 115 Guards Tank Destroyer Regiments 30 Antitank Artillery Brigade 615, 263, 290 Mortar Regiments 5 Antiaircraft Artillery Division (670, 743, 1119, 1181 AA Regiments) 162, 225 Guards AA Regiments 662 Separate Aerial Observation/Air Raid Warning Company
Armored and Mechanized Units	27 Guards Tank Brigade 34, 38 Separate Armored Train Battalions
Engineer Units	175, 329 Separate Engineer Battalions 46, 49, 329 Engineering Battalions
Estimated Combat Strength Reported Ration Strength	33,029 men * 39,874 men and women * (including an estimated 6,845 support personnel.)

(BSSA 44, BSA 44, TsAMO Archive)

*** A TsAMO Archive report for the 2 UF, ID #113723856, lists the ration strength of the 7th Guards Army on 1 March 1944 as 39,874, along with 5,756 horses.**

The spring thaw started in March and continued into April, making dirt roads completely unsuitable for motor vehicles. But before that happened, a blizzard hit northern Ukraine; drifting snow blocked trains and disrupted logistical efforts. Of course, the local population was recruited and mobilized to clear the snow-covered tracks and streets. [408] The snow did not last long as "[t]he coming thaw and the rains that began in early March further worsened the condition of dirt roads, and the movement of vehicles along their entire length was extremely difficult." The 2 UF only had 1,763 trucks in operating condition and in the mud, they could only advance six - nine miles per day, so "all available horse-drawn vehicles and pack units were mobilized." At the rear of one army, "5,400 people [civilians] carried loads manually" to the front. [409]

In preparation for the new offensive, the 27 GTB was reassigned to the 7th Guards Army after its transfer to the Korsun pocket area. The 27 GTB would need several days to arrive by rail.

1 March 1944

Due to General Vatutin's wounds, Marshal Zhukov took over direct command of the 1 UF.

Despite von Manstein's efforts, Hitler refused to shorten the front line and to gather reserves. Hitler told General Zeitzer (Chief of the Army General Staff OKH), "The Russians are bound to stop attacking some time. They have been attacking non-stop since last July and can't go on [forever]." [410]

Axis naval forces operating in the Black Sea transferred to the Crimea 7,000 men; 11,000 horses; and 45,000 tons of supplies. On the return trip, a good number of wounded were evacuated by sea. Also, most of the troops of the 111 Infantry Division were flown to the Crimea.

2 March 1944

The 7th Guards Army continued to regroup troops, artillery and any tanks to its right flank. Thick mud on and off the roads made the movement of troops, weapons and supplies difficult.

3 March 1944

The attack plan called for the 27 and 25 GRCs on the right flank of the Army to make the main attack toward Novo-Ukrainka. The 24 GRC would be covering the center and left flank of the Army, with its rifle divisions holding their positions at first, only advancing when the time was right.

To the north, the 5th GA also would attack and converge upon Novo-Ukrainka. To the south, the 57th A would advance and was expected to keep pace with the 24 GRC when it moved forward.

4 March 1944

Units of the 2 UF conducted reconnaissance along the entire front, gathering valuable intelligence and keeping the Germans guessing what was taking place. During the regrouping, units of the 2 UF used normal *maskirovka* methods, but no major deception effort was made. "German intelligence had not expected such a rapid Soviet resumption of the offensive after the heavy fighting of February. Most of the Soviet concentration went undetected." [411]

5 March 1944

Once again, Marshal Konev stated in his book, *Front Commander's Notes*, Chapter 4:

> On March 5, the 5th Air Army had 129 – Po 2 bombers and 89 – Il-2 attack aircraft. The 4th Fighter Aviation Corps had 95 - Yak and La 5 aircraft. The 7th Fighter Aviation Corps had 150 aircraft, the 312nd Night Light Bomber Aviation Division [had] 74 - Po 2 aircraft, [and] a Separate Reconnaissance Regiment of the front [contained] 14 - Po 2 aircraft, [totaling] only 551 aircraft. The enemy by this time had 500 aircraft. In essence, the number of aircraft was equal, but the enemy still had superiority in daytime bomber aviation. The 5th Air Army was quantitatively inferior to the enemy in bombers by 90 combat aircraft, but surpassed it in fighters by 94 aircraft.

In heavy fog, the right flank of the 2 UF began the Uman Operation, also called the Uman-Botoshanskoy Operation. The German front line was broken in several places as Soviet tank forces advanced through the gaps. Stalin exerted direct control of the operation and coordinated the efforts of both the 1 UF and 2 UF.

Further south, the 5th GA and 7th Guards Armies were scheduled to open their part of the offensive on 8 March with attacks directly west of Kirovograd. This area just happened to be the junction between the German 8th Army to the north, in the 5th GA's attack zone, and the 6th Army to the south, in the 7th Guards Army's attack zone. Also on 8 March, the right flank armies of the 2 UF would continue their attacks using maximum force, while the center armies of the 2 UF would hold back at first. With both flanks of the 2 UF advancing, German forces in the center would be forced to withdraw or risk being surrounded. Unfortunately, during the first three days of the offensive the weather was so bad that aircraft could not fly which reduced the power of the Russian attack.

With the new offensive already underway, the 27 GTB had not yet fully arrived. The only tank support for the entire 7th Guards Army was scattered behind the lines, with 14 – T 34s at one train station, two – T 34s and one – SU 122 at another, with five more T 34s on the way. A TsAMO record ID #100304994 stated, "The combat equipment did not arrive restored [repaired] and ready for loading of combat materiel [;] [the] unit is held back due to the lack of arming supplies." This lack of planning certainly did not please General Shumilov. Regardless, the Army would advance on 8 March as planned even if no tank support were available.

6 March 1944

The right flank armies of the 2 UF continued their attacks, making good progress despite terrible road and weather conditions.

7 March 1944

The 5th and 7th Guards Armies, commanded by Generals A. S. Zhadov and M. S. Shumilov, were tasked to conduct a secondary attack. Their mission was to break through the enemy's defense along an 11-mile section of the front between Shestakovka and Mukhortovka, then advance to the Novo-Ukrainka area and capture the city. An advance of this magnitude was thought to be sufficient to ensure the success of the main offensive. [412]

The overall initial objective of the 2 UF offensive was an attempt by General Konev to surround and to destroy any German forces caught between the two widely separated flanking attacks. As the offensive developed further, the 5th and 7th Guards Armies continued their advance, acting as left flank protection for the main attack of the right flank.

8 March 1944

The 5th GA and 7th Guards Army deployed 1,230 guns and mortars during their initial breakthrough effort. Many heavy artillery pieces were unavailable due to the muddy conditions. The two armies were initially directed toward the towns of Novo-Ukrainka and Pervomaysk. Pervomaysk was located directly on the Southern Bug River. Apparently, T-34 tanks from the 27 GTB were fighting with the 36 GRD and maybe eight more were fighting with the 53 RD. The historical record is not clear on this issue. The dispersal of the few available tanks in this manner indicates they were being employed in an infantry support role rather than attempting an armored breakthrough on their own.

According to Marshal Konev, the logistical supply effort within the 2 UF varied between armies. As for artillery ammunition, the plan was to accumulate 1.5 to 2 Units of Fire (UOF) at the front and then expend 1.3 UOF during the initial bombardment with the remainder being used during the advance. The ammunition goals were mostly achieved, but fuel supplies were lacking in some armies. For example, the 5th GTA had 2.1 standard loads of gasoline and 2.6 of diesel, while the 2nd TA had 0.2 loads of gasoline and 5 loads of diesel. As expected, fuel supplies were critical for the successful operation of mobile troops like tank armies.

Marshal Konev further outlined these operations in his book, *Front Commander's Notes*, Chapter 4:

> The 2nd Ukrainian Front was to strike from the front of Kirovograd — Zvenigorodka — Shpola in the general direction of Uman with the task of breaking up the Uman group of the

enemy and mastering the border: Ladyzhyn, Gayvoron, Novo-Ukrainka. In the future, the front troops were to develop an offensive and go to the Dnestr, on the site Mogilev — Podolsky — Yagorlyk. On March 8, the offensive of the left wing of the front began in the general direction of Novo-Ukrainka. The troops of the 5th and 7th Guards Armies went on the offensive at 7 o'clock, 50 minutes [07:50 hours] after 56 minutes of artillery [bombardment]. On the same day, they broke through the enemy's defenses over a 12 km [7 miles] stretch and advanced to a depth of 7 km [4 miles]. Such a slight advance was due to the low density of the artillery of the armies and the almost complete absence of tanks for direct infantry support. (See Map #35.)

9 March 1944

Along its front, the 7th Guards Army attacked southwest of Kirovograd with two rifle corps, and they managed to make good progress mostly by avoiding the heavy fighting to the north between the 5th GA and the Pz GD *Grossdeutschland.* Five T-34 tanks from the 27 GTB were fighting with the 53 RD with the tank brigade having lost three T-34s so far. At the end of the day, the 27 GTB contained about four T-34s and one SU-85 that were operational. (Note: The reference to a SU-85 might be a typographical error in the records and might be a SU-122 instead.)

10 March 1944

The 7th Guards Army continued to advance southwest of Kirovograd. The 27 GRC continued to make progress, with the 8 GAD on the far-left flank joining in and advancing to the Ingul River.

The right flank armies of the 2 UF had made good progress, pushing German forces back and capturing the cities of Uman, a major road center, and 12 miles northwest of Uman, the city of Khristinovka, a major rail center. The loss of these two major logistical and transportation centers guaranteed AG South would fall back to the lower Southern Bug River. The final liberation of Ukraine was well underway.

Even as the troops of the 5th and 7th Guards Armies achieved success, Marshal Konev thought their advance was not developing fast enough. Part of the reason for the slow advance of the 5th and 7th Guards Armies was the lack of supporting mobile forces, especially tanks. After all, the 7th Guards was an infantry army, supported only by its depleted 27 GTB, listing only eight T-34s and one SU-85/ SU-122 at this time. Note: If the Army did have a SU-85 assigned, the SU-85 was a tank destroyer based upon a T-34 chassis with an 85 mm AT Gun. The 5th GA lacked even that, having no tank support at all. Nevertheless, the advance continued at a slow pace through viscous mud and against determined German resistance.

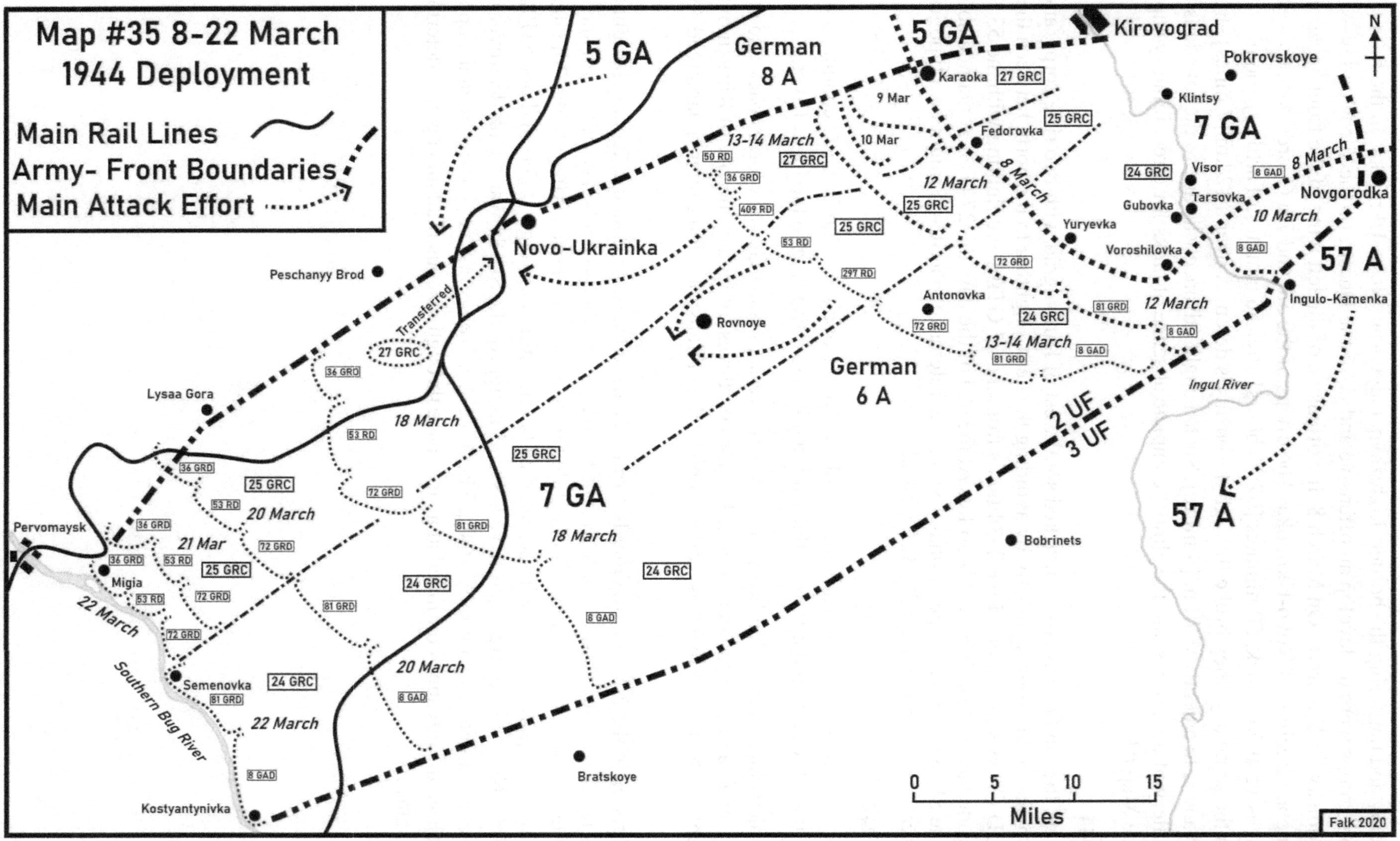
Map #35 8-22 March
1944 Deployment
Main Rail Lines
Army- Front Boundaries
Main Attack Effort
5 GA
German
8 A
5 GA
Kirovograd
Pokrovskoye
Klintsy
7 GA
Karaoka
27 GRC
9 Mar
10 Mar
Fedorovka
25 GRC
13-14 March
50 RD
36 GRD
27 GRC
12 March
8 March
24 GRC
Visor
8 GAD
8 March
Tarsovka
Gubovka
Novgorodka
10 March
409 RD
25 GRC
25 GRC
Yuryevka
Voroshilovka
8 GAD
57 A
Novo-Ukrainka
53 RD
72 GRD
Peschanyy Brod
297 RD
Antonovka
81 GRD
12 March
Ingulo-Kamenka
Transferred
Rovnoye
24 GRC
72 GRD
8 GAD
27 GRC
13-14 March
8 GAD
81 GRD
36 GRD
German
6 A
Ingul River
Lysaa Gora
18 March
53 RD
2 UF
3 UF
25 GRC
36 GRD
25 GRC
72 GRD
7 GA
53 RD
20 March
57 A
Pervomaysk
36 GRD
81 GRD
18 March
21 Mar
72 GRD
36 GRD
53 RD
25 GRC
Bobrinets
Migia
24 GRC
24 GRC
53 RD
72 GRD
81 GRD
8 GAD
22 March
72 GRD
20 March
Semenovka
24 GRC
Southern Bug River
81 GRD
8 GAD
22 March
8 GAD
Bratskoye
0
5
10
15
Miles
Kostyantynivka
Falk 2020

11 March 1944

All the rifle divisions of the 7th Guards Army began a general advance as the German defenders started to fall back. Fighting was heavy in places and sporadic in others. The Germans retreated, but they were doing so at their own pace.

12 March 1944

The 7th Guards Army continued to press forward across its entire front. Progress was slow but steady. German forces that were facing the 7th Guards Army were executing a planned retrograde movement, and they were very skilled in the use of this tactic.

Due to the not-so-secret Hungarian desire to open surrender negotiations with the Soviets, Adolf Hitler decided to launch Operation Margarethe, the plan to militarily occupy Hungary and remove the Hungarian Regent Admiral Miklós Horthy from power. To facilitate this action, Hitler invited Admiral Horthy to Salzburg to discuss military options for Hungary. [413]

13 March 1944

Units of the German 8th Army withdrew at a steady rate, giving up ground as needed to prevent being surrounded. Occasionally, rearguard troops would stand and fight, delaying the advance of the Red Army for a time, so that the main body could retreat unimpeded. Eventually, the German plan was to gain enough time to regroup, reorganize and resupply its forces and to establish a new continuous front line along some defensible position behind the Southern Bug River.

14 March 1944

At this point in the battle, the 5th and 7th GAs were about halfway to Novo-Ukrainka, their first major objective.

15 March 1944

In his book *Front Commander's Notes*, Chapter 4, Marshal Konev commented about the slow advance of the 5th and 7th GAs:

> In my telegram of March 15 sent [to] the commander of the 5th and 7th Guards [A]rmies, it was stated:
>
> The pursuit of the enemy by your armies is slow. Understand that the enemy is running [retreating]. Our front troops are already fighting for Vapnyarka. You [occupy] the left wing of the front; you [lag] 200 km [125 miles] behind the right wing of the front. The main reason for the slow pace of pursuit is:
>
> 1. Lead troops [are] evenly [spaced] along the entire front and comb through the whole area.... There is no courageous maneuver with access to the enemy's retreat path. There is no fist to destroy the enemy with his resistance in the main directions. In this regard, you allow the enemy [the opportunity] to withdraw his manpower and equipment.
>
> 2. The separation of the control of armies, corps, divisions from the troops. The rear is especially behind. Troops always require solid control. Strong control ensures fast pace, and troops do wonders.

In their defense, note that the 5th and 7th Guards Armies were advancing through thick mud with little tank support. The commanders of the 5th and 7th Guards Armies were well aware they lacked that armored fist to use against the determined German rearguard units. Overall, they were advancing about eight-ten miles per day, which was sustainable under these conditions.

16 March 1944

Units of the 27 GRC fought west of Novo-Ukrainka and repelled several German counterattacks. Having cleared part of the city, they pursued German troops to the west.

The 2 UF released a rail report ID #455198953, *The restoration of the railway in the front zone lags far behind the advance of the troops.* Marshal Konev was informing Moscow about the German destruction of rail lines and stations is his area. The resulting supply problems due to the rail disruptions and thick mud added to the slow advance of infantry armies.

17 March 1944

Formations of the 5th and 7th Guards Armies fully occupied Novo-Ukrainka and achieved their first major objective of the offensive. The 5th and 7th Guards Armies at this point were half way to the Southern Bug River. To their front, German forces continued to withdraw toward the Bug.

On the south flank, the 57th A lagged behind somewhat as it captured the road junction city of Bobrinets. (See Map #35 for location.) The 57th A pushed forward as quickly as possible and soon accelerated its advance; the race to the Bug was on.

At this point of the offensive, the armies of the 2 UF operated in two widely separated groups. The main attack force, with three tank armies on the right flank of the front, had already reached and blasted their way over the Southern Bug. These forces then pushed forward another 60 miles, reached the Dnestr River along a 62-mile front, and were preparing to force the river. In turn, the left flank armies were just approaching the Southern Bug River.

Accepting Hitler's invitation, Admiral Horthy boarded a train that left Budapest station, destined for his confrontational meeting with Hitler in Salzburg, Austria. Admiral Horthy must have known or suspected that Hitler knew of his secret desire to surrender Hungary to the Soviets and that he very well could have been travelling to his own arrest, torture or execution.

18 March 1944

At 18:00 hours, the command staff of the 2 UF ordered the 25 GRC to take over the section of the front occupied by the 27 GRC. Additionally, the 25 GRC receive the 36 GRD from the 27 GRC and in turn gave the 297 RD to the 27 GRC. The 25 GRC now contained the 36, 73 GRDs and 53 RD. After this minor reorganization, the 27 GRC then transferred to the 2 UF reserve with the 50, 297 and 409 RDs. [414] Marshal Konev ordered this transfer so that the 27 GRC eventually could reinforce the 52nd A on the right flank of the 2 UF.

This transfer left the 7th Guards Army with only two rifle corps and five rifle divisions. During this transfer, the 7th Guards Army had retained its Guards' Divisions, thus helping to maintain its elite status.

During the night and into the next day, the right flank armies of the 2 UF crossed the 650-1,000 foot-wide Dnestr River. At this point, units of the 2 UF were about 65 miles from the prewar border between the Soviet Union/Ukraine and Romania.

Admiral Horthy arrived at Salzburg, along with several Hungarian government officials in attendance. Hitler met with the Hungarian Regent to discuss a change of leadership within Hungary (i.e., to a government more oriented toward Germany and its war effort against the Soviet Union.) [415] This meeting was deliberately inconclusive as Admiral Horthy avoided Hitler's demands, and he did not sign the protocol that Hitler presented. Hitler kept Admiral Horthy occupied until late in the evening before allowing him to return to Budapest. Hitler knew full well that by this time German troops already were marching into Hungary.

19 March 1944

Due to the recent transfers, the now reduced 7th Guards Army continued to advance toward the Southern Bug River. German units were withdrawing anyway, so this advance was more of a slow pursuit than an offensive charge. Simply put, both the 5th and 7th Guards Armies were too weak to do much more than to keep the pressure on retreating German units. Not letting them settle into a good defensive position, the Armies kept forcing them back, giving them no rest, constantly nipping at their heels.

When Admiral Horthy's train reached Budapest at noon, a German military honor guard was waiting for him at the station. Horthy knew Hitler had outmaneuvered him, both politically and militarily, and that he and Hungary had few options left.

20 March 1944

The Army had reached a line about 10 miles from the Southern Bug River. General Shumilov ordered the right flank 25 GRC with its three rifle divisions to prepare for a final thrust to the river. The left flank 24 GRC was lagging behind somewhat.

To the north and south, both the 5th GA and 57th A also were approaching the Bug in strength. During the evening, advance units of the 5th GA had in fact reached the Bug and were engaged in combat for control of the city of Pervomaysk. The first German line of defense stopped this minor attack. The main body of the 5th GA, 7th Guards Army and 57th A were now aligned as they pushed forward together to the Bug.

To the north, by the end of the day, the right flank of the 2 UF had established several bridgeheads over the Dnestr. Almost immediately, Soviet forces started to operate pontoon ferries on the river and to build bridges across the river.

21 March 1944

The attack by the 25 GRC was successful, but the rifle troops were able to advance only five-six miles. They would reach the Southern Bug River the next day. Marshal Konev was expecting both the 5th and 7th Guards Armies to cross the Bug and to seize bridgeheads on the far shore. Units of both the 5th and 7th Guards Armies were now engaged trying to capture Pervomaysk. This action was most likely the last rearguard action by the Germans on the east side of the Bug. They needed some time to prepare bridges and key structures for demolition.

The weather began to clear somewhat, which allowed the VVS, 5th Air Army, to fly 123 sorties against enemy positions and aircraft. The Luftwaffe, having access to better, all-weather airfields, were able to generate over 600 sorties in the Bug/Dnestr area. The results of all this air activity are unknown.

22 March 1944

The rifle divisions of the 7th Guards Army reached the Southern Bug River along its entire assigned sector and firmly held the left bank of the river. The divisions were aligned north to south: 36 GRD, 53 RD, 73 GRD, 81 GRD and 8 GAD. Likewise, the 5th GA and 57th A also reached the river. Note: By the end of the day, units of the 5th and 7th Guards Armies had captured Pervomaysk. At this time, most likely small advanced units from all three armies had crossed, or were trying to cross, the Southern Bug. Once again, the changeable weather brought a piercing wind, low hanging clouds, and snow mixed with rain that made traveling on roads treacherous. Spring had not yet arrived.

In his book *Front Commander's Notes*, Chapter 5, Marshal Konev noted that he ordered General Shumilov to speed up the crossing of the Southern Bug River and to seize the Razdelnaya station, railway junction area. He also said, "But since M.S. Shumilov had no maneuvering means, except infantry, he did not succeed in performing the maneuver south of Razdelnaya." This was a strange order because at this time Razdelnaya was located 90 miles behind enemy lines to the southwest and just 43 miles north of Odessa on the Black Sea coast. In other words, this was a totally unrealistic order to give an understrength infantry army with no mobile forces. But *Stavka* might have prompted Konev to issue this unworkable order.

Stavka ordered Marshal Konev to turn the left flank armies of the 2 UF to the south in order to assist the advance of the 3 UF and to cut off retreating enemy forces to prevent them from crossing the Dnestr to the west. With this order, the 5th GTA, 53rd A, 5th GA and the 7th Guards Army were supposed to turn toward the south and begin to operate in the areas west of the Bug and Dnestr Rivers. To accomplish this, the 5th and 7th Guards Armies first would need to cross the Southern Bug River in strength.

23 March 1944

After a good deal of political intrigue, the Hungarian Regent Admiral Horthy relented to Hitler's demands and appointed a new president of Hungary. In defiance of Hitler's wishes, Horthy chose Hungarian Field Marshal Döme Sztójay to be the new Minister President. Hitler reluctantly accepted this choice. Now Germany would be able to officially station troops on Hungarian territory, and the Hungarian military would renew its opposition to the Soviet Union. [416] The coming German/Soviet confrontation in the Balkans now could proceed, as Hitler planned, with the full participation of Hungarian forces. Admiral Horthy would remain Regent, and he also would continue scheming to save Hungary from destruction.

General Carl Spaatz, commander of the U.S. Fifteenth Air Force (which was flying from southern Italy), told his boss, General Henry H. Arnold, commander of all U.S Army Air Forces, that he intended to prioritize air attacks on Romania soon. He said, "It is of critical importance to the situation on the Eastern Front and in Romania to act immediately and in the fullest possible strength with the Fifteenth Air Force." [417] Their future efforts would do much to interfere with German and Romanian troop and supply movements by rail. These attacks could only help the Soviet offensive plans against Romania.

24 March 1944

On or about this date, major units of the 7th Guards Army attempted to cross the Southern Bug and to form a bridgehead on the western/right bank of the river. These efforts failed to win a bridgehead due to firm German resistance. General Shumilov's troops would not seize a small bridgehead on the western shore until 27 March. Shumilov then reorganized his forces and prepared for further offensive operations to the southwest.

25 March 1944

A much-reduced 7th Guards Army issued a combat strength report; this was TsAMO record ID #132344492. This historical document showed the combat strength of the five assigned rifle divisions. These divisions contained the following combat troops:

8 GAD – 2,971 men
36 GRD – 2,889
72 GRD – 3,235
81 GRD – 2,797
53 RD – 2,882

These numbers totaled 14,774 men, which was slightly less than the 16,000 troops normally found within only two full-strength rifle divisions at this time. With these five rifle divisions and all supporting combat arms units combined, the entire Army had a combat strength of only 23,963 men.

On this same day, another report was issued; this was TsAMO record ID #132344494, *Information on the course of mobilization of the local population by compounds of 7th Guards Army.* This report listed the number of new conscripts that were, or were going to be, assigned to each rifle division. These totals were as follows:

> 8 GAD – 850 men (conscripts)
> 36 GRD – 4,829
> 72 GRD – 890
> 81 GRD – 2,405
> 53 RD – 1,490

The 190 Army Reserve Training Regiment retained another 1,345 men within its ranks.

By these efforts, a total of 11,809 men were added to the 7th Guards Army.

As can be seen, the influx of these new conscripts increased the combat strength of the Army by almost 50 percent. Together, these two TsAMO reports clearly illustrated just how important these so called "Booty Troops" were to maintaining the fighting strength of the Red Army. This list also shows how guard units like the 36 and 81 GRDs obtained preferential treatment by receiving the bulk of the new recruits.

However, these new conscripts were not real soldiers yet. If time allowed, these recruits would spend several weeks in the Army's 190 Army Reserve Training Regiment for basic combat training with their newly allocated infantry gear: uniforms, boots, and small arms. After their training in the rear, they would be allotted, as needed, to rifle divisions. If they lived long enough at the front, the skilled frontline combat troops would teach them how to really fight and survive.

Field Marshal von Manstein, Commander of Army Group South, flew to Hitler's *Berghof* home near Berchtesgaden, Bavaria, for a conference. Von Manstein had recently established his HQ in the city of Lvov in what was now western Ukraine, so the flight was a relatively short 550 miles. Lvov was actually a Polish city located 200 miles southeast of Warsaw. This area had been annexed by the Soviet Union in accordance with the secret protocols between Hitler and Stalin in 1939 to partition Poland between their two countries. The conference did not go well as Hitler and von Manstein continued to have different points of view about how to fight the war on the Eastern Front.

26 March 1944

The 5th GA continued fighting to expand its bridgehead on the right bank of the Southern Bug and successfully gained a large bridgehead. On the other hand, the 7th Guards Army continued to be thwarted in its efforts to cross the river in strength. The VGK would soon rectify this situation.

On the right flank of the 2 UF, Red Army units had reached and then crossed the Prut River to enter Romania. They now were fighting on enemy territory!

In his book *Front Commander's Notes*, Chapter 5, Marshal Konev stated:

> Our warriors showed a particularly good attitude towards the Romanian units, which independently stopped all resistance. An exciting event in the history of the Great Patriotic War – [our] …front troops were the first to reach the state border of the Soviet Union, and Moscow saluted [them] …with 24 volleys of 324 guns.

Note: Some Romanian units did surrender to the Russians while most bravely continued to fight alongside their German allies.

This achievement on the Prut River appears to have distracted Marshal Konev from the needs of his other armies. Taking stock of the situation, the VGK issued Directive #220057 to Marshal Konev ordering him to support the 7th Guards Army in its efforts to cross the Southern Bug River.

> *March 26. 1944 21 h 20 min*
>
> *According to reports, the 5th Guards Zhadov's army* [Colonel General Aleksey Semenovich Zhadov], [forced the Southern Bug River], *successfully moving towards the Crooked Lake, providing a large bridgehead on the right bank of the river. At the same time, the 7th Guards Shumilov's army continues to remain on the river Southern Bug, unsuccessfully trying to force it on their own.*
>
> *The High Command Headquarters orders:*
>
> *1. Through a section of 5th Guards* [b]*ring to the right bank of the river, Southern Bug major forces of 7th Guards, ...to* [be] *use to strike south in order to collapse the defense* [of the] *enemy in front of the 7th Guards, and the right wing of the 3rd Ukrainian Front, thereby contributing to their westward movement.*
>
> *Supreme High Command Headquarters I. Stalin A. Antonov*

On the diplomatic front, on this day Finnish diplomats left for Moscow to discuss peace terms.

27 March 1944

In his book *Front Commander's Notes*, Chapter 5, Konev stated, "[T]he 5th and 7th Guards Armies continued fighting to expand the bridgehead on the right bank of the Southern Bug."

The above statement is true, but due to the VGK Directive from the previous day, units of the 7th Guards Army, most likely just the 53 RD and 72 GRD, had performed a forced march during the night to cross the Bug in the area of the 5th GA. In the morning, this flanking movement allowed General Shumilov to launch an attack into the left flank of German forces along the river. By the end of the day, the 7th Guards Army had won a small bridgehead from the German defenders. This successful operation apparently allowed the 81 GRD to attack directly across the river and to join the battle on the western side of the Bug. (See Map #36.)

28 March 1944

With the Southern Bug River defensive line broken, the weak German defenders continued to withdraw to the west.

General Shumilov hurriedly moved the remaining rifle divisions of the 7th Guards Army over the Bug and began the pursuit of the defeated enemy. During the day, units of the 25 GRC reached the suburbs of Vradiyevka, a major city in the area. At the end of the day, the Army had advanced over 14 miles in some places. Also, the 72 GRD was transferred to the 24 GRC to strengthen the left flank.

29 March 1944

The pursuit continued as Germans forces lacked the strength to stand and to fight against a superior force. The 7th Guards Army gained another 15 miles during the day.

30 March 1944

As they continued to advance, the Army won another 11-14 miles from the enemy. The 7th Guards Army wheeled to the south and reached the tiny, swampy Tylihul River. To the north and south, the 5th GA and 57th A kept pace as German forces all along the southern section of the Eastern Front fell back toward the Dnestr River.

In the morning, Hitler sent his own Condor aircraft (Fw 200) to transport Field Marshal von Manstein, Commander of Army Group South, and Field Marshal von Kleist, Commander of Army Group A, to Berchtesgaden for a surprise meeting. During the evening, Hitler awarded von Manstein a medal (the Swords to the Knight's Cross) and then Hitler informed both Field Marshals that they would be replaced. Hitler said their commands would be placed "in other hands." Specifically, Field Marshal Walther Model would take over command of AG South on 2 April 1944, while Colonel General Ferdinand Schörner would take over AG A. [418] As being senior commanders, both von Manstein and von Kleist were then most likely transferred to the *Führerreserve* (leaders reserve), but they never held another command during the war.

The dismissal of von Manstein was partly due to his arguments with Hitler over how to conduct the war on the Eastern Front. Their dispute also became known as "the battle of arguments" inside the German high command. [419]

Guido Knopp, in his book *Hitler's Warriors*, on page 129 stated, "General Staff officer (OKH), Johann Von Kielmansegg, said when asked about it: 'There was no Commander-in-Chief who argued with Hitler as much as [von] Manstein did.'"

Knopp also noted in his book one final comment about von Manstein's dismissal. This comment came from General Rodion Malinovsky, commander of the 3 UF and future commander of the 2 UF. Malinovsky said, "We considered the hated von Manstein our most dangerous opponent. His technical mastery of every, and I mean every, situation was unequalled. Things would perhaps have gone much worse for us if every general in the German *Wehrmacht* had possessed his stature."

An interesting side note to the replacement of Field Marshal von Manstein is that Field Marshal Walther Model knew on 28 March of Hitler's wishes for him to replace von Manstein. On 30 March Model flew from AG North to see Hitler and to take over command of AG South. When his aircraft left the ground, a staff officer at the airport phoned AG North HQ and issued the code word "*Schweinfuhrt*. This could have been merely a reference to the German city, but it was not. In its most literal sense, *Schweinfuhrt* expressed the true feelings of Model's subordinates at Army Group North: 'The swine has flown.'" [420]

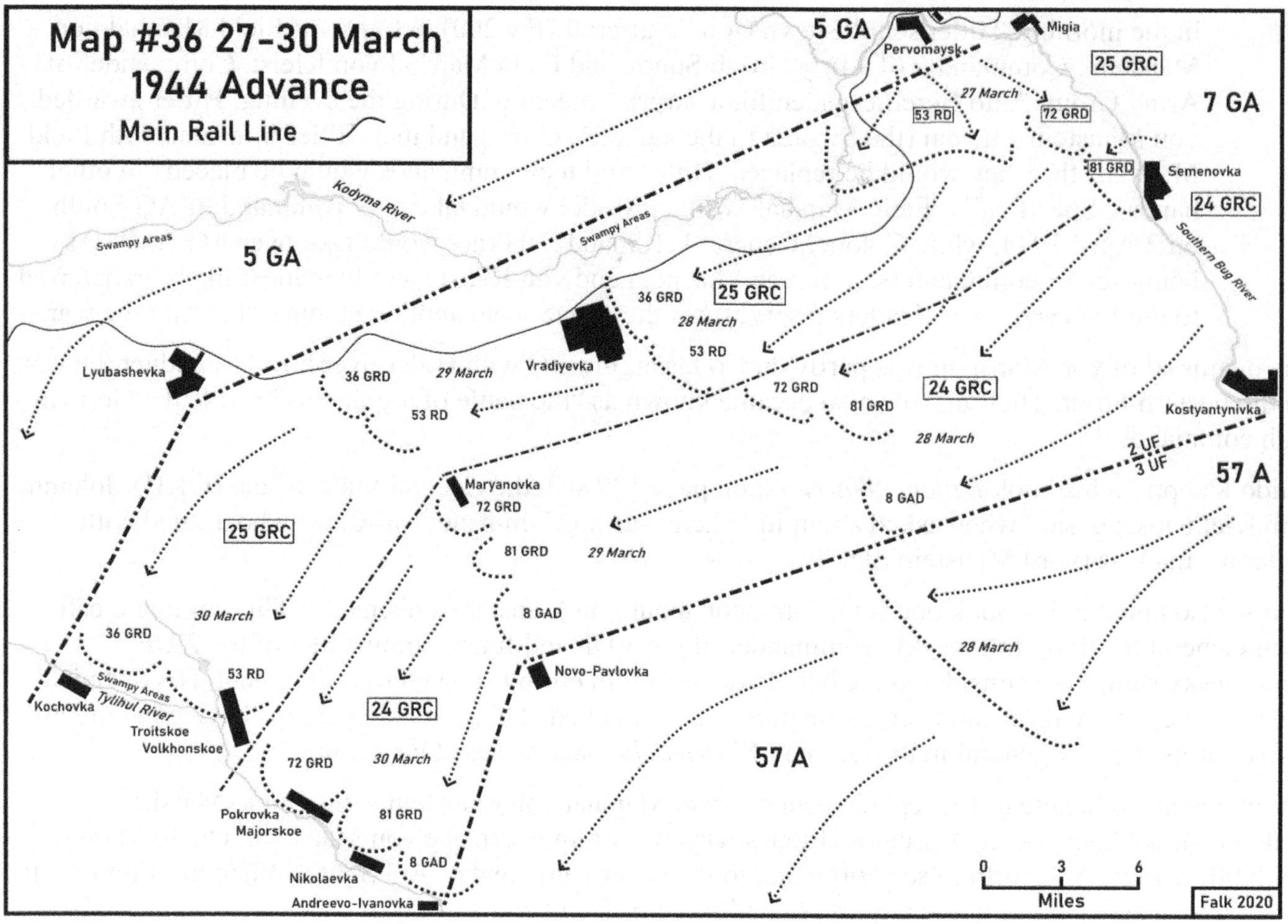

31 March 1944

Some units of the 7th Guards Army advanced a short distance over the swampy Tylihul River, stopping after improving their overall combat position. They spent the remainder of the day resting and resupplying the divisions. After advancing more than 40 miles in four days, the troops needed a rest.

A loss report was issued by 2 UF, TsAMO record ID #113723857, *Report on the loss of personnel in the troops of 2 UkrF from March 1 to March 31, 1944*. During this time, the 7th Guards Army suffered the following losses: 1,193 killed; 4,008 wounded; 623 sick and evacuation to hospitals; 3 with frostbite to hospital; 71 missing; 3 surrendered to the enemy [POW]; and 24 non-combat loss for a total of 5,925 casualties for the month, both men and women.

Losses for the entire 2 UF were 56,850 men and women or 1,833 casualties per day.

Another TsAMO Archive Document, ID #113723859, stated that for the month of March 1944, the 2 UF mobilized 108,000 servicemen in the liberated territory. These were, of course, civilian conscripts. So, with overall losses of 56,850 men and the addition of 108,000 new recruits, the 2 UF ended the month with a surplus of 51,150 men. These extra men could hardly bring the armies up to full strength, but they eventually would add some offensive capability after training.

Marshal Konev summed up the operations of the 2 UF during March 1944. Relevant points were:

- The 5th AA could not be fully employed due to bad weather conditions, lack of paved airfields and limited fuel. He also stated aircraft were diverted to deliver ammunition and fuel to tank units because of the overall poor supply situation due to muddy conditions at the front.

- The front was supplied with adequate amounts of ammunition and fuel by utilizing all types of transport, land vehicles, aircraft and animal transport. Limited food supplies were supplemented with living off the land [taking food from civilian sources] and using captured enemy supplies.
- As to medical support, "There was no serious difficulty in the medical service and in organizing the evacuation of the sick and wounded, since the losses were insignificant. [See losses listed above.] All empty [vehicles], which went back to the rear, [were used] to evacuate the sick and wounded; in addition, since the area was densely populated, our doctors used local hospitals and health facilities."
- For command-and-control issues, Army HQs move forward every 2-3 days; sometimes the HQs were 40-50 Km [24-31 miles] behind the front. Army commanders moved around the front by the use of tracked vehicles and aircraft. Front and Army HQs maintained wired connections as well as radio links. "Our radio stations were mounted on wheeled vehicles. In connection with the thaw, part of the radio stations had to be re-mounted on horse traction so that they could continuously accompany the advancing troops."
- March 1944 was a successful month for the 2 UF. "The victories of the soldiers of the 2nd Ukrainian Front were the result of their high offensive fighting spirit, moral and political [motivation], adamant determination to defeat the enemy at all costs and free the Soviet land from Hitler's invaders." [421]

An alternative summation might include that the Red Army was one month closer to the end of the war, they won a few victories, they suffered hundreds of thousands of casualties, and they were prepared to confront an equally determined enemy to the end.

1 April 1944	
7th Guards Army	
Assigned – 2 Ukrainian Front	
Colonel General M.S. Shumilov Commanding	
Rifle, Airborne and Cavalry Units	24 Guards Rifle Corps (72, 81 GRDs, 8 GAD) 25 Guards Rifle Corps (36 GRD, 53 RD) 3 Separate AT Rifle Battalion 62, 63, 64, 65, 66, 94 Penal Companies
Artillery Units	11 Artillery Division (40 Howitzer Artillery Brigade, 45 Gun Artillery Brigade) 161 Guards Gun Artillery Regiment 114, 115 Guards Tank Destroyer Regiments 114 Tank Destroyer Regiment 30 Antitank Artillery Brigade 263, 290 Mortar Regiments 5 Antiaircraft Artillery Division (670, 743, 1119, 1181 AA Regiments) 162, 225 Guards AA Regiments 662 Separate Aerial Observation/Air Raid Warning Company
Armored and Mechanized Units	27 Guards Tank Brigade 34, 38 Separate Armored Train Battalions
Engineer Units	175, 329 Separate Engineer Battalions 46, 49, 329 Engineering Battalions
Estimated Combat Strength Estimated Ration Strength	30, 247 men 37,062 men and women (including 6,815 estimated support personnel)

(BSSA 44, BSA 44, TsAMO Archive)

At the start of the month, the 615 Mortar Regiment was transferred out of the Army, so it is not listed in the chart or combat strength totals above.

1 April 1944

The 7th Guards Army continued to advance to the southwest as weak defending German troops rapidly withdrew. (See Map #37.)

Finnish diplomats returned from Moscow with harsh Soviet terms that were not acceptable to the Finnish government. The war between Finland and Soviet Russia continued.

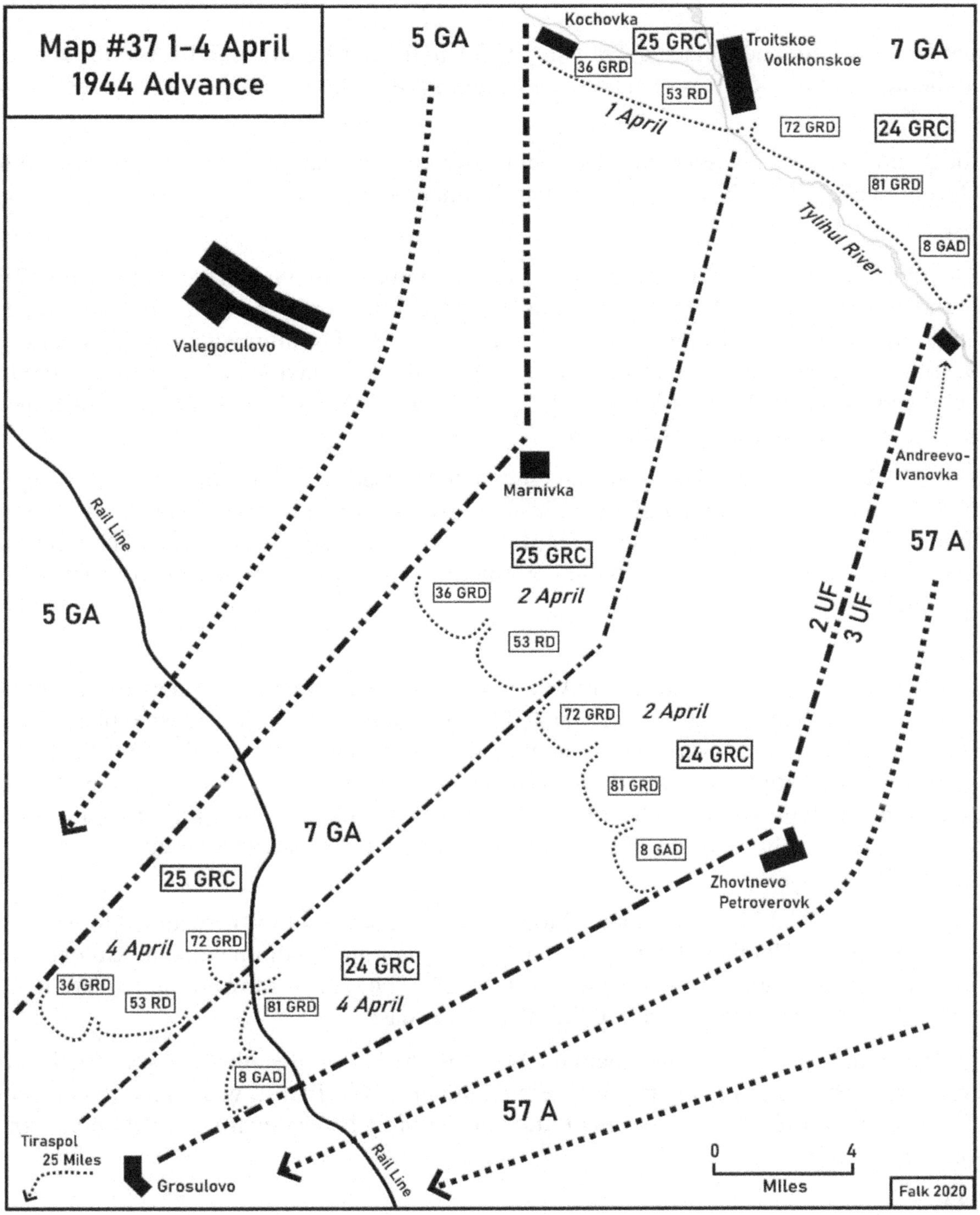

2 April 1944

The 7th Guards Army continued its advance. Weather conditions made movement along dirt roads difficult.

Field Marshal Walter Model and Colonel General Ferdinand Schörner took over command of AG South and AG A.

Marshal Konev continued the 2 UF attack into northern Romania while the 3 UF continued its assaults along the Dnestr River line. Konev was hoping to envelop the defending German and Romanian armies using the combined forces of the 2 UF and 3 UF. Along the way, they would advance to the Prut River, thus capturing key terrain features from the march before German and Romanian forces were ready to defend in strength.

3 April 1944

The 7th Guards Army continued its advance. The 5th GA to the north and the 57th A to the south compressed the frontage of the Army. This was by design as Marshal Konev had other plans for the 7th Guards Army.

Forces of the 2 and 3 UFs continued their attacks but suffered heavy casualties in the process. At this point only formations of the 2 UF were operating on Romanian territory.

4 April 1944

The five rifle divisions of the 7th Guards Army reached a point about 25 miles from the Dnestr River, northeast of the city of Tiraspol. As the 5th GA advanced from the north and the 57th A advanced from the south, they squeezed the 7th Guards Army out of the front line. For all practical purposes, the entire 7th Guards Army was now behind the front. The 57th A, 3 UF, had taken over the section of the front previously covered by the 7th Guards Army. With the 7th Guards Army leaving the front line, the 5th GA now became the left flank army of the 2 UF.

Meanwhile, the U.S. Fifteenth Air Force attacked the main railroad station in Bucharest, dropping 863 tons of bombs. This was the beginning of a U.S. lead air offensive against crude oil and transportation targets within Romania. After the war Romanian officers stated that air attacks against the rail network and supply lines from 4 April to 18 August 1944 crippled the Romanians' ability to move troops and equipment through the country. [422]

5 April 1944

Attacks into northern Romania continued, but Axis forces were defending tenaciously as new infantry and Panzer units were deployed to the front. Marshal Konev submitted a new offensive plan to *Stavka* for his 2 UF, to be launched on 8 April. This plan would shift the main attack to the right flank area in and around the towns of Targu-Frumos and Iasi (also spelled Iasi or Iassy). This new offensive initially would use the 40th and 27th Armies with additional armies being added as they arrived in the area. At 22:10 hours *Stavka* issued Directive #220072 authorizing the proposed advance of the 40th and 27th As on Konev's right flank.

Under the command of Field Marshal Walther Model, Army Group South was renamed to Army Group North Ukraine (AG North Ukraine). On the same day, Army Group A, commanded by Colonel General Ferdinand Schörner, was renamed to Army Group South Ukraine (AG South Ukraine) where it continued to operate in the Crimea and along the Black Sea coast.

Starting on 5 April, then on 15 April and again on 24 April, the U.S. Fifteenth Air force struck hard at Ploesti oil refineries. By the end of the month, German imports from Ploesti were only 56 percent of what they had been in March. [423] This loss in production certainly had a negative effect upon Axis operations all over Europe.

6 April 1944

Stavka ordered the 2 and 3 UFs to go over to the defensive, while the right flank of the 2 UF could continue their attacks. *Stavka* recognized they had lost the chance to quickly and to easily capture critical areas of northern Romania. The German and Romanian defenders had won a defensive victory. David Glantz, in his book *Red Storm over the Balkans*, stated on page 45:

> To their advantage as Germans withdrew into northern Romania, they were able to exploit the excellent and still intact rail and road system to replace lost weaponry and replenish their supplies. In addition, although many Germans questioned their reliability, as many as 200,000 Romanian troops were available to support the defending Germans.

Stavka issued Directive #220073 at 22:45 hours approving of Konev's new plan for a major offensive launched from the right flank of the 2 UF. Major forces needed to be shifted to this flank, including the

7th Guards Army and the 5th GTA. This transfer, of course, took much time and effort. Surprisingly, *Stavka* allowed Konev to continue with the independent advance of the 40th and 27th As and even ordered reinforcements sent to support their attack.

7 April 1944

Without doubt, General Shumilov knew of the developing plan and the reason the 7th Guards Army was currently behind the front. Shumilov had most likely been working with Konev for days and was actively preparing the Army for its transfer to the right flank of the 2 UF. The 7th Guards Army HQ issued Operation plans Document #1091 about the transfer of the Army to a new location. This is from TsAMO record ID #153704459, *Extract from the table of the march of troops 7th Guards Army*. This one-page document shows:

- The 36 GRD and 53 RD would begin their movement on 10 April. They would move to the northwest and in four night marches, they would cross the Dnestr River and concentrate around the city of Rybnitsa, arriving on or about 13 April.
- The 72 and 81 GRDs would begin their march on 11 April, also moving to the northwest. In five night marches, they too would cross the Dnestr River and concentrate around Rybnitsa, arriving on or about 15 April.
- The 8 GAD would follow on 12 April. In six night marches, they also would concentrate around Rybnitsa. Army level artillery units likewise would follow on 12 April, arriving on or about 17 April.

The emphasis of using night marches implied that the Soviet command had implemented full *maskirovka* methods during the transfer. Troops would march at night and sleep and hide from Luftwaffe reconnaissance aircraft during the day. According to the *Regrouping Plan*, TsAMO record ID #132653907, the march would cover 160-170 kilometers [99-105 miles] but due to the r*asputitsa*, they would cover no more than 25-30 Km [15-18 miles] per day. At least this was the plan. Note: This was Phase One of the overall transfer plan. Due to the distance involved, and the weather and ground conditions, the Army required two distinct movements to reach its new area of operations. General Shumilov also would have taken this respite from frontline duties to rest, resupply and restore the strength of the Army as much as possible.

On the German side this day, OKH intelligence maps showed the 7th Guards Army located behind the front line with a question mark next to it signifying an uncertain location. [424]

8 April 1944

From the north, the 2nd GA and 51st A, 4 UF, began the offensive to retake the Crimea from the Germans. From the east came the Independent Costal Army, which also joined in the attack. Both of these forces were aimed at one objective: the fortress city and naval port of Sevastopol. At this time 130,000 Germans and 70,000 Romanian combat troops held the Crimea. On this day, 80,000 redundant men and women along with 2,500 tons of material were evacuated by sea from the Crimea to the port of Constanta, Romania. [425]

The attacking 27th A began fighting for Targu-Frumos; the German 8th Army responded by moving up Pz GD *Grossdeutschland* to counterattack.

9 April 1944

The 27th A, 2 UF, captured Targu-Frumos from defending German troops. Formations along the right flank of the 2 UF felt confident as they advanced and reached the foothills of the Carpathian Mountains in Romania.

In the Crimea, defending German and Romanian forces held their positions against powerful attacks, thus preventing a breakthrough of the Red Army into the isolated peninsula. Simultaneously, the 8th

GA, commanded by General Chuikov (of Stalingrad fame), along with the 5th Shock Army and the 6th Army attacked Odessa on the Black Sea coast. (See Map #32 for the location of Odessa.) The insufficient German and Romanian defenders lost most of the city in one day.

10 April 1944

Sometime after 10 April, the 162 and 225 Guards AA Regiments transferred out of the 7th Guards Army.

The 7th Guards Army started its transfer to the right flank of the 2 UF. During the early evening, the first group, the 36 GRD and 53 RD, began their long march through the mud to the northwest. (See Map #38. Note the staggered start and arrival dates.)

Further south, the great city and port of Odessa fell to the victorious Red Army. In the Crimea, Red Army formations began forcing Axis troops out of their prepared defensive positions. Out in the open, Axis troops were too weak to hold any position for long. The Red Army held more than a 2-1 advantage in troops, and a huge superiority in tanks, artillery and aircraft. Soon they would begin the race to capture Sevastopol.

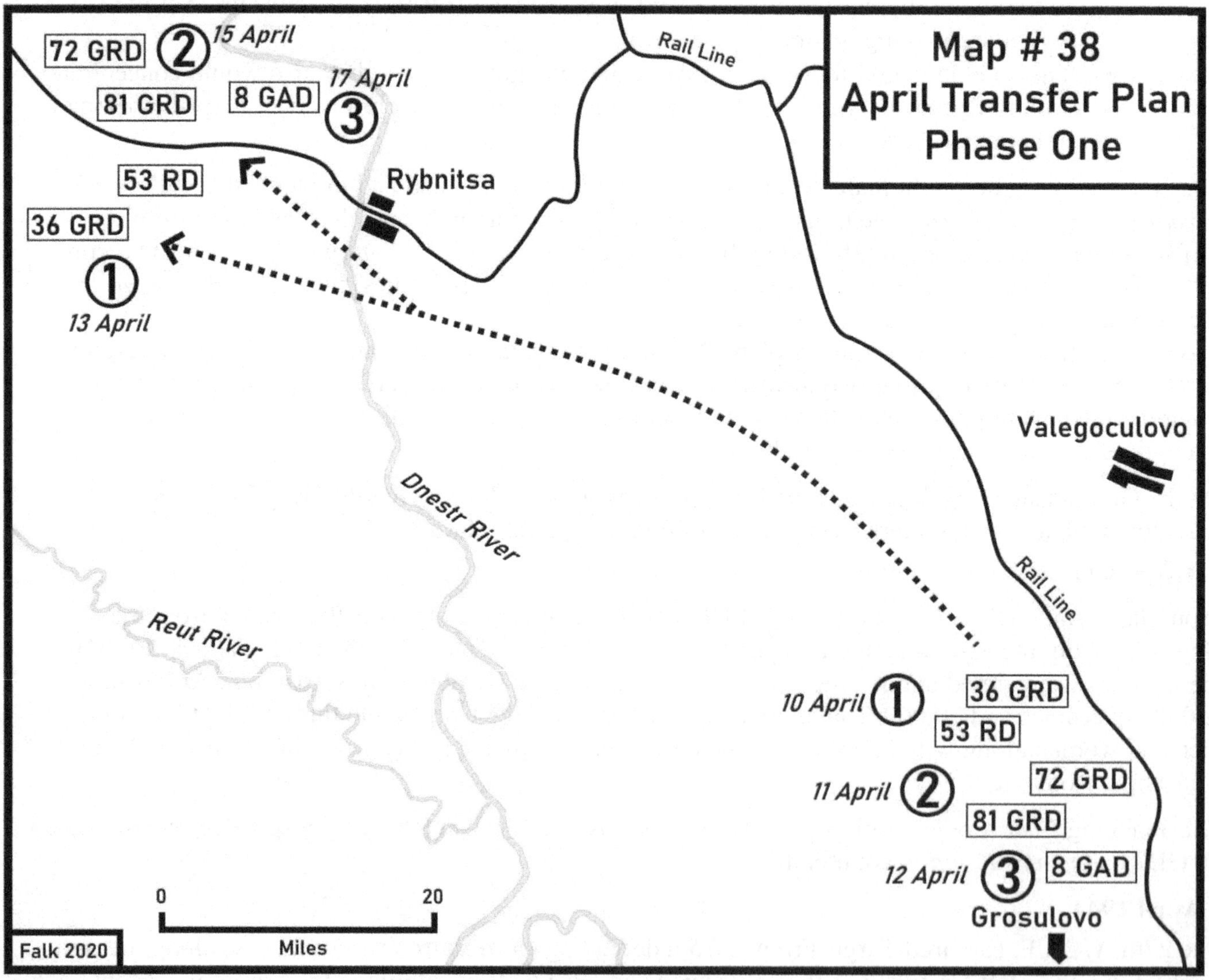

11 April 1944

During the evening, following the 36 GRD and 53 RD, the 72 and 81 GRDs began their march toward Rybnitsa.

Units of Pz GD *Grossdeutschland* counterattacked and recaptured Targu-Frumos. They also forced units of the 27th A several miles away from the city to the north. Targu-Frumos commanded the road network in the area and was the key to controlling the area east of the Seret River. (See Map #39.)

12 April 1944

Bringing up the rear of the Army, the third march group consisting of the 8 GAD and assorted artillery formations followed the same path to Rybnitsa. With a night movement like this, hundreds of things could go wrong and probably did. Units became lost in the dark despite traffic regulators guiding them along the way, scores of vehicles were stuck for hours in the mud, horses died, leaving wagons and artillery pieces without transportation; damaged bridges blocked traffic until engineers performed the required repairs; food for both man and beast went astray; and so forth. In the coming days, the troops could only expect long, cold, exhausting night marches with wind and rain adding to their misery. At lease they were behind the lines and momentarily free from the threat of sudden death. But the war called them forward, and they would do their duty as Guards men and women.

7th Guards Army Senior Sergeant Elsekov (on the left) inspects documents at a checkpoint behind the front, while (on the right) a traffic regulator assists him. The sign in the background says, "ПРОВЕРКА ДОКУМЕНТОВ" which means "VERIFICATION OF DOCUMENTS."

Units of the 5th GA began forcing the Dnestr via improvised means while the entire 7th Guards Army continued their lateral movement behind the front. Note: Despite the recent capture of key rail lines in western Ukraine, rail transportation apparently was unavailable to assist with the transfer of the 7th Guards Army. The troops marched all the way to their new battleground.

Increasing German and Romanian resistance along with repeated counterattacks effectively ended Marshal Konev's right flank offensive. Both Konev and *Stavka* understood they needed more forces to complete a breakthrough in that area. Both the 7th Guards Army and 5th GTA were on the way, but there was little hope they would completely arrive and be fully deployed before early May.

Fighting in and around Targu-Frumos diminished as Pz GD *Grossdeutschland* began constructing defenses around the city. In this position, the *Grossdeutschland* had contact with the Romanian 1 Guards Division on its left flank and with a KG (*Kampfgruppe*) from the 24 Pz D on the right. David Glantz, in his book, *Red Storm over the Balkans*, quoted from the *Grossdeutschland* Divisional history on page 71: "Near Targu-Frumos, meanwhile, contact [had] been established with German and Romanian units attacking from the west. The road from Targu-Frumos east to Jassy [Iasi] was open again. Traffic returned to normal; the supply vehicles rolled."

General Hasso von Manteuffel, commander of Pz GD *Grossdeutschland*.

Over the next several days, German and Romanian forces continued to push the front line further north, but they encountered increased Russian resistance in doing so.

In the Crimea, defending German and Romanian forces in the field began their complete withdraw toward Sevastopol as Soviet forces raced them to the city.

13 April 1944

The 7th Guards Army HQ issued new operation plans in Document #149. Starting on or about 13 April and continuing to 18 April, the 24 and 25 GRCs would move their units to the front, concentrating them for the coming offensive. This is from TsAMO record ID #132653909, *Concentration plan 7th Guards Army*. This was Phase Two of the overall transfer plan. This planning document and the attached map displayed the separate routes and phased positions that each rifle corps needed to reach each day. This plan was hopelessly optimistic. The troops did their best, but their movements already were behind schedule. (See Map #39. Note: Map #39 displays the realistic [estimated] positions and arrival dates for both rifle corps.)

Note: Limited German and Romanian counterattacks in the Targu-Frumos area even now were capturing the planned final concentration areas for the 24 and 25 GRCs. During war, when do detailed plans ever proceed as intended? Marshal Konev and General Shumilov would be occupied over the next several days with making adjustments to the transfer plan.

During this transfer phase and with the Army moving behind the lines, the Army had time and opportunity to acquire additional manpower. Guards Colonel Khairullin and Guards Major of Administrator Service Slesarev issued an intelligence report on this day, *Information on the progress of mobilization by the units of the 7th Guards Army*. This is from TsAMO record ID #132344496. (See Table #9 below.)

Unit	Conscripted Mobilized	Fit for Combat with Training	Fit for Combat without Training	1925 Cohort (19 year olds)	Fit for Service Units	Former Military Personnel (POWs)	Sent to Rifle Divisions	Sent to 190 Replacement Regiment
8 GAD	1,924	508	539	122	132	623	1,074	850
36 GRD	7,427	3,841	1,978	147	483	978	5,527	1,900
72 GRD	1,498	342	470	178	-	508	310	1,188
81 GRD	4,772	990	1,059	718	956	1,049	240	4,532
53 RD	3,230	1,010	1,070	170	495	485	912	2,318
190 RR	2,252	671	878	91	165	447	-	2,252
Total	**21,103**	**7,362**	**5,994**	**1,426**	**2,231**	**4,090**	**8,063**	**13,040**

Table #9: Progress of Mobilization by the Units of the 7th Guards Army – 13 April 1943.

In essence, as the Army moved through newly liberated territory, rifle divisions and the 190 Training Regiment swept up every able-bodied or partially able-bodied person, mostly men, that they could find. After assessing these conscripted troops, commanders sent the rifle divisions 8,063 replacements right away. These people would then have been sent to a divisional replacement battalion, also called a school battalion, for refresher training. The remaining 13,040 people were sent to the 190 Training Regiment for more extensive training and for fitting out with equipment. This report illustrates how a training regiment could contain more troops than an entire rifle division. This report also explains why the Germans forced the evacuation of all military aged men to the rear when they retreated. They knew the Red Army would quickly turn these civilians into combat troops, support personnel, or at least cannon fodder.

14 April 1944

Russian forces conducted powerful attacks for two days against German positions seven miles north of Targu-Frumos. After heavy fighting, the Germans held their positions. German troops and commanders noticed the buildup of Russian troops in this area and knew that further attacks were just a matter of time. After the succession of attacks and counterattacks during early April, the front finally settled down. This is widely considered to be the end of the first battle for Targu-Frumos. At this point, the Germans controlled the entire city and the surrounding area.

In the Crimea, advanced elements of the 51st A reached the outlying areas of Sevastopol, the largest city in the Crimea, and immediately attacked. The battle for Sevastopol had begun.

German OKH intelligence maps from US National Archives, T-78 Roll 136, image #429, shows that the 7th Guards Army was located far behind the front but moving north.

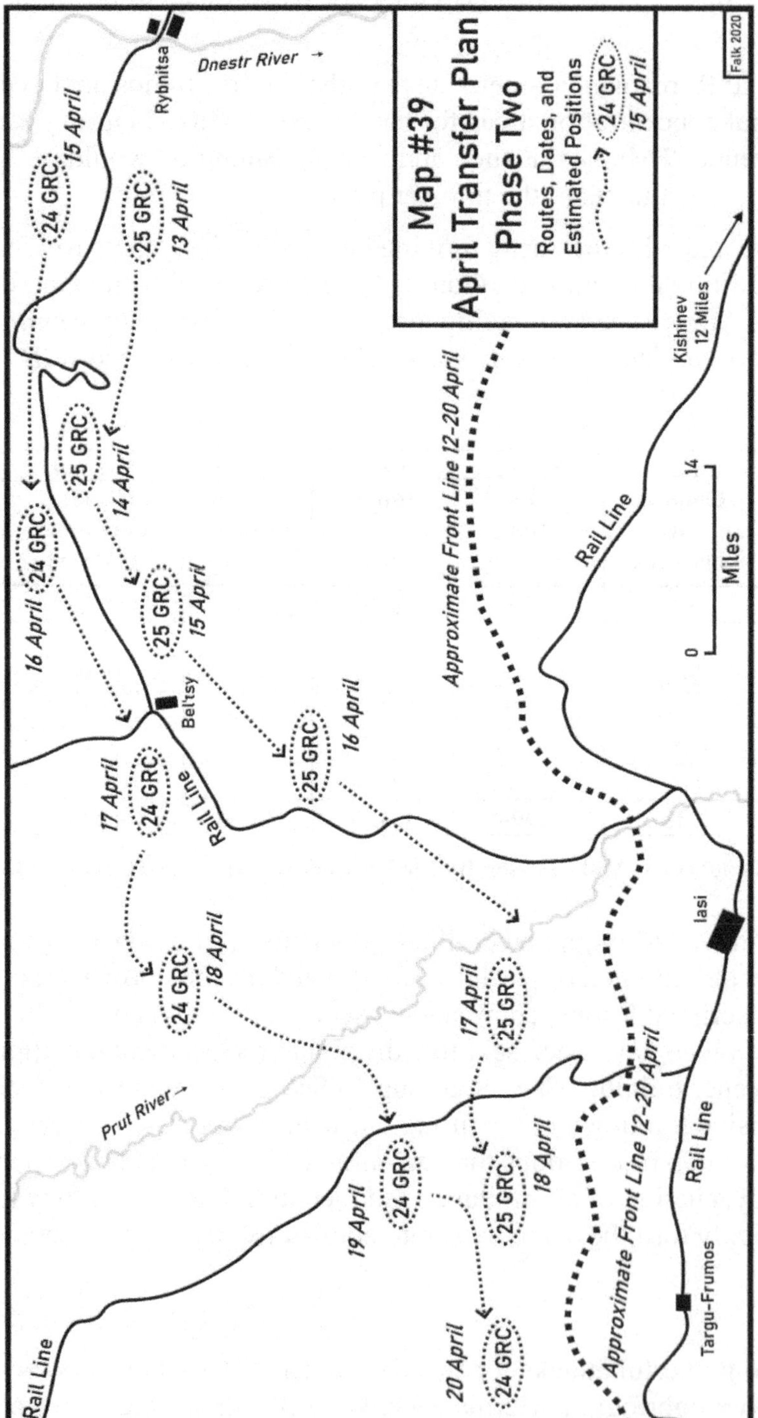

15 April 1944

In February 1944, Russian soldier Athanasius Vatutin died at the front of combat wounds. In March 1944, his brother Semyon Vatutin also died at the front. [426] On this April day, General Nikolai Fyodorovich Vatutin, Commander of the 1 UF, died in Kiev of the wounds he sustained on 28 February 1944.; he was 43 years old. Three brothers were lost in three months. On 17 April 1944, General Vatutin was buried in Kiev, where he rests to this day. [427]

German OKH intelligence maps from US National Archives, T-78 Roll 136, image #432, shows the 5th Guards Tank Army moving into position behind the right flank of the 2 UF.

16 April 1944

By this date, some reports stated that units of the 72 GRD had crossed the Prut River and moved into Romania. This was an overstatement as the transfer plan was at least two days behind schedule. The 7th Guards Army continued to move forward, but the regrouping had already been underway for six days and everyone was exhausted.

Major units of the 5th GTA were concentrating about 15 miles directly north of Targu-Frumos. In turn a small German attack with supporting Panzers pushed back the Russian line somewhat to the northwest of Targu-Frumos. This attack would draw a response from the 5th GTA.

In the Crimea, the German 17th Army withdrew the bulk of Axis troops, approximately 120,000 strong (75,000 German / 45,000 Romanian), to the Sevastopol area and held the city and surrounding fortifications. During the retreat the 17th Army lost approximately 13,000 German and 17,000 Romanian troops who were either killed or captured. [428] The approaching Red Army formations of the 2nd GA, 51st A and the Independent Coastal Army were much larger than the 17th Army, but they needed some time to prepare for their assault upon fortress Sevastopol. Along the way to Sevastopol, troops of the Independent Coastal Army captured the picturesque resort city of Yalta on the south coast of the Crimea.

17 April 1944

Formations of the 18 TC, 5th GTA, moved south and engaged German Panzer forces northwest of Targu-Frumos.

Due to the death of General Vatutin, *Stavka* began planning the reassignment of all of their Ukrainian Front Commanders. Marshal Zhukov would soon transfer command of the 1 UF to Marshal Ivan S. Konev, current Commander of the 2 UF. This move in turn caused *Stavka* to transfer General Rodion Yakovlevich Malinovsky, Commander of the 3 UF, to take over command of the 2 UF in May 1944. Then Marshal Fyodor Tolbukhin, 4 UF Commander, took command of the 3 UF, also in May. Eventually, General Ivan Yefimovich Petrov took command of the 4 UF in August 1944. At this point, the shuffling of front commanders northward finally ended.

18 April 1944

A historical map from TsAMO Archive ID #100736932 showed the counterattack by the Russian 18 TC drove German forces back to their original line in the area of Targu-Frumos. The Germans now understood sizable Red Army tank forces were in the area. That could mean only one thing: a major Soviet offensive was pending.

On this day, some advanced units of the 7th Guards Army should have reached their newly assigned forward assembly areas, but due to terrible road conditions, their march columns were spread out for many miles. How long it took to fully reassemble the combat elements of the Army is unknown, but the best estimated dates and locations are displayed on Map #39. Artillery, medical and logistical support units would follow in the days and weeks to come.

Marshal Konev submitted a new offensive plan for the 2 UF to *Stavka* with operations intended to begin sometime from 27-29 April. This offensive would eventually become the second battle for Targu-Frumos.

In the skies over Germany, more than 2,000 Allied heavy bombers, by both day and night, dropped more than 4,000 tons of bombs on Germany. This was the largest one-day amount to date. The frequency of and devastation caused by these raids would continue to increase in the coming months.

19 April 1944

The 5th GA, acting as the vital left flank of the 2 UF, continued fighting along the Dnestr River and within its two bridgeheads on the western side of the river. Armies of the 3 UF also reached the Dnestr River and put pressure on Axis forces trying to hold the river line.

20 April 1944

On or about this date, the 6 Guards Airborne Division, 6 Rifle Division and the 66 Engineering Battalion were added to the 7th Guards Army. At the same time, the 225 Guards AA Regiment and the 114 Tank Destroyer Regiment were transferred out of the Army. The 7th Guards Army now contained the 24 GRC (72, 81 GRD, 8 GAD), and the 25 GRC (36 GRD, 6 GAD and 53 RD) with the 6 RD as Army reserve. The estimated combat strength of the Army now stood at 40,297 men with an estimated ration strength of 47,092 men and women.

The German 8th Army, which was covering the Targu-Frumos area, also included the 3 SS Pz D *Totenkopf* and the 24 Pz D in reserve. German OKH intelligence maps from US National Archives, T-78 Roll 136, image #435, appear to show the 7th Guards Army moving into position behind the right flank of the 2 UF.

21 April 1944

Fighting along the lower Dnestr River intensified as both the 2 and 3 UFs sought to cross the river in force or to enlarge existing bridgeheads. German and Romanian forces vigorously resisted these efforts.

22 April 1944

TsAMO record ID #132344498, *The calculation of the understaffing of army formations* (7th Guards Army) is a record that shows the strengths of the rifle divisions of the Army as of 21 April and the redistribution of manpower within the Army. (See Table #10.)

Unit	Strength 21 April	Transferred out	Redistribution	Final Strength
8 GAD	2,985	-	1,015	4,000
36 GRD	7,683	2,683	-	5,000
72 GRD	5,770	-	230	4,000
81 GRD	2,745	-	1,438	4,185
53 RD	4,434	-	-	4,434
Total	**21,617 men**		**2,683**	**21,617**

Table #10: Manpower Redistribution within the 7th Guards Army.

This document is interesting because the Army apparently received no reported outside support, new conscripts or returns from hospitals during this time, which is unlikely. These transfers were only a balancing of forces already within the Army with no increase in actual combat strength. This redistribution brought all the divisions up to the 4,000-strength level, which would allow them to be somewhat combat effective.

23 April 1944

The 7th Guards Army issued operation plans Document #1118, *The plan for the replacement of parts 40 and 27 A with parts of 7th Guards Army*. This is from TsAMO record ID #132653904. From 23 to 25 April, the 7th Guards Army started replacing parts of the 40th and 27th As at the front with their own units. This exchange would have taken place as secretly as possible so as not to alert the Germans that a new army, a veteran Guards Army, had arrived at the front.

Marshal Konev issued 2 UF Directive #00324/op. This directive covered overall planning, objectives and coordination between armies for the impending offensive. The 7th Guards Army and the 5th GTA would be working closely together to pierce the enemy line and to advance deeply into their rear. Konev ended the directive with, "Be prepared to begin the offensive on 27 April 1944. The date and hour of the offensive will be in accordance with a special order."

24 April 1944

With the first units of the 7th Guards Army just now reaching the front in strength, it should have been obvious to General Shumilov that major offensive operations would be delayed for some time. The Army was in a disorganized state as stragglers, artillery, trucks and tractors along with much needed supplies lagged far behind the leading elements. Also, the rear services would take days or even weeks to be in a position to accumulate adequate supplies to support offensive combat operations. Supply dumps of all types had to be established, hospitals needed to be moved into position, roads and bridges required attention, and workshops had to be organized to repair and to maintain an army's worth of equipment. The combat troops were reaching their final positions, but the entire Army still lacked adequate logistical support.

Actually, Red Army forces of both the 2 and 3 UFs were exhausted. They required massive amounts of supplies of all types to keep fighting. Equipment needed repair or replacement, and depleted combat formations required rest and replenishment, especially in manpower. All this would take time.

Meanwhile, German and Romanian forces conducted spoiling attacks into the region north and northwest of Targu-Frumos. This was the same area where newly arrived units of the 7th Guards Army were assembling. Combat in this area shifted back and forth until 27 April. Ultimately, Axis forces were able to stabilize the front by forcing the Russian front line back some six miles.

Upon inspecting the numerous destroyed Soviet tanks in the area, the Germans discovered that several of them were the new IS 2 (JS 2) heavy tanks with a 122 mm main gun. These tanks were also known as "Stalin Tanks" with the JS for Joseph Stalin. [429] The Red Army was receiving new and improved equipment at the front. From the German point of view, this was not a good sign as the latest equipment normally was reserved for use in major offensives.

A German solider inspects three destroyed IS-1 (JS-1) Soviet heavy tanks. The IS-2 (JS-2) had a slightly different upper glacis front armor layout.

25 April 1944

Despite all the pressing issues facing the 2 UF, Marshal Konev was determined to launch his breakthrough attack with his right flank armies as soon as possible. To support this offensive, Konev began a series of diversionary attacks against enemy positions around Iasi.

26 April 1944

Regrouping of formations on the right flank of the 2 UF continued, but the planned participation of the 3 UF in the offensives was in doubt. Due to stubborn enemy resistance, units of the 3 UF could not break out from their bridgeheads over the Dnestr River and join the attack. Diversionary attacks against enemy positions by the 2 UF around Iasi continued as Konev struggled with moving his own assault forces into position.

27 April 1944

Marshal Konev was forced to postpone the offensive again, even as formations of the 7th Guards Army hurried to deploy to the front.

28 April 1944

Marshal Konev once again postponed the offensive. He would eventually move the date to 2 May 1944. The recent German and Romanian spoiling attacks also might have caused Konev to delay his offensive.

Konev was a determined commander, but he was fighting against Mother Nature (mud), overwhelming logistical issues, depleted formations, and troop exhaustion. He also faced the skillful use of maneuver engagements and the tactical expertise of a very capable enemy. "The actions of these [German] leaders demonstrated that, when well led, superbly trained but badly outnumbered German divisions and regiments still retained the tactical fighting skills that made them so effective, deadly, and famous during the first two years of the war." [430]

Stalin and *Stavka* recognized the truth of the issue and when Konev called Stalin, on or about this date, Konev said, "A respite is needed… [and he] suggested moving on to the defense."

Stalin in turn replied, "That's right, - Go to the defense and put the troops in order. Further tasks will be given special instructions." [431] These special instructions would arrive on 6 May via a new *Stavka* directive.

29 April 1944

David Glantz, in his book, *Red Storm over the Balkans,* on page 101, summed up Marshal Konev's position:

> As he orchestrated these offensives during the second half of April, Konev consistently believed that if his forces could breech the German partially prepared defenses in multiple sectors across his front, possibly from the march, the German defenses in northern Romania would ultimately collapse.

In this, Konev failed. As Glantz stated, "German defenses bent but did not break."

Marshal Konev had attempted to quickly break through the many natural barriers in northern Romania and to reach the more open Romania plain before German and Romanian forces could properly prepare their defense. Northern Romania was shielded by the Carpathian Mountains to the west, the Dnestr, Prut, Seret and Danube rivers (among others) in the center, and the Black Sea to the east. Mountains, rivers, swampy ground, and rugged coastal terrain made for slow going. All these barriers formed a natural choke point, blocking easy access into central Romania. Konev now knew he would need to plan carefully and to assemble massive forces to achieve his objectives, and that would take time.

30 April 1944

Based upon the best historical evidence available, at the end of the month the 7th Guards Army had successfully deployed both of its rifle corps to the front. The 24 GRC had the 72, 81 GRDs in the front line with the 8 GAD in the second line. The right flank of the corps was anchored on the Seret River, about 11 miles northwest of Targu-Frumos. The 25 GRD was positioned to the left of the 24 GRC, deploying the 36 GRD, 6 GAD and 53 RD in the front line with none in reserve. The newly assigned 6 RD, acting as the Army reserve, also was placed in the second line. Acting upon *Stavka* orders, the 7th Guards Army had five RDs in the front line with two RDs in the second line. This was almost the same defensive configuration that General Shumilov used during the battle of Kursk. But this deployment would not last the night because after dark, Shumilov, acting on orders from Marshal Konev, began moving his first wave assault units 81 GRD and 8 GAD (24 GRC) and 36 GRD and 53 RD (25 GRC) into jump-off positions. The battle would start in two days' time. (See Map #41 for 1 May deployment.)

Because the 7th Guards Army was moving behind the front line during most of the month, and out of combat, total reported casualties were very low. Casualties amounted to 128 men and women, which equals to a loss rate of 4.2 people per day.

Map #40 displays the overall strategic situation along the southern section of the Russian Front at the end of April 1944. Despite the many problems facing his forces and with approval from Stalin to eventually go onto the defense, Marshal Konev decided to forgo the massive buildup that clearly was needed and decided to launch one last, almost desperate attack into Romania. This *ad hoc* operation would only require a few more days to prepare. In this Konev did not act alone. Konev certainly was responding to orders directly from *Stavka. Stavka* wanted another attack along the southern part of the Russian Front to distract the Germans from what they were planned further north: the total destruction of the German Army Group Center.

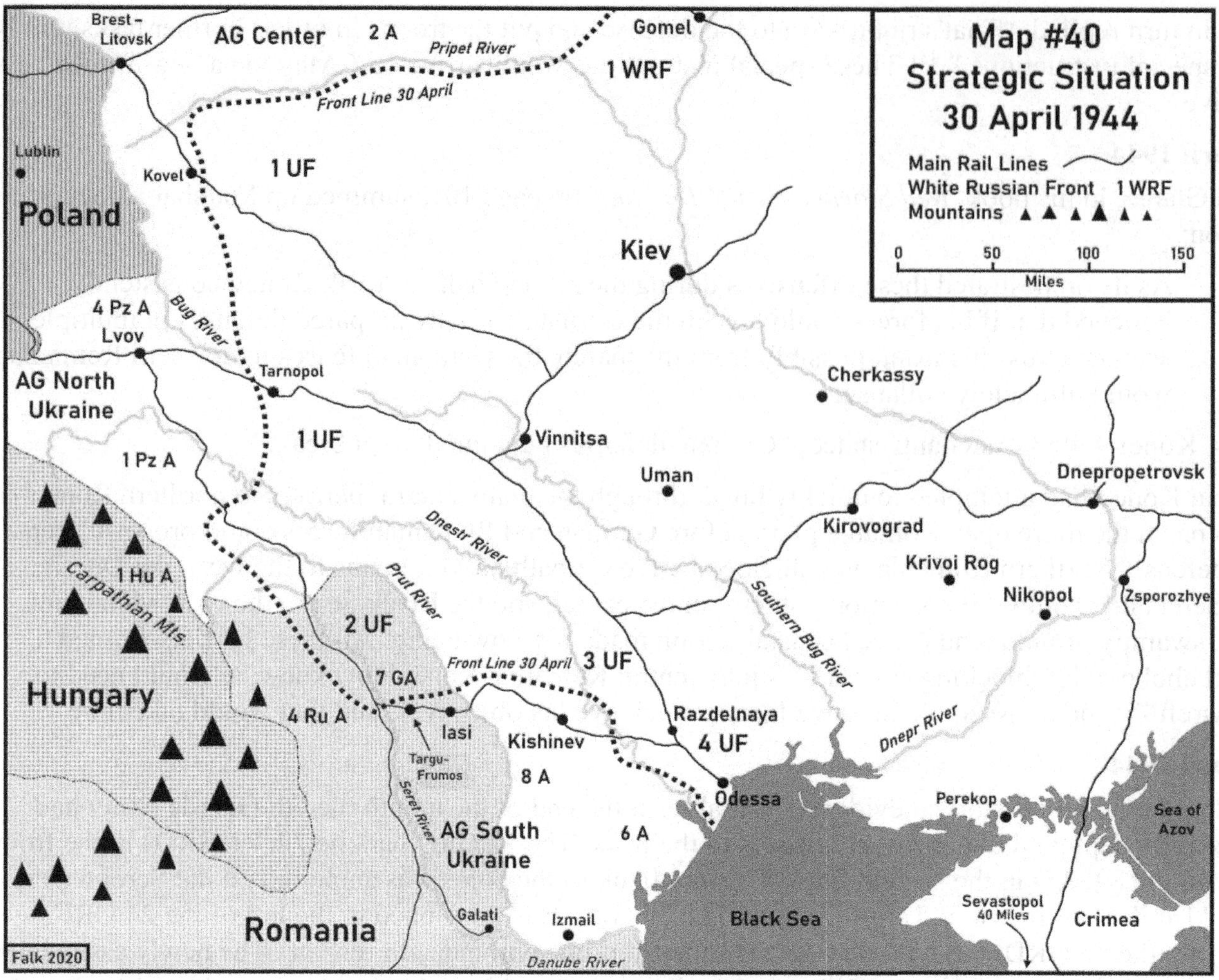

Note: National borders are shown as of 1941. From 1938 to 1941 Hungary annexed large amounts of territory from their neighbors, notably Czechoslovakia in 1938-39, Romania in 1940, and Yugoslavia in 1941.

"Although in the spring German Army strength on the Eastern Front reached a new low (2,242,649 men) and that of the Soviet Army another high (6,077,000 men), in other respects German strength was actually in an upswing." German "synthetic oil production reached its peak in April 1944, and stocks of aviation fuel were larger than any time since 1941." [432] Rising industrial production produced new aircraft and all types of military equipment, including enough Panzers and weapons to equip new divisions and to rebuild depleted ones. But this surge in strength was fleeting, as Allied Forces from both the East and the West were ready to decisively crush Axis armies, navies and air forces wherever the Allies encountered them.

The coming battles would rage for months, but eventually Nazi Germany would face total defeat. The only escape from final and complete devastation would be unconditional surrender. Surrender, however, was not an action Hitler ever would consider. His mentality was, "If you win, you need not have to explain. If you lose, you should not be there to explain!" [433]

1 May 1944	
7th Guards Army	
Assigned – 2 Ukrainian Front	
Colonel General M.S. Shumilov Commanding	
Rifle, Airborne and Cavalry Units	24 Guards Rifle Corps (72, 81 GRDs, 8 GAD) 25 Guards Rifle Corps (36 GRD, 6 GAD, 53 RD) 6 RD 3 Separate AT Rifle Battalion 62, 63, 64, 65, 66, 94 Penal Companies
Artillery Units	11 Artillery Division (40 Howitzer Artillery Brigade, 45 Gun Artillery Brigade) 16 Artillery Penetration Division (49 Light Artillery Brigade, 61 Gun Artillery Brigade, 52 Howitzer Artillery Brigade, 90 Heavy Howitzer Brigade, 109 High-Power Howitzer Brigade, 14 Mortar Brigade) 161 Guards Gun Artillery Regiment 114, 115 Guards Tank Destroyer Regiments 30 Antitank Artillery Brigade 263, 290 Mortar Regiments 5 Antiaircraft Artillery Division (670, 743, 1119, 1181 AA Regiments) 162 Guards AA Regiment 662 Separate Aerial Observation/Air Raid Warning Company
Armored and Mechanized Units	27 Guards Tank Brigade 34, 38 Separate Armored Train Battalions
Engineer Units	175, 329 Separate Engineer Battalions 46, 48, 66 Engineer Battalions
Reported Combat Strength Reported Ration Strength	44, 460 men [434] 53, 475 men and women [435] (including 9,015 support personnel) *

(BSSA 44, BSA 44, TsAMO Archive)

***The reason the number of support personnel increased so much this month is unknown.**

The 162 Guards AA Regiment returned to the 7th Guards Army with a reported strength of 394 men, and the Army also took control of the 16 Artillery Penetration Division with a reported strength of 5,442 men. Note that only a small part of the 16 Artillery Penetration Division would be in position to support the upcoming attack.

1 May 1944

Following orders form Marshal Konev, during the night the 7th Guards Army and adjacent armies finished moving their initial assault units into jump-off positions. This was the final preparation needed for the renewal of offensive operations against German and Romanian positions around Targu-Frumos. The 27 GTB would be employed as the Army's antitank reserve. The offensive would begin early the next day.

To support this attack, the 7th Guards Army had a limited amount of artillery. The Army did have 573 guns and 540 mortars assigned to it, but during the prolonged advance and transfer during March-April, approximately 423 artillery pieces of all types had been left behind due to fuel shortages for prime-movers, being unserviceable, or being immobile (i.e., stuck in the mud). Lack of ammunition also compounded the problem of having limited amounts of artillery support. [436] (See Table #11.) Fuel supplies were equally low for all participating armies. As ordered by *Stavka*, Marshal Konev was attempting one last attack to break into Romania before he was reassigned to the 1 UF. German and Romanian forces were ready and waiting for this attack.

Weapon	Average Available Rounds
82 mm mortars	170
120 mm mortars	110
45 mm AT guns	220
76 mm regimental guns	200
76 divisional guns	100
122 mm howitzers	50
152 mm howitzers	110
152 mm gun- howitzers	40

Table #11: Average Number of Rounds per Gun for the 7th Guards Army and 27th Army on 30 April 1944.

Realistically, the ammunition totals listed above were enough for about one day of modest combat. Despite being a diversionary attack, these totals highlight the desperate and unrealistic nature of this offensive.

2 May 1944

The 7th Guards Army continued its nocturnal activities by moving its second echelon units into assault positions. Joining them were the leading elements of the 5th GTA, which would be supporting the 7th Guards Army. These movements were done as discreetly as possible to maintain the element of surprise, but they had to hurry as the offensive was scheduled to begin at 05:15 hours Moscow time or Decreed Time. (Due to the difference in time standards, this would be 06:15 hours in Germany's records.) Volleys of fire from *Katsuhisa* rocket launchers signaled its start. The Soviet ground assault followed after an hour-long bombardment.

Unknown to the Soviet command, the Germans had expected the attack to begin on 1 May (May Day, or International Labor Day) so they were fully prepared during the early hours of 2 May. The attack to the south was intended to break through the main German line, with the 7th Guards Army advancing between the Seret River on its right flank and Targu-Frumos on the left flank. The 27th A also would attack, supported by the 2nd TA, to capture Targu-Frumos. The 5th GTA with 271 tanks and 87 self-propelled guns (SPG) and the 2nd TA with 77 tanks and 44 SPGs would then exploit the breach, pushing back or bypassing any further German or Romanian resistance as they advanced deeply into the enemy's rear. The 5th AA would support the Red Army attack with 515 aircraft.

On the Axis side of the line, the 4th Romanian Army (4th Ru A) and the German 8th Army stood in their way. Units of the 4th Ru A were deployed astride the Seret River, with the 6 Ru ID on the west side of the river, 1 Ru Guards ID on the east side of the river, and the 1 Ru Pz D with 80 light Romanian tanks (mostly in reserve) also on the east side of the river. The Germans deployed the Pz GD *Grossdeutschland* in an arc around Targu-Frumos with 77 operational tanks and assault guns, and the 24 Pz D had 24 operational tanks and assault guns to its right. The 3 SS Pz D *Totenkopf* was in reserve to the east with no tanks or assault guns operational at the start of the battle. Approximately 600 German and Romanian aircraft would support the Axis defense. (See Map #41 for deployment, Table #12 for strength ratios, and Appendix 13 for further details on tank strengths.)

Forces	Tanks and SPG/Assault Guns	Men	Artillery	Aircraft
2 UF	600+	377,500	8,300	515
4th Ru A - 8th A	250+*	160,000	Unknown	600

Table #12: Approximate Ratio of Forces 1 May 1944. [437] **(* Author's estimate)**

Notably, the *Grossdeutschland* division had received a large shipment of *Panzerfausts* just before the battle. They would put these hand-held anti-tank rocket launchers to good use over the next several days.

The *Panzerfaust* ("armor fist" or "tank fist"), was a secret weapon of sorts, as it was the first operational, mass produced, hand-held, disposable anti-tank weapon in the world that could penetrate any Allied armored vehicle with ease. With over six million produced during the war, Allied tankers learned to fear this simple but effective weapon. The British did have their PIAT launcher and the Americans developed their Bazooka, which were similar, but neither weapon was as effective or as widely used.

A German infantryman aims a *Panzerfaust*.

Despite low levels of ammunition, the assembled Soviet forces had a clear advantage over the defenders in terms of sheer numbers. The offensive began on time at 05:15 hours with a thunderous roar from over a thousand guns and mortars directly north of Targu-Frumos within the 27th A's area of operations. Adding to this barrage, 30 minutes later an additional 800 guns joined in northwest of Targu-Frumos in the 7th Guards Army area. Following behind this barrage, the 7th Guards Army and the 27th A spearheaded the attack with special infantry and sapper teams (combat engineers) that were used to clear obstacles and minefields for the main assault. [438] Konev was quick to add substantial tank support early in the battle to help the infantry advance. This early tank support might have been an attempt to compensate for low levels of artillery ammunition (i.e., blast a path through the German and Romanian defensives with direct tank fire).

Historical maps and documents show that Russian intelligence efforts had located many German bunkers and strong points along the front. The opening artillery barrage would have systematically targeted these "known positions," but after that, artillery observers/spotters would have adjusted the fire as needed to support the infantry and tanks as they advanced.

General Hasso von Manteuffel, commander of Pz GD *Grossdeutschland,* commented about this opening Russian barrage, "Our camouflage must have been very good, for our losses in men and material was very low." [439]

A Russian infantryman sits in a forward dugout position on the telephone in 1944. Note the 1942 model PPSh-41 with a box magazine ready to fire.

For the 24 GRC on the right flank, the attacks on the first day were fairly successful, overrunning German and Romanian positions with the help of the 3 GTC from the 2 UF Reserve. Both German and Soviet sources reported unyielding fighting, with the Germans counterattacking with tanks and assault guns. In this confused fighting, the 24 GRC, with additional tank support from the 29 TC, advanced approximately three to three to one half miles during the first day; however, losses in men and tanks were heavy.

On the left flank, the 25 GRC ran into determined German defenders within strong defensive positions. General von Manteuffel countered the Russian attack with artillery and dug in anti-tank guns including the formidable 88s. The Germans then followed up with Panzer counterattacks into the flanks of the advancing Red Army tank formations. The 24 GRC advanced a maximum of two miles during the day.

Both sides received effective air support, but reports indicate that German air support, led by Luftwaffe pilot Hans Ulrich Rudel (a ground-attack expert), was highly effective in knocking out Russian tanks. Rudel himself is credited with over 500 Soviet tank kills during the war.

At approximately 12:00 hours, elements of the 3 SS Pz D *Totenkopf* with a few tanks and possibly a few assault guns, along with the 24 Pz D with 36 tanks and 18 assault guns, deployed forward to launch counterattacks against advancing Russian tank formations north and northeast of Targu-Frumos. Note: The total number of operational fighting vehicles within these two formations is debated to this day, so these values should be taken as estimates.

While General von Manteuffel committed the reserve Panzer divisions to battle north and northeast of Targu-Frumos, he also commented about the fighting taking place northwest from the city:

> I saw from my battle headquarters the Soviet tanks rolling up to the town. Our own infantry had allowed them to roll past – partly in order to leave the prize to our anti-aircraft battery of 88 mm guns dug in at the northern exit of Targu Frumos. Most of the attacking tanks, of which there were about 25, were shot up in flames; the remainder, about 10 tanks, broke into the concentration area of our Panzer regiment and were destroyed. [440]

At the end of the day, the 2 UF attack had taken little ground and failed to make the expected breakthrough.

This first day, the 7th Guards Army reported losses of 123 dead and 573 wounded with an unknown number missing and captured. The adjacent 27th Army suffered 184 dead and 402 wounded, also with an unknown number of missing or captured. The supporting Soviet 40th A casualties were 89 killed and 379 wounded. [441] Overall tank losses suffered by the 5th GTA and 2nd TA on this first day have been estimated to be about 180. [442] These vehicles should be considered knocked out and not all destroyed (exploded or burned out). By this point in the war, Soviet tank recovery crews were quite skilled and would have worked all night to return some of these "knocked out" tanks to operational status. Regardless, Konev would continue the battle the next day.

3 May 1944

During the night, Marshal Konev regrouped his forces, shifting the focus of the attack more to his right flank (left flank for the Germans) and along narrower sectors.

Following orders, General Shumilov moved all his reserve divisions into the front line which allowed four rifle divisions, the 72, 81, and 36 GRDs and the 6 GAD to focus their attack on the narrow sector where the 29 TC had made its furthest advance the previous day. This was in the area around the village of Dumbrevitsa. The 29 TC and 3 GTC (5th GTA) supported this attack with approximately 180 tanks.

[443] The day's objective for the 7th Guards Army was to penetrate the enemy's defensive positions with a concentrated infantry attack, thus allowing the two tank corps from the 5th GTA to break through and then wheel to the left and capture Targu-Frumos from the west. This was a simple and limited attack plan requiring only an advance of about eight miles.

Likewise, the 27th A, commanded by General S.G. Trofimenko, was ordered to regroup three of its rifle divisions to the west (right flank) and into a four-mile-wide sector directly north of Targu-Frumos. The 3 and 16 TCs (2nd TA) supported this attack with about 70 tanks and self-propelled guns and as much artillery as they could gather. Once again, this was a simple limited attack with the infantry breaking through the enemy line, which would allow the assigned tank forces to advance and to attack Targu-Frumos from the north. [444] The 27th A would need to advance only about five-six miles to capture Targu-Frumos. With the commitment of significant forces along very narrow attack sectors, Konev, Shumilov and Trofimenko must have felt confident of reaching and capturing Targu-Frumos, which was so close.

On the other side of the line, German and Romanian forces also regrouped during the night. The German command expected Marshal Konev to renew his offensive to the west of Targu-Frumos and acted to reinforce their left flank. This deduction might have been influenced by efforts of the German intelligence services along with lax Soviet radio security. Nevertheless, the Germans made the correct decision. By morning reserve elements of the *Grossdeutschland* were deployed to the west and north of Targu-Frumos, while the 24 Pz D and 3 SS Pz D *Totenkopf* had moved into position directly west of Targu-Frumos. The stage now was set for another day of intense combat.

At 06:00 hours, (07:00 hours German time) the Soviet attack opened, again with a powerful artillery barrage. The 7th Guards Army made slow progress and suffered heavy losses in infantry and supporting tanks. David Glantz, in his book, *Red Storm Over the Balkans*, on page 254 recounts the results of the day's fighting:

> However, effective German machine-gun, mortar, artillery, and antitank fire along the immediate front, combined with long-range antitank fire from Manteuffel's 88-mm antiaircraft guns and both long-range tank fire and local counterattacks by his [Manteuffel's] Panzer regiment, repelled each and every Soviet assault. Attacking repeatedly but with increasing futility, Shumilov's rifle divisions were decimated, and about half of Rotmistrov's [Pavel Rotmistrov, commander of the 5th GTA] remaining 180 tanks and self-propelled guns fell victim to the Germans' deadly fire. The attack by the right wing of Konev's shock group collapsed in complete exhaustion by nightfall without registering any significant forward progress.

The adjacent 27th A suffered a similar punishing result at the hands of aggressive German Panzer troops. Fighting was heavy with little ground gained. This attack, too, exhausted itself by the end of the day.

The TsAMO record ID #113984903 shows this second day of combat cost the 7th Guards Army an additional 135 dead and 575 wounded. The 27th A suffered 173 dead and 352 wounded. The 40th A continued its supporting attack and suffered 56 dead and 225 wounded, which pushed the total losses even higher. After two days of combat, the 5th AA (VVS) had flown 1,319 combat sorties over the battlefield while German and Romanian aircraft flew over 2,000 sorties in support of the ground battle. [445] Notably, Romanian air units performed very well and flew constantly during this battle, providing good air support to the defending Axis troops.

4 May 1944

Being a determined commander, Marshal Konev launched his forces once again against the formidable Axis line, only this time with significantly fewer troops and tanks. The attack went in at staggered times of 06:45 and then 08:00 hours, with the same formations in the same general locations. The results also were certain. On this third day of battle, TsAMO record ID #113984903 shows the 7th Guards Army lost 90 killed and 624 wounded. The 27th A lost 119 killed and 443 wounded and the 40th A suffered no reported casualties for the day, most likely due to the army holding its current position. Even the weather turned against Konev as temperatures fell to 7.7-16.6º C [46-62º F] with 100% cloud cover and continuous rain, which turned the roads to mud. [446]

By day's end Konev called off the offensive and started to withdraw the 5th GTA to the rear for refitting, leaving most of its tanks behind in support of the 7th Guards Army. (See Map #41 for final positions.)

At the end of the day, the 7th Guards Army had achieved its maximum advance during the offensive of about four miles southward toward of Targu-Frumos. The confused nature of the back-and-forth fighting prevents the exact identification of the Army's furthest advance. After three days of intense fighting, the main German and Romanian defense line prevented the Army from making any further progress to the south and firmly blocked any meaningful advance directly toward Targu-Frumos. The supporting flanking infantry armies also had been stopped far short of their objectives. The 5th GTA and 2nd TA had lost heavily in tanks and now were depleted. Sporadic fighting continued along the front for several more days, but after making an assessment of the recent fighting, the Germans began to develop their own plans.

David Glantz, in his book *Soviet Military Deception in the Second World War*, page 355-356, puts forth the idea that Konev's Targu-Frumos offensive was really an elaborate deception, and part of an overall *Stavka maskirovka* plan.

> STVKA ordered forces in the eastern and southern Ukraine to conduct limited objective attacks during May, in accordance with earlier operational objectives, to convince the German command that this was a region of continuing Soviet strategic interest….The fact that the operation to this day has been considered a major Soviet effort attests to the success of the Soviet *maskirovka* plan. It, in fact, perpetuated German concerns for their position in Rumania and kept German reserves rooted to Rumania.

The second battle for Targu-Frumos might have been organized as part of an overall Soviet deception plan, but, the use of two tank armies and two or three infantry armies with over 377, 000 men; 600 tanks; 8,300 artillery; and 525 aircraft reveals more about the overwhelming disparity of power between the two antagonists than anything else. At this point in the war, even a Soviet diversionary attack, like this one, posed a real danger for any Axis strategic position and therefore had to be defeated at all costs. On the other hand, if the Soviets took the time and effort to properly plan and launch a major offensive, such an attack had the potential to quickly become a disaster for the defending Axis armies.

Further south, Russian troops in the Crimea attacked the Axis front line north of Sevastopol; however, this attack was a feint. The real attack would be made along the south section of the Axis defense line.

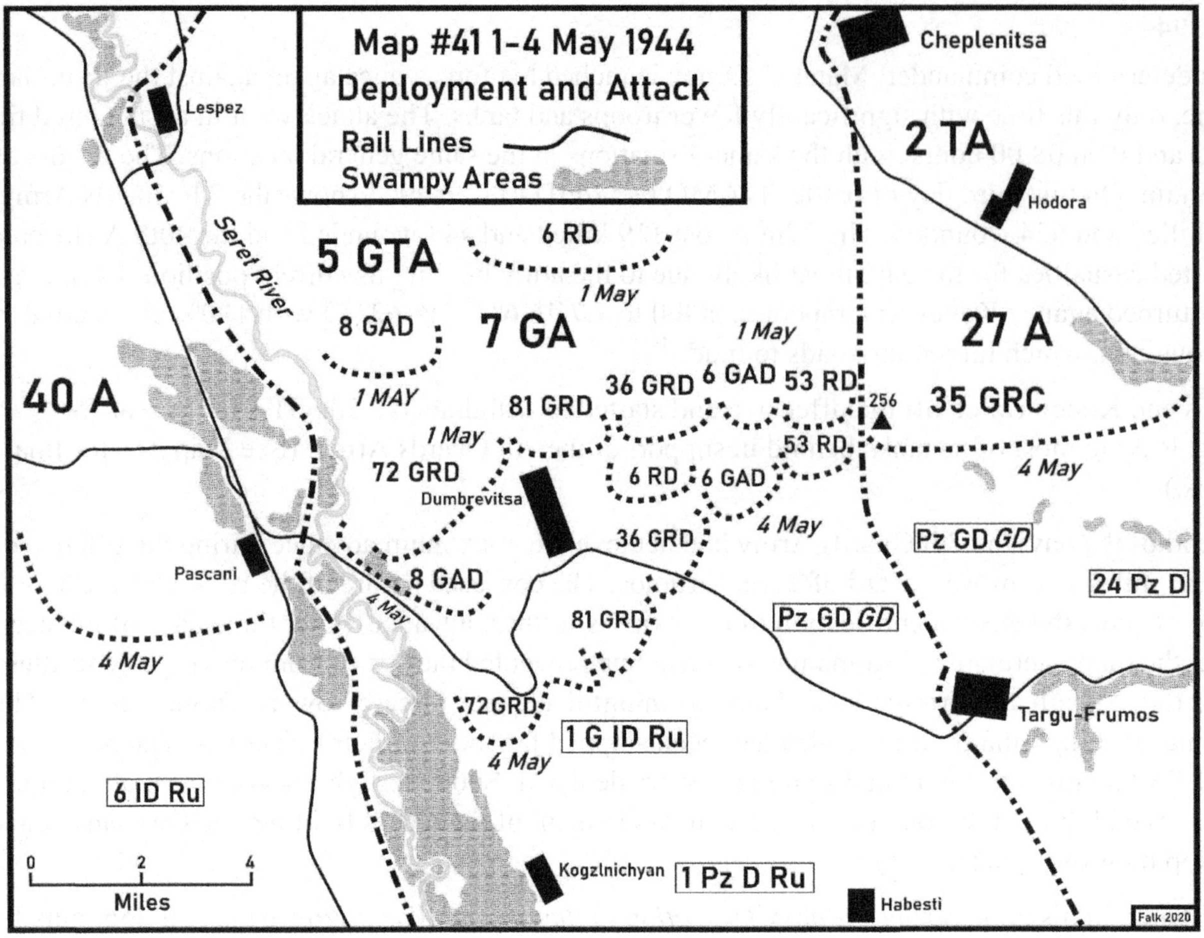

5 May 1944

A historical map, found within TsAMO record ID #100211278, shows the front line in the area of the 7th Guards Army had not changed from the previous day, and it indicated the Army was engaged in small probing attacks. This agrees with historical accounts, but TsAMO record ID #113984903 clearly shows substantial losses for the 7th Guards, 40th and 27th As on this day. These three armies may have been engaged in adjusting their front lines to gain a tactical advantage and/or depriving the enemy of favorable positions. The 7th Guards Army lost 118 killed and 465 wounded. The 40th A becoming active again, suffered 51 dead and 307 wounded, and the 27th A lost 49 killed and 182 wounded.

After the three frontline Soviet armies concluded these low-level actions, combat along the front decreased. For the 7th Guards, 40th, and 27th As, the second battle for Targu-Frumos now was over. The 7th Guards Army was in no shape to do much of anything; they urgently needed a long recovery period to restore the fighting strength of the Army.

Reported personnel losses for the three main attacking infantry armies from 2-5 May are depicted in Table #13.

Army	Killed	Wounded	Total Casualties
7th Guards Army	466	2,236	2,702
40th Army	196	911	1,107
27th Army	525	1,379	1,904
Total	**1,187**	**4,526**	**5,713**

Table #13: Losses from 2-5 May 1944.

Note: Soviet records provide no data on missing men or POWs for the three infantry armies during these four days of battle. However, a conservative estimate would suggest that an additional 50-75 men were lost on the battlefield, never to be found, and that Axis forces had taken between 400-500 POWs.

Soviet tank losses during the battle are difficult to determine. The 2nd TA lost at least 74 tanks and self-propelled guns through 8 May. [447] Data for the 5th GTA is limited, but Soviet and German accounts of the battle indicate the 5th GTA was hit the hardest by German airpower and suffered a larger number of destroyed tanks as a result. As the Germans retained control of most of the battlefield, they would have destroyed any damaged Soviet armored vehicles they found. So, an estimated loss of 150 tanks and self-propelled guns through 8 May is realistic for the 5th GTA. Therefore, both tank armies lost (destroyed) approximately 224 tanks and self-propelled guns during the battle. Several hundred other armored vehicles would have been damaged during the fighting and would have been repaired in the field or sent to the rear for factory repair.

After the battle, the 7th Guards Army issued an intelligence document *Damage to the Enemy,* TsAMO ID #132344638, which covered the Army's assessment of the most recent fighting. (See Table #14.)

Axis Personnel and Equipment	Destroyed/Killed	Captured
Soldiers and Officers	3,260	93
Tanks	50	-
Guns	25	24
Gun Shells	-	2,750
Mortars	28	6
Mortar Shells	-	5,415
Machine Guns	103	68
Rifles	-	461
Rifle Cartridges	-	83,800
Sub-Machine Guns	-	87
Sub-Machine Gun Ammunition	-	1,100
AA Guns	-	2
Food Warehouse	2	-

Table #14: Damage to the Enemy - 5 May 1944.

Researchers and historians recognize that reports like these almost certainly used inflated values, mostly due to the attacker putting a positive spin on the bad news of a lost battle, but "Fog of War" also played its part. This document illustrates just how detailed, even if inflated or fabricated, these army level reports could be.

Documentation of Axis losses during this battle is difficult to find. Estimates suggest that German and Romanian losses of both men and equipment were moderately heavy, but accurate totals are elusive.

A partial total shows the German 8th Army suffered the following losses from 1-10 May 1944: 603 killed; 2,882 wounded; and 274 missing, for a total of 3,759 casualties. Note: These are German losses only. [448]

Needless to say, archives in both Russia and Germany need to be searched to further develop the historical documentation for this largely obscure, but important, second battle for Targu-Frumos.

Meanwhile, German mobile forces once again were regrouping around Targu-Frumos. No records indicate that Soviet forces were aware of these movements.

A column of Panther tanks and other vehicles from the Pz GD *Grossdeutschland* travel down a dirt road in Romania during May 1944.

6 May 1944

Stavka acted to address the issues facing the 2 and 3 UFs by issuing a new Directive. A synopsis is listed below:

VGK Stavka Directive No. 220094 to the Commanding Military of the 2nd Ukrainian Front on the Transition to Defense
May 6, 1944 15 h 15 min
The High Command Headquarters orders:

1. *The troops of the 2nd Ukrainian Front with the receipt of this directive go to* [firm] *defense in the entire front line. In order to create a favorable starting position for a subsequent offensive is permitted carry out private operations with a limited purpose in certain areas* [of the] *front.*

2. *When organizing defense, pay special attention to the Romanesque and Yassky* [Iasi] *directions.*
3. *The defense is deeply layered. In the front line prepare at least three defensive lines with a total depth of defense of 30-40 km.* [18-24 miles].
4. *Front troops to prepare for relocation* [transfer], *immediately proceed to the creation of a stock of ammunition and fuel. Willingness* [readiness] *to resume offensive operations no later than 25 May.* [Note: Eventually this offensive date would be pushed back seven days to 1 June 1944.]
5. *To convey the given orders.*

[For the 2 UF] *The civilian population must be evicted from all settlements in the specified strip, including from Radaut, Solki, Feltichen, Dolgami, Hirlau, Skuleni, Orheev, Dubossar, Grigoriopol, Tashlyk, Buto.*

[For the 3 UF] *The civilian population must be evicted from all settlements in the specified strip, including Malaest, Parkan, Glinnoy, Troitsky, Belyaevka, Mayakov, Ovidiopol, Slobodzeya Russian and Moldavanskaya and the city of Tiraspol.*

Supreme High Command Headquarters

Stalin A. Antonov

Note: See the 20 May 1944 entry about the fate of these displaced, mostly Romanian civilians.

Stavka directive drove home the point that Konev's second attack against Targu-Frumos had failed. The Red Army planned future attacks in this area, but this Axis defensive victory gave the Germans and Romanian forces a respite from major combat operations in eastern Romania. AG South Ukraine desperately needed a break in the fighting to rebuild, to reorganize its forces, and to plan for the future.

The 7th Guards Army and other armies dug in along the new front and conducted local operations, which amounted to small raids and reconnaissance in force missions. In line with the new defensive stance, Marshal Konev began withdrawing the 2nd TA into reserve for refitting. On the other side of the line, German Panzer forces secretly were moving into their final concentration areas north and northeast of Targu-Frumos.

7 May 1944

At 05:50 hours, Panzer units of the German LVII Pz C (*Grossdeutschland,* 24 Pz D and 46 ID) attacked northward from the Targu-Frumos area aiming for the junction between the 7th Guards and 27th As. This surprise attack forced Konev to recall the 16 TC to the front; it also obliged General Shumilov to commit his only reserve force, the 6 RD, to a counterattack in an attempt to regain the positions lost by the 53 RD. At this point in the attack, the 24 Pz D and 46 ID turned their attention to the east and thrust into the flank of the 35 GRC of the 27th A. This sudden and powerful attack forced several rifle divisions of the 35 GRC to fall back, losing a key observation point, Hill 256, which was named Kalmü. (See Map #41 for Hill 256's location.) The final German assault to capture Hill 256 required an artillery barrage, fire from *Nebelwerfers* (smoke mortar - rock launchers) and the use of flame-throwers along with close support from Panther tanks and *Sturmgeschutz* (assault guns). [449] This one-day German counterattack had erased most of the gains the 27th A had achieved over the 2-4 May offensive. Note: The name *Kalmü* is a shortened version of *Kalmück*, which refers to or suggests the Kalmyk Steppe or people, which appears on Soviet historical maps of the area.

Soviet TsAMO record ID #113984903 shows the 7th Guards Army suffered 92 killed and 491 wounded this day. The 27th A suffered 90 killed and 336 wounded. This short but forceful attack must have unsettled Soviet army and front commands. This counterattack also drove home the point that German Panzer formations continued to be dangerous opponents.

The capture of the highest point in the area, Hill 256, allowed the Germans to have good observation into the rear of the 27th A, and that advantage was a significant reversal for the 27th A. The loss of Hill 256 enraged Marshal Konev so much that he ordered an investigation into the matter… "investigate all the circumstances surrounding the surrender of Hill 256 and immediately bring the guilty to trial before the judgment of a military tribunal." He went on to say "…there was no reason whatsoever to justify the surrender of Hill 256." [450] Despite repeated attempts to recover this hill position, Hill 256 was firmly under German control by 11 May, and apparently in the end, Soviet commanders never held the military tribunal.

A German soldier assists two wounded German infantrymen to the rear.

The Western Allies, US and Britain, informed Stalin and the Red Army General Staff that 31 May 1944 was the approximated date of the invasion of Western Europe (D-Day – Operation Overlord). This was done so that Allied armies from the east and west could coordinate offensives during and after Operation Overlord. [451] Alternating offensives would keep German forces off balance and force the German High Command to respond to one widely separated crisis after another.

Far to the south, the main Russian attack in the Crimea against Sevastopol began south of the city near Balaklava. By the end of the day, the Red Army had captured the commanding Sapun Heights. Without control of the Sapun Heights, Axis forces had little chance to mount a successful defense of Sevastopol. At this point the battle for the Crimea was all but over. Fighting would go on for a time, but the outcome was obvious to all.

8 May 1944

Once again, TsAMO record ID #113984903 showed the 7th Guards Army suffered a further 28 killed and 70 wounded on this day. The 27th A likewise suffered 15 killed and 37 wounded. These losses were due to scattered fighting along the front as German forces made adjustments to their newly captured positions. The two-day totals for the 7th Guards and 27th As were 225 killed with 934 wounded. With the conclusion of this last bit of fighting, both sides settled down; the battle was finally over.

General Shumilov and the command staff of the 7th Guards Army finished production of the defensive plan for the Army. This was Operation Plans Document #184, TsAMO record ID #13265391. This plan deployed five rifle divisions to the front line. On the right flank were the 72 and 81 GRDs (24 GRC) and on the left flank were the 36 GRD, 6 GAD, and 53 RD (25 GRC). The 8 GAD was holding the Seret River line in the rear. Numerous supporting combat units were deployed to the rear. By 13 May, the 7th Guards Army had completed the layout for the required three defense lines, but the second and third lines did not contain rifle divisions, most likely due to the weakness of the rifle units. Therefore, at this time, the second and third lines would have been considered fallback positions.

Over the next two days, the main body of the 16 Artillery Penetration Division began to arrive at the front to support the 7th Guards Army. The most recent battle was, of course, over by this date.

Fighting in and around Sevastopol continued, and in an uncharacteristic move, Hitler authorized an evacuation of the troops defending the city.

9 May 1944

Despite no real fighting taking place, the 7th Guards Army issued a routine, *Combat report of the headquarters of the 7th Guards Army, #473,* TsAMO record ID #454034908.

1. The enemy during the night of activity with [the use of] manpower showered methodical artillery fire at the battle formations of the army. Enemy aircraft carried out reconnaissance flights with associated bombing during the night.
2. [The] Army occupied the previous line [front], strengthening it in engineering terms [digging in, building bunkers], conducting a firefight with the enemy and reconnaissance of him.
3. Communication with the troops [via] telephone and with neighboring [units via] telegraph worked smoothly.

One could say just another day at the front. Any losses associated with this normal level of activity are unknown.

Troops of the 4th UF liberated Sevastopol, the major city and naval port of the Crimea, but fighting persisted in the surrounding hills and valleys.

Stavka issued Directive #220096 to the 1st Belarusian Front (1st BF) about the transfer of the 8th GA to its command, making it subordinate to the commander of the newly created 1st Polish Army (1st PA). *Stavka* was secretly building up forces opposite Germany's Army Group Center.

10 May 1944

The 7th Guards Army issued a ten-day long *Plan for counter-artillery training of the 7th Guards Army*, TsAMO record ID #455087681. General Shumilov was using the quiet time to start much-needed training actives for the Army.

The 7th Guards Army issued another routine *Combat report of the headquarters of the 7th Guards Army, #142,* TsAMO record ID #454034919.

1. In the first half of the day nothing significant happened at the army front. The enemy, defending himself at the front line with rare fire, fired at the front lines of our troops and did not take active actions in the morning. [Enemy] [a]viation with single planes conducted reconnaissance to a depth of 20-30 kilometers [12-18 miles].
2. The troops of the army perfected the positions they occupied in engineering terms; they destroyed the targets identified in the [enemy] defense and conducted reconnaissance by observation.
3. Changes in the position of the army did not occur. Communication with troops [by] telephone, with neighbors [by] telegraph.

This report reinforces the sense that within the 2 UFs area matters had gone quiet, with both sides occupying themselves with normal, routine activities.

On 31 May 1944, the 27th A issued a report, *Short Conclusions on the Conduct of Operations from 2 through 8 May 1944,* TsAMO record ID #132882326, page 13-14. A synopsis of this relevant report is listed here to highlight Soviet impressions of why the most recent offensive against Targu-Frumos failed. These same issues also would be valid for the 7th Guards Army's attack.

1. The enemy had a deeply echeloned defense.
2. Poor cooperation between attacking forces, artillery, air, and tanks. As the infantry reached the forward edge of the enemy's defenses, artillery ammunition ran out, VVS aircraft were not available, and even tank support lagged behind the assault.
3. Infantry formations had little AT artillery with them and when enemy tanks attacked, they were forced to go to ground and fight the tanks with what weapons they had or withdraw.
4. The tank armies did not support the infantry attacks with massed tanks, but committed only 5-10 tanks at a time, which were quickly knocked out.
5. The VVA air support bombed our own forces, not only along the forward edge, but up to five kilometers [3 miles] to the rear.
6. The infantry themselves were poorly trained and not fully equipped. Combat experience indicates that if we do not have satisfactorily armed, prepared, and steadfast infantry, then despite the artillery and air support, their effectiveness will be insignificant.
7. The issue of command and control, and communications requires further improvement.
8. The enemy's defense system and his system of fires were not detected to a sufficient degree before the battle began, and as a result, the artillery fire during the artillery preparation had little effect. [452]

As can be seen, this assessment is pragmatic, but it also largely agrees with German accounts of the battle.

But further to the east, in the 3 UF sector, German forces from the 6th Army were once again on the attack. They were attempting to eliminate the Soviet bridgehead over the Dnestr River northeast of Kishinev. Fighting in this area lasted until 20 May, which resulted in a greatly reduced Soviet position on the west side of the river. Once again, a short but powerful German Panzer attack had effectively disrupted *Stavka* plans for future offensive operations from this area. The bridgehead was now too small to support a major attack.

11 May 1944

With the fall of Sevastopol, the Soviet GKO (State Defense Committee) anticipated the complete liberation of the Crimea. The GKO issued a decree on 11 May that required all Tatars (native people who were members of a Turkic ethnic group) to be deported from the Crimea to Uzbek SSR (Soviet Socialist Republic, modern day Uzbekistan), which was some 1,600 miles east of the Crimea on the far side of the Caspian Sea. This was done to punish the indigenous population en masse for collaboration with the Germans. On 20 May the NKVD/NKGB (State Security) issued a report,

> [On] 18 May 180,014 persons have been deported, loaded into 67 trains…6,000 Tatars were conscripted [into the military] …8,000 people have been sent to work for the Moscow Coal Trust [of them] 5,000 are Tatars…. During the deportation 5,989 persons, from amongst anti-Soviet elements, were arrested [which would normally lead to some type of trial. The person could be acquitted of wrongdoing and released, found guilty and executed or, if lucky, sent to Siberia for ten or more years.] [453]

While Stalin's actions were not as heinous as Hitler's attempt to eliminate the Jews, Stalin's treatment of the Tatars and other ethnic groups within the Soviet Union, although not widely known, was just as despicable. In nine days, the deed was done; the state security services were very efficient.

12 May 1944

South of Sevastopol, the final Axis forces in the Crimea surrendered to units of the 51st A and the Independent Coastal Army. The Crimea was now free of Axis forces. During the Crimea battle, the 4 UF reportedly lost 13,000 killed and 50,000 wounded with the Independent Coastal Army losing another 4,000 killed and 16,000 wounded. [454]

Friedrich Ruge, in his book, *The Soviets as Naval Opponents 1941-1945,* stated on page 132, "In one month, 130,000 Germans and Romanian soldiers had been evacuated by sea [from the Crimea], [with] 26,000 soldiers and 6,000 wounded in the last three days…About 78,000 men were killed or made prisoners of war." Due to Hitler's refusal to evacuate troops from the Crimea in a timely manner (that is, before the final battle) over 8,000 men were lost at sea, drowned, as overwhelming Soviet naval and air power assaulted the final groups of evacuation ships. The German 17th Army temporarily had been destroyed, but it would be rebuilt and would reappear in Poland in September 1944.

In the air over Germany, more than 800 heavy U.S. bombers attacked German synthetic oil plants. [455] The resulting damage was sufficient enough that Armaments Minister Albert Speer issued a statement, "The technological war was decided. It meant the end of German armaments production." [456]

On the political front, the U.S, Britain and the Soviet Union issued a joint statement calling upon Hungary, Romania, and Bulgaria to withdraw from the war or to face the consequences of vigorous peace terms. [457]

13 May 1944

The Army issued a *Combat log of the 7th Guards Army*, recorded in TsAMO record ID #2600166.

> Opposing [enemy forces] defended their previous line. He did not fire artillery, but improved his infantry positions and artillery OP [observation posts]. Single enemy aircraft conducted reconnaissance during the day, making 27 aircraft over flights.
>
> Army formations occupied [the same] lines [and] continued to defend their positions [improving] engineering and antitank defenses. [They] carried out a firefight with the enemy and conducted a combat readiness lesson according to the plan.

Once again, this report simply says that nothing much was happening at the front.

14 May 1944

A historical map, TsAMO record ID #100114464, shows the 7th Guards Army had begun to modify its defensive layout with frontline rifle divisions now having two regiments forward with one in a backup position. This map also shows (with no dates given) two rifle divisions, 8 GRD and 6 GAD, occupying the second line of defense. These changes were of course in response to German Panzer attacks in and around Targu-Frumos. A defense in depth was more resilient, and the "reserve regiment" behind the front line could more easily/safely train new recruits and resupply and offer rest to the veterans.

15 May 1944

Stavka (VGK) issued Directive #220097 on 15 May 1944, *On the movements of the commands of the 2nd, 3rd, 4th Ukrainian Fronts.* The reason for this directive was to release Marshal Zhukov from the command of the 1st UF. This would allow the front commanders of the 2nd, 3rd and 4th UFs to assume their new positions. The directive went on to say that "Comrade [Malinovsky would] take command of the 2nd Ukrainian Front no later than May 24 with Comrade Konev to take command of the 1st Ukrainian Front no later than May 27."

16 May 1944

The 2nd and 3rd UF command staffs continued planning for the 25 May offensive. Soviet units were regrouped behind the lines and undertook resupply and training efforts. The Germans could not help but notice all this activity and correctly determined that the Soviets were planning another attack somewhere in the Targu-Frumos - Iasi area. The German command would take the appropriate steps to counter this Soviet threat.

To illustrate what was going on behind the headlines, the entries covering the next few days present several typical reports/orders produced by the 7th Guard Army and the 2 UF. Note: The 7th Guard Army normally would create or review 40-60 major documents each day, with the 2 UF producing at least 50-60 such documents. These average document totals were obtained from a daily search of the historical documents held by the TsAMO archive. Most likely, hundreds of other records also were produced each day by the different HQs.

17 May 1944

General Shumilov issued an order to the troops, TsAMO record ID #133373918. An extract of this order states:

> The liaison officer of the operational service of the 7th Guard Army, Captain Murashnmy Nikolai Pikolaevich, is relieved of his post and is appointed adjutant of the senior rifle battalion of the 235th Guards' Rifle Regiment of the 81st Guards Rifle Division.

This order appears to be a normal transfer, but in fact it illustrates the combing out of fit rear area service personnel for assignment to combat units. Front commanders would routinely order army commanders to look to their rear area service troops for needed replacements for combat units.

18 May 1944

The 5th AA issued a report of air activity for the middle part of May for the 2 UF, recorded in TsAMO record ID #454036309, page 9.

> [T]he period from 10 to 25.5.44 [10-25 May 1944] is characterized by a sharp reduction in the activity of the enemy aircraft, with the exception of reconnaissance, which, as in the first period, continued to carry out intensive reconnaissance to a depth of 500 - 600 km [300 - 370

miles]. During this period, the troops of the 2nd Ukrainian Front went over to the defensive, and the [enemy] ground forces against Nika [bringer of victory] were regrouping and replacing parts of the first line, therefore hence the "lull" at the front - the relative calm and invariability of the ground situation were reflected in the combat activities of the enemy air force. During this period, only 1,366 enemy aircraft over flights were recorded. [The total aircraft over flights covers only 15 days of activity].

19 May 1944

The 7th Guards Army continued to hold its position, to improve defensive works and to accumulate supplies, as did all the other armies of the 2nd and 3rd UFs.

For German and Romanian forces within AG South Ukraine, the pause in the fighting allowed depleted units to rebuild and forces at the front to reorganize. Some of the benefits of fighting on Romania territory were the good road and rail transportation networks, which allowed reliable logistical support with the added benefit of not having bands of Soviet partisans operating in the rear areas. The Luftwaffe and Romania air units also could operate from well-supplied airfields, many of which were all weather capable (hard surfaced runways), which gave them an advantage over the VVS. But not all was well behind the lines.

> However, during the summer of 1944 Army Group South Ukraine began to notice a growth in agitation and a decline in co-operation, both in the oilfields and railways, where it appears that the Communists [instigators] were taking advantage of the destruction wrought by Angle-American air raids to stir up the [Romanian] workforce, and the management was increasingly obstructive. There were even isolated acts of sabotage. [458]

20 May 1944

Once again, the 7th Guards Army HQ sent a personnel request, TsAMO record ID #133373920, "To the Chief of the Operations Department of the Headquarters. For the second time I ask you to send to the personnel department of the 7th Guards Army copies of the combat characteristics for the below-named officers of your department." This list included the names of at least twelve officers suggesting that further transfers from the rear support services to combat units were pending.

The 7th Guards Army HQ also was acting to fulfill *Stavka* Directive #220094 from 6 May, about clearing the combat zone of civilians. "C*ivilian population must be evicted from all settlements in the specified strip."* TsAMO record ID #454034995, from Guards Colonel Kupriyanov, recounted, "I report: from the front-line zone in the 7th Guards Army [that] 5,427 families were subject to eviction on 20.5.44 [20 May 1944]." The paperwork never ends at the HQ. These families were almost certainly Romanian civilians, mostly old men, women and children. In situations like this, they would be told to pack their things, only taking what they could carry, and marched, under guard, miles to the rear. What the Soviet authorities ultimately did with these superfluous populations is unknown.

General Rodion Yakovlevich Malinovsky left the command of the 3 UF for his reassignment as the commander of the 2 UF.

The German 8th Army reported the following losses during 11-20 May 1944: 220 killed; 1,228 wounded; and 88 missing, for a total of 1,536 casualties. Note: These are German losses only. [459]

21 May 1944

The 2nd UF issued an "Operational Summary" for the front, TsAMO record ID #454490265.

> [For the] 40, 7 Guard, 27, 52, 4 Guard, and 53 Armies -There were no changes at the front and in the disposition of army units over the past day. The troops carried out engineering and defensive work, conducted combat reconnaissance and continued to engage in combat training.

So, for six entire armies and along the entire 2 UF, nothing much was taking place, just normal routine activities. This prolonged timeframe of inactivity showed just how much all of the armies on both sides of the conflict needed to rest and to regroup before they could continue fighting.

22 May 1944

The 7th Guards Army issued an *Extract from the combat log of anti-aircraft artillery units of the 7th Guards Army*, TsAMO record ID #152505434.

> The task of the unit and their dislocation are the same. Within 24 hours, [the unit recorded] 90 airplanes, by types: 38 Fw 190, 25 Ju 88, 17 Ju 87, 5 He 123, 3 He 126, [and] 2 He 111. Bombs [were] dropped in districts Bumorovitsa, Migura, Ruzhenos, Buzhka, Bukhalniua, Samodtami, [by] types Ju 88, Ju 87, [and] He 123, in groups of up to 45 aircraft. [They] dropped about 200 FAB [general-purpose bombs]. Reconnaissance was carried out in the Pumbrovitsa, Beicheniya, Ruzhenos districts. In the zone of [AA] fire, 3 group[s] and one single target, [were] fired upon. Consumption [of] 85 mm. shells – 17 pieces, 37 mm shells – 99 pieces. There were no downed planes. During the day, air battles were recorded, as a result of which one Fw 190 was shot down; [two] Fw 190s and 6 of our Airacobra fighters took part in the air combat. Losses of anti-aircraft artillery units: rank-and-file personnel – 2 killed, 4 wounded, 2 vehicles disabled. The morale of the units is good.

Even though these AA units were not the targets of these German air raids, apparently the bombing caused the six anti-aircraft artillery casualties and vehicle damage.

On this date, General Rodion Yakovlevich Malinovsky assumed command of the 2 UF as Marshal Konev took command of the 1 UF. Of course, Marshal Malinovsky continued with the offensive plan begun by Marshal Konev. Time was short, so Malinovsky would have relied upon the 2 UF HQ staff to move the plan along while he settled into his new position.

A 1957 photo of then Marshal Rodion Yakovlevich Malinovsky.

"Beginning with the spring of 1944, [in the newly returned Romanian Moldavia territory, which the Soviets had annexed from Romania in 1940] Soviet authorities extorted 480,000 ton[s] of [wheat] from war-torn [eastern] Moldavia alone, even before the Iasi-Kishinev operation and the liberation of all of Bessarabia." As noted, "Liberation by the Red Army did not engender any particular joy [from the Romanian population]" [460] because the Russians were largely regarded as an invading army intent upon seizing territory from Romania and pillaging the land. Large amounts of this "liberated" wheat and other supplies were sent back to the Soviet Union to feed and to support the home population. In fact, the 7th Guards Army issued a seven-page TsAMO report ID #133391589, covering 1 January to 1 August 1944 titled, *Information about the help of the 7th Guards Army and the national economy*. The seven sections of the report were:

1. Military self-procurement
    a. Prepared and transferred to the planned supply of the army
    b. Received in 1944 from the army subsidiary for the supply of the army
2. Aid to agriculture
3. Assistance to road authorities
4. Assistance to industry and urban households
5. Assistance in the line of defense
6. Veterinary care
7. Other types of assistance to the civilian population

As can be seen, the plundering of captured enemy territories was well underway, with the Red Army receiving first call for all supplies, vehicles, materials, equipment, and food with further food assistance eventually filtering down to the general Soviet civilian population.

23 May 1944

Stavka, realizing their forces were not entirely ready for offensive operations, issued a new Directive delaying the 1st and 2nd UFs offensive that was ordered on 6 May with Directive #220094. The new Directive (number unknown) delayed the two-front offensive until 1 June 1944.

24 May 1944

The 7th Guards Army continued to be deployed defensibly, with five rifle divisions holding the front line with two regiments forward and the third regiment in a reserve position further back. But the left flank of the Army contained a third position to the rear, which was held by the 6 and 8 GADs. This location also secured the boundary line with the 27th A. Overall, the Army was deployed in a firm but flexible layout.

25 May 1944

German planning for a spoiling attack against the anticipated Soviet offensive continued. The Germans had correctly identified the area directly north of Iasi as the most important Soviet force concentration area, so they chose the Russian 52nd A of the 2 UF as the main target for their spoiling attack.

26 May 1944

Stavka sent another directive to further clarify the issue along the southern section of the Russian Front, Directive #220107. The commanders of the 2 and 3 UFs were ordered to postpone further offensive operations in their area.

Directive #220107

May 26, 1944 02 h 35 min

> *The headquarters of the Supreme Command orders:*
>
> *1. To postpone the readiness for the offensive until receiving special instructions of Stavka. The front troops will firmly defend the occupied lines in accordance with the General Headquarters directive of May 6, No. 220094 (for the 3rd Ukrainian Front, No. 220095, dated May 6), having strong second echelons and reserves.*
> *2. Organize combat training of troops with the task of preparing them for subsequent offensive operations. Combat training to cover all troops and headquarters, for which to establish the sequence of withdrawal of troops to the second echelons. When preparing, pay special attention to less trained divisions and headquarters.*
> *3. Defense plan in accordance with the changes in the composition troops and a combat training plan for one month to submit to General Staff by June 1.*
>
> *Headquarters of the Supreme Command*
>
> *I. Stalin A. Antonov* [461]

This *Stavka* directive halted all preparations for the imminent offensive operations that were to have taken place in Romania. The two fronts instead shifted to the defense and began a month-long training program for their troops. This move is significant as it indicates the true situation at the front. *Stavka's* plan devoted over two months of perfect summer offensive weather to rebuild and to reorganize the 2 and 3 UFs depleted forces. Stalin and *Stavka* would have never done this unless they were forced to do so. Stalin wanted a big victory, and that required time to assemble massive forces to achieve that goal.

Of course, General of Infantry Otto Wöhler, commander of the German 8th Army, was unaware of this last-minute change in plans. Preparations for the German spoiling attack against the now canceled Soviet offensive continued.

David Glantz, in his book, *Red Storm over the Balkans,* on page 330, summed up the German position nicely:

> …, the German Eight Army's intelligence organs had already detected his [Malinovsky] fronts' extensive regrouping and, unaware of the Stavka's change of heart, assumed it was associated with Soviet preparations for a fresh offensive. Armed with this knowledge, … Wöhler's Eight Army had no choice but to react to its intelligence and prepare to disrupt the anticipated Soviet offensive.

27 May 1944

Stavka had worked out the summer of 1944 offensive plan for the entire Russian Front, from the Baltic Sea in the north to the Black Sea in the south. The offensives would unfold as five distinct operations launched sequentially, roughly from north to south.

1. Karelian Isthmus – south Karelia – 10 June to 9 August 1944
2. Belorussia – 23 June to 29 August 1944
3. Lvov – Sandomierz – 13 July to 29 August 1944
4. Lublin – Brest – 18 July to 2 August 1944
5. Yassy [Iasi] – Kishinev – 20 August to 7 September 1944 [462]

The Iasi/Kishinev offensive would consist of two converging attacks, the 2nd UF attacking south and east, with the 3rd UF attacking south and west. The plan was to envelope Axis forces holding the line around Kishinev and to force a breakthrough toward Bucharest, the capital of Romania. The 2nd UF would capture Targu-Frumos and Iasi along the way. With this *Stavka* plan in place, the 2nd and 3rd UFs would now have almost three months to rest to assemble and reequip their forces. (See Map #42.)

To carry out the overall summer offensive plan, *Stavka* needed to regroup massive numbers of troops and equipment, and to do it secretly by using *maskirovka* methods and enhanced security measures. Critical deployment orders were written by hand, known to only two or three people, and delivered to and/or reported by commanders in person. The use of radio or telephone was strictly forbidden and powerful radio stations were shut down. *Stavka* was not taking any chances that their real plans for the summer offensive season would be revealed prematurely to the Germans. [463]

In general terms, most historians recognize the main focus of the 1944 Soviet summer offensive season was to destroy German AG Center during *Операция Багратио́н* (Operation Bagration) and to drive enemy forces from Soviet territory. Likewise, the areas occupied by AG North, AG North Ukraine and AG South Ukraine would be major but secondary targets. As for AG South Ukraine, *Stavka* would be happy if the 2 and 3 UFs were able to advance deeply into the Balkans and to remove Romania, Bulgaria, and Hungary from the war, or better yet, to bring them over to the Soviet/Allied side and use them to fight the Nazis. The Soviets also hoped that Finland could be persuaded to break with Germany and to leave the war. These were lofty goals, but it remained to be seen if the Red Army had the strength and stamina to achieve all these objectives during one summer campaigning season.

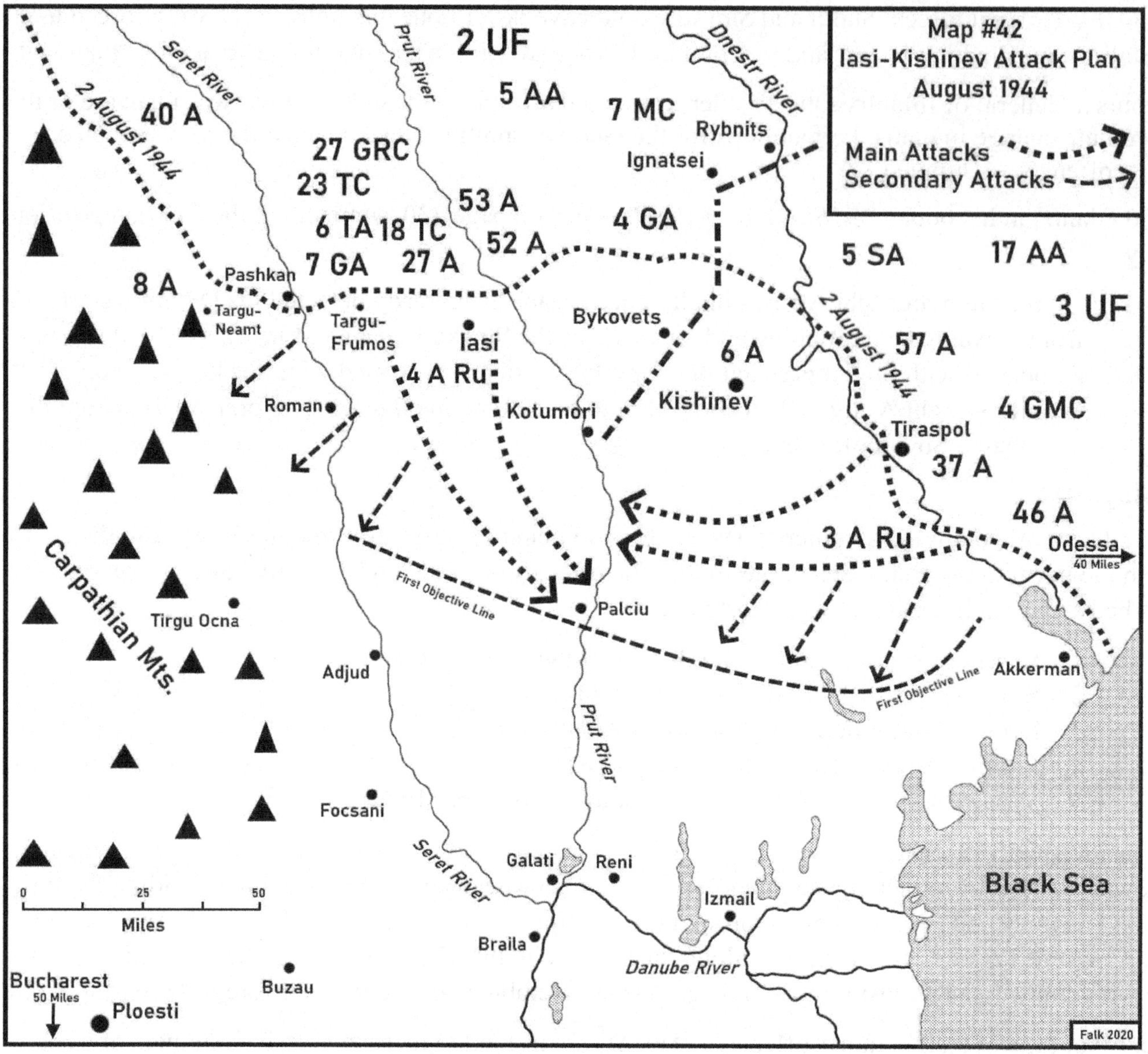

Note: All Armies on the map are shown as they would be deployed for the attack in August 1944.

28 May 1944

Intelligence officers from the German 8th Army noted that major Soviet forces had completed or were in the process of completing their movement into assembly areas north of Iasi. They also calculated these regroupings involved at least eleven RDs, two TDs and one TB. [464]

29 May 1944

The initial German assault forces had completed their own regrouping north of Iasi. This spoiling attack was limited in nature and would consist of two distinct phases. Operation *Sonja* [wisdom] would begin on 30 May and would destroy the assembled Soviet forces and push the remaining Soviet troops back five to six miles and drive them north and over the Jijia River. Once Operation *Sonja* was completed, Operation *Katja* [a female German given name] would be launched on 2 June after the required Panzer formations had been regrouped. This attack would be west of the *Sonja* operation and was designed to push Soviet forces back some ten miles and away from Isai. The German command ultimately wanted to form a new defensive line along the more suitable and more defensible high ground and to link up with the new German positions along the Jijia River. [465] (See Map #43.)

30 May 1944

The German 8th Army launched Operation *Sonja* against the Soviet 52nd A north of Iasi. Initially the 23 and 24 Pz Ds attacked with support from the 79 ID on the left flank. This German surprise attack took place about 30 miles east of the nearest position of the 7th Guards Army, so the 7th Guards Army did not directly participate in the battle.

Operation *Sonja* began at 04:00 hours after a short but intense artillery barrage. After heavy fighting, German assault forces had advanced about five miles along a four-mile-wide front and captured the village of Stanca, which was located south of Carpiti, just over the Jijia River. This attack took Soviet forces by surprise, but the 2 UF reacted quickly and soon reinforcements began flowing into the area.

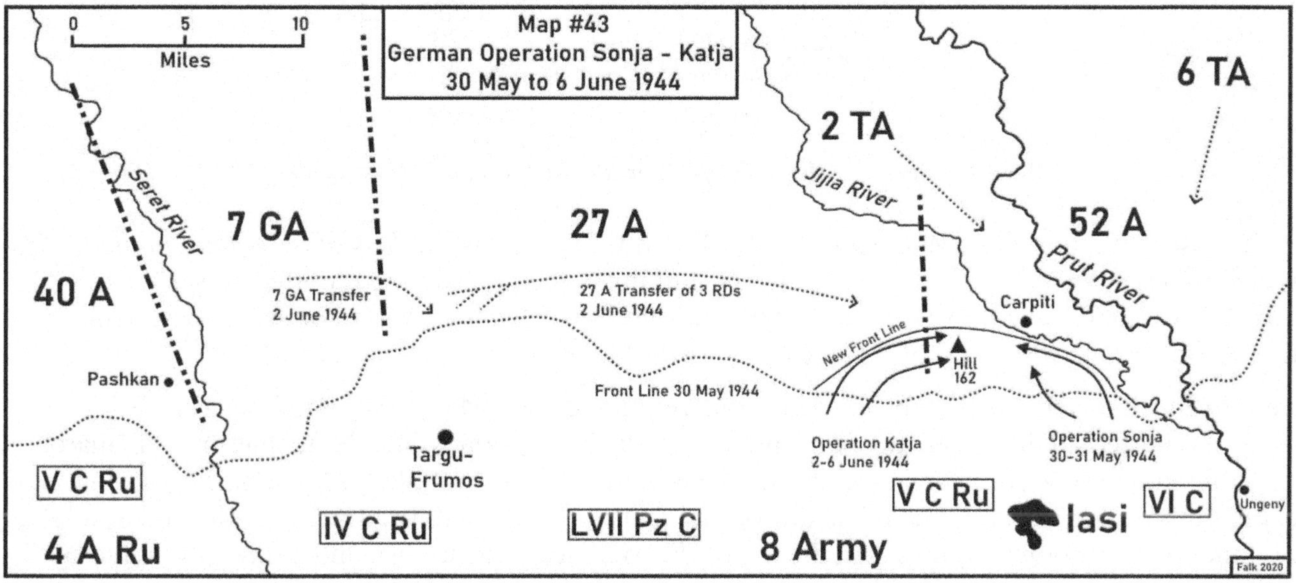

31 May 1944

The German spoiling attack against the 52nd A and the 2 UF continued. Fighting raged as German and Romanian forces (11 Ru ID) widened their penetration and attempted to advance closer to the Jijia River. The Russian defenders fought tenaciously knowing that further reinforcements, including tank formations, were on the way. Having achieved most of their objectives at this point, Operation *Sonja* continued on for another two days with scattered fighting before ending on 2 June.

The 5th AA continued to report air activity for the 2 UF, this time covering the period 30-31 May, TsAMO record ID #454036309, page 9.

> The activity of the enemy's aviation increased again, and… on [30-31 May 1944], when his ground troops launched a counteroffensive in the sector to the north and north west of Iasi, his aviation showed an exceptionally high level of activity; for example: [on 30 May 1944] marked – 2,082 aircraft over flights and [on 31 May 1944] – 1,547 aircraft over flights of the enemy.
>
> [This] fighting took place in a narrow sector of the front: Vulturul, Golzeshtiy; moreover, very intensive and continuous massive strikes were carried out by all types of its aviation, in groups of up to 80 aircraft each. During this period, a total of 4,740 enemy aircraft flights were recorded. [This assumes these 4,740 aircraft were the total for the entire 2 UF during these two days].

Despite its losses, the 5th AA ended the month with 1,338 aircraft, according to TsAMO ID #454036309, page 23.

On this date, the German 4th Air Fleet had the following operational aircraft. (See Table #15.)

Aircraft Type	Number
Single engine fighters	138
Night fighters	9
Bombers	34
Ground attack	302
Night ground attack	70
Strategic reconnaissance	31
Tactical reconnaissance	28
Transports	76
Total	**688**

Table #15: German 4th Air Fleet Operational Aircraft -- 13 May 1944. [466]

The 2 UF issued a report, TsAMO record ID #113723601, *Report on the loss of personnel in the troops of the 2nd UkrF from May 1 to May 31, 1944.* Within this document are the reported losses for the 7th Guards Army: 1,080 killed: 4,410 wounded: 523 missing: 1,494 sick: 40 non-combat losses, none captive or frostbitten, for a total of 7,569 casualties for the month.

This report illustrates the magnitude of the intense fighting during May 1944, with the entire 2 UF suffering a reported 49,443 casualties. This report also lacks important data. Note that the 7th Guards Army suffered no reported POW losses during the month, which is not likely. Most of the missing were probably POWs. The high number of sick troops indicates a weakened, tired Army at the end of a long supply line. Ammunition was always at the top of the logistical priority list, and ammunition during May was in short supply, so food, fodder, and medicine also would have been scarce. Only time and a temporary end to large scale fighting would heal the wounds suffered by the 7th Guards Army.

By this stage of the war, the Red Army medical services provided good support to frontline troops, but treatment still was harsh compared to other armies. For example, the U.S., British and even the German Army had lower death rates because they devoted more resources to medical care.

> [During the war,] the Red Army found that, in offensive operations, it would generally lose one man killed or missing for every three medical casualties (i.e., wounded or sick). [This implies a death/missing rate of about 25%, which the author believes is a low estimate of total Soviet fatalities.]

The wounded right at the front were given only basic first aid. But at the regiment or brigade level,

> only those casualties were treated that could be returned immediately to combat. At each stage [from the front to the rear area], emergency aid would be given to save the soldier's life and stabilize them, but therapeutic treatment was given at any stage only to those who could be expected to recover in a usefully short period of time; such prioritization was considered the greatest contribution the medical services could make to the restoration of combat effectiveness. Those with little prospect of proving useful [service] during the course of the operation were sorted by injury and dispatched to [a] specialized army or, in the most serious cases, [to] front hospitals. [467]

As stated in an earlier chapter, Russian medical reports maintain that sanitary losses (that is wounded who died after reaching medical support,) were about 6% of the wounded. The decision to send the seriously wounded troops to the deep rear of the army or the front, before receiving suitable treatment, certainly contributed to this 6% loss rate. [Once again, 25% plus 6% yields an overall death/loss rate of 31%, which is probably a low estimate of actual fatalities and losses suffered by the Red Army during the war.]

The German 8th Army, in Romania, reported the following losses during 21-31 May 1944: 331 killed; 1,673 wounded; and 34 missing, for a total of 2,038 casualties. In total, the 8th Army suffered 7,313 casualties during all of May 1944. Note: These are German losses only. [468]

Further to the west in Romania, 480 U.S. heavy bombers raided oil refineries and communication targets (rail centers) in the Ploesti area. [469] This massive air raid disrupted both fuel production and rail services for Axis forces.

1 June 1944	
7th Guards Army	
Assigned – 2 Ukrainian Front	
Colonel General M.S. Shumilov Commanding	
Rifle, Airborne and Cavalry Units	24 Guards Rifle Corps (72, 81 GRDs) 25 Guards Rifle Corps (36 GRD, 6 GAD, 53 RD) 8 GAD, 6 RD 3 Separate AT Rifle Battalion 62, 63, 64, 65, 66, 94 Penal Companies
Artillery Units	11 Artillery Division (40 Howitzer Artillery Brigade, 45 Gun Artillery Brigade) 16 Artillery Penetration Division (49 Light Artillery Brigade, 61 Gun Artillery Brigade, 52 Howitzer Artillery Brigade, 90 Heavy Howitzer Brigade, 109 High-Power Howitzer Brigade, 14 Mortar Brigade) 161 Guards Gun Artillery Regiment 114, 115 Guards Tank Destroyer Regiments 41 Guards Gun Artillery Brigade 30 Antitank Artillery Brigade 263, 290 Mortar Regiments 5 Antiaircraft Artillery Division (670, 743, 1119, 1181 AA Regiments) 162 Guards AA Regiment 662 Separate Aerial Observation/Air Raid Warning Company
Armored and Mechanized Units	27 Guards Tank Brigade 34, 38 Separate Armored Train Battalions
Engineer Units	60 Engineer Sapper Brigade 175, 329 Separate Engineer Battalions 46, 48, 66 Engineer Battalions
Reported Combat Strength Estimated Ration Strength	54,993 men 62,993 men and women (including an estimated 9,000 support personnel)

(BSSA 44, BSA 44, TsAMO Archive)

1 June 1944

German and Romanian forces continued with Operation *Sonja*, but having achieved their objectives, they now were more concerned with holding the ground they had gained and keeping Soviet forces distracted and focused upon their section of the front. Meanwhile, the regrouping of Panzer units continued for Operation *Katja*.

2 June 1944

Operation *Katja* began at first light, supported by heavy fire from artillery and rocket launchers. The 24 Pz D, with major elements from the PG D *Grossdeutschland* leading the way along with flank support provided by the 3 Ru ID and 18 Ru MtD, attacked along a six-mile section of front. (See Map #43.) After fierce fighting, the German and Romanian forces advanced between two and a half to five miles. The 2 UF now had to deal with two separate but converging attacks. To counter this new German attack, General Malinovsky ordered General Shumilov to take over a section of the front from the 27th A, thus releasing the entire 35 GRC's three RDs in order to deploy them against the left flank of the German attack. These three RDs would take some time to reach the battle. This order also required General Shumilov to move his two reserve RDs, the 36 GRD and the 6 RD, to the front.

On 2 June, as part of Operation Frantic, U.S. heavy bombers and fighters began using the shuttle bombing technique of flying between Allied air bases in the west to bases in the Ukraine behind the Soviet lines, bombing Axis targets on the way. Then after a few days of rest for the crews, and with the bombers rearmed, they would return to their western bases, once again attacking enemy targets along the way. Overall, the Americans flew seven of these shuttle bombing missions. But after operational mistakes (see 21 June 1944 entry for an example) Soviet and American relations suffered greatly. The entire operation was ultimately considered a failure and was canceled.

3 June 1944

Operation *Katja* continued to unfold with a turn toward the northeast aiming at Hill 162 and the Jijia River. The battle intensified as Soviet reinforcements arrived and launched counterattacks, but the German attack force was not stopped quite yet.

4 June 1944

Units of PG D *Grossdeutschland* finally captured Hill 162, with the 24 Pz D attacking to its right, but strong Russian resistance blocked further progress. Soviet reinforcements had arrived in force.

In Italy, the U.S. 5th Army with supporting Allied forces captured the eternal city, Rome.

5 June 1944

Further German attacks were launched around Hill 162 by units of *Grossdeutschland* and the 24 Pz D, with the 14 Pz D in support. During the day, the Germans captured some new ground, but they failed to break through to the Jijia River. Substantial Soviet forces now firmly blocked any further advance.

6 June 1944

Fighting continued along the front during the day, but Operation *Katja* had in fact ended. "The Eighth Army suspended further offensive operations on 6 June because they had completely occupied the high ground northwest of Jassy [Iasi], and, therefore, achieved their attack objectives." [470]

Completing his task, Guards Major Rymarenko sent a plan to 7th Guards Army HQ for *Counter-Artillery Training* (counter-battery training.) This is recorded in TsAMO record ID #454902294. Despite fighting taking place along the 2 UF front, life and service in the 7th Guards Army continued in a normal day-to-day manner, while fierce combat was taking place nearby, thus revealing the odd contrasted life at the front.

During the early morning hours, on the north coast of France, the U.S. 1st Army and the British 2nd Army along with supporting Allied land, sea and air forces, landed in Normandy; Operation Overlord had begun. By the end of the day approximately 156,000 Allied troops had secured several bridgeheads on the north coast of France. During the first day of the invasion (D-Day), the combined Allied Air

Forces flew 11,085 sorties over the battlefield and dropped 10,395 tons of bombs. Hitler's worst nightmare, a large amphibious landing in Western Europe, was an Allied success! From this point forward, Axis forces in Europe would be progressively crushed between two massive Allied land armies advancing from both the east and the west. With Allied navies controlling the seas and huge air forces dominating the skies over Germany, the end was nigh for Nazi rule.

Joseph Stalin cabled congratulations upon hearing the news of the successful Allied landing in France, saying, "It brings joy to all of us." [471]

7 June 1944

On or about 7 June, Stalin informed Churchill and Roosevelt about Soviet intentions in relation to the D-Day invasion of France:

> The summer offensive of the Soviet troops, to be launched in keeping with the agreement reached at the Teheran Conference, will begin in mid-June in one of the vital sectors of the front. The general offensive will develop by stages, through consecutive engagements of the armies in offensive operations. Between late June and the end of July the operations will run into a general offensive of Soviet troops. I will keep you posted about the course of the operations. [472]

8 June 1944

The 7th Guards Army held its front, as opposing German forces remained quiet. By this date the Army had moved the 8 GAD to take control of part of the front from the 27th A. This process would continue and expand until in late June when the 7th Guards Army would have three RDs in position covering for the 27th A. These movements were in response to the German *Sonja* and *Katja* Operation.

9 June 1944

Deputy Chief of Staff of the Army Guards Colonel Kupriyanov issued, *Combat report of the headquarters of the 7th Guards Army #602/OP,* TsAMO record ID #454035109. A summary of the report is as follows:

1. The situation on the front of the army did not produce anything new. The enemy continued to defend themselves and fired at the positions of our troops with harassing fire. Enemy aircraft conducted reconnaissance with a pair of fighters. [This was a fighter sweep; they were looking for something to shoot down.]
2. The army's troops occupied the former lines and conducted combat training. The position of the army [is] without change.

10 June 1944

The command staff of the 7th Guards Army released a massive 46-page, 11 section plan. This very long plan was entitled, *Defense plan of the 7th Guards Army*, TsAMO record ID #131310317. This proposal almost certainly was meant for the command staff at 2 UF HQ. This document not only covered the placement of combat troops and their supporting elements like artillery and engineers, but also presented one wire and two radio/telegraph diagrams detailing commutation links to the 2 UF and adjacent armies. The report also provided a complete list of medical support units for the Army on page 40, detailed below:

> One HQ Field Evacuation Point – #224.
> One Field Evacuation Point – #24.
> Surgical Hospitals, #869, 899, 2207, 2208, 2209, 2409, 4344, 4611, 5217, 5226.

Lightly Wounded Hospitals – #3247, 3255.
Therapeutic Hospitals – #34, 2712.
Infectious Hospitals – #2241, 2242.
One Separate Medical Reinforcement Company – #51.
One Sanitary Epidemiological Unit – #115.
One Disinfection Company – #111.
One Pathological and Anatomical Laboratory – #112.
One Separate Ambulance Company (Auto) - #97.
One Ambulance Company (Horse-Drawn) - #251.
Field Laundry Squads – #116, 117, 118.
Laundry Bathhouses – #1, 2, 3, 4.
Military Ambulance Trains – #520, 521.

With sixteen hospitals assigned, the Army would have had approximately 5,350 beds to care for the wounded of the entire Army.

The Army continued to receive adequate veterinary services from the following units:

Army Veterinary Hospitals – #958, 060.
Army Evacuation Veterinary Hospitals – #948, 959.
Mobile Veterinary Lab - #16.

Note that the Army's medical and veterinary support structure is similar to previous ones. The total number of hospitals and supporting medical units continued to change somewhat over the course of the war, but now, there was an interesting new addition: the Army had two medical ambulance trains directly assigned to it. These ambulance trains would speed the transfer of critical patients to the deep rear, thus opening space within local hospitals which would increasing the survival rate of these critically wounded troops. To see that medical support within the Red Army was changing for the better was an encouraging sign. For reference, a typical U.S. hospital train during WW II could transport about 360 stretcher patients per trip.

A wounded solider is delivered to a rear area first-aid post on a cart. The inscription on the cart reads: "For the Seriously Wounded."

In the far north, in action against Finland, the Red Army launched the south Karelia offensive. Air and artillery attacks had started the day before. The Red Army committed 270,000 troops; 620 tanks; 1,660 artillery; and 1,500 aircraft, which yielded a 4:1 advantage in troops, 5:1 in tanks, 6:1 in artillery and 15:1 in aircraft over the Finnish defenders. [473] This offensive against Finland was the first stage of *Stavka's* overall summer offensive plan for 1944.

11 June 1944

Guards Lieutenant Colonel Ryzhov issued an intelligence report, documented in TsAMO record ID #454280500, that stated, *According to reports, the German Army received 170 mm guns on a* [howitzer] *carriage*. This is an interesting report because it could only be referring to the German 17cm (170 mm) K 18 MrsLaf heavy gun. This weapon, which was very accurate, was used for long-range counter-battery fire at ranges up to 18 miles. When the 7th Guards Army issued the plan for *Counter-Artillery Training* on 6 June, maybe they did so because of the arrival of this new German weapon.

12 June 1944

Guards Colonel Kupriyanov issued a report entitled, *Combat report of the headquarters of the 7th Guards Army #613/OP,* recorded in TsAMO record ID #454035131. The document reported that

1. Over the past night, no significant changes occurred in the position of the army front. The commander [Army Commander?] defended himself and with methodical fire, combined with short fire raids, he organized the positions of our troops. The enemy's aviation conducted reconnaissance with [a] single aircraft.
2. The troops of the army continued to occupy and fortify the former lines, carried out fires [small arms fire] with the enemy and carried out reconnaissance of the [area].
3. Communication with the troops is [by] telegraph and telephone, with neighbors by telegraph.

Overall, Kupriyanov reported just another quiet day at the front.

Another interesting report, this time by Lieutenant Colonel Leibman, *Armored summary of the headquarters of the BT and MV* [Armored and Mechanized Forces] *7th Guards Army #135,* is found in TsAMO record ID #132309842. This report listed the current status of the 27 GTB. (See Table #16.)

Vehicles Authorized	Currently in Service	Under Repair
64 – T-34 tanks	56 – T-34s	1 – T-34
1 – SU-85 assault gun	1 – SU-85	–
1 – SU -76 assault gun	–	1 – SU -76

Table #16: 27 GTB Strength – 12 June 1944.

So, the 27 TB was currently understrength by 7 – T-34 tanks and 1 – SU-76. Also having only one tank under repair is a strong indicator the 27 GTB had not seen combat recently. The report goes on to say the brigade held the following supplies:

Ammunition 1.8 - combat loads
Diesel fuel - 2.3 loads (for the tanks)
Gasoline B-70 - 3.5 loads (low octane)
Oil/lubricants – 1.3 loads
Gasoline - 2.0 loads (automobile)
Food – four days' supply

And finally, the report states, "The road for all types of transport is passable," thus recording that the roads were dry after the spring *rasputitsa* season. Overall, this nice, neat, one-page report indicated, without saying so, that the 27 GTB was ready for battle. The brigade had on hand sufficient tanks and ammunition for several days of combat and sufficient diesel fuel for approximately 250 miles of off-road travel. The report also noted that the 38 Armored Train Battalion was operating somewhere to the rear of the Army. The 7th Guards Army was gathering and preserving its growing strength.

13 June 1944

Guards Colonel Kupriyanov issued a report entitled, *Combat report of the headquarters of the 7th Guards Army #617/OP,* archived in TsAMO record ID #454035139. He noted that

1. Over the past night, no significant changes took place on the army line. The enemy defended [their] previously occupied line, and with methodological fire, combined with short fire raids, fired at the positions of our troops. Enemy aviation conducted reconnaissance with [a] single aircraft.
2. The army's troops continued to occupy the former lines, to strengthen their positions in engineering terms, and to carry out a firefight with the enemy and small groups of his reconnaissance.
3. Communication with the troops is [by] telegraph and telephone, [and] with neighbors by telegraph.

This report again details another normal day at the front, with an enemy aircraft performing reconnaissance, German artillery pounding some Russian positions and then shifting suddenly to another area [fire raid] for a short time, with small arms combat/fire against enemy troops and infiltrators. All the while communications lines between units and armies remained open. Even the low-level nature of this type of action was enough to cause causalities. No combat activity is insignificant at a personal level.

Adolf Hitler and his propaganda minister Joseph Goebbels had been boasting of *Wunderwaffen* (wonder or miracle weapons) for several years, weapons that would shift the balance of power in Germany's favor. Most people thought this was just propaganda, but the first one of these weapons was now ready for use. (Note: many types of secret *Wunderwaffen* were under development within Germany, and this book will highlight only a few of the most successful ones.)

On the Western Front, Nazi Germany began employing its first secret weapon against Great Britain, the Fieseler Fi 103, also known as the V-1, or *Vergeltungswaffe-1* (Vengeance or Retaliatory Weapon). The V-1 was an unmanned flying bomb capable of carrying a one-ton warhead. Also known by the Allies as a buzz bomb, or doodlebug. The V-1 had a range of about 150 miles and flew at 400 miles per hour, which made it hard to shoot down. In reality this was the world's first operational cruise missile, and it would kill and injure thousands of mostly civilians, and terrorize the populations of major cities like London and Antwerp, Belgium, for many months to come.

14 June 1944

Once again, the 7th Guards Army HQ sent out a personnel request, TsAMO record ID #133373940, *The plan for the movement of personnel of the operational department of the HQ of the 7th Guards Army*. This list included the names of at least 14 officers who were detached and reassigned to different positions, like clerk of the operations department, deputy head of operations department, and so on.

The document also listed eight vacant positions:

- Senior assistant to the head of the logistics department, one position.

- Assistance to the head of the department, two positions.
- Help for the head of the department with the use of war experience, one position.
- Civilian drivers, three positions.
- Civilian draftsman, one position.

Unexciting changes of assignments, yes, but filling these positions was necessary to keep the Army fighting.

15 June 1944

Guards Major Vasiliev issued, *Scheme of the battle order of the air defense units of the 7th Guards A*, TsAMO ID #152502597. This map displayed the placement of four out of five AA Regiments assigned to the Army. Anti-Aircraft Regiments 162, 743, 1119 and 1181 were deployed to protect the frontline rifle divisions in a central area located between the 24 and 25 GRCs. The 670 AA regiment was not shown on the map and was most likely deployed to the rear covering the Army's main supply center or railhead.

16 June 1944

Once again Lieutenant Colonel Leibman issued a report on, *Armored summary of the headquarters of the BT and MV* [Armored and Mechanized Forces] *7th Guards Army #139,* TsAMO record ID #132309846. This report listed the current status of the 27 GTB:

Vehicles Authorized	Currently in Service	Under Repair
64 – T-34 tanks	64 – T-34s	–
1 – SU-85 assault gun	1 – SU-85	–
1 – SU -76 assault gun	–	–

Table #17: 27 GTB Strength – 16 June 1944.

At this point the 27 GTB was at full strength, so it must have received 7 – T-34 tanks and 1 – SU-76 as reinforcements over the previous four days. The Brigade also contained the following supplies:

Ammunition – 76 mm 2.0, 85 mm 1.1 combat loads
Diesel fuel – 2.11 loads (tanks)
Gasoline B-70 – 3.5 loads (low octane)
Oil/lubricants – 1.4 loads
Gasoline – 2.73 loads (automobile)
Food – three days' supply

The 38 Armored Train Battalion continued to operate to the rear of the Army.

17 June 1944

Guards Colonel Kupriyanov issued a report, *Combat report of the headquarters of the 7th Guards Army #634/OP,* TsAMO record ID #454035170. Once again, Kupriyanov seemed to be filling his paperwork obligation without reporting any significant action by the Army. The paperwork never ends at Army HQ.

1. By dawn, no significant changes occurred on the army front. Combat actions continue to be limited to firefights with the enemy.
2. Non-deployment of army troops without deception. [This must be referring to some type of *maskirovka* methods].

3. Communication with the troops is [by] telegraph and telephone, [and] with neighbors by telegraph.

At this point, Kupriyanov appeared to be running out of creative ways of saying the same thing, so his report is both short and efficient.

18 June 1944

Lieutenant Colonel Leibman issued a report entitled, *Work plan of the office of the commander of the BT and MV* [Armored and Mechanized Forces] *7th Guards Army,* recorded in TsAMO record ID #133373946. This plan contained seven activity sections:

1. 16 June - Technical conference with mechanics, drivers and technical workers 27 GTB, 119 Collection Point for Armored Vehicles, and 62 Mobile Repair Shops base.
2. 19 and 23 June - Verification of battle formations 27 GTB and technical condition of tanks on the defensive (according to a special plan).
3. 15-18 June - Selection of tankers [crews] not used by specialty.
4. 20-25, 30 June - Classes with the management [staff] officers.
5. 27 June - Command post exercise 27 GTB.
6. 20-30 - Control and management of combat training units.
7. 18-22 June - Verification of secret training/exercise proceedings.

This plan shows that the command staff of the 27 GTB was seeing to the maintenance and condition of their vehicles. They also were reassigning crews, as needed, and devoting time for training of both crews and the staff. Note: This report appears to be documenting a work plan that was already underway -- i.e., the report was issued on 18 June, but some activities had begun on 15 and 16 June.

19 June 1944

In preparation for Operation Bagration, the offensive against German's AG Center (scheduled to start on 22 June), *Stavka* unleashed a massive wave of partisan attacks against German logistical targets behind AG Center. Numbers vary, but over 100,000 partisans conducted over 10,000 explosive attacks against rail targets and facilities. German supply lines and communications were disrupted for many days. [474]

20 June 1944

After continuous attacks that had begun on 10 June, the major Finnish city of Viipuri, the Karelian capital, north of Leningrad, fell to units of the advancing Red Army. At this point Finland had reached a military and political crisis that would be difficult to resolve. For Finland, the one bright spot was the report that numbers of Red Army units were being transferred south, away from Finland, for use against German forces in the Baltic States. [475]

21 June 1944

After the second successful U.S. shuttle bombing mission targeting Berlin, 184 U.S. bombers and escort fighters landed in and around the city of Poltava using several Soviet airbases. But this time the Luftwaffe was ready to strike back. During the night 75 Luftwaffe bombers attacked the main base and destroyed 47 U.S. heavy bombers (B-17s) and 15 fighters on the ground and damaged many others. Overall, personnel casualties on the ground were low. [476] Accumulated fuel and ammunition supplies also were destroyed. The Luftwaffe suffered no losses due to inadequate Soviet air defenses.

22 June 1944

Exactly three years to the day after Nazi Germany attacked the Soviet Union, four Soviet fronts launched Operation Bagration against Germany's AG Center. Operation Bagration was named after Prince Pyotr Bagration who fought against Napoleon. Soviet forces totaled over 1.2 million troops; 2,700 tanks; 24,000 artillery; 2,300 *Katyusha* rocket launchers; and 5,300 aircraft in four separate air armies. This massive offensive caught the Germans mostly by surprise and soon tore open a 250-mile gap in the German line. The offensive would continue until early August, advancing over 300 miles, destroying over 25 German divisions that stood in their way, and reaching East Prussia and the gates of Warsaw, Poland, before ending. [477]

At the start of the operation, Soviet forces had a huge superiority: 3.7:1 in troops, 9.4:1 in artillery, 23:1 in tanks, 3.6:1 in assault guns, and 10.5:1 in aircraft. As the attack unfolded, and as *Stavka* committed even more forces, the ratios shifted even more against the German defenders. [478] This was the centerpiece operation for the entire Russian Front, and its success greatly influenced other secondary Soviet and German campaigns during the summer.

Regarding Allied bombing attacks against critical targets in Germany, Armaments Minister Albert Speer sent a letter to Hitler saying, "The enemy has succeeded in increasing our losses of aviation gasoline up to 90 percent by June 22. Only through speedy recovery of damaged plants [synthetic fuel plants] has it been possible to regain partly some of the terrible losses." [479]

23 June 1944

In response to peace overtures to the Soviet Union, Finland received what amounted to an unconditional capitulation demand from the Soviets. At the same time, Hitler was pressuring Finnish leaders to publically declare that Finland would remain at Germany's side, continuing the war. [480] But the war did continue, and all the players of this drama contemplated their own ending to what was unfolding in the land of a thousand lakes. For their part, Finnish leaders were desperately trying to find a way out of the war with the fewest losses. Their efforts would require several more months before a mutually agreeable resolution was reached for all parties.

24 June 1944

One of the long-lasting effects of the German Operations *Sonja* and *Katja* can be seen on historical maps. By this date, TsAMO map ID #100114467 shows that the 25 GRC of the 7th Guards Army had taken over a large section of the front previously held by the 27th A. The 8 GAD, 53 RD, and 6 RD now held this section of the front. With this deployment layout, the 7th Guards Army held a much larger section of the front and had only the 81 GRD in reserve.

25 June 1944

The Army issued a *Combat Log of the 72 GRD*, for the month of June, recorded in TsAMO record ID #130269174. Page 13 displays an interesting map showing the 72 GRD as it was deployed on 25 June. (See Map #44.)

The 72 GRD contained three rifle regiments, #222, 224 and 229. Each regiment contained three battalions displayed on the map as 1/222, 2/222 and the missing 3/222 for the 222 RR. The other regiments are displayed in a similar manner. Each battalion contained three companies and so on. This arrangement is called a triangular division. This map shows the 72 GRD deploying all three of its regiments right at the front within a few miles of the front line. But each regiment is missing one of its battalions, in this case all the #3 battalions. We can only assume the missing battalions were deployed to the rear for refitting or were undergoing training. Another map from the 27th A shows a similar arrangement, so the 2 UF might have ordered that each regiment at the front was to deploy one of its battalions to the rear. As can be seen, the 72 GRD Divisional HQ was deployed far back with its regimental HQs further forward. Interestingly, right along the front line were two odd units, the 194 Separate Penal Company (enlisted criminals), with 200-300 men and a small unit labeled BO (or БО in Russian), which is an acronym for flank guard. Behind the Russian lines were several prominent heights (358, 256 and 322), which made good observation posts.

Between the front lines is a fairly wide no man's land, which is preferable to having the enemy within grenade throwing range. As active combat came to an end, both sides would occupy what they considered to be their best defensive positions. Low lying ground, exposed areas and swampy terrain would not normally be held unless they gave the defender/attacker a distinct advantage.

German and Romanian forces were entrenched and firmly defending the entire area from their front line to the area depicted on the bottom of Map #44. They also had two nicely placed heights (305 and 358.1). These two locations allowed German observers to peer deeply behind the Russian front line and to monitor the enemy's activities to the rear. A Russian solider commented about Hill 358.1, "Before us is a ridge of small heights covered with golden grass. The height of 358.1 stands out among them. She dominates the area. Our defense can be seen from its top for many kilometers." [481]

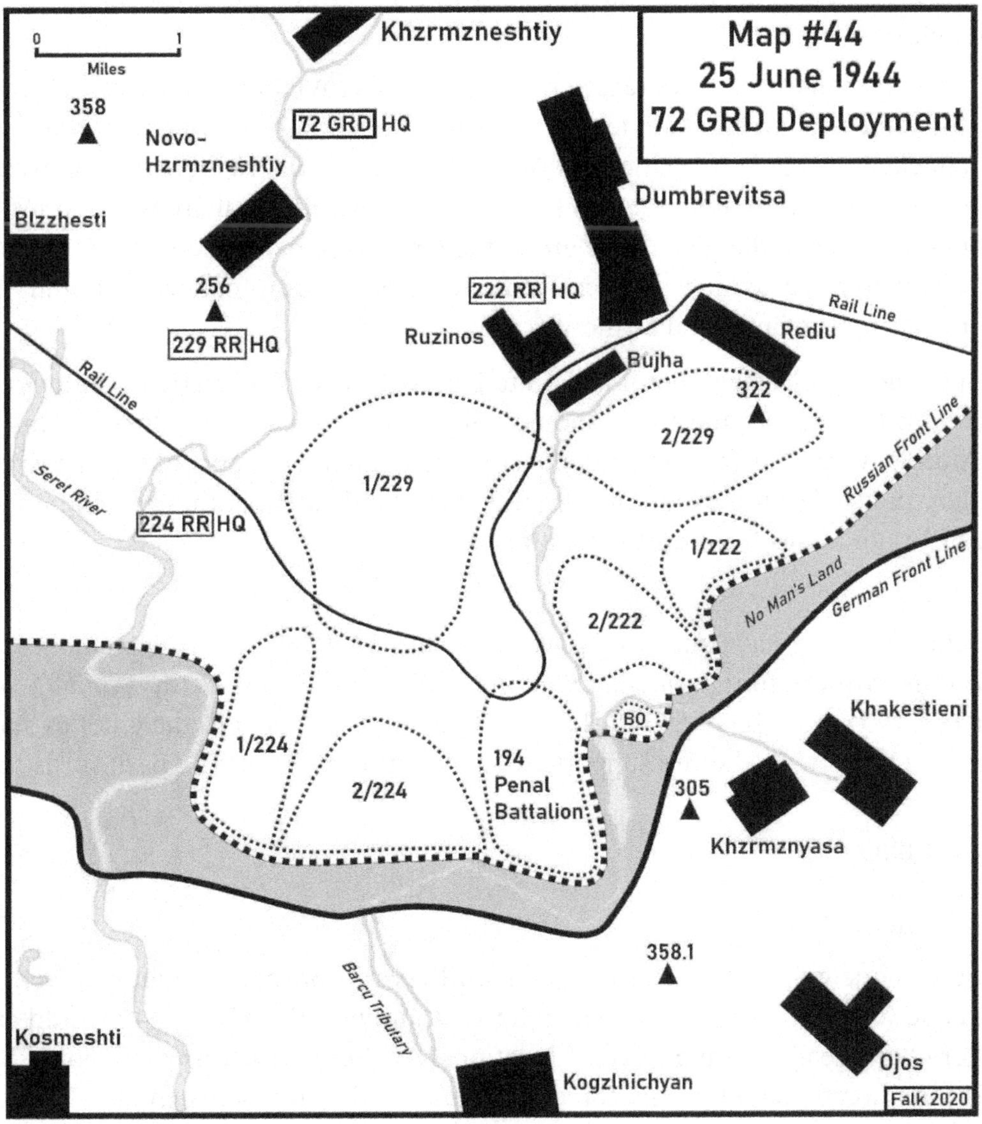

26 June 1944

TsAMO record ID #133373956 shows that Lieutenant Colonel Dolzhishnikov issued a statement to 7th Guards Army HQ. "I inform you that no Romanian population was found on the territory of the zone." The zone he is referring to would be the combat zone that *Stavka* ordered cleared of all civilians, *Stavka* Directive #220094 from 6 May. The removal of civilians was carried out mainly to remove potential spies from the area and to prevent civilians from interfering with combat operations. The removal of civilians also allowed looting, seizing of food supplies, and other living-off-the-land actions performed by the army. The fate of these removed civilians is unknown.

On this same day, another TsAMO record ID #132344625 showed the deployment of one company of ten tanks from the third-tank battalion of the 27 GTB. The 27 GTB supported the rifle divisions holding the newly acquired section of the front from the 27th A. By this date the 27 GTB had replaced its old tanks and might have started to receive the newer T-34/85 tanks with the 85 mm gun. [482] This map also confirmed that the tanks of the 27 GTB were spread out along the front in an infantry support role and not concentrated as an armored assault force.

27 June 1944

The 7th Guards Army issued a report, *Extract from the combat log of anti-aircraft artillery units of the 7th Guards Army* [for 27 June]. The author of the document was Captain Zakharov. This was recorded in TsAMO record ID #437716156. Due to the difficulty of translating this report (fuzzy letters), a synopsis is listed below:

- The mission of the air defense unit and their deployment is the same. During the day, four Ju 88 [bombers], and ten Fw 190s [fighters/bombers] were recorded, which attacked the battle formations of our troops…and construction and supply troops along roads. There were no targets in the area under attack and there were no losses.
- On the basis of the plan of the army, the units were given a combat Order #0027 and #0028 to relocate the anti-aircraft forces to new areas of concentration in the area of the Army's infantry and artillery.

On the surface, this report is clear-cut, but it is hard to believe that 14 Luftwaffe aircraft could make a deliberate attack against the rear of the Army and not hit anything or cause any losses. This report also states that the AA units were ordered to moved closer to the combat elements of the Army and cover infantry and artillery units. This relocation of AA units would indicate that these infantry and artillery formations were suffering losses from Luftwaffe air attacks.

28 June 1944

Guards Captain Kutsevol issued a change of password document, *I am reporting a change to passwords for calling* [artillery fire] *at the junction of 25 GRC* [7th Guards Army] *and 35 GRC* [27th Army], found in TsAMO record ID #454345611. This document was sent to the chief of staff of the 16 Artillery Penetration Division, so the fire support would come from them. The passwords for the next three days were:

28 June – Lighthouse
29 June – Anchor
30 June – Boat

The use of passwords/code words allowed for a quick artillery response to an enemy attack. Transmitting the code word "Lighthouse" and then a target number like #30 or #46 would call down a prearranged artillery barrage against the selected locations. Artillery observers then would be able to adjust the fire to more directly target enemy forces operating in the area between the 25 and 35 GRCs.

29 June 1944

Guards Colonel Kupriyanov issued a report, *#452, Information about the damage inflicted on the enemy and the losses of the 7th Guards Army for the period from 21.6 to 30.6.1944*, TsAMO record ID #454035279. The issue date of the report is 30 June, thus it covered ten days of activity.

> Enemy losses;
> Officers and soldiers killed – 400
> Machine Guns destroyed – 7
> Mortars destroyed – 1
> Captured prisoners – 11
> Light machine guns captured – 1
> Rifles captured – 4
>
> Losses for the 7th Guards Army;
> Killed – 90
> Wounded – 287

The claims made by this report of enemy soldiers killed, exactly 400, a nice round number, are improbable and likely inflated to make the report look better. The other values are within a believable range. The reported losses for the 7th Guards Army revealed 37 casualties per day, which would indicate a moderate amount of low-level combat activity. Based upon the average casualties per day (with 377 casualties reported for ten days, 21-30 June) the estimated casualties for the Army during the entire month are 1,131 men and women. Losses would have included those from scouting missions, trench raids, artillery and mortar bombardments, along with machine gun and sniper fire all along the front. Actual German losses for this period are unknown.

30 June 1944

Gradually entering service during April-June 1944 was the Luftwaffe's second major secret weapon against the Allies: the Messerschmitt Me 262, the world's first operational jet fighter. The Me 262 was called *Schwalbe* or Swallow by the Germans and was more than 110 miles per hour faster than the best piston-engine Allied fighters of the time. The late war introduction of this revolutionary fighter along with production and fuel problems limited its impact on this war, but the advanced technology foretold jet aerial combat of the future.

For the month of June, German V-1 attacks against London, both by day and night, took a heavy toll of the civilian population with 1,935 killed and 5,906 wounded. [483] The Geneva Convention of 1929 only contained treaties regarding the humanitarian treatment of POWs and of the sick and wounded, so little international protections existed at this time regarding civilians in wartime. Because the target area was the entire city of London, numerous public and private businesses, schools, financial institutions, places of worship, and residential homes were destroyed or damaged. The Luftwaffe also undertook intermittent air raids against London for the rest of the war. Losses resulting from these conventional air attacks are included in these totals.

1 July 1944	
7th Guards Army	
Assigned – 2 Ukrainian Front	
Colonel General M.S. Shumilov Commanding	
Rifle, Airborne and Cavalry Units	24 Guards Rifle Corps (72 GRD, 6 RD) 25 Guards Rifle Corps (36 GRD, 6 GAD, 8 GAD) 81 GRD, 53 RD 3 Separate AT Rifle Battalion 15 Penal Battalion 62, 63, 64, 65, 66,194 Penal Companies
Artillery Units	16 Artillery Penetration Division (49 Light Artillery Brigade, 61 Gun Artillery Brigade, 52 Howitzer Artillery Brigade, 90 Heavy Howitzer Brigade, 109 High-Power Howitzer Brigade, 14 Mortar Brigade) 114, 115 Guards Tank Destroyer Regiments 41 Guards Gun Artillery Brigade 30 Antitank Artillery Brigade 263, 290 Mortar Regiments 5 Antiaircraft Artillery Division (670, 743, 1119, 1181 AA Regiments) 162 Guards AA Regiment 662 Separate Aerial Observation/Air Raid Warning Company
Armored and Mechanized Units	27 Guards Tank Brigade 38 Separate Armored Train Battalion
Engineer Units	60 Engineer Sapper Brigade
Reported Combat Strength Estimated Ration Strength	61,600 men 68,600, men and women (including an estimated 7,000 support personnel)

(BSSA 44, BSA 44, TsAMO Archive)

The start of the month saw the confirmation that the Army retained its six Penal Companies and at least one Penal Battalion. Several units were, in turn, transferred out of the Army. The transferred units included the 11 Artillery Division, the 161 Guards Gun Artillery Regiment, the 34 Separate Armored Train Battalion and several engineering units. The next big offensive was scheduled for August, so the Army was preparing for battle in subtle ways.

The frontline layout of the Army did not change much during the month. Four of the five AA regiments covered the left flank of the Army with the remaining AA regiment somewhere in the rear. The 27 GTB ultimately was positioned behind the right flank of the Army. Overall, during this time, the Army was deployed defensively.

Note: Various Army and Front level reports are deliberately used during this month to illustrate and detail the normal day-to-day activities and planning procedures used by the 7th Guards Army and 2 UF. This type of staff work is vital for the proper functioning of the Army.

1 July 1944

Colonel Abalakin, of the 5th AA, issued a report to the 2 UF, *Information about the basing of the enemy's air force in front of the 2nd UkrF.* This report is found within TsAMO record ID #450417471. This report was an assessment by the 5th AA of opposing enemy aircraft facing the 2 UF and not the actual numbers of the opposing Axis aircraft. The report lists:

185 bombers (Ju-87, Ju-88, He-111)
160 fighters (Me-109, Fw-190 and the Romanian IAR 80)
30 other types (Fw-189, He-126, Fi-156 and Me-110)

The above figures totaled 375 aircraft. The true Axis air strength in this area, at this time, is unknown. But this estimate is most likely close to the real total. Regardless of this strength report, the 5th AA was operating in the 2 UF area with a significant superiority in aircraft over the enemy.

The 2 UF issued a *Report on the combat activities of the troops of the 2nd UkrF for the month of June 1944,* from TsAMO record ID #455193156, page 12. This report stated: "During June, the divisions [of the 2 UF] were replenished with personnel. Changes in the number of members can be seen from the following table."

Divisions' Strength up to this Value	On 1 June 1944	On 1 July 1944
3,500	2	-
4,000	2	-
4,500	6	4
5,000	17	4
5,500	4	7
6,000	5	14
6,500	2	3
7,000	12	-
7,500	6	9
8,000	6	4
8,500	-	4
9,000	-	3
9,500	-	1
Total Divisions	**62**	**53**
Total People in Divisions*	**~369,000 People**	**~350,000 People**

Table #18: 2 UF Divisions' Strengths - 1 June to 1 July 1944.

***Author's Addition.**

As can be see, between 1 June and 1 July, the 2 UF lost control of nine divisions, but the 2 UF greatly increased the individual strengths of its remaining divisions. This boosting of strength was necessary to support offensive operations.

2 July 1944

The 7th Guards Army HQ staff produced large numbers of reports throughout the month, which General Shumilov read and used to organize the coming offensive. These reports are too long and involved to include here, but their titles indicate the type of information that was being considered for planning purposes.

Guards Engineer Captain Laschevsky issued a *Climatic information for July on the territory of the 7th Guards Army*, TsAMO ID #133391537. The report covered five factors: wind, temperature, precipitation, cloudiness, and relative humidity. The local weather influenced logistical issues (road conditions), air reconnaissance, and the general health of troops and operations of the Army.

Guards Engineer Captain Laschevsky issued another report, *Submitting reconnaissance material for the Bahlui River*, TsAMO ID #133391558. The Bahlui River was located a few miles behind the Axis lines and the 7th Guards Army would need to cross this small river during the coming battle. The advancing Army needed to know where the bridges and fording places were located, in advance, to plan appropriately.

Guards Major General Lukin and Guards Colonel Kupriyanov issued a four-page document *Counter* [Artillery] *Training Plan for the 7th Guards Army*, TsAMO ID #451310795, which listed several sections:

- Troops (participating in the training)
- Night Strike
- First Fire Raid
- Methodical Suppression and Destruction of Targets and Air Strikes on Accumulations of Tanks
- Second Fire Raid

The last time the 7th Guards Army launched an attack, the opening artillery bombardment failed to sufficiently suppress enemy forces, including enemy artillery, tanks, and dug in infantry. As a result, losses were heavy and the attack failed. General Shumilov was taking no chances of a repeat with this offensive. As can be seen, artillery fire would start before dawn, then it would proceed to hit selected points (fire raid), then move to systematic fire against known targets along with air attacks against known or suspected German tank assembly areas, and finally end with a second fire raid.

At the end of the month, Guards Colonel Kupriyanov produced, *Weather forecast for the month of August 1944 in the area of the 7th Guards Army,* TsAMO ID #454285899. The local weather during an offensive could be critical to success.

Then, Guards Lieutenant Colonel Skorobogatov produced *I inform you that exercises of the 27th* [GTB] *will be held,* TsAMO ID #455140888. The report included the note that the exercise would be using "live ammunition." After resting, training, then repairing tanks and equipment, a live fire exercise would be the last detail to assure the command staff that everything and everyone was ready for the coming battle.

With a little over a month to prepare, planning for the offensive was coming together, but there were many details to work out, and time was in fact short.

3 July 1944

With sufficient time to prepare, *Stavka* ordered a vast quantity of logistical supplies transported to the southern section of the front. The overall logistical plan was to amass enough supplies to support an initial broad offensive operation (tactical) and then to sustain it to a significant depth (operational).

> The railroads along the main feed lines had been rebuilt as close as possible to the front line. Both fronts' [2 and 3 UFs] forces were supplied as follows: with 2-4 combat loads of the basic types of munitions, [and with] up to 8-9 refills of fuel and lubricants for the main types of vehicles. The Second Ukrainian Front, together with its component armies, had up to 3,000 transport vehicles, with an overall lift capacity of more than 6,000 tons; the Third Ukrainian Front had about 2,500 transport vehicles [or about 5,000 tons of capacity]. [484]

General Shumilov (seated on a stump on left) talks with some of his veteran Guardsmen. Note the young age of these "combat veterans."

4 July 1944

An early Soviet assessment of the overall force ratios in Romania are listed in Table #19, Soviet/Axis Strength Ratio - August 1944. As can be seen, the ratios were in favor of the Red Army.

Group of Forces	Artillery Pieces	Tanks and Assault Guns	Aircraft
AG South Ukraine	6,275	549	350 Ger, 300 Ru
2 UF	9,758	1,437	915
3 UF	7,616	425	1,307
Soviet/Axis Ratio	**2.76:1**	**3.39:1**	**3.41:1**

Table #19: Soviet/Axis Strength Ratio – August 1944. [485]

Note: These numbers are early gross values. In the assault areas, the actual Soviet force ratios would eventually be much more favorable to them. One must always remember that the attacker chooses the time and place for battle, and they can and will stack the odds in their favor. The Iasi-Kishinev Operation was intended to deliver a decisive knockout blow to German and Romanian forces operating within Romania.

During the advance of Operation Bagration, forces of the 1 and 2 Belorussian Fronts liberated the capital of Belarus, Minsk, on 4 July. German forces continued to suffer heavy losses as they withdrew from the city.

5 July 1944

Guards Colonel Kupriyanov issued a report at 18:00 hours, *Operational information from the headquarters of the 7th Guards Army,* TsAMO ID #454035316. His report stated the following:

1. On the front of the Army, troops positions were unchanged.
2. Enemy aviation had not been operating since 13.00.
3. Rain started at 15.00, so visibility was limited.

His was a typical short, to-the-point report. Nothing much was going on; it was raining which limited visibility and enemy aircraft activity.

6 July 1944

During the previous few quiet months, German and Romanian armies rebuilt their forces as best as they could in terms of equipment and with large numbers of inadequately trained combat troops. They also attended to logistical issues, but they encountered problems.

> At this stage of the war the Germans no longer had the advantage of efficient logistic support lines, especially rail connections. That advantage was with the Red Army. Forwarding of supply trains and troop transports on the German side over mostly single-tracked branch-lines caused difficulties, especially since the Hungarian and Romanian railroad employees were unfriendly and often delayed taking over and forwarding trains for days at a time. Supply trains required up to three weeks to get from the *Reichs* territory to the supply bases of the troops at the front. [486]

Attacks by U.S heavy bombers against rail targets in Hungary, Romania, and Germany also prevented the running of an efficient Axis rail network.

Logistical and transportation problems for the Germans also manifested themselves in another way on the other side of the front. By this date, the Red Army was operating with hundreds of thousands of trucks, and jeeps plus other vehicles supplied by the Western Allies via Lend Lease. The Soviet concept of "Deep Battle" and "Deep Operation," which was conceived as vast Soviet mobile operations before the war, were now possible. These mobile operations had been proven during the ongoing Operation Bagration to be extremely effective against the largely horse-drawn *Wehrmacht*. Once a fixed German defense line was broken and operations became mobile, the Red Army had all the advantages. Huge numbers of tanks, artillery, and truck-transported infantry could roam at will behind German lines all the while being supplied by truck-borne logistical support. Infantry armies then would follow, at a slower pace, to surround and to destroy the stunned enemy. That was *Stavka's* plan of action for Romania, too.

7 July 1944

The war diary of the German 6th Army noted, "The summer offensive of the Red Army against Army Group 'Center' and the concentration of enemy forces against Army Group 'North Ukraine'…confirms that the enemy has postponed his planned offensive in the Balkans." [487] The extensive *maskirovka* efforts by the 2 and 3 UFs were paying off.

8 July 1944

Because of the Soviet shift in focus of their offensive operations to the north, mainly against AG Center, the summer months were relatively quiet in Romania. But constant small attacks "kept the troops on alert and cost the Romanians a sustainable 3,849 casualties in June; 4,244 in July; and 2,591 up to 19 August." Likewise, the summer months "benefited the Romanians considerably, as they were able to blood their raw troops without undue risk [in other words, allow them to adjust to life and combat at the front without suffering heavy losses]. However, defeatism was rife and morale remained shaky." [488] The German position in the south had improved somewhat in terms of increased numbers of infantry recruits, but the powerful Panzer force that had defeated earlier Soviet attacks had been drawn away to the north to support AG Center, leaving only two weak Panzer divisions (13 Pz and 20 Pz) to cover the front. This denuding of the Panzer force had shifted the balance of power heavily in favor of the Red Army.

9 July 1944

The 7th Guards Army issued a report entitled, *Extract from the journal of combat operations of air defense units of the 7th Guards Army.* This record is found in TsAMO ID #152505448, page 3.

> The deployment and mission of the air defense units is the same. Spotted 25 Ju- 88, 7 Hs-123 and 4 Fw-190. [One] Hs-123 dropped 2 FAB (general-purpose bombs) in the Zapwedelitsa area. Groups of 10 Ju-88 aircraft at night passed through the strip of the army. The rest of the planes were conducting reconnaissance of the combat formations of our troops in the Bedelitsa, Slobozn, Mzgur region and the transport and air transport routes along the Feteshchy, Flemenzi route. Targets that entered the zone of the 1119 AA regiment were fired at, ammunition consumption, 37 mm shells - 60 rounds, and 12.77 mm - 100 rounds. Fire from enemy artillery against the 743 AA regiment killed a single officer and wounded 1 sergeant. There are no downed planes.

Despite relative calm along the front, the Germans were keeping an eye on actives behind the Russian lines. By this date, German intelligence services probably identified all the major Red Army formations deployed against AG South Ukraine, but they would have a more difficult time acquiring detailed information. Air reconnaissance, both visually and photo, radio messages interception, code breaking, captured enemy soldiers and deserters all were ways to fill in the blanks, but that required a constant intelligence effort. Of course, the Red Army displayed false troop deployments and movements and also spread false information to make the enemies' intelligence effort all the more difficult.

10 July 1944

The 2 UF issued an interesting report from the *Combat log of troops of the 2nd UkrF for the month of July 1944.* This is from TsAMO ID #113984879, page 32.

> In the 7th Guards Army, there was a case where the junior commander of the penal company commanded by Gerasims [a Russian surname], was in the trenches, answered the enemy's questions, and then, having agreed, met in the neutral zone and continued the conversations there. Because of this transgression, penal company commander Gerasims has been arrested.
>
> The front commander ordered
> 1. Take all measures to eliminate all possible such cases.
> 2. Establish careful observation [of the front] and in no case allow conversations with the enemy.

Signed Zakharov [Chief of Staff 2 UF, Matvei Vasilevich Zakharov]

From this report it appears the junior commander of the penal company went over to the Germans. In that case, the only thing to do was to arrest the penal company commander Gerasims, instead. The fate of Gerasims is unknown, but in reality, he was destined for one of two endings: to be assigned to a penal battalion (for officers) as a prisoner to redeem himself of his crime in battle, or to stand in front of a firing squad.

11 July 1944

Guards Colonel Kupriyanov issued a *Combat report of the headquarters of the 7th Guards Army*, TsAMO ID #131310883.

> To the commander of the headquarters of the 2 Ukrainian front:
>
> 1. During the past night, nothing significant happened at the army front. Combat operations were limited to a shootout with the enemy and in some areas by the actions of reconnaissance groups.
> 2. Enemy aircraft used [a] single aircraft for reconnaissance. A total of seven sorties were recorded.
> 3. The position of the army's troops is unchanged.
> 4. Communication with troops and neighbors: telegraphic and telephonic.

Overall, this is a boring report, but that was by design. Somewhere on the German side of the line a similar, (theoretical) report was written for this same day, "No change at the front, only an exchange of small arms fire with the enemy, reconnaissance aircraft report nothing significant." The last thing the 2 and 3 UFs wanted to do was to alert the enemy as to what was really going on behind the lines, better to make them complacent through inaction. But the heart of Soviet *maskirovka* methods is "deception" which leads to *vnezapnost* "surprise." A *maskirovka* regulation from 1944 stated:

> *Maskirovka* is a mandatory form of combat support for each action and operation. The objectives of *maskirovka* are to secure concealment of the maneuver and concentration of troops for the purpose of delivering a surprise attack; to mislead the enemy relative to our forces, weapons, actions, and intentions; and thus, force him to make an incorrect decision. [489]

12 July 1944

The 7th Guards Army also collected intelligence about the enemy directly to their front. Guards Colonel Kupriyanov and Guards Lieutenant Colonel Polner detailed their observations in a report, *A brief summary of the generalized combat experience of the troops of the 7th Guards Army, 2 UkrF for the month of July 1944*. P 3-4. This is found within TsAMO ID #113983044.

> Artillery and Mortars Used [by the enemy].
>
> No new types of artillery weapons were noted during July.
> Artillery and rifle troops using reconnaissance established in front of the Army the following:
>
> Artillery batteries 210 mm - 1
> Artillery batteries 150 mm - 9
> Artillery batteries 105 - 27
> Artillery batteries 75 mm – 24
> Anti-aircraft guns - 28
> Anti-tank guns - 53
> Mortar batteries - 45
> Pillbox of artillery – 18
>
> The actions of six-barreled mortars [*Nebelwerfers*- rocket launchers] were not observed.

> [Enemy] artillery counteracted [our artillery fire] by methodical fire and fire raids [which] exposed the combat troops of the army to fire, while [our] fire was conducted over a general area.

In essence, 9-10 artillery batteries, 4-5 mini-batteries [possible deployment of a few guns instead of a full battery] with a consumption of up to 600 shells were noted daily.

This report showed the Germans had assembled a considerable amount of artillery to support this section of their front. This report also explained that Russian artillery fire was answered by return fire, which revealed the positions of enemy artillery batteries.

13 July 1944

The 1 UF, under Marshal Konev, launched the Lvov–Sandomierz offensive against AG North Ukraine right on time according to the *Stavka* plan. This offensive employed 840,000 men; 14,000 artillery; 1,600 tanks; and 2,800 aircraft against Axis forces numbering 500,000 men; 6,000 artillery; 900 tanks; and 700 aircraft. Heavy fighting continued until 29 August after the 1 UF had advanced 280 miles to the west and deeply into southern Poland. [490]

14 July 1944

Hitler moved his headquarters from Berchtesgaden, near Austria, to the Wolf's Lair near Rastenburg in East Prussia. Hitler alternated between these two locations. At this time, Red Army troops were only about 60 miles from the East Prussian border.

15 July 1944

Despite an existing initial overall concept for an offensive against Axis forces in Romania, *Stavka* ordered Malinovsky (2 UF) and Tolbukhin (3 UF) to produce a functional operational plan for the offensive and to present it, in person, to Stalin/ *Stavka* in Moscow by the end of the month.

16 July 1944

During July and into August, German AG South Ukraine was ordered to transfer substantial forces northward in response to the Soviet Operation Bagration and now the Lvov–Sandomierz offensive. Overall, by 20 August, 12 German divisions had been transferred, including six Panzer and one motorized division, which amounted to an overall loss to AG South Ukraine of approximately 245,500 men. [491] Once again, an exhausted Germany was forced to strip forces from one front to reinforce another and in so doing critically weakened AG South Ukraine.

17 July 1944

Behind the lines, Soviet forces gathered the human resources that they would need for the coming offensive. "The factor that up to 80 percent of the enlisted men in the rifle units consisted of reinforcements from those conscripted in the Ukrainian provinces liberated in the spring of 1944 is worthy of mention; also, more than 20,000 conscripts from [newly liberated] Moldavia joined the troops." This influx of recruits leads to the interesting situation within the 53rd A of having accumulated more than 20,000 men in the army's reserve regiment, two full strength rifle divisions' worth! Half of these troops were new conscripts and the other half were returns from the hospital. [492] The pause in offensive operation in the area of 2 and 3 UFs now ensured the gathering of an overwhelming force to shatter any and all opposing Axis armies.

18 July 1944

By mid-July, German intelligence services were aware that *Stavka* had thinned out the front facing AG South Ukraine by transferring several complete armies northward. By this date, the German OKH knew the 5th GA and 8th GA had moved north along with the 2nd TA and the 5th GTA. They knew this because these units currently were advancing deep into Poland as part of Operation Bagration. "The German command, nevertheless, remained confident that Army Group South Ukraine could avoid a disaster similar to those which befell virtually every other German army group on the Eastern Front in the summer of 1944." [493] This was, of course, wishful thinking on the part of the German high command.

By this stage of the war, they understood that Soviet generals were skilled and very dangerous opponents. They also knew the Soviet Union was producing vast amounts of modern weapons, supported by Allied Lend Lease shipments of various supplies. An impartial German assessment of the situation along the Eastern Front would have revealed that the Red Army had achieved near total military dominance and was fully capable of launching and logistically supporting multiple large offensive operations nearly simultaneously. This failure of the German high command and Hitler to correctly evaluate, or perhaps accept, the current state of affairs would condemn a large part of AG South Ukraine to total destruction.

19 July 1944

On the Finnish front, despite the large number of Red Army units that had been transferred south to reinforce the Red Army's attack to the west, recent fighting had seriously weakened the Finnish military, and it continued to face a formidable and determined foe.

> The Finns estimated that the Soviet forces confronting them now consisted of the following: 29 light infantry divisions, two light infantry brigades, ten armored brigades, and a number of fortress units. These forces were certainly capable of carrying out further offensive operations. [494]

The Finns continued to pursue a political settlement to Finland's war with the Soviet Union.

20 July 1944

On 20 July 1944, Colonel Claus von Stauffenberg and other conspirators attempted to assassinate Adolf Hitler by detonating a bomb inside Hitler's Wolf's Lair headquarters. This was *Unternehmen Walküre* (Operation Valkyrie from the Old Norse mythology Valkyrja "Chooser of the Slain"). The bomb exploded, but the plot failed; Hitler suffered only minor injuries. In the aftermath of the assassination attempt, the Gestapo (secret police) arrested more than 7,000 civilians and military officers. Eventually they tortured and executed 4,980 of them. [495] The war and Hitler's madness continued.

21 July 1944

The German high command recalled the Luftwaffe air group that had been supporting the Finnish army back to AG North. More withdraws would follow as the situation on the main front deteriorated for the *Wehrmacht*.

22 July 1944

The German command informed the Finnish government that a promised assault-gun battalion would not be arriving to support them. The battalion had been diverted to fight along the main front that was fast approaching the eastern border of Germany. This lack of support by Germany made seeking negotiations with the Soviets all that more important or enticing.

Meanwhile, taking unilateral action, "Moscow Radio announced the establishment… [of a new Polish government called the] Polish Committee for National Liberation." This new Soviet puppet government was in opposition to the pre-war Polish government, which was in exile in London and receiving support from the Western powers. [496] Unsurprisingly, the Western Allies were not pleased with this political maneuver by Stalin, which lead to the first real diplomatic break between the Allies and ultimately to the division of Europe between East and West and the start of the Cold War.

23 July 1944

The last German forces on Soviet soil were pushed out of the country when the town of Pskov [on Lake Peipus, southwest of Leningrad] was retaken from the invaders. [497] Notably, this event took place three years and thirty-one days after the German invasion of the Soviet Union on 22 June 1941.

24 July 1944

Major General Seleznev issued a report from the 5th AA to the 2 UF, found in TsAMO ID #451001784, *Combat log 5 VA* [AA] *for July 1944*, page 50:

1. General ground and air support
   Noted - 27 enemy aircraft over-flights during the day including 4 He - 111 at night.
2. The task of you and the decision of the commander
   The VA units would continue to fulfill the task by the assignment on 23.7.44. [23 July 1944.] The situation is the same. The readiness of bomber, fighter and assault groups according to the order for 23.7.44. [are set].
3. The actions of the part and the connection
   During the day and the night of 24.07.44, [24 July 1944] VA units conducted reconnaissance with side-by-side bombing of enemy troops and equipment in front of the 2 UF. They escorted spotter aircraft and a particularly important SI-47 aircraft. They were to be ready for combat missions on call.
   They made training sessions and were engaged in training.
   The reconnaissance was carried out in a limited manner due to poor weather conditions.
   Communication with units and subunits is stable.
4. Actions of enemy aircraft
   The activity of enemy aviation was not particularly intense; they conducted reconnaissance with single aircraft, and with bombers and pairs of fighters, by both day and night.
5. The result of combat work for the day
   Aircraft that performed combat missions [were] - 23 aircraft, of which there were 12 la-5, 4 yak-9, 5 Po-2, 2 yak-1.
   The 23 combat sorties totaled a flight time of 32.02 [32 hours 02 minutes].
   Ammunition expended: 7 FAB-50, 4 OAB-15, 15 SAB-8 [bombs], 72 ampoules [containers] with COP [incendiary liquid].
   According to the reports of the crew, four fires were created and one explosion was triggered.
   On the railway, two hits were made [on wagons].

This 5th AA report is interesting because it states that the entire air army only flew 23 sorties during the entire day, while German aircraft flew 27 sorties in the same area. The VVS aircraft dropped only 26 bombs, but interestingly they also dropped 72 containers of incendiary liquid. They must have been targeting a specific object or location that was vulnerable to fire, like a railway marshaling yard.

25 July 1944

On this day, Colonel General Johannes Friessner (Frießner) was transferred from command of AG North to assume command of AG South Ukraine. From the beginning of his arrival in the south, he suspected that the Romanian officers and troops under his command were less than enthusiastic about continuing the war. "As early as my first trips to the front…, we were hearing rumors about the doubtful reliability of the Romanian officers. Although at first there was no concrete proof," he went on to say, "The regular conferences between [Romanian] political figures and generals…seemed very suspicious." And finally, "The crossing of [Romanian] personnel of the front line to the enemy became a daily occurrence." [498]

After receiving briefing from his new headquarters staff, General Friessner came to some conclusions about the position of his AG.

> In the northern sector of the front (Moldavian direction), the terrain was more rugged and partly wooded, which means that it was very favorable for defense. The Trajan position, which served as the second operational line of defense, was especially convenient for this. The key sector of the front of the Army Group was the Yass [Iasi] direction. Here we should under no circumstances allow the enemy to break through with a blow from north to south, as a result of which the troops defending in the [Bessarabia, right flank] direction would be cut off. [499]

In this, General Friessner correctly assessed the situation. The solution was to pull the front back to a more defensible line and less exposed position, but Hitler denied his request.

26 July 1944

After three days and nights of fighting, German forces of AG North Ukraine were forced to give up the major southern Polish city of Lvov. During the night, German troops withdrew out of the city, which allowed Red Army units of 1 UF to advance and to push west, but *Stavka* would change the operational situation the next day. [500] Note: Lvov and this part of southern Poland had been annexed by the Soviet Union in accordance with the secret protocols between Hitler and Stalin in 1939 to partition Poland between their two countries. So, the Soviet Union was just taking back what they considered was theirs, at the expense of the Polish people.

27 July 1944

On or about this date, the twin onslaughts of Operation Bagration and the Lvov–Sandomierz offensive still were advancing, but *Stavka* realized that with growing German resistance and the exhaustion of its own forces, the Red Army needed a new short-term objective. At this point, *Stavka* altered the trajectory of the 1 Ukrainian and 1 Belorussian Fronts to the northwest. The new objective was to reach and to cross the Vistula River, south of Warsaw, and to form operational bridgeheads on the western bank. This they accomplished in due course.

In Moscow, the Polish Committee for National Liberation (also known as the Lublin Committee) signed an agreement with the Soviet Union to form a unified Polish Army, which would be operationally subordinated to the Soviet High command. [501] Russia now would have a Polish army to deploy at will against Nazi Germany.

28 July 1944

During July 1944, the Luftwaffe gradually began operating a third secret weapon: the Messerschmitt Me 163, which was the world's first operational liquid-fueled rocket propelled fighter interceptor. The Me 163 was given the name of *Komet* (Comet) by the Germans and was more than 140 miles per hour faster than the best piston-engine Allied fighters of the time. Once again, the late war introduction of this revolutionary fighter, along with production and fuel problems and its limited range, reduced its impact on the war. Ultimately, however, this German rocket propelled aircraft technology was soon outdated when more advanced jet engine technology achieved dominance after the war.

29 July 1944

Much attention has been given throughout this book to the combat and ration strength of the 7th Guards Army, but what about the command staff? For example, at the end of July, Colonel General Shumilov was commanding the Army with the support from a HQ staff of 405 people (342 officers, 40 sergeants, and 23 enlisted personnel). The 24 GRC HQ staff, commanded by Major General Petr Petrovich Avdeenko, numbered 809 people (135 officers, 172 sergeants, and 504 enlisted personnel). The 25 GRC HQ staff, commanded by Lieutenant General Ganii Bekinovich Safiulin, contained 850 people (136 officers, 180 sergeants, and 534 enlisted personnel). This information was found in TsAMO ID #2600169, p 20. These numbers show that senior Army leadership required 2,064 highly trained officers and staff to control and to direct the formations of the Army. These totals do not include HQs staffs of rifle divisions or support units.

Another interesting order was recorded by the 2 UF in the *Combat log of troops of the 2nd UkrF for the month of July 1944*. This is from TsAMO ID #113984879, page 38.

> The commander of the 40th Army ordered:
> Collect 1,800 horses from [40th] Army units and send them to
>
> > 7th Guards Army - 220 horses
> > 52nd A - 760 horses
> > 53rd A – 300 horses
> > 27 G CC – 240 horses
> > 57 CC – 150 horses
> >
> > [75 CC – 130 horses This entry was missing from the historical list.]
>
> The horses to be collected should be fit for work. Notify me upon completion of this order by 9 August 1944.

Once again, this type of order transferring horses, equipment, and/or troops to another unit was part of the 2 UF balancing what was needed by the offensive verses what was available within the front. The 40th A had extra horses, so the excess was transferred to units in need. A few essential jobs that horse power satisfied mobility reequipments for were mounts for cavalry, transportation of medical units (hospitals, medical personnel and casualties), and the delivery of supplies, all types of weapons, and armaments.

In Finland, "Hitler ordered the 122 Division back to Army Group North [from Finland]. The OKW explained that the quiet situation in Finland was the deciding factor in the division's withdrawal and that the Germans would come to Finland's aid in the future if they were needed." [502] Once again this illustrates Nazi Germany's dire need for combat units to fill breaks in the front line. The unrelenting attacks by the Red Army continued to force the German High Command to denude one section of the front to reinforce another. This practice of course created weak spots, which the much larger Red Army then could exploit.

30 July 1944

On this day, General Shumilov had been in command of the 64th/7th Guards Army for two years, with no end in sight to the war. As the long war continued along its path to an eventual end, Shumilov busily prepared his army for yet another offensive.

31 July 1944

The 2 UF report, *The combat log of the troops of the 2nd UkrF for the month of July 1944*, TsAMO ID #113984879 p 63, 64, detailed the following casualties for the 7th Guards Army:

Killed – 343
Missing – 9
Captured – 0
Non-combat losses – 38
Wounded, Shell-shocked [Post-traumatic stress disorder (PTSD)], Burned – 1,020
Sick – 1,582
Frostbitten – 0

These figures total 2,992 men and women. The report also records the loss of 88 horses during the month.

Guards Lieutenant Colonel Skorobogatov issued a report, *Artillery Combat Log of the 7th Guards Army*, TsAMO ID #150200622 p 12. The report states: "By the end of the month, the army has a large amount of ammunition." Over the previous two months, limits were placed upon the use of artillery ammunition in order to accumulate a large reserve for the coming offensive. The report lists ammunition in Units of Fire.

82 mm (Mortar) – 1.5 Units of Fire
120 mm (Mortar) – 1.8
45 mm (AT Gun) – 1.6
76 mm (Regimental Gun – Infantry Support) – 2.1
76 mm (Divisional Gun) – 1.6
122 mm (Artillery) – 1.7
152 mm (Artillery) – 2.4
25 mm (AA) – 2.5
37 mm (AA) – 2.2
85 mm (AA) – 2.5

Stavka accepted the offensive plan produced by Malinovsky and Tolbukhin for an attack deep into Romania. These two front commanders now would have the responsibility, during the next 19 days, to put the plan into action.

The President of Finland, Risto Ryti, resigned along with most of his cabinet, paving the way for Marshal Carl Gustaf Mannerheim, commander-in-chief of the armed forces and a fierce loyalist of an independent Finland, to take control of the government and the fate of Finland. [503] This transfer of power would take place on 4 August 1944. The resignation of Ryti was a critical step as Mannerheim was ready to break with Hitler and end Finland's war, but only if the Soviet Union would agree to a bearable armistice treaty.

In far off Normandy, France, U.S. General George S. Patton and his 3rd Army captured the key city of Avranches, which would soon lead to the breakout of all the Allied armies from the Normandy bridgehead. Four highly mobile Allied armies now were poised to fan out across France and to engage the largely horse-drawn and badly outnumbered German Army out in the open. Another military disaster was about to befall the German *Heer*, but this time in the West.

For the month of July in Britain, German V-1 attacks killed 2,441 civilians and injured 7,107. Civilian morale eventually started to suffer as these vengeance attacks continued and even increased in intensity.

1 August 1944	
7th Guards Army	
Assigned – 2 Ukrainian Front	
Colonel General M.S. Shumilov Commanding	
Rifle, Airborne and Cavalry Units	24 Guards Rifle Corps (72 GRD, 6 RD) 25 Guards Rifle Corps (36 GRD, 6 GAD, 8 GAD) 81 GRD, 53 RD 5 Guards Cavalry Corps (Attached) 15 Penal Battalion 62, 63, 64, 65, 66,194 Penal Companies
Artillery Units	16 Artillery Penetration Division (49 Light Artillery Brigade, 61 Gun Artillery Brigade, 52 Howitzer Artillery Brigade, 90 Heavy Howitzer Brigade, 109 High-Power Howitzer Brigade, 14 Mortar Brigade) 114, 115 Guards Tank Destroyer Regiments 41 Guards Gun Artillery Brigade 30 Antitank Artillery Brigade 263, 290 Mortar Regiments 5 Antiaircraft Artillery Division (670, 743, 1119, 1181 AA Regiments) 162 Guards AA Regiment 662 Separate Aerial Observation/Air Raid Warning Company
Armored and Mechanized Units	23 Tank Corps (Attached) 27 Guards Tank Brigade 38 Separate Armored Train Battalion
Engineer Units	60 Engineer Sapper Brigade 4 Separate Flamethrower Battalion
Reported Combat Strength Estimated Ration Strength	83,730 men 90,730 men and women (including an estimated 7,000 support personnel)

(BSSA 44, BSA 44, TsAMO Archive)

At the start of the month the 3 Separate AT Rifle Battalion was transferred out of the 7th Guards Army, but at approximately the same time, the 2 UF attached the 5 Guards Cavalry Corps and the 23 Tank Corps to the Army. This move greatly inflated the combat strength of the Army. Other transfers would follow as commanders made adjustments to the main attacking force.

1 August 1944

On or about 1 August, the 2 and 3 UFs began regrouping their forces for the coming offensive. The late date of this movement was intended to deceive German and Romanian commanders as much as possible about any offensive operations. German intelligence services were too good to be misled entirely, but the Soviets did use extensive *maskirovka* methods. The Soviet deception plan was intended to confuse the enemy about the exact concentration and attack locations along the front. The intent was to make the Germans think that an attack would be launched from a certain position, but the real attack would hit the Axis line 20-30 miles away. This would allow the Red Army to concentrate overwhelming forces at the chosen breakthrough location and to quickly pass-through defensive positions and into the rear of the enemy. For the 2 UF, during this regrouping process, 32 rifle divisions completely changed their positions while only seven held their old positions. Additionally, three cavalry divisions and four mechanized corps also completely changed their positions. All these movements would have required many days to accomplish, but they only would have taken place during the night and/or during bad weather. [504]

David Glantz, in his book *Soviet Military Deception in the Second World War*, page 414, stated the overall process; the regrouping was of a varied nature and

> included displacement of forces and equipment from the flanks to the center of the front where they could deliver the main attack. Armies and corps, moved out of the second echelon or front reserve in the process of regrouping, [and] shifted to new regions; divisions and corps transferred from one army to another; and within the armies, forces regrouped into the penetration sectors.

The 2 UF issued a report, *Information on the number of mobilized Soviet citizens in the liberated territory of Ukraine for the period from August 1, 1943, to August 1, 1944*, recorded in TsAMO Archive Document ID #113723646. The one-page report revealed that over a half a million civilians, (528,214 in total), were mobilized during that single year. Most of the people went right into army or divisional level training units and then on to the front. The Red Army needed new troops to fill the depleted ranks of its victorious armies.

In Poland, the Warsaw Uprising began as Polish resistance fighters (Polish Home Army) attacked German forces within Warsaw and quickly were able to capture large sections of the city center. Fighting was intense. At this time, the Red Army's front line was positioned just six to ten miles away from Warsaw, but the Red Army stood fast. Stalin refused Allied calls to assist the Polish freedom fighters. So began the slaughter of thousands of pro-Western Polish soldiers and citizens. The newly installed, pro-Soviet Polish Committee for National Liberation also stood by and waited for the Germans to finish the slaughter of fellow Polish nationals.

In France, U.S. General Patton and his 3rd Army began the breakout from the Normandy bridgehead. Despite repeated attempts to counter this breakout, the German defense position in France finally gave way and the *Wehrmacht* was soon in full retreat back towards Germany. By 19 August the Allied armies had reached the Seine River. By 25 August Paris was liberated and the combined Allied armies were advancing north and northeast, on a broad front, toward the German border. The closest Allied position was only about 180 miles away from the western German border. Most of the Allied soldiers were thinking the war would be over by Christmas, but Hitler had other plans.

2 August 1944

The advancing armies of the 1 and 3 UFs had nearly reached the pre-war borders of Czechoslovakia and Hungary, but their advance was directed to the northwest toward Poland instead. The defending 1st Panzer Army and the 1st Hungarian Army held their positions along the Carpathian Mountains.

3 August 1944

Sometime around this date, corps and divisional level commanders within the 7th Guards Army were briefed about their part in the coming offensive. For the 81 GRD, the 24 GRC commander explained, "We with our guards will break through the fortified positions of the enemy on the Bahlui River, north [of] Targu-Frumos and, bypassing it to the west, we will attack heading south, saddle the Seret River and [cut] off the enemy from the retreat route to the west, to the Carpathians." [505] The overall attack plan was both practical and bold. This plan was not only an attempt to break through the Axis line but also an attempt to surround and to destroy major elements of the German 6th and the 4th Romanian Armies.

In Moscow, the pro-Western Polish Prime Minister Stanislaw Mikolajczyk finally was able to speak with Stalin about the Warsaw Uprising and to ask for Soviet military support for the fighters. Stalin declared that "the Soviet Union did not want two authorities fighting each other in Poland." [The two authorities Stalin was talking about were the official Polish government in London and the Communist puppet government Stalin just appointed.] Stalin finally said that "if the Polish government should prefer the existence in Poland of two different forces… in that event, we shall be forced to continue to assist the Committee for National Liberation. Such is our position." [506] In other words, Stalin was pushing for and supporting the installation of a Communist government in Poland after the war. To make that task easier, Stalin denied the Warsaw Uprising any help, which allowed the Nazis to kill all the pro-Western Poles they wanted.

4 August 1944

Guards Lieutenant Colonel Zherlitsyn and Guards Major Engineer Vasiliev produced a logistical report for the 7th Guards Army, *I report that the supply stations and the unloading stations of our army remain the same.* This is found within TsAMO Archive Document ID #133391584. The report states:

> The supply stations and unloading stations of our army, as well as the technical characteristics of the army railway areas remained the same, as reported by my report No. 01405 dated 15.7.44. [Unfortunately, this earlier report is unavailable.]
> At the same time, I report data on the railway route DORNESHTI-PASHKANI-MZRZSHESHTI-BUZZU. The specified route is single-track; it is the main path for communication to the center of ROMANIA from the BUKOVINO [northwestern Romania] area; [it] is kept in good condition.
> The average technical speed in most sections is 43 km / h [26.7 MPH]. The average distance between the stations is 8 km [4.9 miles]; the greatest is 14 km [8.7 miles].
> The most important stations on the line are:
>
> - Buzzu (which has a rail yard with about 40 tracks, a depot with 20 stalls, a turntable and a turning triangle, and 2 large unloading platforms).
> - Mzrzsheshti ([which] has a rail yard with 18 tracks, a depot for 10-16 stalls, workshops, a turntable, a turning circle, oil tanks and an unloading platform with a length of 277 meters [908 feet]).
> - Adjudnow ([which] has a yard with up to 30 tracks, a depot, a fuel storage [tank/yard?], a turning circle and a triangle, and unloading platform 200 meters long [656 feet]).
> - Roman ([which] has a yard with 16 tracks, a depot, workshops, a turntable and an unloading platform up to 150 meters [492 feet]).
>
> It can be assumed that the capacity on this railway can be brought up to 16-18 pairs of trains per day.

This report is very interesting as it indicates clearly that this single rail line would be used to supply the 7th Guards Army during its advance into Romania. This rail line runs almost directly north/south and crossed the Seret River, among others. With 16-18 pairs of trains per day, the 7th Guards Army would need between 5-7 trains per day, at 600-900 tons per train, to be fully supplied, leaving at least nine trains for adjacent armies' logistical needs, thus insuring all would be well supplied.

To support the offensive at first, trucks and wagons from the 7th Guards Army were used to transport food, fuel, fodder and ammunition into the combat zone from the Army's rear area, while 2 UF trucks and wagons did the same from 2 UF supply depots. When the rail line became available, trains began to deliver bulk supplies as far forward as possible, thus easing the logistical burden upon army and front transportation.

Then, as the advance continued, captured rail lines and facilities would be repaired, as needed, and put back into service to support the 2 UF advance for hundreds of miles further. Initially, the actual rails might not need to be converted to the wider Russian gauge if captured enemy rolling stock, steam engines and rail cars/wagons, could be used instead. If sufficient enemy rolling stock were available, this would temporarily allow for faster movement of supplies to the ever-advancing front. The physical change in the rail gauge could then take place at a later date, as required.

In Finland, Marshal Carl Gustaf Mannerheim took the oath of office as president of Finland. The 77-year-old newly appointed president also retained his position as commander-in-chief of Finland's armed forces. Mannerheim was respected within both Germany and Russia, and because he now wielded both military and political power, the overall state of affairs noticeably improved. This fortuitous change in head of state would now transform relationships and expedite negotiations between Finland and its two powerful warring neighbors.

5 August 1944

Guards Major of the medical service Lev issued a medical report, *Information about the movement of the wounded and sick in medical institutions of the 7th Guards Army*, found in TsAMO Archive ID #133391583. This insightful report covers the period from 30 July to 5 August 1944, and is an assessment of the current medical burden upon the Army's medical establishment. These sick and wounded troops would need to be discharged or sent to the deep rear, as soon as possible, in preparation for the coming battle.

Forward Troop Area – up to divisional medical battalions

- Arrivals:
  - At the start (30 July) – 1,245 men and women
  - Newly arrived – 1,658
  - Total – 2,903
- Departures:
  - Evacuated – 342 (most likely to the Army rear) (moderately wounded)
  - Returned to active service – 1,288
  - Died – 12
  - Total – 1,642
  - In the Forward Troop Area on 5 August – 1,261 (lightly wounded)

In Army Hospitals

- Arrivals:
  - At the start (30 July) – 2,963 men and women
  - Arrived from forward units' medical installations – 378
  - Arrived directly from units – 215
  - Total – 3,556
- Departures:
  - Evacuated – 278 (out of the Army) (severely wounded)
  - Returned to active service – 416

Died – 6
Total – 700
In Army Hospitals on 5 August – 2,856 (moderately wounded) (with 5,350 hospital beds available).

Total in all medical installations on 5 August – 4,117. [507]

As noted in June, the Army held approximately 5,350 beds in all its hospitals, not including divisional level medical support. The Army's medical staff would have attempted to clear all of the 1,261 sick and wounded from the forward area (divisional level) before the battle. That action would concentrate most of the wounded and sick into army level hospitals totaling 4,117 people, and with 5,350 total hospital beds, that left only 1,233 open beds. This could not be called excessive medical support when they were expecting hundreds of wounded per day during the upcoming battle. Most likely, the two medical ambulance trains that were assigned to the Army were busy transporting badly wounded to the deep rear and the Army's medical staff would be working to discharge the sick and wounded from the hospitals as quickly as possible. Hopefully, sufficient medical facilities would be ready in time.

Of special note, during the six-day time period of that report, the Forward Troop Area medical units returned to service 1,245 men and women, while the Army Hospitals only returned to service 416 men and women. This disparity confirms that the forward medical units retained the lightly wounded and returned them to their units fairly quickly. The Army hospitals were reserved for the more seriously wounded, requiring longer treatment and recovery times before they could be expected to return to active duty.

6 August 1944

Providing assistance for the coming offensive, engineer troops of both the 2 and 3 UFs worked to prepare the rear area for unobstructed movement of troops and equipment toward the front. Additionally, this work also allowed for the smooth flow of logistical support as the army advanced and for the evacuation of wounded to the rear. This engineer support also included

- Maintenance of roads and bridges, and preparation for crossing of rivers and streams during the attack.
- Removal of friendly minefields and eventually the clearing of enemy minefields.
- Clearing or filling in anti-tank obstacles like anti-tank ditches or shell craters and infantry obstacles like barbed wire entanglements.
- Coordination of communications and troop control efforts. As for command and control and observation posts, they were moved as close to the front as possible. For example: The front commander had observation posts [OP] 5 Km from the front, army commanders 3-4 Km, corps commanders 3-4 Km, and divisional commanders 2-3 Km. [In total, all these OPs were located between 1-3 miles from the front]. [508]

7 August 1944

Guards Captain Gordeev produced *A summary of the material support of units and formations of the 7th Guards Army as of 20.00 on August 7, 1944*. This report can be found within TsAMO Archive ID #132309977. This report shows stocks of the following ammunition, fuel and food supplies:

82 mm mortar - 1.55 Units of Fire (UoF, all listings are average amounts per rifle division)
120 mm mortar - 1.67
45 mm AT Gun – 1.23
76 mm Regimental and Divisional guns – 2.93

122 mm - .96
Gasoline – 2.83 Standard Loads (Refills are average amounts per rifle division)
Biscuits and bread flour – 10.61
Grains and macaroni – 9.7
Meat – 3.15
Vodka – 0.0 (zero at divisional level but ample supplies at army level of storage)
Oats – .53
Hay – 39.1

And low levels of other types of supplies.

With these amounts of supplies, the Army was ready for extended combat. Note: On 17 August the divisions had about 1.0 units of vodka on hand. By 19 August, the divisions had none, but army level storage showed 1,300 liters of vodka; then the 20 August report showed only 500 liters of vodka at army level and none at divisional level. So, it appears vodka was being distributed to combat units, but it was being consumed just as quickly.

8 August 1944

After taking over command of AG South Ukraine, Colonel General Johannes Friessner became more concerned about the overall situation. His command contained two German armies, the 6th and 8th, and two Romanian armies, the 3rd and 4th. The German armies were in fairly good shape regarding the total number of troops and equipment, but many troops lacked training and combat experience. The Romanian armies also had adequate troop levels, but they lacked training and were armed with mostly outdated equipment. The most troubling issue with Romanian forces was their low morale and apathy concerning the war. Underpinning all these issues with the defending armies was the inadequate AG South Ukraine reserves. Most of all, what they lacked were powerful Panzer forces that could counterattack Red Army breakthroughs and drive the attackers back.

> After evaluating the situation, Friessner feared that the Soviets would undertake an advance from the Jassy [Iasi] area to the south, by which they would easily cut off the German forces on the southern front, around Kishinev, from their rear communications. [509]

General Friessner understood that the German 6th and Romanian 3rd armies, in their current positions, were vulnerable to powerful flanking attacks. Adolf Hitler, true to form, refused to pull back the right flank of AG South Ukraine to less exposed positions. Axis forces in Romania now would stand and fight along the currently held line.

9 August 1944

Russian offensive operations against Finland begin to decrease most likely in an attempt to limit losses and to give Finland time to accept the Soviet surrender terms. By this point, the Russian offensive losses in Finland amounted to 23,000 killed and 71,700 wounded. Finnish forces had suffered at least 60,000 casualties. [510] Soviet forces on the Karelian Isthmus (just north of Leningrad) had been reduced further to just 10 infantry divisions and five tank brigades. [511] This reduction in troops gave the Finnish government hope that further talks with Moscow would be more productive.

10 August 1944

The 2 UF devised a *maskirovka* plan to hide the true regrouping of troops and the true concentration areas from the enemy. These *maskirovka* tasks were to be carried out between 13-19 August, and the 7th Guards Army had a big role to play in this plan.

In part, this plan required the 7th Guards Army to build 400 dummy guns and artillery positions by 14 August. This was to disguise the mass transfer of artillery to the 27th A. Additionally, the Army had to build and emplace 60 dummy tanks. Once again this was done to conceal from the Germans the transfer of the 27 GTB to the 27th A. Then, during several nights before the attack, between 22:00 to 05:00 hours, the Army was to drive vehicles towards the front (on the road to Targu-Frumos) with lights on. This "lights on traffic" was to simulate hurried movement of troops/supplies to the front in a certain area. Finally, during 17-18 August, with the purpose of distracting the enemy's attention, the Army would fire heavy artillery (152 mm and 203 mm), over open sights and with flat trajectory fire directly engaging enemy pillboxes along the front. [512] All this activity was intended to confuse and to mislead the Germans as much as possible.

"By 10 August [the Romanian] 4th Army had detected the arrival of four Soviet corps from Poland, and underestimated that 600 tanks were opposite it." General Gheorghe Avramescu, the commander of the 4th Army, "predicted that a Soviet offensive was probable any time after 15 August. This was highly disturbing, but Hitler and most of his commanders considered the Romanians to be congenitally alarmist and continued to discount their reports." [513]

11 August 1944

On or about this date, the 16 Artillery Penetration Division and other artillery units were reassigned from the 7th Guards Army to the neighboring 27th A. Also included in the transfer was the full strength 27 GTB, so now the 7th Guards Army had lost its tried and true 27 GTB and all its direct tank support. The Army would have the 23 Tank Corps and the 5 Guards Cavalry Corps on hand, but they were to be used to exploit the breach in the enemies' line that the 7th Guards Army hoped to make. Thus, the infantry apparently would go into battle supported mostly by only organic artillery [i.e., artillery from within rifle formations of the 7th Guards Army] and the VVS.

Note: The 27 GTB would remain with the 27th A until November when it would be returned to the 7th Guards Army where it would stay until the end of the war. Additionally, the 227 RD was assigned to the 7th Guards Army around this date, and the 8 GAD transferred from the Army around this same time.

The 2 UF artillery commander stated, "In order to achieve an overwhelming superiority in artillery along the 27th and 52nd Armies' attack axis, subordinate parts of the 40th, 7th Guards, 53rd, and 4th GAs' artillery [would be moved] to these armies." [514] This transfer left only the following 7th Guards Army subordinate units in place: 290 Mortar Regiment and the 5 Antiaircraft Artillery Division (670, 743, 1119, 1181 AA Regiments) and the 162 Guards AA Regiment. Of course, the divisional level artillery still was in place to support the 7th Guards Army attack.

Additionally, starting on 11 August and continuing until 19 August, Soviet forces of the 2 and 3 UFs launched a series of diversionary and probing attacks against Axis positions in Romania. Useful intelligence was gathered about the placement of German vs. Romanian forces along the line, as Romanian units would be the main target at the start of the coming battle. The Soviets knew Romanian formations overall were weaker than their German counterparts. Romanian troops had less equipment, training and determination to keep fighting.

Prit Buttar, in his book, *The Reckoning*, page 437, stated:

> It was now standard practice for the Red Army to make local attacks at company or battalion strength immediately before an offensive; these 'reconnaissance in force' [actions] helped identify strongpoints in the defenses and as has already been described sometimes led to the main attack being committed early to take advantage of an unexpected weakness. In order to

> avoid alerting the Germans to the precise location of their planned attack, the commanders [on] both Fronts ordered a series of such attacks in the weeks preceding the offensive at a wide variety of locations, many of them far from the intended points of assault.

At the same time, far behind the front lines in Bucharest, Romanian loyalists and Communist agents were working together on methods to further weaken the resolve of Romanian armies at the front.

12 August 1944

In Poland, German forces launched powerful concentric attacks against the Polish Home Army, which was entrenched within downtown Warsaw. The Germans would soon be using super heavy artillery within the city including *Mörser Karl* (Mortar Karl), a 600 mm mortar, which fired a shell weighing 4,850 lbs.

13 August 1944

An *Artillery Combat Log of the 7th Guards Army*, from TsAMO ID #150200623, page 7-8, shows normal activity from the 7th Guards Army and the enemy, along with a bit of deception for good measure.

On 12 August, this report stated, in part,

> At the old OP [observation post], mock-ups remained and roaming guns and mortars began to operate, which would simulate normal artillery fire after the bulk of the Army's artillery was transferred to the 27th Army.

Then later in the report,

> Over the past day, the enemy artillery actions were not fixed, with rare methodical and disruptive fire at our troops. In total, the action of 13 artillery batteries was noted with a total consumption of up to 500 rounds.
> In some areas, the enemy had large groups of soldiers in the trenches.
> Other parts of the army continued to firmly hold the occupied positions.
>
> Army artillery destroyed in a day:
> Soldiers – 25, mortars – 2, destroyed OPs – 3, artillery dugout – 1, suppressed machine guns – 2. Ammunition consumption: 82mm – 136 [rounds], 120mm – 56, 76mm – 149, 122mm – 71, 152mm – 102. [A total of 514 rounds].
>
> [The] 16 [Artillery Penetration Division], on the basis of orders from the commander of artillery 2 UF, at 18:00 13 [August] left the [7th Guards Army].
> From the 16 [Artillery Penetration Division], one 152mm howitzer and a 152mm cannon-howitzer were left at the OP to fulfill the plan for firing roaming guns.
> A battery of 203mm howitzers from 109 [High-Power Howitzer Brigade] received 4 units of ammunition, 2 fuel supplies, food supplies for 7 days and were left with the task of firing at bunkers.

This report is full of interesting bits of information. It details the establishment of roaming guns (single artillery guns that would fire from one position, then move to another and fire again) to deceive the Germans into thinking that all the pre-existing artillery batteries were still in place. Also in use was the 203 mm howitzer battery that would be used in direct fire against enemy bunkers to distract the enemy's attention from the real offensive area.

General Friessner, Commander of AG South Ukraine, commented, "…the strength of the army group was constantly decreasing. From the end of June to 13 August, we were forced to transfer as many as 11 divisions to other army groups!" This reduction of strength was directly related to the losses suffered by AG Center and AG North Ukraine during the summer. General Friessner knew his only option to save his AG was to withdraw his forces to a shorter, more defensible line, but Hitler and the High Command refused. The results of this unfortunate choice were predictable. [515]

14 August 1944

Leading up to the offensive, Guards Captain Gordeev produced another report *Information of the troops of the 7th Guards Army, Ammunition, Fuels and Lubricants and Foodstuffs at 18:00 August 14, 1944.* This report can be found within TsAMO Archive ID #132309985. This report shows stocks of the following supplies:

> 82 mm mortar – 1.48 Units of Fire (UoF, all listings are average amounts for the seven rifle divisions)
> 120 mm mortar – 1.70
> 45 mm AT Gun – 1.24
> 76 mm Regimental and Divisional guns – 2.73
> 122 mm – .98
> 85 mm AA – 1.0
> Gasoline – 2.67 Standard Loads (Refills are average amounts for the rifle divisions)
> Biscuits and bread flour – 5.92 (most likely listed in daily usage, i.e., 5.92 days' worth of biscuit and bread flower)
> Grains and macaroni – 12.8
> Meat – 9.71
> Vodka – 0.0 (Zero issued to troops, but 2,600 liters at divisional level supply dumps and ample supplies of 59,100 liters at army level supply bases)
> Oats – 14.50
> Hay/straw– 40.1

And low to moderate levels of other types of supplies.

This report shows that Army level supply bases or supply stations had adequate amounts of ammunition and gasoline, which would be required for any offensive, and good amounts of basic food for the troops and hay/oats for the horses. So, the Army had adequate supplies under current combat conditions, but it still needed more supplies for sustained combat. Note: with approximately 84,000 combat troops within the Army, the total available vodka (61,700 liters) would provide each combat solider with about 2.48 ounces of vodka if it was given to them before going into battle. Two good shots, then off to fight!

15 August 1944

The employment of extensive Soviet *maskirovka* methods within the 2 and 3 UFs worked well. On this day, the German OKH intelligence organization "Foreign Armies East" issued a report covering AG South Ukraine:

> The development of the enemy situation opposite AG "Sud [South] Ukraine" since the beginning of the summer operations has been marked by the withdrawal of enemy forces in strength for employment against the Army Groups "Nord Ukraine" and "Center" (overall 2-3 tank armies, 8-10 mobile groups, 3 infantry armies, 28 rifle divisions). It cannot be determined clearly if this shifting of forces is still going on, but…. it seems probable. An attack operation of larger size with far-reaching objectives against AG "Sud [South] Ukraine" appears therefore to be unlikely. [516]

With Operation Bagration and other massive Soviet summer offensives winding down against German AG North, Center and North Ukraine, AG South Ukraine appeared to be out of danger. But this was a false sense of security as the 2 and 3 UFs were almost ready to destroy the Axis position in the Balkans and to drive Romania out of the war. (See Map #45 for the overall strategic situation.)

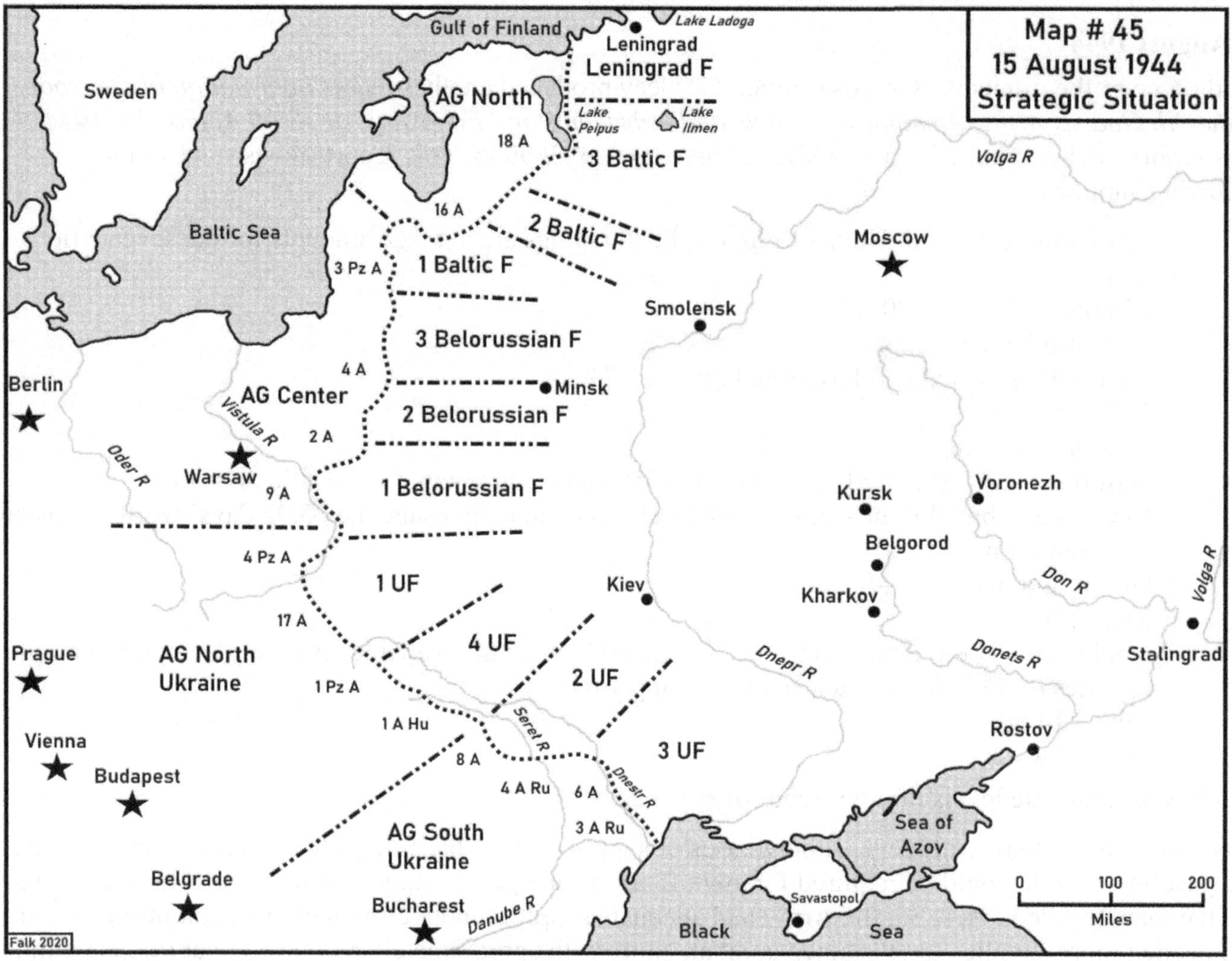

The 2 UF Major General Pavlovsky issued a strength report, *The Balance of Forces for 2 UF*, which is found within TsAMO Archive ID #455199038. This one-page chart details the balance of forces (Red Army vs. Axis) along the entire 2 UF and the ratios along the breakthrough sector, which are listed below. These ratios would only be valid for the 7th Guards Army, 27th A and the 52nd A and only during the initial breakthrough period during the first two days of the offensive. (See Table #20.)

Forces	Troops	Machine Guns	Mortars	Artillery	Total Tubes	SU and Tanks	Aircraft
Red Army	184,400	2,741	1,622	2,255	3,877	973	905
Axis	58,900	1,822	320	208	528	323	300
Ratios	**3.1:1**	**1.5:1**	**5.8:1**	**9.1:1**	**7.3:1**	**3:1**	**3:1**

Table #20: 2 UF Force Ratios - 15 August 1944.
Note: This was the 2 UF's perception of what Axis forces it was facing; the true totals would be somewhat different.

In far off France, the strategic situation for Nazi Germany had become much more despairing. The Western Allies launched Operation Anvil (Dragoon), the second seaborne invasion of France. This amphibious assault force came ashore along France's south Mediterranean coast. The invading U.S. 7th

Army surged inland against ineffective German resistance. This second seaborne invasion of France demonstrated what command of the sea and air permits, essentially, the ability to strike from the sea with overwhelming force at the time and place of one's own choosing. The enemy is therefore caught off guard and forced to respond in a limited and haphazard manner.

16 August 1944

Back on 16 March 1944, Field Marshal Antonescu, the Romanian Chief of State, told the Germans

> that he wished to form [the] 3rd and 4th Armies into a Romanian army group under Romanian command. The Germans were against this, but as they were now on Romanian soil, they had to compromise. They conceded the creation of two new mixed groups of armies, one German-commanded and one Romanian-commanded. Group Wöhler, created on 26 March 1944 [was]…comprised of General Wöhler's [German] 8th Army and [General Ioan Mihail] Racoviță's newly re-fielded Romanian 4th Army. Group [Petre] Dumitrescu consisted of [General] Dumitrescu's own [Romanian] 3rd Army and the German 6th Army. All four armies contained divisions of both nationalities and had integrated command structures. [517]

This mixed group of armies and command arrangements would prove to be disastrous in the near future.

David Glantz, in his book *Soviet Military Deception in the Second World War*, page 420, notes,

> German intelligence began picking up movement in the Soviet rear, in particular in the Tiraspol area and north of Yassy [Iasi] on 16 August but stuck to the prediction of only local attacks. The next day, however, their concern began to increase as they recorded, "The reinforcements by night… have increased in the area of Yassy [Iasi]. One has therefore to anticipate strong attacks, which will have more than local importance."

As the clock ticked down to the start of the Red Army offensive, the Germans still were largely unaware of the scale and scope of the pending attack. The true movements and positions of entire Soviet armies, including the powerful 6th TA, were missing from their intelligence maps. The Axis forces would pay a heavy price for this intelligence failure.

The operations diary from the Romanian 4th Army recorded, "The situation is critically weak on the Basarabian Front. Only the presence of mass German armor justified holding this front, which is naturally vulnerable. The consequences [of the withdrawal of German armor] could be disastrous." [518] Once again, the key point was brought forward: the lack of adequate numbers of Panzer divisions within AG South Ukraine endangered the entire position. With the bulk of German Panzer formations occupied elsewhere, AG South Ukraine would be forced to make due with what forces it had. Hitler thought the Soviets were too weak after the great offensive battles against AG North, Center and North Ukraine to launch yet another major offensive in the south. In a few days' time, *Stavka* would prove him wrong.

17 August 1944

At this point, individual armies of the 2 and 3 UFs began their own regrouping "and occupied their jumping-off positions in one night – the evening of 17 August." [519]

18 August 1944

In the main attack area ("breakthrough sector" as the Soviets called it) of the 2 UF, the artillery concentration reached high levels of superiority. For the 27th A, the ratio was 10.5:1, and for the 52nd A, the ratio reached 5.5:1 against the defending Axis forces. However, in the area of the 7th Guards

Army, the ratio was only 1.2:1. [520] This low ratio was due mostly to the transfer of many artillery units to the 27th A.

Both the 40th A and the 7th Guards Army faced the issue of tough Romanian defense positions. These defensive works covered the front from

> the Targu Neamt area [on the west side of the Seret River] to Targu-Frumos, where the Romanians had a fortified area. Permanent structures were located in two lines and represented semi-caponiers [fortification providing flanking fire] with 37mm and 76mm guns and double aperture machine guns pillboxes in the first line and machine guns pillboxes in the second line, with intervals of 250-300 m [820-980 feet] along the front and 150-300 m [490-980 feet] in depth. There was an anti-tank ditch, three meters deep [9 feet] and 5-8 meters wide [16-26 feet] in front of the machine gun pillboxes, while 3-5 rows of barbed wire were present. The spaces between the pillboxes were mined. [521]

Lacking abundant artillery, the 7th Guards Army would have to use what firepower it had to help the infantry take these positions by storm.

19 August 1944

On or about this date, the 24 GRC commander ordered the 81 GRD to launch a reconnaissance-in-force attack with a reinforced battalion. Following this attack, the corps commander altered the deployment of the division with two rifle regiments in the first echelon and the third regiment in the second echelon behind the left flank of the division. This reconnaissance-in-force attack also confirmed that the German 46 ID and the 1 Ru G ID were deployed to their front. [522]

German intelligence forces finally detected one of the main attack sectors, and during the night they bombed the rear area of the 52nd A with 100 aircraft. But it was far too late for the Germans to make any major adjustment to their defense arrangements. The Soviet Iasi–Kishinev offensive would begin the next day. [523]

General Friessner commented on these failures.

> The results of our air reconnaissance activities were generally very insignificant until the last days before the start of the offensive. This was probably due to the fact that the Russians made the movement of troops secretly and only at night. Since the Russians were good at masking such events, our intelligence agency was able to provide the necessary information also only with a great delay.

He went on to say, "From August 7, the picture changed dramatically. Reconnaissance regularly began to report on more and more intensive movements of enemy troops on both sides of the Prut [River]." [524]

Further north, the massive Operation Bagration offensive against German AG Center had ended with both sides completely exhausted. Nazi German had lost huge amounts of land and equally massive numbers of men (more than 400,000 casualties) and equipment to the advancing Red Army, but the Red Army suffered too. (See Table #21.)

Front	Killed	Wounded	Totals
1 Baltic	41,000	125,000	166,000
3 Belorussian	45,000	155,000	200,000
2 Belorussian	26,000	90,000	116,000
1 Belorussian	65,000	215,000	280,000
Totals	**177,000**	**585,000**	**762,000**

Table #21: Soviet Bagration Offensive Losses. [525]

Even though the Bagration offensive was much larger than the pending Iasi–Kishinev offensive, it might be useful, in due time, to compare its losses to the August Soviet offensive in Romania.

Just before the attack, the 7th Guards Army contained the following seven rifle divisions:

(24 GRC) 72 and 81 GRDs, 6 RD.
(25 GRC) 6 GAD, 36 GRD, 53 and 227 RDs.

The following units were directly attached to the Army:

290 Mortar Regiment
5 Antiaircraft Artillery Division
162 Guards AA Regiment
38 Separate Armored Train Battalion.

All other units had been sent to other armies to support their attacks.

Note: The 8 GAD had earlier been withdrawn from the Army and was far to the rear, positioned behind the 40th A. The 8 GAD was on its way into reserve in the Moscow Military District where it would be reformed into the new 107 GRD. [526]

However, the 2 UF seemed to have had seen the error of stripping the 7th Guards Army of all its artillery assets, or perhaps General Shumilov complained about the issue. So apparently at the last minute, to support the pending attack by the 7th Guards Army, the 2 UF ordered the 27th A to transfer the following units to the left flank of the 7th Guards Army and the 25 GRC:

41 Cannon Artillery Brigade
114 and 115 Anti-Tank Artillery Regiments
263 Mortar Regiment
309 Guards Mortar Regiment (Katyusha rockets).

Additionally, the 7 and 127 Pontoon-Bridge Battalions were assigned to the 23 TC for support in river crossings. [527] Most of these assets were only temporarily assigned, but at least the Army now had a bit more firepower to pound forward enemy positions during the opening phase of the offensive.

20 August 1944

The 2 UF issued a combat strength report of the forces of the 2 UF, *List of the combat strength of the troops of the 2nd UkrF*. This report is found within TsAMO Archive ID #113723709. See Table #22 for 7th Guards Army unit strengths.

Unit/Formation	Combat Strength
24 GRC HQ	814
6 RD	6,127
72 GRD	6,393
81 GRD	6,243
25 GRC HQ	840
6 GAD	6,013
36 GRD	6,070
53 RD	6,337
Army Subordinate Units	
5 AAD	1,915
162 GAAR	400
38 ATB	197
4 Separate Artillery Battalion	510
290 Mortar Regiment	496
60 Engineer Sapper Brigade	1,462
190 Army Reserve Regiment	4,111
227 RD	6,262
Totals	**54,190***

Table #22: 7th Guards Army Combat Strength - 20 August 1944.
***Not including the two attached mobile corps, also known as the cavalry-mechanized group.**

This report is telling as the 7th Guards Army now had 7,400 fewer combat troops than the reported combat strength of 61,600 men on 1 July 1944. This was due to the transfer of major artillery units out of the Army, and it might be one of the reasons why Malinovsky attached the two-additional mobile corps to the 7th Guards Army, to give the Army more striking power for the offensive.

The Soviet Iasi–Kishinev offensive began with a 90-minute barrage along the 2 and 3 UFs. (See Map #42 [page 313] for the overall initial attack plan and army positions.)

Along the 3 UFs breakthrough sector, the 37th and 46th Armies and the supporting 57th A made fair progress along the boundary line between the German 6th Army and the Romanian 4th Army. By the end of the day Soviet forces had advanced about six to eight miles into the Axis positions. Most of the German and Romanian divisions were holding, but some broke and ran or withdrew under heavy pressure. A worrying note for Axis forces was that the German 6th Army had already committed most of its reserves in launching counterattacks.

Along the 2 UFs breakthrough sector, events turned out somewhat better. The 27th and 52nd Armies made fairly good progress with a maximum advanced of about 30 miles in the zone of the 27th A, to the west of Iasi.

During the first 24 hours of the offensive, the 5 Ru ID, 7 Ru ID and the newly reformed 4 Ru MtD collapsed under Soviet assaults. Other Romanian units fought on bravely, only withdrawing when their flanks were exposed. During the breakthrough battle, the Romanian 1 Ru Pz D and the 18 Ru MtD even launched counterattacks in an attempt to hold the line east of Targu-Frumos, but it was all in vain. [528] At the end of the day, the Axis defensive line was badly damaged.

On or about this date the 227 RD was assigned to the 7th Guards Army, but this rifle division was not yet deployed at the front and would not arrive in the combat zone for some time.

21 August 1944

The 3 UF with its 57th, 37th and 46th As had further developed their attack against the Axis line, expanding the breach and penetrating some 25 miles with infantry formations and around 45 miles for mobile units. The Romanian 6th Army line now was broken and the right flank of the German 6th Army was folded back, opening a gap which Russian mobile forces poured through.

On the second day of the attack, General Malinovsky ordered the 7th Guards Army to take part in the offensive along with the 27th and 52nd Armies. To the right of his front, Malinovsky "broadened his offensive by committing Shumilov's 7th Guards Army and the 'cavalry-mechanized group' in a supporting attack to clear the [right] flank." [529] Major General Sergey Gorshkov commanded this Cavalry-Mechanized Group. Note: On related maps, the location and movements of the 23 TC and 5 GCC are approximate. Also, the 40th A, on the right flank of the 7th Guards Army, would delay their own attack for several more days, as part of the overall plan.

With this assignment, the 7th Guards Army once again began offensive maneuvers within Romania. At first, only the Army's left flank 25 GRC attacked, but soon the 81 GRD joined in from the 24 GRC. With help from the 27th A on the left flank, Shumilov's forces broke the Romanian line and by day's end had reached the elusive Targu-Frumos and were fighting within the city. The 27 TC and the 5 GCC were fed into the breach in anticipation of a major breakthrough the following day.

General Friessner once again asked Hitler for permission to fall back and to save his troops; Hitler denied his request. General Friessner later wrote,

> Therefore, I decided to issue such an order at my own peril and risk. On August 21, this order was communicated to the troops. Of course, in the current situation, there could be no question of a systematic separation of troops from the enemy, despite the preliminary preparation. These maneuvers were now completely up to the enemy. They could only be carried out step by step. It was no longer a question of withdrawing troops, but rather of a retreat with battles. [530]

22 August 1944

The 36 GRD, with the help of tanks from the 23 TC, finally completely captured Targu-Frumos and the surrounding area. Units of the 7th Guards Army extended their advance to the south and west and continued to advance into the night. The cavalry-mechanized group continued their advance south, keeping contact with units of the 27th A.

The 81 GRD recounted the status of the attack. "After two days of fighting, …the division in cooperation with a neighbor on the right [6 GAD] and on the left [6 RD], … completely broke through the field fortified defenses [of the] enemy and captured Mount Zhora, the villages of Derechiy, Zhora, the city of Targu-Frumos and [reached] the railway to the west and east of the city." [531] The attached 23 TC and 5 GCC supported all of these attacks.

At the same time, the 52nd A advanced quickly and fully captured Iasi. With the capture of both Targu-Frumos and Iasi, Soviet forces had broken the Axis defense line in front of the 2 UF. Soviet mobile formation began to penetrate into the Axis rear areas, scattering logistical, medical and support troops. No strong German or Romanian reserve forces were available to block their advance or to drive them back. As a precaution before the attack, General Friessner had deployed small combat groups at key bridges along the Seret and Prut Rivers. The hope was these troops would be able to hold these crossings long enough for rear area units and any retreating combat units to cross to safety, and then withdraw to the west or further south.

23 August 1944

The 7th Guards Army continued its advance and by the end of the day the 81 GRD, 6 GAD, and the 72 GRD of the 24 GRC had reached and crossed the Seret River. These units held bridgeheads on the western shore of the Seret as their leading units approached the city and major rail center of Roman. Elements of the cavalry-mechanized group, having forded the river, reached the outskirts of Roman during the evening. Units of the 25 GRC kept pace but were advancing along the east side of the Seret River. After three days of combat, the furthest advance of the Army was about 29 miles as the advance accelerated. (See Map #46.)

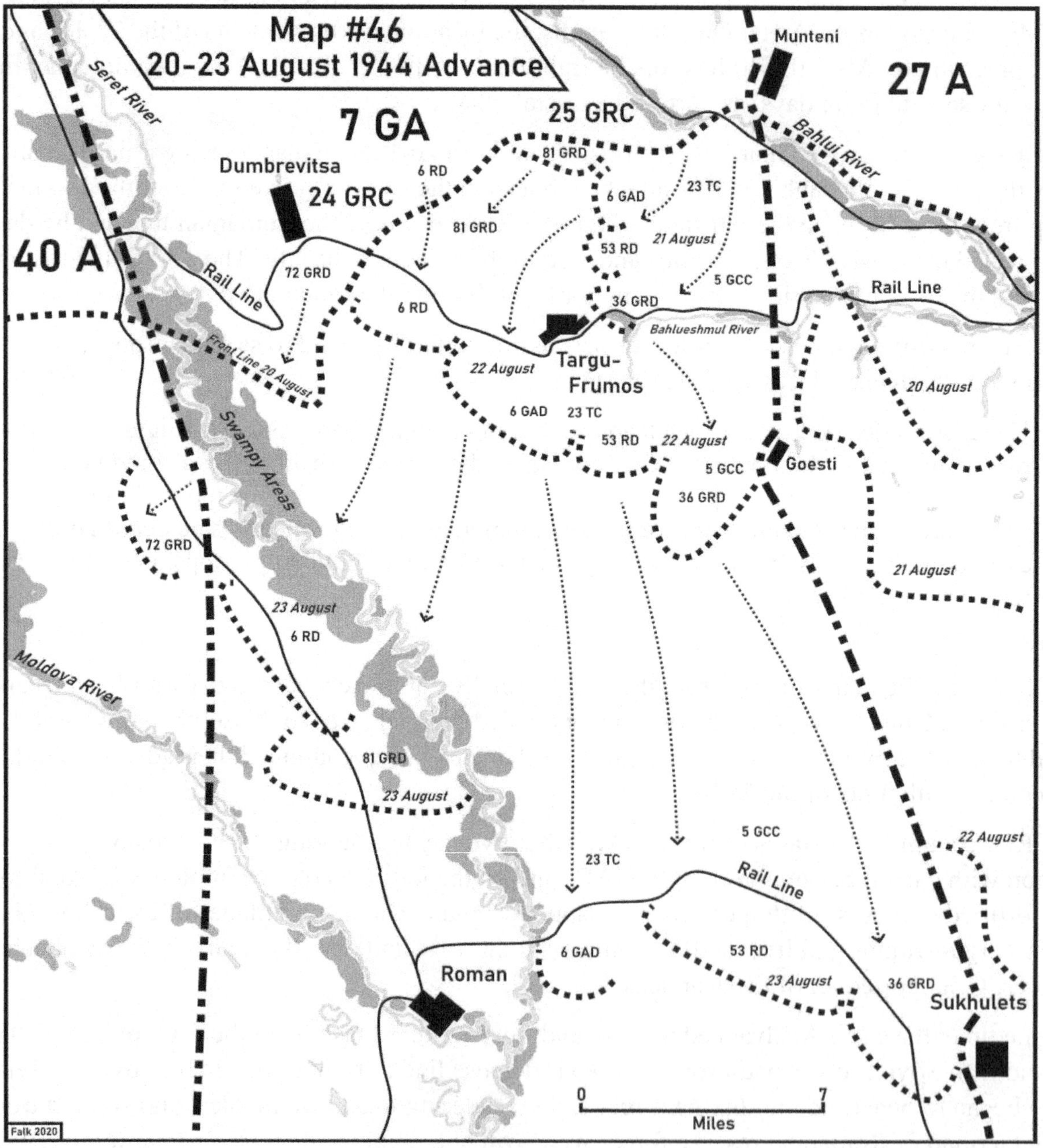

Realizing that the war against the Allies was lost, King Michael I of Romania launched a coup against the Axis sympathizer Field Marshal Antonescu, the Romanian Chief of State, which the king considered a dictator. The king had support from political and military leaders and later that night, King Michael I announced on the radio a cease-fire and an armistice with the Allies. In essence Romania surrendered to Soviet/Allied forces and would soon turn Romanian armies against their former ally, Nazi Germany.

With this surrender announcement, Adolf Hitler, the OKH, and the commander of AG South Ukraine, Colonel General Johannes Friessner, had a real dilemma on their hands. The offensive by the 2 and 3 UFs continued and with the armistice order by King Michael I, Romanian armies largely stopped fighting, which caused the collapse of the entire defensive battle plan of SG South Ukraine. Because of this, German armies were being assaulted by vastly superior Red Army formations while adjacent Romanian units surrendered en masse to advancing Soviet troops, allowing the Red Army free passage to advance deep into their rear.

In response to the cease-fire and an armistice order from the Romanian King, General Friessner immediately ordered Romanian General Dumitrescu, commander of the Romanian 3rd Army, to his HQ.

> Against his better judgment Friessner implored Dumitrescu to support the Army Group *Südukraine* [South Ukraine] in their struggle against the Red Army and ignore instructions from Bucharest. Dumitrescu's simple reply said [it] all: "I cannot break my oath of allegiance to the king." Friessner had no answer to this kind of soldier's logic, but Dumitrescu's stance did not lessen his anger. [532]

The German 8th Army, positioned on the German left flank, had a way out of the approaching debacle by quickly withdrawing westward into the foothills of the Carpathian Mountains. But that maneuver also would swing open the door for the 2 UF to advance to the south. The German 6th Army, defending in the center, had already withdrawn some 25 miles to the south, reaching the fortified Traian Line. The 6th Army was still trying to hold this position around Kishinev, which allowed Russian mobile units to complete their wide flanking movements around both of the 6th Army's flanks. The German 6th Army at this point was in extreme danger of being surrounded.

In an account from a Soviet POW in Romania about the end of his captivity, he stated,

> As of 23 August 1944, 59,856 Soviet prisoners of war remained in the Romanian camps, of which 57,062 were soldiers.
>
> August 23, 1944, when our army was already entering Romania, a popular uprising took place there, the fascist government of Antonescu was overthrown, … the Romanian army turned its weapons against Nazi Germany, and battles began between the former allies. The German units, defeated in Yugoslavia, left along the Danube and fired at the [POW] camp in Calafate; there was a threat of its destruction. Then [sometime in late September 1944] the camp commandant, under cover of darkness, lined up all the prisoners of war in a column and led us towards the Soviet army. We walked all night, and in the morning, there was a long-awaited meeting with units of the 2nd Ukrainian Front. [533]

As a byproduct of Romania changing sides, the new government began the process of forcibly capturing or clearing all German troops and rear area forces from Romanian territory. By 31 August, 56,455 Germans had been captured and another 5,048 were killed. Romanian casualties totaled 8,586 killed and wounded. Even before the Red Army liberated Romania, the Romanians had freed themselves from Nazi entanglements. Of course, all the captured Germans were turned over to the Soviets as POWs. [534]

24 August 1944

The 7th Guards Army advanced quickly now, facing little organized opposition from enemy forces. The rifle divisions of the 24 GRC, along with the 23 TC, captured the city of Roman during the early morning. They then crossed the small Moldova River and continued pushing south, reaching the Bistrita River where they captured the city of Bacau on the far bank. Bacau was where the Romanian 4th Army had recently located its headquarters. This rapid advance might be due to Romanian forces in this area offering no

further resistance to the Red Army after the surrender. At this point, the 7th Guards Army was just reaching the foothills of the Carpathian Mountains. The troops soon would be advancing up winding river valleys with looming hills growing ever higher as they pushed further into the mountains. At the end of the day, advanced units of the rifle divisions were aligned along the Bistrita River and held Bacau. On the 7th Guards Army's right flank, the 40th A began their attacks and also advanced into the foothills of the Carpathians. (See Map #47.)

Wedged between the 2 and 3 UFs, the German 6th Army now was fully, but loosely, surrounded within a fairly large pocket just south of Kishinev. The 2 UFs 52nd A and 4th GA pressed in from the west, while the 3 UFs 5th Shock A, 57th A and 37th A were enveloping the pocket from the north and east. Some units of the German 6th Army attempted a breakout to the west, and were partially successful, while others held out for days within the pocket.

In Finland, sensing the time was right, President Carl Gustaf Mannerheim decided to sue for peace with the Soviet Union. On 25 August, Finland repealed the Ryti-Ribbentrop Pact and contacted the Soviet ambassador in Sweden to open direct talks with Moscow. [535] The Ryti-Ribbentrop Pact was a personal agreement between the former president of Finland, Risto Ryti, and Hitler not to reach a separate peace in the war with the Soviet Union without approval from Germany. Mannerheim felt he was not bound by this agreement by the former president and was thus free to negotiate with Russia without Germany's approval.

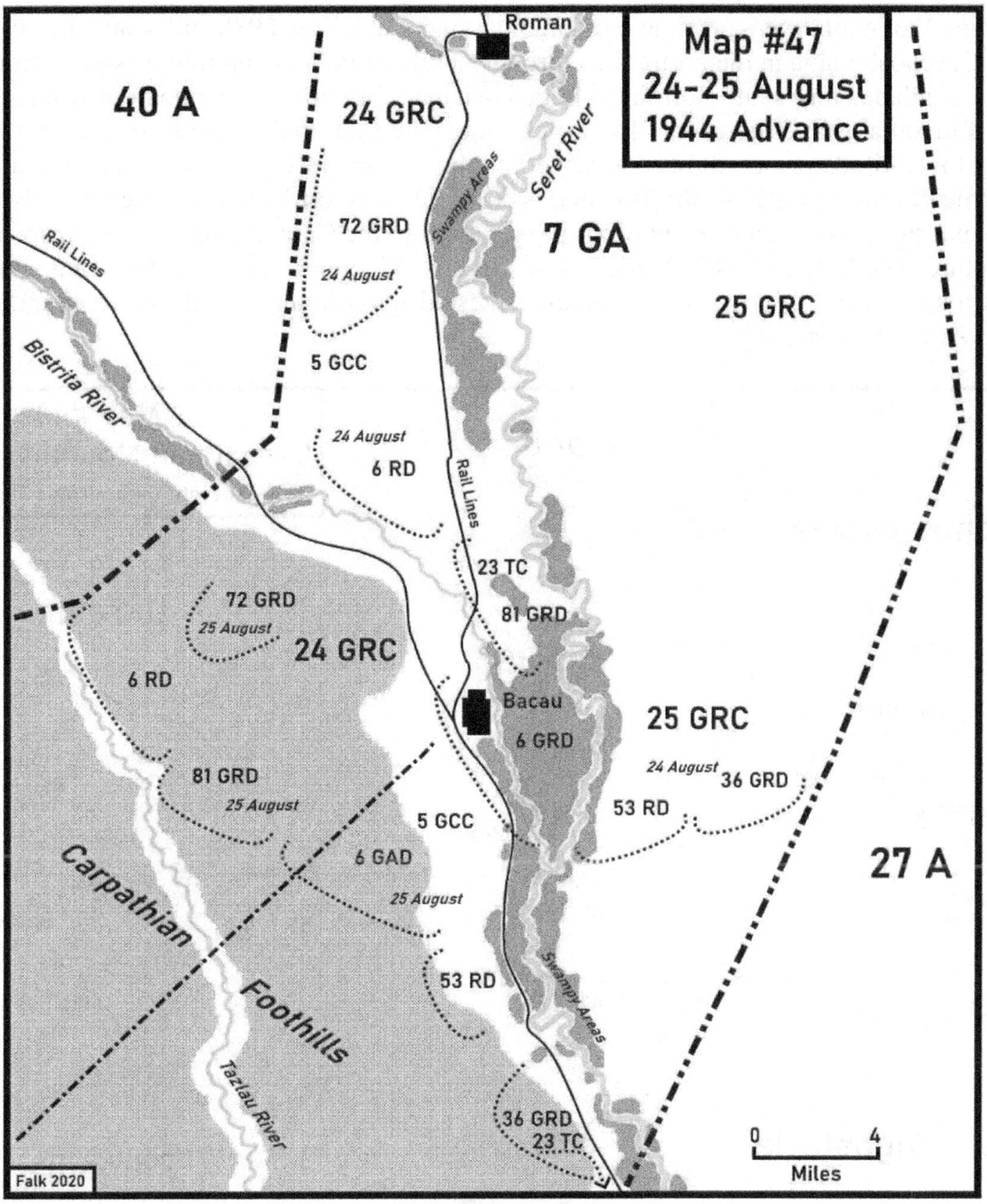

25 August 1944

Taking advantage of weak opposition, the 7th Guards Army pushed forward, but the army boundary line had been shifted by 2 UF HQ forcing the 7th Guards Army to turn to the right and to drive southwest directly into the Carpathians Mountains. General Malinovsky ordered this shift in the direction of the Army's attack in an attempt to quickly break through the Carpathian barrier. The 27th A operating on its left flank also shifted its attack more to the right, thus crowding out the 7th Guards Army and forcing them from the flatlands and into the mountains.

The Carpathians Mountains were no simple mountain range, but a broad 930-mile-long arc sweeping from eastern Czechoslovakia in the north, then through southern Poland and into western Ukraine. At that point, the highlands turn south, forming a huge U-shaped formation in the heart of Romania. The outer bands of mountains averaged some 100 miles wide, with their highest peaks over 8,000 feet. The few relatively low altitude passes allowed passage into the interior valleys (Transylvanian Plateau) and low-lying fertile planes beyond. For the 7th Guards Army, the two critical passes were the Gymes and Oitoz Passes, but there were other less perilous passes through the Carpathians that the Army also needed to capture. The Carpathians had aided in the defense of central Europe from invading hordes from the east for thousands of years. It would now be tested against tanks, artillery and aircraft of the invading Red Army. See Map #48.

Note: National borders displayed on Map #48 are post war.

An artillery battery commissar, (a Communist Party official in charge of political education and organization within a unit) from the 72 GRD described the scene:

> Together with the 7th Guards Army, we reached the Eastern Carpathians. Slowly, slowly climbed to the top. We went down a narrow, winding road. There, making its way between the stones and hiding in a dense bush, a small rivulet gurgled. As if copying it, the road wound along it in the same zigzags. Huge mountains surrounded us on all sides.

He went on to say:

> The Carpathians are good, but the roads are stingy. Therefore, a mass of troops [must] move along one narrow track. This plays into the hands of enemy pilots. Several squadrons suddenly appear from behind the heights and start bombing us.
>
> In the free Don steppes, in the golden expanses of Ukraine, we have learned and adapted to fight in open areas. And here are the mountains, and even covered with forest. Topographic maps provide only a rough idea of the area. It is very difficult to prepare [fire] data [for the artillery] and conduct zeroing [in] of targets due to the lack of landmarks. We need smoke and tracer shells [to identify targets].
>
> It is excruciatingly difficult to climb the mountain. The gun crews have to harness themselves into the straps and drag up the cannons first, and then the cars. [536]

The 7th Guards Army made moderate progress as they advanced into the foothills of the Carpathians. This was truly strange terrain for an army accustomed to conducting war on the open, treeless, rolling countryside of the Steppe. Due to the shift in boundary lines, all the rifle divisions of the Army were now on the right (western) shore of the Seret River. (See Map #47.) During this advance, in a bold move, the 23 TC independently advanced far to the south along the rail line and into the 27th A zone. Then it turned right and followed the rail line into the foothills towards Targu-Ocna in an attempt to outflank German defenders in that area. (See Maps #47 and 49.)

As one of the conditions for peace with the Soviet Union, Romania was forced to declared war on Germany. This act now opened direct hostilities between German and Romanian forces. Because of this change in sides, "One major detachment of the 1st [Romanian] Armored Division [containing 1,058 men,] under lieutenant Colonel Matie was delayed in Roznov by crowded roads, and came to a local understanding with the Soviet 7th Guards Army's 24th Corps on 25 August, subsequently helping it to force the [Gymes] Pass." [537]

During the night,

> loudspeakers [along the front] …broadcast the [recorded] appeal by the Soviet command to the soldiers and officers of the [surrounded] German Sixth Army, [the speaker was the 6th Army's former commander and now POW] Field Marshal Paulus, [along with] news on the events in Romania, and an appeal by [other captured] German generals.

During the following night, "the text of the ultimatum by the commander of the Third Ukrainian Front with a proposal for the capitulation of the encircled German forces was transmitted several times through a powerful radio station." [538] The remaining elements of the 6th Army did not surrender.

The leaders of Finland and the Soviet Union began to come to terms, with Moscow agreeing to a Finnish armistice on the condition that all German troops leave Finnish territory by 15 September. [539] Of course, there were other political Soviet conditions too.

26 August 1944

Units of the 7th Guards Army pushed deeper into the mountains. Due to the increased ruggedness of terrain, individual rifle divisions became concentrated into their own separate river valleys. There simply was no room for divisions to advance side by side. *Stavka* and General Malinovsky attempted to take advantage of the confusion caused by the defection of Romania from the Axis cause. By driving their armies into the Carpathians, *Stavka* anticipated capturing the vital mountain passes before German forces could secure them. The race was now on to see who would reach and control those passes first.

27 August 1944

After six days of combat and marching, the advance continued, but the pace had slowed noticeably. The 7th Guards Army now had reached the real Carpathians. (See Map #49.) At this point, the Army had advanced about 95 miles from the start of the offensive. Along the way the troops crossed several rivers, navigated around swamps, and advanced uphill along narrow river valleys and over foothills to reach the Carpathians. This advance must have been a new experience for most of the Red Army troops. The further the Army advanced into the mountains, the greater the resistance offered by German troops of the 8th Army and the remnants of the 6th Army. The German resistance and the terrain were doing a good job of slowing the Red Army's progress through the Carpathians. The battle for several key mountain passes was about to start.

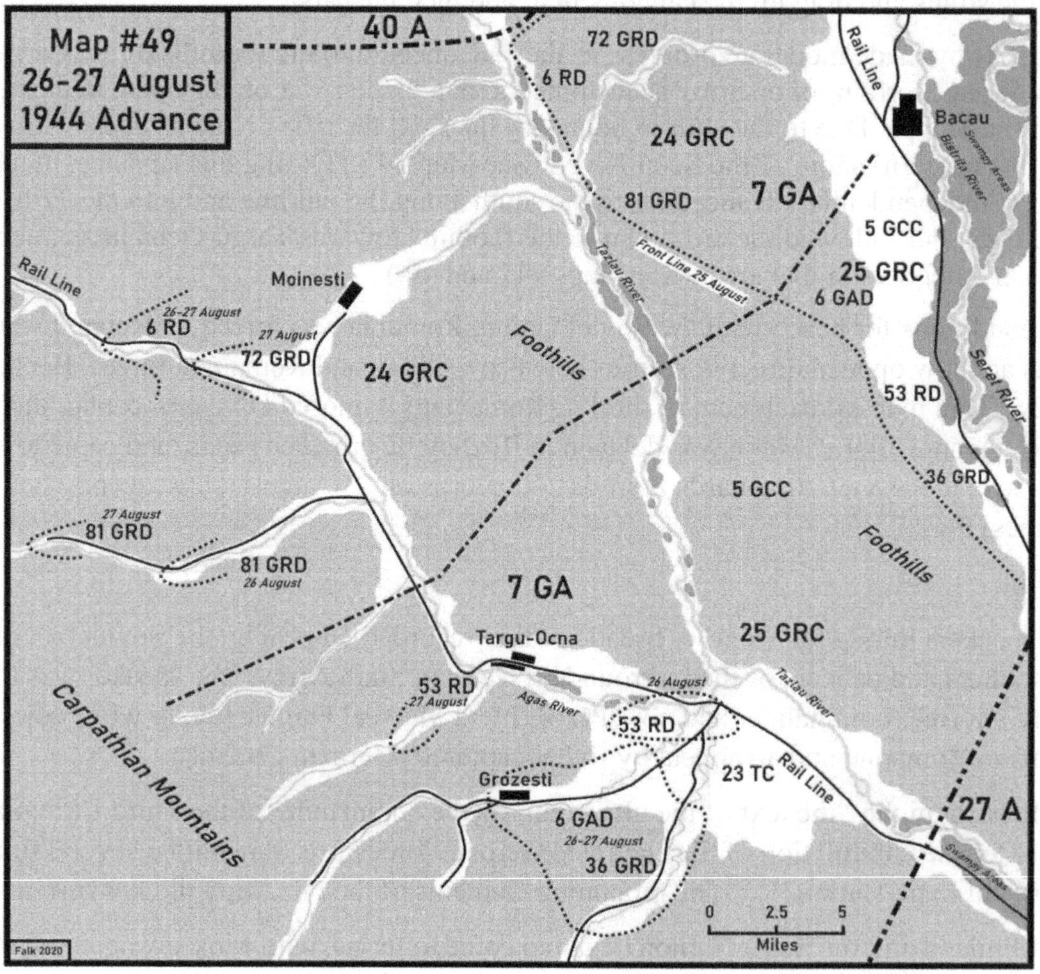

28 August 1944

On this day, 2 UF commander General Rodion Yakovlevich Malinovsky and the deputy chief of the General staff Lieutenant General V.D. Ivanov flew in two Po-2 aircraft (biplanes) on a reconnaissance flight along the frontline area. German ground fire hit Malinovsky's aircraft, and when back on the ground, his pilot said, "There are 12 holes in my plane!"

Malinovsky replied, "I don't know how many holes are in the plane, but I have one in my back." [540] Malinovsky was slightly wounded, but despite the wound, Malinovsky and General Ivanov saw for themselves the situation on the ground. The battle would proceed as planned.

At this point in the offensive, the 7th Guards Army was fighting in three separated groups. The 24 GRC was split into two groups; the northern group followed the rail line headed for the Grymes Pass with the 6 RD, 72 GRD and the 5 GCC in support. Fighting along this narrow valley slowed their progress. The center force consisted of only the 81 GRD, which headed due west toward the town of Banku near the Olt River. The 81 GRD advanced over extremely rough terrain along a single narrow road. The southern force consisted of the three rifle divisions of the 25 GRC. The 53 RD moved southwest up a valley north of the Oitoz Pass. The 6 GAD and two regiments of the 36 GRD marched directly toward the Oitoz Pass. The third regiment of the 36 GRD moved along an adjacent valley to the south. The 23 TC was in reserve near the town of Grozesti, but it did provide some tanks to support the assault against the Oitoz Pass. (See Map #50.)

Further to the east, several German mobile groups, with one under the command of General Friedrich Mieth, broke out of the 6th Army pocket and headed toward the southwest and the Seret River. This was the last desperate attempt to avoid capture by the surrounding Soviet armies. The German attack was directed mostly through the rear area behind the front, scattering Russian support troops along the way, which actually disrupted Russian logistical support efforts.

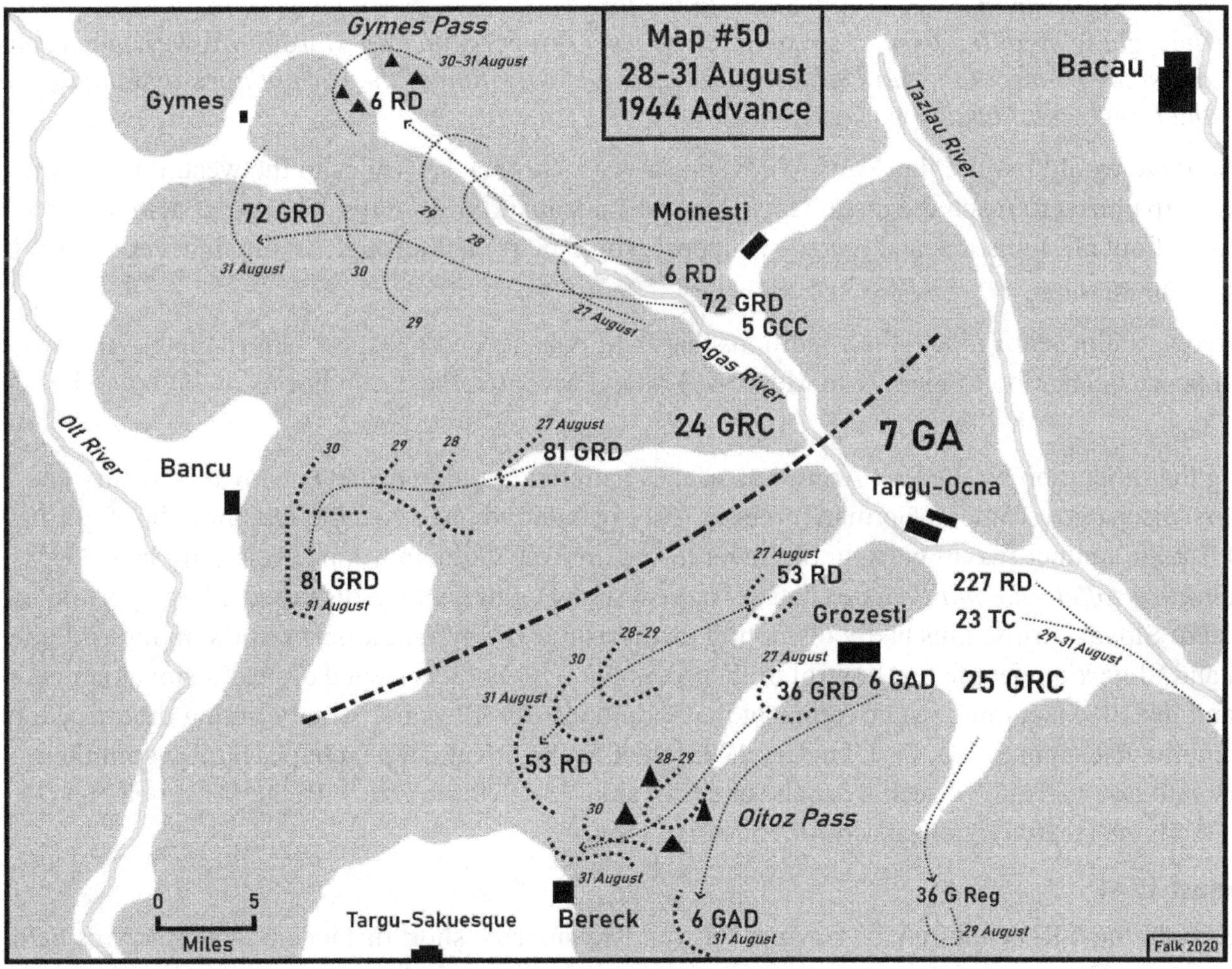

***Note: Rail lines and swampy areas are not shown on this map.**

29 August 1944

The 7th Guards Army continued to fight its way through valleys and towering hills toward the key mountain passes in their area. In the north the 6 RD approached the Grymes Pass while in the south the

36 GRD fought within the Oitoz Pass. Flanking divisions made supporting attacks. (See Map #50.)

Soviet units destroyed several escaping German groups from the encircled 6th Army, but the German group under the command of General Mieth, after heavy fighting during several encounters with Soviet combat troops, finally reached the Seret River and crossed to the western side, supposedly some 10,000 men strong. [541] The 7th Guards Army responded to this threat to its rear supply line by sending the 23 TC and the newly assigned 227 RD eastward to block the Germans' escape. The Russian 27 GRC also arrived from the north along the Seret River.

The 2 UF received Directive #220191 from *Stavka*, regarding capturing the passes through the eastern Carpathians. In part this directive stated:

> *Directive of the VGK Stavka No. 220191 to the commander of the 2nd Ukrainian Front troops, Representative of the Stavka on the development of the offensive in the depth of Rumania.*
>
> *August 29, 1944 01 h 50 min.*
>
> 2 *The right wing of the front - 40, 7th Guards Armies and the mechanized cavalry group of Gorshkov - to seize the passes through the Eastern Carpathians and by September 15 to reach the line of Bistrita, Cluj, Seyud, Sibiu. In the future, covering from the west, the main forces of the right wing of the front to strike in the general direction of Satu Mare* [northwest] *in order to assist the troops of the 4th Ukrainian Front in the transition of the Carpathians and the exit to the Uzhgorod, Mukachevo region.*

This directive would require the Army to advance at least another 130 miles to the west, through the eastern Carpathians to reach the inner Transylvanian Plateau, all in 17 days. This order also required that most, if not all, logistical and rear area support would be routed through mountain river valleys by truck or wagon.

In Moscow, Stalin sent Finland the conditions that the Finns needed to accept before the Soviets would allow a peace delegation to journey to Moscow. Finland accepted these conditions on 2 September 1944. [542]

Sensing that liberation by the Red Army was near, a national uprising began in Slovakia against the pro-Nazi Fascist government. German forces quickly invaded to put down the rebellion, but fighting would last for another two months, until Soviet forces arrived. German troops eventually launched Operation *Kartoffelernte mit Prämie* (Potato Harvest with Bonus), which disarmed a large portion of the eastern Slovak Army, thus badly disrupting the uprising. Despite their best efforts, in the end, the Czech and Slovakian people lost control of their country. The Soviets staged a Czechoslovakian coup in 1948 and installed a communist government that would last for 40 years. After winning their freedom from communist suppression, on 1 January 1993, the Czechoslovakian people peacefully split their country into two parts, the Czech Republic and Slovakia. In 1999 the Czech Republic joined NATO (North Atlantic Treaty Organization) for collective security.

30 August 1944

To the north, the 6 RD captured the Grymes Pass and advanced a short distance beyond but was held back by strong German defensive positions. In the south, by the end of the day, the 36 GRD had captured the Oitoz Pass but could go no further. The news of the capture of these two passes should have pleased General Shumilov.

General Friessner talked about the critical fighting for the Carpathian Passes.

> The German-Hungarian garrisons of strong points and centers of resistance defended the passes in stubborn battles that took place with varying success. Particularly heavy battles were fought for the Oytuz [Oitoz] pass, where the enemy used tanks. For some time, the defenders managed to repel attacks. Everyone understood that everything depends on whether or not it will be possible to keep the passes. [543]

The escaping German 6th Army forces under the command of General Mieth now were trapped between the Seret and the Tazlau Rivers. Fighting lasted until 4 September when the last group was overwhelmed. General Friedrich Mieth died in combat on 2 September 1944. The Soviets had destroyed the last operational formation outside the surrounded German 6th Army. The main pocket of the German 6th Army would surrender on 3 September.

But Prit Buttar in his book, *The Reckoning,* on page 446 stated that some Germans escaped the encirclement of General Mieth's group: "In heavy fighting most of the German units were overwhelmed and destroyed. Groups of soldiers of up to company strength [140 men] managed to slip through to the Carpathian foothills." Once reaching German lines, most of these escaping troops would have continued to fight in *ad hoc* groups within the Carpathians.

In a major accomplishment, advancing Russian forces of the 2 UF 27th A and 6 TA captured the oil fields and facilities in and around the city of Ploesti. Germany had now irreversibly lost its only major remaining source of oil and refined products.

An interesting report concerning Soviet logistics during the Romanian offensive stated:

> The Yassi [Iasi] -Kishinev Operation was an exception. There, the railway lines were taken virtually intact, and in their swift offensive into Romania, [the] 2 and 3 Ukrainian Fronts captured about 2,000 locomotives and 56,000 wagons [train cars], enabling them to exploit standard-gauge railways from the outset; this contributed to the depth of advance achieved by Sixth Tank Army between 20 August and 25 September—about 1,000 km (620 miles).

The report included comments about the Ploiesti oil fields:

> In the Yassi-Kishinev Operation, Sixth Tank Army [along with the 27th A], captured the Ploiesti oil fields and refineries intact and a further advance of more than 500 km (300 miles) by 2 Ukrainian Front was sustained by laying a 225 km (140 mile fuel) pipeline. Of course, such windfalls were only that; they could not be factored into planning. [544]

Without a doubt, the surrender of Romania to the Allies allowed the Red Army to exploit the largely undamaged road and rail transportation network, as well as the food and fuel supplies throughout Romania. These "windfalls" greatly aided the advance and increased the overall negative impact upon the Axis position within the Balkans. The town of Adjud on the Seret River (not shown on maps, but is located 39 miles southeast of Bacau) soon became a major railroad transshipment point from the Russian gauge to the Western gauge of the Romania rail network. Additionally, logistically speaking, the rivers in the Carpathians Mountains were unsuitable to transport supplies, as they were either too small, swift, or flowing in the wrong direction.

General Malinovsky, commander of the 2 UF, began the Bucharest-Arad Operation, which was intended to clear Romania of all Axis forces. The city of Arad is located on the northern border of Romania and is on the way to Budapest, Hungary. [545] The Iasi-Kishinev Operation originally was scheduled to end on 7 September, but with the launching of this new offensive, the Iasi-Kishinev Operation most likely ended on this day; however, some accounts say 3 September. Needless to say, this operation was a complete victory for the Red Army and the 2 and 3 UFs. Stalin soon would express his pleasure in the performance of his two front commanders, General Rodion Malinovsky (2 UF) and General Fyodor Tolbukhin (3 UF).

31 August 1944

Bucharest, the capital of Romania, fell to advancing Red Army troops of the 2 UF.

In the north, units of the 7th Guards Army, including the 5 GCC, could not advance beyond the Grymes Pass, but the 72 GRD and two regiments from the 6 RD were able to cross the mountain to the south and began to outflank the German defenders. In the center, the 81 GRD continued to advance westward but met stiff resistance and German counterattacks, which included the use of armor. The 81 GRD could go no further without assistance. In the south, the 53 RD continued to advance, but it too met strong German resistance. The main attack with the 6 GAD and 36 GRD cleared the Oitoz Pass and advanced downhill, but they were unable to make a clean breakthrough and to seize the town of Bereck.

The 7th Guards Army reported losses for the month of August in a report entitled, *Combat Log of the 7th Guards Army Troops*, recorded in TsAMO record ID #130117628, p 53. The report noted the following casualties:

Killed – 798
Missing – 4
Non-Combat Losses – 22
Wounded – 2,867
Sick – 1,363

Total casualties – 5,054 men and women.

On page 53 of the report was a table listing the transfer of replenishment troops from different sources into various active combat units for the month of August 1944. This table showed the following:

From the 190 Reserve Regiment – 728 men
From Hospitals – 625 (Recovered wounded returning to active duty)
From Penal Units – 1,153 (with 319 of these men transferred to 3 Separate AT Rifle Battalion which the 7th Guards Army no longer controlled).

An interesting note, the Army suffered 2,867 wounded and 1,363 sick for a total of 4,230 people in the hospital during August. The Army received back from the hospitals 625 people during August, which left 3,605 people in some type of hospital (inside or outside the Army) at the end of the month, yielding a 14.8 percent return rate from the hospital for the month. These hospital returns mostly came from lightly wounded troops that required less than a month to recover.

Page 54 of this same report recorded the damage that the 7th Guards Army believed it had inflicted on the enemy during August 1944:

Destroyed/Killed

Soldiers and Officers – 7,570 men
Guns of different calibers – 133
Mortars – 210
Heavy and light machine guns – 360
Submachine guns – 452
Motor vehicles – 141
Carts – 1,020
Destroyed and burned tanks – 10
Pillbox destroyed and blown up – 28
Pillbox captured – 168

Taken Prisoner

56,774, of which 5 were generals, 24 colonels and 1,855 other officers.

Trophies (with the note: Captured from a warehouse with military equipment).

Guns of different calibers – 482
Rifles – 25,249
Mortars – 358
Heavy and light machine guns – 1,537
Submachine guns – 1,330
Tanks and self-propelled guns – 40
Tractors – 12
Horses – 16,490
Carts – 3,608
Motor vehicles – 523
Armored personnel carriers - 9

Despite the large totals listed here, these numbers are believable because the entire Romanian 4th Army had surrendered on the third day of the 7th Guard's Army advance. German units also were fighting in the area of the 7th Guards Army, but after the Romanian surrender, they were forced to withdraw with some haste and disorder, which would only add to the number of captured trophies. One should also notice the large numbers of captured horses -- 16,490 -- which the 7th Guards Army quickly put to use.

German losses during this offensive have been estimated at 256,000 troops killed or captured. Romanian losses amounted to 8,305 killed; 24,989 wounded; and 130,000 troops captured. [546] Equipment losses would have been comparable to these huge personnel losses. An official German source estimated German losses in Romania as 286,000 troops dead or captured. [547]

The true Soviet loses in Romania have been understated in official records, but recent research estimates that Soviet forces, containing two fronts (2 and 3 UF), lost around 40,000 killed or missing and another 94,000 wounded, for a total of 134,000 men and women. [548] Comparing these losses to Operation Bagration losses, containing four fronts, 177,000 killed; 585,000 wounded; for a total of 762,000 (Table #21), one can see that the victories achieved in Romania came at a relatively low cost.

A TsAMO report ID #113723710 lists the August casualties for the 7th Guards Army at 798 killed, 4 missing, 22 non-combat losses; 2,867 wounded; 1,363 sick for a total of 5,054 men and women. The Army also lost 74 horses killed, 3 missing, 17 wounded, 10 sick, for a total of 104.

Fighting the enemy was one thing, but other dangers lurked about the battlefield. In this area of Romania, due to the swampy nature of the surrounding river valleys, mosquito infestation must have been high. One bite of a mosquito infected with the plasmodium parasite, which causes malaria, had devastating results on closely assembled and unprotected troops. During July and August 1944, the "2nd and 3rd Ukrainian Fronts suffered 87,603 cases of malaria which reduced considerably the combat capability of the troops." [549]

For the month of August in Britain, German V-1 attacks killed 1,103 civilians and injured 2,921. Allied counter measures, aircraft interception and anti-aircraft gunfire against the V-1 menace began to take effect, which reduced the damage to London and to its people. [550]

1 September 1944	
7th Guards Army	
Assigned – 2 Ukrainian Front	
Colonel General M.S. Shumilov Commanding	
Rifle, Airborne and Cavalry Units	24 Guards Rifle Corps (72, 81 GRDs, 6 RD) 25 Guards Rifle Corps (36 GRD, 6 GAD, 53 RD) 227 RD 15 Penal Battalion 62, 63, 64, 65, 66,194 Penal Companies
Artillery Units	114, 115 Guards Tank Destroyer Regiments 41 Guards Gun Artillery Brigade 263, 290 Mortar Regiments 5 Antiaircraft Artillery Division (670, 743, 1119, 1181 AA Regiments) 162 Guards AA Regiment 662 Separate Aerial Observation/Air Raid Warning Company
Armored and Mechanized Units	38 Separate Armored Train Battalion
Engineer Units	60 Engineer Sapper Brigade 4 Separate Flamethrower Battalion
Reported Combat Strength Estimated Ration Strength	54,665 men 61,665 men and women (including an estimated 7,000 support personnel)

(BSSA 44, BSA 44, TsAMO Archive)

At the start of the month, the Cavalry-Mechanized Group (23 TC and 5 GCC) commanded by Major General Sergey Gorshkov was still operating in the zone of the 7th Guards Army, but it appears that it was no longer directly attached to the Army. Instead, the 2 UF directed its actions.

Note: The changing weather had begun to affect the progress of the fighting at the beginning of September. If the weather was significant, the author will reference it immediately following the day's date.

1 September 1944 – Weather (high to low) 87° to 62° F, (30.5° to 16.8° C), wet ground

The 7th Guards Army continued to fight its way over the mountains and the critical passes. The fighting was intense, and gains were small. (See Map #51.)

At the insistence of Stalin, Field Marshal Antonescu, the arrested former Romanian Chief of State, was eventually handed over to Soviet forces for safekeeping in Moscow. After the end of the war, he was returned to Romania to stand trial, where he was found guilty of many crimes by the Communist-led government. He and several other defendants were shot by a military firing squad on 1 June 1946.

2 September 1944

The 7th Guards Army continued to press its attacks in the full knowledge that supporting armies, the 6th TA and 27th A, were putting flanking pressure on the Axis defenders.

In the far north, as part of their agreement with the Soviet Union, Finland broke off diplomatic relations with Germany and demanded the exit of German forces from the country by 15 September. This of course was impossible, but Finland was keeping strictly to the terms of the Soviet armistice agreement.

3 September 1944

The 7th Guards Army's 25 GRC made some progress toward the town of Bereck, while the northern 24 GRC attempted to advance beyond the Gymes Pass and/or to outflank it from the south. (See Map #51.)

Due to political events taking place between Finland and the Soviet Union, Germany began its Operation *Birke* (Birch) early. This plan intended to withdraw the entire German 20th Mountain Army from northern Finland into northern Norway. This withdraw/evacuation was intended to keep the army together as a fighting force and to take along all its equipment and as many supplies as possible. The Finnish army would pursue, moving the Germans along as needed, but the real danger would come from a possible sudden Soviet offensive to block the 20th Mountain Army from reaching Norway at all.

4 September 1944

Report #476 from the 2 UF to *Stavka* was in response to the *Stavka* Directive #220191 issued on 29 August 1944 about the great difficulties overcoming passes in the Carpathians. This report is listed in full to illustrate the type of communications passing from front commanders and their staff to *Stavka* in Moscow.

Report of the Representative of the Stavka and the Commander of the troops of the 2nd Ukrainian Front No. 476 to Verkhovny to the Chief of the Operations

Plan to Overcome the Transylvanian Alps and the Carpathians.

September 4, 1944

Fulfilling your directive No. 220191, the left-flank grouping of the front, advancing south of the Transylvanian Alps, is acting according to the following plan - reconnaissance and forward units of the 6th Tank Army today, on September 4, will capture the Craiova area and will cover the exit of the main forces of the 6th Tank Army, which will be able to enter this area on September 7-8 with two corps and one tank corps in the Tirgu-Jiu area. With the approach of the main forces of the 6th Tank Army to this area, reconnaissance units and forward detachments will go to the crossings on the river Danube in the Turnu-Severin, Bychetu section. Managarov's army, advancing behind the 6th Tank Army, will reach the Danube River in the Chetatya sector, capture Giurgiu with its forward units by the end of September 9. Moreover, to the Durju site, Korabia army will be released on September 6. The 27th Army, advancing behind the Tank Corps of Kravchenko, with its left flank will reach the Danube River in the Turnu-Severin, Cetatya sector by September 12. With the right flank, part of the army's forces acts on Brasov from the south, helping the 7th Guards Army in the seizure of this area. The right-flank grouping of the front (as part of the 40th and 7th Guards Armies and the Gorshkov Group) encountered great difficulties in overcoming the passes and great resistance from the Germans and Hungarians, relying on a system of permanent structures in the passages, and is advancing slowly now. [We hope] *that the 40th and 7th Guards Armies will quickly break through with a frontal blow; it is not necessary, therefore, with the arrival of the armies of Managarov and Trofimenko on the Danube River, Kravchenko's army consisting of three corps and the main forces of Trofimenko (two corps) to turn north and seize the area of Cluj, Oradea-Mare, Hatseg. By occupying this area, we thereby create a threat to encircle the Germans and Hungarians operating against the armies of Zhmachenko* [Filipp Feodosevich Zhmachenko, commander of the 40th A] *and Shumilov* [7th GA], *and help them quickly reach the Dezh, Cluj line for subsequent actions on Satu Mare to interact with the 4th Ukrainian Front. After the capture of Brasov, Gorshkov's group* [cavalry-mechanized group] *will operate through Sibiu in the general direction to Cluj. After unloading, the 6th and 4th cavalry corps will concentrate in the Craiova, Pitesti area no earlier than 09/25/1944 and will be used as appropriate.*

Please approve. Zharov Morozov Savin (most likely code names). [551]

In essence, what this long-worded report states is that

- The 6th TA and 27th A of the 2 UF had broken through, further to the south, and were about to turn north and enter the Transylvanian Alps, outflanking the Axis armies blocking the advance of the 40th A and the 7th Guards Army through the mountains. Thus, the 6th TA and 27th A would eliminate the need for the 40th A and the 7th Guards Armies to make costly frontal attacks through narrow mountain passes.
- Once this maneuver had taken place, the combined forces would then advance to the northwest, entering the Transylvanian Plateau and assist in the advance of the 4 UF. Along the way, they hoped to encircle and to destroy the Germans and Hungarian armies defending the Carpathians.
- The ultimate goal of this operation was to break through the formidable Carpathians quickly and to reach the Hungarian border and the Hungarian plains.

Note: This report is full of place names, small towns or villages, that are extremely difficult to locate on maps, even period maps. Many of these locations also have changed names and/or spelling over the last 70 plus years. Despite this issue, identifiable and notable locations and critical towns will be accurately displayed on the maps illustrating this work.

At this time, Axis forces defending within the Carpathians consisted of the German 8th, the remains of the 6th Army, and the Hungarian 1st and 2nd Armies.

The 7th Guards Army issued an internal report by Guards Major Lyamov and Guards Colonel Ryzhov, *Information on the elimination of the encircled enemy grouping by the 7th Guards Army*. This is found within TsAMO ID #131310355.

> On August 31, in the forests and settlements east of Onesti and the Ajud Noua northeast, the remnants of the Chisinau grouping of enemy troops emerging from the encirclement with a total number of up to 8,000 people were concentrated under the command of the commander of the 370th RD, Major General Graf von Hulsen [Lieutenant General Botho Graf von Hülsen]. The grouping included the remnants of the defeated 70, 79, 106, 153, 258, 282, 294, 302, 370, and 376 RD, 10 MD, 236 assault gun brigade and various rear and special units of the 6th and 8th army [All were Germans].
>
> By the actions of the troops of our army, the indicated Hulsen group was again surrounded and eliminated in the battles from 1 to 4.9.44 [1-4 September 1944].
>
> From 31.8 to 4.9.44, [31 August – 4 September 1944] 3,872 prisoners were captured, [one was] major general Graf Hulsen [Count von Hülsen], [who] was the commander of 370 RD, one colonel was the commander of the 370 artillery regiment, [another officer was] Doctor Zaidel [a medical doctor], [another one was] lieutenant colonel Shirscher, who was the chief of staff 370 RD and 102 other officers up to major were included. [In all,] 4,112 enemy soldiers and officers were killed.

Trophies captured:

- Rifles – 1,088
- Machine guns – 13
- Submachine guns – 15
- Pistols – 185
- Horses – 184
- Carts – 102
- Oxen – 8
- Saddles – 14
- Motor vehicles – 3
- Motorcycles – 2

The total losses of the enemy killed and captured were 7,984 people.

This report was written from the viewpoint of the 7th Guards Army. As there were other Red Army units involved in the fighting, they too would have killed and/or captured their share of German troops and gathered trophies.

5 September 1944

The 6 RD finally captured the town of Gymes with help from the 72 GRD on its left flank. Despite this small victory, the Germans still were blocking any movement through the Gymes Pass. To the south, units of the 25 GRC, 53 RD, 36 GRD and 6 GAD had almost surrounded the town of Bereck, with the German defenders blocking the Russian forces from advancing into the river valley to their rear.

On this date, the Red Army finally overcame the final small remnants of the surrounded troops of the German 6th Army, southeast of Kishinev. Once again during the war in the East, the German 6th Army had been surrounded and destroyed by the advancing Red Army. But their stout defense occupied up to five Russian armies, which allowed time for many other German units to fight their way out of the pocket to withdraw south and west to reach safety. Overall, four German corps (30, 52, 44 and 7) were destroyed along with their fifteen infantry divisions (9, 15, 62, 106, 161, 257, 258, 282, 294, 306, 320, 335, 370, 376, 384). Some combat elements of the 76, 79, 358, and 370 ID were able to fight their way to the west and to reach the Carpathian Mountains. Of special note, the victorious Soviet forces also took some 130,000 Romanian troops as prisoners, despite King Michael I announcing a cease-fire and an armistice with Russia and the Allies. These troops were sent to POW camps along with captured Germans, where many of the Romanians died in captivity. Only Romanian units that actually fought against the Germans were spared captivity. [552]

The official *Stavka* reply to the 2 UF report #476 of 4 September was issued as Directive #220207, which corrected the plan for overcoming the Transylvania Alps.

Directive of the Stavka of the VGK #220207 to the Representative of the Stavka,
to the Commander of the 2nd Ukrainian Front
on Refining the Overcoming Operation Plan Transylvanian Alps and Carpathians.
September 5, 1944 18 h 20 min

The headquarters of the Supreme High Command orders the plan of operation submitted by you 4 September No. 4762 to be corrected, guided by the following instructions: [This directive makes minor changes to the 2 UF plan. The pertinent section related to the 7th Guards Army is listed below.]

1. The main task of the troops of the 2nd Ukrainian Front is to strike from the south through Brasov and Sibiu (in the direction of Cluj) and the offensive of the 40th and 7th Guards Armies from the east to overcome the Transylvanian Alps, the southern part of the Carpathian ridge and [to] *reach the front of Satu Mare, Cluj, Deva, Turnu Severin. Entering the Satu Mare area to help the 4th Ukrainian Front cross the Carpathians and* [to] *capture the region of Uzhgorod, Chop, Mukachevo. In the future, develop the offensive with the aim of reaching the main forces on the river Yew in the Nyiregyhaza, Szeged sector, with cover on the Szeged, Turnu Severin front. Please approve.*

Once again, this Directive states that forces of the 2 UF will attack to the northwest, and support the 40th A, 7th Guards Army and 4 UF through the mountains, and advance together toward the Hungarian border.

At the same time but further south, Bulgaria, an Axis partner, had been trying to leave the Axis alliance for several weeks, but they did not want to suffer the wrath of Adolf Hitler for doing so. After much political maneuvering on the part of Bulgaria, the Soviet Union ended the debate by declaring war on them, which prompted Bulgaria to immediately surrender. The Soviets soon would install a puppet Communist government in Bulgaria.

A combined German and Hungarian force of five divisions struck back at the advancing Romanian 4th A in the Cluj-Turda area (55 miles west of the current position of the 7th GA). This surprise attack forced the Romanians back and over the Mures River, which obliged Malinovsky to send the 6th TA to assist in its defense.

6 September 1944

The 7th Guards Army's 24 GRC halted any major attacks, mostly due to exhaustion and/or the lack of supplies. To the south, the 25 GRC kept the pressure on the German defenders.

German OKH intelligence maps from US National Archives, T-78 Roll 136, image #468-469, shows the Germans lost track of the 7th Guards Army and other Red Army units. This might have been due to the confused fighting in the mountains. The German 6th and 8th Armies were also in an exposed salient position trying to hold the Carpathian passes and would soon be forced to withdraw to escape being surrounded.

7 September 1944

Units of the 24 GRC remained in their positions, but units of the 25 GRC reached the valley floor and were fighting just outside of Bereck.

At this point in the battle for the Carpathian passes, General Friessner knew the remnants of the German 6th Army, which were the most exposed, were being outflanked from the south, so he gave the order for the army to begin withdrawing during the night to the northwest. [553]

The new Romanian government had reorganized their military forces and placed their 1st Ru and 4th Ru Armies under Soviet control. As of 7 September, they both were assigned to the 2 UF where Malinovsky put them to use clearing the Germans and the hated Hungarian troops from Romanian lands. The 1st Ru A was subordinate to the 53rd A, and the 4th Ru A was subordinate to the 27th A. [554]

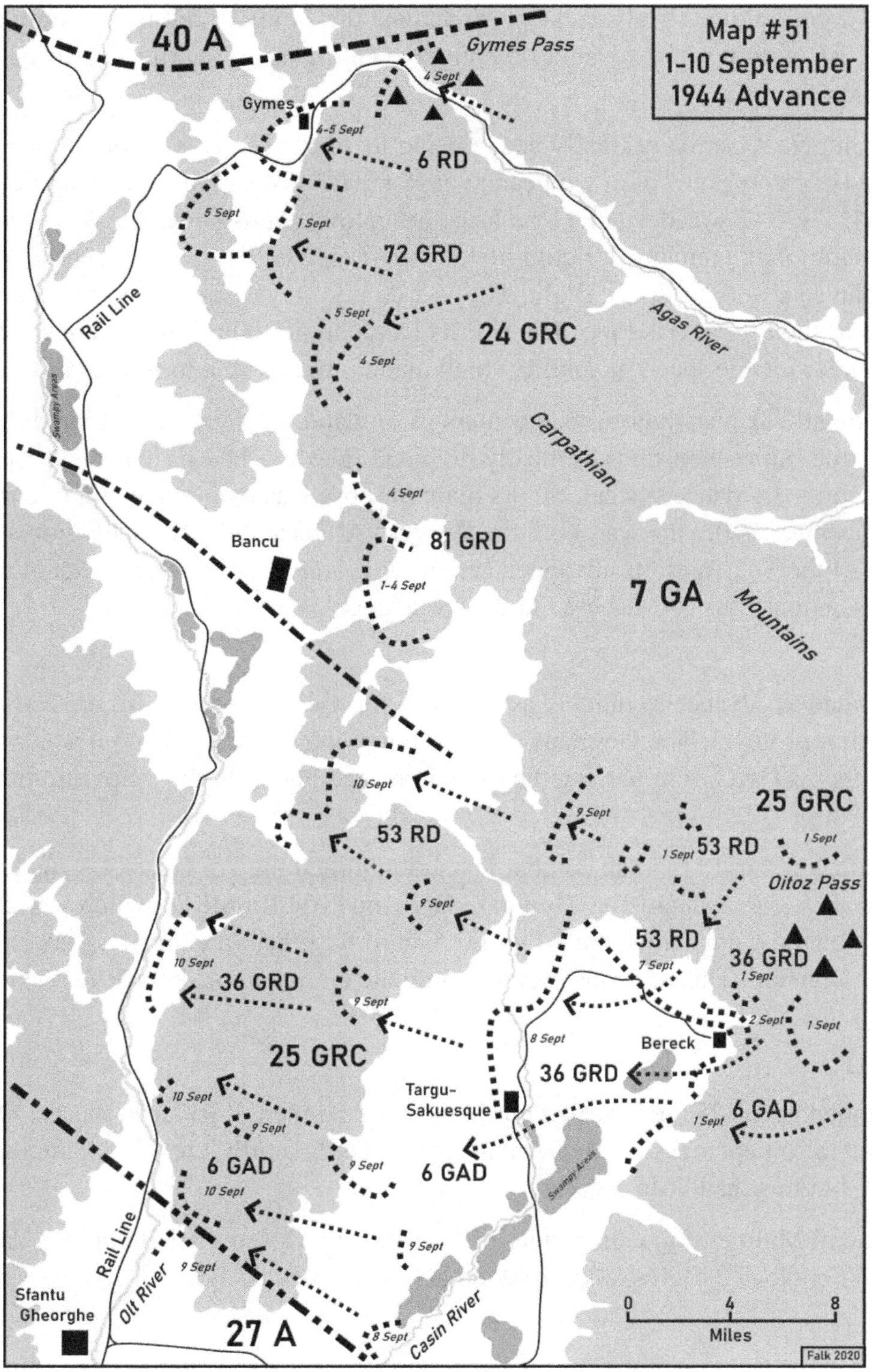

The Finnish peace delegation arrived in Moscow. After waiting days, the Soviet armistice terms were presented to them on 14 September with the Finnish cabinet finally ratified the agreement on 19 September 1944. Events in Finland would unfold quickly in the coming weeks as all sides undertook independent action.

8 September 1944

Sections of the German 6th Army continued to withdraw during the night. Hitler had forbidden any further withdraws but was forced to accept it due to the combat situation. General Friessner was determined to save his forces to rebuild a new shorter defense line further to the west.

German forces still were holding firmly in the north against the 24 GRC, but in the south they started to withdraw. The 25 GRC was quick to advance, keeping the Germans on the run, capturing Targu-Sakuesque along the way.

On the Western Front, Nazi Germany finally began using its fourth secret weapon, in mass, against the Allies. The V-2, or *Vergeltungswaffe-2* (Vengeance or Retaliatory Weapon), had been developed under the designation A-4. The V-2 was designed by a large group of scientists lead by Wernher von Braun and was the first operational, unmanned, liquid fueled long-range ballistic missile and the first human-made object to achieve sub-orbital space flight. The V-2 could carry a one-ton warhead and was known by British civilians as the Flying Gas Pipe. The V-2 had a range of 200 miles and arrived at its target at 1,790 miles per hour, twice the speed of sound, which made it impossible to intercept or to shoot down.

The V-2 was used mostly against major urban centers in England, Belgium, France and the Netherlands where it would kill and injure thousands of mostly civilians. The V-2 had little military application due to the lack of an accurate guidance system. So, its main purpose was to damage and to terrorize large civilian population centers. After the war, both the Western Allies and the Soviet Union studied captured examples of the V-2. With its advanced technology, and help from hundreds of captured or defecting German scientists, the V-2 ushered in the space age.

9 September 1944

The 25 GRC continued to advance as quickly as possible, but German rear guard units always stood in their way to slow their progress. The Germans were withdrawing to the north, so it was critical for them to prevent units of the 24 GRC from leaving the mountains and blocking the roads and rail line along the Olt River.

On or about this date, another Romanian combat formation joined the 7th Guards Army and the 24 GRC. This was the *ad hoc* Romanian 103 Mountain Division (103 Ru Mt D) which was comprised of the 103 and 104 Mountain Commands, the 8 Frontier Guard Regiment and artillery taken from the 6 Romanian Infantry Division. This unit was used to eliminate German pockets in the rear of the 24 GRC. [555]

10 September 1944

The advance continued as the 25 GRC turned north and reached the Olt River in places. On the Army's left flank, the 27th A also kept pressure on German forces. To the north, the 40th A attempted to cut off the German retreat, but they had little success.

Stalin recalled General Malinovsky, commander of the 2 UF, to Moscow where on this day he promoted Rodion Yakovlevich Malinovsky to Marshal of the Soviet Union.

11 September 1944

By 11 September the rifle divisions of the 7th Guards Army were aligned mostly along the Olt River. With the Axis forces withdrawing, the Army was poised for a fast advance to the Mures River and to the city of Targu-Mures, their next major objective, some 80 miles away. The only thing that could slow them down was the lack of supplies, the mountainous terrain, terrible roads, the weather (currently 61-78° F, dry ground with 7 hours of sunshine), utter exhaustion, German demolition teams blowing up vital rail lines, bridges, structures, and equipment along the way, along with determined rearguard troops meant to keep weary footslogging Red Army soldiers on their guard. Other than these few obstacles, it should have been an easy advance.

General Friessner, Commander of AG South Ukraine, expanded upon these issues, but from the German point of view.

> On the narrow mountain roads of the Carpathians, traffic jams were often formed, and this happened in conditions when the enemy applied pressure from both the front and the flanks. The rearguard groups were solving extremely difficult tasks these days, providing from time to time, sometimes by purely improvised methods, the movement of unwieldy and cumbersome columns. Among all the means and methods of maintaining continuous movement, in addition to personal authority, three means turned out to be the most effective: a Fieseler-Storch [Fi 156 liaison aircraft], a Volkswagen car, and an army tractor. [556]

Further northwest, the 6th TA was now in position to launch a counterattack to support the 4th Ru A and repel the Axis attack in the Cluj area. Fighting in this area actually was outflanking Axis forces that were blocking the advance of the 27th A and 7th Guards Army.

12 September 1944

The 7th Guards Army began its final advance through the Carpathian Mountains in pursuit of the withdrawing Axis forces. With winter approaching, the weather would become critical, but the weather was still holding with short but temperate days. The greatest obstacle, of course, was the mountainous terrain. (See Map #52.)

With the Gymes Pass behind them, the 24 GRC took the northern valley route via the towns of Sandominic and Suseni to the Mures River area. The 72 and 81 GRDs started by trying to advance due west, but the terrain was so difficult the 72 GRD diverted north and the 81 GRD diverted south into the 25 GRC zone, which caused some congestion. The shifting of the 72 GRD to the north pushed the 6 RD further north into the 40th A zone. This was necessary because only so many units/formations could advance in an organized manner along narrow mountain valleys. The trailing regiments of the 72 GRD especially were delayed by these needed maneuvers, lagging some four days behind the leading elements of the division.

With the Oitoz Pass behind them, the 25 GRC took the southern valley route west of the town of Bancu. The 81 GRD followed behind the 53 RD, which delayed the 81 GRD by at least a full day. Due to the terrain, the 36 GRD and the 6 GAD were sent in a roundabout route far to the south advancing along the 27th A boundary line. Despite taking this longer route, they were the leading units of the 25 GRC and soon approached the Balta River area.

The 6th TA succeeded in its counterattack in the Cluj area, and *Stavka* awarded them Guards Status the same day.

Romania signed an armistice with the Allies on this date, thus officially ending hostilities between them.

General Fyodor Tolbukhin, commander of the 3 UF, also was promoted to Marshal of the Soviet Union. Marshal Malinovsky stated that General Tolbukhin was promoted two days after him because, "In this way Stalin underscored [Malinovsky's] seniority over Tolbukhin in his marshal's rank." This remark highlights the behind-the-scenes competition among senior officers of the Red Army for rank, glory and prestige.

Allied armies on the Western Front continue their advance as British forces reached the Netherlands and American troops crossed the German border and pushed forward five miles into Germany.

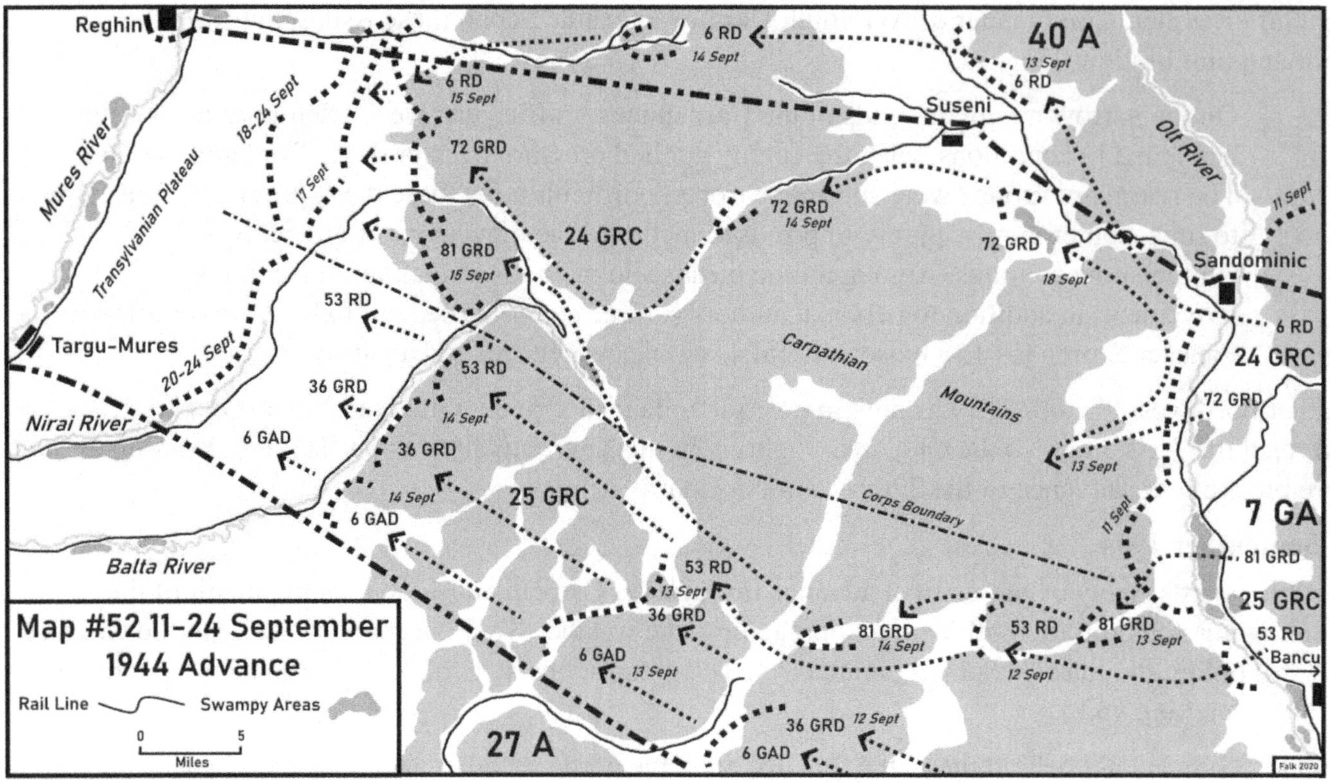

13 September 1944

The 7th Guards Army's advance continued with little or no fighting. The 24 GRC headed toward the city of Reghin and the Mures River. Likewise, the 25 GRC also headed toward the Mures River and the city of Targu-Mures.

The war between Finland and the Soviet Union would be over soon, but as part of the agreement, Finland still needed to remove the German army from Finnish soil. In this, Hitler acted first, withdrawing civilians and staff from Finland as preparations for Operation *Birke*.

"All Germans in southern Finland, including diplomatic and military staff, had left Finland by September 13." [557] This act fulfilled, in part, Finland's obligation to the Soviets to remove German forces from the country. But German operations further north, Operation *Birke,* eventually would lead to open combat among the three combatants. (There was no Finnish/Soviet Alliance. They all were combatants with slightly different points of view.)

14 September 1944

The 7th Guards Army was making good progress. Units of the 25 GRC, which were leading the advance, reached the Balta River and the final great barrier of the Carpathians. The leading elements of the 24 GRC lagged a day behind and reached the western border of the mountains the next day. Below them lay the rolling hills of the Transylvanian Plateau, averaging only 1,000-1,600 feet above sea level.

15 September 1944 – Weather: 62° to 50° F, (17.2° to 10.0° C), dry ground. As the weather began to turn, the 7th Guards Army marched once again into winter.

The six rifle divisions of the 7th Guards Army now were aligned along the western edge of the Carpathians with the much lower Transylvanian Plateau stretched out before them. Reconnaissance efforts discovered that German formations and defensive positions were just a few miles away, aligned roughly along the Nirai River line.

Directive #220215 ordered the 40th A and 7th Guards Army to help with the 4 UF attack by 30 Sept.

Directive of the VGK Stavka № 220215 to the commander of the 2nd Ukrainian Front, the representative of the Stavka on the continuation of the offensive and the exit with the main forces. September 15, 1944 01 h 50 min

The headquarters of the Supreme Command orders:

1. The troops of the front, upon reaching the line of Kirlibaba, Bistrita, Cluj, Lugozh, on 15-19.09, [15-19 September] *continue their offensive in the general direction of Cluj, Debrecen* [in Hungary], *Miskolts with the aim of reaching the main forces of the front on the river Tisa* [in Hungary], *Szolnok sector and help the 4th Ukrainian Front cross the Carpathians and capture the Uzhgorod area* [in far northern Romania]. *The immediate task of the front troops is to capture the border* [with Hungary] *no later than 30.09:44)* [30 September 1944] *40th Army - Siget, Satu Mare, Karei. Gorshkov's horse-mechanized group to advance in the direction of Bistrita, Satu Mare, b) 7th Guards Army* [should remain in the]- *Carey, Oradea Mare* [region on the Hungarian border]. *6th Guards TA to advance in the direction of Cluj, Carey, Nyiregyhaza;*

This Directive basically states that:

- The right flank forces of the 2 UF should advance northwest, clearing the Transylvanian Plateau and reaching the Hungarian border by 30 September. The 40th A should move further north to assist the 4 UF in crossing the Carpathians; the 7th Guards Army would remain deployed to the left of the 40th A with the horse-mechanized group, with the 6th GTA moving independently between them. The goal was to reach the Hungarian Plain as quickly as possible.

The defending German 8th Army and the 1st Hungarian Army attempted to block this movement. Not noted in this directive was that other armies of the 2 UF were advancing in the same general direction but further to the south. *Stavka* also failed to comment on the strung-out nature of these armies and the inadequate logistical support. Nevertheless, *Stavka* ordered these objectives met even in the face of growing German and Hungarian strength.

A Hungarian M 41 Turan tank loaded with troops pulls a disabled truck in 1944.

16 September 1944 – Weather 60° to 48° F, (15.5° to 8.8° C), dry ground

On or about this date, units of the 7th Guards Army begin their attack/advance out of the Carpathians and onto the Transylvanian Plateau, but progress was slow as the Army advanced only about ten miles over the next nine days. German defense efforts had something to do with this sluggish advance, but almost certainly logistical issues and the need to allow units of the Army to regroup and to rest were the main reasons for the slow progress.

For the 7th Guards Army, the first part of the Iasi-Kishinev offensive started on 21 August with the attack against Targu-Frumos, which then continued for another three days until 24 August when the Army approached the foothills of the Carpathian Mountains. Immediately thereafter, the second part of the offensive began which consisted of the successful advance through the Carpathians, which lasted for an additional 23 days of almost continuous combat, from 25 August to 16 September. So, at this point, the 7th Guards Army had been engaged for a total of 27 days of unceasing effort, and the personnel were tired, disorganized, and weakened by losses of both troops and equipment.

17 September 1944

The 7th Guards Army continued its slow advance toward the Nirai River with units of the 24 GRC leading the way. German forces resisted, but they also fell back to prepared positions along the higher ground, along the western side of the Nirai River line.

18 September 1944

The 7th Guards Army continued its advance to the west with most of the fighting taking place on the right flank, in the zone of the 24 GRC.

19 September 1944 – Weather: 64° to 47° F, (17.7° to 8.3° C), dry ground

At this point, the attack by the 24 GRC stalled due to tired troops and the stout defensive efforts by the German 46 ID. In turn, the German 4 Mountain Division (4 Mt D) blocked the quick advance of the 25 GRC on the left flank. Both these German divisions, under the command of the German 8th Army, had vast experience fighting on the Eastern Front. Despite being much reduced in strength; they were able to hold off the relatively weak attacks by the 7th Guards Army.

On this date, Finland's armistice with the Soviet Union was ratified, bringing an end to the war between these two combatants. Germany had lost its northernmost ally. With the end of fighting between Finland and the Soviet Union in the far north, Russia now would be able to transfer redundant forces south to further increase the military pressure that was aimed at the heart of Germany. [558]

20 September 1944 – Weather: 56° to 48° F, (13.3° to 8.8° C), dry ground

Units of the 25 GRC, on the left flank, reached and crossed the Nirai River but could not advance further. The total advance for the Army so far was approximately ten miles. German forces now were defending along the western hills that were 200-300 feet higher than the Nirai River valley, which gave them a distinct advantage.

Guards Major of the medical service Lev issued a medical report, *Information about the movement of the wounded and sick in medical institutions of the 7th Guards Army*, found in TsAMO Archive ID #133391610. This insightful report covers the period from 15 September to 20 September 1944, and is an assessment of the current medical burden upon the Army's medical establishment.

The total medical burden on 20 September was 6,101 people.

Forward Troop Area – up to divisional medical battalions

Arrivals:

At the start (15 September) – 897 men and women

Newly arrived – 1,320

Total – 2,217

Departures:

Evacuated – 886 (most likely to the Army rear) (moderately wounded)

Returned to active service – 447

Died – 36

Total – 1,369

In the Forward Troop Area on 20 September – 848 (lightly wounded)

In Army Hospitals

Arrivals:

At the start (15 September) – 4,748 men and women

Arrived from forward units' medical installations – 1,103

Arrived directly from units – 298 (severely wounded)

Totals – 6,149

Departures:

Evacuated – 429 (out of the Army) (severely wounded)

Returned to active service – 467

Died – 25

Total – 896 (not including the dead)

In Army Hospitals 20 September – 5,253 (moderately wounded)

Total in all medical installations on 20 September – 6,101.

The Army's hospitals contained approximately 5,350 hospital beds for all types of casualties, so the medical units were somewhat overburdened (excluding divisional level medical support).

21 September 1944 – Weather: 48° to 41° F, (5.0° to 19.4° C), dry ground

Overall, units of the 7th Guards Army maintained their current positions while launching probing attacks at a few locations along the line. The entrenched German defenders, on the heights west of the Nirai River line, repelled all these reconnaissance-in-force attacks and drove the tired Russian infantry back.

22 September 1944

Units of the 7th Guards Army continued their attacks but made little progress. Shumilov was looking for weaknesses in the German line, but in truth, his depleted, exhausted and unsupported infantrymen were not up to the task.

23 September 1944

Once again, the 7th Guards Army mostly maintained its position with scattered fighting along the line.

24 September 1944 – Weather: 67° to 45° F, (19.8° to 7.6° C), dry ground

The 7th Guards Army continued with small probing attacks but was unable or unwilling to move forward. This most recent advance of the 7th Guards Army had brought them out of the rugged Carpathian Mountains and onto the Transylvanian Plateau (also called the Carpathian Plateau). This area varies between 1,000 to 1,600 feet in elevation but consists of more rolling hills with fewer high peaks than the Carpathians. Small towns and farmlands dotted the landscape with fast running rivers and streams flowing mostly from northeast to southwest. Winter was coming to this area of Romania, and it was best to have the high mountain passes and valleys behind them.

The German AG South Ukraine was renamed to AG South, which remained under the command of General Friessner and which now contained the 2nd and 3rd Hungarian Armies (2nd Hu A, 3rd Hu A) along with the German 6th and 8th Armies.

25 September 1944

From 25 to 27 September, the 7th Guards Army suspended all attacks. The rifle divisions rested and regrouped as needed. Moving critical artillery forward, and amassing ammunition, food and other supplies closer to the front line was undoubtedly needed at this point. (See Map #53.) Previously, due to its ad hock makeup, the 103 Ru MtD had been only used to hunt for stragglers, clean up the battlefield, and secure rear areas, a secondary but vital task. But now, the 103 Ru MtD was ordered to occupy a section of the front line within the 24 GRC zone.

26 September 1944

The German 46 ID and the 4 Mt D (8th Army) and an assortment of improvised units had stopped the 7th Guards Army from freely advancing without a prepared battle plan and a deliberate attack. General Shumilov went to work to correct this requirement.

27 September 1944

The 7th Guards Army maintained its position. The 227 RD remained deployed to the rear and would stay there for the remainder of the month.

28 September 1944

After resting his troops for three days and preparing a new battle plan, General Shumilov launched his army in a general attack all along the line, using all the rifle divisions of both the 24 and 25 GRCs. By the end of the day the entire Army's front had advanced over eight miles and reached the Mures River. The 6 GAD, 25 GRC, achieved the most success with the capture of Targu-Mures, a long-sought objective and a key transportation hub. Both the 6 GAD and the 81 GRD had some troops over the Mures River, but just barely.

In the north of Finland, the first open clash between German and Finnish forces took place, which led to further fighting on 31 October. [559]

29 September 1944 – Weather: 59° to 46° F, (15° to 7.7° C), wet ground

The 7th Guards Army attack continued for a second day. All divisions were able to cross the Mures River and form bridgeheads of various sizes on the far shore despite German units offering a tenacious defense.

30 September 1944 – Weather: 58° to 56° F, (14.6° to 13.6° C), wet ground

The attack continued for a third day with the 6 RD, 72 GRD, 81 GRD and the 53 RD forming a large and continuous bridgehead on the western shore. The Axis troops in the city of Reghin still held out, but it was only a matter of time before it fell to units of either the 7th Guards Army or the 40th A as Reghin was in the 40th As zone. General Shumilov and Marshal Malinovsky must have been pleased with the Army's success over the previous few days.

Attacks by the 7th Guards Army continued into early October.

By late 1944, the war in the air had decidedly turned against Germany. For the previous few months, the Luftwaffe had been using its last reserve stocks of fuel to engage the enemy, in full force, to the east, west and south, but now they mostly had consumed these reserves. By the end of September only about 180,000 tons of high-octane fuel remained. With fuel production not keeping up with consumption, the Germans had to make drastic cuts in air operations. "Only day fighter operations in defense of the Reich were allowed to continue at their previous level." All other areas were scaled back, which meant, "air operations in support of the Army were permitted only in decisive situations," [560] which signified that the VVS would be free to dominate the air along the Eastern Front. Air dominance by the VVS would become evident in October 1944 when the 5th Air Army, which was supporting the 2 UF and the 7th Guards Army, was able to fly over 8,400 combat sorties.[561] Luftwaffe sorties rates for October are unknown, but they would have been only a fraction of what the Soviets could generate. The Red Army and the Red Army Air Force would continue their advance on all fronts into October 1944.

The 2 UF issued a report, *Report on the loss of personnel in the troops of the 2nd UkrF from September 1 to October 1, 1944.* This was found within TsAMO record ID #113723728. The 7th Guards Army suffered substantial losses for September. The reported total was 12,462 casualties for the month: 2,412 killed; 225 missing; 1 captured; 16 non-combat losses; 7,949 wounded; and 1,859 sick. The report goes on to list 7th Guards Army horse losses: 438 killed, 1 missing, 104 wounded and 24 evacuated for veterinarian care, for a total of 567 equine casualties. The same report shows that the troops of the entire 2 UF suffered 59,892 total human casualties during September 1944.

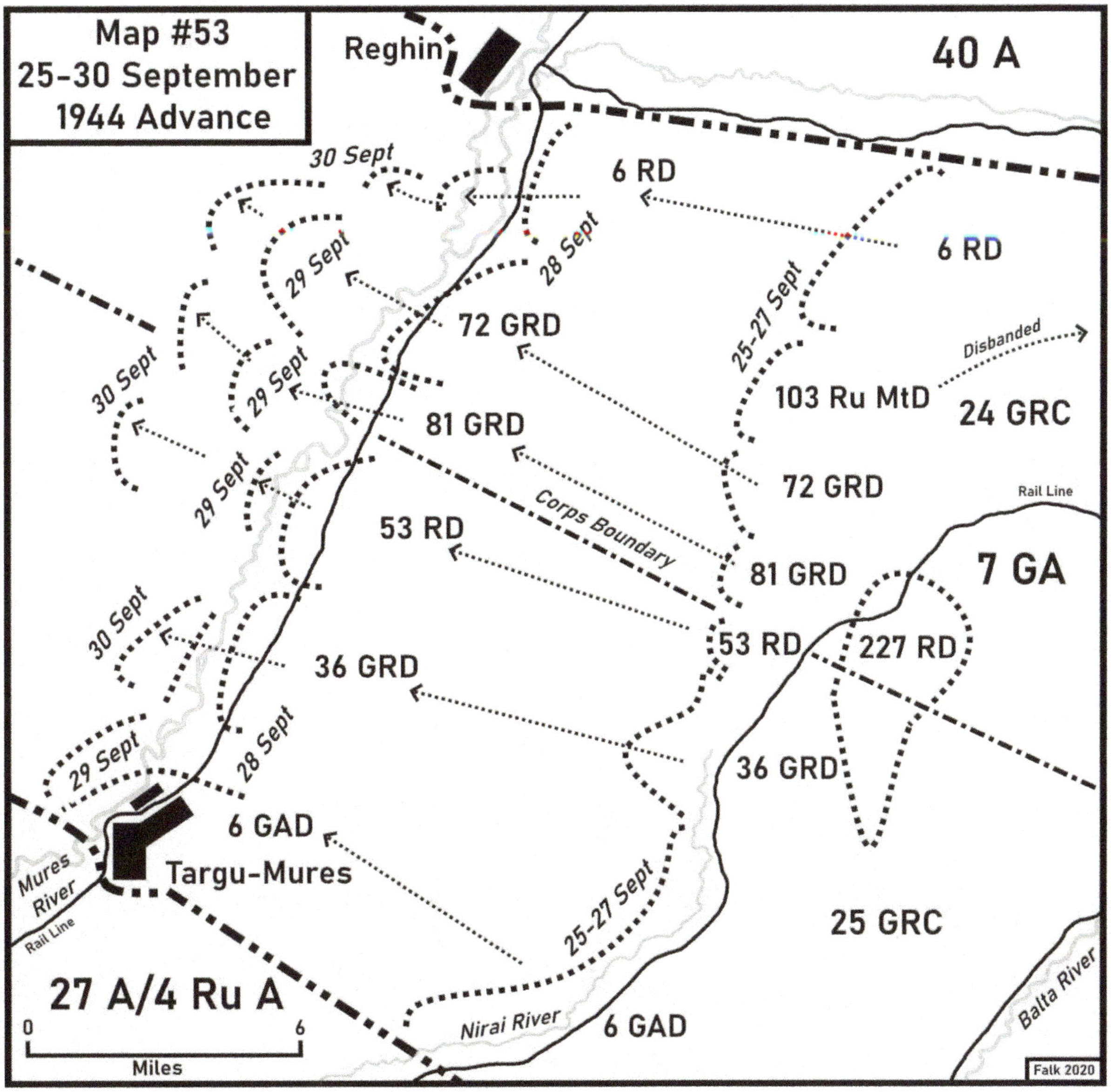

At the end of September, the Soviet command disbanded several *ad hoc* Romanian combat units (like the 103 Ru MtD) that had assisted in conquering the Carpathians. The 1st Ru A and 4th Ru A did remain in the field fighting alongside Russian armies as they advanced toward neighboring Hungary.

Sometime during late September, on the Western Front, Nazi Germany began using its fifth secret weapon against the Allies. This time the weapon was the Arado Ar 234 *Blitz* or Lighting aircraft. This was the world's first operational jet reconnaissance and bomber aircraft. Once more, production and fuel availability problems prevented this aircraft from reaching its full potential. The Ar 234 did operate effectively as a reconnaissance aircraft, which once again gave the Luftwaffe much needed eyes in the sky. But this revolutionary aircraft came too late to change the outcome of the war.

For the month of September in Britain, German V-1 and V-2 attacks killed only 190 people and injured 360. This reduction was due to the Allied advance in the west, which disrupted vengeance attack operations.

1 October 1944	
7th Guards Army	
Assigned – 2 Ukrainian Front	
Colonel General M.S. Shumilov Commanding	
Rifle, Airborne and Cavalry Units	24 Guards Rifle Corps (72, 81 GRDs, 6 RD) 25 Guards Rifle Corps (36 GRD, 6 GAD, 53 RD) 227 RD 15 Penal Battalion 62, 63, 64, 65, 66,194 Penal Companies
Artillery Units	114, 115 Guards Tank Destroyer Regiments 41 Guards Gun Artillery Brigade 263, 290, 493 Mortar Regiments 5 Antiaircraft Artillery Division (670, 743, 1119, 1181 AA Regiments) 162 Guards AA Regiment 662 Separate Aerial Observation/Air Raid Warning Company
Armored and Mechanized Units	38 Separate Armored Train Battalion
Engineer Units	60 Engineer Sapper Brigade 4 Separate Flamethrower Battalion
Reported Combat Strength Estimated Ration Strength	45,624 men 52,624 men and women (including an estimated 7,000 support personnel)

(BSSA 44, BSA 44, TsAMO Archive)

1 October 1944 – Weather: 64° to 47° F, (17.7° to 8.3° C), dry ground

The 7th Guards Army continued its attacks over the Mures River with minor gains of ground mostly on the right flank. Note: The Army had more than 9,000 fewer combat troops from the beginning of the previous month. The freely conscripted civilian "Booty Troops" were not available, as the Army no longer was operating on Soviet territory.

The 2 UF shifted the 27th A's operating zone further to the left, which allowed the 4th Ru A to move to the front and to deploy on the left flank of the 7th Guards Army. Additionally, at some point in October, the 2 UF transferred the 1st Ru A from the 2 UF to the 3 UF.

The political landscape in the Balkans was about to change once again. The Hungarian Regent, Admiral Miklós Horthy, sent a secret armistice delegation to Moscow. He was looking for a way out of the war and to safely break with Hitler. Horthy hoped to make a deal with Stalin to save his country from the destruction that an extended battle for Hungary would cause. [562]

2 October 1944

Once again, the 7th Guards Army continued its attacks. The most notable accomplishment was that the 6 RD, with the help of units of the 40th A, captured the city of Reghin.

The 7th Guards Army issued a report, *Brief Description of Army Highways*, TsAMO record ID #133391623. Among other things, this report listed three separate supply routes —33, 22 and 26 miles long—totaling 81 miles within the zone of the Army. These routes reached back to 2 UF supply points where road and rail transportation would allow the creation of supply dumps, warehouses or depots. From there, it would be up to the 7th Guards Army service units to transport supplies to individual units at the front. Ensuring that a steady stream of munitions and supplies reach the front line area guaranteed the Army's continued success.

Army and/or front engineer units were charged to maintain the gravel roads and bridges and to remove anything that would hinder the movement of vital supplies to the front. While the 7th Guards Army had only one engineer sapper brigade assigned, the 2 UF controlled three engineer brigades, two pontoon bridge brigades, one pontoon bridge regiment, and two other engineer battalions. These front controlled engineer assets were deployed as needed throughout the 2 UF's zone.

In Poland, the Warsaw Uprising was over as the final small groups of Polish fighters were ordered by their commanders to surrender to the Germans. The wholesale slaughter of thousands of Polish civilians and the utter destruction of Warsaw itself now began. Hitler ordered the entire population to be killed or deported to labor or concentration camps, and the city was to be razed to the ground. In no way were these actions a military necessity. These acts were driven by Hitler's need for revenge against the Polish people.

3 October 1944

The 7th Guards Army launched more attacks, but the defending German units were not giving up much ground. The German 46 ID and 4 Mt D were yielding slowly, preventing any breakthroughs, which certainly frustrated General Shumilov greatly.

On 3 October, Malinovsky's 2 UF with support from friendly Romanian forces completed the Bucharest-Arad Operation, which cleared Romania of all Axis troops.

4 October 1944 – Weather: 50° to 45° F, (10° to 7.2° C), wet ground

The attack over the Mures River was dying down. The 7th Guards Army was unable to overcome the German and any scattered Hungarian units that were dug in on the high ground. The 227 RD remained in reserve but was gradually moved closer to the front.

On both flanks, Russian armies were now mostly in line with the position held by the 7th Guards Army. On the left flank, the 4th Ru A moved up and was aligned along the lower Mures River. To the north, on the right flank, leading units of the 40th A finally arrived in the Mures River area. The bulk of the 40th A took several more days to fully approach the river.

In Finland, the German OKW ordered a change from Operation *Birke,* the total withdraw of German forces from northern Finland, to Operation *Nordlicht* (Northern Light), a fighting phased withdraw using scorched earth tactics which denied enemy forces (Finnish or Soviet) resources and shelter. Despite the peace treaty with the Soviets, the war on Finnish soil would continue for some time, with the added twist that Finland was now fighting the Germans.

5 October 1944

A partial report about the 7th Guards Army highlights the weakened condition of the Army. The report from TsAMO record ID #132636291 listed three divisions of the Army and their overall combat strength (all fighting troops) and their bayonet strength (combat infantry). (See Table #23.)

Unit	Combat Strength	Bayonet Strength
72 GRD	3,682 men	1,285 men
81 GRD	3,897 men	3,415 men
53 RD	3,534 men	1,067 men

Table #23: Rifle Division Combat Strength vs. Bayonet Strength.

During the late 1943/44 time period, a full-strength Russian rifle division should have contained roughly 7,000-8,000 men in the combat strength category and 5,000-6,000 men in the bayonet strength category (numbers vary depending upon which sources are used), so this report shows that for the 72 GRD and the 53 RD, the bayonet strengths were critically low.

6 October 1944 – Weather: 56° to 40° F, (13.3° to 4.4° C), wet ground

Despite their weakened state, the rifle divisions of the 7th Guards Army continued to launch attacks. Even with large amounts of artillery providing fire support, gains were negligible. The 24 GRC made the furthermost advance with the 6 RD, 72 and 81 GRDs. Flanking support was provided by the 53 RD and part of the 36 GRD from the 25 GRC. German forces would not give way, and the 7th Guards Army was not making any meaningful progress.

The 2 UF began the Debrecen Operation, which was intended to use the left flank armies of the 2 UF to sweep around the Carpathians and to reach and to cross the Tisza River to "capture the trans-Carpathian area." This was on the direct path to Budapest, the capital of Hungary. To make this possible, the 2 UF had been strengthened with 75,000 reinforcements, which attests to the severity of the fighting during the Bucharest-Arad Operation (30 August to 3 October).[563]

In northern Finland, German Operation *Nordlicht* (Northern Lights), which was the total extraction of the remaining German forces from Finland to Norway, began on or about this date. "It involved the phased withdrawal, under pressure, of over 250,000 men and their equipment and supplies during winter in the arctic." [564] Operation Northern Lights ended in January 1945, but minor numbers of German troops remained on Finnish soil until 28 April 1945. At that point, Finland's participation in World War II had truly ended.

> Finland was the only European ally of the German Reich, and the only state adjoining the Soviet Union, that managed to avoid occupation of its territory. This fortunate circumstance, which was of course due partly to the country's geo-strategically peripheral location, but no less to the government's flair for realpolitik and skillful tactics, is the underlying reason why the Finnish nation was the only one of the powers besieged in the Second World War that was able-despite some initial restrictions-to preserve its political culture and system of government. [565]

7 October 1944

For the 2 UF, the battle to free all of Romania from Axis control, to clear the Transylvanian Plateau, and to advance to the border of Hungary was not developing as desired by Stalin, *Stavka* or Malinovsky. The 7th Guards Army was doing its part, but progress was slow. Sometime around this date, Malinovsky was pondering a change of focus for the 2 UF, which in turn meant a change for General Shumilov and his 7th Guards Army.

8 October 1944 – Weather: 71° to 47° F, (21.6° to 8.3° C), wet ground

Marshal Malinovsky's mobile forces, the 6th GTA and the Mechanized Group Pliyev, largely achieved an unexpected success by breaking through Axis defenders and by quickly approaching the Tisza River west of Debrecen. This progress threatened to outflank German positions on the Transylvanian Plateau and to cut their line of supply. This, in turn, forced General Friessner, commander of AG South, to order (against Hitler's protests) the withdraw of his most exposed forces, namely General Wöehler's army group which was composed of the 8th Army and the 1st Hungarian Army. Overnight, the resistance in front of the 7th Guards Army evaporated. As always, German rearguard formations and combat engineers would do their part in delaying the advance of the 7th Guards Army.

9 October 1944

A historical map from the TsAMO archive, record ID #100114477, shows that units of the 7th Guards Army were able to advance to the northwest with little or no opposition.

Also, on or about this date, the 7th Guards Army received transfer orders from Marshal Malinovsky. As a result of this order, the 7th Guards Army produced document #1213 on 9 October entitled, *March table of troops of the 7th Guards Army to enter the area: Teshad, Kanoshda, Beyush,* TsAMO archive, record ID #132336444. Some confusion exists about this document because it was produced on 9 October, but it shows that the 227 RD should have begun its movement on 8 October. The Army was to conduct a forced march from the Targu-Mures area to the Oradea-Marye area. This involved a road march of between 201-307 miles, depending upon the starting position of individual units.

Perhaps Marshal Malinovsky played chess during the war, because his orders to the 7th Guards Army causing it to shift its position from the right flank to the left flank of the 27th A is reminiscent of a chess move call castling. Nevertheless, Malinovsky was moving a powerful piece, a Guards Army, out of the hills and placing it on the lowlands, where he pointed it directly toward Budapest located some 130 miles away.

In far off Moscow, British Prime Minister Winston Churchill met with Stalin; this was Churchill's second visit to Moscow. As the war now was going very well for the Allies, Churchill's main focus during this summit was with "what to do after the war." What political considerations and agreements should be made at that time so there would be less confusion and disagreements between the Allied powers once the fighting stopped?

A decisive meeting took place in Stalin's office and was attended only by four people-- Stalin, Churchill and two interpreters. At one point Churchill said, "I have this naughty document with some ideas of certain people in London." Churchill handed the document to Stalin who looked it over for some time. The British document showed how Central and Eastern European countries could be divided into Russian and Western zones.

> Romania: 90 per cent Russia, 10 percent Others
> Greece: 90 per cent Great Britain (in accord with USA), 10 per cent Russia
> Yugoslavia: 50-50
> Hungary: 50-50
> Bulgaria: 75 per cent Russian, 25 per cent Others

After a while, Stalin picked up a blue pencil and checked the top left-hand corner of the document and then pushed it back to Churchill. In reply Churchill said, "Might it not be thought rather cynical if it seemed that we had disposed of these issues, so fateful to millions of people, in such an offhand manner? Let us burn the paper."

Stalin replied, "No. You keep it."

Churchill left Moscow on 19 October on much better terms with Stalin. [566] Despite their arrogance, Stalin and Churchill had patently divided Europe between their two countries with a little bit of land left over for "Others." The seeds of the Cold War were taking root.

10 October 1944 – Weather: 63° to 53° F, (17.2° to 11.6° C), wet ground

The bulk of the 7th Guards Army now began, in earnest, their forced march from the region of Targu-Mures to the vicinity of Oradea-Marye in northwestern Romania. Two rifle divisions, the 6 RD and the 6 GAD, were not included on the march table for unknown reasons but possibly because they were bringing up the rear. With this transfer already taking place, the 4th Ru A and the 40th A advanced toward each other and took over the 7th Guards Army position on the front line. (See Map #54, October Transfer Plan Phase One.)

One definition of a forced march is "[a] movement on foot, by soldiers or military prisoners, who must, in order to satisfy a military requirement, travel at a speed or in adverse conditions that would normally tire them excessively." [567]

Forced marches have long been part of military maneuvers worldwide, such as in the United States where Lieutenant General Thomas 'Stonewall' Jackson in 1872 remarked, "The hardships of forced marches are more often painful than the dangers of battle…I would rather lose one man marching than five in fighting." [568] In 1885, Ulysses S. Grant, the Commanding General of the United States Army during the American Civil War (1861-1865), added his own comment about forced marches: "I did not believe this possible because of the distance and the condition of the roads, which was bad; besides, troops after a forced march of twenty miles are not in a good condition for fighting the moment they get through." [569]

The march table indicated that the rifle divisions of the 7th Guards Army were expected to march anywhere between 26-31 miles per day. Units like artillery, which were equipped with trucks or horses, were expected to cover 90-100 miles per day. The combined Army faced a long hard road to their new deployment positions.

The German 1, 23 and 13 Pz Ds launched a counterattack against advancing Soviet units south of Debrecen and managed to briefly surround three Soviet tank and mechanized corps, which ultimately resulted in severe Soviet troop and equipment losses. (This Debrecen battle is not shown on maps.)

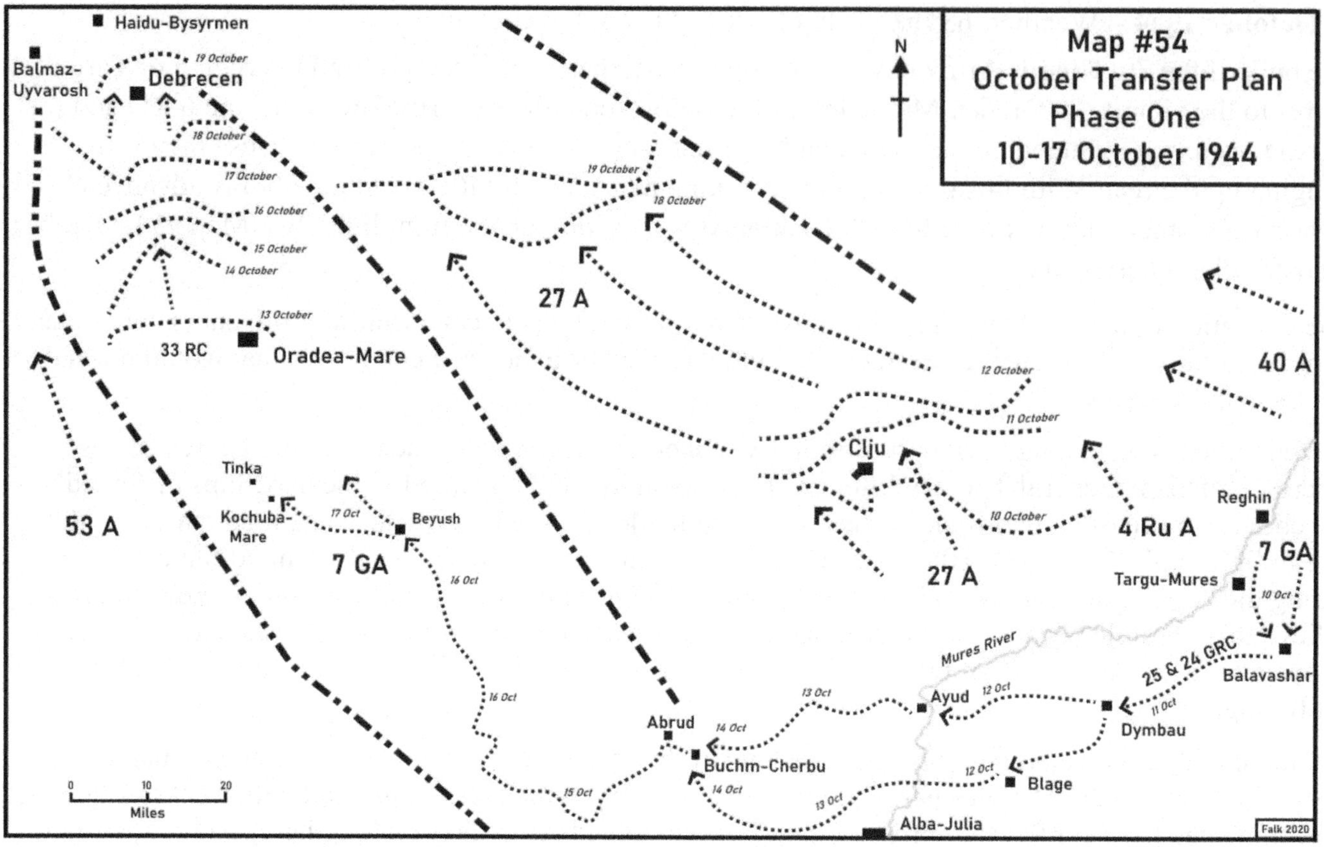

Guards Major of the medical service Lev issued a medical report, *Information about the movement of the wounded and sick in medical institutions of the 7th Guards Army*, found in TsAMO Archive ID #133391270. This insightful report covers the period from 5 - 10 October 1944 and is an assessment of the current medical burden upon the Army's overall medical establishment.

This report is divided into two sections: The forward area, or divisional level, which provided basic first aid and medical care, and the Army level, where more extensive treatments would be provided like surgical operations, x-rays, burn and infectious disease care.

The total medical burden on 10 October was 8,021 people. This was an increase from the 20 September value of 6,101 people.

Forward Troop Area – up to divisional medical battalions

Arrivals:

At the start (5 October) – 1,136 men and women

Newly arrived – 1,207

Total – 3,043

Departures:

Evacuated – 1,298 (most likely to the Army rear) (moderately wounded)

Returned to active service – 638

Died – 56

Total – 1,992 (including dead)

In the Forward Troop Area on 10 October – 1,051 (lightly wounded)

In Army Hospitals

Arrivals:

At the start (5 October) – 7,314 men and women
Arrived from forward units' medical installations – 1,095
Arrived directly from units – 220 (severely wounded)
Totals – 8,629

Departures:

Evacuated – 1,058 (out of the Army) (severely wounded)
Returned to active service – 559
Died – 42
Total – 1,659 (including the dead)
In Army Hospitals 10 October – 6,970 (moderately wounded) with 5,350 Army hospital beds available

Total in all medical installations on 10 October – 8,021.

The overloaded nature of the medical support for the Army was due to constant combat operations. In essence, the available medical services could not keep up with the surge of incoming wounded. Even with transporting 1,058 wounded troops out of the Army, most likely by medical train, the medical load far outstripped the available facilities. In the end, medical personal performed beyond the call of duty, but the war continued regardless of the cost, and as the increased death rates showed, the wounded suffered and died needlessly due to poor planning and inadequate medical resources.

11 October 1944

The 27th A advanced to the northwest and captured the city of Clju, another transportation center. To the right of the 27th A, the 4th Ru A and 40th A continued to advance as the German defenders withdrew. During these attacks, the 27th A also was engaged in screening the movements of the 7th Guards Army behind the lines.

The secret Hungarian delegation in Moscow signed an armistice with the Soviet Union. The Soviet terms were as expected: declare war on Germany, withdraw their troops to Hungarian territory, and assist Soviet Forces in fighting the Axis. But the Germans learned of these treasonous acts and had already reacted to them on 10 October. Hitler had secretly begun Operation *Panzerfaust,* the operation to overthrow the Hungarian government and to arrest Admiral Horthy. Just like in Romania, top Hungarian government and military officials attempted to surrender their entire army and country to the Soviets. But the quick action by Hitler thwarted their efforts and restored, in time, the Hungarian Army as a steady Axis ally.

12 October 1944

As the 7th Guards Army began its long march, the Army relied upon a series of control posts and traffic regulators to direct the Army to its new area of operations. The report *Relying on the organization of control posts on the march of troops of the 7th Guards Army*, is found in TsAMO archive, record ID #132336443. This document lists 21 controlling positions along the march route, ending with Tinak (18 miles) and Kochuba-Mare (23 miles) directly south of Oradea-Mare. When they arrived, the troops were spread out in a fairly large area around these two towns and would have needed several days of rest to recover from their march and to reassemble their separate units. But would they actually receive these needed days of rest?

Of note, on this day units of the Army split their march path by taking two different but parallel routes. This minor change temporarily reduced congestion and accelerated their advance.

A female Traffic Regulator directs a towed 76 mm gun, Zis-3, past an abandoned German Jagdpanzer 38T Tank destroyer (nicknamed *Hetzer* or Baiter).

13 October 1944

Report of the commander of the forces of the 2nd Ukrainian Front No. 00641/op to the supreme commander [Stalin], [about the Debrecen] *Operation development plan October 13, 1944 13 h 00 min*

I am submitting for your approval a plan for the further operation of the 2nd Ukrainian Front for the period from 13 to 20 October 1944.

1. *The main goal of the operation is to cover the Tisza River with parts of the 46th and 53rd Armies, strike by the cavalry-tank forces of the front - the main forces of the 4th, 6th Guards CC and 7 MK in the direction of Nyiregyhaza, Chop to assist the 4th Ukrainian front in the fastest exit to the Hungarian Valley* [Plains] *and part of the forces consisting of 5 Guards CC, 23 TK by a blow from the Oradea-Mare area to Karey, Satu-Mare to* [block] *the withdrawal routes of the enemy's Cluzh-Syget grouping, surround and destroy it.*

2. Stages of the operation:

a) On [13-14 October] *forces of the Pliev group, units of the right flank of the 53rd Army and the 6th Guards Tank Army, destroy the encircled enemy grouping west and south of Bretto-Uyfalu;*

b) [On 15-17 October] *Pliev's group captures Debrecen and goes to the Nyiregyhaza region, capturing bridgeheads* [over the Tisza River]. *Gorshkov's group goes to the Karei region, Satu Mare. 53rd Army with 4 divisions of the Romanian* [Ru 1st A] *goes completely to the* [Tisza River], *with a right flank in the Rakamaz area. The 40th and 27th Armies continue their offensive and reach the line of Petrov, Baia Mara, Margitta. 7th Guards Army continues to concentrate in the area south of Oradea-Mare, and with the forces of* [the 33 RC] *moves to the Sykueni Debrecen area. 6th Guards Tank Army is concentrated in the Debrecen area to put the units in order and makes up the front reserve;*

c) [On 18-20 October] *Pliev's group leaves for the Shotoralia-Uikhel, Chop area. Gorshkov's group, in cooperation with the 27th and 40th Armies, is fighting to destroy the retreating Kluzh-Siget enemy grouping. The 40th and 27th Armies enter the Siget, Karei, Veshad line. 7th Guards Army* [with] *the* [33 RC] *corps enters the line Nyir-Bogat, Nyiregy-khaza. The main forces of the 7th Guards Army go to the area of Oradea Mare. The 6th Guards Tank Army either strengthens the direction of the Pliev group, or the direction of the Gorshkov group, or receives an independent direction to Miskolc. 53rd and 46th armies firmly hold the line along the* [Tisza River]. *In the case of a favorable situation, they may regroup to strengthen the direction* [to] *Szolnok.*

Malinovsky Suslikov Zakharov

This report stated that Malinovsky's 2 UF would develop its Debrecen offensive from the Arad-Oradea line to the northwest reaching the Debrecen-Tisza River line. Along the way it would help the 4 UF breach the last Carpathian barrier and reach the Hungarian Plain. This report also provided direct confirmation concerning the transfer of the 7th Guards Army to its new concentration area south of Oradea-Mare, and it also attached, at least temporally, the 33 RC to the 7th Guards Army.

14 October 1944 – Weather: 62° to 41° F, (16.6° to 5.0° C), wet ground

Units of the 7th Guards Army continued to march, making good progress along its journey despite the terrain and the weather. By the end of the day, the two different march columns of the Army rejoined near the towns of Abrud and Buchm-Cherbu. The convergence of these columns certainly caused a major traffic bottleneck in this area.

The Army logistical situation at this point apparently was in shambles. The historical TsAMO document ID #132313848, page 22-23, illustrates the problem. Despite the Army marching behind the front line, the ammunition situation must have been critical because starting on 14 October, all Army level supply vehicles were ordered to begin transporting ammunition 230 miles through the Carpathians to more forward dumps. Despite working around the clock, with drivers only getting four hours of rest, this effort was unable to transport the required amount of ammunition. Therefore, 100 trucks were diverted from combat units to assist. These trucks were required to make only one trip carrying ammunition and they helped finish the job on time. Because the direct rail line through the Carpathians was damaged, rail deliveries were shifted south to a new route. Starting at the town of Adjud on the Seret River (a major railroad transshipment point) trains soon were arriving at Sighisoara (not shown on maps), but that destination is located about 25 miles southeast of Targu-Mures. This was the same general area the 7th Guards Army just left.

During the 14-25 October time period, when the 7th Guards Army was completing Phase One and Two of its transfer plan and advancing to the Tisza River, the Army supply services performed well. Overall, they transported the following cargo:

Ammunition - 8,432 tons
Fodder - 5,432 tons
Quartermaster/Food - 90.5 tons
Clothing and Equipment - 951 tons
Other cargo - 1,205 tons
Wounded and sick people - 20,637

With Axis mobile units launching counterattacks to cover the withdrawal of their main forces back to the Tisza River line, the front was in flux. The 2 UF also was attacking and shifting forces around, which made the 7th Guards Army vulnerable on the march. What provided an extra layer of protection to the Army was that a bit further ahead of the forced march, the 33 RC, with the help of the 6th GTA, were engaging German and Hungarian units as they moved north toward Debrecen, thus clearing the assembly area between Oradea-Mare and Debrecen for the approaching 7th Guards Army.

15 October 1944

The 7th Guards Army continued to march. The Army at this point was about 65 miles from the nearest fighting.

Hitler sent his favorite commando *Waffen*-SS Lieutenant-Colonel Otto Skorzeny and former *Brandenburger* [special forces unit] Adrian von Fölkersam to Hungary to arrest the Hungarian Regent Admiral Miklós Horthy for secretly negotiating Hungary's surrender to the Red Army. Operation *Panzerfaust,* the code name for the German military operation to occupy Hungary, had begun in earnest. Admiral Horthy and other supporters were arrested, paving way for the Hungarian Arrow Cross Party (Arrow Cross Party-Hungarist Movement, abbreviated NYKP) to seize control of the government and military. Just in case of problems, Skorzeny also kidnapped Admiral Horthy's 37-year-old son Miklos Horthy Jr. (Operation Mickey Mouse), [570] and sent him to Dachau, a German concentration camp, where he remained a prisoner until American troops liberated the camp on 5 May 1945. [571]

Sometime from 10-20 October, Marshal Malinovsky reassigned the entire 16 Artillery Penetration Division back to the 7th Guards Army. This powerful unit would allow the Army to pound stubborn enemy positions into submission or help destroy a large city like Budapest.

German troops, in Budapest, stand among confiscated Hungarian weapons on the ground, after Operation *Panzerfaust.* Note the two soldiers in the foreground are holding two different sizes of anti-tank *Panzerfausts.*

16 October 1944

In Hungary, the ultra-Fascist Arrow Cross Party formed a puppet government called the Government of National Unity. This type of government was more suitable to Hitler's military needs; it also ensured that Budapest and its people would be besieged and that Hungary would be further devastated by the war.

17 October 1944

The leading foot elements of the 7th Guards Army should have started to arrive at their assembly areas in and around the towns of Tinak and Kochuba-Mare, but the historical record is unclear on this fact. Regardless, the troops were exhausted, sore, and ready for some good hot food and a long rest. The more mobile elements of the Army should have arrived several days earlier.

Despite the ongoing World War, Hitler's insane plan to utterly destroy Warsaw as revenge for the Warsaw Uprising was being put into practice. Heinrich Himmler, head of the Nazi SS *Schutzstaffel*, stated, "The city must completely disappear from the surface of the earth and serve only as a transport station for the Wehrmacht. No stone can remain standing. Every building must be razed to its foundation." [572] By January 1945, the Germans had blown up or burned to the ground vast sections of the depopulated city. Libraries, churches, museums and historical building all were destroyed indiscriminately. This mindset is reminiscent of Hitler's plans for Moscow. If the Soviet Union had been defeated, Moscow was to be razed to the ground and beyond. The site of Moscow was to be turned into a vast, deep lake, erasing from the face of the Earth all traces of that ancient city.

18 October 1944

Despite its transfer to a new section of the 2 UF, the work of the Army headquarters never ends. The 7th Guards Army issued a report, S*chedule, the march of units and formations to the areas of concentration according to cypher telegram No. 444/op.* This was found within TsAMO record ID #132336458. This document appears to list the first group of units that were to be concentrated in a new combat area. This was the beginning of Phase Two of the overall transfer plan. (See Map #55.) The report covered a group of artillery units and the 72, 81 GRDs, and 227, 53 RDs. Somewhere, a second document must have listed the remaining units of the Army detailing their concentration area, but that document is not available. The listed units were scheduled to begin their next movement between 15:30 – 24:00 hours, (3:30 p.m.-12:00 a.m.), and were expected to travel anywhere from 30 to 43 miles, if possible. The new concentration area would require the units to cross the Crisul River, march north of Oradea-Mare, and reach the area around the small town of Pochay and the tiny Barcau River. At that point, the 7th Guards Army would have crossed into Hungary.

Also on this day, the 7th Guards Army appeared on German OKH intelligence maps for the first time in days, according to the US National Archives, T-78 Roll 136, image #618. The German intelligence units had correctly located the 7th Guards Army behind the lines and south of Oradea-Mare. German reconnaissance aircraft must have spotted the obvious columns of marching troops. This of course was not by chance. When the 7th Guards Army withdrew behind the front line on 9-10 October, German intelligence units noticed its absence, and eventually they lost track of the Army. An entire army doesn't simply disappear; it was doing something, and their job was to find out what. So, German aircraft were sent out to look for it, and radio reconnaissance/intelligence detachments (*funkaufklärung*) were alerted to listen for radio chatter, call signs or anything that would indicate what the Army was doing and where. Keeping track of all major enemy formations, especially a Guards army, was extremely important.

19 October 1944

Units of the 7th Guards Army began to reach their assigned positions north of the Crisul River. (See Map #55.) The troops had little rest during the march. As different formations trickled into position, the exhausted troops knew the march would continue the next day.

In a disturbing move, General Friessner launched a counterattack from the Szolnok area with the German 24 Pz D, 4 SS Pz GD and the supporting 503 Pz Battalion with Tiger II tanks. This was Operation Gypsybaron, a spoiling attack intended to divert Soviet forces from the Debrecen area. (Not shown on maps.) German forces had been holding a bridgehead on the east side of the Tisza River, and this short attack to the east crushed the resistance put up by the nearby 4 Ru RD and inflected heavy losses on the division. The Panzer attack then continued on to Mesetur where it broke the resistance of a Soviet Guards' cavalry division that was detraining there. After several days of operation, the attacking Panzer force fell back to Szolnok for deployment elsewhere. [573] The Germans simply lacked the strength to do more than launch these short but sharp counterattacks to disrupt Soviet operations and to inflict losses.

20 October 1944

Moving to the west, the rifle divisions of the Army kept to their march schedule as they crossed the Hungarian border. As the Red Army advanced deeper into Hungary, they were no longer on friendly territory. While the Romanians openly greeted and assisted the Red Army by joining in their attack against Hungary, the Hungarians thought of the Russians and especially the Romanians as invaders. The Hungarian civilian population would not willingly collaborate with these "invading armies."

Just to the north of Debrecen, the 33 RC still screened the 7th Guards Army from German interference. The Army also had entered an area that contained a complex rail network. These railways eventually would help ease the logistical burden of supplying the Army.

On or about this date the 1st Ru A was transferred back to the 2 UF and was subordinated to the 7th Guards Army. Also around this date, the 4th Ru A moved to the right flank of the 40th A and was made subordinate to the 40th A. In this position the 4th Ru A was the right flank army for the entire 2 UF.

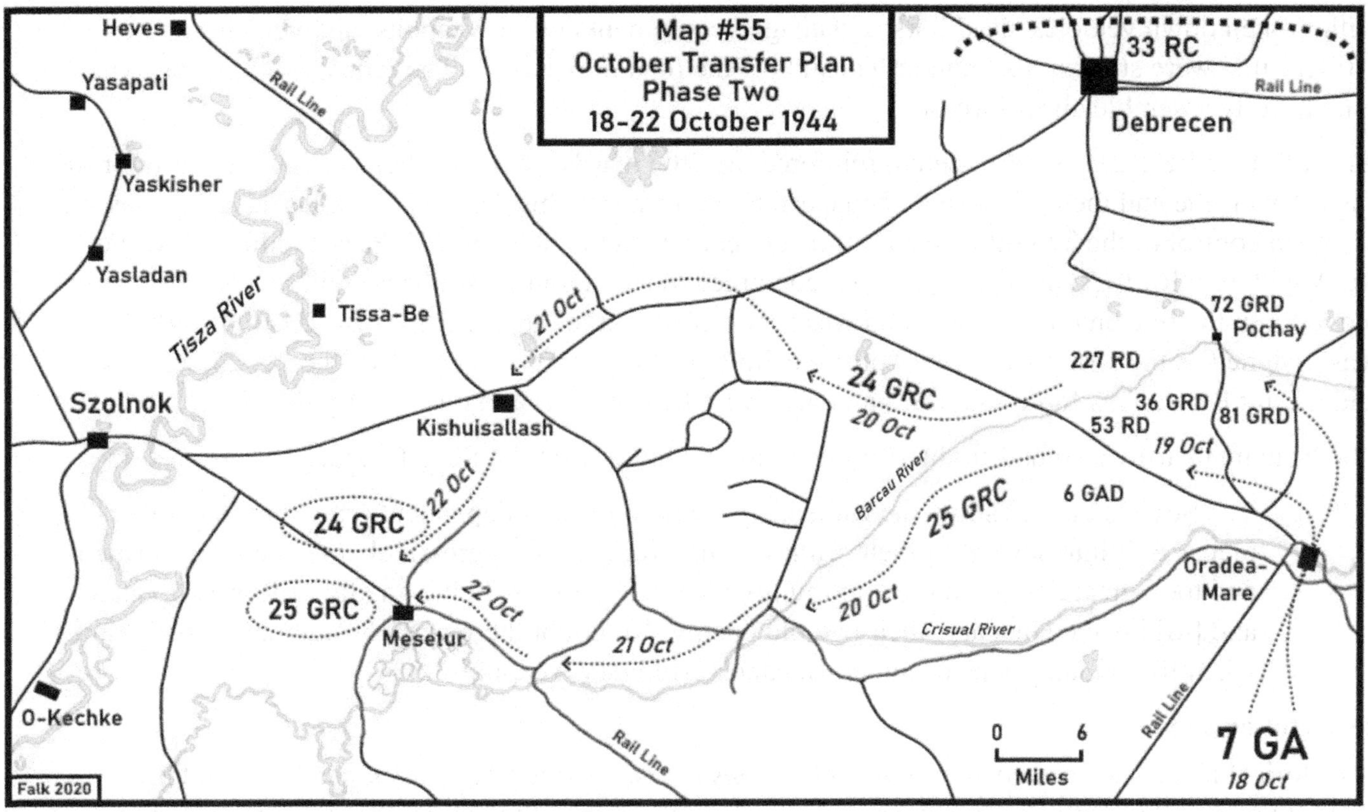

Along the eastern side of the Tisza River south of Szolnok, counterattacking German units of the 24 Pz D, 4 SS Pz GD and a battalion of tiger tanks cut off and destroyed the Romanian 4 Ru RD which had recently crossed to the west side of the river. Overall, Romanian forces were suffering heavy losses at the hands of German and Hungarian troops. Between 1 September and 25 October, Romanian casualties totaled 49,744 men, which was far higher than their Axis adversaries. This disparity most likely was due to lack of equipment, poor training, and their national urge to force Axis forces from Romanian territory as soon as possible.[574]

General Malinovsky had been planning to deploy the 7th Guards Army northward toward Debrecen, but with this latest series of German attacks east of the Tisza River, Malinovsky diverted the 7th Guards Army west toward the Szolnok area instead. [575] The spoiling attack that General Friessner launched worked. Soviet pressure against Debrecen was reduced, as reserves were sent west.

21 October – Weather: 54° to 48° F, (12.8° to 8.8° C), soaked ground

Units of the 24 and 25 GRC slowly approached, through the deepening mud, their final destination just short of the eastern bank of the Tisza River. Care had to be taken at this point because German mobile forces recently had been operating on the same side of the river.

At the direction of the 2 UF, the 7th Guards Army transferred control of the 33 RC to the 27th A.

22 October 1944

The 7th Guards Army reached its concentration area just west and north of the town of Mesetur, but it also engaged scattered and withdrawing German units in heavy fog. The Tisza River was some six to eight miles away. The indicated concentration locations for the 24 and 25 GRC on Map #55 are approximate. Their individual rifle divisions were dispersed around the general area as they battled German rear guards. Eventually, the rifle division of the Army began to take up defensive positions and to collect trailing sub units. With any forced march like this, the Army's route would have been littered with broken down vehicles, dead horses, damaged equipment and/or weapons, and vehicles and/or artillery that were stuck in the mud or had slid off the road into a ditch. The Army would need some time to fully assemble all its forces.

Also of note, the 2 UF saw the need to reinforce the 7th Guards Army for their upcoming battle for the Tisza River line and their advance to Budapest. So on or about this date, the 7th Guards Army was assigned control of the 27 GRC which ultimately contained the 303 and 227 RDs. Additionally, the 1st Ru A with two Ru RCs, the 4 and 7, which contained the 1, 2, and 19 Ru RDs and the 9 Ru CD, was moved into the line on the left flank and was subordinate to the 7th Guards Army. Also, the 409 RD was assigned as the Army's reserve. Thus, the 7th Guards Army now controlled the 24, 25, 27 GRCs and the 1st Ru A. This force contained a total of 11 rifle divisions and one cavalry division.

The Romanian army also had an interest in continuing the war on the Soviet's side.

> The Soviets did not allow the Rumanian Army to be an independent army, but integrated pieces of it into several of their armies.... the Rumanians aggressively engaged the Germans and took great delight in the opportunity afforded them, by being an ally of Russia, to attack and [to] invade Hungary, their age-old enemy. No doubt there was considerable satisfaction when Rumanian troops occupied Budapest, Hungary's capital. [576]

23 October 1944

A historical map, *The scheme of military operations of army troops 22-31 October #3*, found in TsAMO record ID #455178105, shows that all rifle units of the 7th Guards Army were advancing toward the Tisza River. Local Axis forces resisted this slow advance as they withdrew back to the Tisza.

Further to the northeast, German mobile forces attacked south of Nyiregyhaza where they surrounded three corps of Group Pliyev. General Pliyev was able to break out of the encirclement on 26 October, but he lost most of his equipment doing so. This was the second time German mobile forces were able to surround attacking spearheads of the 2 UF and to inflict heavy losses upon them.

24 October 1944

Units of the 7th Guards Army continued to move toward the Tisza River. Possibly in response to the Army's advance, the Germans abandoned their bridgehead at Szolnok and destroyed the local major bridges over the Tisza River.

25 October 1944

The advance to the Tisza River continued with most formations reaching the river during the day. Small formations of the 25 GRC began to cross to the west side (right bank of the river), using big loops in the river to avoid the Axis defenders who were deployed on slightly higher ground away from the river. These were the opening moves in forcing a crossing of the Tisza to support the 46th A to the south and to expand their bridgehead. The near-term objective for the 7th Guards Army was to capture the cities of Szolnok and Abony as soon as possible and then to capture or clear the entire west bank of the Tisza River in their sector. (See Map #56.)

Of note, a historical TsAMO map ID #100155645 shows the 18 TC along with the 5 and 9 MC in the zone of the 7th Guards Army around this time. These three independent mobile units most likely were moved into position by the 2 UF to enable the exploitation of any breakthrough across the Tisza River. These three units are not displayed on Maps #55 and #56 because they were transferred elsewhere soon after this date. The determined Axis defenders along the river prevented any quick breakthrough over the Tisza, thus interfering in *Stavka's* plans to rush to Budapest. Note: the 5 MC is incorrectly displayed on the afore-mentioned historical map because the 5 MC had become the 9 Guards Mechanized Corps (GMC) on 12 September 1944.

On this day, a German OKH intelligence map (T-78 Roll 136, image #613) showed the 7th Guards Army deployed around Oradea-Mare, 90 miles to the rear. In truth, the Army actually was approaching the Tisza River in strength. The German map revealed a slight assessment error on the part of German intelligence agencies. Once again, please note that losing track of a Guards Army is very dangerous.

26 October 1944 – Weather: 62° to 41° F, (16.6° to 5.0° C), wet ground

Rifle units of the Army continued to fight their way over the Tisza River and to slowly expand their bridgehead. The German and Hungarian formations vigorously held the river line and counterattacked with armor.

Continuing to use the big loops in the Tisza River, the 81 GRD reinforced their tiny bridgehead under the cover of smoke screens from 26 to 28 October and then again on 3 and 4 November. Further south, the 53 RD was aided by the use of smoke on the 26, 29, 30, and 31 October and on 1 November. These smoke screens were used to block the Germans from seeing and accurately shooting at the crossing locations, which should have reduced the loss of men and equipment.

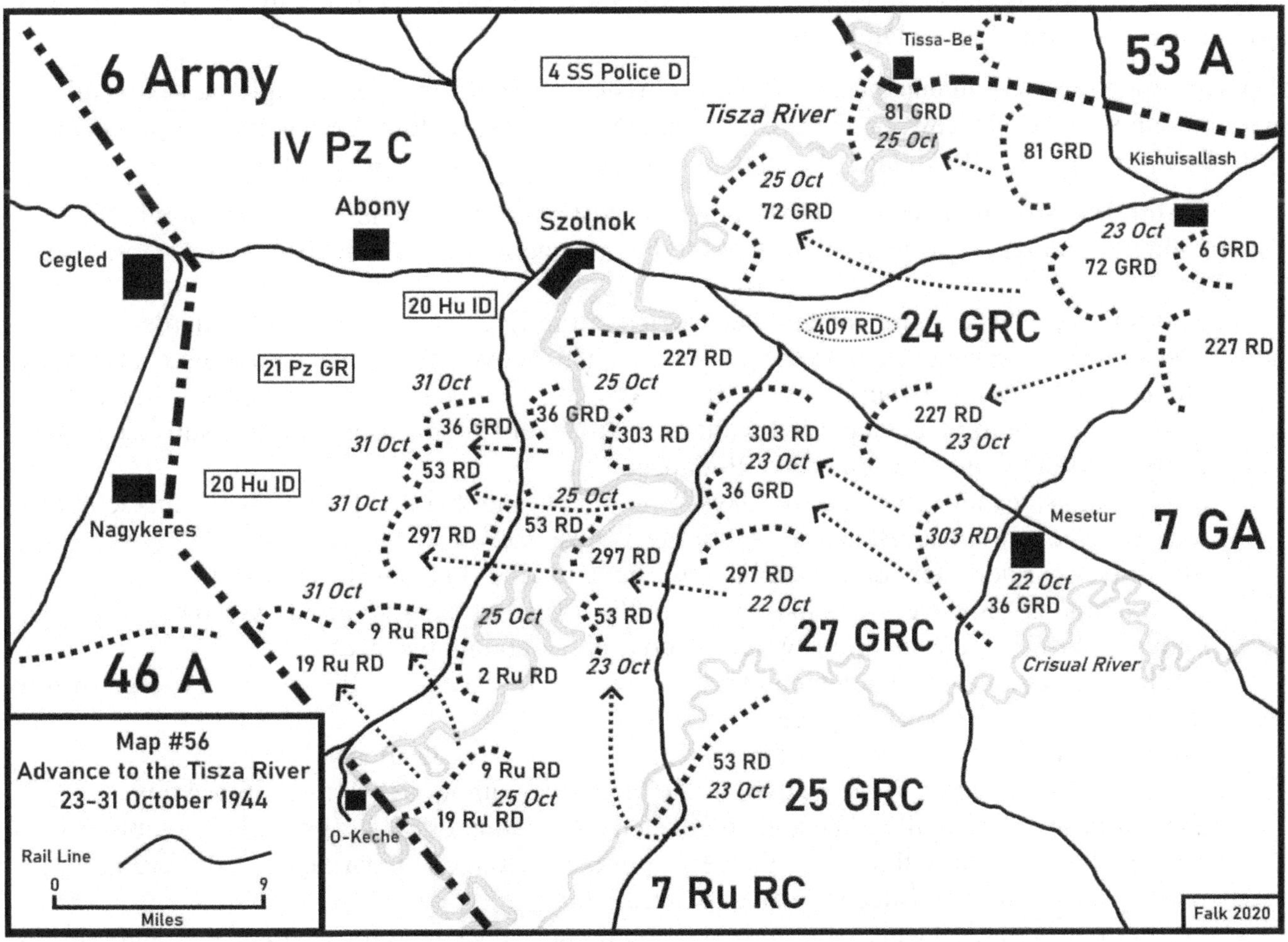

27 October 1944

Stiff fighting continued along the Tisza River. The 46th A, on the west side of the river, was fighting on the 7th Guards Army's left flank, but it also was facing determined enemy resistance.

28 October 1944 – Weather: 63° to 51° F, (17.2° to 10.5° C), soaked ground

Stavka issued Directive #220251 concerning the 2 UF attack over the Tisza River toward Budapest.

Directive of the VGK Stavka No. 220251 to the Commander of the 2nd Ukrainian Front on preparation of the strike to Budapest.

October 28, 1944 22 h 00 min

The headquarters of the Supreme High Command orders the forces of the 46th Army and the 2nd Guards mechanized Corps [on] *October 29* [to] *go on the offensive between the* [Tisza River] *and Danube in order to curtail enemy forces on the western bank of the* [Tisza River] *and* [support] *the 7th Guards Army* [and the 1st Ru A in crossing the Tisza River]. *Later, with the approach of the 4th Guards mechanized Corps, deliver a decisive blow to the enemy group defending Budapest. Report the orders given.*

Stalin A. Antonov

In this directive, the 46th A was to play the main role in forcing a crossing of the Tisza River for the other armies of the 2 UF and to capture Budapest. But Malinovsky knew the 46th A and the supporting 2 GMC were not strong enough to capture Budapest on their own and to clear the west side of the river. Marshal Malinovsky then proposed adding the 7th Guards Army and the 6th GTA to the attack and asked Stalin to give him several days for the transfer of tank formations to the Budapest attack. At this time, the 2 UF was low on tanks and other armored vehicles after suffering heavy losses at the hands of counterattacking German Panzer Divisions.

Then Malinovsky received an unexpected HF (High Frequency (*ВЧ*) Encoded Telephone Line) call from Stalin, who was in Moscow, on the eve of the planned offensive. John Erickson, in his book *The Road to Berlin*, pages 395-396, relates the content of the conversation that took place.

Stalin. It is absolutely essential that in the shortest possible time, in days, you capture the capital of Hungary – Budapest. This has to be done no matter what it costs you. Can you do this?

Malinovsky. This assignment could be carried out in five days, once the 4th Guards Mechanized Corps moves up to the 46th Army. This movement is expected to be complete by November 1. Then the 46th Army, reinforced by two Guards mechanized corps – 2nd and 4th – would be able to mount a powerful attack, which would come as a complete surprise to the enemy and in two to three days take Budapest.

Stalin. Stavka cannot give you five days. You understand that it is for political considerations that we have got to take Budapest as quickly as possible.

Malinovsky. I definitely understand that we have to take Budapest in view of these political considerations. However, we should wait for the arrival of the 4th Guards Mechanized Corps. Only under these considerations will it be possible to count on success.

Stalin. We cannot consider postponing the offensive for five days. It is necessary to go over to the offensive for Budapest at once.

Malinovsky. If you give me, as of now, five days, five days as an absolute maximum, Budapest will be taken. If we go over to the offensive without delay, then 46th Army, for sheer lack of forces, will not be able to develop its blow quickly [and] it will inevitably get bogged down in heavy fighting at the very approaches to the Hungarian capital. Putting it briefly, it cannot seize Budapest off the march.

Stalin. You are arguing all to no purpose. You do not understand the political necessity of mounting an immediate attack on Budapest.

Malinovsky. I understand all the political importance of taking Budapest and for that very reason I am asking five days ...

Stalin. I categorically order you to go on the offensive for Budapest tomorrow.

At this point Stalin abruptly broke the connection. Marshal Malinovsky, in his memoirs relates a similar conversation, but in softer tones. But in the end, the attack would take place no matter what! Orders are orders, and Stalin would not be deprived of his prize.

General Friessner, while in Budapest for a meeting, looked around the city and stated, "I realized that the population had absolutely no idea how serious the situation at the front was." In frustration, Friessner understood that the new Hungarian government wasn't going to mobilize the population for the defense of Budapest or even inform them of the need for such action. [577]

29 October 1944

On the far-left flank of the Army, units of the Romanian 1st A crossed the Tisza and began outflanking the German and Hungarian defenders. This was not a huge breakthrough, but it did allow the Army to deploy more troops on the western side of the river. In line with Stalin's orders, Malinovsky began the 46th A attack towards Budapest, which made some progress, but the Germans were quick to launch powerful counterattacks.

30 October 1944

Romanian and Soviet units continued to make slow but steady progress in the zone of the 7th Guards Army. Even the 46th A kept up and advanced on the Army's left flank, but the Germans brought forward the 1, 23 and 24 Pz Ds to stop the advance of the 46th A and to defend Budapest.

Guards Major of the medical service Lev issued a medical report, *Information about the movement of the wounded and sick in medical institutions of the 7th Guards Army*, found in TsAMO record ID #133391634. This insightful report covers the period from 25 October to 30 October 1944, and is an assessment of the current medical burden upon the Army's medical establishment.

The total medical burden on 30 October was 7,409 people. This was fewer than from the 10 October value of 8,021 people.

Forward Troop Area – up to divisional medical battalions

- Arrivals:
  - At the start (25 October) – 521 men and women
  - Newly arrived – 1,891
  - Total – 2,412
- Departures:
  - Evacuated – 1,004 (most likely to the Army rear) (moderately wounded)
  - Returned to active service – 290
  - Died – 53
  - Total – 1,347 (including dead)
  - In the Forward Troop Area on 30 October – 1,065 (lightly wounded)

In Army Hospitals

- Arrivals:
  - At the start (25 October) – 6,017 men and women
  - Arrived from forward units' medical installations – 1,063
  - Arrived directly from units – 132 (severely wounded)
  - Totals – 7,212
- Departures:
  - Evacuated – 446 (out of the Army) (severely wounded)
  - Returned to active service – 401

Died – 21
Total – 868 (including the dead)
In Army Hospitals 30 October – 6,344 (moderately wounded) (with 5,350 hospital beds available.)

Total in all medical installations on 30 October – 7,409. While this total is better than that of September, the medical facilities still were overloaded with casualties.

31 October 1944 – Weather: 60° to 55° F, (15.5° to 12.7° C), wet ground

For the 7th Guards Army, the divisions on the left flank and in the center continued their slow advance on the other side of the river. The right flank units were having a hard time crossing the Tisza. The now combined 7th Guard Army bridgehead on the west side of the river was about nine miles deep. The 2 Ru and 19 Ru RDs and the 9 Ru CD along with the 36 GRD, 53 and 297 RDs were fully on the western side of the Tisza. The Axis defense line was not broken, but it certainly was dented.

As General Friessner was moving his mobile units around to counter the 2 UF advance toward Budapest, he stated,

> In the zone of the 4th Panzer Corps, battles began with the enemy, who had already crossed the Tisza south of Szolnok and was now advancing north towards the city. … the enemy, even during the tank battle in Puszta, began to systematically expand the bridgeheads he had captured on the Tisza. Only gradually, in the course of heavy fighting, the Russians managed to force the Tisza in the Szeged, Chongrad, Szolnok sector. [578]

In this, General Friessner was talking primarily about the recent activities of the 46th A, 1 Ru A, 7th Guards Army and the 2 and 4 GMC that were operating between the Danube and the Tisza Rivers. (See Map #57, 1 November Strategic Situation.)

The civilian population of Budapest should have begun to understand the situation when posters like the one on the next page started to appear around the city, but few took notice. The vast majority of men, women and children remained within the doomed city, for one reason or another, to the end.

"I Have No Mercy…" Arrow Cross Party - Fascism Poster in Budapest, Hungary.

The 2 UF issued a report, *Report on the loss of personnel in the troops of the 2nd UkrF from October 1 to November 1, 1944.* This was found within TsAMO record ID #113723554. The 7th Guards Army suffered 11,119 total casualties for the month: 1,952 killed; 722 missing; 0 captured; 21 non-combat losses; 7,322 wounded; and 1,102 sick. The report goes on to list the 7th Guards Army's horse losses: 560 killed; 0 missing; 134 wounded; 9 evacuated to veterinarian care; for a total of 703 equine casualties. The reason for the high number of horse deaths is not officially known, but overwork due to the forced march by the Army over hills, through valleys and mud probably played a major role. The same report shows that the troops of the entire 2 UF suffered 84,010 total casualties during October 1944.

For the month of October in Britain, German V-1 and V-2 attacks killed 172 civilians and injured 416. The decrease in losses was due primarily to the Western Allied armies capturing the launching sites of the vengeance weapons on the continent. The Germans eventually made adjustments and shifted their attacks to targets closer to Germany, namely Antwerp and Brussels in Belgium and other targets in the Netherlands.

1 November 1944	
7th Guards Army	
Assigned – 2 Ukrainian Front	
Colonel General M.S. Shumilov Commanding	
Rifle, Airborne and Cavalry Units	24 Guards Rifle Corps (72, 81 GRDs) 25 Guards Rifle Corps (36 GRD, 53, 297 RDs) 27 Guards Rifle Corps (227, 303 RDs) 409 RD 1st Ru A, 7 Ru Rifle Corps (2, 19 RD, 9 CD) (attached) 15 Penal Battalion 62, 63, 64, 65, 66,194 Penal Companies
Artillery Units	16 Artillery Penetration Division (49 Light Artillery Brigade, 61 Gun Artillery Brigade, 52 Howitzer Artillery Brigade, 90 Heavy Howitzer Brigade, 109 High-Power Howitzer Brigade, 14 Mortar Brigade) 2, 11 Antitank Artillery Brigades 114, 115 Guards Tank Destroyer Regiments 41 Guards Gun Artillery Brigade 263, 290, 493 Mortar Regiments 5 Antiaircraft Artillery Division (670, 743, 1119, 1181 AA Regiments) 662 Separate Aerial Observation/Air Raid Warning Company
Armored and Mechanized Units	27 Guards Tank Brigade 38 Separate Armored Train Battalion
Engineer Units	60 Engineer Sapper Brigade
Reported Combat Strength Estimated Ration Strength Reported Combat Strength 1st Ru A (4, 7 Ru RC)	62,131 men 74,820 men and women (including an estimated 7,000 support personnel) 29,178 men

(BSSA 44, BSA 44, TsAMO Archive ID #113723626)

The 2 UF HQ returned the 27 Guards Tank Brigade to the 7th Guards Army after being posted to other armies. At this time the 27 GTB contained 46 – T-34 tanks, 16 – SU-85 self-propelled guns, and 4 – SU-76 self-propelled guns. This data was found within TsAMO record ID #454861851, p 2.

Note that at or around this date, a full-strength tank brigade should have contained 65 – T-34 tanks. Apparently the 27 GTB also had an attached light SU regiment containing 20 – SU-76s (at full *Shat* strength) and a medium SU regiment with 20 – SU-85s (at full *Shat* strength), for a total of 105 armored fighting vehicles.

The 1st Ru A only deployed the 7 RC to the front with three divisions that were within the zone of the 7th Guards Army. The 4 Ru RC was deployed to the rear for unknown reasons, but maybe for refitting. Adding to the confusion about Romanian units, some historical maps occasionally show RDs from the 4 Ru RC on the front line.

Overall, the strengths of the Russian rifle divisions assigned to the 7th Guards Army were surprisingly good, despite all the recent fighting. The Army's medical support services provided a steady stream of treated wounded soldiers back to active duty. Undoubtedly, the 2 UF also supplied reinforcements to keep up the Army's strength. (See Table #24.)

Unit	Strength - Men and Women
24 GRC HQ	794
72 GRD	5,646
81 GRD	5,136
25 GRC HQ	857
36 GRD	5,363
53 RD	4,845
297 RD	5,549
27 GRC HQ	814
227 RD	5,839
303 RD	5,030
409 RD	4,834
Total	**44,707**

Table #24: Rifle Unit Strengths - 1 November 1944.

1 November 1944 – Weather: 57° to 50° F, (13.8° to 10.0° C), dry ground

The 7th Guards Army issued a report, *The scheme of the military road of the 7th Guards Army*, found in TsAMO record ID #133391636. This map showed the Army level supply network extending from around the town of Mesetur, which was located about 27 miles to the rear, and over the Crisual River along the rail lines another 12 miles. This network of roads was good news because Army level transport only needed to move supplies about 40 miles to reach the units at the front line. This map also indicated that the rail lines, controlled by the 2 UF, were operational this close (40 miles) to the fighting, which reinforces the idea that the Russians captured the Romanian railroads largely intact.

With the start of the new month, the 7th Guards Army continued to attack mainly with the 7 Ru RC and 25 GRC on the left flank. (See Map #57 for the overall strategic situation on 1 November 1944.) The 46th A also continued its attacks, and their leading units were about 30 miles from southern Budapest.

With the recent advance of the 2 and 3 UFs, the 5th AA took advantage of the situation and moved its air operations closer to the front.

> The liberation of a significant part of the territory of Hungary created favorable conditions for the redeployment of all regiments and divisions of the 5th Air Army to the airfields of the eastern part of the Hungarian Plain. On November 1, the army's combat aircraft fleet amounted to 925 serviceable units, including 450 fighters, 295 attack aircraft, 145 bombers, 22 reconnaissance aircraft and 13 spotters. [579]

Keeping pace with the advance of ground operations was critical for effective air support during World War II. Generally, most aircraft had a relatively short-range. This was especially true for fighters, tactical bombers and ground attack planes. Ideally, airfields were located as close to the fighting as possible, normally about 25-50 miles behind the lines, which allowed the VVS to provide prompt support to the Red Army.

Despite difficult weather conditions during November, both on the ground and in the air, the 5th AA flew 7,667 sorties. The total Luftwaffe effort during November is unknown, but Soviet reports indicate it was significant, with 2,221 sorties recorded over the period of just nine days. [580] This Luftwaffe effort also would have used significant amounts of increasingly scarce aviation fuel.

At this stage of the war, the Luftwaffe's strength on the Eastern Front was fading fast. Table #25 illustrates the German intelligence's estimate of the disparity between VVS and Luftwaffe forces on the Eastern/Russian Front in November 1944.

Aircraft Type	Soviet VVS	German Luftwaffe
Ground Attack	4,785	Unreported
Fighters	8,162	Unreported
Bombers	4,058	Unreported
Transports	1,733	Unreported
Total Aircraft Deployed	**18,738**	**2,500 of all types**

Table #25: VVS vs. Luftwaffe Aircraft on the Eastern/Russian Front - November 1944. [581]

2 November 1944

The left flank corps of the 7th Guards Army continued to advance, slowly forcing back the German and Hungarian defenders.

In the 46th A area, mobile units of the 2 and 4 GMC broke through the Axis defenders and rapidity approached Budapest from the south. By the end of the day, they had approached to within nine to ten miles of the southern boundary of the city. Krisztian Ungvary, in his book, *The Siege of Budapest, 100 Days in World War II*, on page 14, recounted this event:

> In short intervals between air raids, I can still hear the constant thunder of guns. Are they firing so intensely at enemy planes that have lost their way? Then I realize that these are not antiaircraft guns. Could the battlefield have gotten so close to us? …After the brief lunch break the [air raid] sirens [were still] silent, but the noise of shooting grows ever more threatening. By now there can be no doubt that it is ground artillery that is firing and not antiaircraft guns, or a big tank battle is going on south of us…. We are in the battle zone. How long will it last and are we going to survive?

This comment, and others by the same source, illustrates the lack of preparation by the new Hungarian Government for the coming battle. The capital of Hungary, at this time, contained some 900,000 inhabitants who were basically unaware that they were about to endure a brutal assault by the Red Army and Red Air Force that would last for over three months. This assault would threaten their entire way of life, destroying their economy, their homes, and leaving tens of thousands of dead, both civilian and soldiers, buried in the rubble of their once beautiful city.

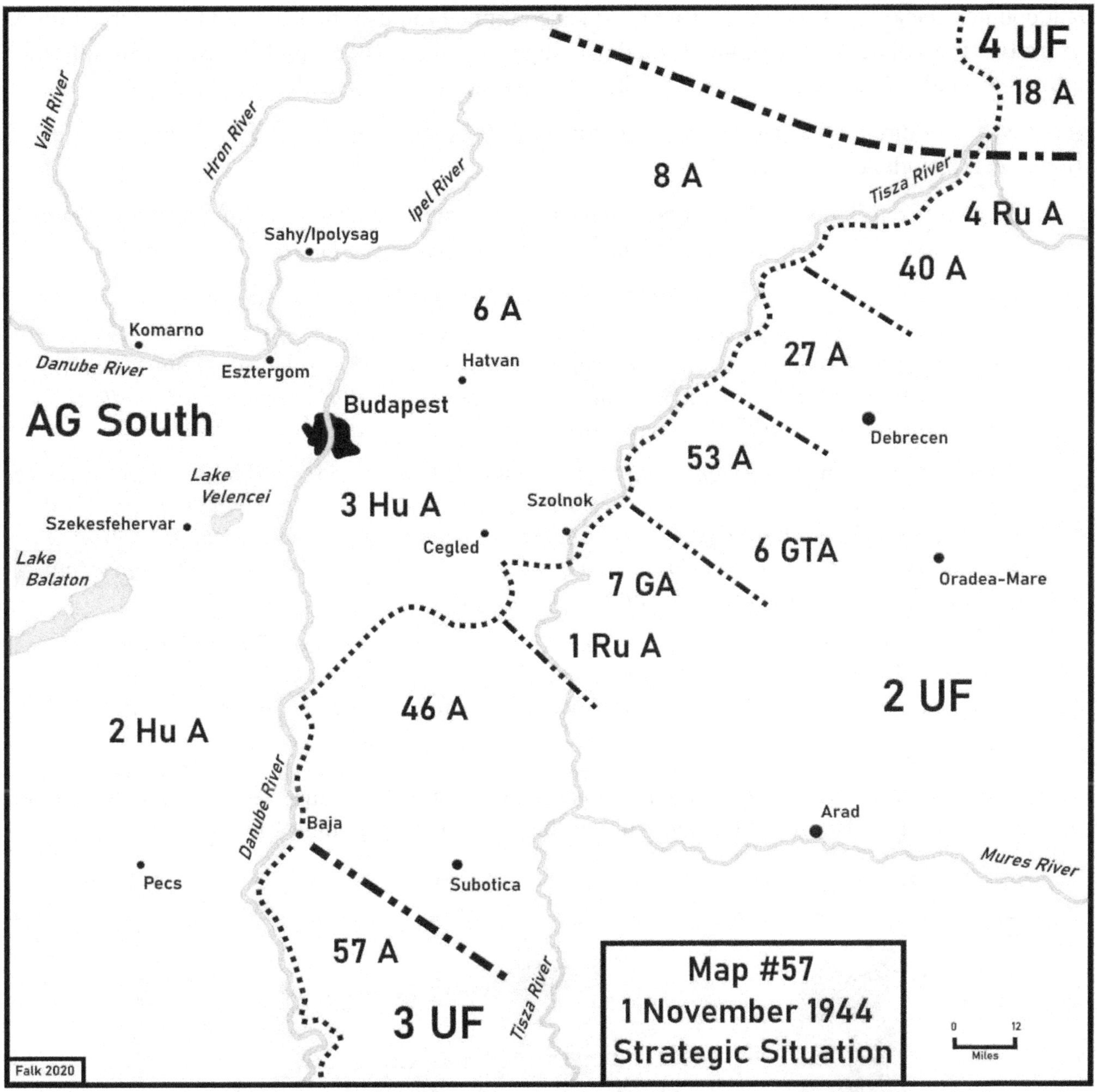

3 November 1944

Fighting against elements of the German LVII and IV Pz Cs, the left flank corps of the Army continued to advance northward, clearing the western side of the Tisza River as they went. They also supported the advance of the 46th A as it surged ahead toward Budapest.

The mobile units of the 46th A, 2 and 4 GMCs, penetrated the Axis "Attila I" defense line south of the city (not shown on maps) and reached a position about four miles east of the Soroksar District of Budapest, but soon they were counterattacked by German and Hungarian troops and forced back.

4 November 1944 – Weather: 54° to 44° F, (12.2° to 6.6° C), soaked ground

Units of the 7th Guards Army captured the towns of Szolnok and Abony (located right next to Szolnok), as the rifle divisions of the 27 GRC crossed the Tisza River and joined in the attack to the northwest. Szolnok is approximately 24 miles from Budapest. (See Map #58.)

Constant counterattacks by Axis forces from all sides continued to push the leading units of the 46th A away from Budapest. The 46th A simply lacked the strength at the tip of its mobile spearhead to hold the captured ground. To do that would require infantry divisions, and those divisions lagged behind.

Stavka issued a new directive clarifying the high command's position on the 2 UF's offensive operations toward Budapest.

Directive of the VGK Stavka no. 220256 to the Commander of the 2nd Ukrainian front on organization of the offensive to Budapest
November 4, 1944 20 h 00 min

> *The headquarters of the Supreme High Command believes that an attack on Budapest on a narrow front by forces of only two mechanized corps with an insignificant number of infantry can lead to unjustified losses and put the troops operating in this direction under a flank attack by the enemy from the northeast. Based on this, the Headquarters orders:*
>
> *1.To* [quickly advance] *the right wing of the front (7th Guards, 53rd, 27th and 40th Armies) to the western bank of the river Tisza in order to launch an offensive on a wide front and* [to] *smash the enemy's Budapest grouping with a strike from the right wing of the front from the north and northeast in cooperation with a strike from the left wing of the front (46th Army, 2nd and 4th Guards MK) from the south.*
>
> *2.To accomplish this task, the forces of the Pliev group, no later than 7.11* [7 November], *strike from the Szolnok area to the north in order to roll up the enemy's defenses on the western bank of the river Tisza and* [advance] *the right wing of the front over the Tisza River.*
>
> *3. Report the orders given.*

This *Stavka* directive was most likely an attempt to correct the problems caused by Stalin's 28 October order to attack toward Budapest right away, without the required regrouping of forces.

A Red Army Traffic Regulator, 80 Km from Budapest, directs troop traffic.

5 November 1944

For the 7th Guards Army, the advance continued to gain ground despite the Germans deploying the 24 Pz D, part of the 46 ID, 5 Hu ID and the 4 SS Police D to block them.

The 2 UF cancelled the futile attacks by the 46th A at the doorstep of Budapest. General Malinovsky and *Stavka* needed to devise a new plan for capturing the city.

6 November 1944 - Weather 45 to 36° F, (7.2 to 2.2° C), wet ground

Once again, General Friessner, Commander of AG South, commented about the current situation in and around Budapest; some of this information is a bit odd.

> The first stage of the battle for Budapest ended, and when it became calm again on the southern outskirts of the capital, the life of the civilian population returned to normal. Trade enterprises were opened, tram traffic resumed, the streets were revived by a dense stream of cars. The population behaved as if there was no war at all… The soldiers reached their positions by tram. When I learned that Hungarian officers were spending the night in their apartments in the city, and not in their positions, I was forced to take drastic measures. [582]

General Friessner did not elaborate on what measures he took, but one would think he prevented Hungarian officers and troops from going home to their warm beds in Budapest at the end of the workday. These comments by General Friessner clearly show that the civilian population and the local government were totally unprepared for the imminent battle for Budapest.

7 November 1944

From 4 to 7 November the 7th Guards Army only gained about six miles, which was not thrilling, but at least it was steady forward progress. The Axis forces were in fact withdrawing at that rate, so they were the ones setting the tempo of the battle. Despite this, the 7th Guards Army had cleared the west bank of the Tisza River in its zone and finally was able to move the 24 GRC over the river.

By this date, the 46th A had lost its most recent gains and had been pushed back behind the "Attila I" defense line all around Budapest. This was a small victory for the defenders of the city. Note: The Attila I defense line, and several other lines, are not shown on maps due to their close proximity to the city.

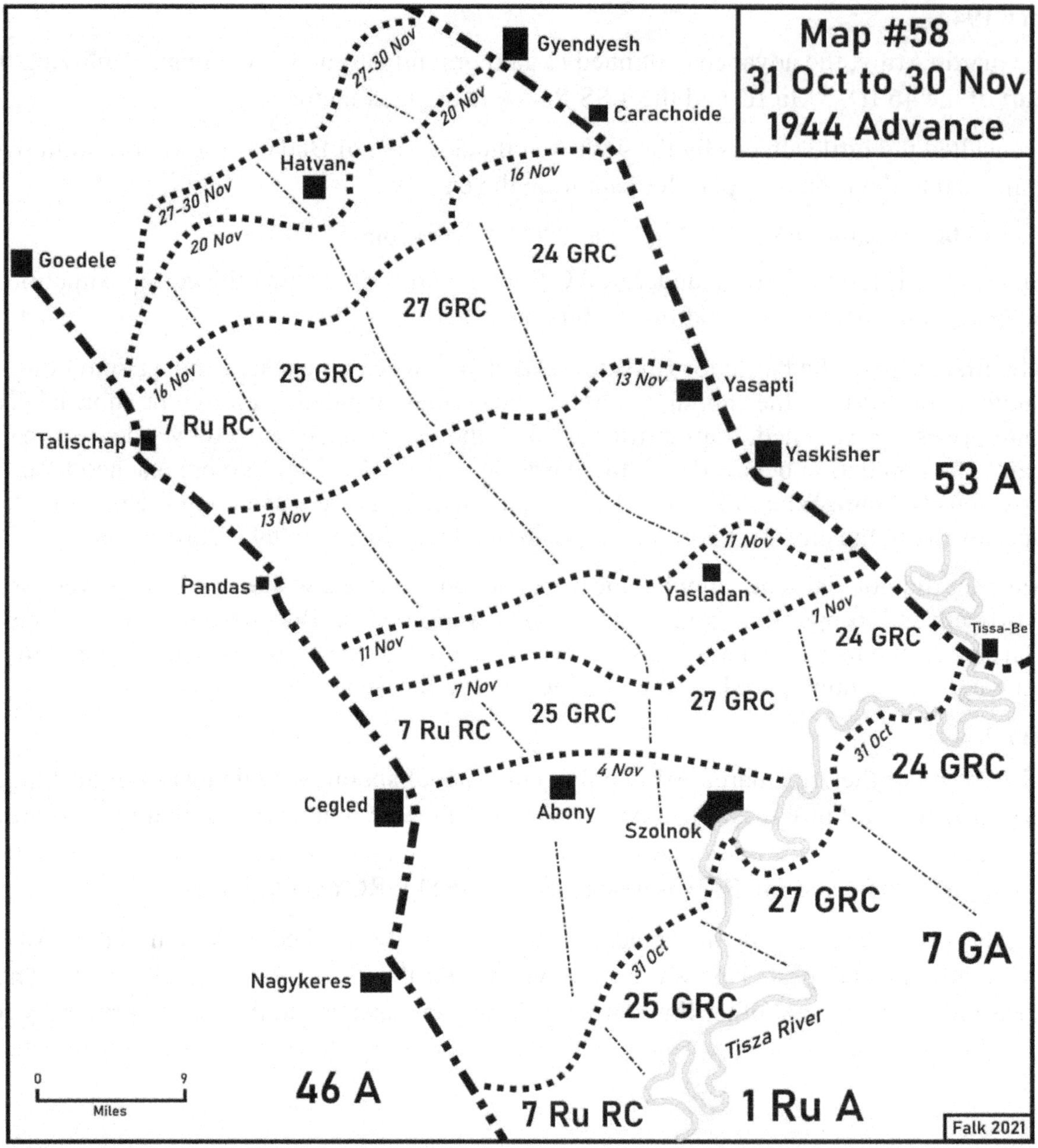

Note: See Map #57 for location reference to Budapest for the towns of Szolnok and Hatvan.

8 November 1944

The 23 and 27 TBs are depicted on historical maps concentrated behind the lines around Szolnok. The 23 TB came from 2 UF reserves and most likely was brought up to add additional mobile firepower to the 7th Guards Army. These two TBs had not yet been committed to battle; General Shumilov was waiting for the opportune moment to do so.

At this point in the battle, 10 of the 7th Guards Army's 11 rifle divisions were on the west side of the Tisza River and were making their way, shoulder to shoulder, to the northwest.

Faced with the overwhelming Soviet military capability that was advancing toward Budapest, namely the 2, 3 and 4 UFs, General Friessner knew his German and Hungarian forces were badly outnumbered. This fact forced him to conduct a fighting withdraw, which was a risky maneuver intended to wear the enemy down to the point of exhaustion by forcing them to constantly overcome new defensive positions. This tactic also traded space for time, all the while keeping one's own forces intact, as much as possible, and denying the enemy the chance to make a decisive breakthrough into the rear of AG South. This type of withdraw was a deadly balancing act, but General Friessner was quite skilled in the use of this tactic.

9 November 1944

Infantry units of the 46th A reached the Axis defense line south of Budapest in force, but there the advance stopped. In turn, Marshal Malinovsky, in response to *Stavka* Directive #220256, regrouped his forces to the east, concentrating a substantial number of mobile units in the zone of the 7th Guards Army, but that transfer would take time. Meanwhile, the 7th Guards Army continued its slow but steady advance to the northwest.

10 November 1944

Guards Major of the medical service Lev issued a medical report, *Information about the movement of the wounded and sick in medical institutions of the 7th Guards Army*, found in TsAMO record ID #133391646. This insightful report covers the period from 5 November to 10 November 1944, and is an assessment of the current medical burden upon the Army's medical establishment.

The total medical burden on 10 November was 10,059 people. These casualties were a substantial increase from the 30 October value of 7,409 people. The total medical burden is a good indicator of the severity of the fighting during that period.

Forward Troop Area – up to divisional medical battalions

- Arrivals:
  - At the start (5 November) – 1,691 men and women
  - Newly arrived – 1,977
  - Total – 3,668
- Departures:
  - Evacuated – 1,624 (most likely to the Army rear) (moderately wounded)
  - Returned to active service – 469
  - Died – 92
  - Total – 1,347 (including dead)
  - In the Forward Troop Area on 10 November – 1,483 (lightly wounded)

In Army Hospitals

- Arrivals:
  - At the start (5 November) – 7,585 men and women
  - Arrived from forward units' medical installations – 1,845
  - Arrived directly from units – 202 (severely wounded)
  - Totals – 9,632
- Departures:
  - Evacuated – 777 (out of the Army) (severely wounded)
  - Returned to active service – 221
  - Died – 58
  - Total – 1,056 (including the dead)
  - In Army Hospitals 10 November – 8,576 (moderately wounded)

Total in all medical installations on 10 November – 10,059.

With approximately 5,350 Army level hospital beds in total for all casualties, the medical establishment was overflowing with wounded. Note, over the period of five days, 5-10 November, the forward medical area returned to service 469 troops and Army hospitals added a further 221, for a total of 690 returns. This type of effort to return soldiers to the front was keeping the 7th Guards Army in the battle. After heavy artillery shelling of positions held by the 23 Pz D, 46 ID and 4 SS Police D, the 7th Guards Army renewed its offensive.

11 November 1944 – Weather: 39° to 34° F, (3.8° to 1.1° C), soaked ground

After heavy artillery shelling of positions held by the 23 Pz D, 46 ID and 4 SS Police D, the 7th Guards Army renewed its offensive. But the Army had been reinforced, so with this new attack the Army now had armor support.

> Having made the march, the 2nd and 4th Guards Mechanized Corps, in cooperation with the 23rd Tank Corps, which was part of the mechanized cavalry group of General I.A. Pliev, and the troops of the 7th Guards Army, resumed the offensive on November 11. Moving to Budapest from the east, the tank and mechanized corps, together with rifle divisions, met fierce enemy resistance. In addition, muddy roads began and the terrain became difficult to pass. The brigades were forced to advance together with the infantry at a pace of 5-6 km [3.1 - 3.7 miles] per day. [583]

12 November 1944 – Weather: 37° to 32° F, (2.7° to 0.0° C), soaked ground

As the 7th Guards Army continued to advance, the 2 GMC supported the left flank, the 4 GMC was in the center and the 23 TC was on the right flank. Spreading out the tank support to different rifle corps was necessary to ensure the enemy line would be broken without delay and without undue losses to the supporting infantry. The weather, which resulted in muddy roads and determined enemy resistance, greatly contributed to the slow advance.

13 November 1944 – Weather: 44° to 35° F, (6.6° to 1.6° C), soaked ground

By the end of the day, the 7th Guards Army and its mechanized support had advanced about 12 miles during the previous two days. If the Army could sustain this advance, Shumilov hoped to capture the city and transportation center of Hatvan quickly, which would cut a major enemy supply line and outflank Budapest from the northeast.

Albert Speer, Minister of Armaments, delivered to Hitler a memorandum, an inventory of current stocks of all critical alloy metals. The report showed that manganese ore would last 19 months, nickel 10, chromium 5.6, wolframite 10.6, molybdenum 7.8 and silicon 6.4 months. Speer went on to state,

> Hence, the element in shortest supply is Chromium. This is especially grave since Chromium is indispensable to a highly developed armaments industry. Should supplies from Turkey be cut off, the stockpile of Chromium is sufficient only for 5.6 months. That meant no more or less than that the war would be over approximately ten months after the loss of the Balkans. [584]

The loss of the Balkans would sever the export supply line of chromium from Turkey to Germany. Hitler took the news in total silence.

14 November 1944

As the 7th Guards Army moved steadily to the northwest, it captured numerous small towns and villages. This farmland area of the Hungarian Plain was densely populated. As the Army advanced, troops would have consumed and/or "liberated" any and all food, livestock or items (booty) in their path. The fate of these liberated communities and populations is unknown, but the war continued nevertheless.

15 November 1944

Once again General Friessner, commander of AG South, commented on the fighting in the zone of the 7th Guards Army. He wrote,

> Likewise, as a result of our counterattack, the enemy's attempt to break through our defenses in the sector south of [Gyendyesh] ended in failure. Combat operations here assumed the character of a mobile defense, that is, a method of conducting combat operations that our troops were excellent at. In these battles, our troops were able to eliminate enemy breakthroughs several times and preserve the integrity of their front. [585]

16 November 1944

The 7th Guards Army, with its four-rifle corps and supporting tanks, continued to advance despite minor setbacks. The city of Hatvan was now about seven miles away. Shumilov must have thought that one more good assault by tanks and infantry would be enough to capture this important city.

17 November 1944

The Russian advance to the northwest continued as the Germans fell back slowly, leaving villages and civilian populations vulnerable to the devastation, debauchery, and destruction brought about by the invading Red Army.

18 November 1944 – Weather: 41° to 28° F, (5.0° to 2.2° C), soaked morning and ice-crusted ground

The 7th Guards Army made minor progress on the flanks with little forward movement in the center, directly against Hatvan.

Stavka issued Directive#297140 transferring the 4th GA, with eight rifle divisions, from *Stavka* reserve to the control of the 3 UF. This information was found in TsAMO record ID #130117570. While this transfer was taking place, Marshal Malinovsky was ordered to transfer control of the 31 GRC from the 46th A to the 3 UF. He did this by shifting the boundary line on the left flank of the 2 UF to the right, thus effectively transferring the 31 GRC and its sector to the 3 UF. This same area was where the 4th GA would be deployed. *Stavka* was planning a little surprise for the German LXVII Corps and the 3rd Hu A, which were located on the other side of the Danube River from the 3 UF.

19 November 1944

The 7th Guards Army made little forward progress in its attack toward Hatvan. The advance stalled just before reaching the city.

20 November 1944

After attacking for four days, the 7th Guards Army had advanced only about five miles toward Hatvan, which lay another two miles away. Determined German resistance, the weather, and muddy roads slowed the Army's advance to a crawl.

21 November 1944

At this point the 7th Guards Army paused in its attack against Hatvan. General Shumilov needed to regroup and to rest his forces before the next round of combat could begin. He also made certain that all required supplies and ammunition were brought forward. The combat troops were able to take a short break while the rear services delivered the required provisions and equipment before the fighting resumed.

22 November 1944

Attacks against Hatvan began again with units of the 7th Guards Army making only minor progress.

23 November 1944 – Weather: 36° to 34° F, (2.2° to 1.1° C), soaked morning and ice-crusted ground
Attacks against Hatvan continued with rifle divisions moving around both flanks of the Germans defending within the city.

On the Western Front, units of 1st French Army captured their first large German city, Strasbourg.

24 November 1944 – Weather: 40° to 28° F, (4.4° to 2.2° C), soaked ground (for the rest of the month)
Fighting within Hatvan and the surrounding area continued.

Acting upon a report of the situation by Marshal Timoshenko, on or about this date *Stavka* ordered more forces transferred into the sector held by the 7th Guards Army. These additions included the entire 6th GTA, along with two additional mechanized corps, one tank corps, two cavalry corps, and two newly formed artillery divisions. [586] With the promise of these additional units, General Shumilov must have believed he would be able to take Hatvan rapidly and to move far beyond it. Because with this new support "[in] the Hatvan sector [the Soviet] infantry and tanks [would outnumber] those of the defenders by a factor of nine and their artillery by a factor of eight." [587]

The 46th A, in a bold attack, captured the 30-mile long and two-mile wide Csepel Island in the Danube River just south of Budapest. This island had strategic significance, which Marshal Malinovsky and the 2 UF would use successfully in the near future.

25 November 1944
The 7th Guards Army finally seized the city of Hatvan. Of course, the retreating Germans regrouped and formed a new defense line just outside the city, once again blocking the way forward.

Marshal Malinovsky issued report No. 00736/op at 16:30 hours. This report/plan was presented to *Stavka* for approval. A synopsis of the report is listed below.

1. Temporarily halt the offensive by the 53rd, 7th Guards and 46th Armies.
2. Launch attacks on the right flank of the front with the 27th A and the left flank with the 40th A.
3. Reopen the offensive with the 7th Guards Army, 6th GTA and General Pliev's group on 4-5 December in the general direction of Sahy, with the goal of capturing the Balassagyarmat, Sahy, Sob, [Felschoged] area. [588]

26 November 1944
With the capture of Hatvan, the 7th Guards Army began to dig in and started to reorganize its forces as it brought up supplies. Losses had been heavy, and the exhausted Army needed a rest.

According to *Stavka*, on this day the first stage of the Budapest Operation had been completed because forces of the 2 UF had reached the outer defensive line surrounding Budapest from the south, southeast and east. Phase Two would complete the total encirclement of the city. Therefore, *Stavka* issued Directive #220273 to regroup 2 UF forces and to set the stage for the start of the second phase of the Budapest Operation. This directive was in reply to Malinovsky's report from the previous day.

Stavka Directive of the VGK No. 220273 to the Commander of the 2nd Ukrainian Front, representative of the VGK on measures to ensure success in the offensive on Budapest
November 26, 1944 03 h 00 min

The headquarters of the Supreme High Command believes that the slow advance of the front's troops is due to the fact that the available sufficient forces and means are scattered over different sectors, instead of creating a decisive superiority over the enemy in the direction of the main attack and concentrating [in this area] *artillery breakthrough divisions and tank formations.*

The VGK orders:

1. To ensure the success of your planned offensive in the sector of the 7th Guards Shumilov's Army to concentrate at least two artillery breakthrough divisions.

2.[To bring] *the 6th Guards* [Tank Army Kravchenko] *to the offensive on the Shumilov sector. To strengthen the 46th Army from the 7th Guards Armies* [Army's] *section, to transfer not two, but only one mechanized corps.*

3. [To use] *Pliev's group to develop an offensive* [following] *the 6th Guards Tank Army* [attack].

4. To consolidate the battle formations of the 7th Guards Infantry Army, for which to expand the 53rd Army's offensive zone to the southwest, thereby making it possible for the 53rd Army to bypass from the west the wooded-mountainous region located to the north of Gyendyesha. [This was part of the Inner Western Carpathian Zone. See Map #48.]

5. [To] *start the offensive no later than December 2-3, 1944.*

6. To inform about the given orders.

Headquarters of the Supreme Command

I. Stalin A. Antonov

This *Stavka* directive caused Marshal Malinovsky to issue two directives of his own.

In accordance with the Stavka of the Supreme High Command's instructions, the front commander issues the following directives to his troops.

Directive #00744/op of 26 November ordered: (A synopsis follows)

a) The 7th Guards Army will turn over to the 53rd Army the 24 GRC HQ with the 81 and 72 GRDs, along with their sector. This is to be completed by 18:00 hours on 28 November.
b) The 53rd Army will receive the 24 GRC, troops and sector.
c) Establish by 18:00 hours on 28 November a new boundary line between the 7th Guards Army and the 53rd Army.

Malinovsky then issued a second Directive #00738/op: (A synopsis follows)

1. The commander of the 7th Guards Army will completely relieve during the night of 26-27 November units of General Pliev's cavalry-mechanized group.
2. Upon relief by the 7th Guards Army, Pliev's cavalry-mechanized group (4 and 6 GCC, 4 GMC) will concentrate to the rear. The 23 TC will be placed into the front reserve.

Also on this same day, the 30 RC, transferred from *Stavka* reserve, was ordered to begin moving during the night of 26-27 November and by 29 November concentrate to the rear of the 7th Guards Army and become subordinated to the 7th GA. [589]

27 November 1944

Marshal Malinovsky issue another Directive #00745/op, assigning his armies their objectives. For the 7th Guards Army, a synopsis is listed below.

> *The commander of the 7th Guards Army* [General Shumilov] *was to break through the enemy's front with the forces of eight rifle divisions (with six divisions in the first echelon) ...It was to secure the commitment into the breach of the 6th Guards Tank Army and General Pliev's cavalry-mechanized group. By the close of the third day, it was to capture the line Ipolszeg-Nograd-Veroce and completely reach the Danube with its left flank.* [590]

So during the upcoming attack, the 7th Guards Army was to rupture the enemy's defense line northeast of Budapest, then introduce the 6th GTA to expand and to deepen the penetration and then follow up with General Pliev's Cavalry-Mechanized Group to exploit the breach to the fullest extent.

If that was not enough force to do the job, the commander of the 5th Air Army was ordered to

> *concentrate the activities of all his assault and bomber aviation along the 7th Guards Army's breakthrough front and, besides this, to allot one assault air division apiece* [up to 124 aircraft] *for cooperating with the cavalry-mechanized group and the 6th Guards Tank Army, to cover the 7th Guards Army's main group of forces and the 46th Army....* [591]

Also on this day, the right flank of the 3 UF launched a major assault over the Danube River using the 57th A and the 4th GA. The weak German LXVII Corps literally did know what hit them as the Soviets had used *maskirovka* methods to secretly move the 4th GA into position along the river. German intelligence units completely failed to identify the arrival of this fresh Guards army at the front. Axis forces along the Danube now paid the price for losing track of a Guard's army. [592]

28 November 1944

The 24 GRC and its two rifle divisions (72 and 81 GRDs), along with their section of the front, were reassigned to the 53rd A. This transfer was done so that the 53rd A could advance more easily by shifting its attack sector to the left, thus avoiding the hilly terrain directly to its front. These hills were part of the Inner Western section of the Carpathian Mountains. (See Map #48.)

General Friessner, commander of AG South, noted at this time a general regrouping of Soviet mobile forces and a likely shift in the focus of further attacks. He noted, "The possible change in the direction of the main enemy attack demands a rapid regrouping of our forces to strengthen the defenses along the Danube River in the third Hungarian Army's sector." [593]

So, General Friessner took action to strengthen the area south of Budapest, but German intelligence failed to identify the buildup within the 7th Guards Army and the area around Hatvan. Marshal Malinovsky was indeed planning another offensive.

29 November 1944

The 7th Guards Army held its current position.

The German High Command (OKH) assessed the actions of the AG South during the recent fighting.

> During the fighting in Central Hungary between Budapest and Miskolc [85 miles northeast of Budapest], the enemy's attempt to seize Budapest with a blow from the southeast and break through into the regions of Western Hungary bordering with Slovakia was thwarted. German troops, together with the Hungarian allies under the general command of Colonel-General Friessner, delayed the almost continuous onslaught of 61 divisions and 7 tank corps of the

> Bolsheviks and Romanians, which began on October 29. The enemy suffered heavy losses in people and equipment. [594]

This author and many other historians would agree with this overall assessment. The Soviet forces of three fronts, 2, 3 and 4 UFs, should have been able to crush the defending Axis forces and to seize Budapest. Maybe Marshal Malinovsky had been correct after all. He did say he needed five days to redeploy troops. In his conversation with Stalin on 28 October, he stated,

> If we go over to the offensive without delay, then 46th Army, for sheer lack of forces, will not be able to develop its blow quickly [and] it will inevitably get bogged down in heavy fighting at the very approaches to the Hungarian capital. Putting it briefly, it cannot seize Budapest off the march.

Marshal Malinovsky wisely did not bring up this conversation with Stalin after the fact, even though he undoubtedly was correct in his original assessment.

30 November 1944

The 7th Guards Army held its current position.

Guards Major of the medical service Lev issued a medical report, *Information about the movement of the wounded and sick in medical institutions of the 7th Guards Army*, found in TsAMO record ID #133391659. This report covered the period from 25 November to 30 November 1944, and was an assessment of the current medical burden upon the Army's medical establishment.

The total medical burden on 30 November was 11,091 people. This was a higher casualty total than from the 10 November value of 10,059 people. This is an indication of very heavy fighting.

Forward Troop Area – up to divisional medical battalions
- Arrivals:
  - At the start (25 November) – 1,266 men and women
  - Newly arrived – 1,523
  - Total – 2,789
- Departures:
  - Evacuated – 1,241 (most likely to the Army rear) (moderately wounded)
  - Returned to active service – 644
  - Died – 74
  - Total – 1,959 (including dead)
  - In the Forward Troop Area on 30 November – 830 (lightly wounded)

In Army Hospitals
- Arrivals:
  - At the start (25 November) – 10,540 men and women
  - Arrived from forward units' medical installations – 2,616
  - Arrived directly from units – 219 (severely wounded)
  - Totals – 13,375
- Departures:
  - Evacuated – 2,247 (out of the Army to the deep rear) (severely wounded)
  - Returned to active service – 760
  - Died – 107
  - Total – 3,114 (including the dead)
  - In Army Hospitals 30 November – 10,261 (moderately wounded) (with 5,350 hospital beds)

Total casualties in all medical installations on 30 November were – 11,091 people. This report shows that the hospitals were trying to cope with almost twice the number of casualties they normally could be expected to handle, which was an impossible medical situation. The increased number of wounded who died while under medical care are indications not only of heavy fighting but also of medical services on overload.

The 2 UF issued a report, *Report on the loss of personnel in the troops of the 2nd UkrF from November 1 to December 1, 1944.* This was found within TsAMO record ID #113723627. The 7th Guards Army suffered 13,322 total casualties for the month: 2,154 killed; 254 missing; 0 captured; 14 non-combat losses; 9,567 wounded; 1,333 sick; and 0 frostbite. The report goes on to list 7th Guards Army horse losses: 414 killed; 4 missing; 47 wounded; 26 evacuated to veterinarian care; for a total of 491 equine casualties. The number of horses that died had dropped from the previous month, but overall, the weather, combat actives and overwork kept overall horse losses high. The same report shows that the troops of the entire 2 UF suffered 90,241 total casualties during November 1944.

In the above report, the 0 captured figure for the 7th Guards Army is interesting and almost certainly a lie to higher headquarters. For reference, German AG South reportedly captured the following number of Soviet/Romanian prisoners: 2,495 in November; 3,943 in December; 5,517 in January 1945; and 2,240 in February 1945. These numbers total 14,195 POWs or almost two rifle divisions' worth of troops. AG South mostly was on the defensive during this time, which would limit the number of people that could be taken prisoner during battle. [595] The number of defectors from the Red Army is unknown, but they would have been added to the reported numbers of POWs.

For the month of November in Britain, air raids and German V-1 and V-2 attacks killed 716 civilians and injured 1,511. Hitler's revenge attacks against Great Britain were not yet finished.

1 December 1944	
7th Guards Army	
Assigned – 2 Ukrainian Front	
Colonel General M.S. Shumilov Commanding	
Rifle, Airborne and Cavalry Units	25 Guards Rifle Corps (6 GAD, 25, 36 GRDs, 53 RD) 27 Guards Rifle Corps (227, 303 RDs) 30 Rifle Corps (141, 151, 155 RDs) 409 RD 1st Ru A, 7 Ru Rifle Corps (2, 19 Ru RDs, 9 Ru CD) (attached) 15 Penal Battalion 62, 63, 64, 65, 66,194 Penal Companies
Artillery Units	5 Guards Heavy Artillery Penetration Division (71 Light Artillery Brigade, 17 Guards Gun Artillery Brigade, 67 Howitzer Artillery Brigade, 95 Heavy Howitzer Artillery Brigade, 18 Guards High-Power Howitzer Artillery Brigade, 27 Mortar Brigade) 16 Artillery Penetration Division (49 Light Artillery Brigade, 61 Gun Artillery Brigade, 52 Howitzer Artillery Brigade, 90 Heavy Howitzer Brigade, 109 High-Power Howitzer Brigade, 14 Mortar Brigade) 2, 34 Antitank Artillery Brigades 114, 115 Guards Tank Destroyer Regiments 220, 228 Gun Artillery Brigades 41 Guards Gun Artillery Brigade 263, 290, 493 Mortar Regiments 5 Antiaircraft Artillery Division (670, 743, 1119, 1181 AA Regiments) 26 Antiaircraft Artillery Division (1352,1357,1363,1369 AA Regiments) 662 Separate Aerial Observation/Air Raid Warning Company
Armored and Mechanized Units	27 Guards Tank Brigade 38 Separate Armored Train Battalion
Engineer Units	60 Engineer Sapper Brigade
Reported Combat Strength	78,218 men
Estimated Ration Strength	90,290 men and women (including an estimated 7,000 support personnel)
Reported Combat Strength 1st Ru A (7 Ru RC)	24,703 men

(BSSA 44, BSA 44, TsAMO Archive)

The start of the month saw the 7th Guards Army massively reinforced. Its four rifle corps now contained 11 rifle divisions and one cavalry division. Two complete artillery divisions and two complete antiaircraft artillery divisions provided enormous firepower and support, plus the normal Army level complement of combat support units were available as needed. With a combat strength of 78,218 and an additional 24,703 men from the 1st Ru A, the 7th Guards Army now had at least 90,000 combat troops ready for battle.

Stavka and the 2 UF were ensuring that the 7th Guards Army was able to lead the northern arm of the offensive successfully, deploying enough firepower and troops on a narrow sector to break through any enemy defense line and to clear the battlefield of the enemy north of Budapest all the way to the Danube River. While the Army was doing this, the 6th GTA and General Pliev's Cavalry-Mechanized Group would enter the breach in the enemy line and fan out into the rear areas of AG South to conduct wide-ranging attacks.

In the south, the 46th A would be conducting the same type of maneuver for the southern arm of the offensive. But the 46th A was helped by launching its attack from Csepel Island, using it as a springboard over the Danube River. The overall aim was for the two arms of the offensive to meet around the town of Esztergom (also known as Gran) west of Budapest to completely encircle all defending Axis troops, the city of Budapest itself, and the surrounding urban area.

1 December 1944 – Weather: 43° to 40° F, (6.0° to 4.6° C), light drizzle

With the transfer of the 24 GRC and its operational zone to the 53th A, the 7th Guards Army frontage had shrunk to 25 miles. The Army now had 11 divisions crammed into this 25-mile front, with the 409 RD in reserve to the rear. Of course, the 6th GTA and General Pliev's Cavalry-Mechanized Group also were deployed a few miles to the rear while preparing for their part in the offensive.

David Glantz, in his book *Soviet Military Deception in the Second World War*, page 455-461, reveals that the German intelligence assessment of the overall situation was partially successful, but they missed the arrival of the 30 RC to the 7th Guards Army and the transfer of the 24 GRC to the 53th A. They also failed to identify the close concentration of the 6th GTA and the Cavalry-Mechanized Group in the forward zone of the 7th Guards Army. Thus, German intelligence agencies had lost track of at least one rifle corps and more importantly, six powerful Soviet mobile corps: two tank corps, two mechanized corps, and two calvary corps. They believed these mobile formations might be in the area of the 7th Guards Army, but they were unsure.

On this date, a soldier from Pz GD *Grossdeutschland* lamented about the secret weapons being developed by Germany. He wrote,

> What is more intended for us? We hear about miracle weapons; we already have the V-1 and V-2 used on the Invasion Front [France] and against southern England. Are we to expect more? V-5-V8-V9…? Here on the Eastern Front, we have a desperate need of such weapons. But the soldier should not think too much; for that reason, there is strict military training and drunken evenings. [596]

This soldier had been fighting on the Eastern Front for a long time and had seen thousands of dead, both friend and foe. He would not think twice about using miracle, wonder, ultimate weapons against his enemy, no matter what that implied.

2 December 1944

The 7th Guards Army held its position and continued to prepare for the pending attack. Of course, all major movements took place at night with supplies, troops and equipment quietly moving into position.

3 December 1944 – Weather: 40° to 37° F, (4.8° to 3.0° C), afternoon rain

The 7th Guards Army held its position.

> …[B]y this time the second Ukrainian Front, not counting the Romanian armies, numbered 528,000 combat troops; 10,867 guns; 3,974 mortars; and 565 tanks and self-propelled guns, while the opposing German and Hungarian formations from Army Group South had 127,000 combat troops; 2,800 guns; 800 mortars; and 140 tanks and assault guns. [597]

General Friessner knew all about the growing disparity of forces facing AG South. He noted,

> All submissions and reports to higher authorities about such an intolerable state of affairs turned out to be useless. We did not receive any German reinforcements and were left on our own, while our forces were constantly decreasing. [598]

Unknown to General Friessner, Adolf Hitler, for a very good reason, deliberately had been starving the Eastern Front of badly needed supplies, weapons and reinforcements. Hitler was preparing a little surprise for the Allies on the Western Front.

4 December 1944 – Weather: 42 to 32° F, (5.8 to 0.2° C), afternoon rain

Once again, the 7th Guards Army held its position, but senior commanders knew the fireworks would begin the next morning.

5 December 1944

Early in the morning, the 7th Guards Army's front line came alive as a 45-minute-long artillery barrage descended upon the unsuspecting defending Axis troops. This barrage included three new additions, the 66, 302 and 309 Guards Mortar Regiments, and if at full strength, each would have contained 24 Katyusha rocket launchers. This meant that with one full salvo from these three Guards Mortar Regiments, over eleven hundred rockets would land on enemy positions within a few minutes.

The plan for this artillery barrage is 19 pages long and was found within TsAMO record ID #455079406, *The plan of sighting targets and benchmarks with artillery of the 7th Guards Army*. This fire plan was designed to saturate the enemy's forward positions with massive amounts of aimed fire and fire covering specific target areas. Another TsAMO document, ID #131311470, showed the amount of artillery that would be deployed for each corps. The artillery units that were assigned to the 27 GRC were to fire against a 3/4 of a mile section of the front line (1.25 Km), the artillery for the 25 GRC would cover a 3-mile section (4.75 Km), while the 30 RCs artillery would cover 1.8 miles (3 Km), for a total of 5.5 miles or 9 Km. In these zones, the Army had amassed a huge number of weapons. (See Table #26.)

The 25, 27 GRC, and the 30 RC launched their attacks directly to their front with support from the 27 GTB. By the end of a day of heavy fighting, they had made a breach in the enemy's line about five miles deep and eight miles wide. The attack would continue the next day. (See Map #59.)

The opposing German IV Pz C attempted to hold the line but was overwhelmed by the power of the attack. The troops and artillery of the 7th Guards Army along with the VVS essentially broke the German line and achieved a tactical victory on the first day, thus insuring deep penetrations into the operational depth the next day. AG South had no substantial ground forces available to counter this breakthrough and the inevitable waves of Soviet tanks that soon followed. The Luftwaffe assisted in the defense, but they too were outnumbered.

Weapons	Available
82 mm mortars	340
120 mm mortars	535
Katyusha rocket launchers	72
45 mm AT guns	64
57 mm AT guns	7
76 mm regimental guns	59
76 divisional guns	602
100 mm gun	20
105 mm German howitzers	11
122 mm howitzers	102
150 mm German howitzers	8
152 mm howitzers	147
203 mm howitzers	48
Totals	**2,015**

Table #26: 7th Guards Army Artillery Strengths - 1 December 1944.

With over 2,000 artillery pieces directed against only 5.5 miles of the enemy's front line, the defending troops never had a chance. Of note, the 7th Guards Army evidently also used captured German artillery pieces against their former owners, including eleven 105 mm and eight 150 mm howitzers. Once again, the Red Army used every resource available to them, and they also demonstrated that they fully understood that it was easier to kill Germans with a howitzer than with a rifle.

The 46th A joined in the offensive when it launched its attacked over the Danube River south of Budapest. The 46th A captured a bridgehead but suffered heavy losses in the process. A quote from a Hungarian Colonel painted a terrible picture:

> Our artillery uninterruptedly shot up the Russians attempting to cross the river. German bombers also pinned them down. Three mortars fired on them from nearby positions. However, despite this the Russians continued to force the river. Groups of Soviet soldiers were ferried over on boats and ferries. I saw how a German bomb fell on a small barge, which was filled to the brim with people and went straight to the bottom. The most horrible things took place along that part of the bank where the Germans were. They fired on the landing troops with machine guns. There was very little chance of surviving. He who had not been cut down by a bullet was forced to fire back, standing up to his chest in the freezing water and mud. Having gotten his fill of all of this, one hussar [a Hungarian solider from the 1 Hussar Division] turned to me and said: "Colonel, what do the Russians do to their enemies if they treat their own soldiers so cruelly?" [Supposedly] 115 Soviet soldiers were awarded the title of Hero of the Soviet Union--many posthumously--for forcing the Danube in this area. [599]

6 December 1944 – Weather: 34° to 30° F, (1.6° to -0.8° C), rain from 14:00-15:00 hours

The offensive continued as the 7th Guards Army greatly expanded its forward positions. Meanwhile, the 6th GTA advanced in the center of the breakthrough area, within the zone of the 30 RC and 25 GRC, while the Cavalry-Mechanized Group advanced on the right flank in the zone of the 27 GRC.

The 7 Ru RC advanced somewhat on the left flank of the Army, but the Romanians lacked the strength and the artillery firepower to make a clean breakthrough.

The main penetration area for the 7th Guards Army rifle units now had grown to over 10 miles wide and 14 miles deep. With this opening, the mobile formations broke into the operational depth and began rampaging, behind enemy lines, in two general directions. The 6th GTA moved west toward the Danube River and to the northwest, while the Cavalry-Mechanized Group advanced in the general direction of Sahy (also known as Ipolysag) and the Ipel River. (See Map #60 for the location of Sahy and the Ipel River.)

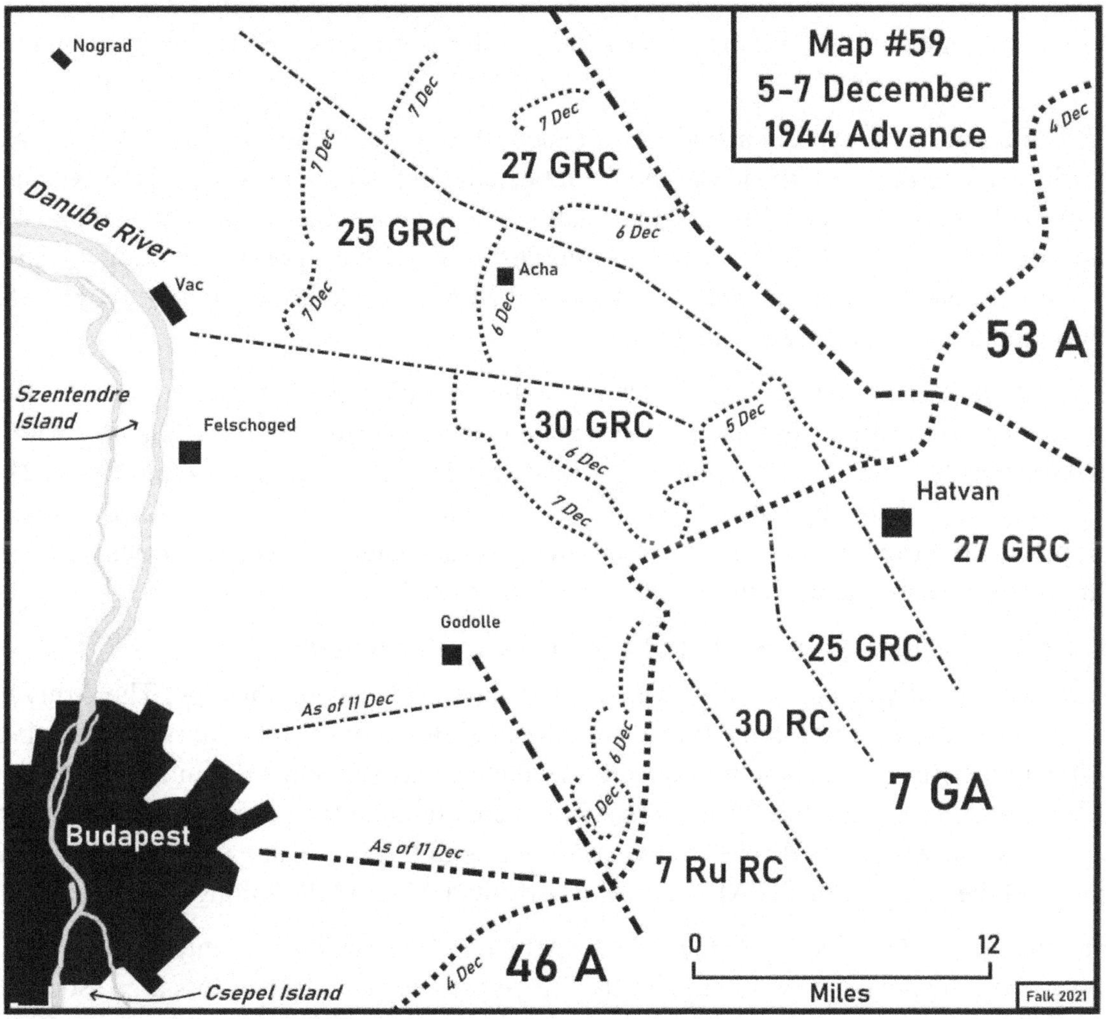

7 December 1944

Units of the 7th Guards Army maintained their sustained advance, securing the ground that had been overrun by the mobile units that preceded them. Because of this and their lack of troops, the enemy was unable to build a new defensive line short of the Ipel River. The forward rifle units of the Army also quickly approached the Danube River near the town of Vac. Meanwhile, the 30 RC turned to the west and made progress toward the town of Felschoged (also located on the Danube), while the 7 Ru RC tried to penetrate the original enemy line with limited success.

On or about this date, the 36 GRD from the 25 GRC was transferred to the 30 RC. Also, the 409 RD was assigned to the 25 GRC from the Army's reserve. General Shumilov was reinforcing his left flank as the Army advanced toward Budapest. General Shumilov knew from personal experience at Stalingrad that fighting within an urban area consumed combat infantrymen at an alarming rate. These adjustments put all of the Army's divisions in the front line. Shumilov also needed to commit every major unit to the front because the Army's frontage had expanded to about 75 miles in length.

8 December 1944

Units of the 7th Guards Army continued their advance, but the front was opening up with Army Corps advancing in different directions, from northwest, west and south toward Budapest. (See Map #60.)

With the 46th A on the left flank of the 7th Guards Army, both armies faced fierce resistance from German and Hungarian units as they advanced into the suburbs of Budapest. On the west side of the Danube, the 46th A made contact with the 4th GA (3 UF) thus forming a continuous front on that side of the river.

Budapest, the capital of Hungary, was actually composed of two ancient cities: Buda, which was located on the hills along the western side of the Danube, and Pest, which occupied the lowlands along the east side of the Danube. Because Budapest had not yet been surrounded, Marshal Malinovsky and General Shumilov attempted to thrust their troops into Pest as quickly as possible, taking advantage of any weakness or confusion on the part of the German and Hungarian defenders before they could receive reinforcements or stabilize their lines.

Also on this day, to add insult to injury, and with a "small amount" of Soviet encouragement, Bulgaria declared war on Germany. Militarily this development was insignificant, but the Bulgarian Air Force claimed that during its brief war with Germany, it shot down 25 German aircraft while losing 21 of their own. [600] With Bulgaria on the Allied side, Turkey also must have felt more inclined to remain neutral. With powerful Allied Armies closing in on all sides, the leaders of Nazi Germany and especially Hitler were becoming even more isolated and desperate.

9 December 1944 – Weather: 51° to 43° F, (10.8° to 6.2° C), afternoon rain

The 7th Guards Army really began its military actions against Budapest on this day. The Army had eight rifle and one cavalry division in the line advancing upon the city from the north and northeast. With the 46th A to the east and the south of the city, Budapest was surrounded on its entire eastern (Pest) side. Additional units from the 7th Guards Army were aligned along the Danube River to the north, which blocked any river traffic to the city. Assisting with the attack toward Budapest, the 27 GTB (7th GA) and the 5 GTC (6th GTA) operated mostly with the 25 GRC along the Danube.

Further north, units of the German IV Pz C held the Ipel River line, which prevented Soviet mobile units from exploiting the breakthrough to the north. Fierce battles took place in this area. South and west of the Ipel River, scattered German units blocked units of the 25 GRC from advancing further to the west. The rolling hills and forested terrain in this area also helped the Germans contain the Soviet advance.

At this time, the 7th Guards Army made no effort to cross the Danube River or even to seize the Island of Szentendre. The Army's efforts and forces were split between trying simultaneously to capture Budapest and to expand its position to the north. Clearly the Army lacked the strength to achieve both of these tasks quickly.

10 December 1944

The positions of the 7th Guards Army and supporting tank units along the Ipel River and at the bend of the Danube mostly were static. German Panzer formations launched counterattacks, which disrupted the Army's attempts to cross the Ipel River. After all, the main effort of the Army was to capture Budapest as soon as possible; these secondary attacks could wait. Despite the Army's best efforts against Budapest, only a slow advance was possible as the divisions fought through towns and villages on their approach to the city.

General Friessner stated that the area of the Ipel River was on the verge of disaster. He wrote,

> The situation at Ipolseg [Ipolysag/Sahy] took on a menacing character. If the enemy launched a strong strike force on the offensive in the western direction, its success would be fully assured. The way to Vienna through [Nove Zamky] and Bratislava was then completely free. The entire area between the Danube and the Slovak border was a vacuum, in which there were almost no German soldiers. The 2nd Ukrainian Front now [had a chance], under the cover of the Danube, to strike in the direction of Vienna. If Malinowski had known how few German forces opposed him here at that time, he would not have long had to puzzle over this decision. However, he did not make such a decision. [601]

General Friessner fears would come true soon enough. Note that in this area, the Ipel River defined the border between Czechoslovakia to the north and Hungary to the south.

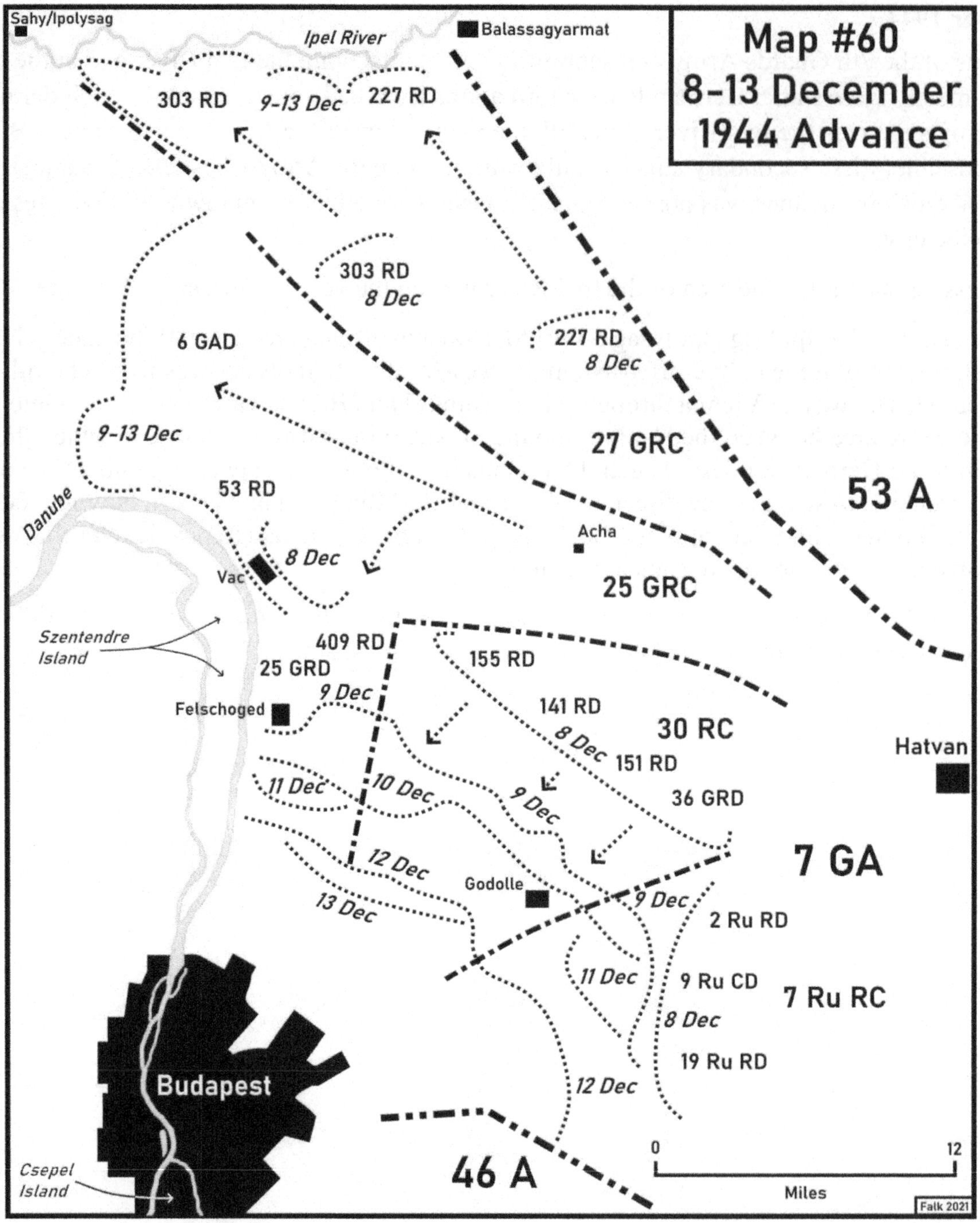

11 December 1944

The 25 GRC and the 7 Ru RC advanced against Budapest on both flanks during the day. This most likely set the stage for a larger advance the next day.

General Shumilov issued Secret Order #0351 to the Army about streamlining the evacuation of prisoners of war from the front to army reception centers. This historical document stated the movement of "prisoners of war [to] army reception points [were] entrusted to unprepared, untrained and accidentally appointed servicemen." He went on to say, there was "no personal responsibility for the maintenance, protection and transportation of prisoners of war." He gave the entire Army four days, until 15 December, to comply with all 10 sections of his order. [602]

A synopsis of General Shumilov's ten points is as follows:

- Divisions and regiments needed to form collection points for POWs with prisoners flowing from regiments to divisions and then to an army level gathering point.
- Prisoners should be guarded by personnel who were "limited fit for non-military service" [i.e., fit for duty, but not for combat].
- The collection and movement of prisoners "shall be guided by the order of the NCO [Central Committee] No. 001 of 1943 and 006 of 1944" and the "instructions of the authorized department of the NKVD for affairs of prisoners of war." [Citing "orders from above" also neatly protects General Shumilov from any criticism of his handling of the POW issue.]
- POWs should be provided with food [and water], and their movements should be communicated along with records/totals of prisoners.
- Lastly, "The control over the implementation of this order shall be entrusted to the head of the rear of the army."

1944/1945 Hungary - German and Hungarian prisoners march into captivity.

Overall, this was a direct and well laid out order, but it lacked any guidance about punishment or mistreatment of POWs. In that regard, item #6 on Shumilov's order was deliberately left blank on the historical document from the archive. We know item #6 did involve something about the handling of POWs, but Soviet authorities did not want to disclose that information. A bit of speculation might clarify this blank space in the historical record. For example; [#6 Prisoners shall be conveyed to the rear of the Army in a timely manner, but escaping, uncooperative or sick/wounded prisoners will be liquated as necessary.]

All during the month, the 27 GTB worked closely with and supported the rifle units of the Army. Listed below, in Table #27, are the evening operational armored vehicles totals for the brigade.

December Dates	T-34s	Su-76s	Su-85s
11	7	1	1
12	13	1	-
16/17	9	1	4
18	12	1	4
19	9	4	4
21	14	-	3
23	17	-	6
24	19	5	-
25	20	-	7
26	18	-	7
27	13	-	8
28	18	-	7
29	17	-	7
30	12	-	6

Table #27: Armored Strengths for 27 GTB – 11-30 December 1944. [603]

As can be seen on the table, armored vehicle strengths for the brigade fluctuated daily. Some vehicles were total losses (burned out) during the fighting, while others were repaired in the field overnight or arrived from repair shops further to the rear. The important thing was that when the infantry divisions went into battle, they almost always had some tanks or self-propelled gun in support.

On or about this date, the 2 UF ordered its front reserve, the 18 GRC, to the south of Budapest where it would operate independently, directly under the control of the 2 UF. The 18 GRC would take several days to arrive and take over the sector from the 46th A.

12 December 1944

All along the 7th Guards Army front line facing Budapest, the troops made a big push this day, which resulted in the capture of the town of Godolle. This was progress, but slow progress.

Further north, units of the 9 GMC, after fighting off German counterattacks, reached the outskirts of Sahy by evening. Then during a nighttime reconnaissance mission with a M4A2 Sherman tank, with which the Russian unit was equipped, Russian tank commander Guards Captain Nikolay Bogdanov captured a *Uhu* (Eagle Owl) device or possibly some other type of infrared device that was attached to a German self-propelled gun. [604] The *Uhu* was a secret night vision infrared apparatus that allowed a vehicle's crew to see, fight and drive in the dark. This was another example of a Nazi miracle weapon, but this one was being developed for the German *Heer* (Army). Guards Captain Bogdanov received an Order of the Red Banner for his exploit, which was awarded for bravery during combat operations. Despite this intelligence success, Soviet forces were not able to deploy infrared devices during the war for various reasons.

13 December 1944 – Weather: 46° to 38° F, (8.2° to 3.6° C), afternoon rain

Another attack by the 25 GRC and the 30 RC pushed a section of the front to less than one mile from Budapest. After eight days of constant marching and combat, the Army was starting to show signs of fatigue.

General Shumilov changed troop positions once again. Both the 27 GTB and the 5 GTC (6th GTA) were pulled out of the line facing Budapest and sent north to rest and to prepare for further operations.

14 December 1944

The right flank of the 7th Guards Army and the 6th GTA became active again. General Friessner stated, "On December 14, the Russians captured Ipolseg [also known as Sahy] and pushed back the 371st Infantry Division and subordinate units of the Dirlewanger Brigade and the 2nd Hungarian Tank Division." [605]

Earl F. Ziemke, in his book, *Stalingrad to Berlin: The German Defeat in the East,* page 382-383, talked about the Dirlewanger Brigade in Hungary. Previously, this brigade had been viciously active in Russia and during the Warsaw Uprising and was notorious for the mass murder of civilians and other war crimes.

> Part of its troops were German Communists, and all were considerably less than dedicated soldiers of the Reich. The officers were roughnecks and sadists, impromptu executioners rather than tacticians. [The day after they failed to hold Sahy] a company of Communists deserted; and in the succeeding days the brigade slid into a state that by normal standards would have been considered mutiny- some troops shot their officers, some deserted-and some did both.

Due to the urgent need for more troops at the front, a desperate General Friessner was forced to put unreliable units like the Dirlewanger Brigade into the line.

On or about this day, General Shumilov transferred the 36 GRD and the 141 RD from the 30 RC to the 27 GRC. He also shifted the corps boundary line between the 25 GRC and the 30 RC to the north, which transferred the 25 GRD to the 30 RC. The 30 RC and the 7 Ru RC now were on their own, with six rifle divisions. This force was required to cover the entire front that was advancing against Budapest, thus simplifying command-and-control issues to the north of Budapest. Shumilov did this to regroup his forces more to his right flank in an attempt to arrange a breakthrough of the German lines between the Danube and the Ipel Rivers so that the 6th GTA and the Cavalry-Mechanized Group could more freely operate in this area.

After receiving *Stavka* Directive #220280 on 12 December, Marshal Malinovsky directed a map planning exercise with his army and air force commanders, trying to satisfy *Stavka* and to develop a plan to complete the encirclement of Budapest and to capture the city. Present were General Ivan Mefodyevich Managarov, Commander of the 53th Army; General Shumilov, Commander of the 7th Guards Army; General Pliev, Commander of the Cavalry-Mechanized Group (4 and 6 GCC, 4 GMC); General Sergei Goryunov, Commander of the 5th Air Army; and General Andrei Kravchenko, Commander of the 6th Guards Tank Army. Time after time the commanders stood before the planning map, one at a time, and presented their strategies, and every time Malinovsky rejected their proposals. Finally, General Kravchenko (6th GTA) stood and said, "Since we need to quickly go to Esztergom, close the ring and not let out the Nazis, please, Comrade Marshal, allow my tank men to advance in the first echelon." After some debate, with the assurance the 6th GTA would receive strong artillery and air support, and with the infantry armies pushing forward forcefully, the plan was approved. [606] The newly arranged offensive would be launched a few days later.

15 December 1944

To counter the loss of the critical defensive position of Sahy, General Friessner transferred the 8 Pz D to the area for a counterattack. Luckily, the Soviet units occupying Sahy did not advance further during the day. Unknown to General Friessner, the Russian units occupying Sahy were waiting for reinforcements: the rifle divisions that General Shumilov already had begun to transfer to his right flank and to the Sahy/Ipel River area.

The 18 GRC, from the 2 UF reserve, arrived south of Budapest and began taking over a section of the front with its two rifle divisions, the 66 GRD and the 317 RD. Additionally, the 18 GRC took control of the 68 GRD from the 46th A. This arrangement allowed the 46th A to conduct operations solely on the western side of the Danube. Once again, this shifting of forces simplified the command-and-control issues around Budapest.

16 December 1944

In the early morning hours, on the western border of Germany, three complete German armies, the 6th Panzer Army [also known as the 6th SS Panzer Army], 5th Panzer and 7th Army launched *Unternehmen Wacht am Rhein* (Operation Watch on the Rhine) against units of the American 1st Army. As conceived by Hitler, this massive surprise attack through the Ardennes Forest area of Luxembourg and Belgium was destined to turn the tide of war in the West in favor of Germany. The secret buildup for this attack was why the Eastern Front had been starved of vital supplies, reinforcements and airpower for the previous few months.

One special unit used during the Ardennes attack was Panzer Brigade 150, commanded by none other than Otto Skorzeny, the same commando who helped rescue Mussolini from captivity and also helped topple the Hungarian government. Skorzeny, the crafty German SS-commando, now was leading a task force of English-speaking German soldiers in Operation *Greif* (Gryphon). These troops would wear mostly American uniforms and use captured equipment in order to operate behind the Allied lines, disrupting troop movements and seizing key bridges needed for the offensive. Hitler's favorite commando was at it again.

On the darker side of this enterprise was the fact that any German soldiers who were captured wearing and fighting in American uniforms were put before a firing squad and shot. The Geneva Convention of 1929 stated that by their actions, (wearing the wrong uniform), they had given up their rights as POWs and were therefore open to execution. At least 14 German soldiers were executed for taking part in Operation *Greif.*

17 December 1944 – Weather: 38° to 32° F, (3.6° to 0.4° C), drizzle between 18:00 - 20:00 hours

For the 7th Guards Army, the 36 GRD and the 141 RD were on the way to join the 27 GRC. The 409 RC and the 27 GTB also were moving into new positions north of the Danube, but they remained within the 25 GRC zone.

Warsaw, or what was left of it, fell to the advancing Red Army. Stalin now was free to install his puppet government, the Polish Committee for National Liberation, in the capital of Poland.

18 December 1944

The German 8 Pz D attacked in an attempt to recapture the town of Sahy. The Russian forces who occupied Sahy held their ground. Soviet forces continued to prepare for their advance over the Ipel River.

The right flank units of the 7th Guards Army, the 25 and 27 GRC, began their attacks to the north and west with support from the 6th GTA. Units of the 27 GRC operated on both sides of the Ipel River. The 4 GMC (Cavalry-Mechanized Group) operated mostly south of the Ipel River and east as far as the town of Balassagyarmat. (See Map #61 18-31 December Advance.) Note: The 243 RD from the 53rd A assisted with the fighting north of Sahy. On 23 December, the 243 RD would be transferred to the 7th Guards Army and the 27 GRC.

Just north of Budapest, units of the 30 RC and the 7 Ru RC defended their positions for the next several days. They were resting, adjusting their lines and bringing forward supplies and probably more artillery and ammunition. Due to the level of enemy resistance so far, Red Army commanders and troops knew that Budapest would not fall quickly or easily. A long, hard battle lay ahead.

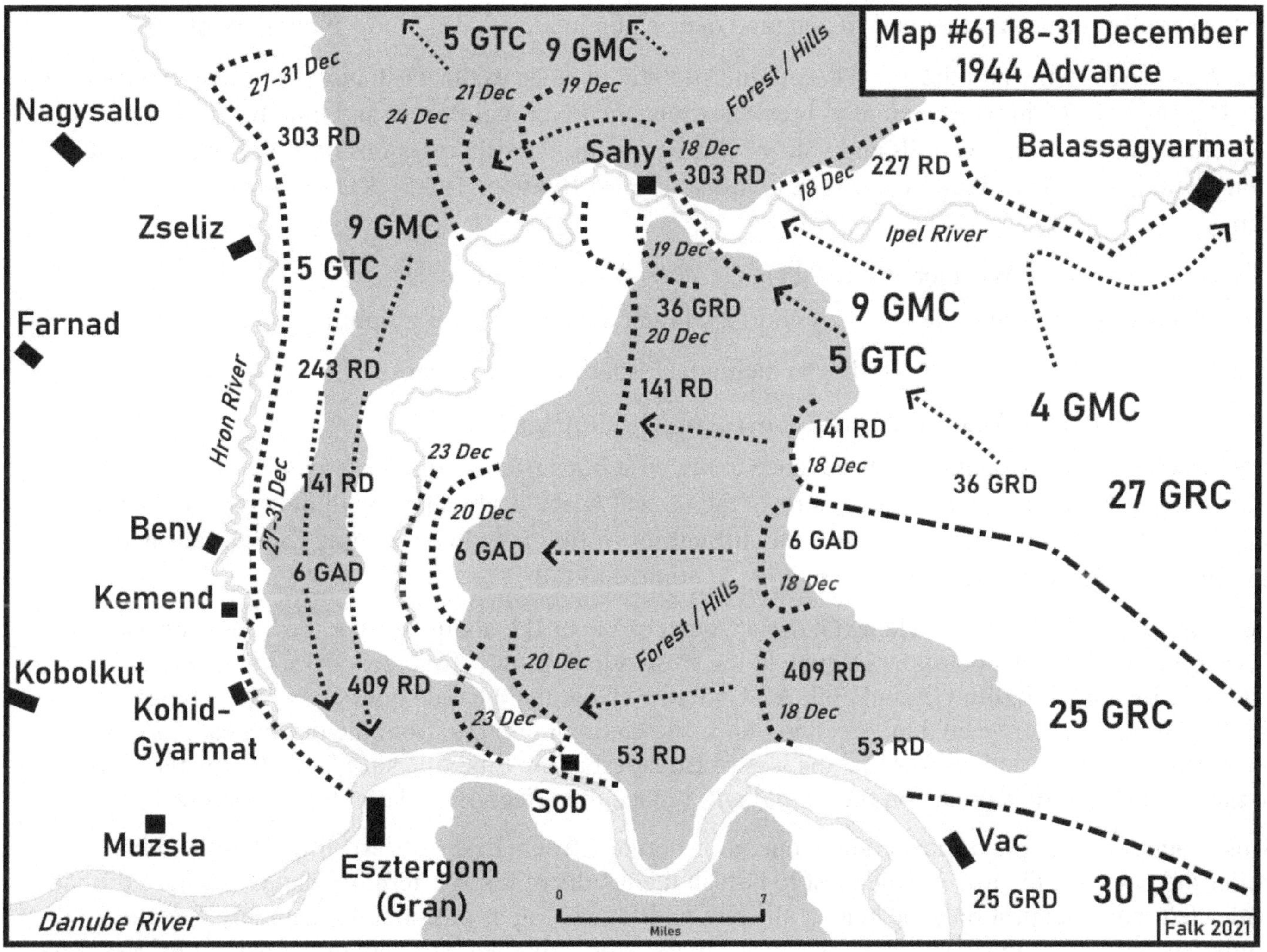

19 December 1944

Launching a sudden attack, the 5 GTC and the 9 GMC broke out of the Sahy bridgehead, and they advanced to the northwest side by side. On the north side of the Ipel River, the tank units had support from the 303 RD (27 GRC). According to plan, the 5th AA supported this attack with 600 aircraft while rifle units from the 7th Guards Army advanced all along the line to the north and west. Meanwhile, the 227 RD covered the right flank of the 6th GTA right on the north side of the Ipel River. South of the Ipel River, the 36 GRD and the 141 RD, also part of the 27 GRC, advanced to the west, covering the mobile corps' left flank.

To the south, units of the 25 GRC, 6 GAD, 409 and 53 RDs along with the 27 GTB advanced due west along the Danube and headed toward the mouth of the Ipel River some eight miles away.

German resistance was scattered, with strong points established at key road junctions. The 3 and 6 Pz Ds (minus all their tanks) and the Hungarian Division St. Laszlo were ordered to join the 8 Pz D north of Sahy for an eventual counterattack.

Southwest of Budapest, the 4th GA and the 46th A (3 UF) developed their own offensive. They planned to advance west of Budapest, driving north toward the Danube River and the town of Esztergom where they would link up with the 2 UF, consequently completing the encirclement of Budapest.

20 December 1944 – Weather: 33° to 26° F, (1.0° to -3.2° C), snow, evening drizzle

The onset of freezing weather and frozen mud once again made off-road movement possible.

The 7th Guards Army and the 6th GTA continued their advance to the west. North of the Ipel River the 5 GTC and 9 GMC, in their attempt to break free, engaged German Panzer and infantry formations in desperate battles among the hills and valleys. Further south, the rifle divisions of the 27 and 25 GRC along with the 27 GTB likewise engaged in hill and forest fighting. The 53 RD captured the town of Sob.

21 December 1944 – Weather: 30° to 22° F, (-1.2° to -5.4° C), snow

The 7th Guards Army and the 6th GTA continued their advance in heavy fighting.

The 4th GA and the 46th A (3 UF) began their attack southwest of Budapest.

22 December 1944 – Weather: 27° to 21° F, (-2.6° to -6.0° C), snow

The 7th Guards Army continued its advance to the west but suffered from heavy and confused fighting, all for negligible progress. However, units of the 6th GTA made a breakthrough of sorts and reached the upper Hron River. One can imagine the difficulties of fighting along winding forest roads and through the hills as the snow and temperature continued to fall.

General Friessner talked with Heinz Guderian, head of the OKH, about shifting the 8 SS Cavalry Division from the east side of the Danube to the west side to protect Budapest from the newly opened offensive by the 3 UF (4th GA and 46th A.) Guderian stated that he did "not believe that Hungary [and/or Budapest] can be held by removing only one German division from the bridgehead." He went on to say, "The Fuehrer believes that the loss of Budapest at the time of a successful German offensive in the West [Operation Watch on the Rhine] will reduce the effect of the latter by 50 percent." [607]

This is another example of how combat operations on one front (East or West) impacted the other. Hitler did not want the loss of Budapest to tarnish the results of his Western offensive. Because of that, Hitler felt that Budapest must be held at all costs and for as long as possible. Presumably General Guderian, as head of the OKH, had other plans in the works that were unknown to General Friessner to hold onto Budapest and Hungary, namely the imminent arrival of the IV SS Pz C.

23 December 1944 – Weather: 28° to 22° F, (-2.2° to -5.2° C), snow

The 6th GTA, having reached the Hron River, turned its 5 GTC and 9 GMC to the south and began to push down the east side of the Hron River. This was an attempt to drive into the enemy's rear area, unhinging their front and reviving the faltering 7th Guards Army's advance. (See the first entry for 24 December)

In the evening, on the right flank of the Army, the 227 RD from the 27 GRC was exchanged with the 243 RD from the 53rd A. This exchange was part of a sector/boundary adjustment between armies.

During a call with the OKH, General Friessner learned that he had been relieved of command of AG South (no reason was given) and that he should report to the *Führerreserve* (leaders reserve). General of Infantry Otto Wöhler would take over command of AG South. General Friessner later stated, "Now I am glad that fate saved me from participating in battles on German soil." [608]

24 December 1944 – Weather: 31° to 19° F, (-0.2° to -7.0° C), snow

The 7th Guards Army was unable to advance because the 6th GTA was engaged in its own swirling tank battle on the east side of the Hron River, directly east of the town of Zseliz (also known as Zhelezovce).

Outside of Budapest, the 30 RC and the 7 Ru RC held their ground, but they would become active again the next day.

In Germany, *Generaloberst* (Colonel General) Guderian, Chief of Staff of OKH, met with Hitler about ending the Ardennes Offensive early and transferring major forces to the Eastern Front. Guderian anticipated a major Soviet offensive against Germany around 12 January 1945 and wanted to build up forces to counter the Soviet attack. Hitler commented, "This [Soviet attack] is the greatest bluff since Genghis Khan. Who dug up this nonsense?" Heinrich Himmler, head of the SS and *Gestapo* (secret police) was present at this meeting and declared, "It's all just a giant bluff…I am quite convinced that nothing is going to happen in the East." [609] Thus, Guderian's legitimate concerns about the defense of Germany were ignored.

25 December 1944 – Weather: 33° to 19° F, (+1.0° to -7.2° C), snow

The units of the 30 RC and the 7 Ru RC launched a combined attack with five divisions against the Axis Pest bridgehead. They made good progress from the start.

On or about this date, German and Hungarian units fighting between the Hron and Ipel Rivers began a fighting withdraw to the west and across the Hron River. The 6th GTA, advancing from the north on the east side of the Hron River, threatened to cut the German supply line and to capture vital river crossings. Units of the 6th GTA attacked German and Hungarian bridgeheads, one after another, along the east side of the Hron, forcing the defenders to eventually cross to the west side of the river and to destroy any remaining bridges.

On the Western Front, the German Operation Watch on the Rhine offensive (also known as the Battle of the Bulge) had reached its maximum extent in the Belgium/Luxemburg Ardennes region. Striking from the north, the U.S. 1st Army under General Courtney Hicks Hodges and from the south, the U.S. 3rd Army under General George S. Patton were beginning to crush the attacking German armies between them. The intent was to drive the German armies back to their starting line. Losses on both sides had been heavy both in men and in equipment, but the Western Allies could replace their losses; Germany could not! Hitler's last chance offensive in the West had failed before reaching any of its objectives and had only delayed the start of the Western Allies' final offensive upon Germany by about six weeks. Suffering from this overwhelming defeat, Hitler once again would turn his attention to the East.

26 December 1944 – Weather: 20° to 14° F, (-6.4° to -9.6° C), snow

The 30 RC and the 7 Ru RC continued their attacks toward Budapest. Also, on or about this date, the 25 GRD conducted a surprise attack against the German-held island of Szentendre in the middle of the Danube River, north of Budapest. The weak defenders were driven back as Red Army infantrymen quickly captured major sections of the island. The 25 GRD also had some help in this effort from the 46th A that was advancing to the north along the western side of the Danube River.

On or about this day, forces of the 2 and 3 UFs officially surrounded Budapest, a belated Christmas gift from Marshal Malinovsky and the Red Army to Stalin. Approximately 100,000 civilians had fled the city before the escape routes had been cut, so roughly 800,000 civilians now were trapped within the city and the surrounding area, along with "approximately 70,000" [610] German and Hungarian defenders.

"At the time of the encirclement there were 450 tons of munitions, 120 tons of fuel and a five-day supply of food for the 800,000 residents and defenders who remained within the city. The garrison alone required 80 tons of food per day, of which only 47 tons [per day] were delivered by air [during the siege]." [611]

Immediately after Budapest being surrounded, German transport aircraft were able to land on the Pest side of the city at emergency airfields located at the racecourse and other locations. These flights delivered supplies and flew out wounded, but the racecourse was operational only until 3 January 1945. After that, only small aircraft were able to land within the city along with cargo gliders in cleared areas. Increasingly, parachute drops of supply containers were used to bring in needed supplies by air.

The desperate need for ammunition within the city forced the Germans to launch an equally desperate supply mission by water. A 40-ton motor barge, loaded with an unknown amount of ammunition (perhaps up to 4,000 tons,) was sent down the Danube at night in an attempt to reach the Budapest garrison. The barge almost made it, grounding on a sand bar about 10.5 miles north of the city near the town of Leanyfalu. Accounts vary, but some or possibly all of the ammunition eventually was salvaged and brought into the city. [612] Another slightly different and expanded account of this same river exploit can be found within the book *Forgotten Legions – Obscure Combat Formations of the Waffen-SS,* by Antonio J. Munoz, page 195-196.

The major military units defending Budapest were the German 13 Pz D, and elements of SS Pz GD *Feldherrnhalle* (named after a building in München where Hitler staged his Beer Hall Putsch) not to be confused with Pz C *Feldherrnhalle* (Pz C F). Also included were the 8 SS Cavalry Division *Florian Geyer* (named after a medieval knight and hero of Germany), and the 22 SS Cavalry Division *Freiwilligen* (a term used for non-German combat volunteers). Major Hungarian units also defended the city; they consisted of the 10 Hu ID and 12 Hu ID, along with remnants of the 1 Hu Pz D and the 1 Hu CD. [613] Note: With three cavalry divisions within the besieged city, (in itself, a strange circumstance) approximately 30,000 horses also were trapped. These horses would prove to be useful later on during the siege.

> After Christmas organized food supplies ceased almost completely. As the population had not been prepared for a prolonged siege, most people began to starve after a few days. Many survived only thanks to the 30,000 or more horses that had been brought to the city by the Hungarian and German cavalry and artillery units.... By the end of January, with only horse meat and carrots (meant for horse feed) or peas available, the soldiers [also] were starving. [614]

Hitler had forbidden, many times before and after the encirclement, any breakout from the city because Budapest had been declared a *Festungen* or Fortress City. In essence, Budapest was to be sacrificed to defend Nazi Germany. But it was even more callous than that. "Thus Hitler's supreme representative in Hungary, Edmund Veesenmayer [*SS Brigadeführer* or Brigade Leader], declared that it mattered not to him if the city was destroyed ten times over as long as Vienna could be defended." [615]

After German and Hungarian units withdrew from the town of Esztergom (Gran) to the north, they destroyed the bridge over the Danube. Later that day Red Army units of the 46th A (3 UF) fully occupied Esztergom, a key town on the Danube located near the mouth of the Hron River. Note: Some reports suggest that units of the 7th Guards Army were the first to capture Esztergom, but this information has not been verified.

During its major advance north of Budapest, General Shumilov recalled the support given by aircraft to clear the way for the Army.

Colonel-General M.S. Shumilov, commander of the 7th Guards Army, recalls the increased role of attack aircraft in defeating the ground enemy: [along the forward] edge and in the nearest tactical depth, [aircraft] inflicted heavy losses on them. By pressing the enemy to the ground with machine-gun and cannon fire, the attack aircraft largely contributed to the infantry's [advance] into the attack, and subsequently to the breakthrough of the entire tactical depth of the enemy's defense. All the commanders of corps and divisions who participated in breaking through the enemy defense [of] northeastern Budapest give a high assessment to the actions of the attack aircraft, [and they] directly felt the enormous help that was provided by the units of the 5th Assault Air Corps to the advancing infantry. [616]

27 December 1944 – Weather: 20° to 19° F, (-6.4° to -7.2° C), snow

As more rifle units of the Army begin to reach the Hron River, the 7th Guards Army now fought on Czechoslovakian territory in strength. This advance to the river was made possible by the daring 6th GTA attack to the south along the east side of the Hron. After reaching the mouth of the Hron River, the 5 GTC spearhead halted its advance while waiting for the 9 GMC to move adjacent.

On the other side of the line, the German IV SS Pz C, consisting of the 3 SS Pz D *Totenkopf* and the 5 SS Pz D Viking, began arriving in the area west of the Hron River. Soviet intelligence services failed to discover the arrival of this Panzer Corps. The weather and lack of air reconnaissance most likely contributed to this intelligence failure. The IV SS Panzer Corps hurriedly had been transferred south from the Vistula River area in Poland to deal with the worsening situation in Hungary. Once they detrained, they moved south over the Danube and quickly assembled for battle along with other units in the hilly, forested terrain northwest of Budapest. This location was not ideal for an attack, but fuel shortages prevented relocation to a more promising area.

Due to Western Allied bombing efforts against German oil refineries and synthetic oil production, the overall German fuel production in December was at 59 percent of capacity with aviation fuel at 56 percent of capacity. These numbers would not improve during the remainder of the war. [617] These limitations translated into fewer sorties by the Luftwaffe and more restricted operations by Panzer formations. Lack of fuel also caused a greater reliance upon rail transportation which by its very nature also was vulnerable to air attack.

During December, in support of the Red Army ground efforts, the 5th AA flew 8,245 combat sorties. [618] Luftwaffe sortie rates for December are unknown.

28 December 1944 – Weather: 20° to 17° F, (-6.4° to -8.0° C), snow

Rifle units of the 7th Guards Army now fully were aligned along the Hron River. Even a few subunits of the 409 RD had crossed to the western shore of the Hron near the mouth of the river and formed a tiny bridgehead. This crossing drew a quick counterattack response from local German forces.

Within Budapest, Axis troops destroyed the southern railroad bridge over the Danube to prevent it from being captured by advancing Soviet troops. The next day, the northern railroad bridge also was destroyed for the same reason.

On this day, General of Infantry Otto Wöhler took over command of Army Group South from Colonel General Johannes Friessner, due to perceived command failings by Friessner.

1945 Hungary – 57mm (Zis-2) AT Gun deployed in Buda or perhaps in the suburbs of Pest.

29 December 1944 – Weather: 25° to 19° F, (-3.8° to -7.2° C), snow

The 9 GMC reached the mouth of the Hron River. With their victory complete, the 6th GTA received orders to withdraw its units to the rear for refitting. The rifle divisions of the 7th Guards Army north of the Danube were busy digging in, consolidating their new positions along the Hron River, and thinking of sleep, clean clothing and most of all hot food.

While the defending German and Hungarian forces had been driven back from Sahy and Ipel River areas and eventually across the Hron River, they did escape to fight another day. Norbert Szamveber, in his book *Days of Battle*, on page 113 summed up the situation well: "However hard they tried, the Soviet Units [7th GA and 6th GTA] failed to cut off the German and Hungarian units between the [Ipel] and the [Hron], although they did inflict heavy losses (mainly in personnel) to them."

30 December 1944 – Weather: 26° to 23° F, (-3.0° to -5.0° C), snow

Despite the need for rest, the 6 GAD and the 409 RD received orders from General Shumilov to cross the Hron in strength and to expand the bridgehead that already existed on the western bank. They accomplished this goal by 31 December with help from the 27 GTB. At the same time, General Shumilov ordered the 36 GRD to withdraw to the rear, within the 27 GRC zone, for rest and replenishment.

By the end of this day, the exhausted German forces stopped their counterattacks against the small 7th Guards Army bridgehead over the Hron River and shifted to the defense. Both sides needed time to rest and to regroup their forces.

The 30 RC and the 7 Ru RC continued their attack against Pest. These two corps finally reached major built-up sections of the city, and because of that, the advance slowed noticeably. Larger buildings were harder to attack and to clear of enemy troops. In total, the city of Budapest covered 80 square miles with most of this made up of a densely populated urban area. The Red Army needed more heavy artillery and air support to batter the enemy into submission.

Guards Major of the medical service Lev issued a medical report, *Information about the movement of the wounded and sick in medical institutions of the 7th Guards Army*, found in TsAMO record ID #133391682. This report covers the period from 25 December to 30 December 1944, and is an assessment of the current medical burden upon the Army's medical establishment.
The total medical burden on 30 December was 7,421 people. This was far less than the 30 November value of 11,091 people, an indication that fighting had tapered off somewhat during December.

Forward Troop Area – up to divisional medical battalions

- Arrivals:
  - At the start (25 December) – 1,437 men and women
  - Newly arrived – 1,908
  - Total – 3,345
- Departures:
  - Evacuated – 1,204 (most likely to the Army rear) (moderately wounded)
  - Returned to active service – 445
  - Died – 84
  - Total – 1,733 (including dead)
  - In the Forward Troop Area on 30 December – 1,612 (lightly wounded)

In Army Hospitals

- Arrivals:
  - At the start (25 December) – 6,141 men and women
  - Arrived from forward units' medical installations – 1,835 (moderately wounded)
  - Arrived directly from units – 120 (severely wounded)
  - Totals – 8,096
- Departures:
  - Evacuated – 1,962 (out of the Army to the deep rear, severely wounded)
  - Returned to active service – 274
  - Died – 51
  - Total – 2,287 (including the dead)
  - In Army Hospitals 30 December – 5,809 (moderately wounded) (with 5,350 hospital beds)

The total in all medical installations on 30 December was 7,421 people.

31 December 1944 – Weather: 27° to 23° F, (-2.4° to -5.0° C), snow

On the last day of a long year, the troops of the 7th Guards Army had been victorious over its enemies. They had reached and crossed the Hron River in the north, and they were on the verge of capturing the capital of Hungary, Budapest, in the south. The 7th Guards Army had performed splendidly during December 1944. See Map #62 which illustrates this impressive textbook-like diverging advance. But the officers and soldiers of the Army probably were thinking about more important things on this New Year's Eve, like, When will the enemy surrender? When will the war be over?

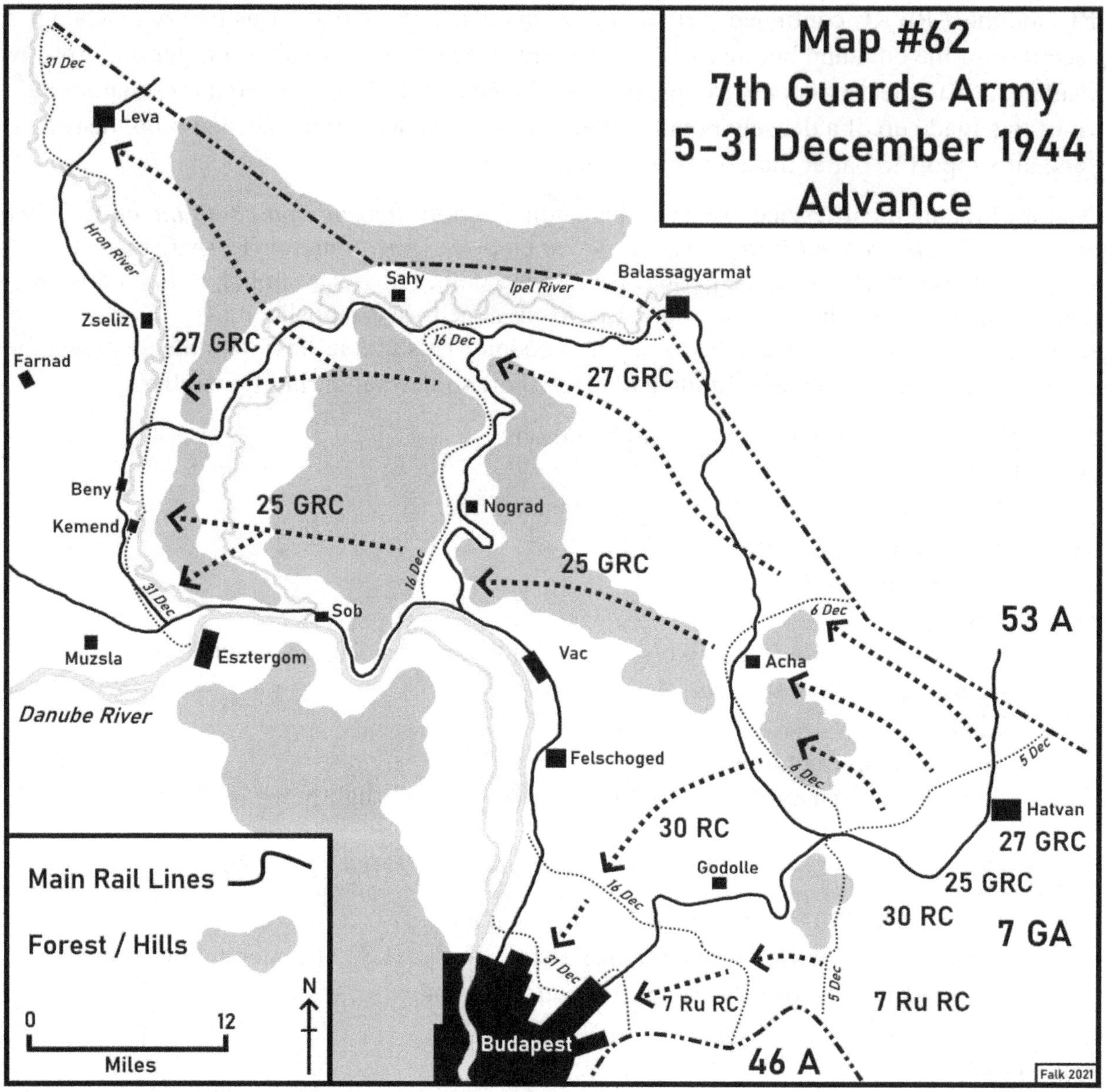

At the end of the year, the German High Command assessed the overall situation and its options for AG South. Axis forces had consolidated their position north of the Danube along the Hron River line. Russian forces in this area would need some time to reorganize and to resupply before resuming offensive operations. To the south, the Soviets had completed their encirclement of Fortress Budapest, but Budapest was strongly held and was unlikely to fall anytime soon. So, the Germans felt that some type of relief or spoiling attack toward Budapest was called for, as it would surprise the Soviets and keep them off guard before they could prepare their own plans. This fleeting opportunity presented two options for offensive operations, a northern (Konrad) or southern (Paula) attack. Ultimately, the Germans chose the northern Konrad option.

The 2 UF issued a report, *Report on the loss of personnel in the troops of the 2nd UkrF from December 1 to December 31, 1944.* This was found within TsAMO record ID #113723739. The 7th Guards Army suffered 13,758 total casualties for the month: 2,562 killed; 424 missing; 0 captured; 20 non-combat losses; 9,526 wounded/burned; 1,226 sick; and 0 frostbite. The report goes on to list 7th Guards Army horse losses: 509 killed; 1 missing; 62 wounded; 13 evacuated to veterinarian care; for a total of 585 equine casualties (in the original document the total was transposed). The number of horses that had died increased from the previous month, probably due to the weather, combat actives and overwork which kept horse losses high. Once again, the 0 people captured figure for the 7th Guards is almost certainly not true. Some of the 424 missing undoubtedly were POWs.

The same report shows that the troops of the entire 2 UF suffered 60,683 total casualties during December.

For the month of December in Britain, air raids and German V-1 and V-2 attacks killed 367 civilians and injured 847.

1 January 1945	
7th Guards Army	
Assigned – 2 Ukrainian Front	
Colonel General M.S. Shumilov Commanding	
Rifle, Airborne and Cavalry Units	25 Guards Rifle Corps (6 GAD, 53 409 RDs) 27 Guards Rifle Corps (36 GRD, 141, 243, 303 RDs) 30 Rifle Corps (25 GRD, 151, 155 RDs) 1st Ru A, 7 Ru Rifle Corps (2, 19 Ru RDs, 9 Ru CD) (attached) 15 Penal Battalion 62, 63, 64, 65, 66,194 Penal Companies
Artillery Units	16 Artillery Penetration Division (49 Light Artillery Brigade, 61 Gun Artillery Brigade, 52 Howitzer Artillery Brigade, 90 Heavy Howitzer Brigade, 109 High-Power Howitzer Brigade, 14 Mortar Brigade) 2 Antitank Artillery Brigade 114, 115 Guards Tank Destroyer Regiments 41 Guards Gun Artillery Brigade 263, 290, 493 Mortar Regiments 5 Antiaircraft Artillery Division (670, 743, 1119, 1181 AA Regiments) 26 Antiaircraft Artillery Division (1352,1357,1363,1369 AA Regiments) 662 Separate Aerial Observation/Air Raid Warning Company
Armored and Mechanized Units	27 Guards Tank Brigade 38 Separate Armored Train Battalion
Engineer Units	60 Engineer Sapper Brigade 21 Pontoon Bridge Battalion
Reported Combat Strength Estimated Ration Strength Reported Combat Strength 1st Ru A (7 Ru RC)	62,986 men 77,089 men and women (including an estimated 7,000 support personnel) 21,940 men (including combat support troops)

(BSSA 45, BSA 45, TsAMO Archive ID #113723524)

A new year, and another battle. They were blending together now so much that the troops had a hard time telling one from another. The war endures, but the soldier succumbs just the same.

The start of the month saw several units withdrawn from the 7th Guards Army: the 5 Guards Heavy Artillery Penetration Division, the 220 and 228 Gun Artillery Brigades, and the 34 Antitank Artillery Brigade. The 21 Pontoon Bridge Battalion was the only unit to join the 7th Guards Army.

1 January 1945 – Weather: 32° to 25° F, (0.0° to -4.0° C), snow

This day saw the start of nearly two simultaneous offensive operations, one Soviet, then one German. On the eastern (Pest) side of Budapest, the 30 RC, 7 Ru RC and the 18 RC once again launched coordinated attacks into the built-up area of Pest. Despite massive artillery support, fighting was heavy and these troops gained little ground. Then at 22:30 hours and deliberately without artillery support, the German IV SS Pz C (at half strength) launched a major surprise attack against Soviet positions of the 4th GA (3 UF) northwest of Budapest. The goal of this attack was to break through the besieging Soviet forces just south of the Danube and to reach the defenders within Budapest; this offensive was codenamed Konrad (also known as Konrad I). Elements of the German 96 ID assisted this attack by conducting a surprise river assault, behind the Russian front line, from the north side of the Danube to the south side of the river.

To support the ground efforts of the 2 UF, the 5th AA deployed 645 aircraft of all types. [619] Opposing them, the German 4th Air Fleet deployed approximately 600 aircraft, but these aircraft had a limited impact due to fuel shortages.

2 January 1945 – Weather: 31° to 18° F, (-0.2° to -7.4° C), snow

Operation Konrad continued with the 3 SS Pz D and 5 SS Pz D leading the way and with parts of the 6 Pz D and 96 ID attacking to the east. The 3 UF command was slow to react to this attack as they were expecting an attack further south. The Germans pushed forward, gaining considerable ground.

1945 – Panzer Grenadiers and a Panther tank from the 5 SS Pz D Viking move into position.

3 January 1945 – Weather: 28° to 12° F, (-2.0° to -11.2° C), snow

The German Konrad attack continued and several small Hungarian supporting units along with the German 711 ID, which was moving up, joined the attack. The 3 UF began moving mobile forces from the south to the north to counter this attack. Marshal Fyodor Tolbukhin, commander of the 3 UF, stopped attacks against the western Buda garrison in order to release forces to use against the German Konrad attack.

On or about this date, Marshal Semyom Timoshenko, acting as a *Stavka* representative, urged that the 2 UF be ordered to launch an attack north of the Danube River to counter and to disrupt the German Konrad attack that was currently making good progress toward Budapest. Supreme Command Headquarters agreed and ordered Marshal Malinovsky to assemble a suitable force for the attack. Ultimately this force would be comprised of the 6th GTA and the 7th Guards Army, and they would go on the offensive in a few days' time. [620]

In January, with the Red Army approaching, Hitler declared the Prussian/Polish/German city of Poznan a Fortress City or *Festungen.* Poznan was a major transportation center and also was on a direct rail line between Warsaw and Berlin. Without capturing Poznan, the advancing Soviet troops could march around the city, but their supplies would have a much harder time reaching their destination at the front. When the Red Army arrived at Poznan, they would find "60,000 men manning the well-prepared and well-stocked armored forts" [621] that had been built during Prussian rule in the 19th century. The resulting battle for the city would last almost a month.

4 January 1945 – Weather: 18° to 11° F, (-7.4° to -11.8° C), snow

The German Konrad attack continued to make progress as Soviet forces of the 3 UF gathered to block them from reaching the Budapest garrison. This sudden German attack compelled *Stavka* to act; it forced them to employ the already weakened formations of the 6th GTA and the 7th Guards Army to deal with the situation.

Directive #11002 regarded the 7th Guards Army and 6th GTA attack north of the Danube from the Esztergom area.

Directive of the VGK Stavka No. 11002 to the Commanders of the Troops 2nd and 3rd Ukrainian Fronts on [deflecting the] *Enemy's* [attack] *from the Komarno Region*

January 4, 1945 02 h 40 min

The headquarters of the Supreme High Command orders the forces of the 2nd and 3rd Ukrainian fronts to defeat the Rotivnik's grouping advancing from the Komarno region to Budapest, for which:

1. To the commander of the 2nd Ukrainian Front to strike with the forces of at least two corps of the 7th Guards Army and 6th Guards Tank Army from the area north of Esztergom along the northern bank of the river Danube in the general direction to Komarno. Master the crossings across the river Danube near Komarno and tank units with motorized infantry to reach the rear of the enemy's Komarnov group.

2. The commander of the troops of the 3rd Ukrainian Front, simply covering the direction to Budapest from the northwest, to strike with the forces of the 4th Guards Army, 5th Guards Cavalry Corps, 18 TK and 1 Guards MK from the area northwest and west of Bichke in the general direction to Nesmey, Komarno with a task in cooperation with the troops of the 2nd Ukrainian Front to defeat the enemy grouping advancing on Budapest, and to seize the border [and capture] *Komarno Mor.*

3. Support the offensive with the bulk of the aviation of both fronts.

4. The offensive to begin on January 6th.

5. To inform about the given orders.

Headquarters of the Supreme Command
I. Stalin A. Antonov

By this order we can see how *Stavka* wanted to outflank the German Konrad attack toward Budapest by advancing along the north side of the Danube River to the west and seizing key bridges and crossing points, mainly in and around the city of Komarno, that were being used to support the German attack. Note that this *Stavka* directive only gave Marshal Malinovsky (2 UF), General Shumilov (7th GA) and Colonel General Andrei Kravchenko (6th GTA) two days to prepare for this offensive. To accomplish this, one tank brigade conducted a 50-mile road march during one night and its commander stated, "In a short period of time the assigned group of sappers prepared the march route. Traffic regulators were posted at isolated difficult sectors and markers that could be easily seen at night at the remaining route segments." [622] This attack would be launched early in the morning on 6 January. This order also disrupted the rest and refitting of both the 6th GTA and the 7th Guards Army after their battles east of the Hron River. Therefore, the German Konrad attack was not only aiming to relieve the Budapest garrison, but it also was functioning as a spoiling attack, disrupting Soviet forces and operations in and around the Budapest area.

Three Red Army traffic regulators install border signs on a bridge in 1945. (Germany to the left, Poland to the right.) The large sign on the right says Germany and the round sign is a 12-ton weight rating for the bridge.

5 January 1945 – Weather: 28° to 14° F, (-2.0° to -10.0° C), snow

On or about this day, the 7th Guards Army was ordered to transfer the 36 GRD to the 26th A of the 3 UF. The 36 GRD would not return to the 7th Guards Army.

Units of the IV SS Pz C launched further attacks toward Budapest, but Soviet resistance had strengthened. The German 711 ID had reached the front just south of the Danube and joined in the attack.

Back in Germany, further secret weapons were under development. Due to the incessant bombing of Germany, three different anti-aircraft missile systems were at an advance stage of development. They were intended to combat the Allied heavy bomber threat. The Henschel Hs 117 *Schmetterling* (Butterfly) was a surface-to-air missile that was visually guided to its target from the ground. Another visually guided surface-to-air missile was the larger and longer ranged *Wasserfall* (Waterfall) missile. Finally, there was the Ruhrstahl Ru X-4, a short range, wire-guided air-to-air missile. [623] Ironically, these advanced anti-aircraft missiles were never used in combat because of production problems due to Allied bombing. After the war, the Allies carefully examined surviving examples, which helped them develop their own anti-aircraft missile systems and wire-guided anti-tank missiles for use in land combat. German engineering advanced the technology of warfare but not necessarily for the benefit of Germany.

6 January 1945 – Weather: 35° to 21° F, (+2.0° to -5.8° C), snow

At 03:00 hours (false dawn), and without artillery preparation, units of the 7th Guards Army and the 6th GTA, with 180 tanks (600 tanks at full strength), struck out from the bridgehead on the west bank of the Hron River. [624] Tanks were in the lead as the offensive attempted to quickly break through the Axis line and to reach the city of Komarno. The German and Hungarian defenders had been alerted to the attack and attempted to repel the advancing Red Army units. The battle to control the Hron River line commenced. One tank battalion commander who participated in the Hron breakout battle recounted,

> The snowstorm howled. We were glad for this. Snow and a cold, gusty wind drove the Germans into their shelters and deep dugouts. The Emachs [an American made M4A2 tank,] was painted white, blended into the white background and were hardly distinguishable in the gusty snowy swirls.... The sappers guided us through the passage lanes in our own minefields. They directed us precisely through the narrow lane prepared in the enemy's explosive obstacle belt [minefield].

The tank commander went on to comment about early fighting on the march:

> The Emach's machine-gun fire was supplemented by sub-machine gun bursts from the *tankodesatniki* [tank riders, assault infantry that rode into battle holding onto handrails that were welded on to the outside of the tank]. When necessary, the motorized riflemen also hurled their grenades. [625]

A Russian solider from the 375 RD recounted an infantry soldier's view of the opening battle.

> With the onset of darkness on [5] January, the personnel were fed a hot dinner, and before the onset they were given 300 grams [10.5 oz] of sausage or boiled meat and 150-200 grams [5.5-7.0 oz] of vodka.... At 3 o'clock in the morning, without artillery preparation, the regiments of the division suddenly began to storm the fortified lines of the Germans with the task of expanding the bridgehead... It is impossible to describe the fierceness of the German counterattacks and the stubbornness of our soldiers to repel them.... In these battles, the penalty battalion [almost certainly the 15 Penal Battalion for convict officers] under the command of the communist senior lieutenant Nikolai Grigorievich Kashpanov performed well.... After the fight, the distinguished [penal battalion members] were [put forward] for the removal of their convictions, and some for [receiving awards]. [626]

The attack by the 6th GTA and the 7th Guards Army was a confused swirling battle, just like the snow storm that enveloped the fighting. From the start, the mobile units of the 6th GTA were separated from the rifle units of the 7th Guards Army that were left behind to deal with pockets of German and

Hungarian resistance. This was the same tactic used in the successful attack by the 6th GTA and 7th Guards Army to clear Axis forces from the Ipel to Hron River area in December. Tank units were in the lead with infantry formations following, but this time the German and Hungarian defenders were better prepared. Gaining ground and capturing Komarno was the key to the battle, but the separation of forces made the offensive vulnerable to counterattacks, and that is what the German were planning to do in a few days' time.

A team of authors in the Soviet book, *Soviet Tank Troops 1941-1945,* Chapter 12, described the attack:

> The 6th Guards Tank Army, in cooperation with the 7th Guards Army, with a surprise night attack on January 6, broke through the enemy defenses and rushed to Komarno. Daring and sudden actions of tankers on the flank and behind enemy lines forced the fascist command to stop the counterblow group advancing on Budapest and begin to transfer it to the north.

Despite the above statement, which is in part true, the direct impact of this attack upon Operation Konrad was minimal due to AG South's shrewd appraisal of the situation. But the area north of the Danube was critical indeed for German operations around Budapest and south of the Danube, so German forces did respond forcibly to this threat. An official German source stated,

> On 6 January Army Group South HQ had toyed with the idea of halting or reducing the offensive south of the Danube [Konrad], but it finally decided to take the risk, in the hope that the newly arrived [20 Pz D] would be able to hold the front [north of the Danube]. [627]

By the end of the first day, rifle units of the 7th Guards Army, supported by the 27 GTB, achieved a maximum advance of about 11 miles while the tank/mechanized units of 6th GTA penetrated about 14 miles into Axis territory. Despite a huge snowstorm in the morning and ground fog later on in the day, the Soviet 5th AA was able to support the attack with 221 sorties, which was a noteworthy achievement owing to the weather. [628] (See Map #63.)

South of the Danube, Operation Konrad continued to gain ground slowly as the 3 UF commander committed his reserves to block the German attack from breaking through to Budapest. [629] The one notable victory was the capture of the town of Esztergom on the south side of the Danube by the German 711 ID. The 711 ID would soon put its artillery to use firing across the Danube River to the north of Esztergom in order to harass Soviet supply lines and Hron River crossing points in the area.

Due to its initial success, German commanders would soon expand Konrad into Konrad II in an attempt to finally break through Soviet lines and to reach Budapest.

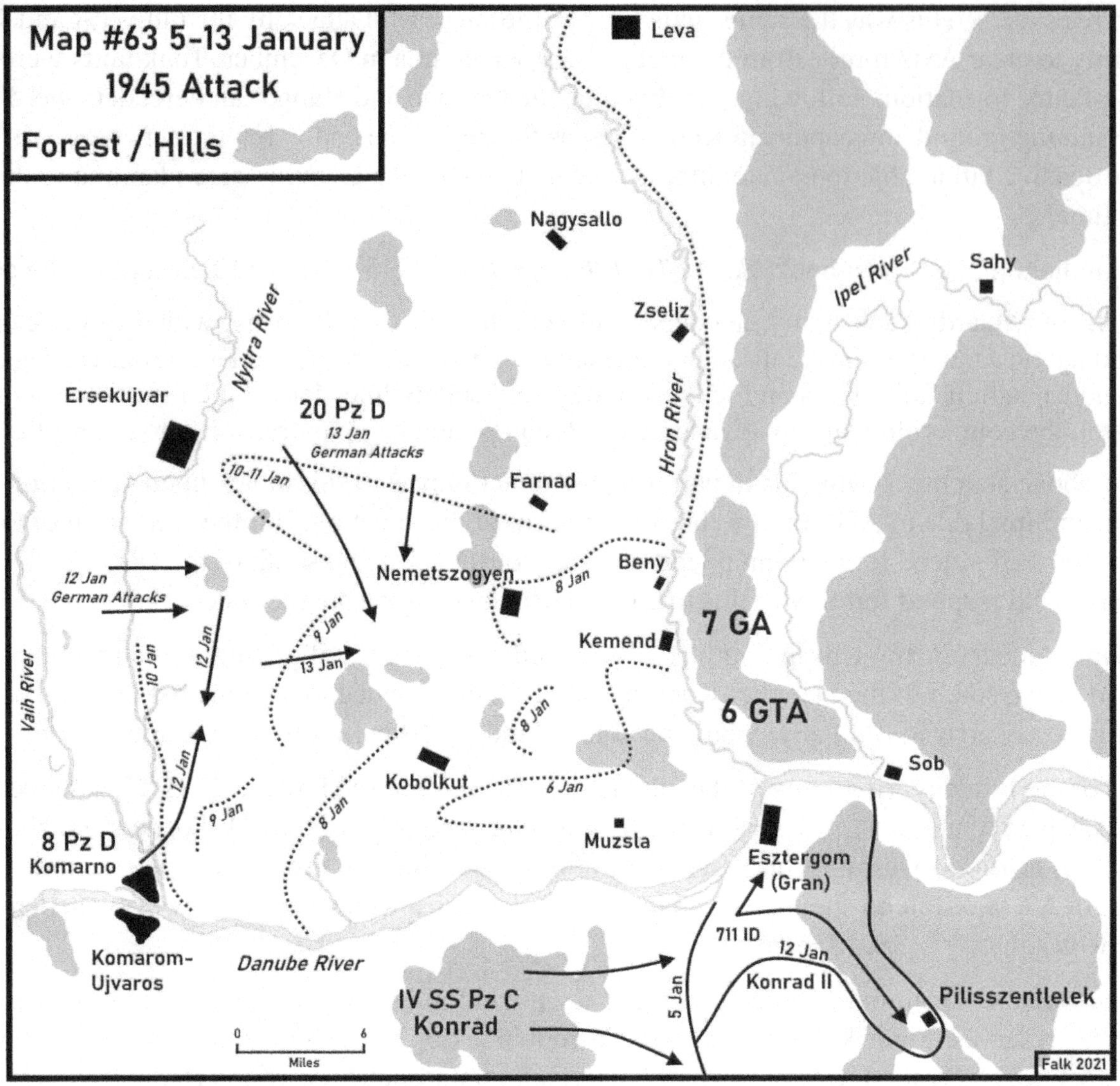

7 January 1945 - Weather 36 to 33° F, (+2.2 to +0.4° C), rain at night and afternoon, wet snow in the evening

The 6th GTA and the 7th Guards Army continued their assault westward toward Komarno and to the north to clear the Hron River line. Soviet forces fought with German and Hungarian blocking positions along the way or drove/marched around them to keep the advance moving. Rifle units of the Army captured the town of Kemend on the Hron, while the German 211 VG D repelled a similar attack against Beny from the east side of the river. Overall, the attack continued to gain ground, but the gap between tank and rifle units widened.

During 6-7 January, the reinforcing German 20 Pz D arrived north of the Danube as it prepared to counterattack the 6th GTA and 7th Guards Army offensive.

Meanwhile, the original Konrad attack now was expanded to Konrad II, with the addition of the 6 Pz D and the entire 96 ID. All of the Konrad II units were engaged in heavy fighting with the newly introduced forces of the 3 UF. Despite Soviet efforts to stop them, the Budapest encirclement was about 17 miles away from the advancing German force. But, only the 711 ID was making good progress just south of the Danube. Reportedly by this time, the IV SS Pz C had already suffered 3,500 casualties and lost 39 tanks and assault guns. [630]

At this point in the battle, the overall situation was somewhat confused which requires some clarification. The 2 UF was attacking westward into Pest on the east side of the Danube with three rifle corps; at the same time, the reinforced German IV SS Pz C was attacking eastward toward Buda on the west side of the Danube. Meanwhile, the 6th GTA and the 7th Guards Army had launched a major attack along the north side of the Danube heading west while simultaneously the 3 UF was counterattacking to the north, on the southwest side of the Danube, to block the advance of the German IV SS Pz C. All this to and fro activity was taking place during a raging snowstorm, which prevented the effective use of artillery fire and grounded most air support, over a period of several days, for both sides.

8 January 1945 – Weather: 31° to 29° F, (-0.2° to -1.8° C), evening snow

Tank and mechanized formations of the 6th GTA reached Komarno and the Vaih River road and rail bridges, but they were unable to force a crossing from the march and were repelled, suffering significant losses. Luftwaffe and Hungarian air units flew approximately 100 unopposed sorties against Soviet forces near Komarno during the day. [631]

The 7th Guards Army continued its now separate attacks, unable to keep up with the advance of 6th GTA mobile units. The rifle divisions fought hard to expand the bridgehead area to the north and northwest side of the Hron, while attempting to support the advance to Komarno. Of growing concern, both the 6th GTA and the 7th Guards Army were beginning to suffer from supply shortages.

During the day, 21 German trains arrived from the north carrying equipment and supplies for the assembling 20 Pz D.

The Konrad II attack continued, with the German 711 ID making steady progress through the hills and forests south of the Danube, and it captured the village of Pilisszentlelek by the end of the day. This town was approximately 14 miles from the Budapest pocket.

9 January 1945 – Weather: 33° to 21° F, (+0.8° to -6.2° C), evening snow with rain

The 24 GRC was reassigned to the 7th Guards Army from the 2 UF reserve. The 24 GRC, which contained the 72 and 81 GRDs, began to move into position between the 25 and the 27 GRCs at the front along the Hron River. The 24 GRC would not be ready for combat operations until 10 January.

Tank formations of the 6th GTA continued to assault Komarno in an attempt to seize the bridges over the Vaih River, but once again they were unable to force a crossing. The 6th GTA also was engaged with scattered German and Hungarian formations further north on the approaches to the town of Ersekujvar. At this point it was unlikely that the two Soviet army commanders, Shumilov and Kravchenko, knew that the German 20 Pz D already was arriving and assembling in the area north of the battlefield. Lagging rifle units of the 7th Guards Army approached Komarno but would not reach it in force until the next day.

In Pest, the 30 RC launched a powerful attack toward the train station in an attempt to split the Pest bridgehead in two. The defending Axis troops counterattacked and then fell back to new positions. [632]

10 January 1945 – Weather: 35° to 33° F, (+2.0° to +0.9° C), light rain

Formations of the 6th GTA were engaged with German counterattacks south and east of Ersekujvar while the 53 RD attempted to capture Komarno. Back at the Hron River, the 24 GRC entered the battle and pushed north along the western back of the Hron with tank support from the 27 GTB. The 27 GTB only had three tanks remaining by the end of the day.

The battle had reached a crisis point. For the most part, formations of the 6th GTA and the 7th Guards Army now were fighting separately within the greater battlefield. The engaged troops had no decisive focus or concentration of forces that they could use to seize the initiative away from the enemy and to achieve overall victory. Colonel General Shumilov and Colonel General Kravchenko (6th GTA) were partially responsible for letting the offensive dissolve into a series of separate almost unrelated battles, but Marshal Malinovsky was really at fault. Marshal Malinovsky failed to control and to guide his two veteran army commanders to success. Malinovsky also failed to adequately supply and to reinforce the offensive when the fighting began to stall. On this account, Marshal Malinovsky and the historical record are silent.

On or about this date, the 11 Antitank Artillery Brigade was assigned to the 7th Guards Army, coming most likely from 2 UF reserve, so its arrival at the front would take some time.

Due to the intense fighting taking place away from Budapest, along with the inclement weather and exhausted troops, the assault upon the city made little headway from 5-10 January. (See Map #64.) Fighting soon would intensify as the surrounding Soviet forces renewed their efforts to capture Pest and then to destroy the remaining trapped Axis troops in the hills of Buda on the west side of the Danube.

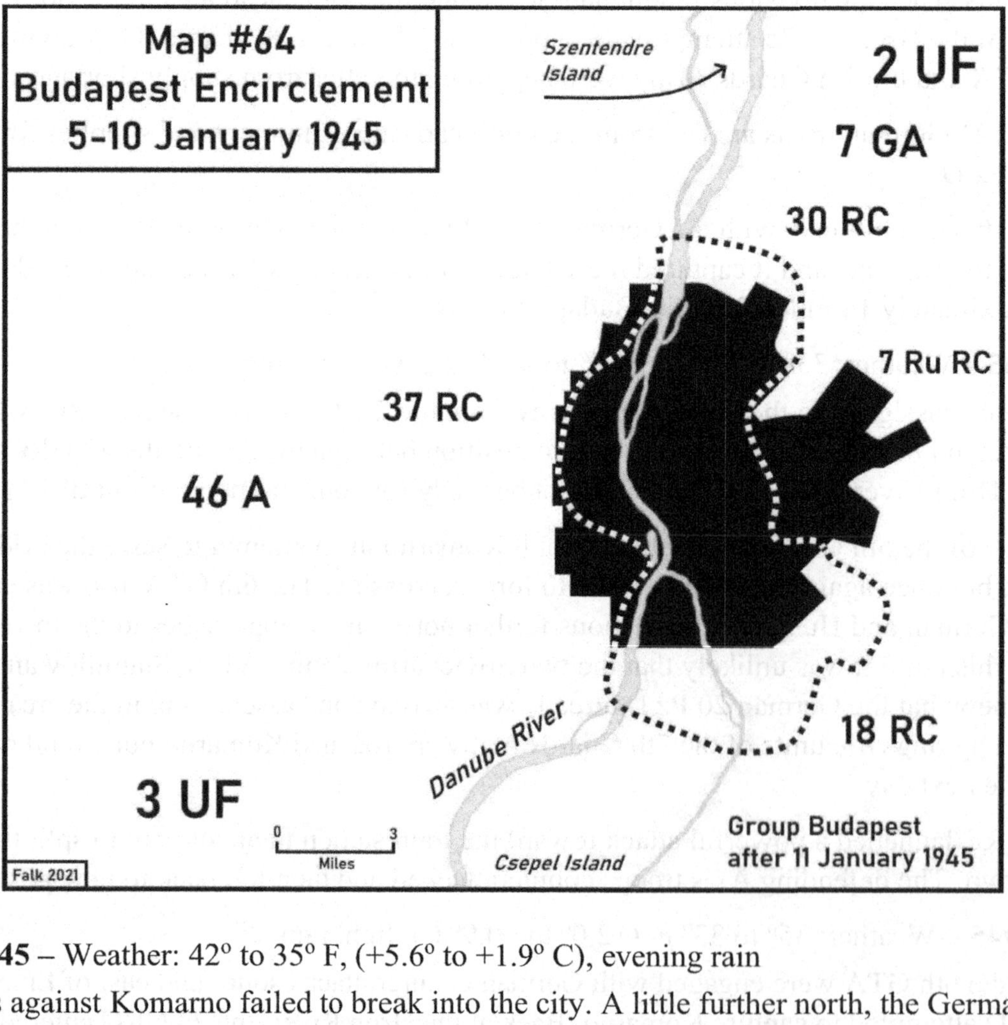

11 January 1945 – Weather: 42° to 35° F, (+5.6° to +1.9° C), evening rain

Russian attacks against Komarno failed to break into the city. A little further north, the German counterattack by the 20 Pz D expanded as it made further progress to the south and east. The fighting in this area was confused as both sides attempted to come to grips with the enemy as they dealt with snow, rain and deepening mud. Later in the day, Russian troops found documents on a dead German soldier. "Only then did the staff of the [2 UF] realize that the counterattack was primarily being carried out by the [20 Pz D.]" During the day the 5th AA only flew 38 sorties to support combat operations along the northern edge of the bridgehead area. [633]

On 10 January, the German 4th Air Fleet, which was supporting AG South, had the following operational aircraft. (See Table #28.)

Aircraft Type	Number
Single engine fighters	78
Night fighters	-
Bombers	56
Ground attack	199
Night ground attack	101
Strategic reconnaissance	38
Tactical reconnaissance	67
Transports	49
Total	**588**

Table #28: German 4th Air Fleet Operational Aircraft – 10 January 1945. [634]

Due to the extreme weather conditions and the lack of fuel, Luftwaffe aircraft operations over Hungary and Czechoslovakia were limited. Bomber and transport sorties were devoted mostly to supplying Budapest by night. Combat operations were equally limited and focused only on critical areas.

The 2 UF troops aligned against Pest -- the 30 RC, 7 Ru RC and the independent 18 RC -- were combined into Group Budapest under the command of Lieutenant General Ivan Mikhailovich Afonin, who was the former commander of just the 18 RC. [635] With this consolidation, the 7th Guards Army lost control of the 30 RC and the 7 Ru RC.

For Konrad II, the German 711 ID reached a point approximately 13 miles from the Budapest pocket, but they could go no further.

1945 Hungary – Soviet T 34/85 tanks and their crews rest while two Hungarian prisoners carry a wounded solider to the rear.

Along with advancing Red Army troops, behind the front came NKVD (Secret Police) troops to "secure" the rear area, i.e., lines of communication and Soviet property. They were also there to control the local people, their movements and actions. Additionally, they apprehended spies, exposed saboteurs, and captured deserters. The NKVD also was charged with guarding POWs and transporting prisoners/criminals and POWs between camps. Moreover, they conducted executions as needed, filled the labor camps with people who were judged by tribunals or courts and found guilty of crimes, oversaw mass deportations of people and suppressed any opposition to Soviet policies or actions. The Red Army liberated the subjugated peoples of Eastern Europe from Nazi or Fascist rule, while the NKVD introduced them to the Soviet Union's equivalent form of tyranny.

On 18 December 1944, the GKO (State Defense Committee) issued a decree raising additional NKVD divisions for security of rear areas in Eastern Europe. By 11 January 1945, it was planned that the 2 UF would have six NKVD Regiments, with 8,528 troops and one new NKVD division, #61, with 5,000 men. The 3 UF would have four NKVD regiments with 6,223 troops. So just for Hungary, a total of 19,751 NKVD troops would be on hand to secure the rear area. Most likely this GKO decree was carried out as ordered.[636]

12 January 1945 – Weather: 40° to 35° F, (+4.6° to +1.6° C), light rain

German attacks north of Komarno, launched from both the north and south, cleared the entire eastern side of the Nyitra River line while further advances by the bulk of the 20 Pz D attempted to split the Soviet bridgehead west of the Hron River.

The last few German trains arrived which carried the remaining troops of the 20 Pz D. The transfer of the 20 Pz D required 70 trains in total. Other German units along with the 16 Hu ID advanced upon the town of Nemetszogyen, threatening to force Soviet troops out of the town.

At 15:00 hours the 24 GRC went over to the defense, and at 17:00 hours the 4 GMC did the same. The German 211 VG D firmly held the line north of Beny, thus preventing the Soviets from expanding their offensive further to the north. [637]

By 12 January, the northern Konrad II offensive toward Budapest could not advance further in its current form. The Germans required further forces to break through to Budapest.

Hitler, never wanting to relieve the garrison anyway, called off the attack. The 711 ID and other units were ordered to pull back from their exposed positions and away from Budapest. So ended the first attempt, (Konrad I & II) to reach and to relieve the Budapest garrison. The engaged German forces were shocked the attack was called off because Budapest was so close. The German combat troops did not understand that Hitler intended for the garrison to remain within Fortress Budapest and to fight to the last man, thus delaying the next major Soviet offensive for as long as possible.

But planning for an even larger offensive, Konrad III, was in the works. Konrad III required the transfer of major forces from the north of Budapest to the area around Lake Balaton, southwest of Budapest. Hitler's overall strategy for the Konrad III offensive was to recapture the entire Danube River line south of Budapest, and in the process to force the 3 UF back over to the east side of the river. Konrad III essentially was a massive spoiling attack to disrupt Soviet offensive operations.

During the night, the IV SS Pz C began moving south some 63 miles to new attack positions for Konrad III. (See Map #66 for Konrad III location.) The Soviets were aware that the German command had withdrawn the IV SS Pz C from the front line. They thought it was headed north to defend Germany, but they were wrong; it was headed south to spearhead another attack in Hungary.

On this day, further north in Poland and Prussia, Marshal Konev's 1 UF and Marshal Zhukov's 1 Belorussian Front (BF) launched the Vistula–Oder offensive into southern Poland and eastern Germany. Along the way, this offensive would capture Krakow, Warsaw, and Poznan and would end only 43 miles from Berlin. Guderian's Soviet offensive estimate, from his conference with Hitler on 24 December 1944, was correct to the location and actual day of this Soviet attack.

On or about this date, Adolf Hitler ordered the 1 and 2 SS Panzer Corps and their attached formations withdrawn from active combat on the Western Front (the Battle of the Bulge). Hitler also ordered these units to be refitted as quickly as possible and made ready for future operations. At this point, most of the German High Command thought these forces would be used to defend Germany, but Hitler had other plans.

13 January 1945 – Weather: 35° to 28° F, (+2.0° to -2.0° C), freezing light rain

The bulk of the 6 GTC and the 5 GMC had been gathered a few miles east of Komarno, most likely to launch a coordinated strike against the city, the prime objective of the entire attack. But attacking Panzer formations operating to their north and rear would make such an attack risky. The 5 GTC also was engaged in attacks against the town of Madar, which was located two miles west of Batorkeszi. Fighting in this area was vicious as both sides attacked and counterattacked.

Meanwhile, Group Budapest, 2 UF, continued their urban assault against Pest.

1945 Hungary – Horse column of 76mm Divisional Guns (Zis-3) advances along a cobblestone road.

14 January 1945 – Weather: 35° to 24° F, (+1.6° to -4.4° C), freezing light rain

German formations from the 20 Pz D along with smaller supporting units struck into the northern flank of the 6th GTA spearhead. At the same time, the 8 Pz D attacked from the Vaih River area to the east. The 53 and 375 RDs from the 7th Guards Army held the line on the western side of the advance for a while, but the Panzer attack from the north was too powerful. The 6th GTA and 7th Guards Army positions began to come apart. (See Map #65.)

A partial losses list for the 6 GAD, 409 RD, 375 RD, and 53 RD between 6-14 January amounted to the following: 144 killed, 618 wounded and 65 missing without a trace, for total casualties of 827 people. This is from TsAMO record ID #132652001. This covered the initial 7th Guards Army attack and part of the German counterattack.

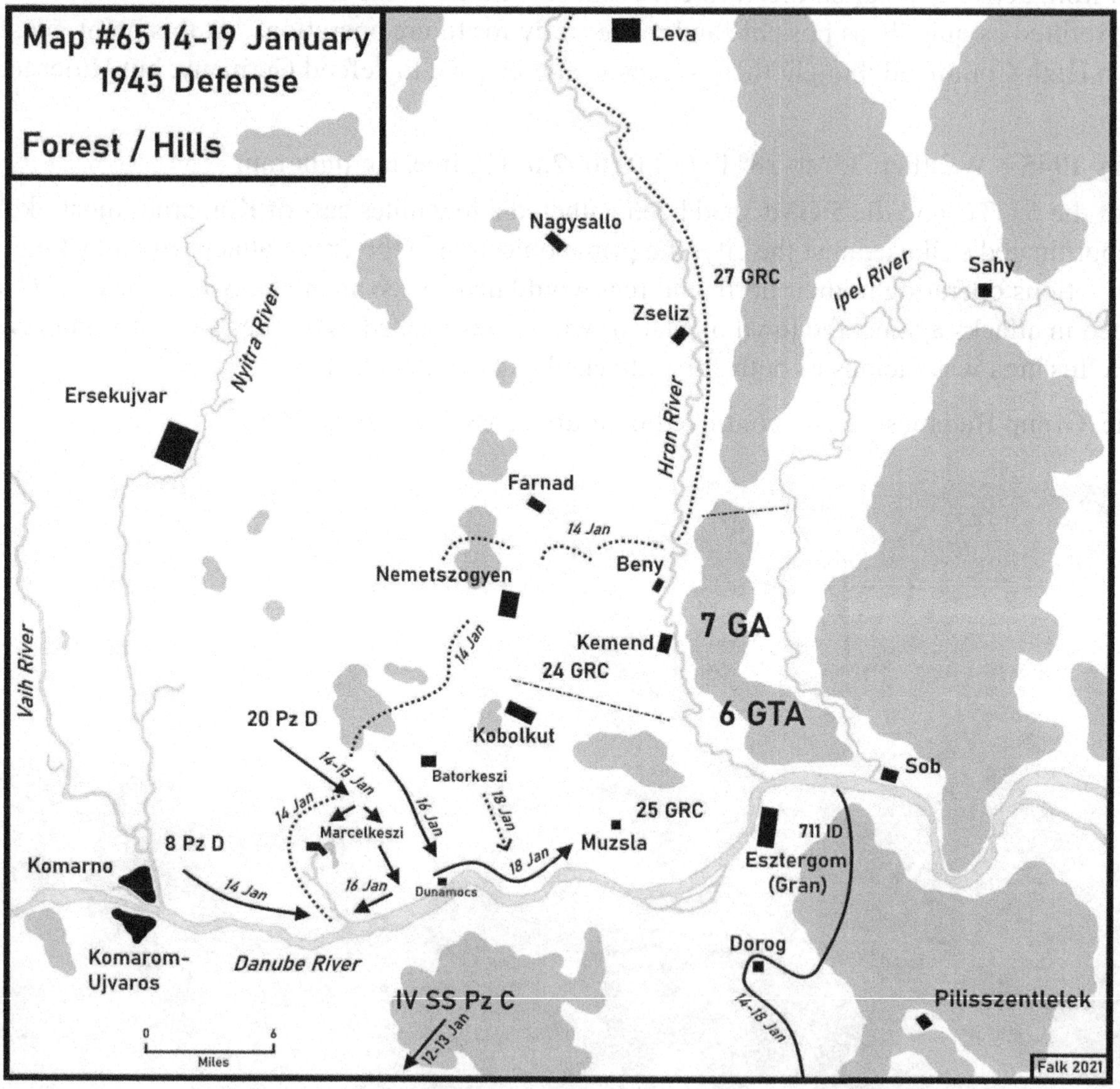

15 January 1945 – Weather: 31° to 26° F, (-0.2° to -3.4° C), freezing light rain

The German 20 Pz D and supporting units continued to attack south toward the Danube as Russian rifle units attempted to pull back and to escape being cut off. Units of the 6th GTA regrouped to the east as rifle divisions of the 7th Guards Army continued to hold positions along the newly established front. The Soviet response to the German counterattack was scattered as individual Red Army units attacked or defended in response to German Panzer thrusts into their rear, especially east of the town of Marcelkeszi.

German and Hungarian defenders within Pest were driven back to the Danube, forcing them to blow up the Horthy Bridge over the Danube to prevent it from falling into enemy hands. The 7 Ru RC, as part of

Group Budapest, was pulled out of the battle by Marshal Malinovsky. The Ru 7 RC had suffered at least 11,000 casualties from November 1944 to January 1945, and they also captured 6,522 prisoners during the fighting. The corps was sent north to Czechoslovakia to rejoin the 4 Ru RC. This transfer also prevented them from taking part in the final assault against Budapest and thus sharing in the glory of capturing the capital of Hungary. They also were most likely in need of both rest and replenishment. [638]

16 January 1945 – Weather: 30° to 24° F, (-0.9° to -4.6° C), morning snow

Attacks by the German 20 Pz D continued until they reached the northern bank of the Danube River, but they were unable to capture Batorkeszi along the way due to tenacious Russian resistance. The 53 RD began retreating to the east to escape encirclement.

During the day, Luftwaffe aircraft numbering up to 100 attacked Russian units of the 6th GTA, 7th Guards Army and the river crossings on the Hron River. Meanwhile, the 5th AA flew 532 sorties in support of the hard-pressed ground troops on the west side of the Hron. [639]

At 17:00 hours, German commanders received orders to transfer the 8 and 20 Pz Ds to AG A in southern Poland. This transfer was scheduled to begin during the evening of 17 January. This order would effectively put an end to the German counterattack against the Russian Hron River bridgehead.

German and Hungarian defenders of Pest destroyed the Ferenc József Bridge (built in 1896) as they withdrew over the Danube and into Buda. The Ferenc József Bridge also might have collapsed due to earlier bomb damage.

In far off Berlin, Hitler and the OKW ordered the I and II SS Pz Cs (6th Panzer Army, then currently located on the Western Front) to be withdrawn from operations and be refit by 30 January. [640] The refitting actually took much longer than anticipated and continued even during the rail transfer to the Eastern Front.

17 January 1945 – Weather: 30° to 22° F, (-1.2° to -5.2° C), morning snow

German and Russian forces continued to maneuver and to perform combat missions north of the Danube but mostly south and east of Batorkeszi. Neither side achieved significant results, but men and equipment losses mounted on both sides. The 53 RD continued to retreat to the east under pressure and in a disorganized manner, perhaps losing as many as 800 soldiers killed, and 16 AT guns, 14 mortars, and 40 machine guns in the process. [641]

Southwest of Budapest the IV SS Pz C discreetly moved into its assigned attack positions. The Soviet command staffs of the 2 and 3 UFs knew the IV SS Pz C had been withdrawn from the front, but "they assumed that it was leaving the Hungarian theatre of war altogether." [642] Operation Konrad III would begin the next day, and once again, the Soviets would be stunned by a sudden Panzer attack for which they were not prepared.

18 January 1945 – Day One of Konrad III. Weather: 28° to 25° F, (-2°.0 to -3.6° C), morning snow

Elements of the German 8 and 20 Pz Ds attacked along the north bank of the Danube, gaining some ground, but they faced growing counterattacks from Russian tank and mechanized forces of the 6th GTA. At 19:00 hours, General Shumilov transferred the 27 GTB from the Beny area south to support the 25 GRC. The 27 GTB had seven T-34s and two SU-85s operational. [643]

The first five trains carrying units of the 20 Pz D departed for Poland. Time was running out for the Germans to complete their counterattack and to destroy the Soviet Hron River Bridgehead. Note: The complete 20 Pz D had only been in action along the Danube since 12 January, or six full days. Now it was off to counter another Soviet attack. This is why German mobile units were called fire brigades, putting out one situational "fire" after another.

The German Konrad III offensive opened to the great surprise of the 3 UF command staff, and especially for Army General Georgiy Zakharov, the commander of the 4th GA, which was the main target. With the IV SS Pz C leading the way, and on the first day the offensive, German forces with four Panzer divisions and supporting infantry units opened an 18-mile wide and 37-mile-deep breach in the Soviet line. Two other smaller supporting attacks on the left flank of the main attack eventually made good progress too. The Luftwaffe flew 270 sorties in support of the attack and claimed 12 enemy aircraft shot down. [644]

The final act within Pest occurred when the remaining German and Hungarian defenders withdrew over the Danube into Buda and then destroyed the Chain and Elisabeth Bridges at 07:00 hours, thus isolating Buda from Pest and the east side of the river. In spite of this, forces defending Buda repelled several Russian ground attacks on the west side of the river and remained under heavy artillery and mortar fire all day. [645] Pest had fallen, and now only the hills of Buda held out. By this time the garrison had been reduced to 34,000 men (German troops), an undefined number of Rumanian troops, approximately 300,000 civilians, and 11,000 wounded soldiers all compressed into a very small area. [646]

The demolition of the Budapest bridges over the Danube River might at first appear to be insensitive and excessive, but it prevented any ground attack over the Danube and into Buda from the Pest side. The German and Hungarian defenders of Buda now could focus their defensive efforts solely against Soviet attacks on the western side of the Danube.

1945 Budapest Hungary – The destroyed Chain Bridge over the Danube River, with Pest in the foreground and the hills of Buda and the Royal Place in the background.

Note that the Pest side of the city occupies about 2/3 of the total capital area. With the Soviet capture of Pest, only 1/3 of the city-- the hills of Buda-- remained to be conquered, but that would require further weeks of intense fighting.

Stavka issued Directive #11099 assigning the task of eliminating the remaining encircled enemy forces (Buda) to Marshal Malinovsky's 2 UF. In part, the directive read,

Directive of the VGK Stavka No. 11099 to the Commanders of the troops 2nd and 3rd Ukrainian Fronts for the Successful Development of the Attack to Lake Balaton

Copy: to the representative of the Stavka January 18, 1945 24 h 00 min

The headquarters of the Supreme Command orders:

1. *The commander of the 2nd Ukrainian Front should be responsible for the management of operations to complete the destruction of the enemy group in the western part of Budapest* [Buda].

a) *Subordination to the commander of the 2nd Ukrainian Front the 37 and 75 RCs and all reinforcing units fighting on the western side of Budapest.*
b) *Material support for the transferred troops remains with the commander of the 3 UF.*

Report the orders given.
Headquarters of the Supreme Command I. Stalin A. Antonov

The transfer of responsibility for the final Budapest attack away from Marshal Tolbukhin would allow him "to devote all of his attention to creating a stable situation along the outer front." [647] In other words, Marshal Tolbukhin and his command staff needed to focus on stopping the Konrad III attack and not worry about capturing the last few Nazi fanatics holding out in Buda. This *Stavka* directive was clear, but interestingly the "b" section left the 3 UF responsible for logistical support of the troops encircling Buda despite Marshal Malinovsky (2 UF) being in command of the operation.

1945 Hungary – SS Panzer Grenadiers rest during an advance. (This photo might actually show SS Panzer Grenadiers taking part in the Konrad III attack.) Note the armored personnel carrier (Sd Kfz 251) on the left with radio antennae and a Volkswagen Schwimmwagen (an amphibious variant of the Kübelwagen) on the right.

19 January 1945 – Day Two of Konrad III. Weather: 31° to 18° F, (-0.2° to -7.8° C), morning snow Just north of the Danube, German and Russian forces once again exchanged attacks with neither side gaining the upper hand. German commanders in the field wanted to regroup their forces and to delay the transfer of the 20 Pz D in order to have sufficient troops to break through the Russian positions in front of Muzsla. After consideration, the transfer of the 8 and 20 Pz Ds "had been given priority over the elimination of the Soviet bridgehead of the [Hron]." [648] German forces north of the Danube therefore were to defend the positions they currently held and to protect the supply line across the Danube and south of the river.

Another six trains carrying units of the 20 Pz D left for Poland while six more trains arrived carrying elements of the 44 Infantry Division *Hoch und Deutschmeister*, an honorary title "High German Master," abbreviated as 44 ID H.u.D. This infantry division was to relieve the two outgoing Panzer divisions at the front.

The Konrad III attack continued to drive Soviet units back in heavy fighting with one Panzer spearhead reaching the Danube far to the south of Budapest at Dunapentele.

20 January 1945 – Day Three of Konrad III. Weather: 32° to 24° F, (-0.0° to -4.5° C), snow at night

During the night a Russian counterattack by the 21 GTB, 30 GMB and parts of the 18 GMB cut the road along the northern bank of the Danube and isolated two German *Kampfgruppe* (combat group) further to the east. Outside attacks had failed to clear the blocking Russian troops, so the two *Kampfgruppe* turned around and attacked westward, breaking through the Russian troops. At this point, the 6th GTA had only 43 armored fighting vehicles in operational condition. Both sides were running short of men and equipment. Later that night, orders arrived from the OKH stating that all armored vehicles of the 20 Pz D were to be withdrawn and loaded onto trains. [649]

Norbert Szamveber, in this book *Days of Battle*, on page 180 points out that as German forces withdrew from the front, AG South issued an order about "standard actions" that should be carried out. "Your attention is called out to the order at all enemy armored fighting vehicles seized by our troops are to be destroyed, unless they can be towed away for repair and reuse by us or have [already] been completely destroyed."

This order was an attempt to prevent damaged but repairable vehicles from falling back into Soviet hands. A good tank recovery/repair shop could restore a disabled vehicle to fighting shape in a few days. Even though fuel was critically low for them, the Germans knew that a few gallons of gasoline or diesel fuel poured down the hatch, a flare tossed in, and the vehicle would burn out, transforming it from a lethal weapon into a hunk of scrap metal. This order also applied to any German vehicles that would be lost to an advancing enemy.

Day Three of the Konrad III operation saw major combat actions taking place between opposing forces, with German forces gaining more ground toward the Danube and Budapest.

The 2 UF issued a *Statement of Spent Ammunition in Budapest* from 25 December 1944 to 20 January 1945, during 27 days of combat. This was found within TsAMO record ID #454819945. Ammunition is grouped by type; a synopsis is listed in Table #29.

Ammunition Type	Number of Rounds Used
7.62 rifle	22,224,122
Heavy MG	158,569
Hand grenades	118,098
AT grenades	11,631
26 mm flare gun	33,904
82 mm mortar	220,251
120 mm mortar	85,348
37 mm AA	23,531
85 mm AA	3,358
45 mm AT gun	46,111
76 mm (all types)	91,697
122 mm howitzer	22,372
152 mm howitzer	26,985
203 mm howitzer	2,482

Table #29: Ammunition Expended against Budapest -- 25 December 1944 to 20 January 1945.

The ammunition consumed during this time totaled 523 truck/wagon loads' worth as shown in the TsAMO report. If the average truck/wagon could carry 1.5 tons, then the 523 truck/wagon loads equaled about 785 tons. The amount of ammunition used by just artillery of 76 mm and above meant that 5,316 artillery shells hit the city every day, or 221 every hour. (Note: Maybe due to low employment of 57 mm AT guns at Budapest, 57 mm rounds are not displayed on this ammunition statement.)

21 January 1945 – Day Four of Konrad III. Weather: 20° to 6° F, (-6.8° to -14.5° C), snow at night

Scattered combat took place along the north bank of the Danube south of Batorkeszi as both sides attempted to stabilize the front and to seize advantageous positions.

All the remaining units of the 20 Pz D were withdrawn from combat, while the 44 H.u.D. ID took control of the vacated sectors of the front.

The Konrad III operation continued as fighting increased in intensity. Soviet forces of the 3 UF were forced back, but they were fighting hard all the way.

In Poland, advancing Soviet troops effectively surrounded the city/fortress of Poznan. But before this took place, the *Deutsche Reichsbahn* (German National Railway) was able to evacuate, before the Red Army arrived, 2 million German civilians from the Poznan railway [district] and another 1.7 million from the Oppeln [district, south of Poznan]. This was a significant achievement because the *Reichsbahn* was under heavy Allied air attack, and it faced a growing shortage of coal to fuel its steam engines. Due to Soviet breakthroughs, the "*Reichsbahn* had to conduct a number of hasty withdraws, frequently under fire, of masses of panicky civilians and large quantities of industrial and military goods." [650] These internal refugees were sent to safe areas, but during a war "safe" was a relative term.

Russian tanks advance into the destroyed city of Poznan in 1945.

22 January 1945 – Day Five of Konrad III. Weather: 19° to -1.5° F, (-7.2° to -18.6° C), snow at night

North of the Danube, during the night and early morning, two separate Russian tank attacks, with tank riders, broke through German lines and almost reached the north bank of the Danube near the town of Dunamocs, south of Batorkeszi. These attacks threatened to cut off German forces along the river. (See Map #65 for Dunamocs' location.) A German counterattack with 25 tanks and a battalion of infantry from the 8 Pz D restored the German front line later in the day.

Adolf Hitler ordered the transfer of the entire 6th Panzer Army (1 and 2 SS Panzer Corps and the 1 Cavalry Corps) from the Western Front to the Eastern Front. But, over objections from High Command staff officers, this Panzer Army went not to the north to protect Berlin, but rather to AG South and to Budapest. Once again, Guderian wanted to use these troops to launch a spoiling attack against Soviets forces east of Berlin, and he maintained that launching a major attack in Hungary would take too long to complete and would only have an indirect impact on the battle further north. Hitler stated, "You want to attack without oil? Good, we'll see what happens when you do that." [651] Working along interior lines, the battered but still operational *Deutsche Reichsbahn* delivered this high priority Panzer formation secretly to Hungary and Czechoslovakia, taking seven to eight weeks to do so. *Stavka* and the 2 and 3 UFs were in for a shock when the arrival of this Panzer Army on the Eastern Front was disclosed during its first commitment to battle in February.

The transfer of the 6th Panzer Army to AG South and not to Germany at first appeared to be an illogical strategic move. But Hitler was not intent on defending; he wanted to attack with his strongest remaining Panzer Army and to impose his will upon the hated Bolshevik enemy. Influencing Hitler's decision was the knowledge that the last major fuel reserves available to the Reich were located in and around Hungary, and these reserves needed to be protected. The Nagykanizsa oil fields southwest of Lake Balaton produced over 900,000 tons of crude oil per year, and they currently had 70,000 tons of crude oil on hand. [652] The nearby major oil refineries located around Komarno and Bratislava, 30 miles east of Vienna, produced the vital fuel, which could more easily be distributed in the AG South area. Because of transportation issues and limited amounts of fuel actually being produced, only a major offensive in this general area could be supported logistically with fuel and lubricants, and Hitler was in an offensive mood.

During the first ten days of February 1945, Hungary produced 925 cubic meters of fuel per day. During the second ten days of February, the refineries produced only 751 cubic meters of fuel per day, and then in March, production dropped to 322 cubic meters of fuel per day. "However, Army Group South consumed only 609 cubic meters altogether in its offensive operations between 6 and 13 March 1945, so other army groups and the Luftwaffe could also be supplied from Hungary [with the surplus]." [653]

To highlight the critical nature of Hungarian fuel supplies, German fuel (gasoline, diesel) production in January 1945 was at 51 percent of total production capacity with aviation fuel (high octane) at only 33 percent of capacity. [654] The Western Allied bombing effort was responsible for this reduction in production output.

Note that when comparing liquid fuel consumption, the German Army [*Heer*] used a logistical measurement called a consumption unit or VS = *Verbrauchssatz.* The VS amount would vary depending upon the size and type of unit being supplied. A VS was how many cubic meters of fuel that a certain military formation needed to travel 100 kilometers or 62 miles. However, the Luftwaffe measured fuel needs in tons. Note: one ton of gasoline = 1.33 cubic meters, while one ton of diesel = 1.16 cubic meters. [655] For example, logistically speaking, a Panzer division would need 2 VS of fuel to launch an attack covering 120 miles, while the Luftwaffe would state they required 48 tons of fuel on hand to fully fuel 20 He-111 aircraft for a bombing mission.

Adding to this gloomy picture was the lack of coal to power factories, steel mills and trains. In February, hard coal production was at 30 percent of normal and what was produced could not be transported easily to end users due to the growing destruction of the railways. [656] For example, the OKW War Diary in February stated that the movement of the 16 SS Pz GD (from the west to the east) would take more than two weeks because "the scarcity of coal for the locomotives also has an obvious hindering and delaying effect." The OKW went on to say that the movement of the 16 SS Pz GD "required 5,000 tons of coal." [657] This was for the movement of only one division from the 6th Panzer Army. The transfer of the entire 6th Panzer Army to the east put a huge strain on the railroad and consumed large amounts of scarce coal. Transferring the 6th Panzer Army across Germany meant that hundreds of trains and thousands of tons of coal were not available to transport civilians fleeing the fighting, to distribute food and other supplies to the population, or even to supply munitions to the fighting front.

Ultimately, the transfer of the 6th Panzer Army to Hungary and Czechoslovakia allowed Hitler and AG South's commander, General of Infantry Otto Wöhler, to launch two offensive operations: the first, *Unternehmen Südwind* (Operation Southwind) and then the much larger *Unternehmen Frühlingserwachen* (Operation Spring Awakening).

While the high-level strategic debate about the 6th Panzer Army was taking place in Germany, the German Konrad III forces still were attacking and had captured the key transportation center of Szekesfehervar. Szekesfehervar was on the direct road toward Budapest, and the Soviets mounted a stiff defense of the city.

23 January 1945 – Day Six of Konrad III. Weather: 6.5° to -4.7° F, (-14.2° to -20.4° C), snow at night

The remaining units of the 8 Pz D withdraw from combat operations along the Danube in preparation for transfer to Poland. With this transfer, armored combat to the west of the Hron River all but ended.

24 January 1945 – Day Seven of Konrad III. Weather: 22° to 1.4° F, (-5.4° to -17.0° C), snow

Attacking formations of Konrad III continued to advance, with the German 3 SS Pz D reaching a point 18 miles from Budapest at the town of Baracska. Soviet resistance grew as they mounted more counterattacks against German spearheads.

25 January 1945 – Day Eight of Konrad III. Weather: 28° to 21° F, (-2.2° to -5.8° C), snow at night

North of the Danube, the decimated formations of the 6th GTA began their transfer to the rear, east of the Hron, where they would become part of the 2 UF reserve force. They needed substantial replenishment of both man and machine before being combat ready again. Only the badly battered 27 GTB remained behind to support the 7th Guards Army west of the Hron. Both the 24 and 25 GRCs fully occupied the Hron River bridgehead, leaving only the 27 GRC to cover the entire extended right flank of the Army.

On this day, the commander of Soviet Group Budapest, Lieutenant General Ivan Mikhailovich Afonin, was severely wounded. He would be out of action for three months and was replaced by Lieutenant General Ivan Mefodevich Managarov.

Major actions against Budapest had been put on hold from 18-25 January due to the German relief attacks and the fall of Pest, but the battle for Buda now intensified once again. The battle for Buda was swiftly reaching its end point. Would the garrison surrender? Would Hitler allow a breakout attempt, or would they fight to the end in the ruins of a once great city?

With the end to the fighting on the west side of the Hron River, an appraisal of the overall cost to both sides is in order. A review of medical data (TsAMO records) [658] for the 7th Guards Army from 5-25 January 1945 shows that medical units of the Army received 9,301 new casualties during this period. This timeframe covers the 7th Guards Army initial attack on the west side of the Hron and the following German counterattack.

A 2 UF report, TsAMO ID #113723525, records losses for the 7th Guards Army from 1-31 January 1945 as 1,569 dead; 491 missing; 0 captured; 25 non-combat losses; 5,398 wounded; 1,637 sick; and 0 frostbitten, for a total of 9,120 casualties. This same report also shows total casualties for the 6th GTA as 3,427.

Support from these two sets of historical data confidently show that the 7th Guards Army suffered approximately 9,000 casualties during the fighting west of the Hron River from 6-25 January, and the 6th GTA suffered approximately 3,000 casualties during this same period. Therefore, the offensive to expand the Hron River bridgehead and the failed attempt to capture the city of Komarno cost the 2 UF about 12,000 casualties.

Norbert Számvéber, in his book, *Days of Battle*, on pages 191-196, presents a detailed review of the available loss information for Axis forces. Records from AG South show German casualties over this period amounted to 4,065, but this total did not include Hungarian losses. Another report, this one by the Russian 9 GMC, states that German/Hungarian losses were 4,200 dead or wounded, along with 973 prisoners. Using this data, an overall estimate of German and Hungarian losses (killed, wounded, missing and POWs) results in a total of about 5,500 people.

Despite some uncertainty, Norbert Számvéber does a good job estimating the overall tank and armored fighting vehicles losses for both sides. German and Hungarian forces lost (damaged or destroyed) 83 tanks and assault guns during the battle while Russian forces lost a minimum of 240 tanks and self-propelled guns. (See Table #12 in Számvéber's book.) [659] Because German and Hungarian forces controlled a large section of the battlefield after the battle, they would have been in a position to recover and to repair many of their damaged vehicles. Russian tank recovery units also would have also been able to recover and to repair at least some of their losses. Overall, Axis forces suffered about 5,500 casualties and had 83 armored fighting vehicles (AFV) that were damaged or destroyed while the Soviet forces suffered approximately 12,000 casualties and had at least 240 armored fighting vehicles damaged or destroyed.

Based upon these reports and estimates, we can see that the 2 UF paid a very high price in men and equipment to gain some useful but not critical ground. In retrospect, the major attack launched by the 6th GTA and the 7th Guards Army from 6-25 January not only failed to capture the important transportation and oil refining center of Komarno, but it also failed to effectively cut the German supply line supporting the Konrad I, II, and III offensives. In contrast, the Red Army salvaged a significant bridgehead west of the Hron River from the fighting which would serve as an ideal staging area for future offensive operations north of the Danube. The road to Austria and southern Germany was spread out before the 2 UF and the 7th Guards Army, but they were prevented at this point from making any meaningful progress toward their ultimate objective, the destruction of Nazi Germany. (See Map #66 for the overall strategic situation in the Budapest area.)

26 January 1945 – Day Nine of Konrad III. Weather: 29° to 28° F, (-1.5° to -2.0° C), snow at night The last trains carrying units of the 8 Pz D left Hungary.

Units of the 25 GRC took over the final positions along the front where the 6th GTA had recently been fighting. Marshal Malinovsky had withdrawn the 6th GTA to an area east of Budapest for a major refit. Now German and Hungarian infantry units only faced Russian rifle units. With the withdraw of the Panzer/Tank formations, German and Russian offensive operations had come to an end north of the Danube.

In the days after the end of the fighting north of the Danube, those that survived the battle had a duty to the dead. One unidentified eyewitness stated,

> The civilians were ordered out to bury the dead. But it was impossible to bury them in the wintry conditions of January. So, they stacked the bodies up like cordwood. Bodies of the men burnt to death, body parts lay everywhere – here a hand, there a leg – bore witness to the horrors of war. But the most horrible sight was the bank of the Danube. They hauled the now useless Soviet tanks to the former [Komarno] market place. In one of them sat the driver. In his right hand he held the control stick, in his left some hair, but he had no head. This sight - for me - marked the end of the January offensive of the Soviets. [660]

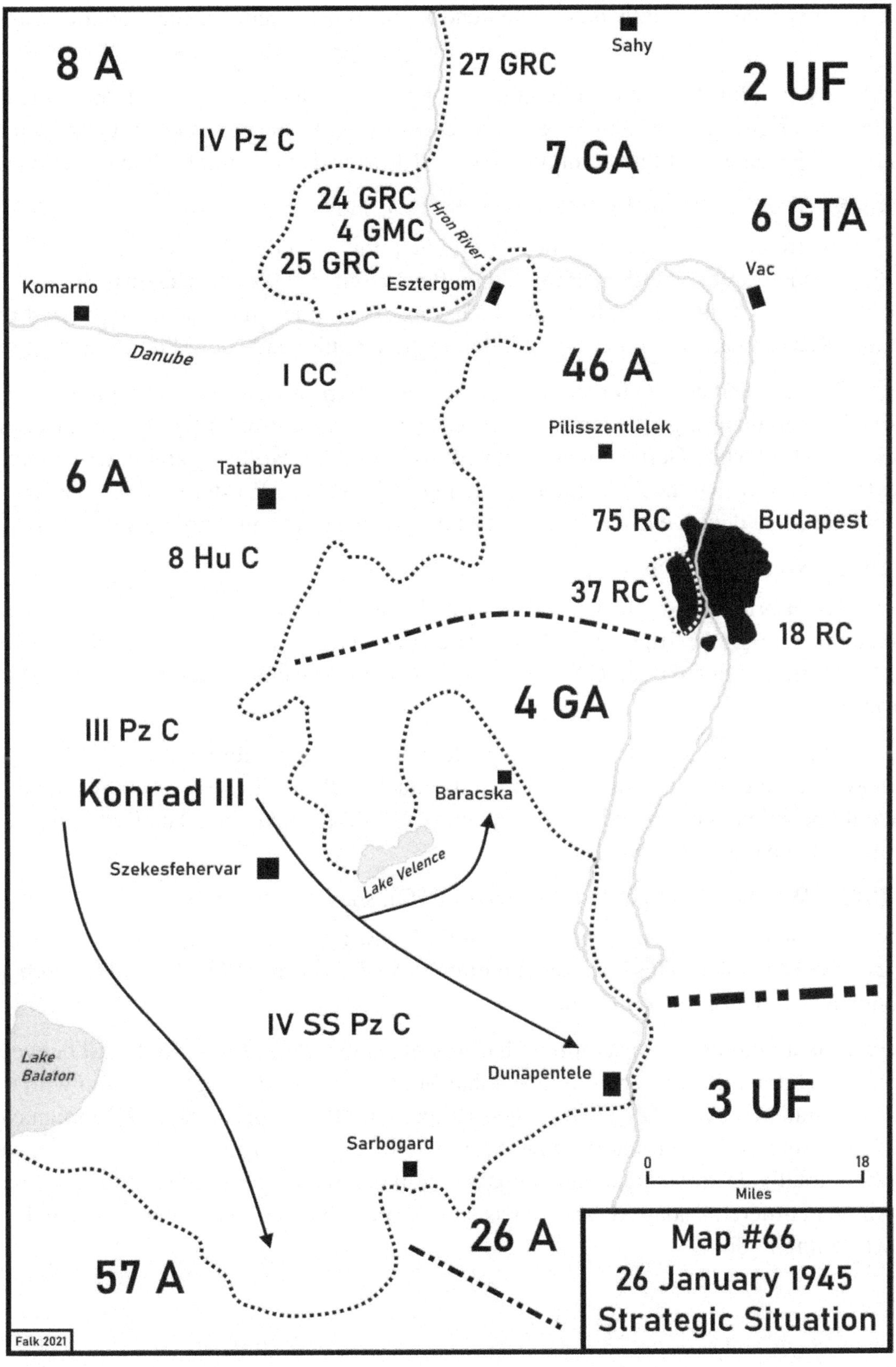

27 January 1945 – Day Ten of Konrad III. Weather: 34° to 21° F, (+1.2° to -5.8° C), snow

The Konrad III offensive advanced slowly, as the Budapest garrison slowly starved. Marshal Tolbukhin's 3 UF, 4th GA, launched counterattacks against the exposed IV SS Pz C around Baracska as *Stavka's* newly assembled 26th A attacked from the south between Sarbogard and Dunapentele.

Stavka hoped that the forces on hand would put an end to this most recent German incursion against the Red Army.

In the face of these powerful Russian counterattacks, supported by large numbers of tanks and aircraft, the attacking German forces of Konrad III were forced to go over to the defense. Time had run out for both Konrad III and Budapest. At this point, the Konrad III offensive essentially had been defeated.

28 January 1945 – Weather: 24° to 16° F, (-4.6° to -8.6° C), snow

At Buda, the Luftwaffe made a major daytime effort to help the defenders by flying 320 combat sorties against attacking Soviet forces, and then at night they flew another 200 sorties against enemy groupings. Additionally, German aircraft dropped 130 tons of supplies into Buda, but most of the canisters fell into Russian hands or into no-man's-land where neither side could retrieve them safely. [661]

On the Western Front, Allied armies forced the last German troops involved in Operation Watch on the Rhine offensive back to their original starting locations. During the entire battle, Allied causalities amounted to over 75,000 while German losses totaled over 100,000. Hitler's gamble was a total failure, which only extended a war that the Axis alliance already had lost. The Western Allies soon would be ready to launch their own decisive offensive, one that Nazi Germany would not be able to stop.

29 January 1945 – Weather: 17° to 8° F, (-8.2° to -13.4° C), snow

Within the Hron River bridgehead, the 4 GMC, detached from General Pliev's Cavalry-Mechanized Group, moved up to support the 7th Guards Army. The 7th Guards Army, 27 GTB, and the 4 GMC were refitting at the front while the 6th GTA and Pliev's Cavalry-Mechanized Group were refitted far behind the front line.

Further to the south, over the previous several days, the 4th GA's attack forced the 3 SS Pz D out of Baracska in heavy fighting, as the 26th A continued its attack to the northwest. *Stavka* expected these two attacks would be able to cut off and to destroy the IV SS Pz C spearhead. The Red Army was in effect reversing all the gains of Konrad III.

30 January 1945 – Weather: 27° to 6° F, (-2.4° to -14.6° C), snow

The 2 UF issued a *Statement of Spent Ammunition in Budapest* from 20 January 1945 to 13 February 1945 during 24 days of combat. This was found within TsAMO record ID #454819944. Ammunition is grouped by type; a synopsis is listed in Table #30.

The huge increase in ammunition consumption that was being directed against Budapest between December 44 and February 45 was a result of Marshal Malinovsky's growing frustration with the slow pace of the attack against the city. Sluggish progress along with the rising number of Russian casualties caused him to substitute overwhelming firepower for manpower in order to capture the city faster. That the city was being totally destroyed and noncombatant civilians were being killed in the process was irrelevant. Stalin had ordered Budapest captured as soon as possible, and that is what Marshal Malinovsky was going to do.

Ammunition Type	Number of Rounds Used
7.62 rifle	6,275,452
Heavy MG	80,810
Hand grenades	44,878
AT grenades	7,087
26 mm flare gun	17,400
82 mm mortar	180,149
120 mm mortar	54,440
37 mm AA	13,200
85 mm AA	1,650
45 mm AT gun	21,651
76 mm (all types)	43,698
122 mm howitzer	29,771
152 mm howitzer	27,557
203 mm howitzer	3,418

Table #30: Ammunition Expended against Budapest -- 20 January to 13 February 1945.

The ammunition expended during this time totaled 400 truck/wagon loads' worth. This table shows a decrease in small arms ammunition consumption and an increase in heavy artillery use from what was shown in Table #29. The fighting was moving into major built up areas of the city, and the Red Army needed more firepower to engage the enemy defending within modern multistory buildings. The 400 truck/wagon loads as shown in the TsAMO report, averaging 1.5 tons each, would equal about 600 tons of munitions. During this period, the amount of ammunition used by just artillery of 76 mm and above meant that 4,351 artillery shells hit Buda every day, or 181 every hour. All these munitions were being directed against only the Buda side of the city now that Pest had fallen, so Buda was literally being pounded into rubble.

1945 Hungary – A Soviet 122 mm howitzer crew uses direct fire during city fighting.

Guards Major of the medical service Lev issued a medical report, *Information about the location of the wounded and sick 7th Guards Army*, found in TsAMO record ID #133393430. This report covers the period from 25 January to 30 January 1945, and was an assessment of the current medical burden upon the Army's medical establishment.

The total medical burden on 30 January 1945 was 5,776 people. This total was far less than the 30 December 1944 value of 7,421 people, an indication that the severity of the fighting continued to decrease during January.

Forward Troop Area – up to divisional medical battalions

- Arrivals:
  - At the start (25 January) – 991 men and women
  - Newly arrived – 799
  - Total – 1,790
- Departures:
  - Evacuated – 347 (most likely to the Army rear) (moderately wounded)
  - Returned to active service – 335
  - Died –20
  - Total – 702 (including dead)
  - In the Forward Troop Area on 30 January – 1,088 (lightly wounded)

In Army Hospitals

- Arrivals:
  - At the start (25 January) – 4,790 men and women
  - Arrived from forward units' medical installations – 309
  - Arrived directly from units – 156 (severely wounded)
  - Totals – 5,255
- Departures:
  - Evacuated – 33 (out of the Army to the deep rear, severely wounded)
  - Returned to active service – 514
  - Died – 30
  - Total – 577 (including the dead)
  - In Army Hospitals 30 January – 4,678 (moderately wounded) (with a newly increased total of approximately 6,700 hospital beds, which results in a surplus of about 2,022 beds)

Total in all medical installations on 30 January – 5,776 people

After months of heavy fighting and even heaver casualties, the Army's medical establishment was at last getting a break. Note the 2,022 vacant Army level hospital beds and the very low number of severely wounded (only 33) sent out of the Army to the deep rear.

31 January 1945 – Weather: 23° to 3.2° F, (-4.8° to -16.0° C), snow in evening

Over the previous two days, fighting in and around Budapest had leveled off as both sides adjusted their positions and resupplied their forces. This short break in the fighting would not last long.

At this point, *General der Panzertruppen* (General of Panzer Troops) Hermann Balck, commander of the 6th Army (also known as Army Group Balck) knew the Konrad III offensive had totally failed with no possibility of resuming offensive operations in the face of the advancing Russian armies. But

General Balck would make the Red Army pay dearly to force his army out of their newly won positions.

For the Budapest garrison, strength estimates come from the War Diary of AG South. At the end of December 1944, German troops were estimated to number 33,000 men with Hungarian formations at 37,000, for a total of 70,000 troops. The many wounded, service troops and non-combat elements should be subtracted from these overall numbers to find out how many of the remaining troops were actually combat effective. With little food or ammunition remaining to supply the surviving combat troops, the prospects of holding out much longer was grim. [662]

The 53rd A and the 7th Guards Army issued a Secret Joint Act about securing the junction between the two Armies. This Act contained four sections detailing just who would cover, defend and fortify the junction and adjacent areas. [663] This type of advanced planning now was standard practice for the Red Army because too many times in the past a sudden German Panzer attack would split the front open right at the boundary line between units. Unknown to these two Army Commanders, this Act would very nearly be put to the test in a few weeks' time.

The 2 UF issued a report, *Report on the loss of personnel in the troops of the 2nd UkrF from January 1 to January 31, 1945.* This was found within TsAMO record ID #113723525. The 7th Guards Army suffered 9,120 total casualties for the month: 1,569 killed; 489 missing; 0 captured; 25 non-combat losses; 5,399 wounded/burned; 1,638 sick; and 0 frostbite. Once again, the 0 people captured figure for the 7th Guards is almost certainly not true as some of the 489 missing probably now were German POWs.

The report goes on to list 7th Guards Army horse losses: 406 killed; 0 missing; 139 wounded; 55 evacuated to the veterinarian; for a total of 600 equine casualties. The number of horses that died decreased from the previous month, but the weather, combat actives and overwork kept the overall horse casualties high.

The same report shows that the troops of the entire 2 UF suffered 45,934 total casualties during January 1945.

For the month of January 1945 in Britain, air raids and German V-1 and V-2 attacks continued, but loss totals are unavailable.

1 February 1945	
7th Guards Army	
Assigned – 2 Ukrainian Front	
Colonel General M.S. Shumilov Commanding	
Rifle, Airborne and Cavalry Units	24 Guards Rifle Corps (6 GAD, 72, 81 GRD) 25 Guards Rifle Corps (53, 375, 409 RDs) 27 Guards Rifle Corps (141, 243, 303 RDs) 93 GRD 15 Penal Battalion 62, 63, 64, 65, 66, 194 Penal Companies
Artillery Units	2, 11 Antitank Artillery Brigades 114, 115 Guards Tank Destroyer Regiments 41 Guards Gun Artillery Brigade 263, 290, 493 Mortar Regiments 5 Antiaircraft Artillery Division (670, 743, 1119, 1181 AA Regiments) 26 Antiaircraft Artillery Division (1352,1357,1363,1369 AA Regiments) 662 Separate Aerial Observation/Air Raid Warning Company
Armored and Mechanized Units	27 Guards Tank Brigade 38 Separate Armored Train Battalion
Engineer Units	60 Engineer Sapper Brigade 176 separate company of backpack flamethrowers
Estimated Combat Strength Reported Ration Strength	51,171 men 70,873 men and women (including an estimated 7,000 support personnel)

(BSSA 45, BSA 45, TsAMO Archive ID #113723570 and 100155614)

The start of the month saw the removal of the 7 Ru RC, 16 Artillery Penetration Division, 26 Antiaircraft Artillery Division, and the 21 Pontoon Bridge Battalion from the 7th Guards Army.

At this time, the 7th Guards Army had the following medical units assigned: 10 KHPPG - Mobile Surgical Field Hospitals, 3 GLR – Hospitals for lightly wounded, 2 IPPG – Infectious Hospitals, 1 EG - Evacuation Hospital, 2 EP - Evacuation Points, and the normal set of medical support companies. These medical units did provide an increase in bed space for the Army, to about 6,700 total beds. But theses medical units also were taking care of 5,776 accumulated wounded from previous battles, so the number of open beds was limited.

1 February 1945

The Deputy Commander of the 72nd GRD issued Secret Order #016 to the division, *"On measures to increase vigilance, strengthen the protection and defense of the rear areas of the division."* This order in part noted, "It has been established that the enemy, with the goal of disabling the soldiers and officers of the Red Army, moves agents into the cities and towns [to be] left behind with the task of undermining vehicles and warehouses, [and] contaminating water sources with poisons and poisoning food." [664] Among other things, this order stated that personnel should

- Increase vigilance in all rear areas.
- Organize and increase night guard posts and patrols in and around warehouses and rear facilities.
- Search the rear areas periodically, detain suspicious persons with or without documents, and turn them over to SMERSH (counter-intelligence agencies).
- Forbid civilians from processing food, and secure any prepared food.
- Check and analyze water sources, and test trophy (captured) products before using.

This order highlighted the fact that the Red Army no longer was fighting on friendly territory and enemy agents, real or imagined, were everywhere! But this level of scrutiny was nothing new, as SMERSH units were "set up to catch spies, [and they] suspected everybody without exception of being an enemy of the Soviet state." [665] In the Soviet Union anyone who had been captured, or even just spent time in occupied territory, carried the stigma for the rest of his or her life. "S.M.E.R.S.H., [was] a powerful institution that trapped real traitors, but also unmasked as spies completely innocent people, in the best traditions of the Stalinist regime." [666] The ultimate fate of so many of these innocent people was arrest, interrogation, conviction and death by firing squad or a single pistol shot to the head.

Guards Major Triumfov issued a report, *List of personnel by army formations*, for 1 February 1945, from TsAMO ID #437714951. This report lists the strength of all the rifle formations of the Army on 1 February and the number and source of replenishment troops that the Army had received during January. (See Table #31.)

Unit	Strength
6 GAD	4,044
72 GRD	4,027
81 GRD	4,206
53 RD	4,042
375 RD	3,857
409 RD	3,780
141 RD	3,785
243 RD	4,261
303 RD	3,722
93 GRD	5,738
Total	**41,462**
All other parts of the Army	29,411
Total Army Personnel as of 1 Feb 45	**70,873**
Replacements from Army commander and front-line reserves (190 Army Reserve Training Regiment)	1,203
From Army hospitals directly to units	1,018
Marching replenishment from military districts	3,983
Total Replacement Personnel in Jan.	**6,204**

Table #31: List of Personnel by Army Formation – 1 February 1945.

The 7th Guards Army suffered 9,120 total casualties during January 1945, and Table #31 shows how those losses were reflected in the understrength nature of the rifle divisions. During the last period of the war a full-strength rifle division, Guards or regular, should have contained between 7,200-8,500 men according to *Shtat* tables. But rifle divisions with strengths of less than 6,000 men were considered strong enough to take part in offensive operations. Because of this, even a Guards Army in 1945 would have been happy to have the strengths of its rifle divisions in the 6,000 range. Having said that, the divisions listed within Table #31 were approximately 45-50 percent under *Shtat* strength, and that is after the listed replacements already had arrived. The replacements delivered to the divisions in January by the Army's training regiment (1,203 men) and from hospital returns (1,018 men and women) and from 2 UF (3,983) totaled 6,204 men and women. If allocated equally, they only would have added 620 men and women to each division. As can be seen in the table, even after receiving these replacements, the rifle divisions still were far under strength. General Shumilov apparently attempted to compensate for the depleted and understrength nature of his Army by ordering extensive field fortifications built, and anti-tank gun zones and minefields prepared, especially within the Hron bridgehead area.

The "All other parts of the Army" section of Table #31 combines all the personnel (29,411) for the remaining combat and support units assigned to the Army. For accounting purposes, this consolidation of personnel totals makes it difficult to separate these people into their separate combat or support roles. Thus, we can only estimate the combat strength of the Army on 1 February 1945.

Also on this date, Captain-Engineer Laschevsky of the 7th Guards Army issued an interesting diagram/report, *Scheme of ice phenomena on the Danube River,* TsAMO ID #133391761. This diagram shows that in the sector between Esztergom-Karva (Karva is located 11 miles west and upriver from Esztergom) the thickness of ice was 10-19 cm (4-7.5 inches) and was considered a "frozen section" of the river. This is interesting because this thickness of ice easily can support the weight of small groups of men or even small vehicles or AT guns. German engineers certainly were recording this same type of information. Note: As part of their regular work, Soviet engineers also monitored the water levels of both the Hron and Ipel Rivers. One never knew when the Army commander might require such information, as it might be critical for planning movements, or maybe it would never be used at all. Also noted was that the lower reaches of the Hron River were 100-130 feet wide and were flowing quickly with the early thaw runoff.

2 February 1945

Major General Lukin, the chief of staff of the 7th Guards Army, produced a planning map, *The scheme of battle formations and counterattacks of the troops of the 7th Guards Army,* TsAMO map ID #100155626. This map displayed three possible enemy attack directions against the Hron bridgehead, as well as counterattack options for the 7th Guards Army. General Shumilov clearly had his staff thinking about what the enemy might do and how the Army could respond. This map also showed possible enemy attacks against the long right flank of the Army and the very exposed 27 GRC.

3 February 1945

During the fighting in January west of the Hron, the 7th Guards Army suffered an average of 294 casualties per day including 53 sick per day. But with no major combat operations taking place this month, losses began to taper off. The number of new casualties, over the next two weeks, fell to an average of just 179 men or women per day, with sickness responsible for 70 of these daily casualties.

A German interception of a Russian radio message from within the Hron Bridgehead area on this date said, "it was intended for work to begin on 8 February." The Germans believed this message indicated that something was being planned on the Russian side for the 8th, and they prepared their forces accordingly. [667] Nothing of note occurred on the 8th, so the meaning of this intercepted message remains unknown.

4 February 1945

President Roosevelt, Prime Minister Churchill and Premier Stalin assembled at the seaside resort of Yalta in the Crimea for the Yalta Conference, code-named Argonaut. This was the last time the "Big Three" would meet during the war. The leaders' conference goals were mostly aimed at post-war security, re-establishment of national borders, and self-determination of European states. Ultimately, the Yalta Conference failed to secure free and fair self-rule for states in the Soviet zone of Europe. The final stage for the cold war between the East and the West almost was established.

5 February 1945

Soviet Forces of Group Budapest launched a powerful attack into the hills of Buda around the Sashegy Hill and the Nematology Cemetery areas. The defenders were hard pressed to repel these attacks. On 6 February, German forces within Buda counterattacked toward Sashegy Hill but were driven back. By the end of the day, Sashegy Hill fell to the Red Army as the Axis troops in that area had run out of food and ammunition. On 7 February, in Buda, Red Army forces captured the southern railway station. [668] With these losses, *SS-Obergruppenführer* (Senior Group Leader) Karl Pfeffer-Wildenbruch, the Commander of the Budapest garrison, knew the end was near, and he began making plans, not for surrender, but for a breakout attempt.

6 February 1945

Along the front guarded by the 7th Guards Army, both sides employed only normal scouting actives, combat patrols, sniper fire, and harassing mortar and artillery fire. In other words, all was quiet. The Army continued to improve its defensive positions and to refit its critically understrength forces.

On the Western Front, advancing British, American, Canadian and French armies had reached the major German defense line called the *Siegfried* Line (also known as the Westwall), and had broken through it in places. This 390-mile-long fortified line stretched from the German/Netherlands border in the north to the German/Switzerland border in the south. Depleted German armies in the West needed the *Siegfried* Line to help guard German's western border, but the line was collapsing under continuous Allied attacks. After the *Siegfried* Line, only the Rhine River would offer a significant barrier to the advancing Western Allied armies.

7 February 1945

With major secret reinforcements already on the way to Hungary, consisting mostly of the 6th Panzer Army, Adolf Hitler, the German high command, and General Otto Wöhler, commander of AG South, met in Berlin to decide how best to employ this powerful Panzer force. They already had decided upon *Unternehmen Frühlingserwachen* (Operation Spring Awakening). Once the entire 6th Panzer Army had assembled, they would launch this offensive from the same general area as the Konrad III operation. But the lingering Soviet Hron River bridgehead represented an ideal assembly area for Soviet forces to threaten German efforts south of the Danube and to jeopardize the safety of Vienna north of the Danube.

So General Wöhler argued that they needed to launch *Unternehmen Südwind* (Operation Southwind) first in order to destroy the Soviet Hron bridgehead and to protect the vulnerable area north of the Danube River. To achieve this goal, Wöhler finally was able to secure the use of the first elements of the 6th Panzer Army to arrive, the I SS Pz C. Wöhler hoped this Panzer Corps would provide the needed offensive firepower to quickly destroy the 7th Guards Army's Hron River bridgehead. Once they achieved that goal, AG South and the 8th Army, commanded by *General der Gebirgstruppe* (General of Mountain Troops) Hans Kreysing, could stabilize their defensive position north of the Danube, and AG South then safely could begin Operation Spring Awakening.

This course of action was a sensible and correct assessment of the situation as it would remove the threat north of the Danube, but it also would sacrifice the element of surprise with the early use of the I SS Pz C, which the Allies thought was deployed along with the rest of the 6th Panzer Army on the Western Front. When the combined 6th Panzer Army finally attacked during Operation Spring Awakening, the Soviets therefore already would know the 6th Panzer Army was in Hungary and would have prepared to counter its attack. If the secret arrival of the 6th Panzer Army could have been kept, the sudden and unexpected appearance of the entire 6th Panzer army on the Eastern Front would have been much more devastating. But the reality was that SG South had no other choice. The Germans needed to remove the Soviet Hron River bridgehead and to destroy or disabled the 7th Guards Army in order to safeguard Vienna and to allow Operation Spring Awakening a chance of success. If the element of surprise was lost for the 6th Panzer Army, so be it. Hitler was determined to attack in Hungary, so both Operation Southwind and then Spring Awakening would proceed regardless.

8 February 1945

The 25 GRD (30 RC) was transferred to the west side of the Danube where it was able to launch powerful attacks against the southern section of Buda. The German/Hungarian line was pushed back as the overall situation became even more desperate in the hills of Buda. The Luftwaffe was able to drop 4.5 tons of supplies over the palace, but that small amount clearly was insufficient. [669]

German armies within AG South were ordered to conserve artillery ammunition in quiet sectors, saving it for more important combat missions.

The German 8th Army reported "no significant engagements to report along the front of Pz C F," [670] the Hron Bridgehead area.

9 February 1945

In the air over the Western Front, 1,500 American heavy bombers entered German airspace determined to bomb seven different targets during one raid. To counter this attack, the Luftwaffe sent up about ten Me 262 jet fighters flown by inexperienced pilots. The resulting air-to-air battle saw one American bomber damaged with six of the attacking Me 262s shot down by escorting American P-51 Mustang fighters. [671] In 1945, the real defense against air attacks upon Germany were the massive numbers of flak guns deployed around key cities and industrial targets. In late 1944, 2,655 heavy and 1,612 light flak batteries, with four guns in each battery, were deployed in Germany. These AA weapons consumed around 3.5 million rounds of heavy and 12.5 million rounds of light ammunition per month! [672] But in early 1945, hundreds of these batteries were transferred to the Eastern Front to provide added firepower for the hard-pressed German Army. The transfer of these AA batteries left German cities and industry even move vulnerable to air attack.

10 February 1945

General Wöhler returned to SG South after the conference in Berlin. Wöhler immediately began preparations for the attack to destroy the Hron bridgehead. The basic plan was that after a short artillery bombardment, the gathered German assault force would attack mainly against the north side of the bridgehead beginning at 05:00 hours. Infantry units of Pz C F would open the attack and clear the way for the introduction of the I SS Pz C to exploit the resulting breach in the Soviet line and to reach the Pariszky Canal on the first day. The Axis forces would use the second day to cross the Pariszky Canal in strength and to prepare for the Panzer thrust to the Esztergom area. The third day's advance would split the bridgehead into two and reach the mouth of the Hron River area. German troops would then begin mopping up operations to eliminate the few remaining pockets of resistance holding out on the west side of the river. The final act would be to cross the Hron and to seize four bridgeheads on the eastern side of the river at Beny, Kemend, Kohidgyarmat and Garamkovesd. [673] Following the end of this successful operation, the 8th Army would withdraw the I SS Pz C from the front at the earliest opportunity to allow it to recover from Operation Southwind before moving south of the Danube to participate in Operation Spring Awakening.

This same night the headquarters of the I SS Pz C was briefed on the general plan for Operation Southwind.

11 February 1945

Karl Pfeffer-Wildenbruch, Commander of the Budapest garrison, ordered a breakout from the doomed city with his remaining troops. Acting on his own and against Hitler's wishes, this was his last attempt to save at least some of the combat troops under his command from certain death or capture. Possible breakout forces totaled 23,900 Germans with 9,600 wounded, and 20,000 Hungarians, including 2,000 wounded. Additionally, 80-100,000 civilians still remained within the city, and an unknown number of these civilians joined in the breakout from Buda. [674] Some accounts state that in the end, Hitler sanctioned a breakout, but this has not been verified.

Commander Pfeffer-Wildenbruch planned two separate routes for the breakout attempt with his approximately 44,000 troops. The largest group would escape to the northwest, totally above ground and the smaller group to the north, initially via the underground drainage system. During the night, some 18,000 soldiers were assembled for the breakout attempt, with 6,000 of them wounded. Most likely these were the first wave of escaping troops. Overall, prepared Soviet forces slaughtered the escaping troops above ground. Only 600-700 of them made it to German lines after several days in the woods.

The underground group managed better at first, but once forced to move above ground, they too were cut down in large numbers; only 785 men reached German lines. [675] The Russians ultimately captured Commander Pfeffer-Wildenbruch and other high-ranking officers during the escape, while less sensible German officers escaped the Soviets with a gunshot to the head. Totals vary from source to source, but overall, about 17,000 Axis troops were killed during the breakout with another 22,350 captured.[676] Soviet losses are unknown.

An unknown number of remaining troops, unwilling or unable to risk escaping, remained with the wounded and civilians within Buda. The end would come soon enough for them.

General Wöhler, commander of AG South, called together the commanders of the I SS Pz C, Pz C F, the chief of staff of the 8th Army, the quartermaster for AG South and other commanders for a more "expansive conference" concerning Operation Southwind. Only verbal orders were given as written orders were prohibited for security reasons. Later that evening, the 8th Army reported that "preparations for Southwind would be put into effect on the evening of 11 February." [677] These preparations were estimated to take five days and would be completed by the evening of 16 February.

12 February 1945

With the desperate breakout attempt from Buda ongoing, Soviet troops were fully occupied with bands of armed and desperate enemy troops wondering through the woods. During the day, Soviet troops did make a cautious advance into Buda, but the final capture of Buda could wait.

North of the Danube, the 7th Guards Army continued to fortify its Hron River bridgehead. A German account stated,

> In the quiet weeks following the stabilization of the front, the Soviets had had time to create an elaborate in-depth defensive system. Villages had been fortified and strong anti-tank gun positions, organized in such a way that they could cross fire and mutually support each other, had been set up. [678]

13 February 1945

North of the Danube, a sudden thaw, which might postpone the attack by a day, delayed the movement of units of the I SS Pz C; however, the 8th Army HQ ultimately told them that "the attack at 05:00 [hours] on 17 February must be adhered to under any and all circumstances." [679]

This day --13 February 1945 -- was the official Soviet date for the fall of Budapest. Looting, rape, murder of wounded enemy soldiers and summary executions of Arrow Cross members filled the ruined city with even more bodies among the thousands of unburied bodies that already existed. Adding to the confusion, some 30,000 male civilians were detained by the Soviets and arrested as enemy soldiers, thus swelling the captured enemy troop totals. [680] Larger numbers of captured enemy troops made for a more impressive report to the high command. By Soviet accounts, the battle for Budapest lasted a total of 108 days, beginning on 29 October 1944. This included actions before they surrounded the city.

After the extended siege of Budapest and with swirling battles all around, Stalin had his victory. Estimated military losses for both sides are listed within Table #32. As can be seen, Soviet armies paid a high price for Stalin's victory, far higher than what the Axis defenders suffered. Civilian casualties are not included in the table below.

Group	Dead	Wounded	Prisoners	Total
Budapest Soviet	44,000	130,000	2,000	176,000
Budapest Romanian	11,000	12,000	1,000	24,000
Budapest German and Hungarian	40,000	Included within prisoner's total	62,000	102,000
Konrad (I, II, III) Soviet	15,000	60,000	5,100	80,000
Konrad (I, II, III) German and Hungarian	8,000	26,000	1,000	35,000
Total Soviet and Romanian	**70,000**	**202,000**	**8,000**	**280,000**
Total German and Hungarian	**48,000**	**26,000**	**63,000**	**137,000**

Table #32: Military Casualties during the Budapest Siege – 3 November 1944 to 11 February 1945. Estimates by John Lukacs. [681]

Unlike Stalin, in the end Marshal Malinovsky had a different feeling about the battle for Budapest. "Malinovsky was furious because the fighting had lasted so long. When General Pfeffer-Wildenbruch was brought to him for interrogation, Malinovsky shouted at him, 'If I didn't have to answer for your head in Moscow, I'd have you hanged in the main square of Buda!'" Despite the threat of hanging, General Pfeffer-Wildenbruch survived the war and Soviet captively, being released in 1955.

Sergei Matveevich Shtemenko, Deputy Chief of the Soviet General Staff, also had thoughts on the issue of Budapest, "…[T]he resistance of Budapest had upset the Soviet war plans: 'The plans of our general staff, according to which our forces were to reach Vienna at the end of December and southern Germany in March, were completely upset, mainly by the long siege of the Hungarian Capital.'" [682]

We also should not forget about the civilian dead, which are much harder to verify. For instance, during the siege of Budapest, an estimated 76,000 civilians died. [683] These included people who died during combat operations, those who starved, died from disease, were executed, raped, or died from forced labor, and those taken away as prisoners of war never to return. We also should include the 180,000 to 600,000 Hungarian Jews who were deported and murdered, starved, or died of other causes. [684] Members of the Fascist Arrow Cross Party, the Nazis, and the Soviets deliberately killed many civilians during the fighting, but most of the civilian deaths occurred because they simply got in the way of the war. During a war, the innocent suffer the most.

On the Western Front between 13 and 14 February, British and American heavy bombers launched several attacks against the German city of Dresden, using mostly firebombs. Up to this point in the war, Dresden had hardly been touched. In the aftermath of these air attacks, approximately 25,000 people were killed and 2.5 square miles of the city center were completely destroyed. The justification for this attack was that it would disrupt German transportation, communications, and war manufacturing efforts in the area and assist in the advance of the Soviet Army. But the city was filled with refugees fleeing the Red Army and largely undefended.[685] Post war, people in Germany and around the world declared this attack as unjustified and even a war crime. Ultimately, the war continued to its turbulent and deadly end.

14 February 1945

General Shumilov had deployed Pliev's 4 GMC within the Hron bridgehead, but he also knew that the remaining parts of General Pliev's Cavalry-Mechanized Group, the 4 and 6 GCC, were refitting to the rear of his army. The 6th GTA was positioned even further back, east of Budapest, and was undergoing a complete refit and rebuilding. *Stavka* was reconstituting these two mobile formations because they were expected to lead the offensive that would capture Vienna!

15 February 1945

On this day, the Soviets considered the Budapest breakout operation to be over, but lone escapees trickled into German lines for weeks to come.

The German 8th Army pointed out for Operation Southwind, "According to aerial photos of the Pariszky Canal, which crossed the bridgehead from west to east northwest of Gran [Hron], it [the canal] formed a major obstacle as a result of the thaw and flooding, which led one to expect delays in carrying out the attack [across the canal]." [686]

At this time the medical service of

> …the 7th Guards Army had 10 HPPG, 3 GLR, 2 IPPG, 1 EG, 2 EP and a reinforcing medical company. All of them were located on the eastern bank of the [Hron]. These medical institutions contained 6,700 beds on February 15, that is, two days before the German offensive. [With 3,419 beds occupied at this time, that left only 3,281 open beds.] [During the coming battle] the chief medical officer, colonel of the medical service, A.N. Grigoriev, did the right thing by sending an operational group to the bridgehead and giving it operational control of 49 ambulances [every motorized ambulance the Army controlled], which were allocated to a separate PMP [Regimental Medical Point] of rifle regiments. The army medical service did everything in order not to leave the wounded on the battlefield. [687]

Despite their best efforts, the 7th Guards Army reported 2,127 missing in February. Compounding the problems of supporting the bridgehead, all of the Army level medical units and rear supply support service units also were located on the east side of the Hron River. With only four crossing points over the river, located at Beny, Kemend, Kohidgyarmat and Garamkovesd, the evacuation of wounded to the rear and the flow of supplies to the front would be endangered if these crossing points were captured or destroyed by the enemy.

16 February 1945 Weather: Average Temperature 34.5° F, (1.6° C), light snow in the evening

For the 7th Guards Army, the situation north of the Danube had been quiet for the previous three weeks. This time was put to good use as all formations needed a rest, and they also required resupply and replenishment of both men and equipment. As General Shumilov put his army in order, troops were tasked with fortifying their positions. Shumilov knew his large bridgehead over the Hron was a threat to the Germans because it was an ideal location to begin an advance, north of the Danube, toward Bratislava and Vienna. With this in mind, both offensively and defensibly, the Army needed to be prepared. (See Map #67.)

General Shumilov (with his hand in his coat) poses with his officers after presenting them with awards.

Along the long right flank of the 7th Guards Army, the 27 GRC covered the Hron River front, while part of the 243 RD held the section of the front around Beny on the western side of the river. For the bridgehead itself, Shumilov positioned the entire 24 GRC with three rifle divisions along the north side of the bridgehead, and the entire 25 GRC with its three rifle divisions along the south side of the bulge. Additionally, Shumilov deployed the entire 4 GMC (7,265 men) around the town of Bart and the entire 27 GTB (1,370 men) [688] to the south, just north of the Danube, west of Muzsla. Also, the Army's reserve rifle division, the 93 GRD, deployed two of its regiments along the north bank of the Danube between Muzsla and Sob, with the third regiment in reserve concentrated around Muzsla.

By the start of the battle, Shumilov had deployed within the bridgehead about 70% of his total rifle troops (30,630 of 43,401 troops). He also deployed within the bridgehead all the tanks and self-propelled guns under his control. The 27 GTB contained 10 T-34s, 5 Su-85s and 2 SU-76s (17 in total) while the 4 GMC added a further 18 T-34/85s, 1 SU-85 and 1 SU-122 (20 in total). These numbers also included three command tanks. At full strength, the 4 GMC alone should have contained well over 200 AFVs so it was operating at less than ten percent of its authorized armored strength. The rifle units within the 4 GMC were operating at about half strength. [689] The depleted nature of these two tank/motorized formations denied Shumilov an effective counterattack force with which to defend the Hron bridgehead. Regardless of their strength, General Shumilov knew these two units would add an additional 8,635 men and 37 AFVs to the force holding the bridgehead. Most of the available artillery and anti-tank formations also would have been deployed within the bridgehead. Shumilov made certain the bridgehead was as secure as possible and that his Army was ready for battle.

But earlier in the morning of 16 February, at 01:00 hours, a regimental assault group from the 271 ID attacked over the Hron River northwest of Leva. These troops advanced about one mile, captured the town of Timace, and reached the rail line north of Leva. This attack was in the zone of the 53rd A but was located at the vulnerable boundary line with the 7th Guards Army. During the day, and with the help of ground attack-aircraft from the Luftwaffe, the Germans repelled all Soviet counterattacks against this tiny bridgehead. [690] A major attack in this area was a Soviet concern, (see the Secret Joint Act referenced on 31 January 1945), but unknown to the Soviet Command, this attack was actually a feint; it was an attempt to divert attention to the upper Hron River area and away from events that were about to take place further south.

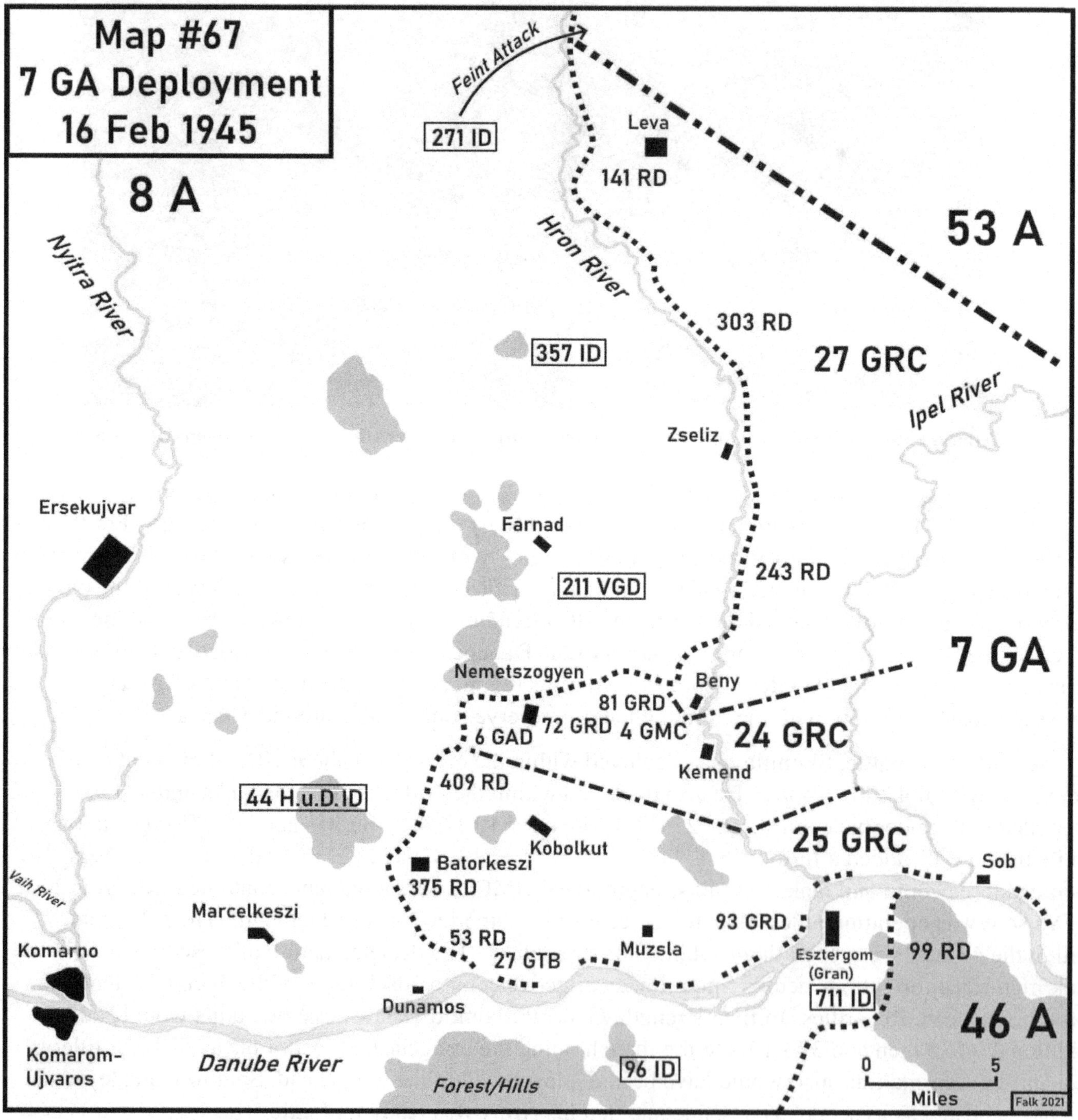

Note: Map #67 shows the deployment of major units of the 7th Guards Army and the Soviets' assessment of German forces facing them along the line. Map #68 displays the true deployment of German forces at this time.

At midnight, 24 miles upriver from Komarno, an 11.5-foot flood crest of water had reached the town of Raab (Gyor). This impending rise in the water level of the Danube, due to melting snow and ice, could only add to the problems of crossing the river during battle.

17 February 1945 - Day One of Operation Southwind. Weather: Average Temperature 33.6° F, (0.9° C), light snow at night

Stavka issued its plan for an attack against Vienna. In part, the directive is listed below.

Directive#11027 7th GA Offensive to take Vienna/Bratislava.

Directive of the VGK Stavka No. 11027 to the Commander of the troops 2nd and 3rd Ukrainian Fronts, representative Stavka for the Operation Vienna [offensive]

February 17, 1945 20 h 15 min

> *The headquarters of the Supreme Command orders:*
>
> *The commander of the 2nd Ukrainian Front to prepare and conduct an offensive operation with the aim of a strike north of the river Danube in the general direction of Nové Zamky, Malacky, Znojmo with a simultaneous offensive of the left wing of the front along the southern bank of the Danube to occupy Bratislava, no later than the 20th day of the operation,* [and to] *capture Brno, Znojmo and, in cooperation with the troops of the 3rd Ukrainian Front, capture Vienna.* [Then] *further develop the offensive in the general direction of Pilsen* [Czechoslovakia].
>
> *To complete this task:*
>
> *a)* [The] *blow north of the river Danube* [will attack] *with the forces of the* [6th Guards Tank], *7th Guards* [Armies] *on the left flank of the 53rd Army, reinforced by three breakthrough artillery divisions, heavy self-propelled brigades and regiments. Create an artillery density of at least 200 barrels (from 76 mm and above) per 1 km of the breakthrough front. To develop success after breaking through the enemy's defense, use along the northern bank of the river Danube* [the] *6th Guards Tank Army and the Mechanized Cavalry Group Pliev.*
>
> *b) Blow south of the river Danube to inflict with the forces of the 46th Army, reinforced by the artillery division of the VGK and the 2nd Guards Mechanized Corps.*
>
> *4. To inform about the given orders.*
>
> *Headquarters of the Supreme Command*
>
> *Stalin A. Vasilevsky*

Stavka's strategic Vienna Operation plan was realistic, but the 2 and 3 UFs required considerable time to rebuild troop strengths and to resupply and to equip depleted combat formations. This offensive was scheduled to begin on 15 March, in 26 days' time. But Hitler and AG South preempted it by launching *Unternehmen Südwind* (Operation Southwind) before dawn on 17 February. The battle had begun to destroy the *Gran-Brückenkopf* (Gran Bridgehead) held by the 7th Guards Army.

A Russian commander recounted the opening of the attack against the 72 GRD, "[A]t dawn, a powerful roar shook the earth - the enemy launched an offensive. Burst flashes illuminated the small foothold. The artillery preparation lasted two hours. Under the roar of artillery cannonade, the fascist tanks reached the line of attack." [691]

Major German forces of the Southwind attack included the I SS Pz C (1 and 12 SS Pz Ds), and Pz C *Feldherrnhalle* (44 H.u.D. ID, 46 and 211 VGDs, with the remnants of the 13 Pz D and Pz D FHH). Also included were three KGs (*Kampfgruppe* - Battle Group): KG *Staubwasser* (Dust Water) from the 357 ID, KG *Hupe* (Horn) from the 96 ID, and an unnamed KG from the 711 ID. Usually, a KG was named for its commander, but in the two KGs listed here, that probably was not the case. Numerous smaller units also took part in the battle, notably the *Schwere Panzer-Abteilung* (Heavy Tank Battalion) *Feldherrnhalle* equipped with 25 Tiger II tanks which supported the 44 H.u.D. ID.

Due to German preparations being completed early, the attack was moved forward one hour to 04:00 hours, thus gaining another hour of darkness to attack. The German attack opened with roughly a two-hour long artillery barrage, but

> [t]he thaw had softened the ground. After the heavy winter snowfall, there were deep pools and puddles of melted snow in every depression in the fields and pastures. Gullies and ditches were brim full of mud and water. The Pariszky Canal, which presented no difficulties at its normal water level, had become a considerable antitank obstacle because of the high water, and could only be crossed after careful preparation. [692]

German infantry divisions led the attack. The 211 VGD attacked Bart, and they failed terribly. They were bogged down outside of the city and were unable to gain any ground. The defending Russian 72 GRD was deployed to the left of Bart, and the 81 GRD was deployed within Bart and to the right of the village. These two rifle divisions repelled all attacks and firmly held the village and the front on both sides of the village. Backing up these two rifle divisions, the entire 4 GMC also was deployed within their positions and to the rear of the 71 and 81 GRDs. The 211 VGD had a combat strength of about 2,200 men (with attachments) [693] while the 72 and 81 GRDs had a combat strength of at least 8,000 men between them. So, the 211 VGD had no chance to make progress against two dug in Guards divisions with the support of the 4 GMC.

This failed attack was due to faulty German intelligence. A little further along the line, the 46 VG D fared better and was able to breach the Russian line and to push forward about 1.8 miles. Then further west, at 04:20 hours, the 44 H.u.D. ID attacked, but it too faced stiff resistance from the 6 GAD forcing the Germans to call in the heavy Tiger II tanks of the heavy tank battalion *Feldherrnhalle.*

Around noon, the 46 VG D and the 1 SS Pz D, having moved up, engaged a Russian antitank gun line that blocked their advance east of Nemetszogyen. Reports say this antitank gun line had a few T-34/85 tanks in support, [694] which would have come from the 4 GMC. The 12 SS Pz D also had moved up and was ready to exploit any breakthrough.

The 211 VGD continued to be completely stalled in front of Bart. Of more concern were air reconnaissance reports indicating that 50 trucks were approaching the Hron river crossings from the east. The Luftwaffe was brought in to attack these soft targets and to disrupt any Soviet plans for a counterattack.

In the evening, after hard fighting, the 46 VG D along with the 1 and 12 SS Pz Ds advanced to the south and reached the Pariszky Canal. The Russians had destroyed all the bridges, so the Germans made preparations to cross the canal after dark. On the right flank of the attack, the 44 H.u.D. ID suffered many causalities from Russian heavy artillery and antitank fire. They reported, "The attacking Tigers [II] came up against an extremely cleverly emplaced battery of large-caliber long-barreled guns. The tanks shot them to pieces or overran them. After that happened the Russian resistance in that sector quickly collapsed." [695] With the help of the Tigers, they were able to overrun the Russian position and to capture the town of Nemetszogyen and the surrounding area.

Sometime during the day, General Shumilov ordered the 93 GRD to launch a counterattack against a German penetration, most likely along the center of the Pariszky Canal, using the division's reserve regiment currently located at Muzsla. But this counterattack weakened the defense along the north bank of the Danube River. Taking advantage of this, during the night and into the early morning hours of 18 February, the German 96 ID launched a preplanned assaulted over the frozen Danube to the north with Battle Group *Hupe,* which was eventually supported by 20 assault guns. This attack over the frozen

Danube came as a complete surprise to the Russians and captured the small town of Ebed, southeast of Muzsla. [696] The Germans now had a battle group in the 7th Guards Army rear area, which disrupted resupply efforts and outflanked the Russian front line to the west.

For many of the small towns and villages north of the Danube, the current battle was the third or fourth time the battlefield had swept over them. Each time the front line passed, the civilian death totals and destruction of their way of life increased. Civilians who could, fled to the west as quickly as possible because they feared the Soviets eventually would return.

In the center of the battle area, at 23:00 hours (German time), the 12 and 1 SS Pz Ds along with the 46 VG D established two small bridgeheads on the south side of the Pariszky Canal. This small but vital achievement allowed for operations the next day to the south of the canal, which threatened to split the Soviet bridgehead in two. [697] Most likely the 44 H.u.D. ID also found a way over the Pariszky Canal during the night or early morning, for they were ready to attack toward the village of Kobolkut the next morning. During the night a Russian infantry counterattack, supported by at least two T-34/85s, was launched against the German positions south of the Pariszky Canal. This attack was repelled with heavy losses. [698] This attack might have been the one General Shumilov ordered the 93 GRD to execute.

For this first day of battle, accounts from Soviet records are limited. We know General Shumilov had few tanks and self-propelled guns available to him, not enough for a proper counterattack. So Shumilov integrated his AFVs (Armored Fighting Vehicles) into the existing antitank and artillery positions or used them to support infantry counterattacks. TsAMO record ID #136074859, *Defense plan of units of the 4th Guards Mechanized Corps*, shows that the 4 GMC had scattered its tanks and SUs along the front line in groups of two or three. Thus, Shumilov was counting upon his fixed defensives to slow and then to stop the German attack. General Shumilov also would have called for assistance from the 2 UF, but would help arrive in time? Acting quickly to this surprise attack, the 2 UF transferred from its reserves the 292 Guards SU Regiment, with 20 SU-76s, to the control of the 4 GMC. This Regiment, however, took several days to arrive at the fighting front.

On or about this date, *Stavka* ordered Marshal Timoshenko, *Stavka's* representative in the south, to visit the HQ of the 7th Guards Army to assess the situation. Additionally, in response to the German attack north of the Danube, attacking Soviet forces in the area of Szekesfehervar (south of the Danube), were ordered to temporarily suspend their own attacks. These attacks were against the remaining German Konrad III force. Notably, Marshal Timoshenko had been retained as *Stavka's* representative for both the 2 UF and 3 UF on 21 January 1945 with *Stavka* Directive #110112. [699] Timoshenko's presence in the south would allow him and front commanders to make faster decisions on the spot, and his unimpeded direct access to Stalin and *Stavka* in Moscow insured a quick response from the highest Soviet command levels.

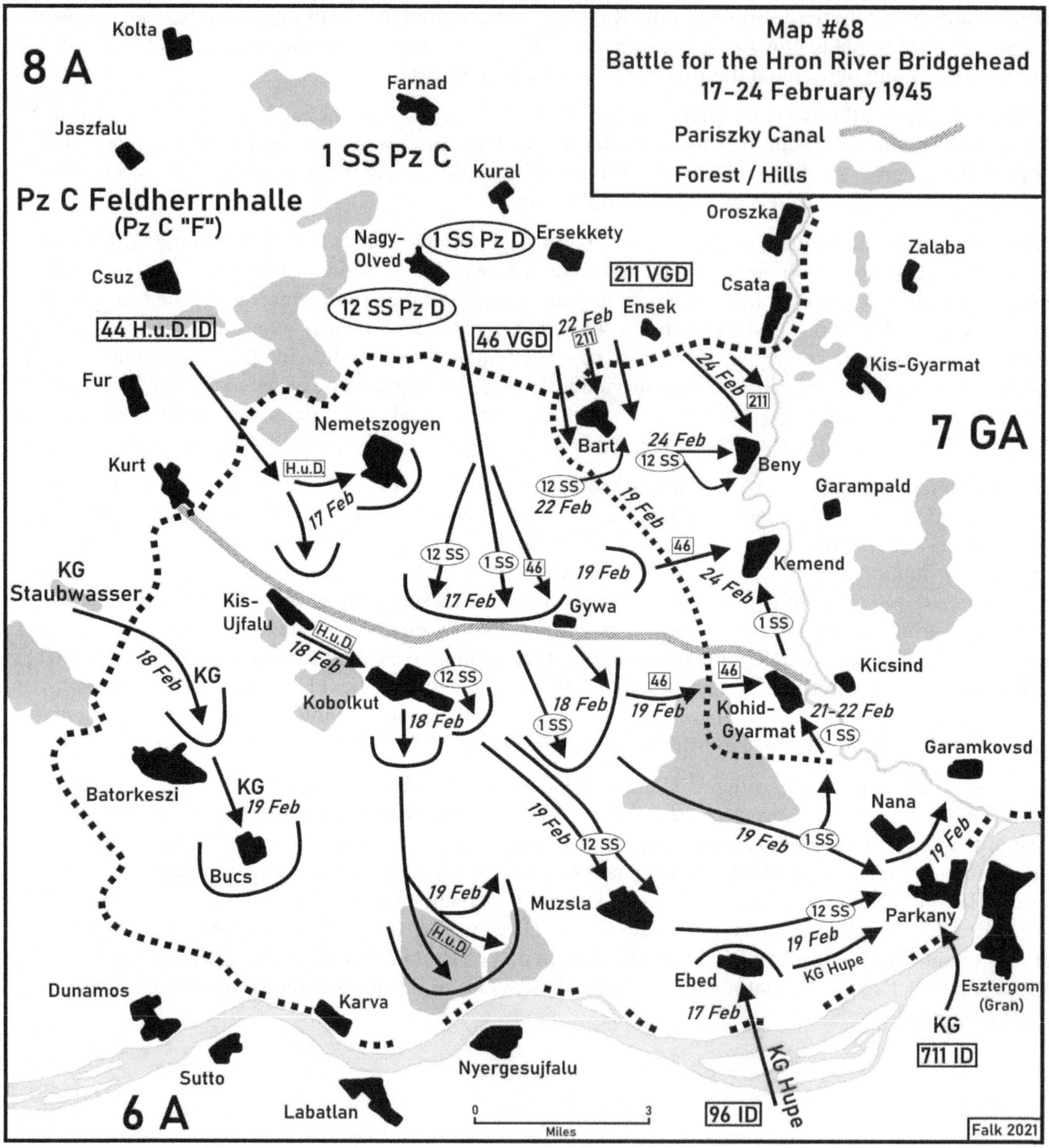

Note: See Map #67 for the deployment of major units of the 7th Guards Army at the beginning of the attack.

18 February 1945 - Day Two of Operation Southwind. Weather: Average Temperature 30.5° F, (-0.8° C), light snow at night

On the left flank of the German attack, the 211 VGD was unable to advance or to capture Bart. The 72 GRD even counterattacked and forced the 211 VGD to assume a defensive stance. German commanders were worried about the lack of progress on the left flank and feared a tank counterattack from this area. Little did they know that General Shumilov had no substantial tank force at his disposal.

In the center of the attack, the 12 and 1 SS Pz Ds along with part of the 46 VG D crossed the Pariszky Canal in force and greatly expanded their bridgehead. They claimed the destruction of four additional T-34s but were held up by minefields and Soviet air attacks. The 44 H.u.D. ID also attacked south of the canal and "after three hours and breaking through a strong position consisting of antitank guns and tanks, we [Tiger II tanks] captured the village" [700] of Kis-Ujfalu. Not content with that success, the 8th Army ordered the 44 H.u.D. to continue its advance that night another 3 miles to assisted the 12 SS Pz D in capturing the town of Kobolkut.

The combined German bridgehead south of the canal was now at least five miles wide and two miles deep. These German advances were met by continuous heavy artillery and antitank gunfire from all sides. Russian positions also were covered by minefields, supported by tanks and self-propelled guns, and were aided by nearly constant air attacks from the VVS.

Part of the 46 VG D also attacked to the east, just north of the canal toward Kemend, but a strong Russian defense repelled this attack. By 12:50 hours, heavy tanks of the I SS Pz C were in the process of crossing the canal.

On the far-right flank, KG *Staubwasser* now launched its attack and broke through Russian positions of the 375 and 409 RDs as it advanced upon the town of Batorkeszi.

In the air, the VVS kept busy bombing and strafing German spearheads while the German 4th Air Fleet bombed bridges and Russian formations along the Hron River.

Luftwaffe reconnaissance aircraft also were airborne, and they found

> surprisingly, only one major group of Russian tanks (30 vehicles) had been discovered up until then. It was north of Nana (northwest of Gran [Esztergom]). The Commander-in-Chief of the 8 *Armee* reported this to *General der Infantries* Wöhler [AG South] at 19:45 hours; "[I]t was quite astonishing that the enemy had not assembled any strong tank force up to now for the bridgehead." [701]

The exact origin of this reported group of 30 tanks is unknown, but during the day reinforcements had arrived by rail for the 27 GTB. Records indicate that "39 T-34s were received at the station." (This information comes from TsAMO record ID #2600181, page 26.) The location of this station is unknown, but it must have been on the east side of the Hron. After detraining, these new tanks were driven over the Hron to reach the battalions of the 27 GTB, and that is most likely when the Luftwaffe reconnaissance aircraft located "the 30 tanks." With this strengthening, the 27 GTB would have on hand 47 T-34s, 4 SU-85s and 2 SU-76s by dawn the next day. So far, the 27 GTB had been lucky. With the main German attack coming down from the north, the 27 GTB saw no combat on the first day and little combat on the second day, with only one T-34 under repair. But with the dawn of 19 February, the soldiers of the 27 GTB would be fighting for their lives.

Despite the small reinforcement for the 27 GTB, the lack of significant tank support for General Shumilov's 7th Guards Army was not an oversight by Marshal Malinovsky or *Stavka*. They knew full well that General Pliev's Cavalry-Mechanized Group and General Kravchenko's 6th GTA were to the rear refitting. They also knew that if they committed these mobile forces to defend the Hron bridgehead, they would not be ready to carry out *Stavka's* offensive, Directive #11027, that had been issued a few days earlier. So, Marshal Malinovsky and *Stavka* waited to see what the enemy would do and what General Shumilov could do with the forces he already had.

With fighting raging within the Hron bridgehead, the German advance south of the Pariszky Canal threatened to outflank and to cut off the 409, 375 and 53 RDs (25 GRC) from the Hron River crossings. To counter this move, at 23:00 hours (Russian time), General Shumilov ordered these isolated forces to withdraw, break through German lines and reach new defensive position along the Hron. A Russian eyewitness stated, "A little dawn broke over the bridgehead; the battle broke out again, everything was blackened by mud and smoke. Everything that could burn burned: cars, earth ... The situation was becoming threatening." [702] The Russian account went on to say these divisions took three days to fight their way through German lines and to reach new positions along the Hron, but that clearly was not true. Most of these troops hurriedly withdrew during the night and early morning of 19 February and with the dawn came fighting with advancing German units of the 44 H.U.D.ID and 12 and 1 SS Pz Ds coming down from the north. They also had to slip past KG *Hupe*, which was holding the town of Ebed, in order to reach Russian positions along the southern Hron. Of course, losses were heavy and these units left most if not all of their heavy equipment and artillery behind; however, most of the men from these rifle divisions avoided death or capture.

Marshal Timoshenko (as *Stavka* representative), on his 50th birthday, went to the command post of General M.S. Shumilov to clarify the situation on the spot. *Stavka* was understandably concerned by this sudden and powerful German attack.

In Poland, the Soviets launched their final assault against the Citadel of Poznan in the center of the city.

19 February 1945 - Day Three of Operation Southwind. Weather: Average Temperature 27.3° F, (-2.6° C), light snow at night

On the left flank of the attack, the 211 VGD remained stuck around Bart unable to advance, but at least they detected no sign of a Russian tank force in that area.

On the far-right flank of the German attack, KG *Staubwasser* advanced against Russian rear guards and mopped up stragglers from the previous night's hurried withdraw. Along the way, KG *Staubwasser* fully occupied the towns of Batorkeszi and Bucs.

In the center, the 44 H.u.D.ID advanced due south and reached the Danube near Muzsla, clearing pockets of Soviet resistance in the woods. At that point a Tiger II tank commander stated, "We would have liked to have a bit of a break at that point, but a new order sent us off [to the northwest] on a new mission. That time the mission led us into hilly terrain. It was a strain for us; we had not shut our eyes for forty-eight hours." [703]

The 46 VG D also attacked to the east along both the north and south sides of the Pariszky Canal. Russian defenders of Kemend repelled the attack along the north side of the canal, but south of the canal, the Germans made good progress toward Kohidgyarmat.

The 12 SS Pz D took the village of Muzsla, linked up with KG *Hupe* in Ebed, and then advanced along the north bank of the Danube toward the Hron and the village of Parkany. Likewise, the 1 SS Pz D advanced to the village of Nana, reached the Hron River, and turned north. At this point, the German 711 ID followed the example of the 96 ID and sent a battle group north from the Esztergom area, over the iced-covered Danube, and captured the town of Parkany on the north bank. Units from the 12 SS Pz D and KG *Hupe* assisted in the attack against Parkany. This second surprise attack over the Danube by the 711 ID insured that the 7th Guards Army would not be able to hold a small bridgehead near the mouth of the Hron River. [704] By the end of the day, German troops were holding the west bank of the Hron from the mouth of the river, three miles north to the town of Kohidgyarmat.

A German account stated, "In general, the day went favorably. The enemy was defeated in the southern part of the bridgehead. He pulled back during the night in the western part, between the Danube and the Pariszky Canal." [705]

In the air, the war took a different turn for the Germans.

> In spite of the employment of every German fighter unit, the Soviet Air Force, which enjoyed great numerical superiority and had no concerns for fuel, finally achieved air superiority over

> the Gran [Hron] bridgehead. It inflicted heavy losses on some elements of the *1 SS-Panzer-Korps* spearhead in particular: "During the entire day, the air superiority that the enemy had managed to gain over the Gran [Hron] Bridgehead, especially in the Muszla [Muzsla] - Parkany-Gran [Esztergom] area, proved to be an obstacle to our movements and caused casualties." [706]

To illustrate what individual units of the 7th Guards Army were doing during this significant day of battle, let us look at the 27 GTB in detail. This account of the 27 GTB is from the historical TsAMO record ID #2600181, page 33.

South of the Pariszky Canal, at 04:00 hours (Russian time), the newly reinforced 27 GTB took up a defense line around the town of Muzsla with one tank and one motorized rifle battalion. They also deployed one tank battalion with nine tanks and SUs around the tiny village of Kaptalok, east of Muzsla. General Shumilov had ordered the rifle divisions from the western side of the bridgehead (25 GRC) to withdraw back to the Hron River, and that is why the 27 GTB was deployed in this area; they were covering the withdrawal of almost four rifle divisions.

From 08:00 to 10:00 hours, the tank battalion at Kaptalok was under attack by enemy tanks, infantry and artillery. The battalion then was ordered to advance north toward Kohidgyarmat and to prevent the enemy from advancing further. This was an attempt to keep open the escape route for Russian infantry units that were retreating through this very area and were trying to reach the Hron River crossings. By 13:00 this battalion was defending southwest of Kohidgyarmat.

Back at Muzsla, by 09:00, the two battalions of the 27 GTB had been surrounded by advancing enemy units (12 SS Pz D), and they were ordered to break out at 11:00 to the east of Muzsla. At that point 30 enemy tanks attacked them with unreported results. Continuing to retreat, they reached the area southwest of Kohidgyarmat by 14:00 where they joined the other battalion of the brigade.

Also at 08:00, another detached tank group from the 27 GTB along with retreating rifle units from the 53 RD and 93 GRD attacked the German KG *Hupe* within the village of Ebed. While this attack was taking place, at 09:00, 25 German tanks with infantry in support (12 SS Pz D) bypassed Ebed to the north and reached the village of Parkany thus surrounding the 27 GTB group and its supporting rifle units at Ebed. For the next six hours this Russian group was engaged in heavy fighting as they attempted to break through the surrounding German forces. They failed during each attempt. Only at 15:00 hours did one tank break through to the Kohidgyarmat area and rejoin the remains of the 27 GTB. Sometime after 15:00, General Shumilov ordered the 27 GTB to withdrawn over the Hron and to deploy along the eastern bank of the river. By nightfall the remains of the 27 GTB were located just north of Garamkovesd, on the east side of the lower Hron River.

So, during this long day of battle, the 27 GTB acted as an armored shield against German forces attacking from the north and west. Undoubtedly, thousands of Russian troops reached safety over the Hron because of their actions, but overall, the 27 GTB had a bad day. Being attacked by two German SS Panzer divisions, along with the 44 H.u.D. ID, had a devastating effect upon the 27 GTB. The troops had fought hard all the way back to the Hron River, and they succeeded in covering retreating rifle units along the way. Despite the tank reinforcements the brigade received the previous day, by 19:00 hours the brigade was down to only 16 operational T-34 tanks and no SUs. This represented a loss of 31 tanks and 6 SUs during one day of combat. All the remaining operational tanks then were consolidated into one battalion.

At 22:00 hours, the reinforcing 292 Guards SU Regiment arrived on the east side of the lower Hron and was turned over to the control of the 4 GMC. This regiment, along with the 27 GTB, currently were deployed on the east side of the Hron River around the village of Garamkovesd. Shumilov was worried the Germans would attack over the lower Hron and seize the village of Garamkovesd, so he concentrated most of his remaining mobile forces in that area. Weather the 292 Guards SU Regiment was at full strength is unknown, but if it was, it would have contained 20 SU-76s, 1 T-34, 1 BA-64 armored car and 440 men.

According to TsAMO record ID #2600181, page 32-34, further north the remaining units of the 4 GMC deployed to the Bart, Beny, Kemend area supporting the 24 GRC. During the day the 4 GMC also saw some action by repelling three enemy attacks in the Kemend area. These German attacks each contained 12-15 tanks supported by infantry in armored personnel carriers. So, 19 February saw the loss of almost the entire bridgehead south of the Pariszky Canal. The soldiers of the 7th Guards Army must have thought, "Maybe tomorrow will be better."

20 February 1945 - Day Four of Operation Southwind. Weather: Average Temperature 25.8° F, (-3.4° C), light snow at night

The 7th Guards Army now only held a narrow bridgehead on the west side of the Hron, centered upon Bart and Beny in the north, Kemend in the center and Kohidgyarmat in the south. (See Map #67.) General Shumilov was determined to hold these outposts, so he covered these areas with his remaining artillery and AA guns. Also, a TsAMO map ID #132313626, *Scheme of flamethrower-anti-tank defense 27 and 3* [separate flamethrower battalions] shows the 27 Separate Flamethrower Battalion was in position around Kohidgyarmat to help hold the town, and the 3 Separate Flamethrower Battalion was deployed about two miles downriver on the east side (left bank) of the Hron protecting the town of Garamkovesd. General Shumilov was attempting to reinforce his small bridgehead with every unit he could find, but most of his available forces were disorganized and had limited combat strength due to heavy losses of both men and equipment. So far, the German attack had done a great deal of damage to the 7th Guards Army.

After four days of combat, the 7th Guards Army finally was able to identify the enemy formations attacking them. [707] Supposedly, this was when Soviet intelligence first identified that elements of the I SS Pz C were leading the assault.

The VVS was very active along the Hron River, which prevented major German operations during daylight hours. Also, the Russians made heavy air and artillery attacks against the 711 ID in the Esztergom area. In the face of Soviet air superiority, German ground forces mostly rested and regrouped as much as they could. They were resupplied in preparation for the final series of night attacks to eliminate the enemy from the west bank of the Hron. In turn, during the day the Luftwaffe was able to mount 220 sorties against the Hron bridgehead, bombing troops, artillery, and tank and vehicle concentrations. Then after dark, the Luftwaffe carried out another 200 night-attack sorties mostly to the east of the Hron River.

The German plan was to launch a night ground attack against Kohidgyarmat with the 46 VG D from the west and the 1 SS Pz D from the south. After dark, they launched both of these attacks and made some progress against increased enemy resistance; however, they were stopped by heavy artillery and mortar fire and in the morning by "waves of [VVS] ground-attack sorties with fighter cover, against the German assault units." [708]

Also during the night, the 12 SS Pz D regrouped to the north to assist with the capture Bart and Beny.

During the evening, Marshal Malinovsky (2 UF) ordered the reforming 6th GTA, with its current 224 tanks and self-propelled guns, to hurriedly move north 75 miles and to assemble around the town of Sahy on the Ipel River. [709] Malinovsky was worried that General Shumilov and the 7th Guards Army would be unable to hold the Hron River line and that powerful enemy units would gain a foothold on the east side of the river. The movement of the 6th GTA was insurance against that prospect.

Additionally, on or about this date, the remaining formations of the 4 GMC were ordered to withdraw over the Hron to the east side of the river. This move would give General Shumilov a few more tanks,

SUs and assorted artillery and AT guns to help hold the river line if the Germans attempted to cross the river in force.

Evidently Luftwaffe reconnaissance aircraft noticed the movement of the 6th GTA and reported, "The enemy was also regrouping. Long enemy columns (3,600 motorized vehicles) were identified moving from the Budapest area toward Balassagyarmat and Ipolysag. The enemy intelligence estimate gave no clear picture of enemy activity. The enemy appears to be bringing in the 6th Guards Tanks Army or the Plijew Army [Pliev's Cavalry-Mechanized Group] …." [710]

On or about this date the 22 and 34 Antitank Artillery Brigades were assigned to the 7th Guards Army. They most likely were coming from the 2 UF reserve.

At the end of the day, on the east side of the Hron, the 27 GTB had 18 operational T-34 tanks and 1 SU-76.

21 February 1945 - Day Five of Operation Southwind. Weather: Average Temperature 26.2° F, (-3.2° C), light snow at night

On or about this date, Marshal Timoshenko, who was still in the area of the 7th Guards Army and reporting directly back to *Stavka* about the current situation along the Hron River, was

> most alarmed by the fact that, according to the testimony of the prisoners, units and formations of the 6th SS Panzer Army [6th Panzer Army] participated in the counterattack [against the 7th Guards Army]. But, as it was known, until recently the 6th SS Panzer Army was in the Ardennes [in Belgium and Luxembourg] and fighting against the Anglo-American troops. 'Apparently, such a regrouping of Hitler's troops is connected with far-reaching plans,' Timoshenko shared his thoughts with Mikhail Stepanovich [Shumilov]. 'It is urgent to double-check the data.' [Many days later, the] information was confirmed. VVS Reconnaissance flights revealed the [new] concentration of a large enemy grouping south-west of Budapest… [This group was assumed to be the entire 6th Panzer Army] [711]

With this small bit of information, Stalin, *Stavka*, Marshal Malinovsky (2 UF) and Marshal Tolbukhin (3 UF) realized they had a real problem on their hands. How and why had the Germans brought the entire 6th Panzer Army secretly to Hungary, and what were they going to do with it? The Soviet command knew for certain that at least one part of the 6th Panzer Army currently was engaged in destroying the Hron River bridgehead, and wrecking the 7th Guards Army, but what further damage could the entire 6th Panzer Army do?

For the 7th Guards Army, on or about this date, the 11 Antiaircraft Artillery Division replaced the 26 Antiaircraft Artillery Division. The reason why these two units were exchanged is unknown. Most likely the 11 Antiaircraft Artillery Division came from 2 UF reserve, and it would reach the front in time to engage in the last part of the ongoing battle.

German forces renewed their effort to capture Kohidgyarmat. Once again, the 46 VG D attacked from the west and the 1 SS Pz D from the south. This attack possibly took place during the late afternoon as they reportedly suffered heavy casualties from artillery and [ground-attack] air attacks. Having forced their way into the village, they completed its capture after dark and in the early morning hours of 22 February.

The German 8th Army reported on the previous day's movement of enemy forces into the area east of the Hron. "It was assumed that the enemy had put these forces into an operational reserve in order to oppose an expected German assault across the Gron [Hron]. …It was concluded from [other information] that [this enemy force] had only a defensive mission." [712]

At 22:10 hours, General Wöhler (AG South) talked with General Guderian (OKH) and confirmed that the enemy force within the bridgehead had been in the range of seven to eight divisions and that retreating Red Army units saved many of their men but left most of their artillery behind. When asked if the Russians had been surprised, General Wöhler replied, "The enemy's surprise had been complete." [713]

At the end of the day, the 27 GTB had 17 operational T-34 tanks, 1 SU-85 and 1 SU-76.

22 February 1945 - Day Six of Operation Southwind. Weather: Average Temperature 31.1° F, (-0.5° C), light snow at night

AG South described the overall situation in its War Dairy as follows:

> The enemy put up fierce resistance in the remaining areas of the bridgehead, especially since he was fighting under the cover of the intense fire of his numerically superior and ammunition-rich artillery. Their long-range guns placed continuous fire on every crossing of the Gran [Hron]. The Red Air Force flew uninterrupted ground-attack missions. It generally controlled the air space on both sides of the Gran [Hron]. [714]

Among other things, this report stressed that units of the 7th Guards Army were fighting under the protection of heavy friendly artillery fire and with continuous air support by the VVS. General Shumilov was desperately trying to hold the remaining bits of his bridgehead and to prevent the Germans from crossing the river and establishing positions on the eastern bank.

During the day, the 46 VG D and the 1 SS Pz D completely cleared Kohidgyarmat of Russian stragglers and then attacked north toward Kemend and the Pariszky Canal. Apparently, they were ignoring Russian artillery fire and air attacks in order to complete their assigned missions. German losses were heavy due to fierce Russian resistance along with constant air and artillery attacks.

Further north, the regrouped 12 SS Pz D with support from the 211 VGD attacked toward Bart and the surrounding area. After fighting all day against determined Russian resistance, this combined German force cleared Bart by the evening. They also repelled all Russian counterattacks from the Beny_area. Once again, both sides suffered heavy losses.

The command staff of AG South, 8th Army and 6th Panzer Army debated the need to continue the attack and to eliminate the entire Soviet bridgehead. Despite the mounting casualties, especially among the troops of the I SS Pz C, the command staff decided to continue the attack until they had removed or eliminated all Russian positions on the west side of the Hron River. They also decided that all future attacks against the bridgehead would be conducted at night to reduce casualties.

Total German casualties for Operation Southwind up to 22 February are listed in Table #33.

Battle Casualties	Dead	Wounded	Missing	Total
I SS Pz C	413	1,923	653	2,989
44 H.u.D.ID	128	739	21	888
46 VG D	288	1,178	127	1,593
211 VD	140	761	100	1,001
Total	**969**	**4,601**	**901**	**6,471**

Table #33: German Battle Casualties, 17-22 February 1945. [715]

At the end of the day, the 27 GTB had 14 operational T-34 tanks, 1 SU-85 and 1 SU-76.

In Poland, the final 12,000 men defending Poznan surrendered to General Chuikov's 8th Guards Army. The direct road from Warsaw to Berlin now was open, but more importantly for the German defenders of Berlin, the main rail line in the Poznan area was not yet operational. Soviet logistic efforts would remain road bound for some time.

23 February 1945 - Day Seven of Operation Southwind. Weather: Average Temperature 33.8° F, (1.0° C), morning drizzle, evening rain

The 8th Army decided to regroup German forces for the attack against Beny and Kemend during the night of 23 February and into the early morning of the 24th. The actual attack would take place on the night of 24 February and then again on the following night, if needed. As AG South stated, "The complete clearing of the bridgehead is required." [716]

So, during the day German forces rested and resupplied; after dark they moved into new assault positions. The 7th Guards Army made efforts to secure the east side of the Hron by distributing supplies and available reinforcements of men and equipment to the depleted units. Both sides were preparing for the final battle of the bridgehead and the Hron River line.

At the end of the day, the 27 GTB had 13 operational T-34 tanks, 1 SU-85 and 1 SU-76. Despite being on the east side of the river, the 27 GTB remained in combat and continued to suffer losses.

24 February 1945 - Day Eight of Operation Southwind. Weather: Average Temperature 36.3° F, (2.4° C), morning drizzle, evening rain

The 1 SS Pz D, along with a KG from the 44 H.u.D. ID and supporting attacks from the 46 VG D, conducted a night assault against the town of Kemend on the Hron River, which was preceded by a reported six-hour artillery barrage (according to a Russian source). But German sources state the attack was to be launched without a long artillery preparation. The confusion on this issue might be due to the Germans keeping the bridges and Soviet concentrations on the east side of the river under continuous artillery attack. The 1 SS Pz D and supporting infantry units had to fight their way through a Russian anti-tank gun belt containing 37 heavy AT guns to reach the town. [717] Nonetheless, fighting was confused as German and Russian formations of the 72 GRD were jumbled together within the town. "By [16:00 hours Russian time] on February 24, [Kemend] fell. [Soviet sources stated that] at the cost of huge losses, the enemy liquidated the bridgehead." [718] Notwithstanding the above Russian statement, German reports talked about house-to-house fighting and that the "Russian artillery fired final protective fire without regard to its own retreating troops in order to prevent the attackers from crossing the Gran [Hron] with them." Of course, the Russians blew up the bridges over the Hron. [719] Losses were heavy on both sides.

Further north, the 12 SS Pz D along with supporting attacks from the 211 VGD, attacked Beny and the area along the riverbank north of the village. Two regiments of the 243 RD held Beny and the west side of the riverbank to the north. By noon the Germans captured Beny and forced Russian troops back over the Hron River. German sources noted that losses for the attack against Beny were less than they suffered for the capture of Kohidgyarmat. With these last attacks and mopping-up operations to remove the last few Russian stragglers from the west side of the river, German forces had eliminated the Red Army's Hron River bridgehead.

At 17:20 hours, the 8th Army reported to AG South, "[t]he complete clearing of the Gran [Hron] Bridgehead." At 17:45 hours AG South reported the same thing to the OKH. The AG South's War Diary highlighted several points during and after the successful Operation Southwind attack. [720]

- Despite heavy losses on both sides, and the stubborn defense by Russian forces, the enemy bridgehead was completely occupied earlier than expected, after eight days of fighting.
- Due to expected losses and the amount of time required, the German High Command dispensed with plans for the capture of four bridgeheads on the east side of the Hron River.

- Soviet commanders (*Stavka*, 2 UF) avoided committing major forces to defend the bridgehead but held reconstituting forces to the rear to help defend the Hron River line, if needed.
- The time needed by the I SS Pz C to recover before the start of Operation Spring Awakening was short and further losses in men and equipment could not be justified in the Hron River area.

25 February 1945 Weather: Average Temperature 36.8° F, (2.7° C), morning drizzle, evening rain

At 00:30 hours, the 7th Guards Army issued a Secret Order #0078/OP. The order stated that by the morning of 27 February, the 4 GMC would be removed from the control of the 7th Guards Army. By 28 February, the corps would concentrate in the Széchenyi area, where it would enter the 2 UF reserve. By this date the 4 GMC had no operational AFVs.

After first defending the bridgehead, then fighting its way through German lines, and then being pushed back over the Hron River, the battered 375 RD finally was ordered to withdraw to the rear for rest and refitting. A historical map TsAMO ID #100114504 actually showed the 72, 81 and 93 GRDs were in the second defense line along with the 375 RD. This was a good indicator that these four rifle divisions were critically low in manpower and equipment after the fighting on the west side of the river. These four divisions actually suffered the most casualties. (See Table #34 for overall losses.) This same map also showed the 4 GAD in the front line along the lower Hron River and that the entire 25 GRD was deployed to cover the north bank of the Danube River to the east of the front line. These last two divisions (4 GAD and 25 GRD) were most likely early reinforcements sent by Marshal Malinovsky from 2 UF reserves. Some accounts state that other reinforcements also were sent to the Hron area.

The first units of the I SS Pz C were taken out of the line so that they could begin to regain some of their strength before moving south of the Danube for further offensive operations in that area.

In the aftermath of the battle, the German 8th Army HQ stated that it "was gravely concerned about the major Soviet formations [6 GTA] located east of the Gran [Hron] and the Eipel [Ipel Rivers], which had remained astonishingly inactive while the divisions of the Russian 7th Guards Army were pushed out of the bridgehead." [721] Basically, Germans forces had successfully eliminated the Russian Hron River bridgehead, but the German 8th Army and AG South had no solid information on what the Soviets would do in response.

Unknown to the German command, *Stavka* and Marshal Malinovsky had in effect allowed Shumilov's 7th Guards Army to absorb the full force of the German attack on its own, with few reinforcements. By doing so, they also accepted the loss of the Hron River bridgehead as an unfortunate but necessary tactical defeat. The Soviet High Command intentionally sacrificed Shumilov's Army so as not to interfere with the refurbishment of General Pliev's Cavalry-Mechanized Group and General Kravchenko's 6th GTA. This deliberate act prevented the premature commitment of these mobile assault formations to battle before the start of the next major Soviet strategic offensive.

In essence, during Operation Southwind Adolf Hitler had committed his forces to achieve a needed, but short-term tactical victory while Josef Stalin and *Stavka* were planning long term to achieve strategically important objectives, like the capture of Vienna and the total defeat of Nazi Germany.

Meanwhile in Berlin, of the four possible courses of action that planners proposed for the pending Operation Spring Awakening, A, B, C1 and C2, Hitler chose the C2 plan, or the southern option. This plan involved a major attack originating from the same general area as the Konrad III attack, and reaching the Danube River south of Budapest. This plan also included a second phase to clear the area directly west of Budapest. This was an extremely ambitious plan for this late in the war. (Operation Spring Awakening and the four attack options are not displayed on maps).

26 February 1945 Weather: Average Temperature 39.9° F, (4.4° C), morning drizzle

With the Hron bridgehead lost, General Shumilov could begin to account for the cost of the battle. Different sources give different totals, but data from historical TsAMO archive document ID #113723574 produces the following two tables. (See Tables #34 and 35.)

Rifle Divisions	Starting Strength – 17 Feb	Human Losses	End Strength – 25 Feb
4 GAD	5,466	172	5,303
6 GAD	4,092	753	3,428
72 GRD	4,108	1,149	3,103
81 GRD	4,259	1,896	2,542
93 GRD	5,774	796	5,143
53 RD	4,167	321	3,917
141 RD	3,975	44	3,970
243 RD	4,199	349	3,952
303 RD	4,106	41	4,165
375 RD	4,010	870	3,295
409 RD	3,983	623	3,437
Totals	**48,139**	**7,014**	**42,267**
Artillery/Tank Units			
16 Artillery Division	7,959	139	7,905
41 Guards Artillery Brigade	1,368	85	1,283
2 Antitank Brigade	1,069	68	1,001
11 Antitank Brigade	1,013	92	921
22 Antitank Brigade	1,109	75	1,039
34 Antitank Brigade	919	77	841
114 Guards Tank Destroyer Regiment	490	37	458
115 Guards Tank Destroyer Regiment	433	49	389
493 Mortar Regiment	573	37	543
290 Mortar Regiment	523	17	521
263 Mortar Regiment	538	9	530
27 GTB	1,283	105	1,408
5 Anti-Aircraft Division	1,819	79	1,770
11 Anti-Aircraft Division	1,944	21	1,940
Remaining Elements	-	276	-
Totals	**21,140**	**1,166**	**20,544**
Grand Total	**69,279**	**8,180**	**62,811**

Table #34: Soviet Human Losses during the Hron Bridgehead Battle 17-25 February 1945.
Note: Units did receive some new personnel during the battle, which is reflected in the totals above.

To further expand on the rough treatment the 7th Guards Army had suffered, the material losses were equally high as displayed in Table #35.

Weapon	Equipment Losses
Machine Guns (all types)	415
120 mm Mortar	44
82 mm Mortar	156
122 mm Artillery	59
76 mm Regimental Gun	44
76 mm Divisional Gun	268 Some under repair.
45 mm AT Gun	62
152 mm Artillery	7
57 mm AT Gun	23
PTR- AT Rifle	164
Tanks and Self-Propelled Guns (AFVs)	54 Eight are subject to evacuation [to the rear].
AA Guns (37 and 85 mm)	12
Vehicles	198

Table #35: Soviet Material Losses during the Hron Bridgehead Battle 17-25 February 1945. Note: Losses for the 4 Guards Mechanized Corps are not included within these two tables.

The human loss for the 7th Guards Army amounted to the pre-battle strength of two rifle divisions and the material losses were sufficient to essentially equip two or three rifle divisions and a tank brigade. Before the Hron River bridgehead battle, the 7th Guards Army was badly understrength, but after the battle, the Army basically needed 24,000 replacements to bring its rifle divisions up to at least the 6,000-troop level. The Army also required a considerable amount of equipment to rebuild its depleted divisions.

By 1 March 1945, five days after the major fighting was over, the 27 GTB only lacked nine men of its pre-battle (1,370) troop strength, but it also had no tanks. Note that different sources list different strength totals for the 27 GTB during this period. Likewise, the 4 GMC only lacked 256 men of its pre-battle (7,265) troop strength, but it too was listed without tanks. This information came from TsAMO record ID #113980918, page 2 and 7. Apparently, it was easier to deliver men than tanks to these worn-out tank units.

Reported overall losses for the 4 Guards Mechanized Corps during the Hron bridgehead battle are listed below. This information comes from the historical TsAMO archive document ID #113723664. (See Table #36.)

People/Equipment	Casualties/Losses
Killed	161
Wounded	492
Missing in action	241
Total	**894**
Tanks - SUs	37
Vehicles/trucks	54
BA-64 Armored Car	1
Guns-Artillery/AT	21
Mortars	39
Machineguns	24

Table #36: 4 Guards Mechanized Corps during the Hron Bridgehead Battle -- 17-25 February 1945.

As can be seen in the above table, the 4 GMC lost about 15 per cent of its starting strength of 5,817 troops. While it lost more tanks and SUs than it started with, the corps must have received some replacements during the battle.

So, accounting for just troop and AFVs losses for the 7th Guards Army and the 4 GMC, casualties/losses during the seven-day bridgehead battle amounted to 9,074 people and 91 tanks and SUs. Equine losses amounted to a little over 1,800.

27 February 1945 Weather: Average Temperature 46.7° F, (8.2° C), morning drizzle

The 7th Guards Army issued a report, *The act to ensure the junction between the 27th and 24th Guards Corps,* TsAMO record #454128884. This report highlights preplanned artillery fire zones in the border area. Then on 16 March 1945, the Army issued another report, *The act to ensure the junction between the 24th and 25th Guards Corps,* TsAMO record #454615389. General Shumilov was making sure the Army was ready to repel any enemy attack over the Hron River.

28 February 1945 Weather: Average Temperature 45.7° F, (7.7° C), morning drizzle

By February, fuel production in Germany had dropped to 40 percent of total capacity with aviation fuel at only 5 percent of capacity. [722] Aircraft fuel production had been hit the hardest because the Western Allied bomber offensive deliberately targeted German synthetic oil and high-octane fuel production.

The total medical burden on the 7th Guards Army for 27 February 1945 was 4,862 people. Despite the battle for the Hron River bridgehead, this total was less than the 30 January 1945 value of 5,776 people, an indication that despite an overall drop in losses, the severity of the fighting had remained high during February.

Forward Troop Area – up to divisional medical battalions

- Arrivals:
  - At the start (25 February) – 1,410 men and women
  - Newly arrived – 891
  - Total – 2,301
- Departures:
  - Evacuated – 686 (most likely to the Army rear) (moderately wounded)
  - Returned to active service – 251

Died –23
Total – 960 (including dead)
In the Forward Troop Area on 27 February – 1,341 (lightly wounded)

In Army Hospitals
Arrivals:
At the start (25 February) – 3,993 men and women
Arrived from forward units' medical installations – 485
Arrived directly from units – 62 (severely wounded)
Totals – 4,540
Departures:
Evacuated – 680 (out of the Army to the deep rear, severely wounded)
Returned to active service – 312
Died – 27
Total – 1,019 (including the dead)
In Army Hospitals 27 February – 3,521 (moderately wounded) (with 6,700 hospital beds)
Total in all medical installations on 27 February – 4,862 people [723]

Note the 680 severely wounded people who were evacuated out of the Army for this period (25-27 February) versus 33 for the end of January (25-30 January).

The 2 UF issued a report, *Report on the loss of personnel in the troops of the 2nd UkrF from February 1 to February 28, 1945.* This was found within TsAMO record ID #113723658. The 7th Guards Army suffered 10,429 total casualties for the month: 1,529 killed; 2,127 missing; 0 captured; 101 non-combat losses; 4,703 wounded/burned; 1,969 sick; and 0 frostbite. The report goes on to list 7th Guards Army horse losses: 1,875 killed; 46 missing; 180 wounded; 32 evacuated to the veterinarian; for a total of 2,133 equine casualties. The number of horses that died greatly increased from the previous month. The weather and overwork kept the overall horse casualties high, but the German attack upon the Hron Bridgehead, with artillery, air attacks and other combat activities pushed these totals to very high levels.

Once again, the 0 people captured figure for the 7th Guards is pure fiction. German sources state they captured about 700 POWs and counted about 4,000 dead (nice round numbers) on the battlefield. [724] These are believable totals. One should also recall that Soviet sources admitted the 7th Guards Army suffered 2,127 missing troops and 1,529 killed during the battle, which totals some 3,656 people. The discrepancy between Soviet and German accounts, in these areas, might indicate that losses for the 7th Guards Army actually might be underreported, as 4,700 (German report) minus 3,656 (Soviet report) = 1,044 overlooked Soviet losses. If this is true, then a more realistic casualty total for all of February 1945 might be closer to 11,473 men and women.

This same report shows that the troops of the entire 2 UF suffered 31,476 total casualties during February 1945.

For the month of February in Britain, air raids and German V-1 and V-2 attacks killed 483 civilians and injured 1,152. These numbers do not begin to reflect the continuing damage done to the infrastructure of major cities or to the quality of life for the civilian population. For example, a single V-2 missile exploding upon contact with the ground would create a crater approximately 66 feet wide, 26 feet deep and eject about 3,000 tons of material into the air. If a house or shop received a direct hit, the entire building and any residents, sheltering in the basement perhaps, would simply be gone, reduced to fragments. [725]

1 March 1945	
7th Guards Army	
Assigned – 2 Ukrainian Front	
Colonel General M.S. Shumilov Commanding	
Rifle, Airborne and Cavalry Units	24 Guards Rifle Corps (72, 81 GRDs, 243 RD) 25 Guards Rifle Corps (4 GAD, 53, 409 RDs) 27 Guards Rifle Corps (93 GRD, 141, 303 RDs) 6 GAD, 375 RD 15 Penal Battalion 62, 63, 64, 65, 66, 194 Penal Companies
Artillery Units	16 Artillery Penetration Division (49 Light Artillery Brigade, 61 Gun Artillery Brigade, 52 Howitzer Artillery Brigade, 90 Heavy Howitzer Brigade, 109 High-Power Howitzer Brigade, 14 Mortar Brigade) 5 Antiaircraft Artillery Division (670, 743, 1119, 1181 AA Regiments) 11 Antiaircraft Artillery Division (804, 976, 987, 996, AA Regiments) 2, 11, 22, 34 Antitank Artillery Brigades 114, 115 Guards Tank Destroyer Regiments 41 Guards Gun Artillery Brigade 263, 290, 493 Mortar Regiments 662 Separate Aerial Observation/Air Raid Warning Company
Armored and Mechanized Units	27 Guards Tank Brigade 28 Separate Armored Train Battalion
Engineer Units	60 Engineer Sapper Brigade 3, 27 Separate Flamethrower Battalions
Reported Combat Strength Estimated Ration Strength	68,073 men 75,073 men and women (including an estimated 7,000 support personnel)

(BSSA 45, BSA 45, TsAMO Archive)

The Army now contained eleven rifle divisions, one artillery division and two antiaircraft artillery divisions plus the normal supporting combat arms regiments and brigades. Note that the 7th Guards Army retained control of the 3 and 27 Separate Flamethrower Battalions. Also, the 28 Separate Armored Train Battalion replaced the 38 Separate Armored Train Battalion. This was most likely due to normal rotation of these steam powered trains for servicing. The 7th Guards Army still was struggling to replace the troops lost during the Hron Bridgehead battle. Because of this, all combat units were far below strength.

1 March 1945 Weather: Average Temperature 44.4° F, (6.88° C), morning fog

Note the displayed weather reports for all of March were issued by the 2 UF HQ and covered the entire 2 UF area.

TsAMO map ID #100230842 shows that the 2 UF provided additional support to the Army. The 18 RC arrived and was deployed to the second defense line of the 7th Guards Army. The 18 RC with two rifle divisions, the 297 and 317 RDs, had been deployed in a line just north of the Danube River between 26 February and 5 March. The 18 RC then was withdrawn on 6 March, because evidently Marshal Malinovsky thought the danger of a German attack over the Hron River had passed by that date.

AG South and 8th Army reported,

> On the Gran [Hron] both sides had to draw back their main battle lines from the riverbanks because of the high water. The appearance of enemy cavalry squadrons east of the [Hron] (seen from the ground!) indicated that the Russian Plijew Army [General Pliev's Cavalry-Mechanized Group] had moved up its cavalry formations into attack positions close to the front lines. [726]

The 7th Guards Army still was recovering from the recent fighting, but this German report was the first indication that on the other side of the line the enemy was very busy preparing for future operations.

Meanwhile, the I SS Pz C hurriedly moved south of the Danube, but it also was refitting and receiving replacements along the way. In front of the 7th Guards Army, the German 8th Army continued to deploy Pz C *Feldherrnhalle* along the Hron River. Its infantry units were deployed from north to south: the 357 ID, 46 VG D and the 211 VGD at the mouth of the Hron River. Behind them and to the rear were the remnants of the 13 Pz D and Pz D FHH. AG South and 8th Army had correctly concluded that the 7th Guards Army and the 2 UF were not in a position to launch a new offensive north of the Danube River for some time. With their flank secure north of the Danube, AG South could launch Operation Spring Awakening according to Hitler's plan.

2 March 1945 Weather: Average Temperature 42° F, (5.5° C), morning fog

After being taken out of the line for rest, refitting and to receive reinforcements, the I SS Pz C mostly was brought up to strength in terms of manpower, but there were issues. Due to extreme need, Nazi Germany was transferring redundant manpower from the Luftwaffe and *Kriegsmarine* (War Navy) to both the *Herr* and *Waffen SS* (Armed SS). They also were enlisting very young or older and inexperienced civilians to fill the manpower needs of frontline combat units. Some of these older men and boys reached these SS assault divisions only hours before being sent into combat for the first time. One SS solider noted,

> We received replacements, more than fifty NCOs and men per company, [but they were] formerly of the Luftwaffe [and] they already wore our uniforms. None of them had any infantry experience. More replacements reached us.... This time they were exclusively Schwabians [a southern German region] ..., aged from seventeen to thirty-five. Over sixty infantrymen. All of them 'young' soldiers that had left their homes for the first time and were correspondingly depressed.... Could the young sons of farmers find a new home with us? [727]

The Iron Cross 2nd Class is awarded to a very young German solider late in the war.

As the 6th Panzer Army assembled, Soviet forces of the 3 UF prepared to receive them. The German command noted that "[w]hile no ground combat activity was reported from the front, except for the usual reconnaissance and patrol activity on both sides and some harassment fire by artillery, both sides were actively engaged in aerial reconnaissance." [728]

The lack of ground combat activity and active aerial reconnaissance by the enemy should have been a warning to the German High Command that the Soviets were aware of the pending German attack. Spring Awakening should have been called off or at least delayed because the element of surprise was absent and the German forces would be forced to fight in terrible weather conditions. Deep mud and poor road conditions greatly hindered the assembly of the entire 6 Panzer Army, the I and II SS Pz C, the I Cavalry Corps and the other units assigned to Operation Spring Awakening. Despite the difficulties, Hitler would not alter the planned attack, with the date fixed for the 5-6 March time frame.

3 March 1945 Weather: Average Temperature 36.3° F, (2.3° C), morning fog

The 7th Guards Army produced a planning document TsAMO record ID #114300194, *Report on the combat activities of the artillery of the 7th Guards Army for the month of March 1945.* In part, this 16-page report shows on page 5 that the Army was taking advantage of the lull at the front to replenished personnel and equipment of its artillery units. Between 1-10 March, the Army received the following artillery replacements: 18 – 120 mm mortars, 17 – 76 mm regimental guns, 269 – 76 mm divisional guns, 14 – 122 mm howitzers and 88 – 45 mm AT guns, totaling 406 weapons. This allotment of new equipment made good the reported losses of 76 mm divisional guns and 45 mm AT guns, but it was still less than what was needed to totally replace the losses of other artillery weapons. At the same time, these units received 600 troop replacements, which amounted to 6 percent of their authorized strength.

4 March 1945 Weather: Average Temperature 31.1° F, (-0.5° C), morning fog

During the day the 7th Guards Army produced a report TsAMO record ID #133375169, *Combat report of the headquarters of the 7th Guards* Army. This typical report contained four parts listing actions/conditions for the Army.

1. During the day, there were no significant changes on the front of the Army. The enemy, continuing to defend himself on the western bank of the GRON [Hron] river, fired rare artillery fire at the positions of our troops and the front-line roads. Military surveillance in the enemy's defensive zone marked the usual movement of people and vehicles. Enemy aviation conducted reconnaissance with single aircraft to a depth of over 30 km [18 miles]. During the day, 6 aircraft over flights were recorded.
2. During the day, the Army troops continued to strengthen the previous lines in engineering respect, conducted reconnaissance of the enemy and destroyed reconnaissance targets with fire, [and] carried out combat training. The position of the troops of the Army [are] without change.
3. During the day, over 100 enemy soldiers and officers, 2 machine guns and 4 wagons with cargo were destroyed. The loss of 7th Guards Army troops: 4 killed and 12 wounded.
4. Daytime weather - changing cloudiness 4-9 points [points most likely indicate the percentage of cloud cover], 600-1000 m high without precipitation, north wind 3-5 m/sec., temperature from 2 degrees of frost to 1 degree of heat. Visibility 6-8 km. Dirt roads during the day are muddy and difficult to drive. The weather is [good for] flying.

Overall, the report was a clear and simple account of the day's events. Section three is interesting, as this account would give the Russian troops a 25 to 1 kill ratio over the Germans, which is not very believable. Under these conditions, with only minor scouting and "rare artillery fire" from the enemy, nice round numbers like the "over 100 enemy … killed" are most likely due to someone at HQ making an estimate of enemy losses or trying to make the report more impressive. The German troops facing the 7th Guards Army probably suffered no more than 20 total casualties during this day. If the Germans really had suffered 100 killed, their artillery would have been blasting the Russian forward positions all day long in retaliation. The most interesting bit of information might be the report of "6 aircraft over flights" during the day. The Germans knew the 7th Guards Army would someday attempt to cross the Hron River again, but the date, time, and location were unknown, so German reconnaissance aircraft were keeping an eye on the situation.

Further south, over 4-5 March, the I SS Pz C moved into their forward assembly area. Other elements of the 6th Panzer Army also moved up to the line as best they could through the thick mud and deep flowing water. Of course, VVS aircraft were watching the German buildup despite the bad weather. The Soviet command knew a major attack was imminent, but they lacked detailed information, so they prepared for the worst.

5 March 1945 Weather: Average Temperature 33.4° F, (0.77° C), morning fog

At this time the 7th Guards Army had ten rifle divisions in the front line with only one in reserve. The Army still retained the two rifle divisions of the 18 RC, but the units of the 18 RC would be transferred the next day, most likely in response to the Spring Awakening attack.

6 March 1945 - Day One of Operation Spring Awakening. Weather: Average Temperature 30.5° F, (-0.83 ° C), morning fog

At 04:00 hours, German forces launched *Unternehmen Frühlingserwachen* (Operation Spring Awakening) with the main attack southwest of Budapest between Lake Balaton and Lake Velence and a smaller attack south of Lake Balaton. The weather and ground conditions were a disaster.

> In the constricted area between Lakes Balaton and Velencei [Velence] the mud became alarming. The closer one came to the … assembly area, the more widespread the land that was under water – impassable for all kinds of vehicles. It looked the same… in the enemy area, as far as the terrain permitted observation. [729]

Commanders of the combat units involved made a last-minute attempt to have AG South, OKH and the High Command delay the attack for several days to allow the ground to dry out, but no changes were made. Hitler refused to alter the start date of the offensive. These severe *rasputitsa* conditions were enough by themselves to doom Spring Awakening to failure. Forces of the Russian 3 UF were ready and waiting for the German attack.

For General Shumilov, the consequences resulting from Operation Southwind and the loss of the Hron Bridgehead due to the German attack had begun. Kamen Nevenkin, in his book *Bloody Vienna,* on page 7 reveals that, "Needless to say, Stalin was furious. The only thing that saved the commander of the 7th Guards Army, General Shumilov, from certain dismissal was his hero's reputation earned during the epic battles of Stalingrad and Kursk."

Stavka also issued a directive on this issue.

Directive of the VGK Stavka No. 11036 to the Commander of the Front troops
on improving the organization of combat actions
Copy: to the representative of the VGK
March 6, 1945 01 h 30 min

> *Recently, on some fronts, there have been cases of carelessness and gullibility, which the enemy has been able to deliver* [to] *us* [surprise attacks] *and sensitive strikes. As a result of these attacks, our troops were forced to withdraw.* [The] *withdrawal in these cases occurred* [in an] *unorganized* [manner]. *The troops suffered heavy losses in manpower and especially in material. For example:*
>
> *7th Guards Army of the 2nd Ukrainian Front, defending east of Komarno, being attacked by the enemy, failed to repulse his offensive, despite a sufficient number of forces and means,* [retreated from] *the operationally important bridgehead occupied by* [the Army] *on the western bank of the river Hron. Having lost 8,194 people, 459 guns of different calibers (of which 374 were 76-mm and above),* [and] *54 tanks and SU.*
>
> *Units of the 26th Army of the 3rd Ukrainian Front, advancing along the Sherviz Canal,* [penetrated] *3-5 km into the enemy's defenses. The enemy, having undertaken a counterattack, easily broke through the battle formations of our advancing units, which did not have serious artillery support, since all their artillery was simultaneously removed from* [firing] *positions and* [ordered to be] *moved forward. As a result of two days of fighting, units of the 133* [RC] *and 135* [RC] *of the 26th Army lost 42 mortars, 90 guns of various calibers and were thrown back to their original position.* [Plus, they also lost an unknown number of troops.]
>
> *The headquarters of the Supreme High Command believes that these cases could have occurred only as a result of criminal carelessness, poor organization of defense, lack of intelligence and control on the part of higher commanders and their headquarters over the position and actions of troops. The commanders of the troops of the 2nd and 3rd Ukrainian fronts did not consider it necessary to inform the headquarters of these shameful facts in time, apparently wishing to conceal them, and the General Staff had to get this information from the* [Army Commanders] *over the heads of the front commanders. The headquarters indicates to the commander of the 2nd Ukrainian Front, Marshal of the Soviet Union Malinovsky and*

the 3rd Ukrainian Front, Marshal of the Soviet Union Tolbukhin, on poor control over the actions of the troops, unsatisfactory organization of reconnaissance and the inadmissibility of failure to report the above losses to Headquarters.

Stavka orders:

a) The commander of the 7th Guards Army, Colonel-General Shumilov for carelessness and poor organization of defense to pronounce a reprimand;

b) The commander of the troops of the 2nd and 3rd Ukrainian fronts, in the indicated cases, conduct a strict investigation and bring the perpetrators to justice. Report the results of the investigation and the measures taken.

Headquarters of the Supreme Command

I. Stalin A. Antonov

The key points from this *Stavka* directive were the following:

- The 7th Guards Army lost 8,194 men, 459 guns (374 above 76mm) and 54 tanks and SUs. They also lost the Hron River Bridgehead.
- The 26th A lost an unknown number of men along with 42 mortars and 90 guns. They also were forced back to their original starting position.
- *Stavka*/VGK were not informed of these reverses by 2 and 3 UFs HQs in a timely manner, and the front intelligence organizations failed to properly identify or to anticipate these enemy attacks and take appropriate action.
- General Shumilov was issued a reprimand, and an investigation into the failures would find "the perpetrators" and bring them to justice!

The reprimand of General Shumilov did not appear to have any lasting effects on his career. Perhaps *Stavka* was a bit harsh with Shumilov as the Hron Bridgehead was fully and properly organized for defense in view of the existing conditions. Apparently, *Stavka* found fault with General Shumilov's deployment of all his rifle divisions within the bridgehead in the front line, i.e., a single echelon. This deployment might not have been the best tactically, but General Shumilov was forced to do so because of the depleted nature of his rifle divisions, the lack of credible tank and SU support, the overall extended length of the front and the awkward deployment of the 7th Guards Army astride the Hron River. Mitigating factors were certainly the suddenness and power of the enemy attack. These factors were due mostly to the secret transfer of the 6th Panzer Army from the west, which the Red Army completely failed to detect, and the number, quality and skill of the enemy forces that were attacking rather than a specific failure on the part of General Shumilov. *Stavka* also failed to take any responsibility for the intelligence failure regarding the transfer of the 6th Panzer Army to the Eastern Front. It's always easier to blame someone else for a battlefield reversal. Once again, even at this late date in the war, German Panzer forces proved to be an ever-present danger to the Red Army.

7 March 1945 - Day Two of Operation Spring Awakening. Weather: Average Temperature 28.2° F, (-2.11 ° C), afternoon rain

The German attack continued, but Russian aircraft appeared in large numbers and "[t]he enemy had recovered by the second day of the fighting. The defense became tougher. Almost everywhere German assaults were met with immediate counterattacks, occasionally supported by tanks." German commanders noted that the mud was the main enemy. [730]

Meanwhile, on the Western Front, having broken through the *Siegfried* Line and advanced to the east, the armies of the Western Allies were poised all along the Rhine River in early March. Units of U.S.

General Hodges' 1st Army captured a key bridge, the Ludendorff Bridge, over the Rhine at the town of Remagen and formed a small bridgehead on the eastern shore. With this, the Allied forces had breached the last great river barrier in western Germany. Other crossings of the Rhine soon followed. But Hitler was enraged by the loss of the Ludendorff Bridge, so he called upon his favorite commando to once again rectify the situation. Otto Skorzeny and his "Danube Frogman Group" were sent to destroy the bridge. Despite a valiant attempt, all the frogmen saboteurs were either killed or captured. Skorzeny personally did not take part in the mission. The damaged bridge finally collapsed on its own on 17 March. [731]

8 March 1945 - Day Three of Operation Spring Awakening. Weather: Average Temperature 26.0° F, (-3.33 ° C), afternoon rain

Overall, the rifle division strengths for the 7th Guards Army were far below the official *Shtat* strengths of 7,000-8,000 men, which was not encouraging. Fortunately, the Army was defending behind the Hron River and not under further attack. See Table #37 for rifle division strengths on 8 March 1945.

Corps HQ/Division	Strength
24 GRC HQ	792
72 GRD	3,205
81 GRD	3,858
243 RD	3,966
Total	**11,821**
25 GRC HQ	842
4 GAD	5,289
53 RD	3,913
409 RD	4,018
Total	**14,062**
27 GRC HQ	805
93 GRD	5,153
141 RD	3,985
303 RD	4,174
Total	**14,117**
Army Subordinate Units	
6 GAD	3,512
375 RD	4,018
Total	**7,530**
Grand Total	**47,530**

Table #37: The 7th Guards Army Rifle Division Strengths - 8 March 1945. [732]

Further south, a German appraisal of the overall Spring Awakening operation was gloomy. "The enemy situation, which had been somewhat murky during the first two days of the attack, has cleared up a bit. The enemy [3 UF] has identified the long-range objectives of the friendly attack and has responded with appropriate long-range countermeasures." [733] At this time, the Germans once again should have called off the attack. The Soviets were defending well, ground and road conditions were terrible and even more Red Army formations were moving into position to counter the offensive. But of course, the generals in the field were not in charge; Hitler was.

For the 7th Guards Army, the 375 RD held a regimental meeting. "On March 8, a solemn meeting of all the girls of the division was held; 84 people were present. [Fifteen] girls were [given awards]. The same [type of] meetings were held in the other regiments." [734] These "girls" were signal operators, mine sweepers, traffic regulators, snipers, medical aids, paramedics, nurses, doctors and other support/combat personnel who were at times right at the front. One photo from the 375 RD showed paramedics of the 454th Medical and Sanitary Battalion. Of the ten people in the photo, four of them were women. Another photo from the same unit showed six different female paramedics. [735] These and other photos should suffice to show how the Red Army integrated women into all aspects of the Army, be it within combat units, headquarters, or support units far to the rear.

Red Army female minesweepers clear a village of mines right behind the front line.

9 March 1945 - Day Four of Operation Spring Awakening. Weather: Average Temperature 27.3° F, (-2.61° C), afternoon rain

The 6th Panzer Army continued to forge ahead, but south of Lake Balaton the 2nd Panzer Army had reached a stalemate against Russian forces as positions were captured only to later be lost again. The 2nd Panzer Army started to regroup its troops in order to advance again.

10 March 1945 - Day Five of Operation Spring Awakening. Weather: Average Temperature 31.2° F, (-0.44° C), blizzard

Despite a lull in the fighting, some type of combat still was taking place at the front. The 7th Guards Army produced a report TsAMO record ID #133375096, *Information about the damage inflicted on the enemy by the troops of the 7th Guards Army,* from 1-10 March. This one-page document reports the Army killed 700 enemy officers and soldiers, destroyed 12 machine guns, and 15 carts/wagons. This report should be considered an estimate; note the nice round number – of 700 killed, which is 70 men killed per day, was unlikely during ten days of combat patrols, harassing artillery fire and the ever-present snipers.

For Operation Spring Awakening, the worsening weather conditions due to the blizzard limited the actives of the VVS, which helped the German effort, but the worsening weather also contributed to bad road and ground conditions. "Even on this fifth day of fighting, the 6 *Panzer-Armee* was denied a breakthrough. The field army's real fighting strength, its armored forces, could not be used effectively in the muddy terrain." [736]

The 7th Guards Army issued a planning document TsAMO record ID #152314321, *Perspective sketch of the front edge of the enemy's defense in front of the front of the 7th Guards A 2 UF.* This document illustrated the "lay of the land" and highlighted significant features, both natural or manmade, which faced the Army over the Hron River. Documents like this were vital in planning the next, and hopefully last, attack over the Hron River.

11 March 1945 - Day Six of Operation Spring Awakening. Weather: Average Temperature 34.8° F, (1.55° C), blizzard

The 2nd Panzer Army continued to have difficulties breaking through the Russian line. Newly arrived Soviet forces supported the few defending Bulgarian units, who were Soviet Allies at the time. AG South ordered the 2nd Panzer Army to regroup and to launch a new assault on 13 March. For the 6th Panzer Army, it reported that it pushed the weakening enemy back as the Red Army units fought a delaying action. For Spring Awakening to succeed, fast and decisive action was needed to keep the Soviet forces off balance and to destroy them. But with the existing blizzard conditions, only a sluggish advance was possible, which allowed the Red Army to trade ground for time, inflict casualties and eventually bring forward enough forces to stop the attack completely.

One must not forget the political/human aspect of the ideological conflict between Nazism and Communism. For most of the combatants on the Eastern/Russian Front, and for many if not all of the political or military leaders involved, the war became an "us versus them" situation. As a result of this mindset, anything was acceptable. As the war wound down to its ultimate conclusion, it also became a war of revenge for the Red Army. For example, the 7th Guards Army reported

> On March 11-12, a two-day seminar was held for deputy commanders for political affairs of regiments and individual battalions, [and] on March 19, an evening of division veterans [375 RD], [was held] which was attended by 150 people of the first formation. The issue of the tasks of veterans in educating young soldiers, in transferring the experience of fighting the fascist invaders was discussed. The veterans accepted the appeal to the young soldiers.

Here are some excerpts from what was distributed.

> "Warrior, you are heading west on the road of victory.
>
> ... You are the judge! Before you are criminals, executioners, be merciless to them ... Kill the cannibal fascist, kill the fascist robbers with a bullet, bayonet, grenade, shovel and buttstock, with everything you can. Don't let [them] go alive! Avenge the vile executioners for blood and tears, for torment and death of innocent people ... " [737]

With official propaganda and encouragement like this being spread to the troops, the stage was set for future mass murders of enemy soldiers and German civilians. This encouragement also included the acceptance of mass rape of any female, no matter how young or old, who was found within the lands of the Nazi/Fascists.

To the north of the Danube, General Hans Kreysing, commander of the 8th Army, was worried about the concentration of the 6th GTA and General Pliev's Cavalry-Mechanized Group (CMG) behind the 7th Guards Army. Unknown to General Kreysing, part of this Soviet deployment was a real threat and the other a deception. Most likely during this same evening, the 6th GTA quietly began pulling out of its current position and moved to the south, crossing the Danube just north of Budapest. The full-strength and completely refurbished 6th GTA contained 406 tanks and self-propelled guns and around 40,000 men, [738] and was relocating to its planned assault position for Phase Two of *Stavka's* Vienna Operation, south of the Danube. General Wöhler, commander of AG South, and General Kreysing, commander of 8th Army, were unaware of this movement at first.

12 March 1945 - Day Seven of Operation Spring Awakening. Weather: Average Temperature 31.6° F, (-0.22° C), blizzard.

This day marked the last day of spring that the troops had to contend with snow. Until the ground dried out, conditions would now be all mud.

The 2nd Panzer Army suspended its attacks during the day and repelled several Russian night attacks. The repositioning of forces continued so as to be ready for the 14 March renewal of offensive operations. The 6th Panzer Army was engaged in heavy fighting and had to repel several Russian counterattacks. The advance continued in this area, but forward progress was met with increased resistance from both the Red Army and from nature.

The American 15th Air Force, flying out of Italy, attacked the Floridsdorf oil refinery outside of Vienna with over 700 heavy bombers, dropping 1,667 tons of bombs. Two days later, on 14 March, they returned to bomb the Hungarian refinery at Komarno. "The German High Command was forced to admit that [as result of] the attack on the Komarno refinery, the Hungarian oil capacity was reduced by 70%... From now on the inescapable fact of severely limited fuel supplies would have a severe effect on the operational decisions of Army Groups South and Center." [739]

13 March 1945 - Day Eight of Operation Spring Awakening. Weather: Average Temperature 34.1° F, (1.16° C), fog

On or about this date, the 25 GRD was assigned to the 7th Guards Army. This addition would bring the Army's strength up to 12 rifle divisions.

The regrouping of 2nd Panzer Army forces went smoothly and the 2nd Panzer Army would be ready to attack by the evening. The 6th Panzer Army once again was engaged in heavy fighting with many reports stating that the enemy has been considerably strengthened and had launched tank-supported counterattacks. The Chief-of-Staff for Operations of the 6th Panzer Army reported:

> In general, friendly attacking formations had made no progress anywhere today. Indeed, the day was marked by deliberate enemy counterattacks as well as successful defense. The enemy had become so strong between the Sarviz Canal and the Danube, that it was only barley possible for the friendly attack to chew its way through in this location. [740]

This report went on to say that losses had been considerable, but the attacking troops were not yet exhausted. George Maier, in his book *Drama Between Budapest and Vienna*, on page 209, presents a table of personnel losses for the 6th Panzer Army from 6-13 March 1945. The table shows 2,451 killed; 11,116 wounded; and 1,251 missing, for a total of 14,818 casualties. Then on page 210 another table listing AFVs for the 6th Panzer Army on 13 March 1945 shows 304 tanks, 279 assault guns and 957 armored personnel carriers operational (APCs). Total losses were 31 tanks, 11 assault guns and 1 APC, with various other vehicles in short or long-term maintenance.

In the air, the VVS continued to make a strong appearance over the battlefield, but in the afternoon the Luftwaffe appeared and flew 280 sorties in an attempt to clear the skies of Russian aircraft and to support the troops on the ground.

14 March 1945 - Day Nine of Operation Spring Awakening. Weather: Average Temperature 32.9° F, (0.50° C), fog

The 2nd Panzer Army launched a new attack and gained 2-4 km (1.2-2.4 miles) of ground after fighting through the Russian defensive positions. For the 6th Panzer Army, the enemy continued to launch numerous counterattacks with heavy air and artillery support and with increasing numbers of tanks. "It was becoming ever clearer that *Unternehmen Frühlingserwachen* had run aground and could no longer be gotten afloat with the forces available." [741]

Intelligence sources indicated that Soviet forces were gathering south of the Danube in the sector opposite of the 6th Army in preparation for a major attack. Hitler and OKH wanted to continue Operation Spring Awakening, but AG South wanted to regroup its forces to counter this new threat. This issue would not be clarified for several days as the ongoing battle hung in the balance.

15 March 1945 - Day Ten of Operation Spring Awakening -The final Day. Weather: Average Temperature 36.5° F, (2.50° C), morning fog

With no clear guidance from Hitler or the OKH about the growing threat to the north, German commanders in the field continued Operation Spring Awakening. The 2nd Panzer Army gained minor amounts of ground in an area that was not critical for the success of Spring Awakening. The 6th Panzer Army also forged ahead in the face of overwhelming Red Army firepower. German forces continued to attack because Hitler could not bring himself to call off a failed offensive.

Later that night at 21:30 hours, AG South learned that Hitler approved limited regrouping of forces.

Meanwhile, the 6th GTA reached its concentration point just west of Budapest, but it was not yet finished with its movement. The next day, acting upon a direct order from Stalin, Stavka subordinated the 6th GTA to the 3 UF to support the 9th GA attack further south, during phase one of *Stavka's* Vienna Operation. A disappointed 2 UF commander Marshal Malinovsky had lost control of his main armored force. By using *maskirovka* methods, the 6th GTA had, up to this point, remained undetected by German intelligence services, but that would change the next day when German intelligence at last noticed the 6th GTA was deploying west of Budapest and the Danube River. [742]

16 March 1945 - First Day of Defensive Fighting, south of the Danube, according to the War Diary of SG South. [743] Weather: Average Temperature 41.1° F, (5.05° C), night and morning fog

Sensing weakness and the growing exhaustion of the most recent German offensive, *Stavka* and the 3 UF commander Marshal Tolbukhin began phase one of *Stavk*a's strategic Vienna Operation. The Russian plan was for a direct attack against the German 6th Army, which if successful would then outflank, encircle and then destroy the entire 6th Panzer Army. A secondary goal would be to reach and to pass through the Vertes Mountains, which was a low mountain range that formed the main line of defense for the German 6th Army.

At 12:38 hours, Soviet forces of 3 UF opened an artillery barrage against positions of the German 6th Army south of the Danube. A general assault soon followed composed of units from the 46th A, 9th GA and 6th GTA. *General der Panzertruppen* (General of Panzer Troops) Hermann Balck, commander of the 6th Army (also known as Army Group Balck), at first reported that his troops were holding the line, but later in the day it was clear to German commanders that the Soviet attack was too powerful to be stopped with the forces on hand.

At this time, Marshal Malinovsky and General Shumilov's 7th Guards Army were waiting for the opportune moment to begin Phase Two of *Stavka's* strategic Vienna Operation. But the 46th A of the 2 UF did begin its attack directly south of the Danube River on this day. This was in direct support of operations by the 3 UF.

With the opening of this new Soviet offensive, Operation Spring Awakening had run its course. Try as they might, the combined power of the German 2nd and 6th Panzer Armies, plus the 6th Army and supporting formations, could not break the 3 UF line. Axis forces had pushed the defending Russian armies back, but the weather, the VVS and the tenacious defense by the 4th GA, 27th, 26th, and the 57th Armies prevented any decisive breakthrough. The 2nd Panzer Army, attacking alone south of Lake Balaton, made little real progress. The most promising advance had been made by the 6th Panzer Army with the 6th Army in support, both attacking north of Lake Balaton, but even here the deepest penetration was only about 26 miles. The defending Russian armies along with the abysmal weather and ground conditions proved to be too much to overcome.

The opposing forces are summed up in Table #38 which clearly shows the Red Army's advantage in terms of divisions, equipment and manpower.

Formations	Divisions	Tanks and Assault Guns	Troops
AG South	25	772	270,000
2 and 3 UFs	77	1,250	1,171,800

Table #38: Relative Forces at the Start of the Soviet Vienna Offensive - 16 March 1945. [744]

17 March 1945 - Second Day of Defensive Fighting. Weather: Average Temperature 45.8° F, (7.66° C), night and morning fog

Once again, Hitler and the OKH hesitated making a decision about officially canceling Spring Awakening. Meanwhile, three Russian armies continued to expand their attack, threatening to completely rupture the German 6th Army line, which would put all of AG South in danger.

In the early morning at 01:55 hours, OKH (General Guderian) ordered several German divisions transferred north to block the Russian offensive, but then at 02:40 hours, Hitler, acting irrationally, prohibited the preparation and/or repositioning of these units from taking place. The 3 UF had already "managed to make a penetration at least 30 kilometers wide [18.6 miles] and 10 kilometers deep [6.2 miles] in the center of *Armeegruppe* Balck in the first assault" and was currently further expanding the penetration. [745]

The day ended with the 2nd Panzer Army being ordered to continue with their original attack, while the 6th Panzer Army was ordered to shorten its lines somewhat, hold the ground taken, and transfer a few units north to resist the Soviet offensive. This plan was doomed to failure. Everyone, including Hitler, should have been aware that the Soviets had anticipated the Spring Awakening attack because they had gathered suitable forces and prepared sufficient defenses to defeat the German attack. Beyond that, *Stavka* had assembled additional armies to launch a strategic offensive of their own at the time and place of their choosing. *Stavka* and the Red Army were shaping and controlling the battle, while Hitler fantasized about continuing to attack, which risked the total destruction of AG South.

18 March 1945 - Third Day of Defensive Fighting. Weather: Average Temperature 44.4° F, (6.88° C), night and morning fog

The 3 UF offensive against the German 6th Army continued as Germans reported,

> "The masses of enemy infantry assaulting the positions suffered heavy, bloody losses in the massed defensive fire of the weapons," but that did not bother the Russian leaders. On the contrary, they were quite indifferent. Moreover, they had huge reserves available on a scale that was no longer even imaginable on the German side. [746]

Some Russian units had already reached the exits from the Vertes Mountains, which were located about 31 miles directly west of Budapest. Once fully through the Vertes Mountains, Russian armies had a direct path northwest to Vienna some 80 miles away. By advancing toward Vienna, they also would outflank most of AG South, which would lead to its quick destruction. Despite these pending disasters, Hitler could not bring himself to officially cancel Operation Spring Awakening or to allow General Wöhler to save his troops. Hitler's indecision on this issue is understandable, because calling off the offensive meant the end of the Third Reich and his end as well.

In response to its transfer to the 3 UF, the commander of the 6th GTA submitted a planning document TsAMO record ID #100567797, *The scheme of the decision of the commander of the army to enter the 6th Guards TA in Breakthrough and Action in Operational Depth.* This plan was to commit the tank army to a wide flanking movement around part of the German 6th Army and all of the 6th Panzer Army and to reach the north shore of Lake Balaton. The successful conclusion of this maneuver would ensure the destruction of most if not all the Spring Awakening assault force.

19 March 1945 - Fourth Day of Defensive Fighting. Weather: Average Temperature 46.2° F, (7.88° C), night and morning fog

AG South was engaged in shifting sectors and zones of control between the 6th Army and 6th Panzer Army. This adjustment added to the growing confusion and bickering within the German command structure. Despite this, counterattacks by the I and II SS Pz C and the III Pz C were able to hold off the Russian advance for another day. [747]

In Berlin, Adolf Hitler issued the *Nerobefehl* Decree (Nero Decree) also known as the "Scorched Earth Order." This decree/order was intended to destroy anything of value or of use to the enemy. Hitler was not concerned that this order also would leave the German people without power, transportation, food or water after the fighting ended. Albert Speer, the Minister of Armaments, quietly and effectively sabotaged this malevolent order, thus saving the civilian population from much further suffering.

The 6th GTA had reached its final assembly position southwest of Budapest in the 3 UF zone. It was concentrated in a tight assault grouping with the 9th GA to the north and the 4th GA to the south.

20 March 1945 - Fifth Day of Defensive Fighting. Weather: Average Temperature 45.8° F, (7.66° C), night and morning fog

On or about this date, the 2 Ru TR (2 Romanian Tank Regiment) was assigned to the 7th Guards Army for the coming offensive.

The 7th Guards Army produced a report TsAMO record ID #133375097, *Information about the damage inflicted on the enemy by the troops of the 7th Guards Army,* from 10-20 March. This one-page document reports the Army killed 650 enemy officers and soldiers, destroyed 1 gun and 16 machine guns, and captured 5 prisoners. The 650 number of enemy officers and soldiers killed over a ten-day period is implausible. The 7th Guards Army only began offensive operations on 21 March, so the person writing this report was confused about the events taking place at the front. The remainder of the report about the number of destroyed guns and captured POWs is believable.

German forces of the 6th Panzer Army and the 6th Army continued to regroup and to launch attacks into the south flank of the 3 UF offensive. Further south the 2nd Panzer Army renewed its attack south of Lake Balaton in an attempt to draw off Russian forces from the main attack. In response to the tenacious Axis defense, the 3 UF committed the 6th GTA to the battle, thus adding its armored weight to the attack south of the Danube. The race was on to see if the 6th Panzer Army could extract itself from the pocket the Soviets were forming around them before the 6th GTA could swing south and reach Lake Balaton.

AG South knew an attack over the Gran (Hron River) aimed at Komarno was imminent. To feed these fears and to support the offensive by the 46th A south of the Danube, the Soviets sent 70 boats of the Danube Flotilla, loaded with 536 men from the 144 Battalion, 83 Naval Rifle Brigade, up the Danube and landed them on the south shore of the river behind the 711 ID. The 711 ID was ordered to destroy this pocket of Russian naval infantry troops in their rear area, which they most likely did over the next several days.

In the air, the VVS launched heavy ground attacks against positions of the 6th Army and the 6th Panzer Army. At the same time, north of the Danube, the VVS made increased reconnaissance flights in the area of the German 8th Army, which should have been a warning to German commanders that the front north of the Danube would soon become active once again. The Luftwaffe also was operating, flying about 300 sorties during the day with about 40 nighttime ground attack missions. [748]

21 March 1945 - Sixth Day of Defensive Fighting. Weather: Average Temperature 49.6° F, (9.77° C), night and morning fog

Units of the 25 GRD, 7th Guards Army, attacked over the Hron River and captured a small bridgehead, 1.2 miles deep and 1.8 miles wide, on the western shore near the mouth of the Hron. This attack should be considered a feint attack to divert attention to the lower Hron River and to the area north of the Danube. German commanders of AG South were undoubtedly waiting for such a ploy, but this did not help them understand from where the expected main Soviet attack would come.

South of the Danube the Soviet attack continued to push back the German 6th Army, but the amassed Soviet troops were unable to make a clean breakthrough of the Axis line. On or about this date, advancing units of the 46th A captured the city of Esztergom (Gran), south of the Danube.

Northeast of Lake Balaton, units of the 3 UF made further advances and reached a point about 15 miles from Lake Balaton. About nine German and Hungarian divisions were within the forming pocket. Time was running out for the forces of SG South to escape disaster.

At this time, units of the threatened 6th Panzer Army "took the appropriate military action of gradually withdrawing north-west. Strangely enough, the officer responsible for this action was not punished, even though he had disobeyed the 'Fuhrer's order'". [749]

22 March 1945 - Seventh Day of Defensive Fighting. Weather: Average Temperature 49.8° F, (9.88° C), night and morning fog

Starting on or about 22 March, the 7th Guards Army began shifting its forces north, greatly compressing the frontage of the 27 GRC, which allowed the 24 GRC to shift its frontage and units north also. These movements would have taken place using *maskirovka* methods, at night or during bad weather to deceive the enemy about just where the 7th Guards Army was amassing its forces. The far-right flank of the Army was being reinforced to serve as the main assault force for Phase Two of *Stavka's* strategic Vienna Operation. Essentially, the 46th A south of the Danube (already attacking) and the 7th Guards Army north of the Danube, both part of the 2 UF, would conduct a coordinated attack toward the city of

Komarno. Additionally, the 7th Guards Army would reach and cross the Nyitra and Vaih River lines in the north.

Around Lake Balaton, the exit corridor for the surrounded German and Hungarian troops had shrunk to only 1.5-2 miles wide. The "9th SS Armored Division managed to keep the corridor open until the bulk of the German [and Hungarian] troops had got back to their own lines, although most of their heavy weapons were lost in the process." [750] With the partial destruction of the 6th Panzer Army, Marshal Tolbukhin and his 3 UF were now free to begin their advance upon Vienna.

Hitler was to blame for the terrible losses suffered by AG South. Instead of ordering a timely withdraw of the Spring Awakening attack force and then conducting a fighting withdrawal in the direction of Vienna, Hitler issued "stand firm" orders until they had no meaning. The defenders of Vienna were left dead on the battlefield or swelled the ranks of POWs columns marching to the east. Their tanks and weapons lay scattered along the roads, out of fuel, or abandoned in the trenches. Vienna now lay virtually defenseless before the invading Red Army.

23 March 1945 - Eight Day of Defensive Fighting. Weather: Average Temperature 49.8° F, (9.88° C), night and morning fog

Striking northward, the 46th A reached the southern bank of the Danube, outflanking and surrounding the German 96 and 711 IDs and the attached 27 Hu ID that were defending in the Esztergom (Gran) area. These Axis troops, some 20,000 strong, eventually evacuated successfully to the north over the Danube with most of them reaching safety on the north side of the river in the German 8th Army zone.

Disorganized elements of the 6th Panzer Army and the 6th Army withdrew in a general northwest direction. The fixed front line had largely disappeared; in turn, it was replaced by isolated holding actions as AG South attempted to delay the Red Army advance. German logistics also were a shambles as American heavy bombers began attacking road and rail targets in the rear areas. With little Luftwaffe interference, the VVS was free to attack anything that moved.

> The Army South divisions lost the bulk of their tanks in the course of their withdrawal--in most cases without any involvement by the enemy. Fuel was in such short supply that the costly weapons had to be blown up or left standing in the road. Whole columns of armored vehicles were left behind on the northern shore of Lake Balaton. [751]

In the air the VVS dominated, launching intensive ground and air attacks in the area north of Lake Balaton, and Soviet aircraft even began attacking ferries working along the Danube. The Luftwaffe was attempting to cover the withdrawing German and Hungarian armies, but it lacked the fuel to intervene decisively in the air battle.

24 March 1945 – (Note the "Days of Defensive Fighting" reference will no longer be included as that entry had lost all meaning. It officially ended on 30 April 1945 with the Forty-Sixth Day of Defensive Fighting.) Weather: Average Temperature 50.7° F, (10.38° C), haze in the evening

AG South commander General Wöhler sent a top secret "Blitz" message to be read only by Hitler and General Guderian. The message, titled "Formation of a Solid Front No Longer Possible," laid out the position of AG South in realistic terms. Hitler's reply: General Wöhler was notified he would be relieved of his command the next day, but because his replacement was not yet available, General Wöhler agreed to stay on until 7 April. Eventually, General Wöhler was relieved of his command on or about 6 April 1945. [752]

The 3 UF advanced steadily, meeting sporadic resistance as Marshal Tolbukhin drove the defeated Axis armies before him.

At this decisive point, Stalin and *Stavka* were ready to increase the pressure on Axis forces to new levels. They were ready to unleash Phase Two of the strategic Vienna Operation north of the Danube. The 7th Guards Army would now return to the fight. General Shumilov had a score to settle with the German 8th Army.

25 March 1945 - Weather: Average Temperature 52.5° F, (11.38° C), night and morning haze

For the 7th Guards Army, the opening of *Stavka's* strategic Vienna Operation, (Phase Two - Bratislava Operation) was ten days late from the original plan. Stalin, *Stavka*, Marshal Malinovsky (2 UF) and General Shumilov certainly could blame the Germans for the delay.

One month after its defeat on the western shore of the Hron River, the 7th Guards Army readied itself to once again cross the Hron. The 375 RD moved up.

> On the night of March 25, 1945, the division reached its starting position to cross the Hron River and attack its western bank. Hron is a fast river with a width of up to 80 meters [262 feet] and a depth of up to 2.5 meters [8 feet]. The boats turned over from the speed of the current and from the blows of ice floes that were still floating on the surface of the water…. The Nazis, well entrenched on the opposite bank of the river, tried to hold the water line. [Their] fire system allowed them to keep [under fire] the coastal strip and the surface of the water…. The silence over the river was broken by the roar of shots, [and] the howl of numerous flying shells and Katyusha [rockets]. The first pillars of explosions soared over the enemy positions. Individual shots are not heard, everything merged into a continuous roar, from which the earth trembled and groaned. The infantrymen watched the fiery whirlwind that rained down on the enemy positions. [753]

In support of this new offensive, Guard's captain Churenkov issued a report, *The list of the availability and technical condition of vehicles in the rifle units of the 7th Guards Army as of March 25, 1945*, from TsAMO record ID #133393484. This one-page report shows that the Army's rifle divisions had on hand the following types of vehicles:

General Type of Vehicle	Number Assigned	Number Available	Excess/Shortfall
Light Vehicles	80	86	+6
Buses	0	9	+9
Cargo Trucks	1,545	936	-609
Specialist Trucks	170	135	-35
Total	**1,795**	**1,166**	**-629**

Table #39: 7th Guards Army Divisional Vehicles Available -- 25 March 1945.

This report showed the Army had a few extra light vehicles on hand, composing of 11 M-1 passenger cars, 17 Jeeps and 58 trophy vehicles, most likely *Kübelwagens*. They also had nine buses that were not officially assigned but were handy for transporting wounded troops. Although they possessed 292 trophy trucks, the Army's rifle divisions were short a significant number of cargo trucks (609) and specialist vehicles (35). Despite the shortfall of logistical vehicles at the front, the 7th Guards Army would attack and advance as best as it could. Note that Army level logistical units would use their own trucks and supply wagons to carry supplies from the rear to the forward zone where divisional assets would pick them up for distribution.

At this time, the overall strategic situation for AG South, in western Hungary and southern Czechoslovak, was moving from bad to catastrophic. South of the Danube the German 2nd and 6th Pz Armies were in full retreat. This retreat also included the 6th Army and supporting Hungarian units. The hope was that they could stabilize the front line somewhere to the west and maybe defend Vienna. Vienna was not only the capital of Austria, but also could be thought of as the gateway into the heart of

Austria and western Czechoslovak. But, with the opening of the 2 UFs offensive north of the Danube, all hope of holding the approaches to Vienna were lost. At this point, the badly weakened Axis forces of AG South could only slow the Soviet advance, not stop it. AG South simply lacked the men, equipment and supplies (food, fuel, ammunition) to fight a protracted battle.

For the German 8th Army north of the Danube, the tactical situation was far from secure. Along the Hron River line in the area facing the 7th Guards Army, the Germans had (from north to south) part of the Pz D FHH, the full 357 ID, the full 46 VG D and part of the 211 VGD. The remaining parts of the 211 VGD, along with the 27 Hu ID, the 96 ID and other units, were still isolated south of the Danube in a small pocket along the river. To the rear of the 8th Army were only a few scattered units that were rebuilding or had just arrived in the area. Overall, the 8th Army contained about 200 tanks and assault guns (105 were combat ready), along with 39,122 men (ration strength) with 18,713 of these as combat strength. They had 31 light and 7 heavy batteries of artillery and a total of 60 75 mm AT guns. [754]

In comparison to the German 8th Army, the entire 2 UF contained 365 operational tanks and self-propelled guns. [755] Additionally, according to TsAMO record ID #113723538, the combat strength of the entire 2 UF on 1 April 1945 was 411,581 people. For the 2 UF formations that were deployed directly against the German 8th Army on 1 April 1945 were the following (from north to south): 53rd A with 43,391 troops; 7th Guards Army with 66,945 troops; 46th A with 86,118 troops; 1 Guards Cavalry Mechanized Corps with 37,065 troops; and an estimated additional 15,000 Romanian troops for a total combat strength of approximately 248,519 men and women. As can be seen by these totals, despite taking huge losses in personnel and equipment over the previous few years, the Red Army still maintained a considerable strength advantage over Axis forces.

So, the German 8th Army deployed 105 tanks and assault guns against 365 tanks and self-propelled guns for the 2 UF. While the 8th Army deployed 18,713 combat troops, the 2 UF outnumbered them with 248,519. The German 8th Army was outnumbered by over 3 to 1 in AFVs and 13 to 1 in combat troops. With ratios like these, once the Red Army broke through the German prepared defenses, there would be no stopping them.

For the 7th Guards Army, the battleplan was for the right flank of the army, 27 and 24 GRCs, to push to the west as fast as possible while the 25 GRC cleared the northern bank of the Danube, capturing Komarno and Bratislava along the way. Once begun, the offensive would push forward as long as supplies and troops lasted.

For direct armored support, the 7th Guards Army was forced to rely upon only the 27 GTB and the attached 2 Rumanian Tank Regiment (2 Ru TR). The 27 GTB had only 9 operations tanks and SU-85s, while the 2 Ru TR had 48 operational AFVs; 7 German Pz IV tanks, 24 Renault R-35/45 tanks, 9 German Pz 38(t) tanks, and 8 German StuG. III Assault guns. [756] General Shumilov would use this armored force to exclusively support the right flank main attack.

The first day of the Bratislava Operation saw the 7th Guards Army cross the Hron River in strength, gaining several bridgeheads in the 27 and 24 GRCs area. Further south the 25 GRC further expanded its existing bridgehead near the mouth of the river. The defending German forces put up a good fight, but they were unable to hold the river line. (See Map #69.)

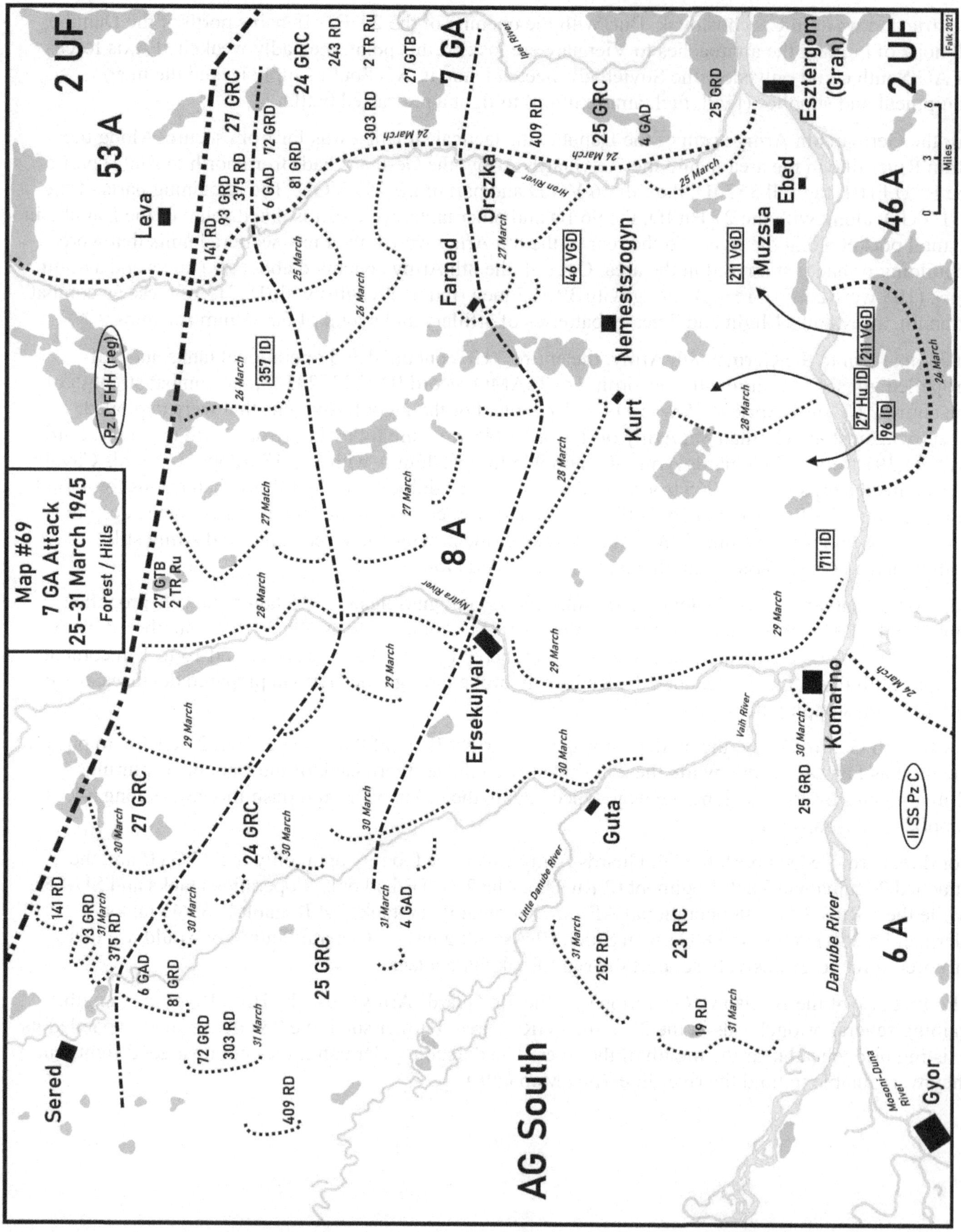

Note: German/Hungarian units are displayed as they were at the start of the offensive.

26 March 1945 - Weather: Average Temperature 56.1° F, (13.38° C), night and morning haze

The 27 and 24 GRCs continued their attack and greatly expanded their bridgehead area. In the south, the much weaker 25 GRC waited for the Germans to abandon the river line. An AG South report stated, "On the lower Gran [Hron], the enemy had beaten down the stubbornly resisting forces of the 8th Army with intense artillery barrages." [757]

Meanwhile, the German 8th Army was engaged with extracting the sizeable number of German and Hungarian troops that were surrounded south of the Danube. Once these troops were on the north side of the Danube, they would be put into the line to hold the north bank of the river.

South of the Danube, Marshal Malinovsky (2 UF – 46th A) asked Marshal Tolbukhin (3 UF – 4th GA) to cooperate with the 46th A to launch a coordinated attack against the heavily fortified city of Gyor. Coming to a quick agreement, these two armies launched their combined attacks toward Gyor.

27 March 1945 - Weather: Average Temperature 57.5° F, (14.16° C), night and morning haze

After a heated exchange, Hitler ordered General Guderian, head of the OKH, to take a six-week medical leave starting the next day. General Hans Krebs, Deputy Chief of Staff of the OKH, replaced Guderian as head of the OKH. [758]

Units of the 27 and 24 GRCs continue their attack and essentially broke out of their bridgehead along the central Hron River. Their advanced units almost reached the Nyitra River during the day. The 25 GRC, 309 RD were able to attack across the Hron and to make a good advance of about nine miles.

In response to the 2 UFs successful attack north of the Danube, AG South ordered the German 356 ID transferred from the 6th Pz Army to the north of the Danube to help the 8th Army defend the north bank of the river. General Wöhler was concerned the Red Army attack north of the Danube (7th Guards Army) would advance too far west and outflank German forces south of the river. "Hitler finally also saw the necessity for pulling back the front. [Later that night] the 8th Army received orders to pull back [its] right wing…." [759]

In Poland during 27-28 March 1945, under the pretext of talks about a post-war Polish government, Soviet NKVD forces arrested 16 senior members of the Polish Underground Government. These leaders were loyal to the Republic of Poland government exiled in London. The 16 leaders were taken to Moscow and charged during a show trial with aiding the Germans. Three of these Polish leaders died in Soviet captivity while the others eventually were returned to Poland between 1945-1949. This act destroyed all hope of an independent Polish government and assured that a Soviet puppet government would be in control of Poland, just like Stalin intended. [760]

This callous act appalled Winston Churchill who

> took it as further proof of the increasingly heavy handed and inflexible approach of Stalin to the postwar governance of Eastern Europe. Since Churchill could not know for certain the limit of the Soviet dictator's ambition, he ordered his chief of staff to draw up a plan-in two versions, one strategically defensive and one strategically offensive-for a possible war against the Soviet Union starting soon after the Germans were finally defeated. [761]

These two plans, which became known as Operation Unthinkable, were thankfully not carried out. As a result, Poland remained under Soviet control until 1989.

28 March 1945 - Weather: Average Temperature 58.1° F, (14.59° C), night and morning haze

As the 7th Guards Army continued to advance, the 27 GTB and the 2 TR Ru appeared for the first time in the 27 GRC zone on a historical map. At this time these tank units were supporting the 27 GRC along the far-right flank of the Army, but they had taken heavy losses the previous day with 20 tanks knocked out, according to German sources.

Units of the 27 and 24 GRCs continued their advance and reached and crossed the Nyitra River north of the city of Ersekujvar. Further south, the 309 RD, 4 GAD and 25 GRD of the 25 GRC were now on the western side of the Hron River. They advanced some 12-14 miles during the day as German forces retreated.

The 46th A (2 UF) and 4th GA (3 UF) finished capturing Gyor. Another major Axis position had fallen.

On the Western Front, General Eisenhower, Supreme Allied Commander in Europe, informed Joseph Stalin that the Western Allied armies would not be advancing against Berlin. Instead, they would occupy western Germany and push into Austria.

29 March 1945 - Weather: Average Temperature 52.8° F, (11.55° C), night and morning haze

On the Western Front, units of the American Third Army captured the major German city and industrial/transportation center of Frankfurt.

The 27 GRC fully crossed the Nyitra River and turned its advance northwest toward the town of Sered on the upper Vaih River.

Meanwhile, the 24 GRC, 81 GRD attacked into Ersekujvar, capturing part of the city.

Units of the 25 GRC reached the Nyitra River line from the Danube in the south to Ersekujvar in the north. The next day they would assault over the Nyitra River and capture Komarno.

At 21:15 hours, the 8th Army received orders to prepare the road and railroad bridges in and around Komarno for demolition. A German General Staff officer, von Gyldenfeldt asked what possible significance the continued defense of the Komarno Bridgehead could have. The answer? “The Fuhrer placed great value on holding the oil refineries. It didn’t matter that the oil refineries only consisted of heaps of rubble and bomb craters, as the aerial photos clearly showed.” [762]

1945 Hungary – Wounded German and Hungarian POWs rest at the roadside.

30 March 1945 - Weather: Average Temperature 52.1° F, (11.16° C), night and morning haze

"On March 30, salutes were fired in Moscow to the valiant troops of the 2nd Ukrainian Front, which broke through the enemy's defenses in the Bratislava direction and captured the cities of Komarno, Nové Zamky [Ersekujvar], Shuravka, Komjatitsa and Vrabla. [The salute was] 20 artillery salvoes from 224 guns." [763] Actually, for the 7th Guards Army, this reference is talking about how the 25 GRD crossed the Nyitra River and captured Komarno, along with the 81 GRD capturing Ersekujvar on this day. This is clearly shown on the historical map found in TsAMO record ID #100114450. The other locations apparently were captured by different units of the 2 UF, but all the rifle divisions of the 7th Guards Army were attacking/advancing and capturing towns and villages.

Overall, units of the 7th Guards Army now were aligned along the Vaih River and ready for further advances.

On this day, the 243 RD was ordered to be transferred from the 7th Guards Army to the 53 A.

Guards Major of the medical service Lev issued a medical report, *Information about the movement of the wounded and sick in medical institutions of the 7th Guards Army*, found in TsAMO record ID #133393477. This report covers the period from 25 January to 30 March 1945, and was an assessment of the current medical burden upon the Army's medical establishment.

The total medical burden on 30 March 1945 was 6,245 people. This total was more than the 27 February 1945 value of 4,682 people, an indication that the severity of the fighting increased in March.

Forward Troop Area – up to divisional medical battalions

Arrivals:

At the start (25-30 March) – 1,461 men and women

Newly arrived – 3,691

Total – 5,752

Departures:

Evacuated – 2,664 (most likely to the Army rear) (moderately wounded)
Returned to active service – 504
Died –120
Total – 3,288 (including dead)
In the Forward Troop Area on 30 January – 1,864 (lightly wounded)

In Army Hospitals

Arrivals:

At the start (30 March) – 2,935 men and women
Arrived from forward units' medical installations – 2,980
Arrived directly from units – 252 (severely wounded)
Totals – 6,167

Departures:

Evacuated – 1,420 (out of the Army to the deep rear, severely wounded)
Returned to active service – 333
Died – 33
Total – 1,786 (including the dead)
In Army Hospitals 30 March – 4,387 (moderately wounded) (with about 6,700 hospital beds)

Total in all medical installations on 30 March – 6,245 people

Despite heavy fighting during an offensive, the Army's medical establishment was managing the casualty load well. Note the large number of those who died (153) after reaching the hospital and the large number of severely wounded (1,420) sent out of the Army to the deep rear.

With the armies of the 2 and 3 UFs advancing upon Vienna, Axis forces were withdrawing all along the line, as there was little else they could do. (See Map #70 for the Stratgic Situtation.) As for the huge numbers of civilians that were caught up in this retreat, the "liberating Red Army" scared them to death. They had good reason to fear because Soviet soliders were seeking revenge against the peoples and lands of the hated enemy. Be they young or old, male or female, they were all Nazies in the Soviet soldiers' eyes and deserved whatever punishment the victorious Red Army could imagine! "The anti-Nazi propaganda of the Red Army was so intense that almost every solider was overwhelmed by an insatiable craving for revenge. As a result, many civilians were indiscriminately killed, women were raped, propery was destroyed, and German military warehouses plundered." [764]

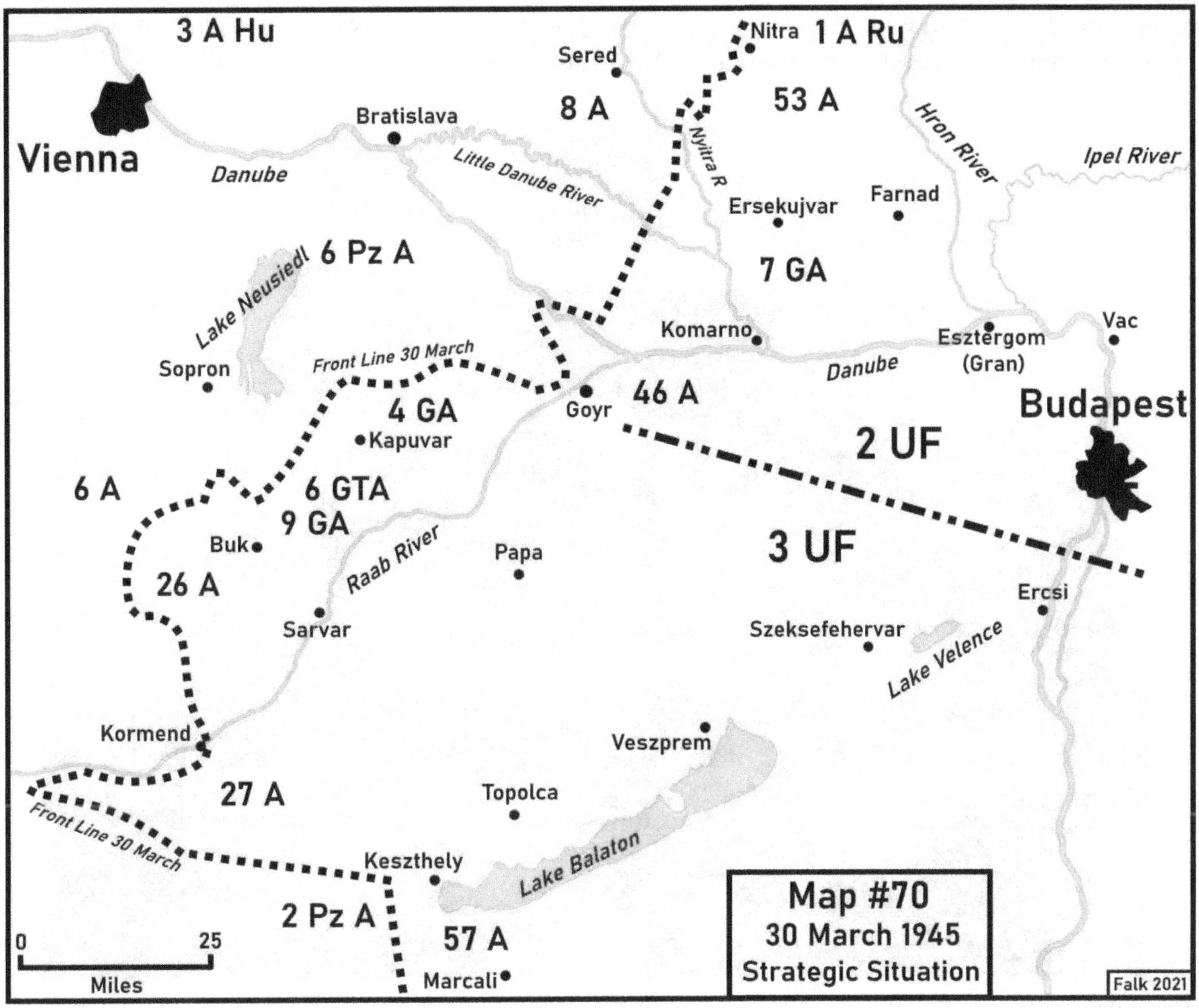

31 March 1945 Weather: Average Temperature 53.6° F, (12.00° C), night and morning haze

The 2 UF issued a report, *Report on the loss of personnel in the troops of the 2nd UkrF from March 1 to March 31, 1945.* This was found within TsAMO record ID #13682461. The 7th Guards Army suffered 8,767 total casualties for the month: 1,197 killed; 189 missing; 0 captured; 105 non-combat losses; 4,677 wounded/burned; 2,599 sick; and 0 frostbite. The report goes on to list 7th Guards Army horse losses: 234 killed; 3 missing; 45 wounded; 42 evacuated to the veterinarian; for a total of 324 equine casualties. The number of horses that died had decreased from the previous month.

At Hitler's direction, AG South ordered all the remaining Hungarian combat formations to be disarmed. Defeatism among the ranks was widespread owing to most of Hungary being under the control of the Red Army. These Hungarian troops had had enough of fighting and dying for a lost cause. [765]

For the liberation of the cities of Nitra and Galant, the 7th Guards Army and the 72 Guards RD by order of the Supreme Command of March 31, 1945, declared gratitude and in Moscow were given a salute with 12 artillery volleys from 124 guns.

On this day, the 23 RC operationally was assigned to the 7th Guards Army from the 46th A. The 23 RC was still part of the 46th A, but the 7th Guards Army had control of its units. At this time the 23 RC contained the 25 GRD from the 7th Guards Army and the 19 RD and 252 RD from the 46th A. The 7th Guards Army now controlled 4 rifle corps and 12 rifle divisions.

General Shumilov (right) with Colonel Philip Kuzmich Scherbak, member of the Army's Military Council. They are taking in the view from a rooftop while having a smoke break in the snow.

The chief of staff of the 23 RC, Sergey Andryushchenko stated:

> On the morning of March 31, our offensive in the area between the Big and Small Danube rivers resumed. Here, in the operational space, the 7th Guards Army of General M.S. Shumilov operated. And since we also ended up in this strip, we were transferred to it on the same day. Not meeting serious enemy resistance, the corps [23 RC] turned its battle formations into columns [marching columns] and, having only advanced detachments [proceeding them], began pursuing the enemy in the direction of Bratislava. [766]

Looking at Map #71, one can see how the 7th Guards Army had shifted it general advance of the 27, 24 and 25 GRC more to the northwest, away from the Danube River and more toward Bratislava and north of Vienna. The newly assigned 23 RC would continue to clear the north bank of Danube River line on the left flank of the Army. German formations were withdrawing so the Army was sure to continue its advance in the days ahead.

The 7th Guards Army produced a report TsAMO record ID #133375099, *Information about the damage inflicted on the enemy by the troops of the 7th Guards Army,* from 20-31 March 1945. This one-page document reports the following as seen in Table #40.

Killed/Destroyed	Number
Officers and Soldiers	5,420
Guns	61
Mortars	26
Machine Guns	127
Destroyed and Burned Tanks	23
Destroyed and Burned Armored Vehicles	5
Wagons/Carts with loads	10
Captured	**Number**
Officers and Soldiers	3,703
Different Caliber Guns	35
Anti-Aircraft Guns	34
Mortars	36
Machine Guns	124
Rifles and Machine Guns (SMGs)	1,300
Tanks and Self-Propelled Guns	20
Armored Personnel Carriers	6
Motor Vehicles	6
Radio Stations	1
Ammunition Warehouse and Food Storage	5

Table #40: The 7th Guards Army Damage Inflicted -- 20-31 March 1945.

Table #40 shows the mixed nature of captured enemy equipment, which represents the remains of a defeated army leaving behind anything that would slow their retreat. Notably, the advancing 7th Guards Army found more enemy dead soldiers than live soldiers to capture. This is a possible indication that in their haste, the retreating German Army was leaving the dead behind on the battlefield or along the road where they fell.

For the month of March in Britain, air raids and German V-1 and V-2 attacks killed 792 civilians and injured 1,426 more men, women and children.

1 April 1945	
7th Guards Army	
Assigned – 2 Ukrainian Front	
Colonel General M.S. Shumilov Commanding	
Rifle, Airborne and Cavalry Units	24 Guards Rifle Corps (6 GAD, 72, 81 GRDs, 303 RD) 25 Guards Rifle Corps (4 GAD, 25 GRD, 409 RD) 27 Guards Rifle Corps (93 GRD, 141, 375 RDs) 15 Penal Battalion 62, 63, 64, 65, 66, 194 Penal Companies
Artillery Units	16 Artillery Penetration Division (49 Light Artillery Brigade, 61 Gun Artillery Brigade, 52 Howitzer Artillery Brigade, 90 Heavy Howitzer Brigade, 109 High-Power Howitzer Brigade, 14 Mortar Brigade) 5 Antiaircraft Artillery Division (670, 743, 1119, 1181 AA Regiments) 26 Antiaircraft Artillery Division (1352, 1357, 1363, 1369, AA Regiments) 2 Antitank Artillery Brigade 114, 115 Guards Tank Destroyer Regiments 41 Guards Gun Artillery Brigade 263, 290, 493 Mortar Regiments 662 Separate Aerial Observation/Air Raid Warning Company
Armored and Mechanized Units	27 Guards Tank Brigade 2 TR Ru 38 Separate Armored Train Battalion
Engineer Units	60 Engineer Sapper Brigade 3, 27 Separate Flamethrower Battalions
Reported Combat Strength Estimated Ration Strength	66,945, men and women 83,245, men and women (including an estimated 7,000 support personnel)

(BSSA 45, BSA 45, TsAMO Archive)

Once again, the 38 Separate Armored Train Battalion returned to the Army while the 28 Separate Armored Train Battalion left. The 11 Antiaircraft Artillery Division also transferred from the Army while the 26 Antiaircraft Artillery Division replaced them.

1 April 1945

The 7th Guards Army continued its advance north of the Danube. (See Map #71.)

Adolf Hitler moved out of the Reich Chancellery and into the nearby *Führerbunker*, a large underground air raid bunker. This was to be Hitler's last headquarters.

For the liberation of Sered and other cities, by order of the Supreme Command on 1 April 1945 declared gratitude to the 7th Guards Army and ordered a salute in Moscow with 12 artillery volleys of 124 guns.

As the 7th Guards Army advanced in the Bratislava area, they also encountered the far western end of the Carpathian Mountains (See Map #48), also known as the Carpathian Minor Range or Little Carpathians. The mountains were not a major barrier with their highest point being only 2,500 feet, but the defending Germans used the winding roads, valleys, forest and hills to good effect to delay their enemy.

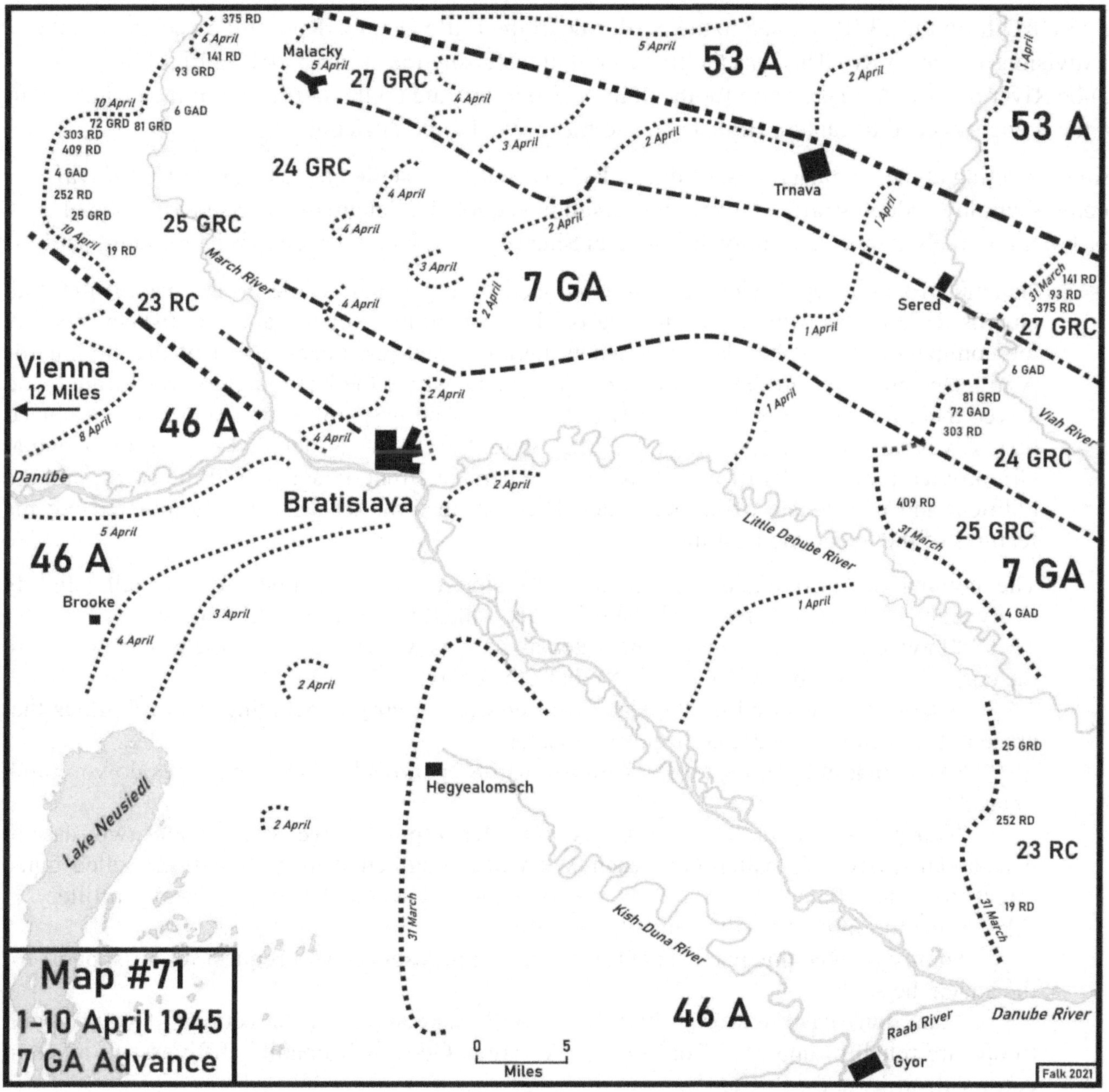

2 April 1945

The critical Hungarian Nagykanizsa oil fields fell to advancing units of the 3 UF. At this point German and Hungarian forces were in full retreat. They desperately were trying to reach what they perceived to be the more humane Western Allied armies to surrender. They would do anything to escape capture by the Soviets.

Units of the 7th Guards Army reached the outskirts of Bratislava.

3 April 1945

Units of the 25 GRC attacked directly into the city of Bratislava with support from the gathering 23 RC.

The 25 GRD from the 23 RC moved to the north side of the Little Danube River followed by the other rifle divisions of the 23 RC. This was the first step in transferring the entire 23 RC north of the Little Danube River in order to make room for the 46th A also to operate on the north side of the Danube. But first the Soviets needed to capture Bratislava, and that proved to be difficult.

During a meeting at the command post of the 23 RC to discuss the stalled attack against Bratislava, General Shumilov suddenly arrived. (This conversation is quoted at length due to its uniqueness of a record of a relatively long conversation by General Shumilov and its relevance to the unfolding battle.)

> General Mikhail Stepanovich Shumilov stopped by for a short while on the way to the 25th Guards Rifle Corps, which was fighting on the northernmost outskirts of Bratislava. We met the commander just a few days ago, when the corps had just become part of the 7th Guards Army. He came to us in the city of Dunaiska-Streda, where the headquarters was located, and stayed there for an hour and a half or two. Everyone who was at the command post introduced themselves to the army commander. MS Shumilov heard a detailed report by General Grigorovich [commander of the 23 RC] on the combat strength, staffing of the corps, and the political and moral state of its personnel. Then he, in turn, set tasks for us, pointed out the features of the upcoming hostilities.
>
> The commander also made a pleasant impression on us this time. He was tall, slightly overweight and good-natured. He behaved easily, talked cheerfully, with a smile.
>
> "So where did you stumble?" he asked as soon as we went down to the basement, where the corps command post was located. "What's the delay?"
>
> Grigorovich reported on the state of affairs in the corps, including the difficulties that arose in the sector of the 252nd Infantry Division.
>
> "And what did you decide?" asked Mikhail Stepanovich. "What measures do you think to take?"
>
> "During the night," the corps commander replied, "we will create two assault detachments, one rifle battalion in each. We will strengthen them with self-propelled guns, direct-fire guns, flamethrower and engineering units[;] we will bring up heavy artillery ... [a]nd in the morning, we will resume the assault."
>
> Shumilov liked this plan. He approved it, and in response to a request for air support for the attack, he said:
>
> "You don't have to worry about that. I will make sure that the actions of your assault troops are reliably supported from the air. At dawn, General Kamanin's fifth assault aviation corps will strike at the fortress."
>
> Leaving, Shumilov remarked with a smile,
>
> "I'm in a hurry to see Ostashenko in the twenty-fifth Guards Corps. You will have to continue the assault on Bratislava with him. Hope you become friends." [767]

This discussion by General Shumilov with his commanders shows a confident and experienced army commander guiding the 23 RC to success while guaranteeing the needed air support would show up for the next attack. The open and friendly nature of General Shumilov when speaking with his subordinates shows his real character, strong, confident, supportive but friendly at the same time. Generals do not need to shout to be heard or obeyed.

4 April 1945

At long last, units of the 7th Guards Army captured Bratislava and pushed beyond the city to the west. Stalin was pleased. For the liberation of Bratislava by the 7th Guards Army, the Supreme Command on 4 April 1945, declared gratitude and in Moscow the Army was given a salute with 20 artillery volleys from 224 guns.

General Shumilov was as good as his word because units of the 5th AA did arrive to add air support to the final attack against Bratislava.

> Despite the extremely tight deadline appointed by the front chief of staff for preparing an air offensive in the new direction [Bratislava], all departments and services of the air army managed to cope with the task in time. Aviators immediately began to actively support the ground forces, which launched an offensive against the capital of Slovakia, Bratislava. On April 4, the formations of the 7th Guards Army of General M.S. Shumilov, in cooperation with the sailors of the Danube military flotilla, seized the city of Bratislava by storm. [768]

Also on this day, the Red Army totally liberated Hungary by removing the last few remaining Axis troops from the country. Four years later, by 1949, the Soviet Union had taken over total control of Hungary. This occupation lasted until the last Soviet troops left Hungary in 1991. Fate created a long road to freedom for the Hungarian people.

5 April 1945

The 7th Guards Army – 141 RD, 27 GTB – for the liberation of Malacky and other cities were commended by the order of the Supreme Command of 5 April 1945, and a salute was given in Moscow with 20 artillery volleys from 224 guns.

Also, on or about this date, the 2 UF shifted the boundary line between the 53rd A and 7th Guards Army to the north. The shift began just east of the town of Malacky. This boundary shift forced the 27 GRC to quickly spread out its rifle units a further nine miles to the north to screen this new frontage area. Fortunately for the 7th Guards Army, the enemy was retreating as these thinned out units would not have been able to hold this line if they had come under attack.

6 April 1945

Units of the 7th Guards Army were for the most part aligned along the March River while some formations forced a crossing of the river and entered into Austrian territory.

"From the political report of the political department of the 7th Army for April 6, 1945: Our troops, having overcome the increasing resistance of the enemy, in a number of places reached the Austrian border." [769]

Adolf Hitler was not pleased about the loss of the Hungarian oil fields, so on or about this date, Hitler finally relieved General Otto Wöhler from command of AG South. His replacement, Colonel General (*Generaloberst*) Lothar Rendulic, was appointed to command AG South on 7 April, and he would last for 23 days in that position, until 30 April 1945 and the end of WW II in Europe.

With the armies of the 3 UF beginning their attack against Vienna, Marshal Tolbukhin had leaflets dropped upon the city. This was a deliberate attempt to avoid a repeat of the Budapest disaster. The leaflets stated, "The retreating German troops want to turn Vienna into a battlefield, as they did in Budapest. Vienna and its inhabitants are thus threatened with the same destruction, the same horrors of war that the Germans inflicted on Budapest and its people." [770]

The defender of Vienna, General of Infantry Rudolf von Buanu, only had some 40,000 troops and 26 tanks and assault guns at his disposal. General Buanu stated that "the idea of mounting a successful defense of the city at all, or even holding it for some time, had to be ruled out from the start." [771] So, Vienna was spared the massive death and destruction that was inflicted upon Budapest due to a lack of troops and the unwillingness of German commanders to destroy a city and its peoples for no reason, as the war was already lost.

In East Prussia, units from the 1st Belorussian Front (1 BF) launched attacks against the newly declared Fortress City and major port of Konigsberg. Soviet land and air forces in this area far outnumbered the German defenders.

7 April 1945

Units of the 7th Guards Army crossed the March River in strength along the south end of the river and pushed to the west.

On or about this date, the right flank units of the 46th A (2 UF) was ordered to cross over to the north bank of the Danube, leaving its left flank formations south of the river, where they were advancing directly toward Vienna. With the splitting of the 46th A in this manner, Marshal Malinovsky was directing the 46th A to approach Vienna on both sides of the Danube. This move also forced a shift in the advance of the 7th Guards Army away from the Danube and Vienna. Because of this, General Shumilov and the 7th Guards Army would not share in the glory of capturing Vienna, the capital of Austria.

8 April 1945

German defenders largely blocked units from the 27 and 24 GRC from crossing the March River. Meanwhile, the 25 GRC and 23 RC did push over the March River and continued to gain ground on the western bank. Also, the 46th A, which was now in part north of the Danube, provided left flank support for the advance of the 7th Guards Army.

9 April 1945

Axis forces continued to block the advance of the 27 and 24 GRC along the March River. Further south, the 25 GRC and 23 RC slowly pushed the German defenders back.

On 9 April, the German 4th Air Fleet, supporting AG South, had the following operational aircraft (See Table #41.) These operational numbers are in spite of losses of men and machines and the lack of fuel and spare parts.

Aircraft Type	Number
Single engine fighters	62
Night fighters	-
Bombers	-
Ground attack	93
Night ground attack	61
Strategic reconnaissance	44
Tactical reconnaissance	68
Transports	-
Total	**328**

Table #41: German 4th Air Fleet Operational Aircraft – 9 April 1945. [772]

Amazingly, the Luftwaffe still existed as a fighting force. At this time, the entire Luftwaffe contained some 3,331 operational aircraft, but night fighters, bombers, and transport types had all but disappeared. Single engine fighters and ground attack aircraft formed the bulk of the remaining force. One German fighter pilot recalled that

> …if a fighter were damaged it was simpler to get a new one than to make repairs. We simply went to the depot nearby, where they had hundreds of brand new [Me] 109s – G-10s, G-14s and even the very latest K models. There was no proper organization any more. The depot staff just said, 'There are the aircraft, take what you want and go away.' But getting fuel – that was more difficult…. [773]

On the other hand, the VVS had no shortage of aircraft, fuel or pilots. During 1945, the VVS received a total of 19, 228 combat aircraft from their factories along with another 6,876 non-combat aircraft for a total of 26,104 aircraft. [774] Lend Lease deliveries only swelled the number of aircraft further.

In East Prussia, units from the 1 BF finished capturing the Fortress City of Konigsberg. After only four days of combat, General Lasch surrendered the city. The defenders suffered 42,000 military and 25,000 civilian deaths and another 92,000 troops were captured. (These values differ depending on the source.) In response to the quick loss of Konigsberg, "Hitler sentence[d] Lasch to death *in absentia* and ordered the arrest of his family and relatives." [775] General Lasch actually survived the war and Soviet captivity, dying in Berlin in 1971. Of note, the occupation of Konigsberg, modern day Kaliningrad, by the Soviets (and currently by the Russians) along with Germany's total defeat in World War II also marked the total dissolution of the German state of Prussia.

10 April 1945

For the 7th Guards Army, the days of easy gains now were over as German defenders all along the line dug in and slowed the Army's advance.

Colonel General M.S. Shumilov at the front with his staff in 1945.

On or about this day, the 38 Separate Armored Train Battalion once again was removed from the 7th Guards Army. The reason for this transfer is unknown.

On the Western Front, units of the U.S. 9th Army captured the city of Essen in the heart of the major Ruhr industrial area in western Germany.

11 April 1945

Fighting inside Vienna continued, with Soviet forces gaining ground and capturing key buildings.

Formations of the 7th Guards Army pushed forward against increased German resistance. The 23 RC along with the 25 and 24 GRC made a good advance with the 27 GRC struggling to move forward over the March River. (See Map #72 for the 7th Guards Army Advance.)

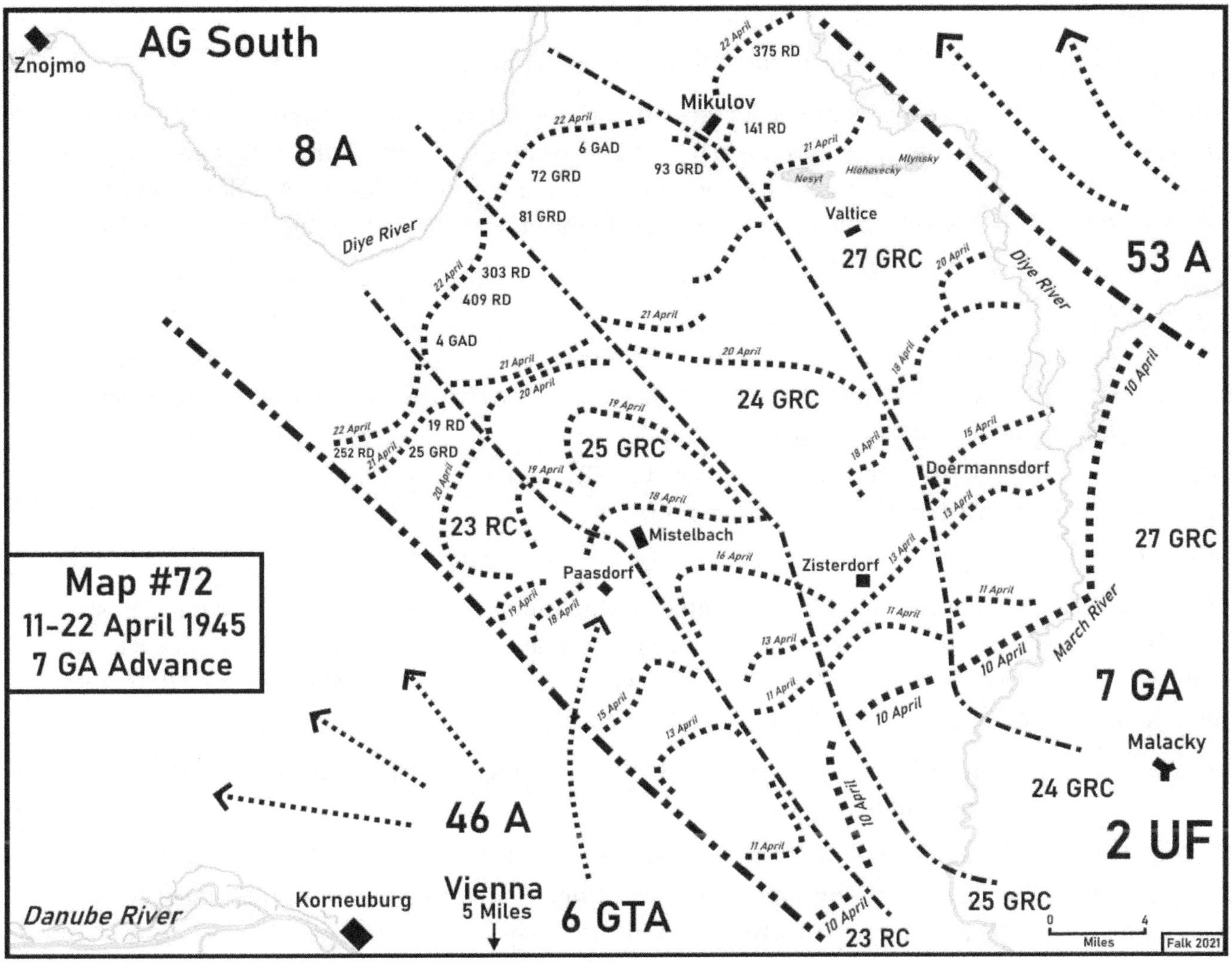

Map #72
11-22 April 1945
7 GA Advance

12 April 1945

Formations of the 7th Guards Army slowly advanced, barley making any headway at all against determined German defenders. Unknown to General Shumilov, the Germans were conducting a fighting withdraw to gain time to prepare defensive earthworks further to the rear.

The American President, Franklin D. Roosevelt, died after suffering a stroke. The Vice President, Harry S. Truman, immediately was inaugurated as President of the United States at the White House in Washington, DC.

13 April 1945

Over the previous two days, the 7th Guards Army gained only 2-3 miles against stubborn German resistance.

Units from the 2 UF and 3 UF finish capturing the city of Vienna and the surrounding suburbs as fighting continued after dark. Despite the relatively brief battle, damage was extensive in areas and losses were heavy but nothing like the carnage of Budapest.

To shift the focus of the advance in the Vienna area, *Stavka* transferred the 6th GTA to the 2 UF north of the Danube. This tank force would take several days and much effort to cross the Danube River and to reassemble in the 2 UF zone.

14 April 1945

Despite heavy fighting, the rifle units of the 7th Guards Army hardly advanced.

AG South reported, "The enemy continued attacks of varying intensity all along the front of *Heeresgruppe Süd.* The main effort was made by the 2nd Ukrainian Front north of the Danube." The Luftwaffe responded with 270 sorties of fighters and ground-attack aircraft. The AG South reported, "In the sector of the *8. Armee*, all enemy attacks were in the area south-west of and in Zistersdorf were repelled by hard-fighting friendly forces." [776]

15 April 1945

The 23 RC and the 27 GRC, on both flanks, made some progress, but the center of the Army was stalled. General Shumilov was shifting his attacks from one area to the next, looking for a soft spot in the German line to force a breakthrough.

On the Western Front, the 1st French Army crossed the Rhine River and captured the city of Kehl, which was directly across the Rhine from Strasbourg.

16 April 1945

Soviet forces launched the Berlin Offensive with overwhelming strength. Stalin believed the Western Allies still might want to secure Berlin for themselves, so Stalin wanted a fast-paced offensive to reach the capital first. Stalin did not trust General Eisenhower's message from 28 March, and he desperately wanted the Red Army to capture the final, ultimate prize: Berlin!

During the day, in the area of the 6th Pz Army and 8th Army, the Luftwaffe flew 180 sorties, using both German and Hungarian aircraft in support of the defenders. [777]

A historical map covering this date, TsAMO map ID #100114451, showed the 7th Guards Army making no attacks this day. Shumilov apparently was bringing up artillery and supplies and repositioning his troops for a more powerful attack the next day.

17 April 1945

The attacking 7th Guards Army seized the oil producing territories of Austria and the town of Zistersdorf.

In response to this breakthrough, Marshal Malinovsky ordered the 7th Guards Army and the 6th GTA to form a "shock group" to exploit this success and to drive into southern Czechoslovakia at the city of Znojmo. In doing so, they would outflank the German defenders of Brno further to the north. Unfortunately, this improvised shock group ran into *Kampf* Groups from the 6 and 25 Pz Ds, supported by King Tiger tanks, about 15 miles north of the Austrian town of Mistelbach. Counterattacking German Panzers surrounded the Soviet spearheads with some Russian formations being decimated in the attack. [778] The historical TsAMO map ID #114203722 shows a swirling tank battle beginning on 19 April and lasting until 21 April. At best, Malinovsky's plan was ill conceived, at worst, foolhardy. The city of Znojmo was at least 45 miles away through enemy territory and the 7th Guards Army and the 6th GTA could hardly be expected to support an advance at that distance with strong forces.

At this point in the battle for Austria and Czechoslovakia, both sides were regularly launching local attacks and counterattacks all along the front line. These attacks made sorting out the flow of battle in any one area extremely difficult. But the much stronger Red Army kept advancing while the weakened German defenders could only withdraw to save themselves from being surrounded and captured.

In spite of its fading strength, the Luftwaffe flew 70 sorties during the day in support of the 8th Army, which was under heavy attack.

18 April 1945

Units of the 7th Guards Army made a moderately good advance of some three to four miles over the previous two days.

The Luftwaffe could only get a few aircraft off the ground to support the army because "the air wings had no more fuel and American fighter-bombers were patrolling the air bases." [779]

On the Western Front, U.S. forces capture 325,000 German troops with the surrender of the Ruhr pocket. Field Marshal Walter Model, who commanded the Ruhr pocket force, did not surrender with his troops. Instead, he escaped capture to determine his own fate a few days later.

19 April 1945

AG South reported, "North of the Danube, the first formations of the 6th Guards Tank Army were identified on both sides of Mistelbach." This sighting report is most likely related to the account from 17 April about combat taking place around Mistelbach. AG South then went on to state, "As *Heeresgruppe Süd* had no more reserves, it could only release forces from the quiet front sectors south of the Danube and send them to the hard-pressed *8. Armee.*" [780]

20 April 1945

Units of the 7th Guards Army and 6th GTA continued to advance together to the northwest. The 6th GTA was coordinating its movements and attacks with the 7th Guards Army, but it was operating under the direct control of the 2 UF. Therefore, for dedicated armor support, the 7th Guards Army had control only of its 27 GTB and the 2 Ru TR. The 2 Ru TR also was the only Romanian combat unit to see action on Austrian soil. [781]

Further north, Soviet forces were close enough to Berlin to start shelling the city with artillery. This day also was Adolf Hitler's 56 birthday, which he spent in his underground bunker.

Meanwhile in East Prussia, a Soviet officer stated "…we drove into Neidenburg, fires lit up the city, entire blocks were ablaze… This is Germany. So: destroy everything, burn everything. Take revenge." [782] After the war a Soviet war correspondent reflected that, "For more than three years, the Soviet soldier was told one thing: 'revenge.' This word was repeated in various ways in the conversation of political instructors, leaflets, and newspapers." [783]

The resulting counterproductive rape and murder of civilians and the destruction of material goods forced a change from the top. On this date *Stavka* issued Directive No. 11072, which in part sought to change the attitude of the troops towards POWs and civilians. "A more humane treatment of the Germans will make it easier for us to conduct combat operations…and will undoubtedly reduce the German's tenacity in defense." Strangely enough, even Stalin understood that with the imminent occupation of Germany, "the amount of terror and destruction the Red Army inflicted on Germany needed to be limited." [784] Gradually, discipline was restored to the Red Army as the need for revenge faded away. The debt had been repaid; the pound of flesh taken.

On the Western Front, U.S forces finished capturing Nuremberg while French forces began attacking Stuttgart, a major urban and transportation center in southwest Germany.

21 April 1945

All four of the rifle corps of the 7th Guards Army were able to advance abreast of each other, but the largest gains were in the right flank area. Unknown to General Shumilov, the Germans were deliberately withdrawing to the Diye River line where they had built defensive works to make a determined stand.

On the Western Front, not wanting to be taken prisoner by the Allies, as he was wanted by the Soviets for war crimes, Field Marshal Walter Model walked off into the woods and shot himself in the head.

22 April 1945

Units of the 7th Guards Army continued to advance against weak German resistance, but the Army was exhausted after 22 days of continuous combat and an advance of over 88 miles. The most notable development for the day was that the 27 GRC almost completely surrounded the town of Mikulov.

Adolf Hitler decided against fleeing Berlin; instead, he chose to await the end in the *Führerbunker* along with Eva Braun, his mistress of 16 years.

23 April 1945

In the zone of the 7th Guards Army, the German 8th Army had built a new defensive line along the Diye River. This line was intended to temporarily stop or at least to slow down the Soviet advance to the north or northwest, thus blocking a Soviet drive into southern Czechoslovakia, while gaining time for auxiliary troops and civilians to flee to the west.

Not deterred by the Germans defending behind the Diye River, divisions of the 27 GRC conducted a military crossing (assault crossing) of the river northwest of Mikulov. The approach and first crossing of the river began on 23 April and the fighting continued until 27 April when a modest bridgehead was consolidated on the western bank. At the same time, the remainder of the Army made small advances and reached the Diye River line. Despite repeated attacks, the 25 GRC was unable to cross the river.

Also, on or about this date, the 24 GRC was withdrawn from the front, most likely for rest and refitting. There is some evidence the 24 GRC was temporally assigned to the 53rd A, but it would return to the front during the first part of May 1945 working with the 2 Guards Mechanized Corps (2 GMC), 6th GTA. Apparently the 24 GRC still was assigned to the 7th Guards Army at that time.

Back in Germany, Hitler stripped Hermann Göring, the second highest-ranking Nazi official, of all his authority because Hitler believed Göring was trying to usurp his power. Hitler was definitely still in charge, but privately Hitler admitted to his closest advisers that the war was lost.

24 April 1945

The 27 GRC continued to expand its small bridgehead in heavy fighting on the west bank of the Diye River. The remaining units of the 7th Guards Army held their positions and dug in.

Adolf Hitler ordered Hermann Göring to be arrested, but Göring was able to temporarily escape justice.

25 April 1945

The 27 GRC continued to expand its small bridgehead on the west bank of the Diye River.

The US 1st Army made first contact with the Red Army, 1 UF, at several locations in the area between the city of Torgau and Riesa on the Elbe River. With American troops advancing from the west and Soviet troops from the east, the remaining German controlled territory now was split in half. The small town of Torgau is located about 60 miles southwest of Berlin.

26 April 1945

Units of the 7th Guards Army were aligned along the Diye River but were unable to advance further. Fighting continued in the 27 GRC bridgehead.

American Second Lt. William D. Robertson (L) embraces Russian Lt. Alexander Selvashko (R). This meeting took place at the Elba River on 26 April 1945, but many more such greetings would soon follow. Robertson and Selvashko stayed friends for the rest of their lives, meeting several times long after the war.

27 April 1945

The fighting abated in the small bridgehead held by the 27 GRC, as General Shumilov and the 7th Guards Army were content to firmly hold the Diye River line. Now that political events were taking on an increased importance, there was little need to fight a defeated enemy that might surrender in a matter of hours. (See Map #73 for 7th Guards Army deployment.) During its most recent fighting, the 27 GRC had inserted parts from at least four different rifle divisions into the bridgehead (72 GRD, 141, 409 and 375 RDs), which resulted in a very confused, intense battle.

Heeresgruppe Süd received orders to "ruthlessly weaken its right wing and center to release forces for the reinforcement of the left wing of the *8. Armee* at Brunn [Brno]." [785]

Italian National Fascist Party Leader and dictator Benito Mussolini and his mistress, Claretta Petacci, were captured by local partisans in northern Italy. They were attempting to flee to Switzerland.

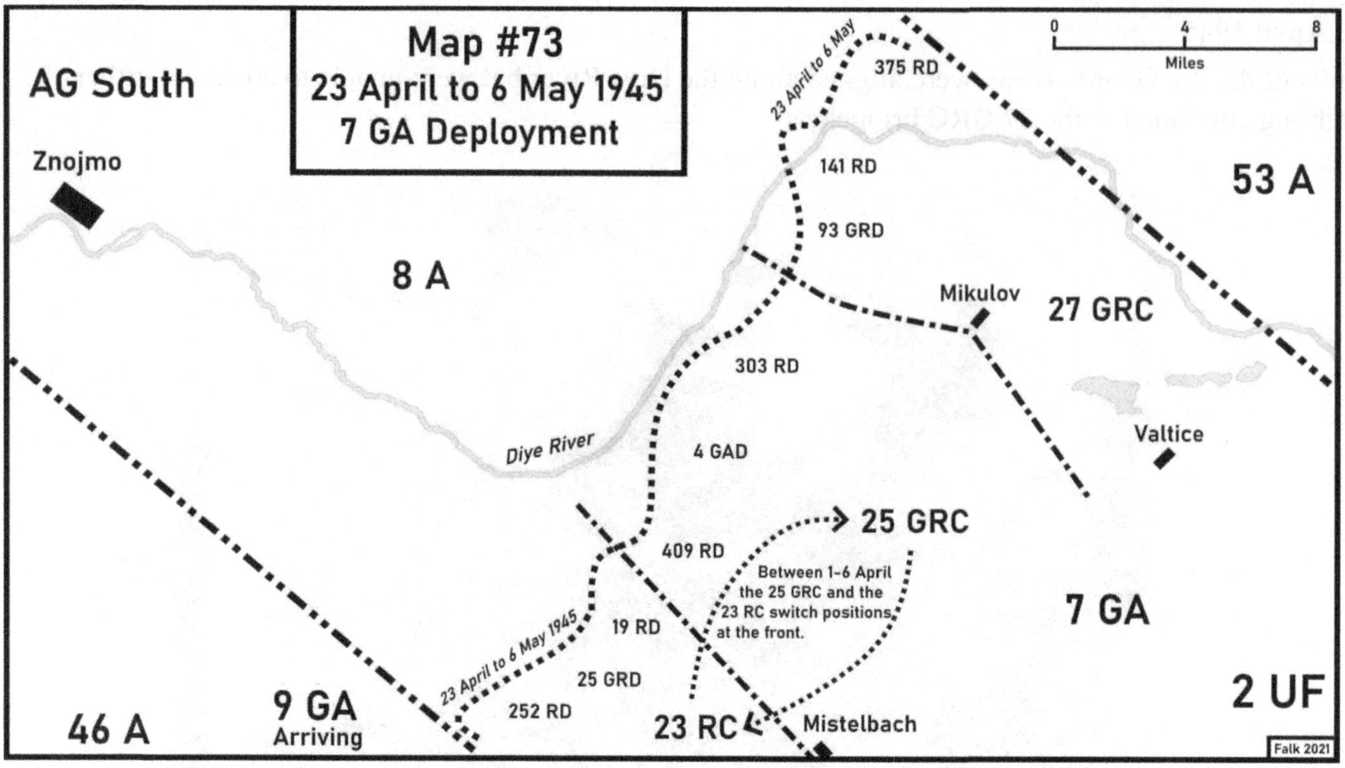

Note: During the 23 April to 6 May time period, the rifle divisions of the Army frequently shifted positions to meet military needs. Map #73 displays their approximate positions on this date.

28 April 1945

The 7th Guards Army held its position and resupplied the troops.

In Berlin, Soviet troops were within one mile of Hitler's *Führerbunker* as the fighting raged all around and within the *Reichstag* (old Parliament building) and the Reich Chancellery.

The partisans in Italy took great joy in the capture of Mussolini and Petacci, who were shot in the afternoon. Afterwards, they were strung up by their feet in the Piazzale Loreto in Milan for crowds to view and to jeer at. Some reports said the bodies were beaten with hammers and shot (again). So ended one Fascist dictator. Mussolini was 62, Claretta Petacci was 33.

29 April 1945

The 7th Guards Army continued to hold its position along the Diye as the troops rested and dug in. Any needed supplies, equipment or replacements also were brought forward to replenish the depleted ranks.

German Army Group Commander General Vietinghoff signed unconditional surrender documents in northern Italy. Despite the surrender to the Allies, the fighting was scheduled not to end until 2 May 1945 as Italian partisans were still in control of sections of Italy and were seeking revenge against German troops.

In Berlin, an ever-distrustful Hitler had a cyanide capsule tested on his beloved female German Shepherd Blondi. Blondi, of course, died.

Later that day, deep in the *Führerbunker*, Adolf Hitler and Eva Braun married during a short civil ceremony. At 33, Eva was still a young and attractive woman. On the other hand, Hitler, who turned 56 nine days before, was an ill and broken man. Hitler suffered from insomnia, drug addiction, and was showing symptoms of advanced Parkinson disease, among other ailments. The marriage would not last very long as both of the newlyweds had cyanide capsules of their own. Instead of wedding bells, Russian artillery fire was raining down on the Reich Chancellery and the *Führerbunker*.

As Hitler began to face reality, he appointed Grand Admiral Karl Dönitz to be his designated successor. Dönitz was currently the Commander-in-Chief of the German Navy (*Kriegsmarine*). This appointment was in lieu of Hermann Göring who was Hitler's previous successor. Göring currently was unavailable as he was on the run from the *SS*. In a final callous act, Hitler also took the time to order the defenders of Berlin to fight to the last bullet.

30 April 1945

The 7th Guards Army held its current position.

Because Western Allied armies were overrunning Germany and the V-1 and V-2 launching sites, German military forces had little incentive to attack England. For the first time in years, "There were no civilian air raid casualties in Britain during the month of April. A total of 60,585 [people had been] killed and 86,175 seriously wounded [in Britain] since September 1939." [786]

Around 15:25 hours, deep beneath Berlin in his *Führerbunker* near the Reich Chancellery, Adolf Hitler, an Austrian by birth and the leader of Nazi Germany, along with his new wife Eva Braun Hitler, committed suicide together. Most sources say they both took cyanide, but some say that Hitler also shot himself in the head while Eva bit into her cyanide capsule. Regardless, as ordered, their bodies then were taken out of the bunker and burned in a nearby shell crater. Ironically, the bodies were not totally cremated due to the lack of gasoline.

After the fall of Berlin, the Soviets dug up and recovered the charred bodies of Hitler and Eva, holding their remains in secret for decades. In 1968 the Soviets revealed the autopsy report for Hitler, which continues to be a source of speculation. Then on 4 April 1970, acting upon orders from Moscow to destroy their bodies, KGB officers dug up the remains of Hitler, Eva, and the Goebbels family (see 1 May 1945 entry for further information) from their hidden location near the city of Megdeburg, southwest of Berlin on the Elba River. The final KGB report stated, "The destruction of the remains was carried out by incineration on a pyre. The remains were completely incinerated and ground together with pieces of charcoal to pulverized ash and thrown in the river." [787] So, Hitler and his associates' final act was to pollute the Elba River.

Units of the U.S. 7th Army captured the southern German city of Munich.

The 2 UF issued a report, *Report on the loss of personnel in the troops of the 2nd UkrF from April 1 to April 30, 1945*. This was found within TsAMO record ID #113723699. The 7th Guards Army suffered an amazing 19,781 total casualties for the month: 3,202 killed; 624 missing; 0 captured; 63 non-combat losses; 13,964 wounded/burned; 1,930 sick; and 0 frostbite. This was the fourth highest monthly total for the 7th Guards Army during the war! The report goes on to list 7th Guards Army horse losses: 572 killed; 0 missing; 137 wounded; 28 evacuated to the veterinarian; for a total of 737 equine casualties. The number of horses that died once again increased from the previous month. These casualties attest to the brutal fighting that continued to the very the end of the war, with too many men, women, children, and yes, even animals, being killed during the closing stages -- a saddening finish to a terrible war!

1 May 1945	
7th Guards Army	
Assigned – 2 Ukrainian Front	
Colonel General M.S. Shumilov Commanding	
Rifle, Airborne and Cavalry Units	24 Guards Rifle Corps (25 GRD, 4, 6 GADs) 25 Guards Rifle Corps (81, 93 GRDs, 303 RD) 27 Guards Rifle Corps (72 GRD, 141, 375, 409 RDs) 23 Rifle Corps (19, 252 RDs) 15 Penal Battalion 62, 63, 64, 65, 66, 194 Penal Companies
Artillery Units	16 Artillery Penetration Division (49 Light Artillery Brigade, 61 Gun Artillery Brigade, 52 Howitzer Artillery Brigade, 90 Heavy Howitzer Brigade, 109 High-Power Howitzer Brigade, 14 Mortar Brigade) 5 Antiaircraft Artillery Division (670, 743, 1119, 1181 AA Regiments) 26 Antiaircraft Artillery Division (1352, 1357, 1363, 1369, AA Regiments) 2 Antitank Artillery Brigade 114, 115 Guards Tank Destroyer Regiments 41 Guards Gun Artillery Brigade 263, 290, 493 Mortar Regiments 662 Separate Aerial Observation/Air Raid Warning Company
Armored and Mechanized Units	27 Guards Tank Brigade 2 Ru TR 38 Separate Armored Train Battalion
Engineer Units	60 Engineer Sapper Brigade 3, 27 Separate Flamethrower Battalions
Estimated Combat Strength Estimated Ration Strength	59,968 men and women 76,268 men and women (including an estimated 7,000 support personnel)

(BSSA 45, BSA 45, TsAMO Archive)

With the pending end of the war in Europe, *Stavka* and front commanders adjusted the locations and assignments of military units and formations to meet Soviet military and political needs. Because events were changing rapidly, *Stavka* switched around the combat units of the 7th Guards Army and within the Red Army in general. Therefore, the BAS and BSSA reports do not agree with each other. The data in the above table is the most likely organizational layout for the 7th Guards Army on 1 May 1945. Also, the normal level and quality of Army and 2 UF staff work (paperwork) had declined, so unit strength reports are largely missing or incomplete, which forces the author to list estimated combat and rations strengths.

On this day, Colonel General M.S. Shumilov has been commanding the 64th/7th Guards Army for two years ten months and two days. With Hitler dead and Berlin all but captured, the war was now nearly over. Much was achieved during the years of fighting, but duties remained as the Red Army needed to eliminate the final scattered pockets of Nazi resistance.

The 28 Separate Armored Train Battalion once again was assigned to the 7th Guards Army.

While the war in Europe was near its end, fighting in the Pacific continued to rage over vast areas. Imperial Militaristic Japan refused to surrender despite being defeated on all fronts.

1 May 1945

The 7th Guards Army continued to hold its position along the Diye River. On or about this date, General Shumilov began shifting around the units of his Army. The 24 GRC withdrew into reserve, and the 23 RC and 25 GRC switched positions at the front. The distribution of rifle divisions within the Army/Corps also started to change.

At 03:50 hours, German emissaries sent by Joseph Goebbels reached the Berlin headquarters of General Chuikov, commander of the 8th Guards Army, to begin negotiating the surrender of Berlin. During the conversation, the Germans announced that Hitler had committed suicide the day before. Thinking quickly, Chuikov stated, “We already know that,” but he was lying. Chuikov was the first Red Army commander to learn of Hitler’s death. Chuikov soon called Marshal Zhukov, commander of the 1 Belarussian Front, and informed him of the arrival of the emissaries and of Hitler’s death.

Then around 05:05 hours, Marshal Zhukov sent a telegram to Stalin in Moscow informing him of the news. A short time later, Zhukov followed up with a telephone call directly to Stalin at his Dacha house [a seasonal second home called the “Near House”] outside of Moscow. “The Soviet dictator was still asleep and was irritated at being awakened.” Then, after hearing of Hitler’s death, Stalin supposedly said, “So that’s the end of the bastard,” and then, “Too bad it was impossible to take him alive.” [788]

A new German government was announced under the leadership of Admiral Karl Dönitz. As his successor, Hitler deliberately had named Dönitz “Reich president” and not “Führer.” The short-lived Dönitz government/regime established its first capital/headquarters in the two northern German cities of Plön and Eutin. Both of these small cities are located south of Kiel, Germany. Then due to circumstances (advancing British forces) Dönitz hurriedly moved the government to Flensburg, 53 miles north of Kiel, where it ultimately surrendered to Allied forces. [789]

The Dönitz Nazi government lasted eight days from 30 April to 7 May 1945. But, after the official surrender on 7 May, the Dönitz government was allowed to exist for an additional fifteen days as an interim, administrative government under Allied control. This arrangement ended on 23 May 1945 when Allied military representatives, sent by Supreme Allied Commander in Western Europe General Eisenhower, placed Dönitz and his ministers under arrest and took them into custody as POWs. [790]

With his official Nazi duties concluded, the day after Adolf and Eva Hitler committed suicide, Dr. Paul Joseph Goebbels, the Nazi propaganda minister, and his wife Magda also killed themselves, but only after poisoning their six children, ages 4-12. Thus, six more innocent children became victims of Nazi Germany’s terror.

The senseless killing continued in Hitler’s inner circle. As ordered, the five pups of Hitler’s dog Blondi were shot outside the Reich Chancellery bunker, along with two dogs belonging to Eva. Erna Flegel, a nurse who worked in the bunker, stated that the death of Blondi affected people in the bunker more than the death of Eva Braun [Hitler]. [791]

2 May 1945

The 7th Guards Army continued to hold its position at the front while shifting units around.

Fighting in Italy officially ended as over one million Axis troops surrendered to the Allied armies. The growing numbers of POWs were carefully screened for Nazi leaders, individuals with SS connections and other wanted criminals. In the coming weeks and months, many of these former soldiers were released back into civilian life, while others languished in makeshift POW camps throughout Europe for much longer.

The RAF launched the last major bombing raid of the European war against the city of Kiel.

General Chuikov accepted the unconditional surrender of Berlin's garrison. A few remaining Axis troops in the city continued to hold out, with scattered fighting in the ruins lasting several days.

Further south, at the foot of the Alps, General Walter Dornberger (director of the Peenemunde research center, V-1, V-2 weapons), Wernher von Braun (lead rocket scientist) and about 1,600 other German rocket scientists surrendered to the American Army. The war was over for them, but they would soon be put to work building rockets for the American military. [792]

3 May 1945

The 7th Guards Army continued to hold its position.

British forces of the 2nd Army, commanded by General Miles Dempsey, captured the major port city of Hamburg in northern Germany.

4 May 1945

The 7th Guards Army continued to hold its position, but the strain must have been terrible as the soldiers wondered whether the war really was over.

The 2 UF issued Order #7982 transferring the 1505 and 1897 Self-Propelled Artillery Regiments (SUR) from the 46th A to the 7th Guards Army. These two SURs arrive in the 7th Guards Army zone on 6 May and were assigned to work together and along with the 27 GTB. The 1505 SUR contained 14 SU-76, 2 SU-85, and 1 BA-64 light armored car. The 1897 SUR contained just 13 SU-76 self-propelled guns according to TsAMO record ID #132313700, page 2. With this transfer the 7th Guards Army gained 30 AFVs, which were needed as the 27 GTB had only a mixed group of 9 operational AFVs.

Over one million German troops in northwest Germany, the Netherlands and Denmark surrendered to British Field Marshal Bernard Montgomery, Commander of the 21st Army Group.

5 May 1945

The Prague Uprising began innocently enough when at 06:00 hours "a speaker on Czech radio opened the daily broadcast by announcing the time of day in the usual mix of Czech and German [as required]. As the day progressed, the hours were [eventually] given in Czech only." This is how the uprising that would free Prague from Nazi rule began. After this signal, rioting and armed combat soon followed in the streets of this great city. [793]

The 7th Guards Army held its position.

German forces in Bavaria and southwest Germany surrendered to the Western Allies.

Directive #11081 arrived from *Stavka*, which transferred the 9th GA to 2 UF and put it between the 7th Guards Army and the 46th A. The Directive read:

> [Regarding] *the VGK Stavka* Directive *No. 11081 to the Commanding troops 2nd and 3rd Ukrainian Fronts, representative Stavka for submission of the 9th Guards Army and development attacks on Pilsen and Ceske Budeevice*
>
> *May 5, 1945 02 h 00 min*
>
> *The headquarters of the Supreme Command orders:*
>
> *1. Commander of the 3rd Ukrainian Front, 9th Guards Army* [commanded by General Vasily Glagolev] *consisting of 37, 38 and 39 Guards* [RCs] *(a total of nine divisions) with all army reinforcement units, rear units, institutions and cash reserves no later than 24.00 on May 7 this year. Transfer to the left bank of the Danube, concentrate in the area of Korneiburg, Floridsdorf, and transfer to the troops of the 2nd Ukrainian Front.*
>
> *2. To the Commander of the 2nd Ukrainian Front:*
>
> *a) 9th Guards Army Glagolev to enter between the 7th Guards and the 46th Armies and used for the offensive in the direction of Neubistritz, Tyn and further Pilsen;*
>
> *b) 46th A* [commanded by General Alexander Petrushevsky] *set the task to reach the river Camp, and under favorable conditions - to advance in the general direction of Ceske Budejovice.*
>
> *3. To establish from 24.00 on May 5 the next dividing line between 4th and 2nd Ukrainian fronts: before Gulin - the former, then Boskovice, Zdirets, Choteborzh, Ricany (all points for the 2nd Ukrainian Front inclusive). The responsibility for ensuring the junction between the 4th and 2nd Ukrainian Fronts shall be left to the commander of the 2nd Ukrainian Front.*
>
> *4. To inform about the given orders.*
>
> *Headquarters of the Supreme Command I. Stalin A. Antonov*

This directive signaled a major shift in focus to the north and highlighted *Stavka's* desire to push quickly into Czechoslovakia before the Western Allies could do the same. That is why *Stavka* transferred both the 6th GTA and the 9th GA north of the Danube. Although the war to conquer Germany and destroy Nazism was nearly finished, the contest to control the fate of central Europe had just begun. The 9th GA soon would appear on the left flank of the 7th Guards Army.

At 17:00 hours, in Flensburg, Reich president Dönitz informed Heinrich Himmler that he was "'forgoing his services' and that he considered 'all ties between him and the current government dissolved.'" Up to this point, President Dönitz felt "that a break would be inadvisable as long as Himmler controlled the police, Gestapo and the SS." [794]

6 May 1945

The 7th Guards Army continued to hold its position, but during the night and into the early morning hours of 7 May, assault troops moved into position. Offensive operations were about to resume as *Stavka* launched its Prague offensive. Soon the entire 1, 2, 3 and 4 UFs would advance into Czechoslovakia.

Hermann Göring, the former commander in chief of the Luftwaffe among other military posts, was captured by American troops. Göring would eventually stand trial at Nuremberg for his many crimes, where he was found guilty and sentenced to hang.

Units of the U.S. 3rd Army, commanded by General George S. Patton, captured the city of Pilsen, which was a major Czechoslovakian arms production center. Pilsen was also a point on the Pilsen-Budweis-Karlsbad boundary/stop line that General Patton was ordered not to cross. General Eisenhower was concerned about starting an accidental conflict with the Russians and did not want an international incident. The area around Pilsen later was shifted into the U.S. 1st Army zone. (See Map #76.)

Fighting within Prague continued and intensified. When news of the uprising reached the new Reich president Admiral Dönitz at 10:00 hours, he ordered AG Center [and AG South] to "withdraw as far west as possible as quickly as possible to save the greatest number of German soldiers [from being captured by the Red Army]." [795]

7 May 1945

At 02:41 hours (German time) "*Heeresgruppe Süd* [HQ ordered its formations] to break contact with the Russians as quietly and inconspicuously as possible behind a weak screen of security forces and [to move] to the west [and surrender] to approaching American forces, but without engaging them in combat." [796] The end of the struggle quickly was approaching for AG South. Forces from an "American armored reconnaissance group under the command of a friendly young lieutenant had reached the field-army group's headquarters" [797] and accepted its surrender. Before its capture, *Heeresgruppe Süd* HQ, under the command of General Lothar Rendulic, transferred command authority to the 6th Panzer Army.[798] (See Map #74 for the 7th Guards Army advance and Map #75 for the overall Strategic Situation on 7 May 1945.)

Early in the morning, General Shumilov called his corps HQs and asked about their mood and the mood of the troops. After verifying the timing of the attack, and being in a good mood himself, Shumilov wished them all good luck! At 07:45 hours [Russian time], a massed artillery barrage opened against German positions along the Diya River. At 07:50 hours all the anti-aircraft artillery of the army and AA machine guns joined the barrage, firing upon the most forward enemy positions. Then at 08:00, supported by air attacks, the right flank of the 7th Guards Army, 27 GRC, attacked out of its bridgehead position. Soon the 23 RC also become active as its rifle units assaulted across the Diye River. After breaking the resistance that was offered, units of the 7th Guards Army achieved an eight-mile advance on the first day: the advance continued into the night. [799]

Despite orders just to screen the front and then withdraw, the German defenders of the Diye River line put up a vigorous fight. The 7th Guards Army achieved a breakthrough, but the cost was paid in blood. During the evening, the 23 RC replenished its rifle divisions. The 19 RD received 850 replacements, the 25 GRD added 680 people, and the 252 RD received 480 people. The 27 GRC must have suffered similar losses. [800]

Reich president Admiral Dönitz, as Hitler's chosen successor, officially surrendered Germany to the Allied powers in Reims, France, at General Eisenhower's (SHAEF) headquarters. Nazi German had been utterly defeated. The War in Europe was over. VE Day (Victory in Europe) was celebrated all over the world. As agreed upon, the fighting would officially end at 23:01 hours 8 May 1945. But a full Soviet delegation was not present at the surrender ceremony, so the fighting along the Russian Front continued for another day.

During May 1945, a mass expulsion of ethnic Sudeten Germans from Czechoslovakia began. Various parties (internal and external) were responsible for this ethnic cleansing, which ultimately removed over 2.3 million Germans from their homes and resulted in the deaths/murders of hundreds of thousands of these refugees. [801] The human cost of World War II continued to rise.

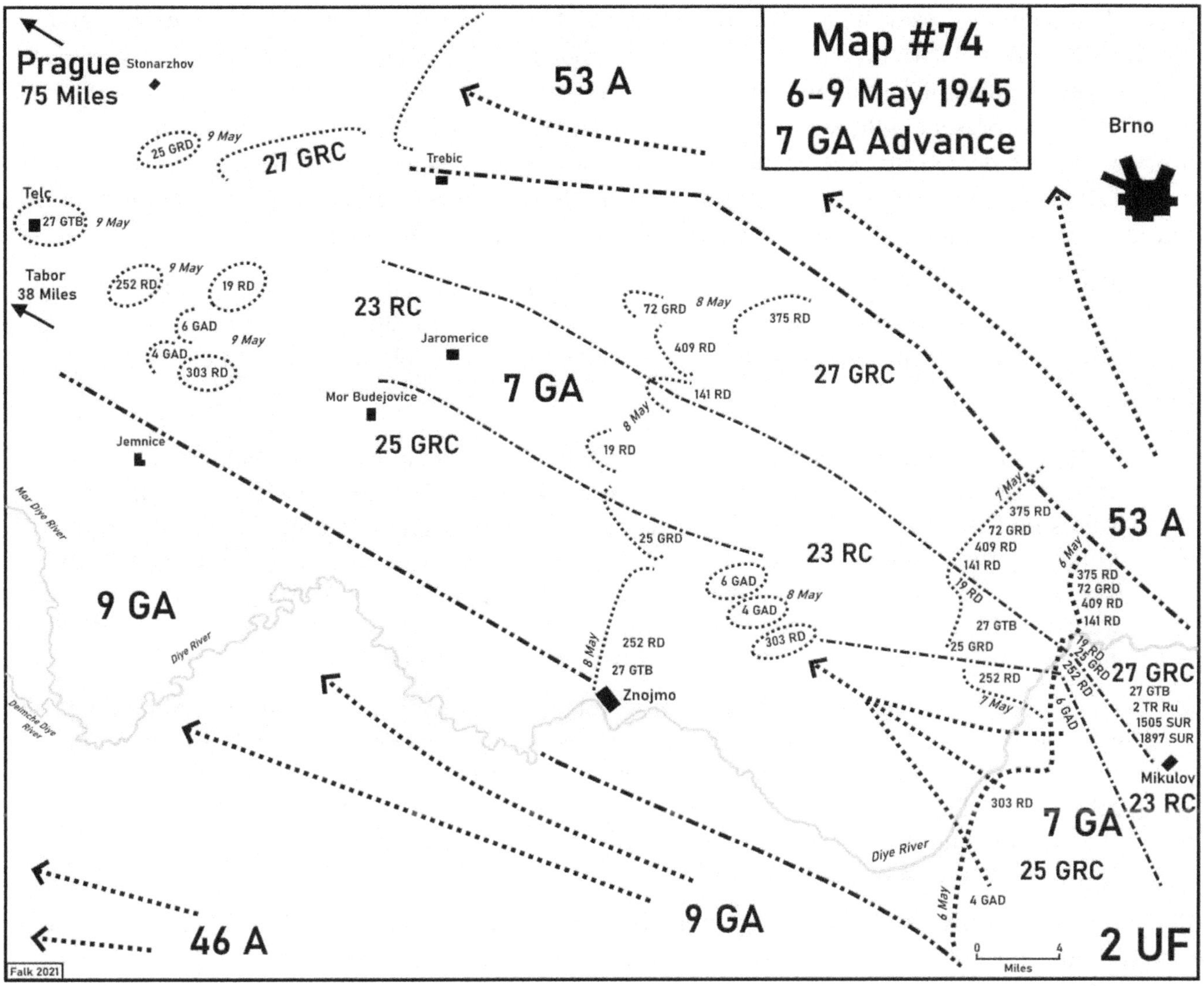

8 May 1945

The 25 GRC joined the offensive and together with the 23 RC and the 27 GRC advanced a further 19 miles against fading resistance. The rifle divisions of the 25 GRC moved in behind the advancing formations of the 23 RC and concentrated to the east of Znojmo. Axis forces withdrew as quickly as possible, hoping to reach and to surrender to Western Allied armies and not fall under the control of the Soviets. The 9th GA, 46th A and 53rd A also began their attacks, thus supporting both the left and right flanks of the 7th Guards Army. Also, on or about 8 May, by order of the 2 UF, the boundary between the 9th GA and the 7th Guards Army shifted north. With the disintegration of the Axis front line, friendly and enemy units were becoming entangled in the resulting chaos. The question then became one of telling who was resisting, who was trying to escape, and who just wanted to quit the war and surrender.

As the 9th GA advanced, it gradually moved further to the south, thus pulling away from the left flank of the 7th Guards Army. Normally this would lead to problems, but with the enemy fleeing there was little chance the advancing victorious Red Army could be stopped or even interfered with in a meaningful way.

The 6th GTA also was operating freely within the zones of both the 7th Guards Army and the 57th A. (The movements of the 6th GTA are not displayed on maps.)

Annoyed that Admiral/President Dönitz had surrendered Germany to the Western Allied armies with no official Soviet representative present, Stalin insisted upon another ceremony in Berlin, before a proper Soviet delegation. General Eisenhower did point out to Stalin that Soviet General Susloparov, a Soviet representative, was present at the surrender ceremony and had signed the surrender document. But Stalin would not be denied, so once again the highest military representatives of Nazi Germany assembled and surrendered, this time to a proper Soviet delegation in bombed out Berlin. With the conclusion of this ceremony, the Great Patriotic War of the Soviet Union ended.

The 7th Guards Army and the 27 GTB were given honors for the liberation of Znoymo and other cities. By order of the Supreme Command on 8 May 1945, they were given gratitude and a salute in Moscow with 20 artillery salvoes from 224 guns. Stalin personally approved each Moscow artillery salute.

The 23 RC reported that

> In the afternoon we received a message signed by the commander of the 2nd Ukrainian Front, Marshal of the Soviet Union R. Ya. Malinovsky. It read: "On May 7, 1945, in the city of Reims [at Eisenhower's SHAEF HQ], the Nazis signed an act of unconditional surrender, which entered into force on May 9 at 1.00 Moscow time. Further, after 23.00 on May 8, we were instructed to demand from the command of the opposing fascist troops to lay down their arms and surrender. If this requirement was rejected, it was necessary to continue performing the previously assigned task until the enemy was completely destroyed.
>
> The Hitlerite command refused our offer of surrender. Therefore, General M.S. Shumilov gave the order to the army to continue the offensive. Not encountering much resistance, the corps formations turned the battle formations into marching columns and, sending reinforced detachments forward, began to pursue the enemy. During May 9, we moved 60–65 kilometers [37-40 miles] northwestward. The troops of our front, including, of course, the 7th Guards Army, were now aimed at Prague. [802]

For the 7th Guards Army, the final military breakthrough of defenses along the Diye River lead to the liberation of the cities of Znojmo, Yaromyerzhitsye, Moravian, Krumlov and Miroslav. Most of these locations fell without a fight. Note: Most of these cities are small and are not located on maps within this book.

Within Prague, a ceasefire agreement between the Wehrmacht, SS and Czech combatants was reached at 16:00 hours. German soldiers and civilians were now free to leave the city and the surrounding area. They, of course, hurriedly fled to the west and away from the advancing Soviets. [803]

9 May 1945

After a brief hope that the fighting really was over, units of the 7th Guards Army resumed their advance and gained another 30-40 miles to the west. German forces were surrendering en masse to Allied armies in the west while still resisting, as best they could, the Soviet advance from the east.

The 7th Guards Army now was conducting a military pursuit against a defeated enemy army, sweeping through Czechoslovakia while meeting limited opposition. Combat units were busy gathering surrendering enemy troops or destroying any isolated pockets of resistance.

Prague, the capital of Czechoslovakia, was captured/liberated/occupied by the advancing Red Army. But the reality was the popular Czech uprising had driven most German forces from the city the day before Russian troops arrived. German troops had fled the city but many vulnerable German civilians remained behind. These noncombatants were now at the mercy of a vengeful population and the changing political situation.

Before the surrender of Germany went into effect at midnight on 9 May 1945, Admiral Dönitz had in the end succeeded in saving large numbers of German soldiers from capture by the Soviets. Approximately 1,850,000 German soldiers on the Eastern Front were able to reach the Western Allied lines. "In total more than half of [the] German troops on the Eastern Front were able to surrender to the western allies." [804]

Map #75 - 7 May 1945 Strategic Situation

Note: The displayed national borders are pre-war.

As can be seen on Map #75, German forces at the time of the surrender held a large bridgehead in Latvia, which was called the Courland Pocket. They also occupied a large part of western

Czechoslovakia, the central part of Austria and a bit of Yugoslavia. Excluding Norway, Allied Armies had overrun and occupied all other territories.

10-14 May 1945

Once again, the 28 Separate Armored Train Battalion transferred out of 7th Guards Army control, but the 38 Separate Armored Train Battalion replaced it. The reason for these constant reassignments is unknown. Also, on or about this date, the 180 RD was assigned to the 7th Guards Army.

Overall, the 7th Guards Army was busy occupying different key areas of Czechoslovakia within their zone of operation. This included setting up roadblocks to help capture enemy troops, destroying any resistance that they encountered, liberating anything of value, and advancing in the general direction of Western Allied armies.

11 May 1945

In Flensburg, after the official surrender of Germany to the Allies, Heinrich Himmler, former head of the SS and a major proponent of the Holocaust, came out of hiding wearing shabby clothing along with a fake paybook identifying him as "Sergeant Henirich Hitzinger." He slipped out of town heading south with several companions. [805]

15 May 1945

During their advance, units of the 7th Guards Army finally encountered American troops from the U.S. 3rd Army (General Patton) east of Pilsen in the area around the town of Rokochaniy. East meets West! This historic meeting took place on or about 15 May 1945; the meeting also might have included units from the U.S. 1st Army (General Hodges). (See Map #76.)

Similar first meetings took place all along the front, with similar outcomes: smiles, hugs, dancing, drinking and the realization that the war was really over, and they had survived!

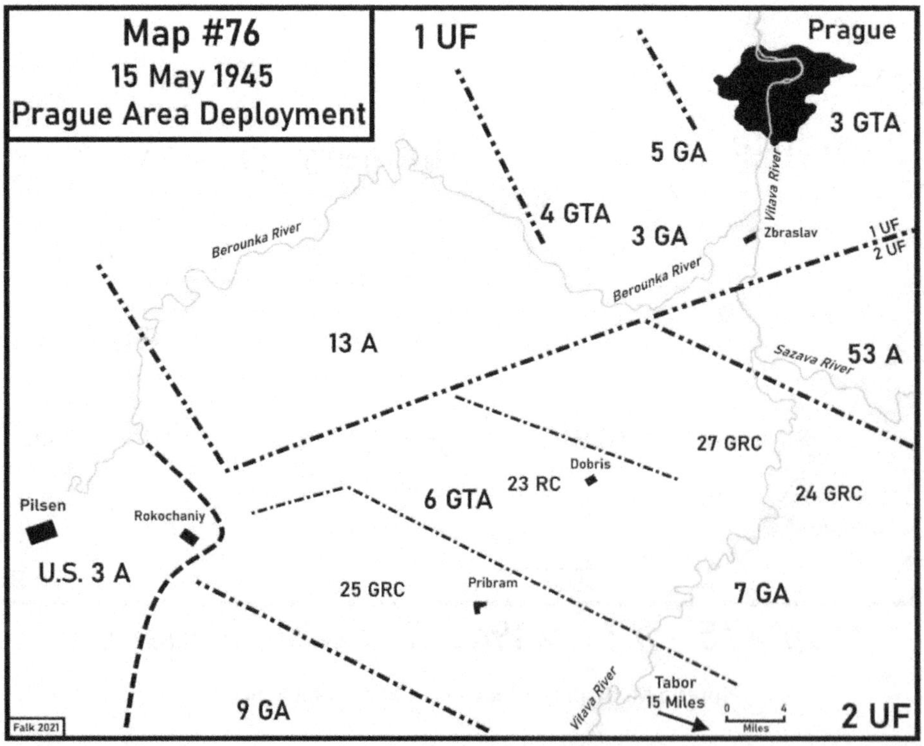

Note: The deployment of Soviet units on Map #76 are approximate as *Stavka* and the 1 and 2 UFs were making changes almost daily.

16 May 1945

The official statement said that the 7th Guards Army helped in the removal of German troops from the territories of Czechoslovakia, but many Axis troops fled to the west and surrendered to the Americans or other Western Allied force while Soviet forces captured others before they could escape. These unfortunate ones disappeared into the Soviet Union for many years, serving as slave labor to rebuild all the destruction caused by the war. Countless POWs would never return home, but many of the survivors had to wait until the 1950's to be released.

At this point, the 7th Guards Army terminated its military actions against the enemy and assumed a more reduced/relaxed military posture.

The 7th Guards Army's Guards Colonel Dolin issued an intelligence report, *Information about prisoners and trophies of the 7th Guards Army, 9-16 May 1945* reported in TsAMO document ID #133375147. The report lists that the Army captured/liberated the following:

Officers – 1,135
NCOs – 11,490
Soldiers – 79,193
Rifles and SMGs – 15,000 with 50% serviceable
Machine Guns – 1,200 with 30% serviceable
Mortars – 42 with 40% serviceable
Guns [of all types] – 181 with 40% serviceable
Tanks and self-propelled guns 45 with 20% serviceable
AFVs and Armored cars – 112 with 20% serviceable
Cars – 3,000 with 5% serviceable
Tractors – 27 with 3% serviceable
Motorcycles – 560 with 10% serviceable
Different warehouses – 26
Combat horses – 177 with 16 artillery horses
Draft horses– 614 (for a total of 807 horses)

During these seven days the Army captured 91,818 enemy troops and a sizeable amount of military equipment. Note the low serviceability rate for captured vehicles, the rounding off of numbers and the keen interest in horses (combat vs. artillery vs. draft horses). Ironically, after a war, a good draft horse was far more valuable and useful than an armored car or a tank.

19 May 1945

Soviet forces overwhelmed the last German pocket of resistance holding out in central Czechoslovakia.

With the surrender of Germany, Table #42 shows the massive number of tanks and SUs the Red Army had on hand to continue the war if that had been necessary, i.e., See Operation Unthinkable, 27 March 1945.

Tanks	Number
Heavy Tanks	5,300
Medium Tanks	11,000
Light Tanks	8,800
Total Tanks	**25,100**
SUs	**Number**
Heavy SUs	2,700
Medium SUs	1,900
Light SUs	5,400
Total SUs	**10,000**
Grand Total	**35,100**

Table #42: Soviet Tank and SU Totals at the End of the War – 1945. [806]

20 May 1945

The 38 Separate Armored Train Battalion was removed from the 7th Guards Army for unknown reasons.

The Army issued a *Summary information on the damage inflicted on the enemy by the troops of the 7th Guard Army during the entire period of participation in the Patriotic War,*

Axis Forces and Equipment	Killed/Destroyed	Captured	Total
Officers and Soldiers	253,300	276,082	529,382
Tanks and SP Guns	1,748	336	2,084
APCs and Armored Vehicles	498	182	680
Different Caliber Guns	1,604	2,365	3,969
Different Caliber Mortars	1,386	1,005	2,391
Six-Barreled Mortars	-	24	24
Machine Guns	5,068	7,804	12,872
Armored Trains	1	1	2
Motor Vehicles	3,476	10,160	13,636
Aircraft	206	28	234
Tractors	37	309	346
Motorcycles	318	1,784	2,102
Radios	3	140	143
Searchlights	-	6	6
Rubber Boats	-	100	100
Metal Pontoons	-	102	102
Wooden Pontoons	-	40	40
Boat Motors	-	14	14
Aviation Engines	-	900	900
Steam Locomotives	10	94	104
Armed RR Trains	-	16	16
Railway Wagons with Military Equipment	-	375	375
Empty Railway Wagons	-	300	300
Steam Boats	-	15	15
River Boats	2	-	2
Different Types of Warehouses	-	138	138

Table #43: Damage Inflicted by the 7th Guards Army on the Enemy between 20 July 1942 – 20 May 1945.

Table #43, TsAMO record ID #132636324, covers 1,035 days of combat from 20 July 1942 to 20 May 1945, which also includes the 64th Army. This Damage Inflicted report is the 7th Guards Army's opinion on what they achieved during the war, but over such a long period of time accuracy is obviously an issue. The Army also claimed to have captured 44 generals and 11,938 other officers. This is also an abbreviated report which omits many common items like rifles, pistols, ammunition, food, horses and so on. Notably, during this entire 1,035 - day period, the Army only destroyed (shot down) 206 aircraft and did not destroy any German six-barreled mortars (Nebelwerfer rocket launchers), but only captured 24 of these highly mobile weapons. Overall, the report is not only interesting but also revealing of what a Guards Army could accomplish.

In contrast to the 7th Guards Army's damage inflicted report, historical monthly casualty reports, partial monthly reports and estimates reveal the approximate number of casualties suffered by the 7th Guards Army between March 1943 and April 1945, or during 26 months of combat. The total casualties, i.e., killed, wounded, sick/injured, and missing during these 26 months were 266,198 people, for an average of 10,238 casualties per month.

The Presidium of the Supreme Council awarded [the Army] orders and medals to 134,603 people, including 220 Hero of the Soviet Union awards, according to TsAMO record ID #131305098, pages 24 and 25.

To facilitate further historical research, the total reported and estimated casualties suffered by the 7th Guards Army are broken out into individual months in the following table.

Month	Casualties		Month	Casualties
March 1943	654		May 1944	7,569
April 1943	1,944		June 1944	1,131
May 1943	1,528		July 1944	2,992
June 1943	2,560		August 1944	5,054
July 1943	38,300		September 1944	12,462
August 1943	25,185		October 1944	11,119
September 1943	11,683		November 1944	13,322
October 1943	29,547		December 1944	13,758
November 1943	12,877		**1944 Total**	**86,499**
December 1943	6,280		January1945	9,120
1943 Total	**130,558**		February 1945	11,473
January 1944	12,059		March 1945	8,767
February 1944	980		April 1945	19,781
March 1944	5,925		**1945 Total**	**49,141**
April 1944	128		**Grand Total**	**266,198 Troops**

Table #44: Monthly Casualties for the 7th Guards Army between March 1943 and April 1945.

Note that as the war developed between 1943-1945, Axis forces became weaker on the Eastern Front and Soviet forces became more skilled and much better equipped. Therefore, the average monthly casualties for the 7th Guards Army in 1943 totaled 13,000 per month, then in 1944 fell to 7,000 per month before finally rising again in 1945 to 12,000 per month.

Many factors lead to the defeat of the Nazi and Fascist regimes in Europe. The combined Allied superiority in logistics, resources, armaments, and food production along with the total work and fighting force that was employed against Axis forces, in the end, was overwhelming.

21-22 May 1945

Disguised as a common soldier, Heinrich Himmler, former head of the SS and *Gestapo*, was captured by a British Army patrol as he attempted to flee Germany. Himmler and his associates were then moved to several different British camps for interrogation. After three days of captivity, Heinrich Himmler, the man that sent millions of people to slave labor camps and murdered millions more in concentration camps, committed suicide by biting down on a hidden cyanide pill during a physical examination. He was 45. [807] Another leading Nazi escaped the hangman's noose.

On the other hand, Otto Johann Anton Skorzeny, Hitler's favorite commando, also was captured at the end of the war. But after being acquitted of war crimes during the Battle of the Bulge, he escaped from an internment camp in 1948 and enjoyed an interesting post war career. Eventually he settled in Spain, dying of lung cancer there at the age of 67 in July 1975.

23 May 1945

Meanwhile, the 7th Guards Army took care of its troops.

> So the war ended. Starting from Prague - through Czechoslovakia, Austria, Hungary - we walked home to the accompaniment of a brass band, marching, proud and cheerful. At the final point of our route, in the area of the town of Monor, not far from Lake Balaton, we set up a camp in the forest and arranged an official send-off of the fighters ...[808]

During the war, the editorial board of the 7th Guards Army newspaper "For the Motherland" produced 612 issues with a total circulation of 743,000 copies. They also wrote and published 20 leaflets dedicated to the exploits of soldiers. In addition to articles by trained journalists, "on its pages were [2,431 military] authors; 2,511 military letters were published, [authored] by 1,255 soldiers and sergeants [and] 1,176 officers." [809]

The 1st Guards Mechanized Cavalry Group (5 RC, 4 GCC, 6 GCC, 7 MC) moved up and were deployed in the 7th Guards Army and 53rd A zone just southwest of Prague.

29 May 1945

Directive of the VGK Stavka No. 11096 to the Commander of the 1st Ukrainian Front on the Rename of the Front to the Central Group of Forces and its Composition
May 29, 1945 03 h 30 min

> *The headquarters of the Supreme Command orders:*
>
> *1. Rename from 24.00 on June 10th. d. 1st Ukrainian Front in the Central Group of Forces. The headquarters of the group from the same time to be in the Vienna area. The commander of the 1st Ukrainian Front, Marshal Konev, is called the commander-in-chief of the Central Group of Forces.*
>
> *2. The central group of troops should be composed of: 5th Guards. Army - 32 Guards. RC (13 Guards, 95 Guards, 97 Guards Rifles), 33 Guards. RC (9th Guards Airborne Division, 14th Guards, 78th Guards Rifle Division), 34 Guards. RC (15 Guards, 58 Guards, 6 Guards Rifle); 7th Guards. Army - 25th Guards RC (6th Guards Airborne Division, 4th Guards Airborne Division, 25th Guards Division), 27th Guards. RC (72 Guards, 42 Guards, 81 Guards Rifle), 23 RC (19, 252, 303 Rifle Division).*

30-31 May 1945

TsAMO record ID #136672255 shows that on or about 31 May 1945 the 7th Guards Army gained control of the 83 Naval Rifle Brigade with 1,377 people. Even at this late date, *Stavka* and/or Front HQs were switching around units and formations.

Epilogue

1 June 1945 and beyond

In the aftermath of this great and terrible war, the Soviet Union undertook the burden of rebuilding its devastated lands. But not all were welcomed.

> For example, … from major cities disappeared [the] disabled [soldiers of] World [War] Service, almost all [of them] and almost overnight. In order not to spoil the appearance of the socialist country, not to undermine the faith in the bright tomorrow and not to overshadow the memory of the great Victory. According to sources, the mass withdrawal of disabled people outside the city line [of Moscow and other cities] happened in 1949, on the 70th anniversary of Stalin's [birthday]. In fact, they were caught from 1946 until Khrushchev's time. You can find reports [from] Khrushchev himself about how many legless and armless beggars … [were] filmed, for example, on the railway and the numbers there are thousands. Yes, not everyone was taken out. They took those who had no relatives, who did not want to burden their relatives with the care of themselves or from whom these relatives because of the injury refused. Those who lived in families were afraid to appear on the street unaccompanied by relatives, so that they would not be taken away. Those who could - moved from the capital to the outskirts of the USSR, because, despite the disability, [they] could and wanted to work, to lead a full life. The island of Valaam, 200 kilometers north of Svetlana in 1952-1984, is the site of one of the most inhumane experiments to form the largest human "factory." Here, in order not to spoil the urban landscape, [were transferred the] disabled [and the diseased]. It was believed that disabled people spoil the appearance of Soviet cities. Valaam was [just] one [of these locations] but [it was] the most famous of the dozens of places linked [to] the disabled [of the] war. [810]

See the following link to read and see the art work from the above article. https://fishki.net/2109354-istorii-invalidov-vov-kotoryh-soslali-na-valaam-za-ubogij-vid.html. (Best viewed with Google Chrome for instant translation of text).

10 June 1945

Following orders from *Stavka*, the 2nd Ukrainian Front disbanded. The 7th Guards Army now became part of the Central Group of Forces commanded by Marshal Ivan Konev.

11 June to 9 July 1945

From 11 June to 9 July (29 days), the 7th Guards Army transferred its troops and operations from Czechoslovakia to Hungary, moving the Army HQ to the city of Veszprem, just north of Lake Balaton. After a 250-mile road march, the 7th Guards Army, with three rifle corps, concentrated in the area around Lake Balaton. Soon after arriving, units of the Army were deployed in a defensive line facing Yugoslavia. The Army would spend the coming winter in this new location. (See Map #77.)

The transfer of the Army south and the formation of a defensive line facing Yugoslavia was strange. The reason for this action might be related to Stalin learning about Operation Unthinkable from Soviet spies in Great Britain. [811] (See 27 March 1945 entry for Operation Unthinkable.) The real cause for concern was most likely the Communist Party of Yugoslavia (CPY) partisan forces under the command of Josip Broz Tito. At this time, Tito's Yugoslav partisans were challenging British forces for control of the Italian city and port of Trieste on the Adriatic Sea. Stalin and or Stavka might have thought that any open conflict between these two forces might trigger Operation Unthinkable. It was best to be prepared for any eventually.

Note: The displayed national borders are post-war.

13 June 1945

While the 7th Guards Army moved into Hungary, *Stavka* and the new Central Group of Forces made other changes. On 13 June 1945, the 7th Guards Army issued order No. 0123.op accepting the 4 Bulgarian Infantry Corps with the 12 and 16 Bulgarian Infantry Divisions into the Army. The 4 Bulgarian Infantry Corps was being transferred from the 3 UF. This is found in TsAMO record ID #132313922 of 35 pages. This union would not last long because on 29 June 1945, the 7th Guards Army issued order No. 0128.op sending the 4 Bulgarian Infantry Corps and attached units from Hungary back to Bulgaria, according to TsAMO record ID #135745563, of 8 pages. This action, at first, might appear odd, but the 7th Guards Army was apparently acting as a holding or controlling Army. As nearby fronts, armies and corps disbanded, a high-level headquarters needed to take temporary command and/or control of random units while they processed out of the active army. Another record from 16 June, TsAMO ID #132313998, *Certificate of Arriving in the 7th Guards Army of Armored Units*, documents the temporary assignment of nine different armored units to the Army from all over, including some from the 4 UF. These units, too, soon would be disbanded and the troops and equipment redistributed as needed.

25 June 1945

The United Nations Charter was unanimously adopted in San Francisco, California. With this the United Nations was formed. Its mission is

> *to save succeeding generations from the scourge of war,...to reaffirm faith in fundamental human rights,...to establish conditions under which justice and respect for the obligations arising from treaties and other sources of international law can be maintained, and to promote social progress and better standards of life in larger freedom.* [812]

1-6 July 1945

British, French, American and Soviet troops moved into their assigned zones within greater Berlin. The victorious Allied forces would now jointly garrison Berlin for the next 45 years until 1990, with Western Allied forces remaining in Berlin (at Germany's request) until all Soviet forces withdrew from German soil in 1994.

16 July 1945

The United States of America, in the state of New Mexico, at the Alamogordo Bombing Range at 5:29 a.m. successfully detonated a small implosion-type plutonium fission atomic bomb. The bomb released approximately 18.6 kilotons of energy, which was equivalent to 24,250 tons (2,000 pounds/ton) of high explosives. The Atomic Age had begun.

26 July 1945

Winston Churchill lost the British General Election and was ousted as Prime Minister. Clement Attlee took over as Prime Minister and as Head of Government.

6 August 1945

An atomic bomb called Little Boy, a small gun-type uranium fission bomb, was dropped by the United States on the Japanese city of Hiroshima. This single bomb destroyed a large portion of the city and killed tens of thousands of its people.

8 August 1945

Per agreement with the Western Allies, the Soviet Union officially declared war on Japan and immediately attacked Japanese forces in Asia.

9 August 1945

An Atomic bomb called Fat Man, a small implosion-type plutonium fission bomb, was dropped by the United States on the Japanese city of Nagasaki. Once again, this one bomb destroyed a large portion of the city and killed tens of thousands of its people.

15 August 1945

Imperial Militaristic Japan surrendered to the combined Allied nations with only one provision, that Emperor Hirohito remains on the throne. People all over the world celebrated VJ Day [Victory over Japan]. World War II was over!

September 1945

In September 1945, on the basis of the order of the Supreme Commander-in-Chief, among other formations of the 7th Guards Army, the Ural 375th [RD] was disbanded. The personnel were demobilized. The younger soldiers were transferred to the 81 Guards Rifle Division. [813]

October 1945

The 7th Guards Army HQ began logistical planning for the transportation of units and formations of the Army by rail (destinations unknown). This planning would continue into December 1945.

January 1946

Units and formations of the 7th Guards Army began leaving Hungary by rail for various other deployments. Some units were disbanded with others transferred to other armies and locations, but the Army HQ itself was destined to survive far into the future.

February 1946

In February 1946, General Shumilov was transferred out of the 7th Guards Army to assume other commands. The 7th Guards Army's combat units were dispersed or disbanded, and the 7th Guards Army HQ was transferred to the Transcaucasus Military District where it would stay an active Army HQ for the next 46 years. The 7th Guards Army and its HQ were officially disbanded on 14 August 1992 after the breakup of the Soviet Union. [814]

15 October 1946

Hermann Göring, former leader of the Nazi Party, creator of the Gestapo, and commander of the Luftwaffe, died by suicide after taking a hidden cyanide pill on the night before he was to be hanged for war crimes. He was 53.

5 March 1953

At 09:50 hours, Joseph Vissarionovich Stalin, Communist dictator of the Soviet Union and murderer of millions of Soviet citizens, died of complications from a stroke in Moscow. He was 74.

24 January 1965

Sir Winston Leonard Spencer Churchill, solider, statesman and twice Prime Minister of Great Britain, died in London following a stroke. He was 91.

28 June 1975

Colonel-General Mikhail Stepanovich Shumilov, commander of the 64th Army and the 7th Guards Army during World War II, continued to serve in the Red Army after the war. In 1946, he became Commanding Officer of the 13th Army, in 1948 Commander-in-Chief of the White Sea Military District and in 1949 Commander-in-Chief of the Voronezh Military District until 1956 when he retired. In 1958 he was recalled to service and served until 1975 as Inspector-Counselor, Group of Inspector-Generals, Ministry of Defense.[815] General Shumilov died on 28 June 1975 at the age of 79 in Moscow. He was buried on Mamayev Kurgan hill, site of intense fighting during the battle for Stalingrad. Stalingrad had meanwhile been renamed Volgograd in 1961 after Stalin's death.

18 March 1982

Marshal of the Soviet Union (as of 1955) Vasily Ivanovich Chuikov, commander of the 1st Reserve Army, the 64th and 62nd Armies and the 8th Guards Army during World War II, died on 18 March 1982 at age 82. To this day, General Chuikov lies next to his old friend and former Commander General Shumilov on Mamayev Kurgan hill, Stalingrad/Volgograd, where this story began.

7 January 1989

Emperor Shōwa (a.k.a. Emperor Hirohito), Emperor of Imperial Japan, died in Tokyo. He was 97.

Appendix

Appendix 1

To use the ID # method, call up any TsAMO record and replace the ID number in the command line (at the top of the page) with the ID number of the record you are trying to view. For example, the ID # is shown in **Bold**: https://pamyat-naroda.ru/documents/view/?id=**133391761**&backurl....

Appendix 2

COMMANDING OF STALINGRAD GROUP OF FORCES, BY FORCES OF VORONEZH FRONT, ABOUT THE REASSIGNMENT OF THE ARMY

On February 28, 1943. 05 h of 30 min
The People Commissioner of Defense ordered:

1. To direct 64 Army at the disposal of the Voronezh front. In army to include 15 GRD, 36 GRD, 204, 29, 38 and 422 RD, 27 Guards Tank Brigade, 1111 and 156 Cannon Artillery Regiment, 186, 493, 496 and 500 Antitank Artillery Regiment and 838 Artillery Observation Battalion.

2. Sending 64 Army and reinforcing means to produce by the following order:

a) Administration of army (HQ) - 20 echelons (trains), loading at Station Beketovskaya, starting at 06.00 hours on 2.3. 1943, loading at the rate of 2 trains/day, trains № 25201 – 25221.

b) 15th GRD, 10 echelons, loading – Station Sarepta, at 06.00 hours on 2.3.1943, rate – 3 trains/day, trains № 25221 – 25230.

c) 36 GRD, 7 echelons, loading – Station Zaplavnaya, at 06.00 hours on 5.3.194, rate – 3, trains № 25281 – 25287.

d) 29 RD, 6 echelons, loading – Station Voroponovo, at 12.00 hours on 8.3.1943, rate – 3, trains № 25241 – 25246.

e) 38 RD, 7 echelons, loading – Station Elshanka, at 18.00 hours on 5.3.1943, rate - 3, trains № 25233 – 25239.

f) 204 RD, 6 echelons, loading – Station Zaplavnaya, at 06.00 hours on 3.3. 1943, rate - 3, trains № 25274 – 25279.

g) 422 RD, 6 echelons, loading – Station Zaplavnaya, at 12.00 hours on 7.3.1943, rate - 3, trains № 25291 – 25296.

h) 1111 Cannon Artillery Regiment, 2 echelons, loading – Station Gumrak, at 06.00 hours on 2.3.1943, rate - 2, trains № 25251 – 25252.

i) 156 Cannon Artillery Regiment, 2 echelons, loading – Station Gumrak, at 06.00 hours on 2.3.1943, rate - 2, trains № 25254 – 25255.

j) 186 AT Regiment, 1 echelon, loading – Station Gumrak, at 06.00 hours on 7.3.1943, train № 25259.

k) 500 AT Regiment, 1 echelon, loading – Station Gumrak, at 06.00 hours on 6.3.1943, Train № 25257.

l) 493 AT Regiment, 1 echelon, loading – Station Voroponovo, at 06.00 hours on 8.3.1943, train № 25261.

m) 496 AT Regiment, 1 echelon, loading – Station Elshanka, at 06.00 on 2.3.1943, train № 25263.

n) 838 Artillery Observation Battalion, 1 echelon, loading – Station Sarepta, at 06. 00 on 2.3.1943, train № 25265.

o) 27 Guards Tank Brigade, 3 echelons, loading – Station Gumrak, at 06.00 on 7.3.1943, rate - 3. train № 25248 - 25250.

Region of unloading for all parts - Valuyki. (Which is east of Belgorod)

3. *From the general composition of the administration 64 of Army to send 20 echelons, where to immerse the parts of control, connection, of engineer support and strictly necessary the part of army rear. Transportation of the remaining part of the army apparatus and establishments will be realized tentatively in the month.*
4. *Those sent to supply: by ammunition - 1 fire unit, by fuel - 1 servicing, by food and forage - on the route of 7 days and, furthermore, 3- days unloading reserve.*
5. *Komsomol of Voronezh Front to organize meeting, reception, accommodation of those arriving and to enroll them to all forms of allowance.*
6. *Commander of the troops of the Stalingrad Group of Forces prior to dispatch of the 64th Army to finish sending the 157th and 169th Divisions and the 11th Artillery division. Stop sending the 68th and 69th GRD. The loaded echelons of these divisions are to be returned to the loading station and unloaded. The empty cars to transport parts of 157th, 169th divisions and 11th artillery division.*
7. *On the progress of sending, arriving to report to the General Staff every day by 18.00. Deputy chief of the General Staff.*

Appendix 3

During World War II most trains were powered by a single steam engine because of control issues when linking more engines together. With two engines, the two different crews had to coordinate their efforts very closely or the engines would not pull together as one unit. In other words, if the crews were not careful, one engine would start to pull or push the other; at that point, the underperforming engine would become a load and not actually help move the train. Modern trains circumvent this problem with the use of interlinking communication systems and computers. These systems allow a single crew member to smoothly control multiple engines, even if they are dispersed throughout the length of the train.

Appendix 4

Boevoi Sostav Sovetskoi Armii – Combat Composition of the Red Army or BSSA tables list all designated units of the active fronts. Data was compiled the first day of each month; therefore, it provides a snapshot of information once a month for the entire Red Army. Further monthly references from this source are listed as BSSA - 42, 43, 44, 45 to designate the years of the records.

The links below provide access to these invaluable records:
http://www.soldat.ru/files/f/boevojsostavsa1941.pdf
http://www.soldat.ru/files/f/boevojsostavsa1942.pdf
http://www.soldat.ru/files/f/boevojsostavsa1943.pdf
http://www.soldat.ru/files/f/boevojsostavsa1944.pdf
http://www.soldat.ru/files/f/boevojsostavsa1945.pdf

Boyevoy Sostav Obshchevojskovykh Armiy - Composition of the Combined Arms Combat Armies or BSA data was compiled three times each month so it gives a bit more detail than BSSA data. Further monthly references from this source are listed as BSA - 42, 43, 44, 45 to designate the years of these records. These historical BSA files are difficult to find, and the author knows of no public source. Please note that BSSA and BSA records do not always agree with each other, which forces historians to reconcile the discrepancies.

Memory of the People, https://pamyat-naroda.ru/, is a Russian web site whose main goal is to "provide users the opportunity to get the most complete information about participants of the Great Patriotic War."

The TsAMO Archive, https://pamyat-naroda.ru/documents/, is a Russian language database that allows researchers to search for historical documents from World War II.

Appendix 5

When solid information from Soviet/Russian archives about combat and/or ration strengths is lacking or inconclusive, the author uses estimated strengths for the army, and these estimates are noted as such in the text. These estimates are based on research data that is then used to approximate the unknown values for units assigned. Listing these combat and ration strengths gives the reader an idea of the forces involved and the losses suffered. These values are not perfect. They are estimates, based upon known and estimated starting strengths, and then known and estimated losses during battle. These values are subject to change if confirmed data becomes available.

Appendix 6

At this time the SU weapon system was new to the Red Army. SU or *Samohodnaya Ustanovka* translates as Self-propelled installation. This was a direct copy of the German *Sturmgeschutz* assault gun design. The idea behind this type of vehicle was that taking the turret off a tank chassis allowed the mounting of a much larger weapon. The resulting mount had limited traverse capabilities, but the large caliber AT gun or howitzer provided a powerful direct support weapon for combat formations.

Appendix 7

Major General Mikhail Stepanovich Shumilov, who took over command of the 64th Army at the end of July 1942, would continue as its commander until 16 April 1943 when the 64th Army was awarded guards status becoming the 7th Guards Army. Shumilov continued to command this newly appointed 7th Guards Army. In October 1943, Shumilov was promoted to Colonel General and would remain in command of the 7th Guards Army until after the end of the war.

Appendix 8

Several Historical OOB charts for 15 April, 29 April, and 19 August 1943 from the TsAMO Archive show the 7th Guards Army controlling the 62, 63, 64, 65, and 66 Penal Companies and the 82, 83 84, 85, and 86 Blocking Detachments. The losses for all these units during the battle for Kursk appear to be unavailable, but we can assume they did take part in the fighting.

Of special note, the 66 Penal Company was listed as having 193 men missing in action. This large number of missing soldiers suggests that either most of these men surrendered (en masse) to the Germans on the first day of combat or they were surrounded and quickly captured. In either case, they became German POWs.

Appendix 9

Russian historian Boris V. Sokolov described the Soviet attitude about losses:

> The Stalinist system created not only the habit of carrying out any order without objection, but also and more frighteningly [the habit] of issuing orders without thinking. The tragic paradox was that the pursuit of victory at all costs with soldiers falling in the belief that they were bringing victory closer only further delayed the end of the war. Had our losses been fewer, and the ratio to those of the Germans more favorable […] our superiority in human resources would have enabled us to smash the German war machine and [to] achieve victory quicker than [what] actually happened. [816]

Appendix 10

Subsequently, the General Staff developed and the Supreme Commander-in-Chief approved three categories of salutes: 1st category - 24 volleys of 324 guns, 2nd - 20 volleys of 224 guns, 3rd - 12 volleys of 124 guns. Permission for each salute was given personally by the Supreme [Stalin]. With rare exceptions, Moscow saluted to the winners on the day of the expulsion of the enemy from one point or another. [817]

Appendix 11

The death of Hans Jeschonnek was not the first time a leading Luftwaffe commander shot himself. On 17 November 1941, Ernst Udet, Director-General of Equipment for the Luftwaffe, shot himself in the head. Alcoholism, issues with unfulfilled needs of the Luftwaffe in Russia (low aircraft production), and disagreements with Hitler and the Nazi party caused his suicide.

Appendix 12

Various bits of information about Operation *Rumyantsev* was obtained from the Axis History Forum Contributor, Der Alte Fritz, https://forum.axishistory.com// and RKKA in World War 2 Forum Contributors, Art (Artyom A.), and Alex (amvas) http://www.armchairgeneral.com/rkkaww2/. Much thanks to these dedicated researchers and historians.

Appendix 13

The best available information indicates that on 1 May 1944, Pz GD *Grossdeutschland* had on hand approximately 128 tanks and assault guns with 77 tanks and assault guns operational. They were the following: 40 Pz IV (with 25 operational), 28 Panthers (19 operational), 20 Tiger I (8 operational), and 40 assault guns (25 operational). The 24 Pz D had 24 tanks and assault guns operational (types unknown) and the 3 SS Pz D *Totenkopf* had no operational tanks or assault guns and were weak in infantry.

On the Soviet side, the 5th GTA had 271 tanks (including 40 of the new IS-2 [JS 2] heavy tanks) and 87 SU self-propelled guns (SU). The 2nd TA had 77 tanks and 44 SUs. So, the armored strengths were approximately 101 operational German tanks and assault guns verses 478 Soviet tanks and SUs, which is more than a 4 to 1 ratio in favor of the Red Army.

Selected Bibliography

Abdulin, Mansur. *Red Road from Stalingrad,* Stackpole Books, 2008.

Ammon, Sella. *The Value of Human Life in Soviet Warfare,* Routledge, 1992.

Andryushchenko, S. A. *We Started on Slavutich,* Voenizdat, 1979.

Armstrong, Richard N. *Red Army Tank Commanders the Armored Guards*, Schiffer Military History, 1994.

Axworthy, Mark. *Third Axis Fourth Ally*, Arms And Armour, 1995.

Bellamy, Chris. *Absolute War*. Alfred A. Knopf, New York, 2007.

Bergstrom, Christer. *Kursk the Air Battle: July 1943,* Classic Publications, 2007.

Bergstrom, Christer. *Black Cross Red Star air war over the Eastern Front Vol 5*, Vanktel Books, 2020.

Buttar, Prit. Th*e Reckoning, The Defeat of Army Group South, 1944,* Osprey, 2020.

Buttar, Prit. *Retribution, The Soviet Reconquest of Central Ukraine 1943,* Osprey, 2019.

Citino, M. Robert. *The Wehrmacht Retreats*, University Press of Kansas, 2012.

Corum, James S. *Wolfram von Richthofen, Master of the German Air War*, University of Kansas Press, 2008.

Crawford, Steve. *The Eastern Front Day by Day, A Photographic Chronology*, Potomac Books, 2006.

Davie, H.G.W. The Journal of Slavic Military Studies, 2020, Vol. 33, No 4, *The Logistics of the Combined-Arms-Army-the Rear: High Mobility Through Limited Means*.

Davie, H.G.W. The Journal of Slavic Military Studies, 2021, Vol. 34, No 2, *Mechanized Corps – a Study in Mobility and Transport.*

Davtyan, S. M. *Fifth Aerial, Military historical outline of the combat path of the 5th Air Army during the Great Patriotic War,* Military Publishing House, 1990.

Dunn JR, Walter S. *Hitler's Nemesis the Red Army 1930-1945*, Praeger, London, 1994.

Erickson, John. *The Road to Berlin,* Weidenfeld and Nicolson, 1983.

Filipescu, Mihai Tone. *Reluctant Axis, the Romanian Army in Russia 1941-1944,* FTM, 2006.

Forczyk, Robert. *Kursk 1943 the Southern Front,* Osprey Publishing, 2017.

Frieser, Karl-Heinz. *Germany and the Second World War, Volume VIII,* Clarendon Press Oxford, 2017.

Friessner, Johannes Hans. *Lost Battles (or Loyal Battles),* Holsten Verlag, 1956.

Glantz, David M. *Red Storm Over the Balkans*, University Press Kansas, 2007.

Glantz, David M. *Colossus Reborn The Red Army at War, 1941 -1943*, University Press Kansas, 2005.

Glantz, David M. *Soviet Military Deception in the Second World War,* Frank Cass, 1989.

Glantz, David M., and Harold S. Orenstein. *The Battle For Kursk, The Soviet General Staff Study,* Frank Class, London, 1999.

Glantz, David M., and Jonathan M. House. *The Battle of Kursk,* University Press Kansas, 1999.

Glantz, David M., and Jonathan M. House. *When Titans Clashed*, University Press Kansas, 1995.

Goralski, Robert. *World War II Almanac 1931-1945*, Hamish Hamilton, London, 1981.

Gudkova, Galina D. *They Will Live,* Moscow: Mol. Guard, 1986.

Guillemot, Philippe. *Hungary 1944-45 The Panzers' Last Stand*, Histoire & Collections, 2010.

Hardesty, Von, and Ilya Grinberg. *Red Phoenix Rising*, University Press of Kansas, 2012.

Harrison, Richard W., editor. *The Battle for the Dnepr The Red Army's Forcing of the East Wall, September-December 1943*, Helion & Company, 2018.

Harrison, Richard W., editor. *The Iasi-Kishinev Operation*, Helion & Company, 2017.

Harrison, Richard W., editor. *The Budapest Operation 1945*, Helion & Company, 2017.

Harrison, Richard W., editor. *The Battle of Kursk, The Red Army's Defensive Operations and Counter-Offensive, July-August 1943,* Helion and Company, England, 2016.

Harutyunian S.S. *Under the banner of the guard: The combat path of the 15th Guards Rifle Kharkov-Prague Order of Lenin, twice the Red Banner, Orders of Suvorov and Kutuzov Division,* Hayastan, 1982.

Haupt, Werner. *Army Group South, The Wehrmacht in Russia 1941-1945*, Schiffer Military History, 1998.

Hill, Alexander. *The Great Patriotic War of the Soviet Union, 1942-45,* Routledge, 2009.

Hinze, Rolf. *With the Courage of Desperation,* Helion & Company Ltd, 2013.

Hinze, Rolf. *Crucible of Combat,* Helion & Company Ltd, 2009.

Kavalerchik, Boris, and Lev Lopukhovsky. *The Price of Victory*, Pen & Sword, 2017.

Khusainovich, Aynutdinov Sergey. *In Memory and Heart,* Yekaterinburg: State Research and Production Enterprise, Chapter "Ural Worker," 2000.

Konev, Ivan Stepanovich. *Front Commander's Notes*, Militera.lib.ru, 1972.

Krivosheev, Colonel-General G.F., editor. *Soviet Casualties and Combat Losses in the Twentieth Century,* Greenhill Books, 1997.

Kurowski, Franz. *Operation Zitadelle July 1943 The Decisive Battle of World War II,* J.J. Fedorowicz Publishing Inc., Canada, 2003.

Lawrence, Christopher A. *Kursk The Battle of Prokhorovka*, Aberdeen Books, 2015.

Lochmann, Franz-Wilhelm, Richard Freiherr von Rosen., and Alfred Rubbel. *The Combat History of German Tiger Tank Battalion 503 in World War II,* J.J Fedorowicz Publishing, 2000.

Lodieu, Didier. *III. Pz. Korps at Kursk*, Histoire & Collections, Paris, 2007.

Loza, Dmitriy. *Commanding the Red Army's Sherman Tanks,* University of Nebraska Press, 1996.

Lukacs, John. *The Siege of Budapest, 100 Days in World War II*, Yale University Press, 2005.

Lunde, Henrik O. *Finland's War of Choice*, Casemate, 2011.

Maier, Georg. *Drama Between Budapest and Vienna*, J.J. Fedorowicz Publishing, Inc, 2004.

Maxim, Zagorulko., and Boris Usyk. *Stalingradskaya Gruppa Voisk (Stalingrad Group of Forces, February – May 1943).* n.d. http://www.marshals-victory.senat.org/marshalvictory/Stalingrad/stalingradgrupp.html.

Melvin, Major General Mungo. *Manstein, Hitler's Greatest General,* St. Martin's Press, 2010.

Moore, Perry. *Panzerschlacht, Armoured Operations on the Hungarian Plains September-November 1944,* Helion & Company, 2008.

Morozov, Ivan. *From Stalingrad To Prague,* Volgograd: Nizhne-Volzhskoye Publishing House, 1976.

Murry, Williamson. *Strategy for Defeat, The Luftwaffe 1933-1945,* Air University Press, Maxwell Air Force Base, 1983.

Nash Sr, Douglas E. *From the Realm of a Dying Sun Vol III*, Casemate, 2021.

Newton, Steven H. *Hitler's Commander, Field Marshal Walter Model – Hitler's Favorite Commander,* Da Capo Press, 2006.

Newton, Steven H. *Kursk The German View*, De Capo Press, 2002.

Nevenkin, Kamen. *Bloody Vienna.* PeKo Publishing Kft., 2020.

Nevenkin, Kamen. *Fortress Budapest*, Vol 1 & 2, PeKo Publishing Kft., 2020.

Piekalkiewicz, Janusz. *Operation Citadel*, Presidio Press, 1987.

Pierik, Perry. *Hungary 1944-1945 The Forgotten Tragedy,* ASPEKT, 1998.

Plocher, Generalleutnant Hermann. *The German Air Force Versus Russia, 1943*, USAF Historical Division, Aerospace Studies Institute, Air University June 1967, USAF Historical Studies #155.

Price, Alfred. *The Last Year of the Luftwaffe, May 1944 to May 1945*, Arms and Armour, 1991.

Raus, General Erhard. *Panzers on the Eastern Front,* London: Greenhill Books, 2006.

Raus, Erhard. *Panzer Operations*, Da Capo Press, USA, 2003.

Reese, Roger R. *Why Stalin's Soldiers Fought*, University Press of Kansas, 2011.

Rehfeldt, Dr. Hans Heinz. *Mortar Gunner on the Eastern Front, Vol 1 and Vol 2*, Greenhill Books, 2019.

Reynolds, Michael. *Men of Steel, I SS Panzer Corps*, Sarpedon, New York, 1999.

Ruge, Friedrich. *The Soviets as Naval Opponents 1941-1945*, Patrick Stephens, Cambridge, 1979.

Schranck, David. *Thunder at Prokhorovka*, Helion & Company, 2013.

Sharp, Charles C. *Soviet Order of Battle WW II vol #1-12,* West Chester: George F. Nafziger, 1995-1998.

Sokolov, Boris. *Marshal Malinovskii: Hero of the Soviet Union*, Helion & Company, 2017.

Stein, Marcel. *Field Marshal von Manstein, The Janus Head – A Portrait,* Helion & Company, 2007.

Stout, Jay A. *Fortress Ploesti the Campaign to Destroy Hitler's Oil*, Casemate, 2011.

Szamveber, Norbert. *Armoured Warfare in the battle for Budapest*, PeKo, 2018.

Szamveber, Norbert. *Days of Battle*, Helion & Company, 2013.

Taylor, Brian. *Barbarossa to Berlin A Chronology of the Eastern Front 1941 to 1945, Vol 1,* Spellmount, 2003.

Taylor, Brian. *Barbarossa to Berlin A Chronology of the Eastern Front 1941 to 1945, Vol 2,* Spellmount, 2004.

Tsouras, Peter G. *Panzers on the Eastern Front*, Greenhill Books, 2002.

Tsouras, Peter G., editor. *The Anvil of War,* Greenhill Books, 1994.

Ullrich, Volker. *Eight Days in May*, Liveright Publishing Corporation, 2020.

Ungvary, Krisztian. *The Siege of Budapest, 100 Days in World War II*, Yale University Press, 2005.

Viktorovich, Efimiev Alexander, Alexander Manzhosov., and Sidorov Petr Fedorovich. *Armored trains in the Great Patriotic War 1941-1945,* Military Publishing, Moscow, 1992.

Vinogradova, Lyuba. *Defending the Motherland,* MacLehose Press, 2015.

Vinogradov, Timofey Zakharovich. *Expensive-Forever,* Moscow: Military Publishing, 1977.

Vladimirovich, Kovalev Ivan. *Transport in the Great Patriotic War (1941-1945),* Moscow Nauka, 1981.

Voytenno, M.F. *Organization of medical support for the 64th army during its advance from its Stalingrad position to a new combat operations position,* Voenno-Meditsinskii Zhurnal, 1984 Feb.

Vuitenko, Colonel M.F. *Organization of the Medical Support of the 7th Guards Army in Belgorod Kharkov Offensive Operation*, Military-Medical Journal #8, August 1983, 40th Anniversary of the Battle of Kursk.

Zaloga, Steven J., and Leland S. Ness. *Red Army Handbook 1939-1945*, Sutton Publishing, 1998.

Zamulin, Valerii Nikolaevic. *On the Role of Soviet Intelligence During the Preparation of the Red Army for the Summer Campaign of 1943*, The Journal of Slavic Military Studies, Vol 32, Number 2, April-June 2019.

Zamulin, Valeriy. *The Forgotten battle of the Kursk Salient*, Helion & Company, 2018.

Zamulin, Valeriy. *The Battle of Kursk, Controversial and Neglected Aspects,* Helion and Company, 2017.

Zamulin, Valerii, Nikolaevic. *To Defeat the Enemy Was Less a Problem Than the Laziness and Indolence of Our Own Commanders*, The Journal of Slavic Military Studies, 29:4, 707-726, DOI: 10.1080/13518046.2016.1232563, 2016.

Zamulin, Valeriy. *Demolishing the Myth,* Helion and Company Ltd, 2011.

Zetterling, Niklas, and Anders Frankson. *The Korsun Pocket*, Philadelphia, Casemate, 2008.

Zetterling, Niklas, and Anders Frankson. *Kursk 1943 A statistical analysis,* Frank Class, London, 2004.

Zhagal, Viktor Makarovich. *Clearing the Path of the Infantry - Chapter: Fiery Arc,* Moscow: Military Publishing, 1985.

Zhukov, Georgy. *Marshal of the Soviet Union G Zhukov, Reminiscences and Reflections*, Progress Publishers, Moscow, 1985.

Ziemke, Earl F. *Stalingrad to Berlin: The German Defeat in the East*, Office of the Chief of Military History, Washington D.C., 1969.

Websites

English language:

Soviet Army (RKKA) in World War II. http://www.armchairgeneral.com/rkkaww2/.
(The RKKA web site is not being supported, but the links at the bottom of the page work.)

Axis and Allied forces in World War II. https://forum.axishistory.com//.

Maps and Aerial Photos from World War II. http://www.wwii-photos-maps.com/index.html.

Site devoted to the Battle for Stalingrad. www.staldata.com.

Eastern Front Research Forum and The Battler for Stalingrad Forum. https://www.facebook.com/.

Russian language:

Memory of the People, TsAMO Archive. https://pamyat-naroda.ru/documents/.

Raw materials about the Red Army in the World War II and before. http://www.soldat.ru/.

Soviet/Russian Military Books and Memoirs. http://militera.lib.ru/.

Military Maps of the Second World War and information about the Red Army. http://www.samsv.narod.ru/.

Note: Using Google Chrome will automatically translate some sections of web sites.

Notes

[1] Historical Document, *Registration card on the numerical and combat composition of parts 64 A*, Retrieved from the TsAMO Archive, ID #133373251.

[2] Historical Document, *Army Strength Report for 1 March 1943*, Retrieved from the TsAMO Archive, ID # unknown.

[3] Kovalev Ivan Vladimirovich, *Transport in the Great Patriotic War (1941-1945).* Moscow: Nauka, 1981.

[4] Valeriy Zamulin, *The Forgotten battle of the Kursk Salient*, 2018, Helion & Company, 84.

[5] Charles C. Sharp, *Red Hammers Volume XII*, George F. Nafziger, 1998, 94.

[6] Maxim Zagorulko, Boris Usyk, From the paper *Stalingradskaya Gruppa Voisk* (Stalingrad Group of Forces, February – May 1943) Volgograd State University and State Museum-panorama "The Battle of Stalingrad". http://www.marshals-victory.senat.org/marshalvictory/Stalingrad/stalingrad-grupp.html. This link graciously provided to me by Aryom in Moscow. Document #30. Note: at this time the chief of staff or the 64th Army was General Ivan Andreevich Laskin, but this document appears to have been signed by staff officers from the Operations Division of the 64th Army and the Chief of Staff for the Stalingrad group of forces.

[7] Historical Document, Relocation Schedule for March 1943, Retrieved from the TsAMO Archive, ID # unknown.

[8] Historical Document, it is possible these leading elements of the 64th Army numbered some 1,343 men. This total was taken from an obscure strength table with the date of 14 March 43. (from Pamyat Naroda web site).

[9] M. F. Voytenno, Organization of medical support for the 64th army during its advance from its Stalingrad position to a new combat operations position. Voenno-Meditsinskii Zhurnal, 1984 Feb, 69-70.

[10] Team of Authors, *Railway Troops of Russia - Book 3 On the fronts of the Great Patriotic War: 1941-1945,* The bridges over the Don, west of Stalingrad.

[11] Valerii Nikolaevich Zamulin (translated by Stuart Britton) (2016) '*To Defeat the Enemy Was Less a Problem Than the Laziness and Indolence of Our Own Commanders*', The Journal of Slavic Military Studies, 29:4, 707-726, DOI: 10.1080/13518046.2016.1232563. 719.

[12] From: http://forum.axishistory.com and postings by forum member Hugh Davie (aka, Der Alte Fritz).

[13] My map is based upon the Historical Document, *Map of the decision of the commander of the 64 A on the deployment of formations arriving by rail on 15.3.43.* Retrieved from the TsAMO Archive, ID #100210818.

[14] Vinogradov Timofey Zakharovich, *Expensive-Forever*, Moscow: Military Publishing, 1977 Ch. On The Kursk Bulge.

[15] David M. Glantz, Jonathan M. House, *The Battle of Kursk,* University Press Kansas. 1999, Appendix E, 254-355.

[16] Semyonovich, Harutyunyan Samvel, *Under the banner of the guard* [15 GRD], Hayastan 1982, Chapter 4.

[17] David M. Glantz, Jonathan M. House, *The Battle of Kursk,* University Press Kansas. 1999, 27.

[18] Valerii Nikolaevic Zamulin, *On the Role of Soviet Intelligence During the Preparation of the Red Army for the Summer Campaign of 1943*, The Journal of Slavic Military Studies (JSMS), Vol 32, Number 2, April-June 2019, 239.

[19] This section of the northern Donets River is also called the Severskii Donets in Russian accounts of the battle.

[20] Vinogradov Timofey Zakharovich, *Expensive-Forever*, Moscow: Military Publishing, 1977 Ch. On The Kursk Bulge.

[21] Valerii Nikolaevic Zamulin, *On the Role of Soviet Intelligence During the Preparation of the Red Army for the Summer Campaign of 1943*, The Journal of Slavic Military Studies, Vol 32, Number 2, April-June 2019, 239-247.

[22] Christopher Lawrence, A, Kursk The Battle of Prokhorovka, Aberdeen Books, 2015, 35.

[23] Valerii Nikolaevic Zamulin, *On the Role of Soviet Intelligence During the Preparation of the Red Army for the Summer Campaign of 1943*, The Journal of Slavic Military Studies, Vol 32, Number 2, April-June 2019, 239.

[24] David Glantz, *Soviet Military Deception in the Second World War,* Frank Cass 1989, 152-153.

[25] Historical Document, *Registration card on the numerical and combat composition of parts 64 A*, Retrieved from the TsAMO Archive, ID #13265405.

[26] Viktor Makarovich Zhagal, *Clearing the path of the infantry,* Military Publishing, Moscow, 1985, Chapter Fiery Arc. http://militera.lib.ru/memo/russian/zhagala_vm/index.html.

[27] Valery Zamulin, (paper) *The forgotten battles fiery arc*, 2009, 18.

[28] M. F. Voytenno, *Organization of medical support for the 64th Army during its advance from its Stalingrad position to a new combat operations position.* Voenno-Meditsinskii Zhurnal, 1984 Feb.

[29] Maxim Zagorulko, Boris Usyk, From the paper *Stalingradskaya Gruppa Voisk* (Stalingrad Group of Forces, February – May 1943) Volgograd State University and State Museum-panorama "The Battle of Stalingrad". http://www.marshals-victory.senat.org/marshalvictory/Stalingrad/stalingrad-grupp.html. Graciously provided to me by Aryom from Moscow. Document #52.

[30] Based upon the Historical Document, Army Strength Report for 1 April 1943, Retrieved from the TsAMO Archive, ID #132654055 and BSSA Part 3 and BSA April 1 43. Due to the rapidly changing order of battle of the 64th Army at this time, it is very difficult to present the combat strength for the army, this is my best estimate as several major units would soon be leaving the army.

[31] Based upon the Historical Document, Army Strength Report for 1 April 1943, TsAMO Archive, ID #132654055.

[32] Christopher A. Lawrence, *Kursk the Battle of Prokhorovka,* Aberdeen Books 2015, 53-57.

[33] David M. Glantz, Jonathan M. House, *The Battle of Kursk,* University Press of Kansas 1999. 28.
[34] Christopher A. Lawrence, *Kursk the Battle of Prokhorovka,* Aberdeen Books 2015, 58-60.
[35] David Glantz, *Soviet Military Deception in the Second World War,* Cass 1989, p.152. Note: in the original text the month of June was listed as the month of deployment of the 7th Guards Army into the 69th Army's sector. This is obviously a typo.
[36] Valerii Nikolaevich Zamulin (translated by Stuart Britton) (2016) *To Defeat the Enemy Was Less a Problem Than the Laziness and Indolence of Our Own Commanders*, The Journal of Slavic Military Studies, 29:4, 707-726, DOI: 10.1080/13518046.2016.1232563. 708.
[37] Ibid., 720.
[38] Ammon Sella, *The Value of Human Life in Soviet Warfare,* Routledge, 1992, 54.
[39] Found at: http://rkka.kiev.ua/node/489, which has an account by Drozhcha Ivan (Drozhchi grandfather A.N.).
[40] Generalleutnant Hermann Plocher, *The German Air Force Versus Russia, 1943*, USAF Historical Division, Aerospace Studies Institute, Air University June 1967, USAF Historical Studies #155, 56-57.
[41] Based upon the Historical Document, Army Strength Report for 5 April 1943, Retrieved from the TsAMO Archive, ID #132654078, TsAMO Archive (Архив ЦАМО), Fund (Фонд) 341, Inventory (Опись) 5312, File (Дело) 194, Document (Документ) 170.
[42] Valeriy Zamulin in his book *The Forgotten Battle of the Kursk Salient,* page 88, states that on 5 April the 64th Army had a ration strength of 49,729 men and a combat strength of 42,464. This is less than what my research shows, but Valeriy is quoting from a TsAMO doc that I do not have access to. F 203, op 2870, d 37.l.
[43] Christopher A. Lawrence, *Kursk the Battle of Prokhorovka,* Aberdeen Books, 2015, 58-60.
[44] Based upon the Historical Document, Army Strength Report for 5 or 6 April 1943, Retrieved from the TsAMO Archive, ID #454068463, and my estimates.
[45] Richard W. Harrison, Editor and Translator, *The Battle of Kursk,* Soviet general staff, Helion & Company, 2016, 83.
[46] A. M. Vasilevsky, *A Lifelong Cause,* Progress Publishers, 1973, 263-267.
[47] Janusz Piekalkiewicz, *Operation Citadel*, Presidio Press, 1987. 39-40.
[48] Marcel Stein, *Field Marshal von Manstein, The Janus Head – A Portrait,* Helion 2007. 192.
[49] Based upon BSA Data from April 10 1943.
[50] David M. Glantz, Jonathan M. House, *The Battle of Kursk,* University Press of Kansas 1999. 362-363.
[51] Valerii Nikolaevich Zamulin (translated by Stuart Britton) (2016) *To Defeat the Enemy Was Less a Problem Than the Laziness and Indolence of Our Own Commanders*, The Journal of Slavic Military Studies, 29:4, 707-726, DOI: 10.1080/13518046.2016.1232563. 708, 709.
[52] David M. Glantz, Jonathan M. House, *The Battle of Kursk,* University Press of Kansas 1999. 364-365.
[53] John Erickson, *The Road to Berlin,* Weidenfeld and Nicolson, 1983, 64-65.
[54] The Journal of Slavic Military Studies, Vol 32, Number 2, April-June 2019, 239-244.
[55] Based upon the Historical Document, *Intelligence - On the number of artillery left on the Firing Positions (OP) and tanks in the concentration areas,* 12 April 1943, Retrieved from the TsAMO Archive, ID #132654071.
[56] Kovalev Ivan Vladimirovich, *Transport in the Great Patriotic War (1941-1945).* Moscow: Nauka, 1981. Chapter 9 -Transport in the defeat of the enemy on the Kursk Bulge. http://militera.lib.ru/h/kovalev_iv/index.html.
[57] David Porter, *Visual Battle Guide – Fifth Guards Tank Army at Kursk 12 July 1943,* Amber Books, 2011, 39.
[58] Taken from the web site: https://en.wikipedia.org/wiki/Russian Guards.

1st Guards Army	1st Guards Tank Army
2nd Guards Army	2nd Guards Tank Army
3rd Guards Army	3rd Guards Tank Army
4th Guards Army	4th Guards Tank Army
5th Guards Army	5th Guards Tank Army
6th Guards Army	6th Guards Tank Army
7th Guards Army	
8th Guards Army	
9th Guards Army	
10th Guards Army	
11th Guards Army	

[59] Valeriy Zamulin, *The Forgotten Battle of the Kursk Salient – 7th Guards Army Stand Against Army Detachment Kempf,* Helion & Company, 2018, 31.
[60] Mansur Abdulin, *Red Road from Stalingrad,* Stackpole Books, 2008, 85.
[61] The Journal of Slavic Military Studies, Vol 32, Number 2, April-June 2019, 243.
[62] Valeriy Zamulin, *The Forgotten battle of the Kursk Salient*, 2018, Helion & Company, 42.
[63] Marcel Stein, *Field Marshal von Manstein, The Janus Head – A Portrait,* Helion 2007. 192.
[64] Info from various Wikipedia web sites.
[65] Chris Bellamy, *Absolute War.* Alfred A. Knopf, New York, 2007, 489.

[66] Amnon Sella, *The Value of Human Life in Soviet Warfare,* Routledge, 1992, 38.

[67] This is per the memorandum from Deputy Defense Commissar E. A. Shchadenko on May 28. Maxim Zagorulko, Boris Usyk, From the paper *Stalingradskaya Gruppa Voisk* (Stalingrad Group of Forces), *February – May 1943.* Volgograd State University and State Museum-panorama "The Battle of Stalingrad". http://www.marshals-victory.senat.org/marshalvictory/Stalingrad/stalingrad-grupp.html. Graciously provided to me by Aryom in Moscow. Document #52.

[68] Valerii Nikolaevich Zamulin (translated by Stuart Britton) (2016) *To Defeat the Enemy Was Less a Problem Than the Laziness and Indolence of Our Own Commanders*, The Journal of Slavic Military Studies, 29:4, 707-726, DOI: 10.1080/13518046.2016.1232563. 720, 721.

[69] David M. Glantz, Jonathan M. House, *The Battle of Kursk,* University Press of Kansas, 1999. 30-31.

[70] Ibid., 365-373.

[71] Von Hardesty, Ilya Crinberg, *Red Phoenix Rising, the Soviet Air Force in World War II,* University Press of Kansas, 2012, 232.

[72] David M. Glantz, Jonathan M. House, *The Battle of Kursk,* University Press of Kansas, 1999. 31-32.

[73] Galina D. Gudkova, *They Will Live,* Moscow, 1986, Chapter 18.

[74] Vinogradov Timofey Zakharovich, *Expensive-Forever*, Moscow: Military Publishing, 1977 Ch. On The Kursk Bulge.

[75] Based upon the Historical Document, Army Strength Report for 23 April 1943, Several Records from TsAMO Archive, ID #132654078. Note; this is the average strength as two different report totals are listed along with an additional pencil entry. So, the strength totals range from 61,924 to 62,303, to 62,830. These varying amounts might reflect totals for the morning, afternoon, and evening entries on the same day.

[76] Janusz Piekalkiewicz, *Operation Citadel*, Presidio Press, 1987. 43-44.

[77] Galina D. Gudkova, *They Will Live,* Moscow, 1986, Chapter 19.

[78] Major General Mungo Melvin, *Manstein, Hitler's Greatest General,* St. Martin's Press, 2010. 358.

[79] David M. Glantz, Jonathan M. House, *The Battle of Kursk,* University Press of Kansas, 1999. 26.

[80] From TsAMO Archive, *Information about the loss of connections in parts of 7 Guards*, F 203, I 2843, C 317 ID #132654086.

[81] Based upon the Historical Document, *Army Strength Report for 1 May 1943*, From TsAMO Archive, ID #132654095 and BSSA and BSA data.

[82] David M. Glantz, Harold S. Orenstein, *The Battle for Kursk 1943 The Soviet General Staff Study,* Frank Class, 1999, 45.

[83] Von Hardesty, Ilia Crinberg, *Red Phoenix Rising, the Soviet Air Force in World War II,* University Press Kansas, 2012, 238.

[84] Steven H. Newton, *Hitler's Commander*, Da Capo Press, 2006, 219.

[85] Christer Bergstrom, *Kursk the Air Battle: July 1943,* Classic Publications, 2007, 19.

[86] Valeriy Zamulin, *The Planning and Preparations for the Battle of Kursk, Vol 1,* Helion and Company, England, 2021, 451.

[87] Christer Bergstrom, *Kursk the Air Battle: July 1943,* Classic Publications, 2007, 18-21.

[88] David M. Glantz, Harold S. Orenstein, *The Battle for Kursk1943 The Soviet General Staff Study,* Frank Class 1999, 45, 47.

[89] Major General Mungo Melvin, *Manstein Hitler's Greatest General*, Thomas Dunne Books, New York, 2010, 357, 358.

[90] Valerii Nikolaevich Zamulin (translated by Stuart Britton) (2016) *To Defeat the Enemy Was Less a Problem Than the Laziness and Indolence of Our Own Commanders,* The Journal of Slavic Military Studies, 29:4, 707-726, DOI: 10.1080/13518046.2016.1232563. 720, 721.

[91] David M. Glantz, Jonathan M. House, *The Battle of Kursk,* University Press Kansas, 1999, 77.

[92] David M. Glantz, Harold S. Orenstein, *The Battle for Kursk 1943 The Soviet General Staff Study,* Frank Class, 1999, 45, 47, 48.

[93] Steven J. Zaloga and Leland S. Ness, *Red Army Handbook 1939-1945*, Sutton Publishing, 1998, Table #1 p x. By the end of the war the Red Army deployed some 174 rifle corps.

[94] Based upon the Historical Document, *Army Strength Report for 10 May 1943*, Retrieved from the TsAMO Archive, ID #132636481.

[95] Christer Bergstrom, *Kursk the Air Battle: July 1943,* Classic Publications, 2007, 19.

[96] Geoffrey P. Megargee, *Inside Hitler's High Command,* University Press of Kansas, 2000, 19.

[97] David M. Glantz, Harold S. Orenstein, *The Battle for Kursk 1943 The Soviet General Staff Study,* Frank Class 1999, 49.

[98] Ibid., 48, 49.

[99] Ibid., 134-146.

[100] Valeriy Zamulin, *The Battle of Kursk Controversial and Neglected Aspects,* Helion and Company, England, 2017, 138.

[101] Walter S. Dunn JR, *Hitler's Nemesis the Red Army 1930-1945*, Praeger, London, 1994, 89-103.

[102] Efimiev Alexander Viktorovich; Alexander Manzhosov; Sidorov Petr Fedorovich, *Armored trains in the Great Patriotic War 1941-1945,* Moscow 1992, 189, On a Fiery Arc. http://militera.lib.ru/h/efimev_manzhosov_sidorov/index.html.

[103] The 7th Guards Army would eventually have the 34 and 38 Separate Armored Train Battalions assigned to it during the Kursk Battle. The 34 contained the armored train named "Krasnoufimsky Railwayman" - #712 and the train named "Death to Fascism" - #667, while the 38 contained trains named after the newspaper "Pravda" - #754 and the newspaper "Krasnaya Zvezda "- #730. It was typical to give armored trains patriotic names.

[104] Valerii Nikolaevich Zamulin (translated by Stuart Britton) (2016) *To Defeat the Enemy Was Less a Problem Than the Laziness and Indolence of Our Own Commanders,* The Journal of Slavic Military Studies, 29:4, 707-726, DOI: 10.1080/13518046.2016.1232563. 722.

[105] Efimiev Alexander Viktorovich; Alexander Manzhosov; Sidorov Petr Fedorovich, *Armored trains in the Great Patriotic War 1941-1945, Moscow 1992, p 189 On a Fiery Arc.* http://militera.lib.ru/h/efimev_manzhosov_sidorov/index.html.

[106] Malikin Lev Samoilovich, the regimental intelligence officer from the 72 GRD, found on the site https://iremember.ru/memoirs/razvedchiki/malikin-lev-samoylovich/.

[107] Christer Bergstrom, *Kursk the Air Battle: July 1943*, Classic Publications, 2007, 21.

[108] Valeriy Zamulin, *The Battle of Kursk, Controversial and Neglected Aspects*, Helion and Company, England, 2017, 139.

[109] David M. Glantz, Jonathan M. House, *The Battle of Kursk,* University Press Kansas, 1999, 77.

[110] Based upon the Historical Document, *Map, 21 April 1943*, From TsAMO Archive, f 203, op 2843, d 588, ID #100186690.

[111] Lyuba Vinogradova, *Defending the Motherland,* MacLehose Press, 2015, 102.

[112] Maxim Zagorulko, Boris Usyk, From the paper *Stalingradskaya Gruppa Voisk* (Stalingrad Group of Forces, February – May 1943) Volgograd State University and State Museum-panorama "The Battle of Stalingrad". http://www.marshals-victory.senat.org/marshalvictory/Stalingrad/stalingrad-grupp.html Graciously provided to me by Aryom in Moscow. Document #52.

[113] Based upon the Historical Document, *Summary of material support for parts and formations 7 Guards, 29 May 1943*, From TsAMO Archive, ID #132636523.

[114] Steven H. Newton, *Kursk The German View*, De Capo Press, 2002, 366-367.

[115] James S. Corum, *Wolfram von Richthofen, Master of the German Air War*, University of Kansas Press, 2008, 318.

[116] Christer Bergstrom, *Kursk the Air Battle: July 1943, Classic Publications*, 2007, 25.

[117] *Information about the loss of parts and formations 7 Guards*, 31 May 43, TsAMO Archive, ID #133373284, F341, I5312, C237.

[118] Based upon the Historical Document, *Information on the combat strength and security of the tank units 7 Guards A. May 31 1943*, From TsAMO Archive, ID #136526877.

[119] Valeriy Zamulin, *The forgotten Battle of the Kursk Salient,* Helion & Company, 2018, 71,72.

[120] Based upon the Historical Document, *The diagram of the flights over Volokonovkoy and the number of raids on the railway objects of the army railway section, May 1-31*, From TsAMO Archive, ID #133390055.

[121] Christer Bergstrom, *Black Cross Red Star air war over the Eastern Front Vol 5*, Vanktel Books, 2020, 182.

[122] Based upon the Historical Document, Army Strength Report for 1 June 1943. From TsAMO Archive, ID #133373300, and BSSA Part 3 and BSA June 1 43.

[123] Based upon an estimated 7,265 men and women in support and non-combat units. Also see note #70 in, *The Forgotten Battle of the Kursk Salient,* 88.

[124] Charles C. Sharp, *Red Hammers, Soviet Self-Propelled Artillery and Lend Lease Armor 1941-1945,* Published by George F. Nafziger 1998, 77.

[125] David Porter, *Fifth Guards Tank Army at Kursk 12 July 1943,* Amber Books, London 2011, 118.

[126] Based upon the Historical Document, *Statement of the combat strength of units and formations 7 Guards*, for 1 June 1943, From TsAMO Archive ID #133373300.

[127] Christer Bergstrom, *Black Cross Red Star air war over the Eastern Front Vol 5,* Vanktel Books, 2020, 175.

[128] Kovalev Ivan Vladimirovich, *Transport in the Great Patriotic War (1941-1945).* Moscow: Nauka, 1981. Chapter 9 - Transport in the defeat of the enemy on the Kursk Bulge, 565.

[129] Valeriy Zamulin, *The forgotten Battle of the Kursk Salient,* Helion & Company, 2018, 441, 442.

[130] Steven H. Newton, *Kursk The German View*, De Capo Press, 2002, 366-367.

[131] Kovalev Ivan Vladimirovich, *Transport in the Great Patriotic War (1941-1945).* Moscow: Nauka, 1981. Chapter 9 - Transport in the defeat of the enemy on the Kursk Bulge.

[132] Based upon the Historical Document, *Summary of material support of parts and formations 7 Guards.* 3 June 1943, From TsAMO Archive, ID #133390052.

[133] Niklas Zetterling, Anders Frankson, *Kursk 1943 A statistical analysis,* London, 2004, 22.

[134] David M. Glantz, Harold S. Orenstein, *The Battle For Kursk, The Soviet General Staff Study,* Frank Class 1999, London, 33-34.

[135] David M. Glantz, Jonathan M. House, *The Battle of Kursk,* University Press of Kansas, United States 1999, 65-68.

[136] Christopher A. Lawrence, *Kursk the Battle of Prokhorovka*, Aberdeen Books, USA, 2015, 200-203.

[137] Valeriy Zamulin, *Demolishing the Myth,* Helion and Company Ltd, England, 2011, 43.

[138] Richard W. Harrison, Editor and Translator, Soviet General Staff, *The Battle of Kursk, The Red Army's Defensive Operations and Counter-Offensive, July-August 1943,* Helion and Company, England, 2016. 52.

[139] Christopher A. Lawrence, *Kursk the Battle of Prokhorovka*, Aberdeen Books, USA, 2015, 1230.

[140] Historical Document, *Information about the losses of personnel blown up at their minefields in the formations of the 24 Guards Corps*, From TsAMO Archive, ID #133373348.

[141] Christer Bergstrom, *Kursk the Air Battle July 1943,* Classic Publications, England, 2007, 20-24.
[142] Edgar M. Howell, *The Soviet Partisan Movement 1941-1944,* Department of the Army pamphlet No 20-244, 1956, 154-161.
[143] James S. Corum, *Wolfram von Richthofen, Master of the German Air War*, University of Kansas Press, 2008, 319.
[144] Based upon the Historical Document, *Registration cars on the numerical and combat composition of units of the 7 Guards, for* 10 June 1943, From TsAMO Archive ID #133373288. All the following strength reports are from the same TsAMO source.
[145] James S. Corum, *Wolfram von Richthofen, Master of the German Air War,* University of Kansas Press, 2008, 319-320.
[146] Ibid., 324.
[147] Brian Taylor, *Barbarossa to Berlin A Chronology of the Eastern Front 1941 to 1945,* Vol 2, Spellmount, 2004, 71.
[148] Valeriy Zamulin, *The forgotten Battle of the Kursk Salient,* Helion & Company, 2018, 98.
[149] Valerii Nikolaevich Zamulin (translated by Stuart Britton) (2016) *To Defeat the Enemy Was Less a Problem Than the Laziness and Indolence of Our Own Commanders,* The Journal of Slavic Military Studies, 29:4, 707-726, DOI: 10.1080/13518046.2016.1232563. 722, 723.
[150] Valeriy Zamulin, *The Forgotten Battle of the Kursk Salient – 7th Guards Army Stand Against Army Detachment Kempf,* Helion & Company, 2018, 68.
[151] David M. Glantz, Jonathan M. House, *The Battle of Kursk* University Press of Kansas, United States 1999, 55.
[152] Various authors, Edited by Major General Ivan Parotkin, *The Battle of Kursk,* Progress Publishers, Moscow, 1974, 244.
[153] David M. Glantz, Jonathan M. House, *The Battle of Kursk* University Press of Kansas, United States 1999, 55.
[154] Various authors, Edited by Major General Ivan Parotkin, *The Battle of Kursk*, Progress Publishers, Moscow, 1974, 242-243.
[155] H. G. W. Davie, The Journal of Slavic Military Studies, 2020, Vol 33 No 4, *The Logistics of the Combined-Arms-Army-the Rear: High Mobility Through Limited Means*, 590.
[156] Valerii Nikolaevich Zamulin (translated by Stuart Britton) (2016) *To Defeat the Enemy Was Less a Problem Than the Laziness and Indolence of Our Own Commanders,* The Journal of Slavic Military Studies, 29:4, 707-726, DOI: 10.1080/13518046.2016.1232563. 710.
[157] Christer Bergstrom, *Kursk the Air Battle July 1943,* Classic Publications, England, 2007, 24.
[158] Kovalev Ivan Vladimirovich, *Transport in the Great Patriotic War (1941-1945).* Moscow: Nauka, 1981. Chapter 9 - Transport in the defeat of the enemy on the Kursk Bulge. http://militera.lib.ru/h/kovalev_iv/index.html.
[159] Valeriy Zamulin, *Demolishing the Myth,* Helion and Company Ltd, England, 2011, 43-44.
[160] Archaeology breakthrough: British WW2 shipwreck found with £173million of gold revealed | World | News | Express.co.uk.
[161] Robin Cross, *Citadel The Battle of Kursk,* Michael O'Mara Books Limited, London, 1993, 143.
[162] Steven H. Newton, *Kursk the German View*, Da Capo Press, USA, 2002, 17.
[163] Christer Bergstrom, *Kursk the Air Battle: July 1943,* Classic Publications, 2007, 25.
[164] Ibid.,
[165] Christer Bergstrom, *Black Cross Red Star air war over the Eastern Front Vol 5,* Vanktel Books, 2020, 178.
[166] General Gehlen was chief of the Wehrmacht Foreign Armies East (FHO) military-intelligence unit.
[167] Karl-Heinz Frieser, *Germany and the Second World War, Volume VIII*, Clarendon Press Oxford, 2017, 77-78.
[168] Christopher A. Lawrence, *Kursk the Battle of Prokhorovka*, Aberdeen Books, USA, 2015, 214-215.
[169] David M. Glantz, Jonathan M. House, *The Battle of Kursk,* University Press of Kansas, 1999, 310.
[170] Based upon the Historical Document, *Registration card on the numerical and combat composition of units of the 7 Guards,* for 3 July 1943, From TsAMO Archive ID #133373312, and BSSA Part 3 and BSA July 1 43.
[171] Jean-Luc Marchand, *Order of Battle Soviet Army WW 2, 1943 July and August, Operation Zitadelle*, *Vol XII*, The Nafziger Collection, 2010, 25.
[172] These weather reports are all from Valeriy Zamulin, *The Forgotten Battle of the Kursk Salient – 7th Guards Army Stand Against Army Detachment Kempf,* Helion & Company, 2018, 418-419, Table 12.
[173] Christopher A. Lawrence, *Kursk the Battle of Prokhorovka*, Aberdeen Books, USA, 2015, 1224.
[174] Alexander Viktorovich Efimiev; Alexander Manzhosov; Sidorov Petr Fedorovich, *Armored trains in the Great Patriotic War 1941-1945, Moscow 1992, 189 On a Fiery Arc.* http://militera.lib.ru/h/efimev_manzhosov_sidorov/index.html.
[175] Based upon the Historical Document, *Material security summary 7 Guards*, 1 July 1943, From TsAMO Archive, ID #133390094.
[176] Franz Kurowski, *Operation Zitadelle July 1943 The Decisive Battle of World War II,* J.J. Fedorowicz Publishing Inc. Canada, 2003, 74.
[177] Valeriy Zamulin, *The Battle of Kursk, Controversial and Neglected Aspects*, Helion and Company, England, 2017, 227.
[178] David M. Glantz, Jonathan M. House, *The Battle of Kursk* University Press of Kansas, United States 1999, 81.
[179] Richard W. Harrison, Editor and Translator, *The Battle of Kursk,* Soviet general staff, Helion & Company, 2016, 129.
[180] Valeriy Zamulin, *The Battle of Kursk, Controversial and Neglected Aspects,* Helion and Company, England, 2017, 201.
[181] Christer Bergstrom, *Kursk The Air Battle July 1943*, Classic Publications, 2007, p 26-34, 120. Note: Figures from the 17th AA are missing totals for fighter losses. So, these additional losses would push the real totals far above 150 aircraft.
[182] G. A. Koltunov, B. G. Solovyov, *The Battle of Kursk,* Moscow, Military Publishing, 1970, Chapter 4.

[183] This information is based upon comments from - https://forum.axishistory.com/viewtopic.php?t=224930.
[184] Based upon the Historical Document, *Characteristics of fords across the river. Norther Donets on a section of 7 Guards A May 20*, From TsAMO Archive ID #454068504, Map on page 4.
[185] Karl-Heinz Frieser, *Germany and the Second World War, Volume VIII,* Clarendon Press Oxford, 2017, 104-105, 113-114.
[186] Some accounts state the German attack was delayed by hours due to the Russian bombardment.
[187] Erhard Raus, *Panzer Operations*, Da Capo Press, USA, 2003, 202-203.
[188] Christer Bergstrom, *Black Cross Red Star air war over the Eastern Front Vol 5,* Vanktel Books, 2020, 212-213.
[189] Ibid., 225-227.
[190] Valeriy Zamulin, *The Battle of Kursk, Controversial and Neglected Aspects,* Helion and Company, England, 2017, 253.
[191] Christer Bergstrom, *Black Cross Red Star air war over the Eastern Front Vol 5,* Vanktel Books, 2020, 230.
[192] Valeriy Zamulin, *The Forgotten Battle of the Kursk Salient – 7th Guards Army Stand Against Army Detachment Kempf,* Helion & Company, 2018, 395-396.
[193] Christer Bergstrom, *Black Cross Red Star air war over the Eastern Front Vol 5,* Vanktel Books, 2020, 231.
[194] Khazanov Dmitry Borisovich; Gorbach Vitaly Grigorievich, *Aviation in the battle over the Oryol-Kursk Bulge*, 2004.
[195] Valeriy Zamulin, *The Forgotten Battle of the Kursk Salient – 7th Guards Army Stand Against Army Detachment Kempf,* Helion & Company, 2018, 393-394.
[196] Ibid., 388.
[197] It was officially called the 92 Red Banner Border Regiment of the NKVD. Zamulin, Valeriy, *The Forgotten Battle of the Kursk Salient,* 400 note 159.
[198] Christer Bergstrom, *Black Cross Red Star air war over the Eastern Front Vol 5*, Vanktel Books, 2020, p 233-235.
[199] Valeriy Zamulin, *The Forgotten Battle of the Kursk Salient – 7th Guards Army Stand Against Army Detachment Kempf,* Helion & Company, 2018, 432.
[200] Ibid., 429.
[201] Christer Bergstrom, *Black Cross Red Star air war over the Eastern Front Vol 5*, Vanktel Books, 2020, 240.
[202] Valeriy Zamulin, *The Forgotten Battle of the Kursk Salient – 7th Guards Army Stand Against Army Detachment Kempf* Helion & Company, 2018, 165.
[203] Ibid., 456-457, table 14.
[204] Ibid., 568, table 16.
[205] Christer Bergstrom, *Black Cross Red Star air war over the Eastern Front Vol 5,* Vanktel Books, 2020, p 242.
[206] Valeriy Zamulin, *The Forgotten Battle of the Kursk Salient – 7th Guards Army Stand Against Army Detachment Kempf,* Helion & Company, 2018, 484-485.
[207] Christer Bergstrom, *Black Cross Red Star air war over the Eastern Front Vol 5*, Vanktel Books, 2020, 244-245.
[208] Named after General Mikhail Kutuzov who fought against Napoleons invasion of Russia in 1812.
[209] Steven H. Newton, *Hitler's Commander, Field Marshal Walter Model – Hitler's Favorite Commander,* Da Capo Press, 2006, 254.
[210] Christopher A. Lawrence, *Kursk the Battle of Prokhorovka*, Aberdeen Books, USA, 2015, p 952-956. Note: These values are based upon the newest research which significantly alters the old hearsay accounts about this battle.
[211] Valeriy Zamulin, *The Forgotten Battle of the Kursk Salient – 7th Guards Army Stand Against Army Detachment Kempf* Helion & Company, 2018, 510.
[212] Ibid., 531-532.
[213] David M. Glantz, Jonathan M. House, *The Battle of Kursk,* University Press of Kansas, United States 1999, 218, 419n.34.
[214] Christer Bergstrom, *Black Cross Red Star air war over the Eastern Front Vol 5,* Vanktel Books, 2020, 248-250.
[215] Roland was a paladin in the court of the Roman Emperor Charlemagne.
[216] Valeriy Zamulin, *The Forgotten Battle of the Kursk Salient – 7th Guards Army Stand Against Army Detachment Kempf,* Helion & Company, 2018, 559.
[217] Robert Forczyk, *Kursk 1943 the Southern Front,* Osprey Publishing 2017, 88.
[218] Christer Bergstrom, *Kursk The Air Battle July 1943*, Classic Publications, 2007, 97-99.
[219] Valeriy Zamulin, *The Forgotten Battle of the Kursk Salient – 7th Guards Army Stand Against Army Detachment Kempf,* Helion & Company, 2018, 588.
[220] Christer Bergstrom, *Black Cross Red Star air war over the Eastern Front Vol 5*, Vanktel Books, 2020, 289-290.
[221] Christopher A. Lawrence, *Kursk the Battle of Prokhorovka*, Aberdeen Books, USA, 2015, 1117.
[222] Ibid., 1106.
[223] Ibid., 1155.
[224] Christer Bergstrom, *Black Cross Red Star air war over the Eastern Front Vol 5*, Vanktel Books, 2020, 290-291.
[225] From the Russian www.istmira.Ru historical library web site: http://www.istmira.ru/knigi-vtoraya-mirovaya-vojna *The Supreme High Command Documents and Materials 1943 year*. Moscow 1999, (Document 279 and 280).
[226] Valeriy Zamulin, *The Forgotten Battle of the Kursk Salient – 7th Guards Army Stand Against Army Detachment Kempf* Helion & Company, 2018, 567, table 16, 17, 18.

[227] Didier Lodieu, *III. Pz. Korps at Kursk*, Histoire & Collections, Paris, 2007, 142.
[228] David M. Glantz, Jonathan M. House, *The Battle of Kursk,* University Press of Kansas, United States 1999, 338.
[229] Steven H. Newton, translator, *Panzer Operations*, Da Capo Press, 2003, 212.
[230] David M. Glantz, Jonathan M. House, *The Battle of Kursk* University Press of Kansas, United States 1999, 223.
[231] Marcel Stein, *Field Marshal von Manstein The Janus Head, A portrait,* Helion & Company Ltd, 2007, 203.
[232] Christopher A. Lawrence, *Kursk the Battle of Prokhorovka*, Aberdeen Books, USA, 2015, 1155.
[233] Christer Bergstrom, *Black Cross Red Star air war over the Eastern Front Vol 5,* Vanktel Books, 2020, 291.
[234] David M. Glantz, Jonathan M. House, *The Battle of Kursk,* University Press of Kansas, United States 1999, 229.
[235] *Kursk operation simulation and validation exercise – Phase II (Kosave II),* September 1998, US Army Concepts Analysis Agency, Office of the Chief of Staff.
[236] Christopher A. Lawrence, *Kursk the Battle of Prokhorovka*, Aberdeen Books, USA, 2015, 1189.
[237] Christer Bergstrom, *Kursk The Air Battle July 1943*, Classic Publications, 2007, 121.
[238] David M. Glantz, Jonathan M. House, *The Battle of Kursk,* University Press of Kansas, United States 1999, 238.
[239] Christopher A. Lawrence, *Kursk the Battle of Prokhorovka*, Aberdeen Books, USA, 2015, 1164.
[240] David M. Glantz, Jonathan M. House, *The Battle of Kursk,* University Press of Kansas, United States 1999, 238.
[241] Found on the following web site (accessed on 21 Sept 2019) https://www.revolutionarydemocracy.org/Stalin od0724to092543.htm.
[242] Christopher A. Lawrence, *Kursk the Battle of Prokhorovka*, Aberdeen Books, USA, 2015, 1195-1197.
[243] This information is based upon comments provided by the Axis History Forum member DirkAH at: https://forum.axishistory.com/viewtopic.php?f=50&t=251366, accessed August 2020.
[244] Dr. Hans Heinz Rehfeldt, *Mortar Gunner on the Eastern Front, Vol II*, Greenhill Books, 2019. 9.
[245] This information is based upon the following web site: https://history.wikireading.ru/413782, accessed August 2020.
[246] This information is based upon the following (Russian fatherland) web site: http://xn--b1afa8admefdd2c.xn--p1ai/norm/003/010.htm, accessed August 2020.
[247] Robert Goralski, *World War II Almanac 1931-1945*, Hamish Hamilton, London, 1981, 275.
[248] Historical Document, *Report on personnel losses of 7 Guards*, 5-25 July 1943, From TsAMO Archive ID #131308701 or 133373328.
[249] Dr. Hans Heinz Rehfeldt, *Mortar Gunner on the Eastern Front, Vol 1*, Greenhill Books, 2019. 329-330.
[250] David M. Glantz, Jonathan M. House, *The Battle of Kursk,* University Press of Kansas, United States 1999, 238.
[251] Robert M. Citino, *The Wehrmacht Retreats*, University Press of Kansas, 2012, 218.
[252] David Schranck, *Thunder at Prokhorovka*, Helion & Company, 2013, 451.
[253] Based upon the Historical Document, *Summary of material support of parts and formations 7 Guards A, 29 July 1943,* From TsAMO Archive ID #133390144.
[254] Based upon the Historical Document, *Settlement-application for transportation by railway 270 RD, 29 July 1943,* From TsAMO Archive ID #454032070.
[255] Ivan Stepanovich Konev, *Front Commander's Notes*, Militera.lib.ru (website) 1972, Chapter 1.
[256] David M. Glantz, Jonathan M. House, *The Battle of Kursk,* University Press of Kansas, United States 1999, 240.
[257] Lev Lopukhovsky, Boris Kavalerchik, *The Price of Victory*, Pen & Sword, 2017, 61.
[258] The Operation Rumianysev plan was also known as the General Rumianysev plan.
[259] Colonel M.F. Vuitenko, *Organization of the Medical Support of the 7th Guards Army in Belgorod Kharkov Offensive Operation*, Military-Medical Journal #8, August 1983, 40th Anniversary of the Battle of Kursk, 1-3.
[260] Richard W. Harrison, Editor and Translator, *The Battle of Kursk,* Soviet General Staff, Helion & Company, 2016, 322.
[261] Peter G. Tsouras, *Panzers on the Eastern Front*, Greenhill Books, 2002, 176-180.
[262] H. G. W. Davie, The Journal of Slavic Military Studies, 2020, Vol 33 No 4, *The Logistics of the Combined-Arms-Army-the Rear: High Mobility Through Limited Means*, 595.
[263] Based upon the Historical Document, *Summary of material support of parts and formations 7 Guards A, 2 August 1943,* From TsAMO Archive ID #133390148.
[264] Based upon the Historical Document, *Calculation of the weight of one army daily supply of food stuffs and one army fueling of fuel and lubricants 7 Guards A, 2 August 1943,* From TsAMO Archive ID #133390876.
[265] Karl-Heinz Frieser, *Germany and the Second World War, Volume VIII,* Clarendon Press Oxford, 2017, 189.
[266] Von Hardesty, Ilya Grinberg, *Red Phoenix Rising*, University Press of Kansas, 2012, 267.
[267] David M. Glantz, Jonathan M. House, *The Battle of Kursk,* University Press of Kansas, United States 1999, 246.
[268] Karl-Heinz Frieser, *Germany and the Second World War, Volume VIII,* Clarendon Press Oxford, 2017, 195.
[269] Karl-Heinz Frieser, *Germany and the Second World War, Volume VIII,* Clarendon Press Oxford, 2017, 191.
[270] Peter G. Tsouras, *Panzers on the Eastern Front*, Greenhill Books, 2002, 180-184.
[271] Based upon the Historical Document, *Information on the number and combat strength of 7 Guards A, 4 August 1943,* From TsAMO Archive ID #133373372.
[272] Richard W. Harrison, Editor and Translator, *The Battle of Kursk,* Soviet General Staff, Helion & Company, 2016, 326.

[273] S. M. Davtyan, *Fifth Aerial, Military historical outline of the combat path of the 5th Air Army during the Great Patriotic War,* Military Publishing House, 1990, Chapter "On the Stepp Front."
[274] Earl F. Ziemke, *Stalingrad to Berlin The German Defeat in the East*, Office of the Chief of Military History, Washington D.C., 1969, 151.
[275] Peter G. Tsouras, *Panzers on the Eastern Front*, Greenhill Books, 2002, 186-193.
[276] Based upon the Historical Document, *Registration card on the numerical and combat composition of units of the 7 Guards A, 6 August 1943,* From TsAMO Archive ID #133373361.
[277] Based upon the Historical Document, *Information on the headcount of the personnel of the 7 Guards A, 6 August 1943,* From TsAMO Archive ID #133373351.
[278] Georgy Zhukov, *Marshal of the Soviet Union G Zhukov Reminiscences and Reflections*, Progress Publishers, Moscow, 1985, 202-204.
[279] Kabanov Pavel Alekseevich, *Steel Spans*, Military Publishing House, 1973, militera.lib.ru, Chapter "The Road of Courage."
[280] Ibid.,
[281] Based upon the Historical Document, *Vehicle Maneuvering (for the Belgorod-Kharkov Operation),* From TsAMO Archive ID #133390875.
[282] Based upon the Historical Document, *Registration card on the numerical and combat composition of units of the 7 Guards A, 7 August 1943,* From TsAMO Archive ID #133373361.
[283] Colonel M.F. Vuitenko, *Organization of the Medical Support of the 7th Guards Army in Belgorod Kharkov Offensive Operation*, Military-Medical Journal #8, August 1983, 40th Anniversary of the Battle of Kursk, 4.
[284] Based upon the Historical Document, *Registration card on the numerical and combat composition of units of the 7 Guards A, 8 August 1943,* From TsAMO Archive ID #133373361.
[285] S. M. Davtyan, *Fifth Aerial, Military historical outline of the combat path of the 5th Air Army during the Great Patriotic War,* Military Publishing House, 1990, Chapter "On the Stepp Front.
[286] Colonel M. F. Vuitenko, *Organization of the Medical Support of the 7th Guards Army in Belgorod Kharkov Offensive Operation*, Military-Medical Journal #8, August 1983, 40th Anniversary of the Battle of Kursk, 4.
[287] Richard W. Harrison, Editor and Translator, *The Battle of Kursk* Soviet general staff, Helion & Company, 2016, 352.
[288] Ibid., 351.
[289] Karl-Heinz Frieser, *Germany and the Second World War, Volume VIII,* Clarendon Press Oxford, 2017, 357.
[290] H. G. W. Davie, The Journal of Slavic Military Studies, 2020, Vol 33 No 4, *The Logistics of the Combined-Arms-Army-the Rear: High Mobility Through Limited Means*, 599-600.
[291] Earl F. Ziemke, *Stalingrad to Berlin: The German Defeat in the East*, Office of the Chief of Military History, Washington D.C., 1969, 174.
[292] Karl-Heinz Frieser, *Germany and the Second World War, Volume VIII,* Clarendon Press Oxford, 2017, 186.
[293] Based upon the Historical Document, *Summary of material support of parts and formations 7 Guards A, 17 August 1943,* From TsAMO Archive ID #133390169.
[294] Hugh Davie, *Steppe Front Rear Units 17th July 1943* From TsAMO Archive ID # unknown. And from TsAMO Archive ID #133390884, *Report on the work of the road troops 7 Guards A, 23 August 1943*.
[295] Robert Forczyk, *Kursk 1943 the Northern Front,* Osprey Publishing 2014, 89.
[296] Karl-Heinz Frieser, *Germany and the Second World War, Volume VIII*, Clarendon Press Oxford, 2017, 197, 196.
[297] Murray Williamson, *Strategy For Defeat, The Luftwaffe1933-1945*, Air University Press, 1983, 182-183.
[298] Earl F. Ziemke, *Stalingrad to Berlin The German Defeat in the East*, Office of the Chief of Military History, Washington D.C., 1969, 156.
[299] S. M. Davtyan, *Fifth Aerial, Military historical outline of the combat path of the 5th Air Army during the Great Patriotic War,* Military Publishing House, 1990, Chapter "On the Stepp Front."
[300] Alexey Grigorievich, Found within https://iremember.ru/memoirs/artilleristi/chicherov-aleksey-grigorevich/.
[301] Colonel M. F. Vuitenko, *Organization of the Medical Support of the 7th Guards Army in Belgorod Offensive Operation*, Military-Medical Journal #8, August 1983, 40th Anniversary of the Battle of Kursk, 3-5.
[302] David M. Glantz, Jonathan M. House, *When Titans Clashed*, University Press Kansas, 1995, 296.
[303] Karl-Heinz Frieser, *Germany and the Second World War, Volume VIII*, Clarendon Press Oxford, 2017, 197, 198.
[304] Based upon the Historical Document, *Sanitary service during the Belgorod Kharkov operation*, p 11, From TsAMO Archive ID #133390890.
[305] Richard W. Harrison, Editor and Translator, *The Battle of Kursk,* Soviet general staff, Helion & Company, 2016, 356.
[306] Prit Buttar, *Retribution The Soviet Reconquest of Central Ukraine 1943,* Osprey, 2019, 195.
[307] Colonel M. F. Vuitenko, *Organization of the Medical Support of the 7th Guards Army in Belgorod Kharkov Offensive Operation*, Military-Medical Journal #8, August 1983, 40th Anniversary of the Battle of Kursk, 5.
[308] Based upon the Historical Document, *Report of Operations 7 Guards A, 3-24 August 43,* p 34, From TsAMO Archive ID #13233621.

[309] Georgy Zhukov, *Marshal of the Soviet Union G Zhukov Reminiscences and Reflections*, Progress Publishers, Moscow, 1985, 216.
[310] Von Hardesty, Ilya Grinberg, *Red Phoenix Rising*, University Press of Kansas, 2012, 268.
[311] Data provided by Hugh Davie based upon his own research using, *A brief overview of the operation of the rear of the Steppe Front during the preparation and conduct of the Belgorod Kharkov operation from August 3rd to 23rd, 1943.* From TsAMO Archive ID #113984469 Fond:240 Opus:0002779 Doc:0453. Also - *Steppe Front Rear Units 17th July 1943* From TsAMO Archive ID # unknown, Rear of the Steppe Front during Operation Rumyantsev.
[312] The city of Kharkov, the fourth largest city in the Soviet Union, was lost to the Germans for the first time during the invasion of the USSR in 1941. Then it was recaptured by the Red Army in February 1942, during the Operation Star winter counteroffensive. Then it was lost yet again in March 1942 to von Manstein's backhand blow attack. The final battle for Kharkov was successfully concluded by the Red Army, Operation Rumianysev, with its final capture in late August 1943.
[313] Ivan Stepanovich Konev, *Front Commander's Notes*, Militera.lib.ru (website) 1972, Chapter 2.
[314] Prit Buttar, *Retribution The Soviet Reconquest of Central Ukraine 1943,* Osprey, 2019, 216-217.
[315] Based upon the Historical Document, *The Schedule of the Combat Personnel 7 Guards A, 4 September 43,* p 34, From TsAMO Archive ID #132344467.
[316] S. S. Harutyunian, *Under the banner of the guard: The combat path of the 15th Guards Rifle Kharkov-Prague Order of Lenin, twice the Red Banner, Orders of Suvorov and Kutuzov Division,* Hayastan, 1982, Chapter 5.
[317] Based upon the Historical Document, *Data on the State of the Railway in the Zone 7 Guards A,* From TsAMO Archive ID #133390181, 3.
[318] Igor Vitukhin, *Soviet Generals Recall World War II,* Sphinx Press Inc, 1981, 278.
[319] S.S Harutyunian, *Under the banner of the guard: The combat path of the 15th Guards Rifle Kharkov-Prague Order of Lenin, twice the Red Banner, Orders of Suvorov and Kutuzov Division,* Hayastan, 1982, Chapter 5.
[320] Ibid.,
[321] Based upon the Historical Document, *Statement of Combat and Strength of Formations 7 Guards A, 10 September 43,* From TsAMO Archive ID #114646064.
[322] Based upon the Historical Document, *Calculation of the Weight of one Army Daily Supply of Food Stuffs and one Army Fueling of Fuel and Lubricants 7 Guards A, 10 September 43,* From TsAMO Archive ID #133390877.
[323] Based upon research by Douglas E. Nash, via Facebook - Eastern Front Research Forum, 12 November 2021.
[324] Erhard Raus, *Panzer Operations The Eastern Front Memoir of General Raus 1941-1945*, Da Capo Press, 2003, 148.
[325] Ivan Morozov, *From Stalingrad To Prague.* Volgograd: Nizhne-Volzhskoye Publishing House, 1976, 150.
[326] Major General Mungo Melvin, *Manstein Hitler's Greatest General*, Thomas Dunne books, 2010, 396.
[327] Dr. Hans Heinz Rehfeldt, *Mortar Gunner on the Eastern Front, Vol II*, Greenhill Books, 2019. 10.
[328] Ibid., 13.
[329] Erhard Raus, *Panzer Operations The Eastern Front Memoir of General Raus 1941-1945*, Da Capo Press, 2003, 252-254.
[330] Charles C. Sharp, *Soviet Order of Battle in World War II, Volume IV, Red Guards,* Published by George F. Nafziger, 1995.
[331] Team of Authors, *Railway Troops of Russia*, Militera.lib.ru, 2002, Book 3 Chapter 5.
[332] Boris Ilyich Mutovin, *Through All the Trials*, Military Publishing House, 1986, Chapter 5.
[333] A. K. Morozov, N. I. Yashchenko, P. S. Belan, H. P. Kalita, *73 Guards,* Alma-Ata: Kazakhstan, 1986, 12.
[334] Richard W. Harrison, Editor and Translator, *The Battle for the Dnepr The Red Army's Forcing of the East Wall, September-December 1943*, Helion & Company, 2018, 246.
[335] Based upon the Historical Document, *Information about the number of mobilized local population in the ranks of the Red Army in the territory liberated from the enemy units, 7 guards A, 25 September 43,* From TsAMO Archive ID #114646069.
[336] Paul Carell, *Scorched Earth The Russian-German War, 1943-1944*, Little Brown and Company, 1966, 328-329.
[337] Richard W. Harrison, Editor and Translator, *The Battle for the Dnepr The Red Army's Forcing of the East Wall, September-December 1943*, Helion & Company, 2018, 261.
[338] Based upon the Historical Document, *The schedule of the combat personnel 7 Guards A, 29 September 43,* From TsAMO Archive ID #132344470.
[339] Based upon the Historical Document, *Information on the progress of the crossing of the Dnepr River troops 7 Guards A, 29 September 43,* From TsAMO Archive ID #455185538.
[340] Werner Haupt, *Army Group South The Wehrmacht in Russia 1941-1945*, Schiffer Military History, 1998, 295-296.
[341] Based upon the Historical Document, *Certificate of personnel losses 7 guards A, 1- 30 September 43,* From TsAMO Archive ID #114646074.
[342] S.M. Davtyan, *Fifth Aerial*, Military Publishing House, 1990, Chapter "In the battle for the Dnepr."
[343] Dr. Alfred Price, *Spearhead for Blitzkrieg*, Greenhill Books, 1996, 149.
[344] Paul Carell, *Scorched Earth, The Russian-German War, 1941-1944*, Little Brown and Company, 1966, 345.
[345] Karl-Heinz Frieser, *Germany and the Second World War, Volume VIII*, Clarendon Press Oxford, 2017, 244.
[346] David M. Glantz, *Colossus Reborn The Red Army at War, 1941 -1943*, University Press Kansas, 2005, 155.
[347] Ivan Stepanovich Konev, *Front Commander's Notes*, Militera.lib.ru (website) 1972, Chapter 2.

[348] Based upon the Historical Document, *Information on the combat composition of the troops 7 Guards A, and crossed to the southern bank of the river Dnepr,* From TsAMO Archive ID #114646091.
[349] Charles C. Sharp, *Soviet Order of Battle in World War II, Volume VI, Red Thunder,* Published by George F. Nafziger, 1995, 4.
[350] David Glantz, *Soviet Military Deception in the Second World War,* Frank Class, 1989, 274.
[351] Richard W. Harrison, Editor and Translator, *The Battle for the Dnepr The Red Army's Forcing of the East Wall, September-December 1943*, Helion & Company, 2018, 281, 289.
[352] Richard N. Armstrong, *Red Army Tank Commanders the Armored Guards*, Schiffer Military History, 1994, 359.
[353] Boris Ilyich Mutovin, *Through All the Trials*, Military Publishing House, 1986, Chapter 5.
[354] Earl F. Ziemke, *Stalingrad to Berlin: The German Defeat in the East*, Office of the Chief of Military History, Washington D.C., 1969, 182-183.
[355] Thomas L. Jentz, Editor, *Panzertruppen, Vol 2*, Schiffer Military History,1996, 47-64.
[356] Team of Authors, *Railway Troops of Russia*, Militera.lib.ru, 2002, Book 3 Chapter 8.
[357] From the web site: *http://www.warheroes.ru/hero/hero.asp?Hero_id=612,* (Retrieved on 22 February 2020).
[358] Based upon the Historical Document, *Report on personnel losses in units and formations of 7 Guards A, from 1 October to 29 October 1943*, From TsAMO Archive ID #114646101.
[359] Roger R. Reese, *Why Stalin's Soldiers Fought*, University Press of Kansas, 2011, 168.
[360] Jay A. Stout, *Fortress Ploesti the Campaign to Destroy Hitler's Oil,* Casemate, 2011, 7-8.
[361] Andrew Brookes, *Air War Over Russia,* Ian Allan Publishing, 2003, 132.
[362] Karl-Heinz Frieser, *Germany and the Second World War, Volume VIII,* Clarendon Press Oxford, 2017, 246.
[363] Field Marshal Erich von Manstein, *Lost Victories*, Presidio Press, 1982, 483.
[364] Richard W. Harrison, Editor and Translator, *The Battle for the Dnepr The Red Army's Forcing of the East Wall, September-December 1943*, Helion & Company, 2018, 281, 206-307.
[365] Team of Authors, *Railway Troops of Russia*, Militera.lib.ru, 2002, Book 3 Chapter 8.
[366] From the Historical Document, *The Reduced Statement of Fighting Personnel of 7th Guards Army,* From TsAMO Archive ID #131308929.
[367] From the Historical Document, *Henceforth special order 37 A and 57 A go on to Strong Defense, 30 November 43,* From TsAMO Archive ID #455185330.
[368] David M. Glantz, *Forgotten Battle of the German-Soviet War 1941-1945, Vol V, Part two*, 2000, 739-742.
[369] Ibid., 741-742.
[370] Graham H. Turbiville Jr, *Soviet Operational Logistics, 1939-1990*, 2014, page 297. Accessed via the RKKA web site.
[371] Team of Authors, *Railway Troops of Russia*, Militera.lib.ru, 2002, Book 3 Chapter 8.
[372] David Glantz, *Soviet Military Deception in the Second World War,* Frank Class, 1989, 306.
[373] Geoffrey P. Megargee, *Inside Hitler's High Command*, University Press Kansas, 2000, 207.
[374] David Glantz, *Soviet Military Deception in the Second World War,* Frank Class, 1989, 306.
[375] Kabanov Pavel Alekseevich, *Steel Spans*, Military Publishing House, 1973, militera.lib.ru, Chapter "Dnepr Bridges."
[376] Steve Crawford, *The Easter Front Day by Day, A Photographic Chronology*, Potomac Books, 2006, 143.
[377] Ivan Vladimirovich Kovalev, *Transport in the Great Patriotic War 1941–1945*, military.lib.ru, 1981, Chapter 11.
[378] Guy Sajer, *The Forgotten Soldier*, Harper & Row Publishers, 1971, 24-37.
[379] Friedrich Ruge, *The Soviets as Naval Opponents 1941-1945*, Patrick Stephens, Cambridge, 1979, 125.
[380] Grylev Anatoly Nikolaevich, *Dnepr - Carpathians – Crimea* - Military Science, 1970, Appendix 3, Army Command.
[381] Pavel Alekseevich Rotmistrov, *Steel Guard*, Moscow, Military Publishing, 1984. 256.
[382] Field Marshal Erich von Manstein, *Lost Victories*, Presidio Press, 1982, 501-505.
[383] S. M. Davtyan, *Fifth Aerial*, Military Publishing House, 1990, Chapter "Over the Right-Bank Ukraine."
[384] Karl-Heinz Frieser, *Germany and the Second World War, Volume VIII*, Clarendon Press Oxford, 2017, 383-384.
[385] Pavel Alekseevich Rotmistrov, *Steel Guard*, Moscow, Military Publishing, 1984. 257.
[386] Ibid.,
[387] Peter G. Tsouras, Editor, *The Anvil of War*, Greenhill Books, 1994, 199.
[388] Interview with S. Otroschenkov, from the following site, https://iremember.ru/memoirs/tankisti/otroschenkov-sergey-andreevich/.
[389] Peter G. Tsouras, Editor, *The Anvil of War,* Greenhill Books, 1994, 200.
[390] Yaroshenko Alexey Andreevich, *The 41st Guards went into battle,* Military Publishing, Moscow, 1982, 90.
[391] Karl-Heinz Frieser, *Germany and the Second World War, Volume VIII,* Clarendon Press Oxford, 2017, 389.
[392] Based upon the Historical Document, *Armored summary of the BT and MV headquarters 7th Guards A*, 24 January 1944, From TsAMO Archive ID #132309778.
[393] Friedrich Ruge, *The Soviets as Naval Opponents 1941-1945*, Patrick Stephens, Cambridge, 1979, 126.
[394] Yaroshenko Alexey Andreevich, *The 41st Guards went into battle,* Military Publishing, Moscow, 1982, 91.
[395] Niklas Zetterling, Anders Frankson, *The Korsun Pocket*, Casemate, 2008, 187.

[396] Based upon the Historical Document, *Conclusion in the case of the explosion of captured ammunition*, 12 February 1944, TsAMO Archive ID #133373860.
[397] Grylev Anatoly Nikolaevich, *Dnepr- Carpathians – Crimea,* Military Science, 1970, Chapter 2.
[398] Rolf Hinze, *Crucible of Combat,* Helion & Company Ltd, 2009, 191.
[399] Niklas Zetterling, Anders Frankson, *The Korsun Pocket*, Casemate, 2008, 277.
[400] Alexander Werth, *Russia At War 1941-1945*, Carroll & Graf, 1984, 782.
[401] Karl-Heinz Frieser, *Germany and the Second World War, Volume VIII,* Clarendon Press Oxford, 2017, 409.
[402] This order was retrieved from the following web site, *Collection of combat documents of the Great Patriotic War, Issue 12*, Moscow, 1954, http://www.teatrskazka.com/Raznoe/SbornikBoevyhDokumentov/Issue12/Issue12.html.
[403] Chanturiya, Kazbek, *After 73 years, the memory of Stalin's deportation of Chechens and Ingush still haunts the survivors,*2017, https://oc-media.org/features/after-73-years-the-memory-of-stalins-deportation-of-chechens-and-ingush-still-haunts-the-survivors/, Accessed 23 March 2020.
[404] Henrik O. Lunde, *Finland's War of Choice*, Casemate 2011, 259-260.
[405] David Glantz, *Soviet Military Deception in the Second World War,* Frank Class, 1989, 341-343.
[406] Based upon the Historical Document, *Report on aviation actions in the 7th Guards Army area,* 1-28 February 1944, TsAMO Archive ID #133672977.
[407] Kelly Bell, *The Ukrainian Underground Army, Strategy & Tactics Magazine #324 Sep-Oct 2020*, 73.
[408] Kovalev Ivan Vladimirovich, *Transport in the Great Patriotic War 1941–1945*, military.lib.ru, 1981, Chapter 11.
[409] Ivan Stepanovich Konev, *Front Commander's Notes*, Militera.lib.ru (website) 1972, Chapter 4.
[410] Karl-Heinz Frieser, *Germany and the Second World War, Volume VIII,* Clarendon Press Oxford, 2017, 416.
[411] David Glantz, *Soviet Military Deception in the Second World War,* Frank Class, 1989, 341.
[412] Ivan Stepanovich Konev, *Front Commander's Notes*, Militera.lib.ru (website) 1972, Chapter 4.
[413] Perry Pierik, *Hungary1944-1945 The Forgotten Tragedy,* ASPEKT, 1998, 18-34.
[414] Based upon the Historical Document, TsAMO Archive ID #113985200.
[415] Earl F. Ziemke, *Stalingrad to Berlin: The German Defeat in the East*, Office of the Chief of Military History, Washington D.C., 1969, 287.
[416] Ibid.,
[417] From the web site, *From Balloons to Drones,* https://balloonstodrones.com/, From the article *Attacking Refugees for Military Effect During the First Iasi-Kishinev Offensive,* by Dr. Luke Truxal, accessed August 2020.
[418] Karl-Heinz Frieser, *Germany and the Second World War, Volume VIII,* Clarendon Press Oxford, 2017, 442-443.
[419] This reference was retrieved from the following web site, *military.wikia.org/wiki/Hitler–Manstein dispute*, Accessed March 2020.
[420] Steven H. Newton, *Hitler's Commander*, DA Capo Press, 2006, 275-276.
[421] Ivan Stepanovich Konev, *Front Commander's Notes*, Militera.lib.ru (website) 1972, Chapter 5.
[422] From the web site, *From Balloons to Drones,* https://balloonstodrones.com/, From the article *Attacking Refugees for Military Effect During the First Iasi-Kishinev Offensive,* by Dr. Luke Truxal, accessed August 2020.
[423] Jay A. Stout, *Fortress Ploesti the Campaign to Destroy Hitler's Oil*. Casemate, 2011, 102-103.
[424] U.S. National Archives, T-78 Roll 136, *Captured German Records, Headquarters OKH*, image #426.
[425] Friedrich Ruge, *The Soviets as Naval Opponents 1941-1945*, Patrick Stephens, Cambridge, 1979, 128.
[426] Kamanin Nikolay Petrovich, *Pilots and Astronauts,* Politizdat, 1971, 269.
[427] Harold Shukman, Editor, *Stalin's Generals,* Grove Press, 1993, 297.
[428] Brian Taylor, *Barbarossa to Berlin A Chronology of the Eastern Front 1941 to 1945,* Vol 2, Spellmount, 2004, 175.
[429] David M. Glantz, *Red Storm Over the Balkans*, University Press Kansas, 2007, 184-188.
[430] Ibid., 102.
[431] Ivan Stepanovich Konev, *Front Commander's Notes*, Militera.lib.ru (website) 1972, Chapter 5.
[432] Earl F. Ziemke, *Stalingrad to Berlin The German Defeat in the East*, Office of the Chief of Military History, Washington D.C., 1969, 311-312.
[433] This Goodreads quote came from https://www.goodreads.com and was retrieved during May 2020.
[434] Based upon the Historical Document, TsAMO Archive ID #113723600, page 4 and 17.
[435] Based upon the Historical Document, TsAMO Archive ID #132344640.
[436] Axis History Forum Staff, Art, *Let's Build The Battle of Targul Frumos,* forum.axishistory.com, accessed June 2020.
[437] David M. Glantz, *Red Storm Over the Balkans* University Press Kansas, 2007, 214.
[438] Ibid., p 215.
[439] Ibid., p 217.
[440] Prit Buttar, *The Reckoning, The Defeat of Army Group South 1944,* Osprey, 2020, 350.
[441] Based upon the Historical Document, TsAMO Archive ID #113984903 *Extract from the combat operations log of the 2nd UkrF for the month of May 1944.*
[442] David M. Glantz, *Red Storm Over the Balkans*, University Press Kansas, 2007, 249.

[443] Ibid., 250.
[444] Ibid., 251.
[445] Ibid., 253-257.
[446] Ibid., 258.
[447] Ibid., 273.
[448] The original web site, http//ww2stats,com is inactive, but an archive copy can be found here: https://web.archive.org/web/20161110010821/http://ww2stats.com/cas_ger_okh_dec44.html. These are 10 -Day Casualty Reports per Army/Group, 1944.
[449] Rolf Hinze, *Crucible of Combat,* Helion & Company Ltd, 2009, 296.
[450] David M. Glantz, *Red Storm Over the Balkans*, University Press Kansas, 2007, 264-268.
[451] John Erickson, *The Road to Berlin*, Weidenfeld and Nicolson, London, 1983, 191.
[452] David M. Glantz, *Red Storm Over the Balkans*, University Press Kansas, 2007, 270-271.
[453] Alexander Hill, *The Great Patriotic War of the Soviet Union, 1942-45,* Routledge, 2009, 227.
[454] Colonel-General G.F. Krivosheev, Editor, *Soviet Casualties and Combat Losses in the Twentieth Century,* Greenhill Books, 1997, Table 75.
[455] Robert Goralski, *World War II Almanac 1931-1945*, Hamish Hamilton, London, 1981, 315.
[456] Edward Davidson, Dale Manning, *Timeline of World War II*, Cassell & Co., London, 1999, 190.
[457] Robert Goralski, *World War II Almanac 1931-1945*, Hamish Hamilton, London, 1981, 315.
[458] Mark Axworthy, *Third Axis Fourth Ally*, Arms And Armour, 1995, 157.
[459] The original web site, http//ww2stats,com is inactive, but an archive copy can be found here: https://web.archive.org/web/20161110010821/http://ww2stats.com/cas_ger_okh_dec44.html. These are 10 -Day Casualty Reports per Army/Group, 1944.
[460] Boris Sokolov, *Marshal Malinovskii: Hero of the Soviet Union*, Helion & Company, 2017, 270.
[461] Russian Archives: The Great Patriotic War, *Stavka VGK: Documents and Materials 1944-1945*, Document #110, 92.
[462] David Glantz, *Soviet Military Deception in the Second World War,* Frank Class, 1989, 351-353.
[463] Ibid., 356.
[464] David M. Glantz, *Red Storm Over the Balkans*, University Press Kansas, 2007, 333.
[465] Ibid., 334-339.
[466] Alfred Price, *The Last Year of the Luftwaffe, May 1944 to May 1945*, Arms and Armour, 1991, 22.
[467] Mitch Williamson, *Soviet Operational Level Logistics in 1944 part-II,* Weapons and War website, 2019, Accessed August 2020. (https://weaponsandwarfare.com/2019/10/13/soviet-operational-level-logistics-in-1944-part-ii/).
[468] The original web site, http//ww2stats,com is inactive, but an archive copy can be found here: https://web.archive.org/web/20161110010821/http://ww2stats.com/cas_ger_okh_dec44.html. These are 10 -Day Casualty Reports per Army/Group, 1944.
[469] Steve Crawford, *The Easter Front Day by Day, A Photographic Chronology*, Potomac Books, 2006, 151.
[470] David M. Glantz, *Red Storm Over the Balkans*, University Press Kansas, 2007, 362.
[471] Michael Korda, *Ike, An American Hero*, Harper Perennial, 2007, 480.
[472] John Erickson, *The Road to Berlin*, Weidenfeld and Nicolson, London, 1983, 210.
[473] Henrik O. Lunde, *Finland's War of Choice*, Casemate 2011, 272.
[474] Dale Manning, Edward Davidson. *Timeline of World War II*, Cassell, 1999, 198.
[475] Henrik O. Lunde, *Finland's War of Choice*, Casemate 2011, 312.
[476] Bekker, Cajus, *The Luftwaffe war Diaries*, De Capo Press, 1994, 355.
[477] John Erickson, *The Road to Berlin*, Weidenfeld and Nicolson, London, 1983, 214-225.
[478] Karl-Heinz Frieser, *Germany and the Second World War, Volume VIII,* Clarendon Press Oxford, 2017, 526-530.
[479] Jay A. Stout, *Fortress Ploesti the Campaign to Destroy Hitler's Oil*. Casemate, 2011, 190.
[480] Henrik O. Lunde, *Finland's War of Choice*, Casemate 2011, 301-304.
[481] Vinogradov Timofey Zakharovich, *Expensive-Forever,* Moscow: Military Publishing, 1977, 126.
[482] Charles C. Sharp, *Soviet Order of Battle in World War II, Volume III, Red Storm,* Published by George F. Nafziger, 1995, 70.
[483] Robert Goralski, *World War II Almanac 1931-1945*, Hamish Hamilton, London, 1981, 329.
[484] Richard W. Harrison, Editor and Translator, *The Iasi-Kishinev Operation*, Helion & Company, 2017, 19.
[485] Ibid., 20-21.
[486] Rolf Hinze, *With the Courage of Desperation,* Helion & Company Ltd, 2013, 141.
[487] David Glantz, *Soviet Military Deception in the Second World War,* Frank Class, 1989, 419.
[488] Mark Axworthy, *Third Axis Fourth Ally*, Arms and Armour, 1995, 159-160.
[489] David Glantz, *Soviet Military Deception in the Second World War,* Frank Class, 1989, p xxxvii, 33.
[490] Steve Crawford, *The Easter Front Day by Day, A Photographic Chronology*, Potomac Books, 2006, 156-162.
[491] Boris Sokolov, *Marshal Malinovskii: Hero of the Soviet Union*, Helion & Company, 2017, 276.
[492] Ibid., 270, 280.

[493] David Glantz, *Soviet Military Deception in the Second World War,* Frank Class, 1989, 412.
[494] Henrik O. Lunde, *Finland's War of Choice*, Casemate 2011, 301-312.
[495] William L Shirer, *Rise and Fall of the Third Reich,* Simon and Schuster, 1960, Chapter 29, 1033-1072.
[496] John Erickson, *The Road to Berlin,* Weidenfeld and Nicolson, 1983, 267.
[497] Edward Davidson, Dale Manning, *Timeline of World War II*, Cassell & Co., London, 1999, 205.
[498] Boris Sokolov, *Marshal Malinovskii: Hero of the Soviet Union*, Helion & Company, 2017, 279.
[499] Johannes Hans Friessner, *Lost Battles (or Loyal Battles),* Holsten Verlag, 1956, 53.
[500] John Erickson, *The Road to Berlin,* Weidenfeld and Nicolson, 1983, 242.
[501] Ibid., 267.
[502] Henrik O. Lunde, *Finland's War of Choice*, Casemate 2011, 313.
[503] Ibid., 314.
[504] Richard W. Harrison, Editor and Translator, *The Iasi-Kishinev Operation*, Helion & Company, 2017, 75.
[505] I. K. Morozov, *From Stalingrad to Prague.* Volgograd, Lower-Volzh book Publishing House, 1976, 185.
[506] John Erickson, *The Road to Berlin,* Weidenfeld and Nicolson, 1983, 275.
[507] This medical table was translated by Art who lives in Moscow, and brilliantly supports the Soviet Union section of the Axis History Forum, https://forum.axishistory.com/index.php. The table was translated on 1 October 2020.
[508] Richard W. Harrison, Editor and Translator, *The Iasi-Kishinev Operation*, Helion & Company, 2017, 79-81.
[509] Rolf Hinze, *With the Courage of Desperation,* Helion & Company Ltd, 2013, 148.
[510] Brian Taylor, *Barbarossa to Berlin A Chronology of the Eastern Front 1941 to 1945,* Vol 2, Spellmount, 2004, 208, 213.
[511] Henrik O. Lunde, *Finland's War of Choice*, Casemate 2011, 315.
[512] Richard W. Harrison, Editor and Translator, *The Iasi-Kishinev Operation*, Helion & Company, 2017, 77.
[513] Mark Axworthy, *Third Axis Fourth Ally*, Arms and Armour, 1995, 162.
[514] Richard W. Harrison, Editor and Translator, *The Iasi-Kishinev Operation*, Helion & Company, 2017, 36.
[515] Johannes Hans Friessner, *Lost Battles (or Loyal Battles),* Holsten Verlag, 1956, 66.
[516] David Glantz, *Soviet Military Deception in the Second World War,* Frank Class, 1989, 419-420.
[517] Mark Axworthy, *Third Axis Fourth Ally*, Arms and Armour, 1995, 157-158.
[518] Ibid., 162.
[519] David Glantz, *Soviet Military Deception in the Second World War,* Frank Class, 1989, 415.
[520] Richard W. Harrison, Editor and Translator, *The Iasi-Kishinev Operation*, Helion & Company, 2017, 39, 41.
[521] Ibid., 75-76.
[522] I. K. Morozov, *From Stalingrad to Prague.* Volgograd, Lower-Volzh book Publishing House, 1976, 186.
[523] David Glantz, *Soviet Military Deception in the Second World War,* Frank Class, 1989, 417.
[524] Johannes Hans Friessner, *Lost Battles (or Loyal Battles),* Holsten Verlag, 1956, 67.
[525] Brian Taylor, *Barbarossa to Berlin A Chronology of the Eastern Front 1941 to 1945,* Vol 2, Spellmount, 2004, 208.
[526] Charles C. Sharp, *Soviet Order of Battle in World War II, Volume IV, Red Guards,* Published by George F. Nafziger, 1995. 25.
[527] Richard W. Harrison, Editor and Translator, *The Iasi-Kishinev Operation* Helion & Company, 2017, 80, 99.
[528] Mihai Tone Filipescu, *Reluctant Axis, the Romanian Army in Russia 1941-1944,* FTM, 2006, 96-97.
[529] John Erickson, *The Road to Berlin,* Weidenfeld and Nicolson, 1983, 356.
[530] Johannes Hans Friessner, *Lost Battles (or Loyal Battles),* Holsten Verlag, 1956, 76.
[531] I. K. Morozov, *From Stalingrad to Prague.* Volgograd, Lower-Volzh book Publishing House, 1976, 187-188.
[532] Perry Pierik, *Hungary 1944-1945 The Forgotten Tragedy*, 1998, Aspekt, 55.
[533] From the web site, https://mil-history.livejournal.com/, *Memories of a Soviet soldier about being taken prisoner in Romania, 1943-44,* Accessed August 2020. Taken from the book: *S.Z. Yatrupolo. "Memories of War".* Military literature, 2004.
[534] Mark worthy, *Third Axis Fourth Ally*, Arms and Armour, 1995, 193.
[535] Henrik O. Lunde, *Finland's War of Choice*, Casemate 2011, 317.
[536] Vinogradov Timofey Zakharovich, *Expensive-Forever*, Moscow, Military Publishing, 1977, 128-129.
[537] Mark Axworthy, *Third Axis Fourth Ally*, Arms and Armour, 1995, 186.
[538] Richard W. Harrison, Editor and Translator, *The Iasi Kishinev Operation*, Helion & Company, 2017, 285.
[539] Brian Taylor, *Barbarossa to Berlin A Chronology of the Eastern Front 1941 to 1945,* Vol 2, Spellmount, 2004, 220.
[540] Boris Sokolov, *Marshal Malinovskii: Hero of the Soviet Union*, Helion & Company, 2017, 272.
[541] Richard W. Harrison, Editor and Translator, *The Iasi-Kishinev Operation*, Helion & Company, 2017, 145.
[542] Henrik O. Lunde, *Finland's War of Choice*, Casemate 2011, 317.
[543] Johannes Hans Friessner, *Lost Battles (or Loyal Battles),* Holsten Verlag, 1956, 114.
[544] Mitch Williamson, *Soviet Operational-Level Logistics in 1944, Part I*, Retrieved November 2020 from - https://weaponsandwarfare.com/2019/10/13/soviet-operational-level-logistics-in-1944-part-i/.
[545] Boris Sokolov, *Marshal Malinovskii: Hero of the Soviet Union*, Helion & Company, 2017, 282.
[546] Mihai Tone Filipescu, *Reluctant Axis, the Romanian Army in Russia 1941-1944,* FTM, 2006, 97.

[547] Karl-Heinz Frieser, *Germany and the Second World War, Volume VIII,* Clarendon Press Oxford, 2017, 816.
[548] Boris Sokolov, *Marshal Malinovskii: Hero of the Soviet Union*, Helion & Company, 2017, 280.
[549] Ammon Sella, *The Value of Human Life in Soviet Warfare,* Routledge, 1992, 65.
[550] Robert Goralski, *World War II Almanac 1931-1945*, Hamish Hamilton, London, 1981, 342.
[551] *Institute of Military History of the Ministry of Defense of the Russian Federation, Russian Archive 16 (5-4) headquarters of the Supreme Command 1944-1945*, Moscow, 1999, #25, 295-296.
[552] Rolf Hinze, *With the Courage of Desperation,* Helion & Company Ltd, 2013, 196.
[553] Karl-Heinz Frieser, *Germany and the Second World War, Volume VIII,* Clarendon Press Oxford, 2017, 824-825.
[554] Mark Axworthy, *Third Axis Fourth Ally*, Arms and Armour, 1995, 194.
[555] Mark Axworthy, *Third Axis Fourth Ally*, Arms and Armour, 1995, 186.
[556] Johannes Hans Friessner, *Lost Battles (or Loyal Battles),* Holsten Verlag, 1956, 99.
[557] Henrik O. Lunde, *Finland's War of Choice*, Casemate 2011, 332.
[558] Henrik O. Lunde, *Finland's War of Choice*, Casemate 2011, 319.
[559] Karl-Heinz Frieser, *Germany and the Second World War, Volume VIII,* Clarendon Press Oxford, 2017, 997.
[560] Alfred Price, *The Last Year of the Luftwaffe, May 1944 to May 1945*, Arms and Armour, 1991, 96.
[561] S. M. Davtyan, *Fifth air*, Moscow: Military Publishing, 1990, 211.
[562] John Erickson, *The Road to Berlin,* Weidenfeld and Nicolson, 1983, 391.
[563] Boris Sokolov, *Marshal Malinovskii: Hero of the Soviet Union*, Helion & Company, 2017, 282, 283.
[564] Henrik O. Lunde, *Finland's War of Choice*, Casemate 2011, 349.
[565] Karl-Heinz Frieser, *Germany and the Second World War, Volume VIII,* Clarendon Press Oxford, 2017, 999-1000.
[566] Chris Bellamy, *Absolute War*. Alfred A. Knopf, New York, 2007, 627-629.
[567] Found on the web site, forced march - Wiktionary.
[568] Peter G. Tsouras, *Warrior's Words*, Cassell Arms and Armour, 1992, 246.
[569] Found on the web site, forced march - Wiktionary.
[570] Charles Whiting, *Skorzeny Ballantine's Illustrated History of a Violent Century, Book #11,* 1972, 82.
[571] Perry Pierik, *Hungary 1944-1945 The Forgotten Tragedy,* ASPEKT, 1998, 69-70.
[572] Krystyna Wituska, *Inside a Gestapo Prison: The Letters of Krystyna Wituska, 1942-1944,* Wayne State University Press, 2006, xxii.
[573] Rolf Hinze, *With the Courage of Desperation,* Helion & Company Ltd, 2013, 262.
[574] Mark Axworthy, *Third Axis Fourth Ally*, Arms and Armour, 1995, 201-202.
[575] Perry Moore, *Panzerschlacht, Armoured Operations on the Hungarian Plains September-November 1944,* Helion & Company, 2008, 127.
[576] George F. Nafziger, *Rumanian Order of Battle World War II*, privately published, 1995, 6.
[577] Johannes Hans Friessner, *Lost Battles (or Loyal Battles),* Holsten Verlag, 1956, 159.
[578] Ibid., 163.
[579] S. M. Davtyan, *Fifth Air,* Moscow: Military Publishing, 1990, 212-213.
[580] Ibid., 212-216.
[581] Richard Muller, *The German Air War in Russia*, N&A Publishing, 1992, 225.
[582] Johannes Hans Friessner, *Lost Battles (or Loyal Battles),* Holsten Verlag, 1956, 170.
[583] Team of Authors, *Soviet Tank Troops 1941-1945*, Moscow: Military Publishing, 1973. Chapter 12, 235.
[584] Albert Speer, *Inside the Third Reich*, Macmillan Company, 1970, 316-317.
[585] Johannes Hans Friessner, *Lost Battles (or Loyal Battles),* Holsten Verlag, 1956, 176.
[586] Ungvary, Krisztian, *The Siege of Budapest, 100 Days in World War II*, 2005, 23-25.
[587] Krisztian Ungvary, *The Siege of Budapest100 Days in World War II*, 2005, 24.
[588] Richard W. Harrison, Editor and Translator, *The Budapest Operation 1945*, Helion & Company, 2017, 36.
[589] Ibid., 41-42.
[590] Ibid., 42.
[591] Ibid.
[592] David Glantz, *Soviet Military Deception in the Second World War,* Frank Class, 1989, 454-455.
[593] Ibid., 458.
[594] Johannes Hans Friessner, *Lost Battles (or Loyal Battles),* Holsten Verlag, 1956, 179.
[595] Boris Sokolov, *Marshal Malinovskii: Hero of the Soviet Union*, Helion & Company, 2017, 291.
[596] Dr. Hans Heinz Rehfeldt, *Mortar Gunner on the Eastern Front, Vol II*, Greenhill Books, 2019. 160.
[597] Boris Sokolov, *Marshal Malinovskii: Hero of the Soviet Union*, Helion & Company, 2017, 293.
[598] Johannes Hans Friessner, *Lost Battles (or Loyal Battles),* Holsten Verlag, 1956, 174.
[599] Boris Sokolov, *Marshal Malinovskii: Hero of the Soviet Union*, Helion & Company, 2017, 293.
[600] Eduardo M. Gil Martinez, *The Bulgarian Air Force in World War II,* 2017, Kagero Publishing, 63.
[601] Johannes Hans Friessner, *Lost Battles (or Loyal Battles),* Holsten Verlag, 1956, 196.

[602] This order was retrieved from the following web site, *Collection of combat documents of the Great Patriotic War, Issue 22*, Moscow, 1954, http://www.teatrskazka.com/Raznoe/SbornikBoevyhDokumentov/Issue22/Issue22.html.
[603] Norbert Szamveber, *Days of Battle,* Helion & Company, 2013, 11-113.
[604] Dmitriy Loza, *Commanding the Red Army's Sherman Tanks,* University of Nebraska Press, 1996, 56-60.
[605] Johannes Hans Friessner, *Lost Battles (or Loyal Battles),* Holsten Verlag, 1956, 198.
[606] Stepan Naumovich Grechko, *Decisions were made on the Ground*, Moscow: Military Publishing, 1984, 259-260.
[607] Ibid., 208.
[608] Ibid.
[609] Georg Maier, *Drama Between Budapest and Vienna*, J.J. Fedorowicz Publishing, Inc, 2004, 112.
[610] Georg Maier, *Drama Between Budapest and Vienna*, J.J. Fedorowicz Publishing, Inc, 2004, 402.
[611] Boris Sokolov, *Marshal Malinovskii: Hero of the Soviet Union*, Helion & Company, 2017, 296.
[612] John Lukacs, *The Siege of Budapest, 100 Days in World War II*, Yale University Press, 2005, 128.
[613] Georg Maier, *Drama Between Budapest and Vienna*, J.J. Fedorowicz Publishing, Inc, 2004, 402.
[614] John Lukacs, *The Siege of Budapest, 100 Days in World War II*, Yale University Press, 2005, 263.
[615] Philippe Guillemot, *Hungary 1944-45 The Panzers Last Stand*, Histoire & Collections, 2010, 32.
[616] S. M. Davtyan, *Fifth Air,* Moscow: Military Publishing, 1990, 222.
[617] Williamson Murry, *Strategy for Defeat, The Luftwaffe 1933-1945,* Air University Press, Maxwell Air Force Base, 1983, 275.
[618] S. M. Davtyan, *Fifth Air,* Moscow: Military Publishing, 1990, 221.
[619] Von Hardesty, *Red Phoenix, The Rise of Soviet Air Power 1941-1945,* Smithsonian Institution Press, 1982, 257.
[620] R. M. Portuguese Domank, A. S., Kovalenko A. P., *Marshal S.K. Timoshenko,* Publishing House of the Moscow Foundation "Victory-1945", 1994, 317-315.
[621] John Erickson, *The Road to Berlin* Weidenfeld and Nicolson, 1983, 470.
[622] Dmitriy Loza, *Commanding the Red Army's Sherman Tanks,* University of Nebraska Press, 1996, 66.
[623] Alfred Price, *The Last Year of the Luftwaffe, May 1944 to May 1945*, Arms and Armour, 1991, 145.
[624] Norbert Szamveber, *Days of Battle*, Helion & Company, 2013, 125.
[625] Dmitriy Loza, *Commanding the Red Army's Sherman Tanks,* University of Nebraska Press, 1996, 67-68.
[626] Aynutdinov Sergey Khusainovich, *In Memory and Heart,* Yekaterinburg: State Research and Production Enterprise, 2000, Chapter "Ural Worker," 313.
[627] Karl-Heinz Frieser, *Germany and the Second World War, Volume VIII,* Clarendon Press Oxford, 2017, 908.
[628] Norbert Szamveber, *Days of Battle*, Helion & Company, 2013, 138-139.
[629] Boris Sokolov, *Marshal Malinovskii: Hero of the Soviet Union*, Helion & Company, 2017, 299.
[630] John Lukacs, *The Siege of Budapest, 100 Days in World War II*, Yale University Press, 2005, 192.
[631] Norbert Szamveber, *Days of Battle*, Helion & Company, 2013, 152-155.
[632] John Lukacs, *The Siege of Budapest, 100 Days in World War II*, Yale University Press, 2005, 136.
[633] Norbert Szamveber, *Days of Battle*, Helion & Company, 2013, 165-166.
[634] Alfred Price, *The Last Year of the Luftwaffe, May 1944 to May 1945*, Arms and Armour, 1991, 121.
[635] Boris Sokolov, *Marshal Malinovskii: Hero of the Soviet Union*, Helion & Company, 2017, 298.
[636] Taken From the Axis History Forum, *Soviet Fronts strengths: Overview - Page 5 - Axis History Forum.*
[637] Norbert Szamveber, *Days of Battle*, Helion & Company, 2013, 168-169.
[638] Mark Axworthy, *Third Axis Fourth Ally*, Arms and Armour, 1995, 209.
[639] Norbert Szamveber, *Days of Battle*, Helion & Company, 2013, 175-176.
[640] Michael Reynolds, *Men of Steel, I SS Panzer Corps*, Sarpedon, New York, 1999, 176.
[641] Norbert Szamveber, *Days of Battle*, Helion & Company, 2013, 176, note 246.
[642] Karl-Heinz Frieser, *Germany and the Second World War, Volume VIII,* Clarendon Press Oxford, 2017, 913.
[643] Norbert Szamveber, *Days of Battle*, Helion & Company, 2013, 180-181.
[644] Georg Maier, *Drama Between Budapest and Vienna*, J.J. Fedorowicz Publishing, Inc, 2004, 47.
[645] Ibid.
[646] Perry Pierik, *Hungary1944-1945 The Forgotten Tragedy,* ASPEKT, 1998, 144.
[647] Richard W. Harrison, Editor and Translator, *The Budapest Operation 1945*, Helion & Company, 2017, 89.
[648] Norbert Szamveber, *Days of Battle*, Helion & Company, 2013, 181.
[649] Ibid., 182-183.
[650] Alfred C. Mierzejewski, *The Most Valuable Asset of the Reich vol 2*, The University of North Carolina Press, 2000, 159.
[651] Georg Maier, *Drama Between Budapest and Vienna* J.J. Fedorowicz Publishing, Inc, 2004, 113.
[652] Ibid., 120.
[653] Karl-Heinz Frieser, *Germany and the Second World War, Volume VIII,* Clarendon Press Oxford, 2017, 925-926.
[654] Williamson Murry, *Strategy for Defeat, The Luftwaffe 1933-1945,* Air University Press, Maxwell Air Force Base, 1983, 275.
[655] Georg Maier, *Drama Between Budapest and Vienna*, J.J. Fedorowicz Publishing, Inc, 2004, 121.

[656] Alfred C. Mierzejewski, *The Collapse of the German Economy 1944-1945*, The University of North Carolina Press, 1988, 171-172.
[657] Georg Maier, *Drama Between Budapest and Vienna* J.J. Fedorowicz Publishing, Inc, 2004, 124, 144.
[658] TsAMO records ID #133393421, 133393422, 133393427, 133393428.
[659] Norbert Szamveber, *Days of Battle*, Helion & Company, 2013, 193-196.
[660] Ibid., 188.
[661] Georg Maier, *Drama Between Budapest and Vienna*, J.J. Fedorowicz Publishing, Inc, 2004, 73, 107.
[662] Ibid., 402.
[663] This order was retrieved from the following web site, *Collection of combat documents of the Great Patriotic War, Issue 20*, Moscow, 1954, http://www.teatrskazka.com/Raznoe/SbornikBoevyhDokumentov/Issue20/Issue20.html.
[664] This order was retrieved from the following web site, *Collection of combat documents of the Great Patriotic War, Issue 22*, Moscow, 1954, http://www.teatrskazka.com/Raznoe/SbornikBoevyhDokumentov/Issue22/Issue22.html.
[665] Lyuba Vinogradova, *Defending the Motherland,* MacLehose Press, 2015, 148.
[666] Ibid., 144-145.
[667] Georg Maier, *Drama Between Budapest and Vienna*, J.J. Fedorowicz Publishing, Inc, 2004, 127.
[668] Brian Taylor, *Barbarossa to Berlin A Chronology of the Eastern Front 1941 to 1945,* Vol 2, Spellmount, 2004, 276.
[669] Georg Maier, *Drama Between Budapest and Vienna*, J.J. Fedorowicz Publishing, Inc, 2004, 82.
[670] Ibid, 127.
[671] Alfred Price, *The Last Year of the Luftwaffe, May 1944 to May 1945*, Arms and Armour, 1991, 133.
[672] Edward B. Westermann. *Flak German Anti-Aircraft Defenses 1914-1945*. University Press Kansas, 2001, 272-273.
[673] Norbert Szamveber, *Days of Battle*, Helion & Company, 2013, 198-199.
[674] John Lukacs, *The Siege of Budapest100 Days in World War II*, Yale University Press, 2005, 201-204.
[675] Perry Pierik, *Hungary1944-1945 The Forgotten Tragedy,* ASPEKT, 1998, 147-149.
[676] John Lukacs, *The Siege of Budapest100 Days in World War II*, Yale University Press, 2005, 255.
[677] Georg Maier, *Drama Between Budapest and Vienna*, J.J. Fedorowicz Publishing, Inc, 2004, 126.
[678] Stefano Di Giusto, *Panzer-Abteilung 208 – I./ Panzer-Regiment Feldherrnhalle*, Tankograd Publishing-Verlag Jochen Vollert, 2010, 149.
[679] Georg Maier, *Drama Between Budapest and Vienna*, J.J. Fedorowicz Publishing, Inc, 2004, 129-130.
[680] Perry Pierik, *Hungary 1944-1945 The Forgotten Tragedy,* ASPEKT, 1998, 149.
[681] John Lukacs, *The Siege of Budapest, 100 Days in World War II*, Yale University Press, 2005, 429-431, Table #18.
[682] Karl-Heinz Frieser, *Germany and the Second World War, Volume VIII,* Clarendon Press Oxford, 2017, 923.
[683] John Lukacs, *The Siege of Budapest, 100 Days in World War II*, Yale University Press, 2005, 428, Table #15.
[684] Perry Pierik, *Hungary 1944-1945 The Forgotten Tragedy,* ASPEKT, 1998, 109, 274.
[685] Barrie Pitt, Editor, *The History of the Second World War, part 97*, 1973, 2198-2199, Statistics compiled by the Imperial War Museum.
[686] Georg Maier, *Drama Between Budapest and Vienna* J.J. Fedorowicz Publishing, Inc, 2004, 130.
[687] Efim Ivanovich Smirnov, *Frontline Mercy*, Moscow. Military publishing, 1991, 348.
[688] TsAMO record ID #113723570, pages 2, 7, 8.
[689] H.G.W. Davie, *The Journal of Slavic Military Studies*, 2021, Vol. 34, No 2, p 240-242, *"Mechanized Corps – a Study in Mobility and Transport."*
[690] Georg Maier, *Drama Between Budapest and Vienna* J.J. Fedorowicz Publishing, Inc, 2004, 130.
[691] Vinogradov Timofey Zakharovich, *Expensive-Forever,* Moscow: Military Publishing, 1977, 134.
[692] Georg Maier, *Drama Between Budapest and Vienna* J.J. Fedorowicz Publishing, Inc, 2004, 131.
[693] Norbert Szamveber, *Days of Battle*, Helion & Company, 2013, 202, 228.
[694] Michael Reynolds, *Men of Steel – 1 SS Panzer Corps,* Sarpedon, New York, 1999, 190.
[695] Dr. Franz-Wilhelm Lochmann, Richard Freiherr von Rosen, Alfred Rubbel, *The Combat History of German Tiger Tank Battalion 503 in World War II,* J.J Fedorowicz Publishing, 2000, 332.
[696] Norbert Szamveber, *Days of Battle*, Helion & Company, 2013, 210.
[697] Ibid., 211.
[698] Michael Reynolds, *Men of Steel – 1 SS Panzer Corps,* Sarpedon, New York, 1999, 191.
[699] General S. M. Shtemenko, *The Last Six Months*, Doubleday & Company, Inc., 1977, 32.
[700] Dr. Franz-Wilhelm Lochmann Richard Freiherr von Rosen, Alfred Rubbel, *The Combat History of German Tiger Tank Battalion 503 in World War II,* J.J Fedorowicz Publishing, 2000, 332, 333.
[701] Georg Maier, *Drama Between Budapest and Vienna* J.J. Fedorowicz Publishing, Inc, 2004, 133.
[702] Aynutdinov Sergey Khusainovich, *In Memory and Heart,* Yekaterinburg: State Research and Production Enterprise, 2000, Chapter "Ural Worker," 324.
[703] Dr. Franz-Wilhelm Lochmann Richard Freiherr von Rosen, Alfred Rubbel, *The Combat History of German Tiger Tank Battalion 503 in World War II,* J.J Fedorowicz Publishing, 2000, 333.

704 Douglas E. Nash Sr, *From the Realm of a Dying Sun Vol III*, Casemate, 2021, 14.
705 Georg Maier, *Drama Between Budapest and Vienna*, J.J. Fedorowicz Publishing, Inc, 2004, 134.
706 Ibid.
707 Norbert Szamveber, *Days of Battle*, Helion & Company, 2013, 202.
708 Georg Maier, *Drama Between Budapest and Vienna*, J.J. Fedorowicz Publishing, Inc, 2004, 135.
709 Richard N. Armstrong, *Red Army Tank Commanders – The Armored Guards,* Schiffer Military/Aviation History, 1994, 439.
710 Georg Maier, *Drama Between Budapest and Vienna*, J.J. Fedorowicz Publishing, Inc, 2004, 135.
711 R. M. Portuguese. Domank A. S., Kovalenko A. P., *Marshal S.K. Timoshenko,* Publishing House of the Moscow Foundation "Victory -1945", 1994, 317-318.
712 Georg Maier, *Drama Between Budapest and Vienna*, J.J. Fedorowicz Publishing, Inc, 2004, 136.
713 Ibid.
714 Ibid.
715 Norbert Szamveber, *Days of Battle*, Helion & Company, 2013, 228.
716 Georg Maier, *Drama Between Budapest and Vienna*, J.J. Fedorowicz Publishing, Inc, 2004, 138.
717 Ibid., 139.
718 Vinogradov Timofey Zakharovich, *Expensive-Forever,* Moscow: Military Publishing, 1977, 138.
719 Georg Maier, *Drama Between Budapest and Vienna*, J.J. Fedorowicz Publishing, Inc, 2004, 139.
720 Ibid., 139, 140.
721 Ibid., 139, 141.
722 Williamson Murry, *Strategy for Defeat, The Luftwaffe 1933-1945,* Air University Press, Maxwell Air Force Base, 1983, 275.
723 (TsAMO ID #133393465).
724 Karl-Heinz Frieser, *Germany and the Second World War, Volume VIII,* Clarendon Press Oxford, 2017, 928.
725 Blitz Street (Channel 4 2010, Tony Robinson) | Memorable TV.
726 Georg Maier, *Drama Between Budapest and Vienna*, J.J. Fedorowicz Publishing, Inc, 2004, 166.
727 Michael Reynolds, *Men of Steel, I SS Panzer Corps*, Sarpedon, New York, 1999, 201.
728 Georg Maier, *Drama Between Budapest and Vienna*, J.J. Fedorowicz Publishing, Inc, 2004, 166.
729 Michael Reynolds, *Men of Steel, I SS Panzer Corps*, Sarpedon, New York, 1999, 202.
730 Georg Maier, *Drama Between Budapest and Vienna*, J.J. Fedorowicz Publishing, Inc, 2004, 187.
731 Charles Whiting, *Skorzeny*, Ballantine Books, 1972, 128-129.
732 TsAMO ID #136824610.
733 Georg Maier, *Drama Between Budapest and Vienna*, J.J. Fedorowicz Publishing, Inc, 2004, 190.
734 Aynutdinov Sergey Khusainovich, *In Memory and Heart,* Yekaterinburg: State Research and Production Enterprise, 2000, Chapter "Ural Worker, 333.
735 Ibid., List of Illustrations.
736 Georg Maier, *Drama Between Budapest and Vienna*, J.J. Fedorowicz Publishing, Inc, 2004, 196.
737 Aynutdinov Sergey Khusainovich, *In Memory and Heart,* Yekaterinburg: State Research and Production Enterprise, 2000, Chapter "Ural Worker," 338.
738 Boris Sokolov, *Marshal Malinovskii: Hero of the Soviet Union*, Helion & Company, 2017, 311.
739 Kamen Nevenkin, *Bloody Vienna.* PeKo Publishing Kft., 2020, 24.
740 Georg Maier, *Drama Between Budapest and Vienna*, J.J. Fedorowicz Publishing, Inc, 2004, 207.
741 Ibid., 211.
742 David Glantz, *Soviet Military Deception in the Second World War,* Frank Class, 1989, 517-520.
743 Georg Maier, *Drama Between Budapest and Vienna*, J.J. Fedorowicz Publishing, Inc, 2004, 222.
744 Karl-Heinz Frieser, *Germany and the Second World War, Volume VIII,* Clarendon Press Oxford, 2017, 943.
745 Georg Maier, *Drama Between Budapest and Vienna*, J.J. Fedorowicz Publishing, Inc, 2004, 228-231.
746 Ibid., 235.
747 Ibid., 240-245.
748 Ibid., 246-251.
749 Karl-Heinz Frieser, *Germany and the Second World War, Volume VIII,* Clarendon Press Oxford, 2017, 946.
750 Ibid.
751 Ibid., 948.
752 Douglas E. Nash Sr, *From the Realm of a Dying Sun Vol III*, Casemate, 2021, 146.
753 Aynutdinov Sergey Khusainovich, *In Memory and Heart,* Yekaterinburg: State Research and Production Enterprise, 2000, Chapter "Ural Worker," 336-337.
754 Norbert Szamveber, *Days of Battle*, Helion & Company, 2013, 237-238.
755 Ibid., 236.
756 Ibid.
757 Georg Maier, *Drama Between Budapest and Vienna*, J.J. Fedorowicz Publishing, Inc, 2004, 301.

[758] Douglas E. Nash Sr, *From the Realm of a Dying Sun Vol III*, Casemate, 2021, 146.
[759] Georg Maier, *Drama Between Budapest and Vienna*, J.J. Fedorowicz Publishing, Inc, 2004, 307.
[760] From the web site, *Remembering the 16: the Soviet betrayal and arrest of 16 Poles of the underground state* – The First News, Accessed January 2022.
[761] Ty Bomba, *Operation Unthinkable-Churchill's Plan for World War III in 1945, Strategy & Tactics Magazine #333 Mar-Apr 2022*, 6.
[762] Georg Maier, *Drama Between Budapest and Vienna*, J.J. Fedorowicz Publishing, Inc, 2004, 309.
[763] Aynutdinov Sergey Khusainovich, *In Memory and Heart,* Yekaterinburg: State Research and Production Enterprise, 2000, Chapter "Ural Worker," Chapter "Further West," 342.
[764] Alemzhan Arinov, *To Seek Vengeance or Not? How the Evolution of Revenge Propaganda Occurred in the Soviet Military Periodical Press,* The Journal of Slavic Military Studies, Vol 34, Number 3, July-September 2021, 394.
[765] Kamen Nevenkin, *Bloody Vienna.* PeKo Publishing Kft., 2020, 106-107.
[766] S. A. Andryushchenko, *We Started on Slavutich,* Voenizdat, 1979, Chapter "Danube Again," 269.
[767] Ibid., 273-274.
[768] S. N. Grechko, *Decisions Were Made on the Ground,* Voenizdat, 1984, 278.
[769] Aynutdinov Sergey Khusainovich, *In Memory and Heart,* Yekaterinburg: State Research and Production Enterprise, 2000, Chapter "Ural Worker," 347.
[770] Karl-Heinz Frieser, *Germany and the Second World War, Volume VIII,* Clarendon Press Oxford, 2017, 951.
[771] Ibid., 951.
[772] Alfred Price, *The Last Year of the Luftwaffe, May 1944 to May 1945*, Arms and Armour, 1991, 151.
[773] Ibid., 129-130.
[774] Von Hardesty, Ilya Grinberg, *Red Phoenix Rising, The Soviet Airforce in World War II*, University Press Kansas, 2012, 368.
[775] Steve Crawford, *The Easter Front Day by Day, A Photographic Chronology*, Potomac Books, 2006, 183.
[776] Georg Maier, *Drama Between Budapest and Vienna*, J.J. Fedorowicz Publishing, Inc, 2004, 362.
[777] Ibid., 364.
[778] Kamen Nevenkin, *Bloody Vienna.* PeKo Publishing Kft., 2020, 129.
[779] Georg Maier, *Drama Between Budapest and Vienna*, J.J. Fedorowicz Publishing, Inc, 2004, 366.
[780] Ibid., 367.
[781] Kamen Nevenkin, *Bloody Vienna.* PeKo Publishing Kft., 2020, 137.
[782] Alemzhan Arinov, *To Seek Vengeance or Not? How the Evolution of Revenge Propaganda Occurred in the Soviet Military Periodical Press,* The Journal of Slavic Military Studies, Vol 34, Number 3, July-September 2021, 394.
[783] Ibid.
[784] Ibid., 399-400.
[785] Georg Maier, *Drama Between Budapest and Vienna*, J.J. Fedorowicz Publishing, Inc, 2004, 370.
[786] Robert Goralski, *World War II Almanac 1931-1945 A Political and Military Record,* Hamish Hamilton, London, 1981 402.
[787] Volker Ullrich, *Eight Days in May*, Liveright Publishing Corporation, 2020, xi, 15, 74-75.
[788] Ibid., xi, 15, 23-26.
[789] Ibid., xi, 15, 82-83.
[790] Ibid., 239, 258-260.
[791] Luke Harding, *Hitler's nurse breaks 60 years of silence, The Guardian*, May 2, 2005.
[792] Volker Ullrich, *Eight Days in May*, Liveright Publishing Corporation, 2020, xi, 15, 84-86.
[793] Ibid., 159-163.
[794] Ibid., 185.
[795] Ibid., 162.
[796] Georg Maier, *Drama Between Budapest and Vienna*, J.J. Fedorowicz Publishing, Inc, 2004, 371.
[797] Ibid.
[798] Karl-Heinz Frieser, *Germany and the Second World War, Volume VIII,* Clarendon Press Oxford, 2017, 954.
[799] S. A. Andryushchenko, *We Started on Slavutich,* Voenizdat, 1979, Chapter "Danube Again," 281.
[800] Ibid.
[801] Hans Dollinger, *The Decline and Fall of Nazi Germany and Imperial Japan*, Bonanza Books, 1967, 297.
[802] S. A. Andryushchenko, *We Started on Slavutich,* Voenizdat, 1979, Chapter "Danube Again," 282.
[803] Volker Ullrich, *Eight Days in May*, Liveright Publishing Corporation, 2020, 164.
[804] Ibid., 240.
[805] Ibid., 186.
[806] Charles C. Sharp, *Soviet Order of Battle in World War II, Volume III, Red Storm,* Published by George F. Nafziger, 1995, 98.
[807] Volker Ullrich, *Eight Days in May*, Liveright Publishing Corporation, 2020, 186.
[808] Vinogradov Timofey Zakharovich, *Expensive-Forever,* Moscow: Military Publishing, 1977, Chapter "Abroad," 141.
[809] Sergey Ainutdinov, *In Memory and in the Heart*, 2000, Chapter "Ural Worker" GIPP, 373.

[810] Retrieved from the following web site on March 2021, https://fishki.net/2109354-istorii-invalidov-vov-kotoryh-soslali-na-valaam-za-ubogij-vid.html. The drawings of the old and wounded soldiers are worth viewing.

[811] Ty Bomba, *Operation Unthinkable-Churchill's Plan for World War III in 1945, Strategy & Tactics Magazine #333 Mar-Apr 2022*, 10.

[812] Web site, *United Nations | Definition, History, Founders, Flag, & Facts | Britannica.*

[813] http://militera.lib.ru/memo/russian/ainutdinov_sh/05.html.

[814] http://www.ww2.dk/new/army/armies/7gvoa.htm.

[815] Biography of Colonel-General Mikhail Stepanovich Shumilov - (Михаил Степанович Шумилов) (1896 – 1975), Soviet Union (generals.dk).

[816] Karl-Heinz Frieser, *Germany and the Second World War, Volume VIII,* Clarendon Press Oxford, 2017, p 204.

[817] Sergey Matveevich Shtemenko, *General Staff during the war*, Military Publishing, 1989, Book 1, Chapter 16. Retrieved from militera.lib.ru.

INDEX

1

2

3

4

5

6

7

8

9

A

B

C

D

E

F

G

H

I

J

K

L

M

N

O

P

R

S

T

U

V

W

X

Y

Z

Made in the USA
Monee, IL
05 November 2026

67858284R00302